ADVANCED KALMAN FILTERING, LEAST-SQUARES AND MODELING

About the Cover

The watercolor portrait of Carl Friedrich Gauss was made by J.C.A. Schwartz in 1803 (photo courtesy of Axel Wittmann, copyright owner). This is the decade when Gauss first used the least squares method to determine orbits of asteroids and comets. The 1963 picture of Rudolf Kalman in his office at the Research Institute for Advanced Studies was taken after he first published papers describing the Kalman filter (photo courtesy of Rudolf Kalman, copyright owner). The GPS IIF spacecraft is the latest operational version of the spacecraft series (picture courtesy of The Boeing Company, copyright owner). The GPS ground system uses a Kalman filter to track the spacecraft orbits and clock errors of both the spacecraft and ground monitor stations. A least squares fit is used to compute the navigation message parameters that are uplinked to the spacecraft and then broadcast to user receivers. GPS receivers typically use a Kalman filter to track motion and clock errors of the receiver. The plot shows the root-mean-squared *user range error* (URE) for 29 GPS satellites operational in 2005. Those URE values were computed using smoothed GPS orbit and clock estimates as the "truth" reference (see Example 9.6 of Chapter 9).

ADVANCED KALMAN FILTERING, LEAST-SQUARES AND MODELING

A Practical Handbook

BRUCE P. GIBBS
Carr Astronautics, Inc.

WILEY

A JOHN WILEY & SONS, INC., PUBLICATION

Published by John Wiley & Sons, Inc., Hoboken, New Jersey
Published simultaneously in Canada

Library of Congress Cataloging-in-Publication Data:

Gibbs, Bruce. P., 1946–
Advanced kalman filtering, least-squares and modeling : a practical handbook / Bruce. P. Gibbs.
p. cm.
Includes bibliographical references.
ISBN 978-0-470-52970-6 (cloth)
1. Engineering–Statistical methods. 2. Kalman filtering. 3. Least squares. 4. Mathematical models. I. Title.
TA340.G49 2010
620.0072′7–dc22

2010013944

Printed in Singapore

oBook ISBN: 978-0-470-89004-2
ePDF ISBN: 978-0-470-89003-5

10 9 8 7 6 5 4 3 2 1

This book is dedicated to the memory of Gerald Bierman,
Larry Levy, and James Vandergraft.
They contributed much to the field, and are greatly missed.

CONTENTS

PREFACE xv

1 INTRODUCTION 1

1.1 The Forward and Inverse Modeling Problem / 2
1.2 A Brief History of Estimation / 4
1.3 Filtering, Smoothing, and Prediction / 8
1.4 Prerequisites / 9
1.5 Notation / 9
1.6 Summary / 11

2 SYSTEM DYNAMICS AND MODELS 13

2.1 Discrete-Time Models / 14
2.2 Continuous-Time Dynamic Models / 17
2.2.1 State Transition and Process Noise Covariance Matrices / 19
2.2.2 Dynamic Models Using Basic Function Expansions / 22
2.2.3 Dynamic Models Derived from First Principles / 25
2.2.4 Stochastic (Random) Process Models / 31
2.2.5 Linear Regression Models / 42
2.2.6 Reduced-Order Modeling / 44
2.3 Computation of State Transition and Process Noise Matrices / 45
2.3.1 Numeric Computation of $\mathbf{\Phi}$ / 45
2.3.2 Numeric Computation of $\mathbf{Q}_D$ / 57
2.4 Measurement Models / 58
2.5 Simulating Stochastic Systems / 60
2.6 Common Modeling Errors and System Biases / 62
2.7 Summary / 65

3 MODELING EXAMPLES **67**

3.1 Angle-Only Tracking of Linear Target Motion / 67
3.2 Maneuvering Vehicle Tracking / 69
3.2.1 Maneuvering Tank Tracking Using Multiple Models / 69
3.2.2 Aircraft Tracking / 73
3.3 Strapdown Inertial Navigation System (INS) Error Model / 74
3.4 Spacecraft Orbit Determination (OD) / 80
3.4.1 Geopotential Forces / 83
3.4.2 Other Gravitational Attractions / 86
3.4.3 Solar Radiation Pressure / 87
3.4.4 Aerodynamic Drag / 88
3.4.5 Thrust Forces / 89
3.4.6 Earth Motion / 89
3.4.7 Numerical Integration and Computation of $\mathbf{\Phi}$ / 90
3.4.8 Measurements / 92
3.4.9 GOES I-P Satellites / 96
3.4.10 Global Positioning System (GPS) / 97
3.5 Fossil-Fueled Power Plant / 99
3.6 Summary / 99

4 LINEAR LEAST-SQUARES ESTIMATION: FUNDAMENTALS **101**

4.1 Least-Squares Data Fitting / 101
4.2 Weighted Least Squares / 108
4.3 Bayesian Estimation / 115
4.3.1 Bayesian Least Squares / 115
4.3.2 Bayes' Theorem / 117
4.3.3 Minimum Variance or Minimum Mean-Squared Error (MMSE) / 121
4.3.4 Orthogonal Projections / 124
4.4 Probabilistic Approaches—Maximum Likelihood and Maximum *A Posteriori* / 125
4.4.1 Gaussian Random Variables / 126
4.4.2 Maximum Likelihood Estimation / 128
4.4.3 Maximum *A Posteriori* / 133
4.5 Summary of Linear Estimation Approaches / 137

5 LINEAR LEAST-SQUARES ESTIMATION: SOLUTION TECHNIQUES **139**

5.1 Matrix Norms, Condition Number, Observability, and the Pseudo-Inverse / 139
5.1.1 Vector-Matrix Norms / 139

5.1.2 Matrix Pseudo-Inverse / 141
5.1.3 Condition Number / 141
5.1.4 Observability / 145
5.2 Normal Equation Formation and Solution / 145
5.2.1 Computation of the Normal Equations / 145
5.2.2 Cholesky Decomposition of the Normal Equations / 149
5.3 Orthogonal Transformations and the QR Method / 156
5.3.1 Givens Rotations / 158
5.3.2 Householder Transformations / 159
5.3.3 Modified Gram-Schmidt (MGS) Orthogonalization / 162
5.3.4 QR Numerical Accuracy / 165
5.4 Least-Squares Solution Using the SVD / 165
5.5 Iterative Techniques / 167
5.5.1 Sparse Array Storage / 167
5.5.2 Linear Iteration / 168
5.5.3 Least-Squares Solution for Large Sparse Problems Using Krylov Space Methods / 169
5.6 Comparison of Methods / 175
5.6.1 Solution Accuracy for Polynomial Problem / 175
5.6.2 Algorithm Timing / 181
5.7 Solution Uniqueness, Observability, and Condition Number / 183
5.8 Pseudo-Inverses and the Singular Value Transformation (SVD) / 185
5.9 Summary / 190

6 LEAST-SQUARES ESTIMATION: MODEL ERRORS AND MODEL ORDER 193

6.1 Assessing the Validity of the Solution / 194
6.1.1 Residual Sum-of-Squares (SOS) / 194
6.1.2 Residual Patterns / 195
6.1.3 Subsets of Residuals / 196
6.1.4 Measurement Prediction / 196
6.1.5 Estimate Comparison / 197
6.2 Solution Error Analysis / 208
6.2.1 State Error Covariance and Confidence Bounds / 208
6.2.2 Model Error Analysis / 212
6.3 Regression Analysis for Weighted Least Squares / 237
6.3.1 Analysis of Variance / 238
6.3.2 Stepwise Regression / 239
6.3.3 Prediction and Optimal Data Span / 244
6.4 Summary / 245

7 LEAST-SQUARES ESTIMATION: CONSTRAINTS, NONLINEAR MODELS, AND ROBUST TECHNIQUES **249**

7.1 Constrained Estimates / 249
7.1.1 Least-Squares with Linear Equality Constraints (Problem LSE) / 249
7.1.2 Least-Squares with Linear Inequality Constraints (Problem LSI) / 256
7.2 Recursive Least Squares / 257
7.3 Nonlinear Least Squares / 259
7.3.1 1-D Nonlinear Least-Squares Solutions / 263
7.3.2 Optimization for Multidimensional Unconstrained Nonlinear Least Squares / 264
7.3.3 Stopping Criteria and Convergence Tests / 269
7.4 Robust Estimation / 282
7.4.1 De-Weighting Large Residuals / 282
7.4.2 Data Editing / 283
7.5 Measurement Preprocessing / 285
7.6 Summary / 286

8 KALMAN FILTERING **289**

8.1 Discrete-Time Kalman Filter / 290
8.1.1 Truth Model / 290
8.1.2 Discrete-Time Kalman Filter Algorithm / 291
8.2 Extensions of the Discrete Filter / 303
8.2.1 Correlation between Measurement and Process Noise / 303
8.2.2 Time-Correlated (Colored) Measurement Noise / 305
8.2.3 Innovations, Model Validation, and Editing / 311
8.3 Continous-Time Kalman-Bucy Filter / 314
8.4 Modifications of the Discrete Kalman Filter / 321
8.4.1 Friedland Bias-Free/Bias-Restoring Filter / 321
8.4.2 Kalman-Schmidt Consider Filter / 325
8.5 Steady-State Solution / 328
8.6 Wiener Filter / 332
8.6.1 Wiener-Hopf Equation / 333
8.6.2 Solution for the Optimal Weighting Function / 335
8.6.3 Filter Input Covariances / 336
8.6.4 Equivalence of Weiner and Steady-State Kalman-Bucy Filters / 337
8.7 Summary / 341

9 FILTERING FOR NONLINEAR SYSTEMS, SMOOTHING, ERROR ANALYSIS/MODEL DESIGN, AND MEASUREMENT PREPROCESSING **343**

9.1 Nonlinear Filtering / 344
9.1.1 Linearized and Extended Kalman Filters / 344
9.1.2 Iterated Extended Kalman Filter / 349
9.2 Smoothing / 352
9.2.1 Fixed-Point Smoother / 353
9.2.2 Fixed-Lag Smoother / 356
9.2.3 Fixed-Interval Smoother / 357
9.3 Filter Error Analysis and Reduced-Order Modeling / 370
9.3.1 Linear Analysis of Independent Error Sources / 372
9.3.2 Error Analysis for ROM Defined as a Transformed Detailed Model / 380
9.3.3 Error Analysis for Different Truth and Filter Models / 382
9.4 Measurement Preprocessing / 385
9.5 Summary / 385

10 FACTORED (SQUARE-ROOT) FILTERING **389**

10.1 Filter Numerical Accuracy / 390
10.2 U-D Filter / 392
10.2.1 U-D Filter Measurement Update / 394
10.2.2 U-D Filter Time Update / 396
10.2.3 RTS Smoother for U-D Filter / 401
10.2.4 U-D Error Analysis / 403
10.3 Square Root Information Filter (SRIF) / 404
10.3.1 SRIF Time Update / 405
10.3.2 SRIF Measurement Update / 407
10.3.3 Square Root Information Smoother (SRIS) / 408
10.3.4 Dyer-McReynolds Covariance Smoother (DMCS) / 410
10.3.5 SRIF Error Analysis / 410
10.4 Inertial Navigation System (INS) Example Using Factored Filters / 412
10.5 Large Sparse Systems and the SRIF / 417
10.6 Spatial Continuity Constraints and the SRIF Data Equation / 419
10.6.1 Flow Model / 421
10.6.2 Log Conductivity Spatial Continuity Model / 422
10.6.3 Measurement Models / 424

10.6.4 SRIF Processing / 424
10.6.5 Steady-State Flow Constrained Iterative Solution / 425
10.7 Summary / 427

11 ADVANCED FILTERING TOPICS 431

11.1 Maximum Likelihood Parameter Estimation / 432
11.1.1 Calculation of the State Transition Partial Derivatives / 434
11.1.2 Derivatives of the Filter Time Update / 438
11.1.3 Derivatives of the Filter Measurement Update / 439
11.1.4 Partial Derivatives for Initial Condition Errors / 440
11.1.5 Computation of the Log Likelihood and Scoring Step / 441
11.2 Adaptive Filtering / 449
11.3 Jump Detection and Estimation / 450
11.3.1 Jump-Free Filter Equations / 452
11.3.2 Stepwise Regression / 454
11.3.3 Correction of Jump-Free Filter State / 455
11.3.4 Real-Time Jump Detection Using Stepwise Regression / 456
11.4 Adaptive Target Tracking Using Multiple Model Hypotheses / 461
11.4.1 Weighted Sum of Filter Estimates / 462
11.4.2 Maximum Likelihood Filter Selection / 463
11.4.3 Dynamic and Interactive Multiple Models / 464
11.5 Constrained Estimation / 471
11.6 Robust Estimation: *H*-Infinity Filters / 471
11.7 Unscented Kalman Filter (UKF) / 474
11.7.1 Unscented Transform / 475
11.7.2 UKF Algorithm / 478
11.8 Particle Filters / 485
11.9 Summary / 490

12 EMPIRICAL MODELING 493

12.1 Exploratory Time Series Analysis and System Identification / 494
12.2 Spectral Analysis Based on the Fourier Transform / 495
12.2.1 Fourier Series for Periodic Functions / 497
12.2.2 Fourier Transform of Continuous Energy Signals / 498
12.2.3 Fourier Transform of Power Signals / 502

12.2.4 Power Spectrum of Stochastic Signals / 504
12.2.5 Time-Limiting Window Functions / 506
12.2.6 Discrete Fourier Transform / 509
12.2.7 Periodogram Computation of Power Spectra / 512
12.2.8 Blackman-Tukey (Correlogram) Computation of Power Spectra / 514
12.3 Autoregressive Modeling / 522
12.3.1 Maximum Entropy Method (MEM) / 524
12.3.2 Burg MEM / 525
12.3.3 Final Prediction Error (FPE) and Akaike Information Criteria (AIC) / 526
12.3.4 Marple AR Spectral Analysis / 528
12.3.5 Summary of MEM Modeling Approaches / 529
12.4 ARMA Modeling / 531
12.4.1 ARMA Parameter Estimation / 532
12.5 Canonical Variate Analysis / 534
12.5.1 CVA Derivation and Overview / 536
12.5.2 Summary of CVA Steps / 539
12.5.3 Sample Correlation Matrices / 540
12.5.4 Order Selection Using the AIC / 541
12.5.5 State-Space Model / 543
12.5.6 Measurement Power Spectrum Using the State-Space Model / 544
12.6 Conversion from Discrete to Continuous Models / 548
12.7 Summary / 551

APPENDIX A SUMMARY OF VECTOR/MATRIX OPERATIONS 555

A.1 Definition / 555
A.1.1 Vectors / 555
A.1.2 Matrices / 555
A.2 Elementary Vector/Matrix Operations / 557
A.2.1 Transpose / 557
A.2.2 Addition / 557
A.2.3 Inner (Dot) Product of Vectors / 557
A.2.4 Outer Product of Vectors / 558
A.2.5 Multiplication / 558
A.3 Matrix Functions / 558
A.3.1 Matrix Inverse / 558
A.3.2 Partitioned Matrix Inversion / 559
A.3.3 Matrix Inversion Identity / 560
A.3.4 Determinant / 561

A.3.5 Matrix Trace / 562
A.3.6 Derivatives of Matrix Functions / 563
A.3.7 Norms / 564
A.4 Matrix Transformations and Factorization / 565
A.4.1 LU Decomposition / 565
A.4.2 Cholesky Factorization / 565
A.4.3 Similarity Transformation / 566
A.4.4 Eigen Decomposition / 566
A.4.5 Singular Value Decomposition (SVD) / 566
A.4.6 Pseudo-Inverse / 567
A.4.7 Condition Number / 568

APPENDIX B PROBABILITY AND RANDOM VARIABLES **569**

B.1 Probability / 569
B.1.1 Definitions / 569
B.1.2 Joint and Conditional Probability, and Independence / 570
B.2 Random Variable / 571
B.2.1 Distribution and Density Functions / 571
B.2.2 Bayes' Theorem for Density Functions / 572
B.2.3 Moments of Random Variables / 573
B.2.4 Gaussian Distribution / 574
B.2.5 Chi-Squared Distribution / 574
B.3 Stochastic Processes / 575
B.3.1 Wiener or Brownian Motion Process / 576
B.3.2 Markov Process / 576
B.3.3 Differential and Integral Equations with White Noise Inputs / 577

BIBLIOGRAPHY **579**

INDEX **599**

Most of the software used for the examples in this book is available online at ftp://ftp.wiley.com/public/sci_tech_med/least_squares/. Chapter 13, "Software Documentation," Appendix C, "Selected First-Principle Concepts," and Appendix D, "Discrete-Time (ARMAX-type) Models" are also available at this location.

PREFACE

Another book on Kalman filtering and least-squares estimation—are there not enough of them? Well, yes and no. Numerous books on the topic have been published, and many of them, both old and recent, are excellent. However, many practitioners of the field do not appear to fully understand the implications of theory and are not aware of lessons learned decades ago. It often appears that model structure was not carefully analyzed and that options for improving performance of the estimation were not considered. Available books present information on optimal estimation theory, standard implementations, methods to minimize numerical errors, and application examples, but information on model development, practical considerations, and useful extensions is limited.

Hence the first goal of this book is to discuss model development in sufficient detail so that the reader may design an estimator that meets all application requirements and is somewhat insensitive to modeling errors. Since it is sometimes difficult to *a priori* determine the best model structure, use of *exploratory data analysis* to define model structure is discussed. Methods for deciding on the "best" model are also presented. The second goal is to present little known extensions of least-squares estimation or Kalman filtering that provide guidance on model structure and parameters, or that reduce sensitivity to changes in system behavior. The third goal is discussion of implementation issues that make the estimator more accurate or efficient, or that make it flexible so that model alternatives can be easily compared. The fourth goal is to provide the designer/analyst with guidance in evaluating estimator performance and in determining/correcting problems. The final goal is to provide a subroutine library that simplifies implementation, and flexible general purpose high-level drivers that allow both easy analysis of alternative models and access to extensions of the basic estimation.

It is unlikely that a book limited to the above goals would be widely read. To be useful, it must also include fundamental information for readers with limited knowledge of the subject. This book is intended primarily as a handbook for engineers who must design practical systems. Although it could be used as a textbook, it is not intended for that purpose since many excellent texts already exist. When discussions of theory are brief, alternate texts are mentioned and readers are encouraged to consult them for further information. Most chapters include real-world examples of topics discussed. I have tried to approach the topics from the practical implementation point of view used in the popular *Applied Optimal Estimation* (Gelb 1974) and *Numerical Recipes* (Press 2007) books.

Why am I qualified to offer advice on optimal estimation? I was very lucky in my career to have been in the right place at the right time, and had the great fortune of working with a number of very talented people. The reputation of these people allowed our organizations to consistently obtain contracts involving R&D for innovative state-of-the-art estimation. In several cases our implementations were—to our knowledge—the first for a particular application. Some of the approximately 40 estimation applications I have worked on include:

1. Spacecraft orbit determination (batch least squares, Kalman filtering, and smoothing) and error analysis
2. Spacecraft attitude determination
3. Determination of imaging instrument optical bias and thermal deformation
4. Combined spacecraft orbit and optical misalignment determination
5. Monitoring of crustal movement using spaceborne laser ranging
6. Ship and submarine tracking (active and passive)
7. Aircraft tracking (active and passive)
8. Missile tracking using radar
9. Real-time tank maneuvering target tracking using multiple hypothesis testing
10. Tank gun tube flexure prediction
11. Ship inertial navigation error modeling
12. Missile inertial guidance error modeling
13. Geomagnetic core field modeling
14. Nonlinear model predictive control for fossil power plants
15. Subsurface ground water flow modeling and calibration
16. Geophysical modeling
17. Optical image landmarking using shoreline correlation
18. Spacecraft bi-propellant propulsion modeling
19. Spacecraft thruster calibration using in-orbit data
20. Global Positioning System constellation orbit/clock/monitor station tracking (filtering, smoothing, and maximum likelihood estimation process noise optimization)
21. Atomic clock calibration

The wide range of applications has included filters with as few as three states and as many as 1000. The ground water flow modeling software has been used on problems with more than 200,000 nodes (400,000 unknowns), although the solution technique does not compute an error covariance matrix. Some models were based on first-principles, some models were empirically derived, and others were both. Because of this wide range in applications and model types, it was necessary to develop software and methods that could easily adapt to different models, and could be configured to easily change the selection of states to be estimated. The need to execute large scale problems on computers that were 200 times slower than today's PCs made it necessary to use reduced-order models of systems and to implement efficient code. This forced the development of tools and methods for simplifying the dynamic models, and for determining which states are important to the estimation. It also encouraged development of software that could easily adapt to different models and to different selections of estimated states. The lessons learned from these experiences are the focus of this book.

This book starts with introductory material and gradually expands on advanced topics. The first chapter briefly describes the estimation problem and the history of optimal estimation from Gauss to Kalman. Notation used in the book is discussed. Other background material on matrix properties, probability theory, and stochastic processes appears in Appendices.

Chapter 2 discusses different types of dynamic models, their use as the basis of estimation, and methods for computing the state transition and process noise matrices. Use of first-principles models, reduced-order models, and models for dynamic effects that are poorly known are addressed. Chapter 3 describes several real-world problems that demonstrate various modeling principles and are used as examples in later chapters.

Chapter 4 derives least-squares estimation from several different points of view (weighted least squares, Bayesian least squares, minimum mean-squared error, minimum variance, maximum likelihood, maximum *a posteriori*) and discusses various implementations. Chapter 5 discusses least-squares solution techniques such as Cholesky decomposition of the normal equations, QR decomposition, Singular Value Decomposition, and iterative Krylov space methods. Also addressed is the theory of orthogonal transformations, solution uniqueness, observability, condition number, and the pseudo-inverse. Numerous examples demonstrate the issues and performance.

Chapter 6 discusses methods to evaluate the validity of least-squares solutions, error analysis, selection of model order, and regression analysis for parameter estimation. Chapter 7 addresses the important topics of least-squares estimation for nonlinear systems, constrained estimation, robust estimation, data editing, and measurement preprocessing. Real-world examples are given.

Chapter 8 presents the basics of Kalman filtering, and shows that it is based on fundamental concepts presented in earlier chapters. Discrete and continuous versions of the Kalman filter are derived, and extensions to handle data correlations and certain types of model errors are presented. Other topics include steady-state filtering, outlier editing, model divergence, and model validation. The relationship to the Wiener filter is also discussed.

Since real-world problems are frequently nonlinear, methods for nonlinear filtering are discussed in Chapter 9. Also discussed are smoothing (fixed point, fixed interval, and fixed lag), analysis of various modeling errors, design of reduced-order models, and measurement preprocessing.

Chapter 10 discusses numerical issues and shows how use of factorized (square root) estimation can minimize the growth of numerical errors. The factorized U-D and SRIF algorithms and their various implementations are discussed at length, and smoothers designed to work with the factored filters are presented. An example based on an inertial navigation system error model is used to compare properties of the covariance and factored filters. Usefulness of the square-root information filter (SRIF) data equation concept as a general approach to estimation is explained, and a hydrological flow problem with soft spatial continuity constraints on hydraulic conductivity demonstrates application of the data equation to two-dimensional and three-dimensional spatial problems.

Chapter 11 presents several advanced topics. It shows how properties of the filter innovations (one-step filter measurement residuals) allow model jump detection/estimation, and calculation of the log likelihood function. The log likelihood is useful when determining which of several models is best, and in determining the "best" values of model dynamic parameters and magnitudes of process and measurement noise. Adaptive filtering techniques such as jump detection and multiple-model approaches are discussed. Other topics include constrained estimation, robust estimation, and newer nonlinear filtering approaches (unscented and particle filters).

Chapter 12 discusses empirical model development in cases where it is not possible to develop a complete model from first principles. It may seem odd that this chapter appears at the end of the book, but the methods used for exploratory analysis of stochastic time series data depend on model concepts discussed in Chapter 2, and on least-squares techniques presented in Chapters 4 through 6. Use of spectral analysis for determining model dynamics and order is explained, and methods for computing parameters of autoregressive (AR) or autoregressive moving average (ARMA) models are presented. Accurate determination of the model order and states is discussed. The theory is presented at a level where the reader can understand the implications of assumptions and limitations of the methods. Applications of the theory for real-world examples are mentioned, and the performance of the different methods is demonstrated using data generated from a fourth-order autoregressive moving average with exogenous input (ARMAX) model.

Most of the advanced methods presented in this book have appeared in previous literature. Unfortunately some methods are rarely used in practice because of difficulty in implementing flexible general-purpose algorithms that can be applied to different problems. Chapter 13 presents a framework for doing this and also describes software for that purpose. The goal is to structure code so that alternative models can be easily compared and enhancements can easily be implemented. Many algorithms and high-level drivers are available as Fortran 90/95 code (compatible with Fortran 2003), downloadable from the web site ftp://ftp.wiley.com/public/sci_tech_med/least_squares/. Software used for many examples described in various chapters, and drivers for advanced algorithms are also included. The various algorithms are implemented in software that I have used successfully for many years. Where the original code was written in Fortran 77, it has been upgraded to Fortran 90/95 usage and standards. In some cases the library functions and subroutines are enhanced versions of codes written by others. In particular, some modules for factorized estimation are an enhancement of Estimation Subroutine Package or Library subroutines written by Dr. G. Bierman and associates. (The original software is still available on the web at sources listed in Chapter 13.) Other libraries and algorithms are used by some drivers. One particularly useful library is the Linear Algebra PACKage (LAPACK). Others packages implement the LSQR and NL2SOL algorithms.

There were several reasons for choosing Fortran 90/95/2003—these are very different languages than the Fortran 77 familiar to many. The Fortran 90 enhancements most relevant to estimation problems are the matrix/vector notation used in

MATLAB® (registered trademark of The Mathworks, Inc.), and inclusion of matrix/vector operators such as multiply, dot product, and element-by-element operations. Other important enhancements include dynamic array allocation, code modules (encapsulation), limited variable scope, argument usage validation, and flexible string handling. Fortran 90 compilers can validate usage of variables and catch many bugs that would have previously been caught at execution time. Fortran 95 extends the capabilities and Fortran 2003 adds many object oriented capabilities. Fortran 90/95/2003 is a much more suitable language than C or C++ for linear algebra applications. Use of MATLAB was also considered, but rejected because it does not easily integrate into some production applications, and many previous estimation books have included MATLAB code. The reader should have little difficulty in converting Fortran 90/95 code to MATLAB because vector/matrix syntax is similar. Use of global variables and module features in the supplied software was deliberately limited so that code could be more easily ported to MATLAB.

After reading this book you may get the impression that development of "good" estimation models and algorithms is less of a science than expected—in some respects it is an art. Creation of universal "rules" for developing models and implementing estimation algorithms is a desirable, but probably unreachable, goal. After four decades of trying to find such rules, I have come to the conclusion that the best one can hope for is a set of "guidelines" rather than rules. A highly respected colleague once announced that "after years of trying to find rules for characterizing the effects of model errors on the estimation," he came to the conclusion that there are no rules. Every time he discovered a possible rule, he eventually discovered an exception. Mis-modeling in the estimation will, of course, usually increase estimation errors, but it is also possible that model errors can partially offset the effects of a particular noise sequence, and hence reduce errors in serendipitous cases. To determine bounds on estimation errors due to particular types of model errors, it is usually necessary to simulate both the estimation algorithm and the true system. This can be done using Monte Carlo or covariance methods, both of which are time-consuming.

With regard to algorithm accuracy, numerical analysts have extensively studied the growth of truncation and round-off errors (due to finite computer word length) for various algorithms, and certain algorithms have been identified as more stable than others. Convergence properties of iterative algorithms (for nonlinear problems) have also been studied and some algorithms have been found to converge faster or more reliably than others for most problems. However, there are always exceptions. Sometimes the algorithm that works best on a simple problem does not work well on a large-scale problem. One fault of many estimation books is the tendency to present "simple" low-order examples, leaving the reader with the impression that the described behavior is a general characteristic. My suggestion is to implement as many "reasonable" algorithms as you can, and to then test them thoroughly on simulated and real data. For example, try to include a "factorized" estimation algorithm (U-D or SRIF) option when a covariance formulation is the prime algorithm because you never know when numerical errors will cause problems. Also try to implement models and algorithms in multiple ways so that results can be compared. If it is important to develop an estimation algorithm that meets requirements under all possible conditions, expect much hard work—there are few shortcuts.

You may be tempted to give up and turn over development to "experts" in the field. I do not want to discourage potential consulting business, but please do not get discouraged. The estimation field needs new bright engineers, scientists, and mathematicians. One goal of this book is to pass on the lessons learned over several decades to a new generation of practitioners.

To provide feedback on the book or associated software, or to report errors, please send email to bgibbs00@ieee.org.

ACKNOWLEDGMENTS

I greatly appreciate the support, suggestions, and review from others. Thomas Englar and David Uetrecht reviewed the early chapters and offered suggestions. William Levine reviewed several chapters, suggested many improvements (particularly in Appendix B), and provided good publishing advice. James Carr reviewed most chapters and made many helpful suggestions. David Porter reviewed material on maximum likelihood parameter estimation and multiple model approaches, provided additional background, and suggested a number of improvements. Dan Simon reviewed several chapters, made many good suggestions, and provided helpful feedback. Catherine Thornton reviewed sections on error analysis and factorized estimation, discussed historical background, and provided additional papers by herself and Gerald Bierman. Wallace Larimore reviewed Chapter 12 discussions of canonical variate analysis (CVA), and provided additional background and references. Mark Pittelkau also reviewed Chapter 12 and made many suggestions. Axel Wittmann provided a picture of Gauss, while Rudolf Kalman provided a picture of himself and further explained filter history. Finally, I greatly appreciate the support and tolerance of my wife Edwina during the long book-writing process.

Bruce P. Gibbs

CHAPTER 1

INTRODUCTION

Applications of estimation theory were limited primarily to astronomy, geodesy, and regression analysis up to the first four decades of the twentieth century. However, during World War II and in the following decades, there was an explosive growth in the number and types of estimation applications. At least four reasons were responsible for this growth. First, development of the new radar, sonar, and communication technology greatly expanded the interest in signal processing theory. Second, development of digital computers provided a means to implement complex math-based algorithms. Third, the start of space exploration and associated expansion in military technology provided a critical need for estimation and control, and also increased interest in state-space approaches. Finally, papers by Kalman (1960, 1961), Kalman and Bucy (1961), and others provided practical algorithms that were sufficiently general to handle a wide variety of problems, and that could be easily implemented on digital computers.

Today applications of least-squares estimation and Kalman filtering techniques can be found everywhere. Nearly every branch of science or engineering uses estimation theory for some purpose. Space and military applications are numerous, and implementations are even found in common consumer products such as Global Positioning System (GPS) receivers and automotive electronics. In fact, the GPS system could not function properly without the Kalman filter. Internet searches for "least squares" produce millions of links, and searches for "Kalman filter" produce nearly a million at the time of this writing. Kalman filters are found in applications as diverse as process control, surveying, earthquake prediction, communications, economic modeling, groundwater flow and contaminant transport modeling, transportation planning, and biomedical research. Least-squares estimation and Kalman filtering can also be used as the basis for other analysis, such as error budgeting and risk assessment. Finally, the Kalman filter can be used as a unit Jacobian transformation that enables maximum likelihood system parameter identification.

With all this interest in estimation, it is hard to believe that a truly new material could be written on the subject. This book presents the theory, but sometimes limits detailed derivations. It emphasizes the various methods used to support

Advanced Kalman Filtering, Least-Squares and Modeling: A Practical Handbook, by Bruce P. Gibbs

batch and recursive estimation, practical approaches for implementing designs that meet requirements, and methods for evaluating performance. It focuses on model development, since it is generally the most difficult part of estimator design. Much of this material has been previously published in various papers and books, but it has not all been collected in a form that is particularly helpful to engineers, scientists, or mathematicians responsible for implementing practical algorithms.

Before presenting details, we start with a general explanation of the estimation problem and a brief history of estimation theory.

1.1 THE FORWARD AND INVERSE MODELING PROBLEM

Modeling of physical systems is often referred to as either *forward modeling* or *inverse modeling*. In forward modeling a set of known parameters and external inputs are used to model (predict) the measured output of a system. A forward model is one that can be used for simulation purposes. In inverse modeling (a term used by Gauss) a set of measured values are used to infer (estimate) the model states that best approximate the measured behavior of the true system. Hence "inverse modeling" is a good description of the estimation process.

Figure 1.1 shows a generic forward model: a set of j constant parameters $\mathbf{p}$, a deterministic time-varying set of l input variables $\mathbf{u}(\tau)$ defined over the time interval $t_0 \le \tau \le t$, and an unknown set of k random process inputs $\mathbf{q}(\tau)$—also defined over the time interval $t_0 \le \tau \le t$—are operated on by a linear or nonlinear operator $\mathbf{f}_t(\mathbf{p},\mathbf{u},\mathbf{q},t)$ to compute the set of n states $\mathbf{x}(t)$ at each measurement time t. (Bold lower case letters are used to represent vectors, for example, $\mathbf{p} = [p_1 \;\; p_2 \;\; \cdots \;\; p_j]^T$. Bold upper case letters are used later to denote matrices. The subscript "t" on $\mathbf{f}$ denotes that it is a "truth" model.) The states included in vector $\mathbf{x}(t)$ are assumed to completely define the system at the given time. In control applications $\mathbf{u}(t)$ is often referred to as a *control* input, while in biological systems it is referred to as an *exogenous* input.

Noise-free measurements of the system output, $\mathbf{y}_t(t)$, are obtained from a linear or nonlinear transformation on the state $\mathbf{x}(t)$. Finally it is assumed that the actual

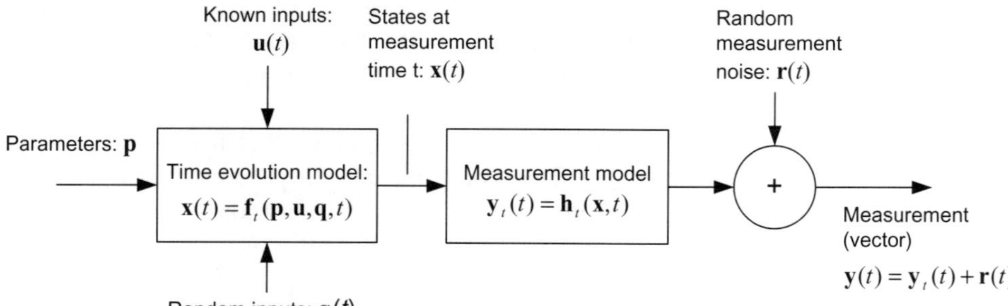

FIGURE 1.1: Generic forward model.

measurements are corrupted by additive random noise $\mathbf{r}(t)$, although measurement noise is often not considered part of the forward model.

A polynomial in time is a simple example of a forward model. For example, the linear position of an object might be modeled as $x(t) = p_1 + p_2 t + p_3 t^2$ where t represents the time difference from a fixed epoch. A sensor may record the position of the object as a function of time and random noise may corrupt the measurements, that is, $y(t) = x(t) + r(t)$. This is a one-dimensional example where neither process noise ($\mathbf{q}$) nor forcing inputs ($\mathbf{u}$) affect the measurements. The state will be multidimensional for most real-world problems.

The inputs $\mathbf{p}$, $\mathbf{u}$, $\mathbf{q}$, and $\mathbf{r}$ may not exist in all models. If the random inputs represented by $\mathbf{q}(t)$ are present, the model is called *stochastic*; otherwise it is *deterministic*. In some cases $\mathbf{q}(t)$ may not be considered part of the forward model since it is unknown to us. Although the model of Figure 1.1 is shown to be a function of time, some models are time-invariant or are a function of one, two, or three spatial dimensions. These special cases will be discussed in later chapters. It is generally assumed in this book that the problem is time-dependent.

Figure 1.2 graphically shows the inverse modeling problem for a deterministic model. We are given the time series (or possibly a spatially distributed set) of "noisy" measurements $\mathbf{y}(t)$, known system inputs $\mathbf{u}(t)$, and models (time evolution and measurement) of the system. These models, $\mathbf{f}_m(\mathbf{p},\mathbf{u},t)$ and $\mathbf{h}_m(\mathbf{x})$, are unlikely to exactly match the true system behavior (represented by $\mathbf{f}_t(\mathbf{p},\mathbf{u},t)$ and $\mathbf{h}_t(\mathbf{x})$), which are generally unknown to us. To perform the estimation, actual measurements $\mathbf{y}(t)$ are differenced with model-based predictions of the measurements $\mathbf{y}_m(t)$ to compute measurement residuals. The set of measurement residuals for the entire data span is processed by an optimization algorithm to compute a new set of parameter values that minimize some function of the measurement residuals. In least-squares estimation the "cost" or "loss" function to be minimized is the sum-of-squares, possibly weighted, of all residuals. Other optimization criteria will be discussed later. The new parameter values are passed to the time evolution and measurement models

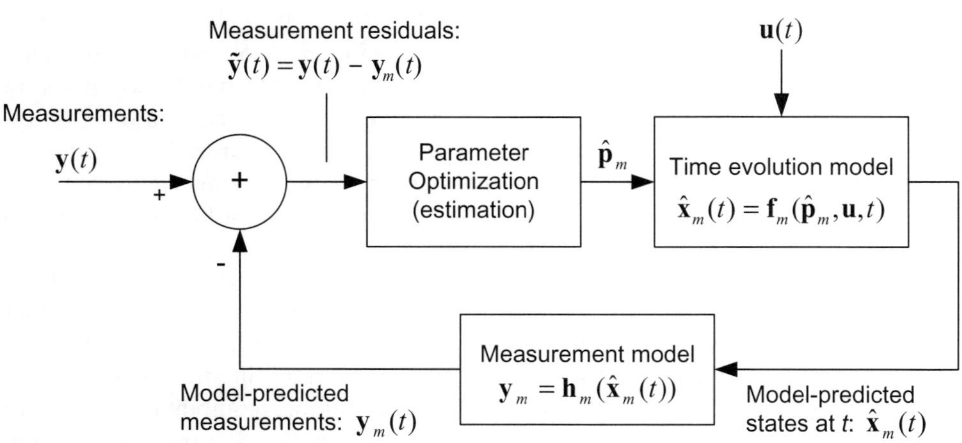

FIGURE 1.2: Deterministic inverse modeling.

to compute another set of model-based predicted measurements. A new set of measurement residuals is computed for the entire data span and a new cost function is computed. If the models are linear, only one iteration is normally required to converge on parameters that minimize the cost function. If the models are nonlinear, multiple iterations will be required to compute the optimum.

The optimization process may update the estimates of **p** after each new measurement is received. This is called recursive estimation and is particularly useful when process noise $\mathbf{q}(t)$ is present in the system. This topic is discussed in Chapter 8 (Kalman filtering) and later chapters.

This inverse modeling summary was intended as a high-level description of estimation. It intentionally avoided mathematical rigor so that readers unfamiliar with estimation theory could understand the concepts before being swamped with mathematics. Those readers with significant estimation experience should not be discouraged: the math will quickly follow.

1.2 A BRIEF HISTORY OF ESTIMATION

Estimation theory started with the least-squares method, and earliest applications modeled motion of the moon, planets, and comets. Work by Johannes Kepler (1619) established the geometric laws governing motion of heavenly bodies, and Sir Isaac Newton (1687) demonstrated that universal gravitation caused these bodies to move in conic sections. However, determination of orbits using astronomical observations required long spans of data and results were not as accurate as desired—particularly for comets. In the mid-1700s it was recognized that measurement errors and hypothetical assumptions about orbits were partly responsible for the problem. Carl Friedrich Gauss claims to have first used the least-squares technique in 1795, when he was only 18, but he did not initially consider it very important. Gauss achieved wide recognition in 1801 when his predicted return of the asteroid Ceres proved to be much more accurate than the predictions of others. Several astronomers urged him to publish the methods employed in these calculations, but Gauss felt that more development was needed. Furthermore, he had "other occupations." Although Gauss's notes on the Ceres calculations appear contradictory, he probably employed an early version of the least-squares method. Adrien-Marie Legendre independently invented the method—also for modeling planetary motion—and published the first description of the technique in a book printed in 1806. Gauss continued to refine the method, and in 1809 published a book (*Theoria Motus*) on orbit determination that included a detailed description of least squares. He mentioned Legendre's work, but also referred to his earlier work. A controversy between Gauss and Legendre ensued, but historians eventually found sufficient evidence to substantiate Gauss's claim as the first inventor.

Gauss's (1809) explanation of least squares showed great insight and may have been another reason that he is credited with the discovery. Gauss recognized that observation errors could significantly affect the solution, and he devised a method for determining the orbit to "satisfy all the observations in the most accurate manner possible." This was accomplished "by a suitable combination of more observations than the number absolutely requisite for the determination of the unknown quantities." He further recognized that "… the most probable system of

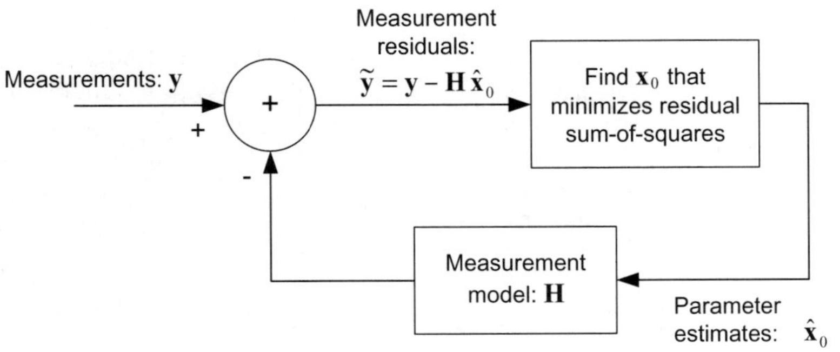

FIGURE 1.3: Least-squares solution to inverse modeling.

values of the quantities ... will be that in which the sum of the squares of the differences between the actually observed and computed values multiplied by numbers that measure the degree of precision is a minimum."

Gauss's work may have been influenced by the work of others, but he was the first to put all the pieces together to develop a practical method. He recognized the need for redundancy of observations to eliminate the influence of measurement errors, and also recognized that determining the most probable values implied minimizing observation residuals. He computed these measurement residuals $\tilde{y}_1, \tilde{y}_2, \ldots \tilde{y}_m$ using equations (models) of planetary motion and measurements that were based on an "approximate knowledge of the orbit." This approach allowed iterative solution of nonlinear problems. Gauss reasoned that measurement errors would be independent of each other so that the joint probability density function of the measurement residuals would be equal to the product of individual density functions. Further he claimed that the normal density function would be

$$p_Y(\tilde{y}_1, \tilde{y}_2, \ldots, \tilde{y}_m) = \frac{\prod_{i=1}^{m} w_i}{(2\pi)^{m/2}} \exp\left[-\frac{1}{2}\sum_{i=1}^{m} \tilde{y}_i^2 w_i^2\right],$$

where w_i are weighting factors that take into account measurement errors. Gauss noted that the maximum of the probability density function is determined by maximizing the logarithm of the density function. For the assumed conditions, this is equivalent to minimizing the measurement residual sum-of-squares. Figure 1.3 shows the general structure of the least-squares method devised by Gauss, where vector $\mathbf{y}$ includes all measurements accumulated over a fixed period of time (if time is an independent variable of the problem).

Improvements in the least-squares computational approach, statistical and probability foundations, and extensions to other applications were made by Pierre-Simon Laplace, Andrey Markov, and Friedrich Helmert after 1809. In the early 1900s Sir Ronald Fisher (1918, 1925) developed the maximum likelihood and analysis of variance methods for parameter estimation. However, no fundamental extensions of estimation theory were made until the 1940s. Until that time applications generally involved parameter estimation using deterministic models. Andrey Kolmogorov and Norbert Wiener were concerned with modeling unknown stochastic signals corrupted by additive measurement noise. The presence of random dynamic noise made the stochastic problem significantly different from least-squares

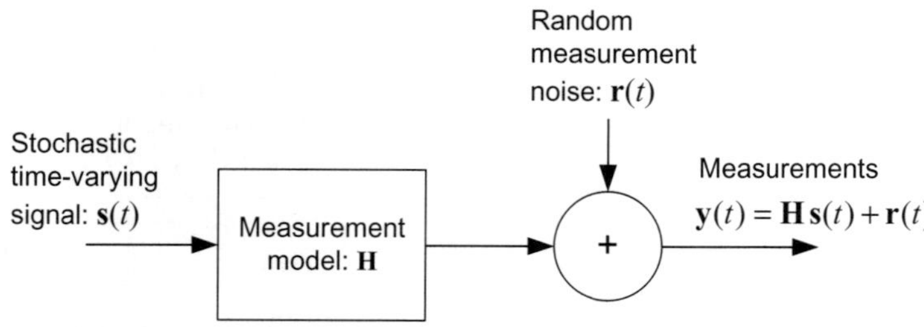

FIGURE 1.4: Simplified forward system model for Wiener filtering.

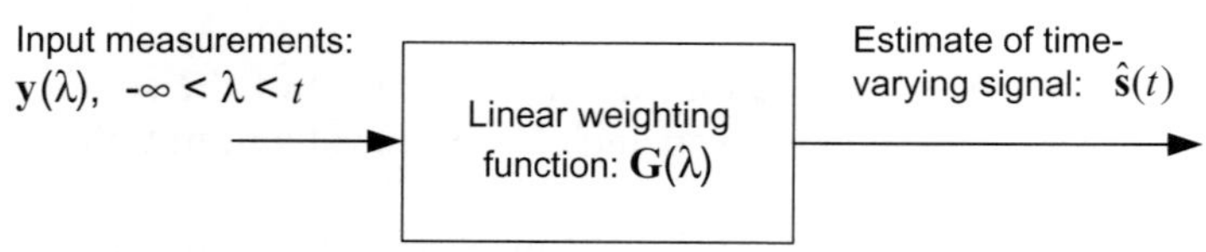

FIGURE 1.5: Wiener filter.

parameter estimation. Kolmogorov (1941) analyzed transition probabilities for a Markov process, and the discrete-time linear least-squares smoothing and prediction problem for stationary random processes. Wiener (1949) independently examined a similar prediction problem for continuous systems. Wiener's work was motivated by physical applications in a variety of fields, but the need to solve fire control and signal processing problems was a driving factor. This led him to study filtering and fixed-lag smoothing. Figure 1.4 shows a simplified version of the general problem addressed by Wiener filtering for multidimensional $\mathbf{y}(t)$ and $\mathbf{s}(t)$.

Message and measurement error processes are described by correlation functions or equivalently *power spectral densities* (PSD), and the minimum mean-squared error solution for the optimal weighting matrix $\mathbf{G}(t)$ is computed using the Wiener-Hopf integral equation in the time domain:

$$\mathbf{R}_{sy}(\tau) = \int_0^\infty \mathbf{G}(\lambda)\mathbf{R}_{ss}(\tau-\lambda)\,d\lambda \qquad 0<\tau<\infty \tag{1.2-1}$$

where $\mathbf{R}_{sy}(\tau) = E[\mathbf{s}(t)\mathbf{y}^T(t-\tau)]$, $\mathbf{R}_{ss}(\tau) = E[\mathbf{s}(t)\mathbf{s}^T(t-\tau)]$, $E[\cdot]$ denotes the expected value of random variables and it is assumed that either $\mathbf{s}(t)$, $\mathbf{y}(t)$, or both are zero mean. The correlation functions $\mathbf{R}_{sy}(\tau)$ and $\mathbf{R}_{ss}(\tau)$ are empirically obtained from sampled data, or computed analytically if the signal characteristics are known. The steady-state filter gain, $\mathbf{G}(t)$, is computed by factorization of the power spectral density function—a frequency domain approach. (Details of the method are presented later in Chapter 8.) The resulting filter can be implemented as either *infinite impulse response* (IIR)—where a recursive implementation gives the filter an "infinite" memory to all past inputs—or as *finite impulse response* (FIR) where the filter operates on a sliding window of data, and data outside the window have no effect on the output. Figure 1.5 shows the IIR implementation. Unfortunately the spectral factorization approach assumes an infinite data span, so the solution is not realizable. Wiener avoided this problem by assuming a finite delay in the filter, where the delay was chosen sufficiently long to make approximation errors small. Other extensions of Wiener's work were made by Bode and Shannon (1950), and Zadeh and Ragazzini (1950). They both assumed that the dynamic system could be modeled as a shaping filter excited by white noise, which was a powerful modeling concept.

While Kolmogorov's and Wiener's works were a significant extension of estimation technology, there were few practical applications of the methods. The assumptions of stationary random processes and steady-state solutions limited the usefulness of the technique. Various people attempted to extend the theory to nonstationary random processes using time-domain methods. Interest in state-space descriptions, rather than covariance, changed the focus of research, and led to recursive least-squares designs that were closely related to the present Kalman filter. Motivated by satellite orbit determination problems, Peter Swerling (1959) developed a discrete-time filter that was essentially a Kalman filter for the special case of noise-free dynamics; that is, it still addressed deterministic rather than stochastic problems. Events having the greatest impact on technology occurred in 1960 and 1961 when Rudolf Kalman published one paper on discrete filtering (1960), another on continuous filtering (1961), and a joint paper (Kalman and Bucy 1961) on continuous filtering. The papers used the state-space approach, computed the solution as minimum mean-squared error, discussed the duality between control and estimation, discussed observability and controllability issues, and presented the material in a form that was easy to understand and implement. The design allowed for nonstationary stochastic signals and resulted in a solution with time-varying gains. The Kalman filter was quickly recognized as a very important tool. Stanley Schmidt realized that Kalman's approach could be extended to nonlinear problems. This led to its use for midcourse navigation on the National Aeronautics and Space Administration (NASA) Apollo moon program in 1961 (Schmidt 1981; McGee and Schmidt 1985).

Figure 1.6 shows the generic model structure used as the basis of the Kalman filter design: $\mathbf{q}(t)$ and $\mathbf{r}(t)$ are white random noise and $\mathbf{\Phi}(\Delta t)$ is the state transition matrix for the time interval Δt. Notice that the dynamic and measurement models

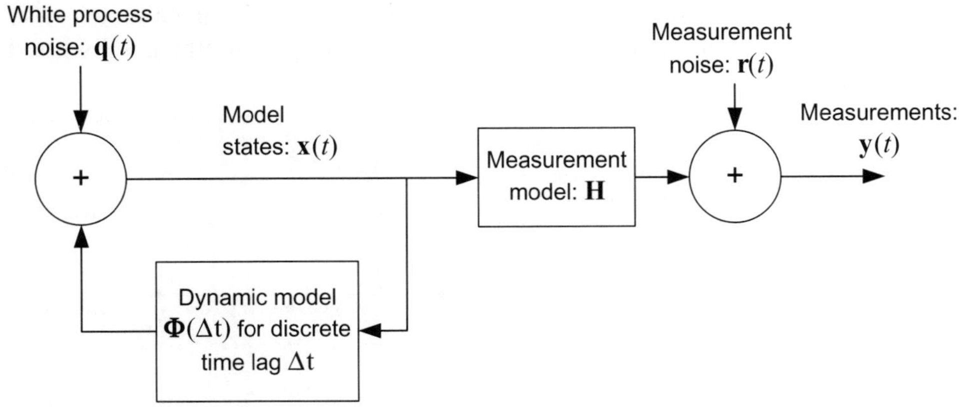

FIGURE 1.6: Forward model structure assumed by discrete Kalman filter.

are similar to those used in least-squares estimation, but they allow for the presence of random process noise, and the state vector **x** is defined using a moving epoch (t) rather than a fixed epoch (t_0).

More information on Gauss's work and subsequent estimation history can be found in Gauss (1809), Bühler (1981), Sorenson (1970), Meditch (1973), Anderson and Moore (1979), Åström (1980), Tapley et al. (2004), Simon (2006), Grewal and Andrews (2001), Kailath (1968), Bennett (1993), and Schmidt (1981).

1.3 FILTERING, SMOOTHING, AND PREDICTION

The estimation problem is further classified as either filtering, smoothing, or prediction. Figure 1.7 graphically shows the differences. In the filtering problem the goal is to continuously provide the "best" estimate of the system state at the time of the last measurement, shown as t_2. When a new measurement becomes available, the filter processes the measurement and provides an improved estimate of the state $\hat{\mathbf{x}}(t)$ at the new measurement time. In many systems—such as target tracking for collision avoidance or fire control—the goal is to provide an estimate of the state at some future time t_3. This is the prediction problem, and provided that the linear filter state estimate at t_2 is optimal, the predicted state is obtained by simply integrating the state vector at t_2:

$$\hat{\mathbf{x}}(t_3) = \hat{\mathbf{x}}(t_2) + \int_{t_2}^{t_3} \dot{\hat{\mathbf{x}}}(\lambda)d\lambda,$$

that is, the optimal linear predictor is the integral of the optimal linear filter. If the dynamic model is deterministic, the same approach can be applied to obtain the optimal state at any time prior to t_2. That is, integrate $\hat{\mathbf{x}}(t_2)$ backward in time from t_2 to t_1. This assumption is implicit when using batch least-squares estimation: the *epoch time* at which the state is defined is arbitrary as the state at any other time may be obtained by analytic or numerical integration. However, when the dynamic model is subject to random perturbations—the stochastic case handled by Wiener or Kalman filtering—the optimal smoothed estimate at times in the past cannot be obtained by simple integration. As you may guess, the optimal smoothed estimate is obtained by weighting measurements near the desired smoothed time t_1 more than those far from t_1. Smoothed estimates must be computed using a time history of filter estimates, or information that is equivalent. Optimal smoothing is discussed in Chapter 9.

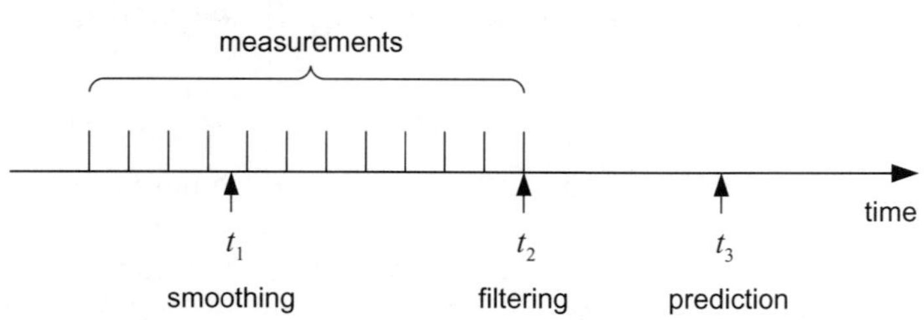

FIGURE 1.7: Filtering, smoothing, and prediction.

1.4 PREREQUISITES

Many early books on Kalman filtering included introductory chapters on linear and matrix algebra, state-space methods, probability, and random (stochastic) processes. This was desirable since few engineers at that time had good background in these areas. That situation has changed somewhat. We assume that the reader has been exposed to these topics, but may need a refresher course. Hence, the first appendix summarizes matrix algebra and the second summarizes probability and random process theory. Numerous references provide more background for those who want it. Particularly recommended are Papoulis and Pillai (2002), Kay (1988), Marple (1987), Cooper and McGillem (1967), Jazwinski (1970), Åström (1970), and Parzen (1960) for information on probability, stochastic processes, and spectral properties; DeRusso et al. (1965) for state-space methods and matrix operations; Lawson and Hanson (1974), Björck (1996), and Golub and Van Loan (1996) for least-squares estimation and matrix theory; and Kay (1988), Marple (1987), Gelb (1974), and Simon (2006) for material on several of these topics.

It is more important that the reader be familiar with matrix algebra than probability theory. Sufficient background is provided within the text to understand discussions involving stochastic processes, but it will be difficult to follow the material if the reader is not familiar with matrix algebra. Although somewhat less important, we also assume that the reader is familiar with Fourier, Laplace, and Z transforms. These topics are covered in many texts for various fields. Of the books previously mentioned, Cooper and McGillem (1967) and DeRusso et al. (1965) provide good background on transforms.

1.5 NOTATION

Unfortunately the notation used in estimation literature is not entirely consistent. This is partly due to the limited number of letters in the combined English and Greek alphabets, and the application of estimation to fields that each have their own notation. Even Kalman and Bucy used slightly different notations in different publications.

We tend to use Kalman's and Bucy's notation, but with modifications. Vectors are represented as bold lower case letters, and matrices are bold upper case letters. Dots above variables denote time derivative ($\dot{x} = dx/dt$, $\ddot{x} = d^2x/dt^2$), an overbar denotes the unconditional mean value of a random variable ($\bar{x} = E[x]$), a "hat" denotes the mean value conditioned on knowledge of other random variables ($\hat{x} = E[x \mid y]$), and a tilde denotes the error in a conditional mean ($\tilde{x} = x - \hat{x}$).

Variable **x** is generally used to denote the state vector. Unless otherwise stated, vectors are assumed to be column vectors. Functional dependence of one variable on another is expressed using parentheses $[x(t)]$, discrete-time samples of a variable are denoted using subscripts $[x_i = x(t_i)]$, and the state estimate at one time conditioned on measurement data (y) up to another time is denoted as $\hat{x}_{i/j} = E[x(t_i \mid y_j, y_{j-1}, \ldots y_1)]$. (Other authors sometimes use (–) to denote *a priori* estimates, $\hat{x}_i(-) = E[x(t_i \mid y_{i-1}, \ldots y_1)]$, and (+) to denote *a posteriori* estimates, $\hat{x}_i(+) = E[x(t_i \mid y_i, y_{i-1}, \ldots y_1)]$.)

Use of subscripts for time dependence may occasionally become confusing because subscripts are also used to denote elements of a vector or matrix. For example, A_{jk} usually denotes the (j,k) element of matrix $\mathbf{A}$. If $\mathbf{A}$ is also a function of time t_i, one might use $(\mathbf{A}_i)_{jk}$. Another alternative is to list the subscripts in parentheses, but this could also be ambiguous in that it implies functional dependence. We only show subscripts in parentheses when listing pseudo-code for algorithms. We also use the MATLAB/Fortran 90 "colon" notation to denote a range of a vector or array, that is, $\mathbf{A}(1:n,i)$ denotes elements 1 to n of column i of matrix $\mathbf{A}$, and $\mathbf{A}(:,i)$ denotes the entire column i. If rows or columns of $\mathbf{A}$ are treated as vectors in algorithms, $\mathbf{a}_{:i}$ denotes column i and $\mathbf{a}_{i:}$ denotes row i.

Even more confusing, we occasionally use subscripts separated by an underscore to denote a condition under which the variable is defined. The text should clarify the notation in these cases. Superscripts generally denote exponents of a variable, but there are a few exceptions. For example, s^* is the complex conjugate of complex variable s. Exponents on matrices have the following meanings for matrix $\mathbf{A}$: $\mathbf{A}^T$ is the transpose, $\mathbf{A}^H$ is the Hermitian (complex conjugate transpose), $\mathbf{A}^{-1}$ is the inverse and $\mathbf{A}^{\#}$ is the pseudo-inverse. These terms are defined in Appendix A.

Set notation is often used to define variables such as vectors and matrices. For example, $\mathbf{A} \in R^{n \times m}$ is sometimes used to denote that $\mathbf{A}$ is a real-valued matrix having n rows and m columns, or $\mathbf{x} \in R^n$ is a real vector with n elements. Another example is $x(t)$, $t \in [0,\mathrm{T})$ to indicate that variable x is defined over the time interval $0 \le t < T$. Set notation is cumbersome for the purpose of this book, and

TABLE 1.1: Comparison of Least-Squares Estimation and Kalman Filtering

Attribute	Least Squares (LS)	Kalman Filter (KF)
Batch versus recursive solution	Batch LS on finite data span (FIR). Recursive LS (IIR)	Recursive (IIR) on unlimited data span. Fixed interval smoother can provide "batch" solution on finite data span.
Real-time processing	Batch LS requires periodic execution. Recursive LS will eventually ignore new data unless re-initialized. Can also de-weight older data.	Ideal for real-time processing
Dynamic model	Deterministic	Stochastic: may be time-varying and nonstationary
Solution for nonlinear models	Iterative Gauss-Newton solution uses linear expansion at each iteration	Extended KF linearizes about current estimate at each measurement. Other approaches are available.
Inclusion of prior information	Yes: Bayesian solution No: Maximum likelihood or weighted least-squares solution	Standard KF requires prior estimate. Information KF can operate without prior.

potentially confusing. Hence we describe **A** as an $n \times m$ matrix, and **x** as an n-element vector.

Also, many probability texts use upper case variables to denote the domain of a random variable with lower case variables used to denote realizations of the variable, for example, $\hat{x} = E[X \mid Y = y]$. While mathematically more precise than $\hat{x} = E[x \mid y]$, it is also cumbersome and usage is limited.

1.6 SUMMARY

We have shown how estimation can be viewed as an inverse modeling problem that accounts for the effects of measurement noise. Gauss and Legendre developed the least-squares method for deterministic models while Wiener and Kolmorgov developed a steady-state filtering approach for stationary stochastic models. Kalman, Bucy, and others extended the method to time-varying nonstationary stochastic systems using a state-space model. Table 1.1 compares various attributes of least-squares estimation and Kalman filtering. This table is partially a summary of Chapter 1, and partly an introduction to later chapters. Some terms have not yet been defined, but you can probably guess the meaning—details will be provided in later chapters. You may want to look at this table again after reading subsequent chapters.

The most important distinction between the least-squares and Kalman filtering methods is the allowance for process noise in the Kalman filter dynamic model; that is, the Kalman filter can handle stochastic problems while the least-squares method cannot. Other differences, such as batch versus recursive implementation, tend to be less important because either method can be made to perform similarly to the other.

CHAPTER 2

SYSTEM DYNAMICS AND MODELS

The performance of estimation algorithms is highly dependent on the accuracy of models. Model development is generally the most difficult task in designing an estimator. The equations for least squares and Kalman filtering are well documented, but proper selection of system models and parameters requires a good understanding of the system and of various trade-offs in selecting model structure and states. We start by discussing dynamic and measurement models. This provides the reader with a better understanding of the problem when estimation algorithms are presented in later chapters.

Chapter 1 was deliberately vague about the structure of models because the intent was to present concepts. The dynamic (time evolution) model of the true system used in Figure 1.1 was a generic form. The system state vector $\mathbf{x}(t)$ was an n-element linear or nonlinear function of various constant parameters $\mathbf{p}$, known time-varying inputs $\mathbf{u}(\tau)$ defined over the time interval $t_0 \le \tau \le t$, random process noise $\mathbf{q}(\tau)$ also defined over the time interval $t_0 \le \tau \le t$, and possibly time, t:

$$\mathbf{x}(t) = \mathbf{f}_t[\mathbf{p}, \mathbf{u}(\tau), \mathbf{q}(\tau), t] \quad t_0 \le \tau \le t. \tag{2.0-1}$$

Function $\mathbf{f}_t$ was defined as a time integral of the independent variables. The m-element measurement vector $\mathbf{y}(t)$ was assumed to be a function of $\mathbf{x}(t)$ with additive random noise:

$$\mathbf{y}(t) = \mathbf{h}(\mathbf{x}) + \mathbf{r}(t) \tag{2.0-2}$$

However, these models are too general to be useful for estimation. We now consider the options. Models are first divided into two categories: *continuous* and *discrete*. Nearly all practical implementations of the Kalman filter use the discrete filter form. The continuous filter developed by Kalman and Bucy is useful for analytic purposes, but it is rarely implemented because most physical systems are discretely sampled. Nonetheless, continuous models are important because the majority of discrete least-squares and Kalman filter applications are based on integration of a continuous system model. Continuous time models generally take one of the following forms:

Advanced Kalman Filtering, Least-Squares and Modeling: A Practical Handbook, by Bruce P. Gibbs

1. An expansion in time using basis functions such as polynomials, Fourier series, or wavelets.
2. A derivation from first-principle (e.g., physical, thermodynamic, chemical, biological) concepts. This may lead to distributed models based on partial differential equations, or lumped-parameter (linear or nonlinear) models based on ordinary differential equations.
3. A stochastic random process model (e.g., random walk, Markov process) to model effects that are poorly understood or behave randomly.
4. Combinations of the above.
5. Linear regression models.

Kalman (1960) and Kalman and Bucy (1961) assumed that discrete dynamic system models could be obtained as the time integral of a continuous system model, represented by a set of first-order ordinary differential equations. If continuous system models are obtained as high-order ordinary differential equations, they can be converted to a set of first-order ordinary differential equations. Hence Kalman's assumption is not restrictive.

The first category of continuous models—basis function expansion—multiplies the basis functions by coefficients that are included in the state vector of estimated parameters. The model itself is not a function of other parameters and hence is *nonparametric*. The second and third categories (except for random walk) depend on internal parameters and are thus *parametric*. The model parameters must be determined before the model can be used for estimation purposes.

Discrete models are based on the assumption that the sampling interval is fixed and constant. Discrete process models are parametric, and may take the following forms:

1. Autoregressive (AR)
2. Moving average (MA)
3. Autoregressive moving average (ARMA)
4. Autoregressive moving average with exogenous inputs (ARMAX)
5. Autoregressive integrated moving average (ARIMA)

Each of these model types is discussed below and in Appendix D (available online at ftp://ftp.wiley.com/public/sci_tech_med/least_squares/). We start the discussion with discrete dynamic models since they are somewhat simpler and the ultimate goal is to develop a model that can be used to process discretely sampled data.

2.1 DISCRETE-TIME MODELS

Discrete models can be viewed as a special case of continuous system models where measurements are sampled at discrete times, and system inputs are held constant over the sampling intervals. Discrete models assume that the sampling interval T is constant with no missing measurements, and that the system is a stationary random process such that the means and autocorrelation functions do not change with time shifts (see Appendix B). These conditions are often found in process control and biomedical applications (at least for limited time spans), and thus ARMAX models are used when it is difficult to derive models from first principles. Empirical computation of these models from sampled data is discussed in Chapter 12.

The general form of a single-input, single-output, n–th order ARMAX process model is

$$y_i = -\sum_{j=1}^{n} \alpha_j y_{i-j} + \sum_{j=0}^{n} \beta_j u_{i-j} + \sum_{j=1}^{n} \gamma_j q_{i-j} + q_i \tag{2.1-1}$$

where

y_i is the sampled measurement at time t_i,

u_i is the exogenous input at time t_i,

q_i is zero-mean random white (uncorrelated) noise input at time t_i, and $t_i - t_{i-1} = T$ for all i.

The summations for y_{i-j}, u_{i-j}, and q_{i-j} in equation (2.1-1) may be separately truncated at lower order than n. Alternate forms of the general model are defined as follows.

1. If all $\alpha_j = 0$ and $\beta_j = 0$, the model is MA because the output measurement is a weighted average of past noise inputs q_{i-j}.
2. If all $\gamma_j = 0$ and $\beta_j = 0$, the model is AR because the output measurement is a weighted average of past measurements and the current noise input q_i.
3. If all $\beta_j = 0$, the model is ARMA because exogenous inputs are not included.

The z-transform transfer function of an ARMA model is

$$\frac{Y(z)}{Q(z)} = \frac{1 + \gamma_1 z^{-1} + \gamma_2 z^{-2} + \ldots + \gamma_l z^{-l}}{1 + \alpha_1 z^{-1} + \alpha_2 z^{-2} + \ldots + \alpha_n z^{-n}} \tag{2.1-2}$$

where z^{-1} is the unit delay operator,

$$\begin{aligned} z^{-1} &= e^{-j2\pi f/T} \\ &= \cos(j2\pi f/T) - j\sin(j2\pi f/T)' \end{aligned}$$

$j = \sqrt{-1}$ and f is frequency in Hz. The transfer function can be factored to obtain

$$\frac{Y(z)}{Q(z)} = \frac{\kappa(z - r_1)(z - r_2)\ldots(z - r_l)\, z^{n-l}}{(z - p_1)(z - p_2)\ldots(z - p_n)} \tag{2.1-3}$$

where

$$\kappa = \left(\frac{\prod_{i=1}^{n}(1 - p_i)}{1 + \sum_{i=1}^{n} \alpha_i} \right) \left(\frac{1 + \sum_{i=1}^{l} \gamma_i}{\prod_{i=1}^{l}(1 - r_i)} \right). \tag{2.1-4}$$

The zeroes, r_i, and poles, p_i, may be real or complex. Complex poles or zeroes must occur in complex conjugate pairs for a real system. Hence equation (2.1-3) can be written in terms of first-order factors for real poles and zeroes, and second-order factors for complex conjugate poles or zeroes. For example, the transfer function for complex conjugate pole (p, p^*) and zero (r, r^*) pairs can be written as

$$\frac{(z-r)(z-r^*)}{(z-p)(z-p^*)} = \frac{z^2 - 2\,\mathrm{Re}(r)z + |r|^2}{z^2 - 2\,\mathrm{Re}(p)z + |p|^2} \tag{2.1-5}$$

where Re(·) and Im(·) are the real and imaginary parts of the complex number, and $|\cdot|^2 = \mathrm{Re}(\cdot)^2 + \mathrm{Im}(\cdot)^2$. If the roots of the denominator (poles) all lie within the unit circle in the complex plane, the model is *stable* (output remains bounded) and *causal* (output only depends on past inputs, not future). If all poles and zeroes (roots of the numerator) are inside the unit circle it is called a *minimum phase* or *invertible* system, because the input noise q_i can be reconstructed from knowledge of the past outputs y_j for $j \le i$.

Since q_i is assumed to be white (uncorrelated) noise, the *power spectral density* (PSD) of q_i is the same at all frequencies:

$$S_q(f) = S_{q0}, \quad |f| < \frac{1}{2T}.$$

The PSD versus frequency of the real output y_i can be computed as

$$S_y(f) = |Y(z)|^2_{z=\exp(j2\pi f/T)} = Y(z)Y(z^{-1})\big|_{z=\exp(j2\pi f/T)} \tag{2.1-6}$$

where $Y(z)$ is the z-transform of y_i. Since an AR process does not have zeroes, it is called an all-pole model and is best applied to systems that have spectral peaks. An MA process is an all-zero model best applied to systems with spectral nulls. ARMA models can handle both spectral nulls and peaks.

Multi-input, multi-output ARMAX models are usually represented using state-space models of the form

$$\begin{aligned} \mathbf{x}_i &= \mathbf{\Phi}\mathbf{x}_{i-1} + \mathbf{G}\mathbf{u}_{i-1} + \mathbf{B}\mathbf{q}_{i-1} \\ \mathbf{y}_i &= \mathbf{H}\mathbf{x}_i + \mathbf{r}_i \end{aligned} \tag{2.1-7}$$

or

$$\begin{aligned} \mathbf{y}_i &= \mathbf{H}\mathbf{x}_i + \mathbf{A}\mathbf{u}_i + \mathbf{B}\mathbf{q}_i + \mathbf{r}_i \\ \mathbf{x}_{i+1} &= \mathbf{\Phi}\mathbf{x}_i + \mathbf{G}\mathbf{u}_i + \mathbf{q}_i \end{aligned} \tag{2.1-8}$$

where $\mathbf{x}_i$ is an n-element state vector, $\mathbf{q}_i$ is a p-element process noise vector, $\mathbf{u}_i$ is an l-element input vector, $\mathbf{y}_i$ is an m-element measurement vector, and $\mathbf{r}_i$ is an m-element measurement noise vector. The various arrays are defined accordingly.

Measured outputs of many systems have nonzero means, trends, or other types of systematic long-term behavior. When successive differences of measurements result in a stationary process, the system may be treated as an ARIMA process. Box et al. (2008, chapter 4) discuss different approaches for handling nonstationary processes.

Appendix D explains AR, MA, ARMA, ARMAX, and ARIMA models in greater detail and includes examples. It also shows how discrete models can be derived from continuous, linear state space models that are discretely sampled.

However, it is usually not practical to compute the model parameters α_i, β_i, and λ_i directly from a continuous model. Chapter 12 discusses methods for determining the parameters empirically from measured data.

To summarize, discrete models can be used when the sampling interval and types of measurements are fixed, and the system is statistically stationary and linear. With extensions they can also be used for certain types of nonstationary or nonlinear systems. Discrete models are often used in applications when it is difficult to develop a model from first principles. For example, they are sometimes used in process control (Åström 1980; Levine 1996) and biomedical (Lu et al. 2001; Guler et al. 2002; Bronzino 2006) applications.

2.2 CONTINUOUS-TIME DYNAMIC MODELS

Continuous time dynamic models are usually defined by a set of linear or nonlinear first-order differential state equations. Differential equations are linear if coefficients multiplying derivatives of the dependent variable x are not functions of x. The assumption of first-order differential equations is not restrictive since higher order differential equations can be written as a set of first-order equations. For example, the stochastic linear second-order system

$$\ddot{x}_1(t) = -k_2\dot{x}_1(t) - k_1 x_1(t) + q(t)$$

can be written as

$$\begin{aligned} \dot{x}_1(t) &= x_2(t) \\ \dot{x}_2(t) &= -k_2 x_2(t) - k_1 x_1(t) + q(t) \end{aligned}.$$

A block diagram of the model is shown in Figure 2.1.

In the general case the state differential equations may be nonlinear:

$$\dot{\mathbf{x}}(t) = \mathbf{f}(\mathbf{x}, \mathbf{u}, \mathbf{q}_c, t), \tag{2.2-1}$$

where $\mathbf{x}$, $\mathbf{u}$, and $\mathbf{q}_c$ are all a function of time. As before, $\mathbf{x}$ and $\mathbf{f}$ are n-element vectors, $\mathbf{u}$ is an l-element vector of control inputs, and $\mathbf{q}_c$ is a p-element *white noise* vector. Notice that function $\mathbf{f}$ in equation (2.2-1) is assumed to be a function of the state $\mathbf{x}$ rather than the parameter vector $\mathbf{p}$ used in equation (2.0-1). We have implicitly assumed that all *unknown* parameters in $\mathbf{p}$ are included in the system state vector $\mathbf{x}$. Other parameters in $\mathbf{p}$ that are known with a high degree of accuracy are treated as part of the model and are thus ignored in our generic model equation.

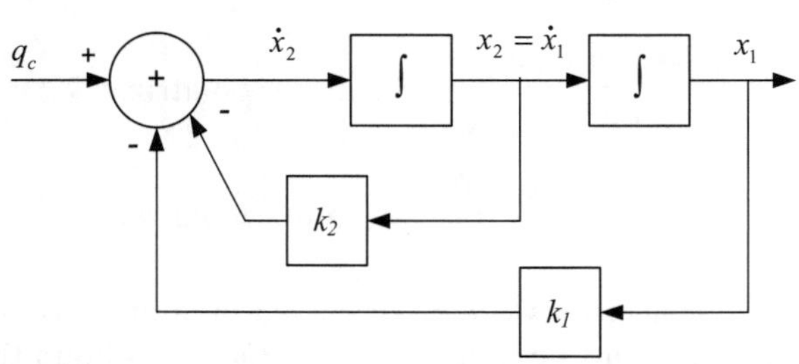

FIGURE 2.1: Stochastic second-order dynamic model.

It should be noted that white noise has infinite power, so inclusion of white noise input in a differential equation is not physically realistic. Furthermore, integration of the differential equation is not defined in the usual sense, even when $\mathbf{f}(\mathbf{x},\mathbf{u},\mathbf{q}_c,t)$ is linear. The problem is briefly described in Appendix B. Jazwinski (1970), Schweppe (1973), Åström (1970), and Levine (1996, chapters 34, 60) provide more extended discussions of calculus for stochastic processes. For estimation purposes we are usually interested in expectations of stochastic integrals, and for most continuous functions of interest, these integrals can be treated as ordinary integrals. This assumption is used throughout the book.

In many cases the function $\mathbf{f}$ is time-invariant, but we will not yet apply this restriction. Since process noise $\mathbf{q}_c(t)$ is assumed to be zero mean and a small perturbation to the model, it is customary to assume that *superposition* applies. Thus the effect can be modeled as an additive linear term

$$\boxed{\dot{\mathbf{x}}(t) = \mathbf{f}(\mathbf{x},\mathbf{u},\mathbf{0},t) + \mathbf{G}(\mathbf{x},\mathbf{u},t)\,\mathbf{q}_c(t)} \tag{2.2-2}$$

where the $n \times p$ matrix

$$\mathbf{G}(\mathbf{x},\mathbf{u},t) = \left.\frac{\partial \mathbf{f}(\mathbf{x},\mathbf{u},\mathbf{q}_c,t)}{\partial \mathbf{q}_c}\right|_{\mathbf{q}=\mathbf{0}},$$

is possibly a nonlinear function of $\mathbf{x}$. For nonlinear models, most least-squares or Kalman estimation techniques numerically integrate $\dot{\mathbf{x}}(t) = \mathbf{f}(\mathbf{x},\mathbf{u},t)$ (without $\mathbf{q}_c$ since it is zero mean) to obtain the state vector $\mathbf{x}$ at the measurement times. However, least-squares and Kalman estimation techniques also require the sensitivity of $\mathbf{x}(t_i)$, where t_i is a measurement time, with respect to $\mathbf{x}(t_e)$, where t_e is the epoch time. This sensitivity is required when computing the optimal weighting of the data, and it is normally computed as a linearization of the nonlinear equations. That is, equation (2.2-1) is expanded in a Taylor series about a reference trajectory and only the linear terms are retained:

$$\boxed{\delta\dot{\mathbf{x}}(t) = \mathbf{F}(\mathbf{x}_{ref},\mathbf{u}_{ref},t)\,\delta\mathbf{x}(t) + \mathbf{B}(\mathbf{x}_{ref},\mathbf{u}_{ref},t)\,\delta\mathbf{u}(t) + \mathbf{G}(\mathbf{x}_{ref},\mathbf{u}_{ref},t)\,\mathbf{q}_c(t)} \tag{2.2-3}$$

where

$\delta\mathbf{x} = \mathbf{x}(t) - \mathbf{x}_{ref}(t)$,

$\delta u = \mathbf{u}(t) - \mathbf{u}_{ref}(t)$

$\mathbf{x}_{ref}(t)$ and $\mathbf{u}_{ref}(t)$ are the reference trajectory,

$$\mathbf{F}(\mathbf{x}_{ref},\mathbf{u}_{ref},t) = \left.\frac{\partial \mathbf{f}(\mathbf{x},\mathbf{u}_{ref},\mathbf{0},t)}{\partial \mathbf{x}}\right|_{\mathbf{x}=\mathbf{x}\_ref} \quad \text{is an } n \times n \text{ matrix,}$$

$$\mathbf{B}(\mathbf{x}_{ref},\mathbf{u}_{ref},t) = \left.\frac{\partial \mathbf{f}(\mathbf{x}_{ref},\mathbf{u},\mathbf{0},t)}{\partial \mathbf{u}}\right|_{\mathbf{u}=\mathbf{u}\_ref} \quad \text{is an } n \times l \text{ matrix,}$$

$$\mathbf{G}(\mathbf{x}_{ref},\mathbf{u}_{ref},t) = \left.\frac{\partial \mathbf{f}(\mathbf{x}_{ref},\mathbf{u}_{ref},\mathbf{q}_c,t)}{\partial \mathbf{q}_c}\right|_{\mathbf{q}=\mathbf{0}} \quad \text{is an } n \times p \text{ matrix.}$$

Sources for reference trajectories will be discussed in later chapters. The perturbation equation (2.2-3), without the $\mathbf{q}_c$ term, is integrated to obtain the desired sensitivities. If the model is entirely *linear*, the model becomes:

$$\boxed{\dot{\mathbf{x}}(t) = \mathbf{F}(t)\,\mathbf{x}(t) + \mathbf{B}(t)\,\mathbf{u}(t) + \mathbf{G}(t)\,\mathbf{q}_c(t).} \tag{2.2-4}$$

This linear form is assumed in the next section, but it is recognized that the methods can be applied to equation (2.2-3) with suitable redefinition of terms.

2.2.1 State Transition and Process Noise Covariance Matrices

To obtain a state model at discrete measurement times t_i, $i = 1, 2, \ldots$, the continuous dynamic models of equation (2.2-2), (2.2-3), or (2.2-4) must be time-integrated over intervals t_i to t_{i+1}. For a deterministic linear system, the solution can be represented as the sum of the response to an initial condition on $\mathbf{x}(t_i)$, called the *homogeneous solution*, and the response due to the driving (forcing) terms $\mathbf{u}(t)$ and $\mathbf{q}(t)$ for $t_i < t \le t_{i+1}$, called the *particular* or *forced solution*, that is,

$$\mathbf{x}(t_{i+1}) = \mathbf{x}_H(t_{i+1}) + \mathbf{x}_P(t_{i+1}). \tag{2.2-5}$$

The initial condition solution is defined as the response of the homogeneous differential equation $\dot{\mathbf{x}}(t) = \mathbf{f}(\mathbf{x}, \mathbf{0}, t)$ or $\dot{\mathbf{x}}(t) = \mathbf{F}(t)\mathbf{x}(t)$:

$$\mathbf{x}_H(t_{i+1}) = \mathbf{x}(t_i) + \int_{t_i}^{t_{i+1}} \mathbf{f}(\mathbf{x}(\lambda), \mathbf{0}, \lambda)\, d\lambda. \tag{2.2-6}$$

In many applications equation (2.2-6) is numerically integrated using truncated Taylor series, Runge-Kutta, or other methods. Numerical integration can also be used to compute effects of the control $\mathbf{u}(t)$ if it is included in $\mathbf{f}(\mathbf{x}(\lambda),\mathbf{u}(\lambda),\lambda)$ of equation (2.2-6). However, a more analytic method is needed to compute the sensitivity matrix (either linear sensitivity or partial derivatives) of $\mathbf{x}(t_{i+1})$ with respect to $\mathbf{x}(t_i)$. It will be seen later that this sensitivity matrix is required to calculate the optimal weighting of measurement data for either least-squares or Kalman estimation. It is also necessary to characterize—in a covariance sense—the effect of the random process noise $\mathbf{q}_c(t)$ over the time interval t_i to t_{i+1} when computing optimal weighting in a Kalman filter.

The state sensitivity is provided by the *state transition matrix* $\boldsymbol{\Phi}(t_{i+1},t_i)$, implicitly defined from

$$\boxed{\mathbf{x}_H(t_{i+1}) = \boldsymbol{\Phi}(t_{i+1}, t_i)\,\mathbf{x}(t_i)} \tag{2.2-7}$$

for a linear system. A similar equation applies for linearized (first-order) state perturbations in a nonlinear system,

$$\delta\mathbf{x}_H(t_{i+1}) = \boldsymbol{\Phi}(t_{i+1}, t_i)\,\delta\mathbf{x}(t_i),$$

but $\boldsymbol{\Phi}$ is then a function of the actual state $\mathbf{x}(t)$ over the time interval t_i to t_{i+1}. This nonlinear case will be addressed later. For the moment we assume that the linear equation (2.2-7) applies. From equation (2.2-7), it can be seen that

$$\boxed{\begin{aligned} &\boldsymbol{\Phi}(t_i, t_i) = \mathbf{I} \\ &\boldsymbol{\Phi}(t_i, t_k) = \boldsymbol{\Phi}(t_i, t_j)\boldsymbol{\Phi}(t_j, t_k). \\ &\boldsymbol{\Phi}(t_i, t_j) = \boldsymbol{\Phi}^{-1}(t_j, t_i) \end{aligned}} \tag{2.2-8}$$

When $\mathbf{F}(t)$ is *time-invariant* ($\mathbf{F}(t) = \mathbf{F}$), the solution is obtained as for the scalar problem $\dot{x}(t) = f$, which has the solution $x(t_{i+1}) = e^{fT}\,x(t_i)$ with $T = t_{i+1} - t_i$. Hence

$$\boxed{\boldsymbol{\Phi}(t_{i+1},t_i)=\boldsymbol{\Phi}(t_{i+1}-t_i)=e^{\mathbf{F}T}.} \tag{2.2-9}$$

The exponential can be represented as an infinite Taylor series

$$e^{\mathbf{F}\Delta t}=\mathbf{I}+\mathbf{F}T+(\mathbf{F}T)^2/2+(\mathbf{F}T)^3/6+\ldots \tag{2.2-10}$$

which is sometimes useful when evaluating $\boldsymbol{\Phi}(T)$. Other methods will be presented in Section 2.3.

In the general linear case where $\mathbf{F}(t)$ is *time-varying*, the solution is more complicated. If $\mathbf{F}(t)$ satisfies the commutativity condition

$$\mathbf{F}(t_1)\mathbf{F}(t_2)=\mathbf{F}(t_2)\mathbf{F}(t_1), \tag{2.2-11}$$

then it can be shown (DeRusso et al. 1965, p. 363) that

$$\boldsymbol{\Phi}(t_{i+1},t_i)=\exp\int_{t_i}^{t_{i+1}}\mathbf{F}(\lambda)\,d\lambda. \tag{2.2-12}$$

Unfortunately equation (2.2-11) is rarely satisfied when systems are time-varying (including cases where the system is nonlinear and $\boldsymbol{\Phi}$ has been computed as a linearization about the reference trajectory). Analytic techniques have been developed to compute $\boldsymbol{\Phi}$ when commutativity does not exist (DeRusso et al. 1965, pp. 362–366), but the methods are difficult to implement for realistic problems. More generally, the relationship

$$\boxed{\frac{d}{dt}\boldsymbol{\Phi}(t,\tau)=\mathbf{F}(t)\,\boldsymbol{\Phi}(t,\tau)} \tag{2.2-13}$$

is true for both time-varying and time-invariant cases. This can be derived by differentiating equation (2.2-7) with respect to time, substituting the homogeneous part of equation (2.2-4) for $\dot{\mathbf{x}}_H(t_{i+1})$, and substituting equation (2.2-7) for $\mathbf{x}_H(t_{i+1})$. Thus equation (2.2-13) may be numerically integrated to obtain $\boldsymbol{\Phi}(t_{i+1},t_i)$. More will be said about this in Section 2.3.

Assuming that $\boldsymbol{\Phi}(t_{i+1},t_i)$ can be computed, the total solution $\mathbf{x}(t_{i+1})$ for the model equation (2.2-4) is

$$\boxed{\mathbf{x}(t_{i+1})=\boldsymbol{\Phi}(t_{i+1},t_i)\,\mathbf{x}(t_i)+\mathbf{u}_D(t_{i+1},t_i)+\mathbf{q}_D(t_{i+1},t_i)} \tag{2.2-14}$$

where

$$\boxed{\mathbf{u}_D(t_{i+1},t_i)=\int_{t_i}^{t_{i+1}}\boldsymbol{\Phi}(t_{i+1},\lambda)\,\mathbf{B}(\lambda)\,\mathbf{u}(\lambda)\,d\lambda} \tag{2.2-15}$$

and

$$\boxed{\mathbf{q}_D(t_{i+1},t_i)=\int_{t_i}^{t_{i+1}}\boldsymbol{\Phi}(t_{i+1},\lambda)\,\mathbf{G}(\lambda)\mathbf{q}_c(\lambda)\,d\lambda.} \tag{2.2-16}$$

Notice that the particular solution (including the effects of $\mathbf{u}(t)$ and $\mathbf{q}_c(t)$) is computed as convolution integrals involving $\boldsymbol{\Phi}(t_{i+1},\lambda)$. If the system dynamics represented by $\mathbf{F}(t)$ are time-invariant, then equations (2.2-14) to (2.2-16) will use $\boldsymbol{\Phi}(t_{i+1}-t_i)$ and $\boldsymbol{\Phi}(t_{i+1}-\lambda)$. As noted before, most nonlinear applications compute the portion of $\mathbf{x}(t_{i+1})$ due to $\mathbf{x}(t_i)$ and $\mathbf{u}(t)$ by direct numerical integration, but even in these cases equation (2.2-16) is used when calculating the covariance of $\mathbf{q}_c(t)$.

Some elements of $\mathbf{u}_D$ and $\mathbf{q}_D$ may be zero if the indicated integrals do not affect all states, but that detail is ignored for the moment. We now concentrate on $\mathbf{q}_D$. The continuous process noise $\mathbf{q}_c(t)$ is assumed to be random: for modeling purposes it is treated as unknown and cannot be directly integrated to compute $\mathbf{q}_D$. It is also assumed that $\mathbf{q}_c(t)$ is zero-mean ($E[\mathbf{q}_c(t)] = \mathbf{0}$) white noise with known PSD *matrix* $\mathbf{Q}_s$:

$$E\left[\mathbf{q}_c(t)\mathbf{q}_c^T(\tau)\right] = \mathbf{Q}_s\delta(t-\tau) \tag{2.2-17}$$

where $E[\cdot]$ denotes expected value and $\delta(t - \tau)$ is the Dirac delta function. Since

$$\int_{-\infty}^{\infty} \delta(t)\, dt = 1 \tag{2.2-18}$$

is unitless, $\delta(t)$ must have units of inverse time, such as 1/s. Hence $\mathbf{Q}_s$ must have units of (magnitude)^2·s. For example, if the i-th element of $\mathbf{q}_c(t)$ directly drives a state having units of volts, then the diagonal (i,i) element of $\mathbf{Q}_s$ must have units of volts^2·s or volts^2/Hz. Thus $\mathbf{Q}_s$ is a PSD.

Using equation (2.2-16), the covariance of $\mathbf{q}_D$ is computed as:

$$\begin{aligned} &E\left[\mathbf{q}_D(t_{i+1},t_i)\mathbf{q}_D^T(t_{i+1},t_i)\right] \\ &= E\left\{\left[\int_{t_i}^{t_{i+1}} \mathbf{\Phi}(t_{i+1},\lambda)\,\mathbf{G}(\lambda)\mathbf{q}_c(\lambda)d\lambda\right]\left[\int_{t_i}^{t_{i+1}} \mathbf{q}_c^T(\lambda)\mathbf{G}^T(\lambda)\mathbf{\Phi}^T(t_{i+1},\lambda)d\lambda\right]\right\} \\ &= E\left\{\left[\int_{t_i}^{t_{i+1}} \mathbf{\Phi}(t_{i+1},\lambda)\,\mathbf{G}(\lambda)\mathbf{q}_c(\lambda)\left[\int_{t_i}^{t_{i+1}} \mathbf{q}_c^T(\tau)\mathbf{G}^T(\tau)\mathbf{\Phi}^T(t_{i+1},\tau)d\tau\right]d\lambda\right]\right\} \end{aligned}. \tag{2.2-19}$$

Since $E\left[\mathbf{q}_c(t)\mathbf{q}_c^T(\tau)\right] = \mathbf{0}$ for $t \neq \tau$ and $\mathbf{q}_c(t)$ is the only variable within the integral that is random, the expectation may be moved within a single integral:

$$\boxed{\begin{aligned} \mathbf{Q}_D(t_{i+1},t_i) &= \int_{t_i}^{t_{i+1}} \mathbf{\Phi}(t_{i+1},\lambda)\,\mathbf{G}(\lambda)E\left[\mathbf{q}_c(\lambda)\mathbf{q}_c^T(\lambda)\right]\mathbf{G}^T(\lambda)\mathbf{\Phi}^T(t_{i+1},\lambda)d\lambda \\ &= \int_{t_i}^{t_{i+1}} \mathbf{\Phi}(t_{i+1},\lambda)\,\mathbf{G}(\lambda)\mathbf{Q}_s\,\mathbf{G}^T(\lambda)\mathbf{\Phi}^T(t_{i+1},\lambda)d\lambda \end{aligned}} \tag{2.2-20}$$

where we have defined the discrete process noise covariance

$$\mathbf{Q}_D(t_{i+1},t_i) \triangleq E\left[\mathbf{q}_D(t_{i+1},t_i)\mathbf{q}_D^T(t_{i+1},t_i)\right]. \tag{2.2-21}$$

Notice that

$$E\left[\mathbf{q}_D(t_i,t_{i-1})\mathbf{q}_D^T(t_j,t_{j-1})\right] = \mathbf{0}$$

for $t_i \neq t_j$ because $E\left[\mathbf{q}_c(t)\mathbf{q}_c^T(\tau)\right] = \mathbf{0}$ for $t \neq \tau$, and no sample of $\mathbf{q}_c(t)$ is common to both intervals $t_{i-1} < t \leq t_i$ and $t_{j-1} < t \leq t_j$. Furthermore $E[\mathbf{q}_D(t_i,t_i)] = \mathbf{0}$ for all t_i because $E[\mathbf{q}_c(t)] = \mathbf{0}$.

At first glance equation (2.2-20) may not appear very useful because the "messy" convolution integral involves products of the state transition matrix, and we have indicated that computation of $\mathbf{\Phi}(t,\tau)$ may not be trivial. You may wonder why $\mathbf{Q}_D$ is needed at all. For least-squares applications $\mathbf{Q}_D$ is zero because the model is deterministic. However, $\mathbf{Q}_D$ is used in the Kalman filter to compute optimal weighting of measurement data. Fortunately $\mathbf{Q}_D$ can be approximated because the covariance equations used to compute weighting need not be as accurate as the state model. This will be discussed again in Chapter 8.

Before considering various methods for computing the state transition and process noise covariance matrices, we first discuss the types of dynamic models that may be used in estimation problems.

2.2.2 Dynamic Models Using Basic Function Expansions

The simplest type of dynamic model treats the measurement data as an expansion in basis functions. This is really just curve fitting using a constraint on the type of fit. Polynomials are probably the most commonly used basis function. For example, a scalar measurement could be represented as a third-order polynomial in time

$$y_m(t) = c_0 + c_1 t + c_2 t^2 + c_3 t^3 \tag{2.2-22}$$

where c_0, c_1, c_2, and c_3 are the polynomial coefficients. If the deterministic model is used in least-squares estimation, the epoch model states can be set equal to the coefficients:

$$\mathbf{x}(0) = [c_0 \quad c_1 \quad c_2 \quad c_3]^T . \tag{2.2-23}$$

Since the states are constant, the dynamic model is simply $\mathbf{x}(t) = \mathbf{x}(0)$ and the measurement equation is

$$y_m(t) = [1 \quad t \quad t^2 \quad t^3]\,\mathbf{x}(0). \tag{2.2-24}$$

However, if the model is to be used in a Kalman filter with process noise driving the states, it is usually better to define the measurement using a moving epoch for the state:

$$y_m(t) = [1 \quad 0 \quad 0 \quad 0]\,\mathbf{x}(t)$$

where

$$\dot{\mathbf{x}}(t) = \begin{bmatrix} 0 & 1 & 0 & 0 \\ 0 & 0 & 1 & 0 \\ 0 & 0 & 0 & 1 \\ 0 & 0 & 0 & 0 \end{bmatrix} \mathbf{x}(t) + \begin{bmatrix} 0 \\ 0 \\ 0 \\ q_4(t) \end{bmatrix} \tag{2.2-25}$$

and the initial state is

$$\mathbf{x}(0) = [c_0 \quad c_1 \quad 2c_2 \quad 6c_3]^T .$$

Equation (2.2-25), without the effect of $q_4(t)$, is integrated over the time interval between measurements, T, to obtain:

$$\mathbf{x}(t+T) = \begin{bmatrix} 1 & T & T^2/2 & T^3/6 \\ 0 & 1 & T & T^2/2 \\ 0 & 0 & 1 & T \\ 0 & 0 & 0 & 1 \end{bmatrix} \mathbf{x}(t). \tag{2.2-26}$$

The extension to include the effect of process noise ($q_4(t)$ in this example) will be addressed later when discussing random process models.

One problem in using polynomials for least-squares fitting is the tendency of the solution to become numerically indeterminate when high-order terms are included. The problem of numerical singularity and poor observability is discussed further in Chapter 5. To minimize numerical problems, it is often advisable to switch from ordinary polynomials to polynomials that are orthogonal over a given data span (see Press et al. 2007, section 4.5). Two functions are considered *orthogonal* over the interval $a \le x \le b$ with respect to a given weight function $W(x)$ if

$$\int_a^b W(x) f(x) g(x) dx = 0. \tag{2.2-27}$$

A set of mutually orthogonal functions $f_i(x)$, $i = 1, 2, \ldots, n$ is called *orthonormal* if $\int_a^b W(x) f_i^2(x) dx = 1$ for all i. A set of orthogonal polynomials $p_i(x)$ for $j = 0, 1, 2, \ldots$ can be constructed using the recursions

$$\begin{aligned} p_{-1}(x) &\equiv 0 \\ p_0(x) &\equiv 1 \\ p_{j+1}(x) &= (x - a_j) p_j(x) - b_j p_{j-1}(x) \qquad j = 0, 1, 2, \ldots \end{aligned} \tag{2.2-28}$$

where

$$\begin{aligned} a_j &= \frac{\int_a^b W(x) x\, p_j(x) p_j(x) dx}{\int_a^b W(x) p_j(x) p_j(x) dx} \\ b_j &= \frac{\int_a^b W(x) p_j(x) p_j(x) dx}{\int_a^b W(x) p_{j-1}(x) p_{j-1}(x) dx} \end{aligned}. \tag{2.2-29}$$

As one example, Chebyshev polynomials are orthogonal over the interval −1 to 1 for a weight of $1/\sqrt{1-x^2}$. They are defined by

$$T_n(x) = \cos(n \arccos x) \tag{2.2-30}$$

or explicitly

$$\begin{aligned} T_0(x) &= 1 \\ T_1(x) &= x \\ T_2(x) &= 2x^2 - 1 \\ T_3(x) &= 4x^3 - 3x \\ T_4(x) &= 8x^4 - 8x^2 + 1 \\ &\ldots \\ T_{n+1}(x) &= 2xT_n(x) - T_{n-1}(x) \quad n \ge 1 \end{aligned}. \tag{2.2-31}$$

Notice that to use Chebyshev polynomials over an arbitrary interval $a \le y \le b$, the affine transformation

$$x = 2\left(\frac{y-a}{b-a}\right) - 1$$

must be applied. Unlike ordinary polynomials, Chebyshev polynomials are bounded by −1 to +1 for all orders. Hence the numerical problems of ordinary polynomials, caused by large values of high-order terms, are avoided when using orthogonal polynomials. Use of orthonormal transformations to minimize the effects of numerical errors is discussed in later chapters, and there is a connection between these orthonormal transformations and orthonormal polynomials.

Other possibilities for basis functions include Fourier series and wavelets. In deciding which function to use, consider the characteristics. Polynomials can accurately model smooth functions and trends, but high-order models are needed to follow abrupt changes. Fourier expansions are ideal for functions that are periodic over a finite data span, but do not model trends well unless linear and quadratic terms are added. Wavelets are useful for modeling nonperiodic functions that have limited time extent (Press et al. 2007, section 13.10). Notice that the measurement data span used for the modeling should be limited so that the basis function expansion can accurately model data within the span. If the span is long, it may be necessary to carry many terms in the expansion, and this could cause problems if the model is used to predict data outside the measurement data span: the prediction will tend to rapidly diverge when high-order terms are included in the data fits.

Example 2.1: GOES Instrument "Attitude"

Basis function modeling can be used not only for curve fitting, but also for modeling a physical system when it is difficult to develop a first-principles model. For example, a combination of a polynomial and Fourier series expansions is used to model imaging instrument misalignments on the GOES I-P spacecraft (Gibbs 2008). The GOES geosynchronous weather satellites provide continuous images of the western hemisphere from a nearly equatorial position and fixed longitude. Because imaging instruments have small internal misalignments, the optical boresight at various scan angles is modeled using five misalignment parameters: three Euler attitude rotations (roll, pitch, and yaw) and two lumped-parameter misalignments. The five misalignments vary with time because instrument thermal deformation is driven by solar heating. Since the spacecraft has a fixed position with respect to the earth, the angle of the sun relative to the spacecraft has a pattern that nearly repeats every day. Thus misalignments have a fundamental period equal to the solar day. The misalignment profile is not exactly sinusoidal so it is necessary to include harmonics of the 24-h period to model the fine structure. For GOES, each of the five misalignment parameters is modeled as the sum of a linear or quadratic term plus a Fourier series of 4 to 13 harmonics, depending on the parameter. For example, instrument roll $\phi(t)$ is modeled as

$$\phi(t) = x_1 + x_2 t + x_3 \cos\theta + x_4 \sin\theta + x_5 \cos 2\theta + x_6 \sin 2\theta + \ldots + x_n \sin(n/2-1)\theta$$

where $\theta = 2\pi\, t/1440$ and t is minutes of the solar day. The other four misalignment parameters are modeled similarly using different x coefficients. Then the instrument-observed scan angles to various stars and landmarks are modeled using a nonlinear function of the five misalignment angles. Least-squares estimation is used to determine the epoch misalignment states, so the state dynamic

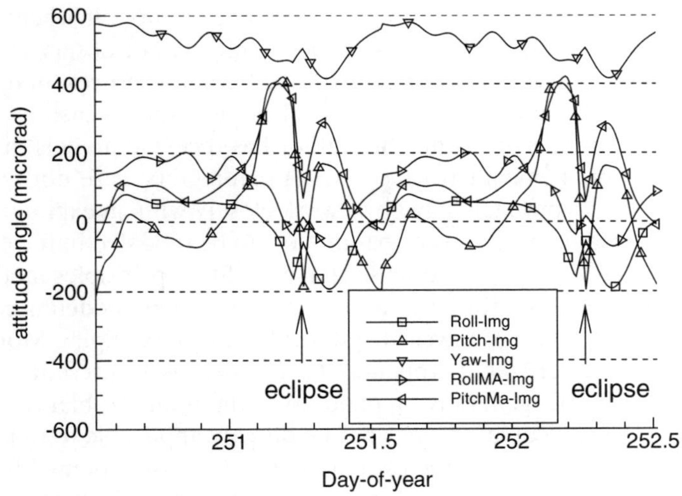

FIGURE 2.2: Estimated Imager misalignments for GOES-13 during eclipse season 2006.

model is simply $\mathbf{x}(t) = \mathbf{x}(0)$ with the time variation of the five misalignment parameters included in the measurement model.

Figure 2.2 shows the estimated GOES-13 Imager misalignment angles over a 48-h period that includes eclipses at approximately Greenwich Mean Time (GMT) days 251.25 and 252.25 in 2006. Eclipse periods lasting a maximum of 72 min at the spring and autumnal equinoxes have a significant effect on instrument temperatures and thus misalignment angles. Instrument thermal recovery can take several hours after eclipse end, so "eclipse periods" of 4 h are modeled separately from the normal 24-h model.

2.2.3 Dynamic Models Derived from First Principles

When working with physical systems, it is usually better to develop models from first principles, rather than computing empirical models from measured data only. The first-principles approach has the potential advantage of greater accuracy and the ability to model an extended (nontested) range of operating conditions. It also has the potential disadvantages of poor estimation accuracy if the model structure or parameters are incorrect, and development of good first-principle models may be difficult for some systems. A standard approach uses first-principles models for parts of the system that are well modeled, and then includes random walk or Markov process (colored noise) models for parts of the system (usually higher order derivatives) that appear to be random or have uncertain characteristics. If it is difficult to develop a first-principles model, or if the model uncertainty is great, then a totally empirical approach may be better. This is discussed in Chapter 12.

It is important to understand all relevant physical laws, the assumptions on which those laws are based, and the actual conditions existing in the system to be modeled when developing first-principles models. You will need to either work with an expert or become an expert yourself. However, beware—system experts tend to focus on the subject that they know best and sometimes create detailed models that include effects irrelevant for estimation purposes. For example, it is generally unnecessary and undesirable to model high-frequency effects not observed at the measurement sampling rate. These effects can often be treated as process or measurement noise. Calibration of simulation models based on an expert's instinct rather than measured data is another problem. This author has received models that exhibited physically impossible behavior because model parameters were not properly calibrated. Thus the best approach may be to work closely with an expert, but consider carefully when deciding on effects or parameters to be modeled in the estimation.

This book does not attempt to discuss all types of first-principles modeling—that would be a formidable task. Rather, we summarize the main modeling concepts and demonstrate the modeling of physical systems by a few examples. More examples are provided in Chapter 3 and Appendix C. The goal is to present concepts and guidelines that can be generally applied to estimation problems. Be warned, however, that the listed equations may not be directly applicable to other problems, so consult references and check the assumptions before using them. More information on model building and identification may be found in Fasol and Jörgl (1980), Levine (1996), Balchen and Mummé (1988), Close et al. (2001), Isermann (1980), Ljung and Glad (1994), Ljung (1999) and Åström (1980).

Typical "first-principles" concepts used in model building include

1. Conservation of mass: for example, the continuity equation
2. Conservation of momentum: for example, Newton's laws of motion
3. Conservation of energy: for example, first law of thermodynamics
4. Second law of thermodynamics and entropy relationships
5. Device input/output relationships: for example, pump, fan, motor, thruster, resistor, capacitor, inductor, diode, transistor
6. Flow/pressure relationships: for example, pipes, ducts, aerodynamics, porous media
7. Heat transfer models: for example, conductive, convective, radiant
8. Chemical reactions: for example, combustion thermodynamics
9. Optical properties and relationships
10. Special and general relativity.

This list is obviously not exhaustive as many other "first principles" exist. Use of these principles may lead to distributed models based on *partial differential equations* (PDE), or lumped-parameter (linear or nonlinear) models based on *ordinary differential equations* (ODE). When PDEs are used, boundary conditions must be specified. To simplify solutions, it is sometimes desirable to discretize a distributed model so that PDEs are replaced with ODEs. Models may be static, dynamic, or both. Other classifications include deterministic or stochastic, and parametric or nonparametric. Time delays—such as those resulting from mass transport—cannot be exactly represented as differential equations, so it may be necessary to explicitly model the delay or to approximate it using ODEs.

After structure of a parametric model is defined, the parameter values must be determined. In some cases component or subsystem test data may be available, and it may be possible to determine parameters directly from the data. In other cases data from a fully instrumented system must be used to identify parameters. Much has been written about this subject, and discussions may be found in previously listed references. Note that parameter observability is affected by the types of perturbing signals and the presence of control feedback.

The remainder of this section summarizes a few basic concepts that are often used in first-principles models. Other useful concepts are described in Appendix C and in listed references.

2.2.3.1 Linear Motion Constant velocity motion is one of the most commonly used models for tracking applications. This often appears as the default for tracking of ships and aircraft since both vehicles move in a more-or-less straight line for extended periods of time. However, any tracker based on a constant velocity assumption must also include some means for detecting acceleration or sudden changes in velocity (when the interval between measurements is longer than the applied acceleration). In two lateral dimensions (x and y) where $\mathbf{r}$ denotes position and $\mathbf{v}$ denotes velocity, the state vector is $\mathbf{x}^T = [r_x \quad r_y \quad v_x \quad v_y]$, which has time derivative

$$\dot{\mathbf{x}}(t) = \begin{bmatrix} 0 & 0 & 1 & 0 \\ 0 & 0 & 0 & 1 \\ 0 & 0 & 0 & 0 \\ 0 & 0 & 0 & 0 \end{bmatrix} \mathbf{x}(t). \tag{2.2-32}$$

This can be integrated to obtain

$$\mathbf{x}(t) = \begin{bmatrix} 1 & 0 & t & 0 \\ 0 & 1 & 0 & t \\ 0 & 0 & 1 & 0 \\ 0 & 0 & 0 & 1 \end{bmatrix} \mathbf{x}(0).$$

If used in a Kalman filter, process noise on the two velocity states should be modeled to account for occasional changes in velocity.

2.2.3.2 Linear Momentum Change Newton's second law is

$$\mathbf{f} = \frac{d(m\mathbf{v})}{dt} = \dot{m}\mathbf{v} + m\dot{\mathbf{v}},$$

where $\mathbf{f}$ is the vector of applied force, m is body mass, and $\mathbf{v}$ is the velocity vector. This equation can be added to the linear motion model above. If mass does not change and a known applied force is from an external source, the two-dimensional model becomes

$$\dot{\mathbf{x}}(t) = \begin{bmatrix} 0 & 0 & 1 & 0 \\ 0 & 0 & 0 & 1 \\ 0 & 0 & 0 & 0 \\ 0 & 0 & 0 & 0 \end{bmatrix} \mathbf{x}(t) + \begin{bmatrix} 0 \\ 0 \\ f_x(t)/m \\ f_y(t)/m \end{bmatrix}. \tag{2.2-33}$$

In other words, the applied force is treated as an exogenous input: $\mathbf{u}$ in equation (2.2-4). If no information on applied force is available, x and y acceleration should be included as "constant" states. If added at the bottom of the state vector, that is, $\mathbf{x} = [r_x \quad r_y \quad v_x \quad v_y \quad a_x \quad a_y]^T$, the dynamic model becomes

$$\dot{\mathbf{x}}(t) = \begin{bmatrix} 0 & 0 & 1 & 0 & 0 & 0 \\ 0 & 0 & 0 & 1 & 0 & 0 \\ 0 & 0 & 0 & 0 & 1 & 0 \\ 0 & 0 & 0 & 0 & 0 & 1 \\ 0 & 0 & 0 & 0 & 0 & 0 \\ 0 & 0 & 0 & 0 & 0 & 0 \end{bmatrix} \mathbf{x}(t). \tag{2.2-34}$$

In this case acceleration will not remain constant for long periods of time, so a Kalman filter should be used and process noise on the acceleration should be modeled. Alternately least-squares estimation could be used, but it would be necessary to include some mechanism for detecting changes in acceleration and re-initializing the estimate (Willsky and Jones 1974, 1976; Basseville and Benveniste 1986; Bar-Shalom and Fortmann 1988; Gibbs 1992).

A different choice of acceleration states may be appropriate in some cases. For example, both ships and aircraft tend to turn at an approximately constant angular rate. They also tend to accelerate along the current velocity vector when additional thrust is applied (or aircraft pitch is changed). When multiple measurements are available during the maneuvers, the estimator may perform better using constant crosstrack acceleration (a_c) as state #5 and constant alongtrack acceleration (a_a) as state #6. This leads to

$$\begin{aligned} \dot{v}_x &= a_c v_y / v + a_a v_x / v \\ \dot{v}_y &= -a_c v_x / v + a_a v_y / v \end{aligned} \tag{2.2-35}$$

where $v = \sqrt{v_x^2 + v_y^2}$. Notice that the model is now nonlinear, but the problems introduced by the nonlinearity may be offset by the improved performance due to the better model. Chapter 9 includes an example that uses this model in a Kalman filter. To implement this model in a filter, the nonlinear equations are usually linearized about reference velocities to obtain first-order perturbation equations, that is,

$$\delta\dot{\mathbf{x}} = \left.\frac{\partial \dot{\mathbf{x}}}{\partial \mathbf{x}}\right|_{\mathbf{x}\_ref} \delta\mathbf{x}.$$

The perturbation form of this model is:

$$\delta\dot{\mathbf{x}}(t) = \begin{bmatrix} 0 & 0 & 1 & 0 & 0 & 0 \\ 0 & 0 & 0 & 1 & 0 & 0 \\ 0 & 0 & (-a_c v_x v_y + a_a v_y^2)/v^3 & (-a_a v_x v_y + a_c v_x^2)/v^3 & v_y / v & v_x / v \\ 0 & 0 & (-a_a v_x v_y - a_c v_y^2)/v^3 & (a_c v_x v_y + a_a v_x^2)/v^3 & -v_x / v & v_y / v \\ 0 & 0 & 0 & 0 & 0 & 0 \\ 0 & 0 & 0 & 0 & 0 & 0 \end{bmatrix} \delta\mathbf{x}(t). \tag{2.2-36}$$

2.2.3.3 Rotational Motion

Rotational motion of a rigid body is described by

$$\boldsymbol{\tau} = \frac{d\mathbf{h}}{dt}$$

where τ is the applied torque and $\mathbf{h}$ is the angular momentum vector measured in a nonrotating coordinate system. If τ and $\mathbf{h}$ are referenced to the coordinate system of the rotating body (the b-system), then $\mathbf{h}_b = \mathbf{I}_b\boldsymbol{\omega}_b$ where $\mathbf{I}_b$ is the inertia tensor of the body and $\boldsymbol{\omega}_b$ is the rotational rate. Since the reference frame is rotating, the change in angular momentum of the rotating coordinate system must be accounted for as part of the angular acceleration (Goldstein 1950, section 4-8; Housner and Hudson 1959, p. 203), that is,

$$\begin{aligned}\boldsymbol{\tau}_b &= \dot{\mathbf{h}}_b + \boldsymbol{\omega}_b \times \mathbf{h}_b \\ &= \mathbf{I}_b\dot{\boldsymbol{\omega}}_b + \boldsymbol{\omega}_b \times (\mathbf{I}_b\boldsymbol{\omega}_b)\end{aligned}.$$

Rearranging yields:

$$\begin{bmatrix}\dot{h}_{b1}\\ \dot{h}_{b2}\\ \dot{h}_{b3}\end{bmatrix} = \begin{bmatrix}\tau_{b1}\\ \tau_{b2}\\ \tau_{b3}\end{bmatrix} - \begin{bmatrix}\omega_{b2}h_{b3}-\omega_{b3}h_{b2}\\ \omega_{b3}h_{b1}-\omega_{b1}h_{b3}\\ \omega_{b1}h_{b2}-\omega_{b2}h_{b1}\end{bmatrix}. \tag{2.2-37}$$

If the inertia tensor is assumed to be diagonal (generally not true), the body angular accelerations are computed as

$$\begin{bmatrix}\dot{\omega}_{b1}\\ \dot{\omega}_{b2}\\ \dot{\omega}_{b3}\end{bmatrix} = \begin{bmatrix}\tau_{b1}/I_{b1}\\ \tau_{b2}/I_{b2}\\ \tau_{b3}/I_{b3}\end{bmatrix} + \begin{bmatrix}-(\omega_{b2}\omega_{b3}I_{b3}-\omega_{b3}\omega_{b2}I_{b2})/I_{b1}\\ -(\omega_{b3}\omega_{b1}I_{b1}-\omega_{b1}\omega_{b3}I_{b3})/I_{b2}\\ -(\omega_{b1}\omega_{b2}I_{b2}-\omega_{b2}\omega_{b1}I_{b1})/I_{b3}\end{bmatrix} \tag{2.2-38}$$

Notice that this model is nonlinear in $\boldsymbol{\omega}_b$. If the body of interest contains other rotating bodies, such as momentum or reaction wheels, the separate angular momentum of those rotating bodies must be included as part of the total and additional states modeling the separate angular rates (or equivalently angular momentum) must be included. Depending upon how the model is structured, the "external torques" may include internal torques applied by motors connected to the momentum wheels. If the body of interest is an aircraft, aerodynamic forces (see Appendix C) and torques about the center-of-mass are computed for each section of the airframe and summed.

There are several choices for defining attitude. Euler angles are commonly used for representing aircraft attitude, and are sometimes used to define spacecraft attitude. If the Euler angles are defined in 3-2-1 order (as used for aircraft), the first rotation is about axis-3 (yaw or ψ), the next rotation is about axis-2 (pitch or θ), and the final rotation is about axis-1 (roll or ϕ). The rotation from the reference coordinates to the body coordinates can be represented as a direction cosine matrix:

$$\begin{aligned}\begin{bmatrix}\omega_{b1}\\ \omega_{b2}\\ \omega_{b3}\end{bmatrix} &= \begin{bmatrix}1 & 0 & 0\\ 0 & C_\phi & S_\phi\\ 0 & -S_\phi & C_\phi\end{bmatrix}\begin{bmatrix}C_\theta & 0 & -S_\theta\\ 0 & 1 & 0\\ S_\theta & 0 & C_\theta\end{bmatrix}\begin{bmatrix}C_\psi & S_\psi & 0\\ -S_\psi & C_\psi & 0\\ 0 & 0 & 1\end{bmatrix}\begin{bmatrix}\omega_{r1}\\ \omega_{r2}\\ \omega_{r3}\end{bmatrix} \\ &= \begin{bmatrix}C_\theta C_\psi & C_\theta S_\psi & -S_\theta\\ S_\phi S_\theta C_\psi - S_\psi C_\phi & S_\phi S_\theta S_\psi + C_\psi C_\phi & S_\phi C_\theta\\ C_\phi S_\theta C_\psi + S_\psi S_\phi & C_\phi S_\theta S_\psi - C_\psi S_\phi & C_\phi C_\theta\end{bmatrix}\begin{bmatrix}\omega_{r1}\\ \omega_{r2}\\ \omega_{r3}\end{bmatrix}\end{aligned} \tag{2.2-39}$$

where $C_\theta = \cos\theta$, $S_\theta = \sin\theta$, etc. The rotation from body to the reference system is the transpose:

$$\begin{bmatrix}\omega_{r1}\\ \omega_{r2}\\ \omega_{r3}\end{bmatrix} = \begin{bmatrix} C_\theta C_\psi & S_\phi S_\theta C_\psi - S_\psi C_\phi & C_\phi S_\theta C_\psi + S_\psi S_\phi \\ C_\theta S_\psi & S_\phi S_\theta S_\psi + C_\psi C_\phi & C_\phi S_\theta S_\psi - C_\psi S_\phi \\ -S_\theta & S_\phi C_\theta & C_\phi C_\theta \end{bmatrix} \begin{bmatrix}\omega_{b1}\\ \omega_{b2}\\ \omega_{b3}\end{bmatrix}. \tag{2.2-40}$$

The body rates can be written in terms of the Euler angles rates by direct resolution for the 3-2-1 rotation order. In the reference frame the body rates are

$$\boldsymbol{\omega}_r = \dot{\phi}\hat{\mathbf{x}}'' + \dot{\theta}\hat{\mathbf{y}}' + \dot{\psi}\hat{\mathbf{z}}_r,$$

which map to the body frame as

$$\boldsymbol{\omega}_b = \begin{bmatrix} \dot{\phi}\hat{\mathbf{x}}''\cdot\hat{\mathbf{x}}_b + \dot{\theta}\hat{\mathbf{y}}'\cdot\hat{\mathbf{x}}_b + \dot{\psi}\hat{\mathbf{z}}_r\cdot\hat{\mathbf{x}}_b \\ \dot{\phi}\hat{\mathbf{x}}''\cdot\hat{\mathbf{y}}_b + \dot{\theta}\hat{\mathbf{y}}'\cdot\hat{\mathbf{y}}_b + \dot{\psi}\hat{\mathbf{z}}_r\cdot\hat{\mathbf{y}}_b \\ \dot{\phi}\hat{\mathbf{x}}''\cdot\hat{\mathbf{z}}_b + \dot{\theta}\hat{\mathbf{y}}'\cdot\hat{\mathbf{z}}_b + \dot{\psi}\hat{\mathbf{z}}_r\cdot\hat{\mathbf{z}}_b \end{bmatrix}$$

where

$$[\hat{\mathbf{x}}' \quad \hat{\mathbf{y}}' \quad \hat{\mathbf{z}}'] = \begin{bmatrix} C_\psi & S_\psi & 0 \\ -S_\psi & C_\psi & 0 \\ 0 & 0 & 1 \end{bmatrix} [\hat{\mathbf{x}}_r \quad \hat{\mathbf{y}}_r \quad \hat{\mathbf{z}}_r]$$

$$[\hat{\mathbf{x}}'' \quad \hat{\mathbf{y}}'' \quad \hat{\mathbf{z}}''] = \begin{bmatrix} C_\theta & 0 & -S_\theta \\ 0 & 1 & 0 \\ S_\theta & 0 & C_\theta \end{bmatrix} [\hat{\mathbf{x}}' \quad \hat{\mathbf{y}}' \quad \hat{\mathbf{z}}']$$

$$[\hat{\mathbf{x}}_b \quad \hat{\mathbf{y}}_b \quad \hat{\mathbf{z}}_b] = \begin{bmatrix} 1 & 0 & 0 \\ 0 & C_\phi & S_\phi \\ 0 & -S_\phi & C_\phi \end{bmatrix} [\hat{\mathbf{x}}'' \quad \hat{\mathbf{y}}'' \quad \hat{\mathbf{z}}''] = \begin{bmatrix} C_\theta & 0 & -S_\theta \\ S_\phi S_\theta & C_\phi & S_\phi C_\theta \\ C_\phi S_\theta & -S_\phi & C_\phi C_\theta \end{bmatrix} [\hat{\mathbf{x}}' \quad \hat{\mathbf{y}}' \quad \hat{\mathbf{z}}']$$

and $\hat{\mathbf{x}}_r, \hat{\mathbf{y}}_r, \hat{\mathbf{z}}_r$ are unit vectors defining the reference axes. Computation of the vector dot products yields

$$\begin{bmatrix}\omega_{b1}\\ \omega_{b2}\\ \omega_{b3}\end{bmatrix} = \begin{bmatrix} \dot{\phi} - \dot{\psi}\sin\theta \\ \dot{\theta}\cos\phi + \dot{\psi}\sin\phi\cos\theta \\ -\dot{\theta}\sin\phi + \dot{\psi}\cos\phi\cos\theta \end{bmatrix}. \tag{2.2-41}$$

Equation (2.2-41) can be inverted to compute the Euler angle rates:

$$\begin{bmatrix}\dot{\phi}\\ \dot{\theta}\\ \dot{\psi}\end{bmatrix} = \begin{bmatrix} \omega_{b1} + \tan\theta(\omega_{b2}\sin\phi + \omega_{b3}\cos\phi) \\ \omega_{b2}\cos\phi - \omega_{b3}\sin\phi \\ (\omega_{b3}\cos\phi + \omega_{b2}\sin\phi)/\cos\theta \end{bmatrix}. \tag{2.2-42}$$

For computational reasons it is often preferable to order the states representing the highest derivatives at the end of the state vector. This tends to make the state transition matrix somewhat block upper triangular. Therefore we use ϕ, θ, ψ as the first three states and ω_{b1}, ω_{b2}, ω_{b3} as the last three states in the model.

There are several problems with the use of Euler angles as states. First, the representation becomes singular when individual rotations equal 90 degrees. Hence they are mainly used for systems in which rotations are always less than 90 degrees (such as for non-aerobatic aircraft). Further, the direction cosine matrix and rotation require evaluation of six trigonometric functions and 25 multiplications; the computational burden makes that unattractive for evaluation onboard spacecraft. For these reasons, most spacecraft use a quaternion (4-parameter) representation for attitude rotations (Wertz 1978), or a similar rotation vector model of the form $\mathbf{x}_{rot} = \alpha \hat{\mathbf{x}}_{rot}$ where $\hat{\mathbf{x}}_{rot}$ is the unit vector axis of rotation and α is the angle of rotation. Then the rotation from the body-to-reference system in equation (2.2-40) can be implemented as

$$\begin{aligned}
\alpha &= \sqrt{x_{rot_1}^2 + x_{rot_2}^2 + x_{rot_3}^2} \\
\hat{\mathbf{x}}_{rot} &= \mathbf{x}_{rot} / \alpha \\
\boldsymbol{\omega}_r &= \boldsymbol{\omega}_b \cos\alpha + \hat{\mathbf{x}}_{rot}(\boldsymbol{\omega}_b \cdot \hat{\mathbf{x}}_{rot})(1-\cos\alpha) + (\hat{x}_{rot} \times \boldsymbol{\omega}_b)\sin\alpha
\end{aligned} \quad . \tag{2.2-43}$$

Notice that the rotation is implemented by summing vector components in three directions where the third vector is orthogonal to the first two. Two trigonometric functions, a square root, and 25 multiplications are still required in this implementation. The trig functions and square root are avoided when using an Euler-symmetric parameter (quaternion) representation (Wertz 1978, p. 414),

$$\mathbf{q} = \begin{bmatrix} (\sin\alpha/2)\hat{x}_{rot\_1} \\ (\sin\alpha/2)\hat{x}_{rot\_2} \\ (\sin\alpha/2)\hat{x}_{rot\_3} \\ \cos\alpha/2 \end{bmatrix}, \tag{2.2-44}$$

but extra computations are required to use the 4-parameter quaternion with a 3-state (first three quaternion parameters) model in a Kalman filter.

Appendix C summarizes other common first-principles models used for examples in this text.

2.2.4 Stochastic (Random) Process Models

Stochastic process models are frequently combined with other model types in Kalman filter applications to account for effects that are difficult to model or subject to random disturbances. These random models are typically used as driving terms for higher-order derivatives of the system model, although they can be used to drive any state.

2.2.4.1 Random Walk The most commonly used random process model is the *random walk*, which is the discrete version of *Brownian motion* (also called a *Wiener process*). This is often used as a default model to account for time-varying errors when little information is available. The Brownian motion model is

$$\dot{x}(t) = q_c(t) \tag{2.2-45}$$

where $q_c(t)$ is scalar white noise with

$$\begin{aligned}
E[q_c(t)] &= 0 \\
E[q_c(t)q_c(\tau)] &= Q_s\delta(t-\tau)
\end{aligned}.$$

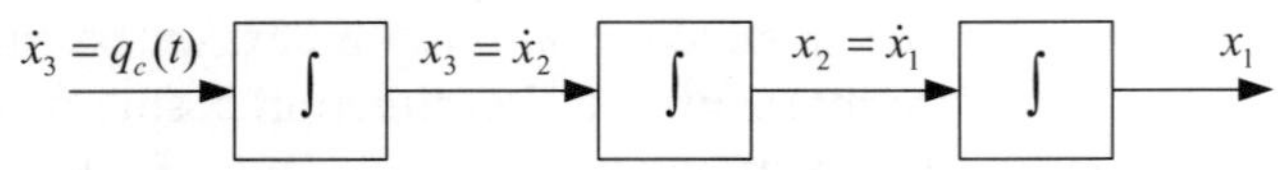

FIGURE 2.3: Integrated Brownian motion.

Since the discrete state transition function (scalar) for equation (2.2-45) is $\Phi(T) = 1$, the discrete process noise variance from equation (2.2-20) is

$$Q_D(T) = \int_t^{t+T} Q_s \, d\lambda = Q_s T \tag{2.2-46}$$

In other words, the variance of x increases linearly with the time interval for a random walk. A random walk model is often used as a driving term for the highest derivatives of the system, such as acceleration in a position/velocity/acceleration model. Hence we are interested in the integrated effect of the random walk on other states. Consider the three-state (position, velocity, acceleration) model

$$\begin{bmatrix} \dot{x}_1 \\ \dot{x}_2 \\ \dot{x}_3 \end{bmatrix} = \begin{bmatrix} 0 & 1 & 0 \\ 0 & 0 & 1 \\ 0 & 0 & 0 \end{bmatrix} \begin{bmatrix} x_1 \\ x_2 \\ x_3 \end{bmatrix} + \begin{bmatrix} 0 \\ 0 \\ q \end{bmatrix} \tag{2.2-47}$$

as shown in Figure 2.3.

The state transition matrix for this system is

$$\mathbf{\Phi}(T) = \begin{bmatrix} 1 & T & T^2/2 \\ 0 & 1 & T \\ 0 & 0 & 1 \end{bmatrix}.$$

Hence the discrete state noise covariance of $\mathbf{x}(t+T)$ from equation (2.2-19) is

$$\begin{aligned} \mathbf{Q}_D(T) &= \int_{\lambda=0}^{T} \begin{bmatrix} 1 & \lambda & \lambda^2/2 \\ 0 & 1 & \lambda \\ 0 & 0 & 1 \end{bmatrix} \begin{bmatrix} 0 & 0 & 0 \\ 0 & 0 & 0 \\ 0 & 0 & Q_s \end{bmatrix} \begin{bmatrix} 1 & 0 & 0 \\ \lambda & 1 & 0 \\ \lambda^2/2 & \lambda & 1 \end{bmatrix} d\lambda \\ &= Q_s \begin{bmatrix} T^5/20 & T^4/8 & T^3/6 \\ T^4/8 & T^3/3 & T^2/2 \\ T^3/6 & T^2/2 & T \end{bmatrix} \end{aligned} \tag{2.2-48}$$

Equation (2.2-48) is very useful when trying to determine appropriate values for Q_s since it is often easier to define appropriate values for the integrated effect of the process noise on other states. For example, we may have general knowledge that the integrated effect of acceleration process noise on position is about 10 m $1-\sigma$ after a period of 1000 s. Hence we set $Q_s(1000)^5/20 = 10^2$ or $Q_s = 2 \times 10^{-12}\,\text{m}^2/\text{s}^5$.

Although this example included white process noise on just the highest order state, this is not a general restriction: process noise can be included on any or all states. For example, one commonly used second-order clock error model includes process noise on both states:

$$\begin{bmatrix} \dot{x}_1 \\ \dot{x}_2 \end{bmatrix} = \begin{bmatrix} 0 & 1 \\ 0 & 0 \end{bmatrix} \begin{bmatrix} x_1 \\ x_2 \end{bmatrix} + \begin{bmatrix} q_1 \\ q_2 \end{bmatrix} \tag{2.2-49}$$

where q_1 and q_2 are white noise and x_2 is the clock timing error. Two separate q terms are used because clock errors have characteristics that can be approximated as both random walk and integrated random walk.

2.2.4.2 First-Order Markov Process (Colored Noise) A *first-order Markov process* model is another commonly used random process model. A discrete or continuous stochastic process is called a Markov process if the probability distribution of a future state (vector) depends only on the current state, not past history leading to that state. This definition also includes random walk models. As with random walk models, low-order Markov process models are primarily used to drive the highest derivatives of the system dynamics, thus modeling correlated disturbing forces. In practice most Markov process models used for Kalman filters are first-order models, with second-order used occasionally. These low-order Markov process models are also called *colored noise* models because, unlike white noise, the PSD of the output is not constant with frequency. Most low-order Markov process implementations have "low-pass" characteristics because most of the power appears in the lowest frequencies.

A first-order Markov process is one in which the probability distribution of the scalar output depends on only one point immediately in the past, that is,

$$F_X[x(t_k) \mid x(t_{k-1}), \ldots, x(t_1)] = F_X[x(t_k) \mid x(t_{k-1})]$$

for every $t_1 < t_2 < \ldots < t_k$. The time-invariant stochastic differential equation defining a first-order Markov process model is

$$\dot{x}(t) = -\frac{1}{\tau}x(t) + q_c(t) \tag{2.2-50}$$

where τ is the model time constant and $q_c(t)$ is white noise. The state transition for equation (2.2-50) over time interval T is

$$\Phi(T) = e^{-T/\tau} \tag{2.2-51}$$

and the discrete process noise variance is

$$\begin{aligned} Q_D(T) &= \int_0^T Q_s\,(e^{-\lambda/\tau})^2\,d\lambda \\ &= \frac{Q_s\tau}{2}\left(1 - e^{-2T/\tau}\right) \end{aligned} \tag{2.2-52}$$

Unlike a random walk, a first-order Markov process has a steady-state variance σ_x^2. Taking the limit of $Q_D(T)$ as $T \to \infty$ in equation (2.2-52),

$$\sigma_x^2 = \frac{Q_s\tau}{2}. \tag{2.2-53}$$

The inverse of this relationship, $Q_s = 2\sigma_x^2/\tau$, is used when determining appropriate values of Q_s given an approximate value of σ_x^2. The autocorrelation function of the steady-state process is

$$\begin{aligned} R_x(T) &= E[x(t+T)x(t)] \\ &= E\{[\Phi(T)x(t) + q_d(T)]x(t)\} \\ &= \sigma_x^2 e^{-|T|/\tau} \end{aligned} \tag{2.2-54}$$

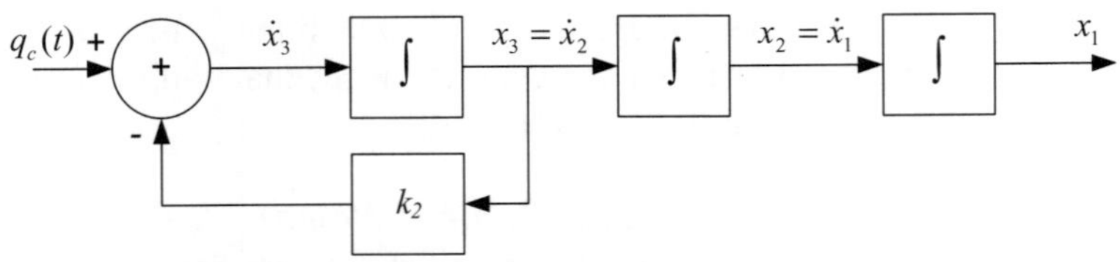

FIGURE 2.4: Integrated first-order Markov process.

where $q_d(T)$ is the integrated effect of the process noise over the interval t to $t + T$ and $E[q_d(T)x(t)] = 0$. (See Appendix B for definitions of random process properties.) The PSD of a stationary random process can be computed as the Fourier transform of the autocorrelation function:

$$\begin{aligned} S_x(\omega) &= \sigma_x^2\left[\int_{-\infty}^{0} e^{\lambda/\tau}e^{-j\omega\lambda}\,d\lambda + \int_{0}^{-\infty} e^{-\lambda/\tau}e^{-j\omega\lambda}\,d\lambda\right] \\ &= \sigma_x^2\left[\left.\frac{e^{(1/\tau - j\omega)\lambda}}{1/\tau - j\omega}\right|_{-\infty}^{0} - \left.\frac{e^{-(1/\tau + j\omega)\lambda}}{1/\tau + j\omega}\right|_{0}^{\infty}\right], \\ &= \frac{2\sigma_x^2\tau}{1+(\omega\tau)^2} = \frac{Q_s\tau^2}{1+(\omega\tau)^2} \end{aligned} \qquad (2.2\text{-}55)$$

where $\omega = 2\pi f$ is frequency in radians/second. This obviously has the low-pass characteristic mentioned previously. By moving the model output from the integrator output to the integrator input, the PSD becomes

$$S_x(\omega) = \frac{Q_s\tau^2\omega^2}{1+(\omega\tau)^2},$$

which has a high-pass characteristic. This alternate model is seldom helpful in Kalman filter implementations because the Markov process output is generally passed through additional integrators in the system model. Hence the same effect can be obtained by moving the Markov process output to the input of an alternate system model state.

We now compute the state transition matrix and discrete state noise covariance for the same three-state system model used for the random walk model, but replace the third state random walk with a first-order Markov process, as shown in Figure 2.4.

Using the property that $\mathbf{\Phi}(t) = \mathscr{L}^{-1}[s\mathbf{I} - \mathbf{F}]^{-1}$ (discussed later in Section 2.3) where s is a complex variable, $\mathscr{L}^{-1}$ is the inverse Laplace transform and

$$\mathbf{F} = \begin{bmatrix} 0 & 1 & 0 \\ 0 & 0 & 1 \\ 0 & 0 & -1/\tau \end{bmatrix},$$

the state transition matrix is computed as

$$\mathbf{\Phi}(T) = \begin{bmatrix} 1 & T & \tau T + \tau^2\left(e^{-T/\tau} - 1\right) \\ 0 & 1 & \tau\left(1 - e^{-T/\tau}\right) \\ 0 & 0 & e^{-T/\tau} \end{bmatrix} \qquad (2.2\text{-}56)$$

and the discrete process noise covariance from equation (2.2-18) is

$$\mathbf{Q}_D(T) = Q_s \int_{\lambda=0}^{T} \begin{bmatrix} \tau\lambda + \tau^2\left(e^{-\lambda/\tau} - 1\right) \\ \tau\left(1 - e^{-\lambda/\tau}\right) \\ e^{-\lambda/\tau} \end{bmatrix} \left[\tau\lambda + \tau^2\left(e^{-\lambda/\tau} - 1\right) \quad \tau\left(1 - e^{-\lambda/\tau}\right) \quad e^{-\lambda/\tau}\right] d\lambda. \tag{2.2-57}$$

The (3,3) element of $\mathbf{Q}_D(T)$ is equal to

$$\frac{Q_s\tau}{2}\left(1 - e^{-2T/\tau}\right),$$

which matches equation (2.2-52). Carrying out the indicated integration, the (1,1) and (2,2) elements are

$$\begin{aligned} Q_D(T)_{11} &= Q_s\tau^2 \int_0^T \lambda^2 + 2\tau\lambda\left(e^{-\lambda/\tau} - 1\right) + \tau^2\left(e^{-2\lambda/\tau} - 2e^{-\lambda/\tau} + 1\right) d\lambda \\ &= Q_s\tau^2\left[\lambda^3/3 - \tau\lambda^2 + \tau^2\lambda - (\tau^3/2)e^{-2\lambda/\tau} + 2\tau^3 e^{-\lambda/\tau} - 2\tau^2 e^{-\lambda/\tau}(\lambda + \tau)\right]_{\lambda=0}^{T} \\ &= Q_s\tau^2\left[T^3/3 - \tau T^2 + \tau^2 T - (\tau^3/2)(e^{-2T/\tau} - 1) - 2\tau^2 T e^{-T/\tau}\right] \end{aligned} \tag{2.2-58}$$

and

$$\begin{aligned} Q_D(T)_{22} &= Q_s\tau^2 \int_0^T \left(1 - 2e^{-\lambda/\tau} + e^{-2\lambda/\tau}\right) d\lambda \\ &= Q_s\tau^2\left[\lambda + 2\tau e^{-2\lambda/\tau} - (\tau/2)e^{-2\lambda/\tau}\right]_{\lambda=0}^{T} \\ &= Q_s\tau^2\left[T + 2\tau(e^{-T/\tau} - 1) - (\tau/2)(e^{-2T/\tau} - 1)\right] \end{aligned} \tag{2.2-59}$$

Notice that while the variance of the Markov process state x_3 is constant, the variance of integrated Markov process states will increase with T, as for the random walk. However, the dominant exponent of T for state x_1 is 3 (not 5) and the exponent of T for state x_2 is 1 (not 3).

We leave integration of remaining terms as an exercise for the reader since most Kalman filter implementations that use first-order Markov process models only evaluate $\mathbf{Q}_D(T)$ for the Markov state. As discussed later in Section 2.3, elements of $\mathbf{Q}_D(T)$ for states that are integrals of the Markov state are usually approximations.

To summarize, the primary differences between the random walk model and the first-order Markov process model are

1. The variance of a first-order Markov process is constant in time. It increases linearly with time for a random walk.
2. The PSD of a first-order Markov process is nearly constant up to frequency $1/(2\pi\tau)$ in Hz. Above this break frequency it falls off at −20 dB per decade. The PSD of white noise is constant at all frequencies. The PSD of a random walk process falls off at −20 dB per decade starting at 0 frequency.
3. The autocorrelation function of a first-order Markov process decays exponentially with the time shift. The autocorrelation function of white noise is zero for any nonzero time shift. The autocorrelation function of a random walk process is equal to 1 for all time shifts.

4. While the mean value of both processes is zero, a first-order Markov process will drive an initial condition towards zero. The initial state of a random walk is "remembered" indefinitely.
5. Because of the last property, physical states of a system model can be easily converted to random walk states by simply modeling the effect on the covariance of white noise added to the state differential equation. Use of a colored noise model, represented as a first-order Markov process, generally involves adding a state to the system model.

More on the pros and cons of using random walk and Markov processes for modeling system randomness appears after the discussion of second-order Markov processes.

2.2.4.3 Second-Order Markov Process

A *second-order Markov process* is a random process in which the probability distribution of the scalar output depends on only two output points immediately in the past, that is,

$$F_x[x(t_k) \mid x(t_{k-1}), \ldots, x(t_1)] = F_x[x(t_k) \mid x(t_{k-1}), x(t_{k-2})]$$

for every $t_1 < t_2 < \ldots < t_k$. Equivalently, the probability distribution of a two-state vector depends on only a single two-state vector immediately in the past. A time-invariant stochastic differential equation defining a second-order Markov process model is

$$\ddot{x}(t) = -2\varsigma\omega_0\,\dot{x}(t) - \omega_0^2 x(t) + q_c(t) \qquad (2.2\text{-}60)$$

where ω_0 is the undamped natural frequency in radians/s and ς is the damping ratio ($\varsigma > 0$). This can also be written as the two-state model:

$$\begin{bmatrix} \dot{x}_1 \\ \dot{x}_2 \end{bmatrix} = \begin{bmatrix} 0 & 1 \\ -\omega_0^2 & -2\varsigma\omega_0 \end{bmatrix} \begin{bmatrix} x_1(t) \\ x_2(t) \end{bmatrix} + \begin{bmatrix} 0 \\ q_c(t) \end{bmatrix} \qquad (2.2\text{-}61)$$

with state transition matrix

$$\mathbf{\Phi}(T) = \frac{e^{-\varsigma\omega_0 T}}{\sqrt{1-\varsigma^2}} \begin{bmatrix} \sqrt{1-\varsigma^2}\cos\theta + \varsigma\sin\theta & \dfrac{1}{\omega_0}\sin\theta \\ -\omega_0\sin\theta & \sqrt{1-\varsigma^2}\cos\theta - \varsigma\sin\theta \end{bmatrix} \quad \text{for } 0 < \varsigma < 1 \qquad (2.2\text{-}62)$$

where $\theta = \omega_0 T\sqrt{1-\varsigma^2}$, or

$$\mathbf{\Phi}(T) = \frac{1}{a+b} \begin{bmatrix} be^{-aT} + ae^{bT} & -e^{-aT} + e^{bT} \\ -ab(e^{-aT} - e^{bT}) & ae^{-aT} + be^{bT} \end{bmatrix} \quad \text{for } \varsigma > 1 \qquad (2.2\text{-}63)$$

where

$$a = \omega_0\left(\varsigma + \sqrt{\varsigma^2 - 1}\right), \quad b = \omega_0\left(-\varsigma + \sqrt{\varsigma^2 - 1}\right).$$

When $0 < \varsigma < 1$ the model is called *underdamped* because it tends to oscillate, and it is called *overdamped* when $\varsigma > 1$. The *critically damped* case $\varsigma = 1$ is handled differently since it has repeated roots. The results are listed in Gelb (1974, p. 44) and in Table 2.1 at the end of this section.

The discrete process noise covariance is

$$\mathbf{Q}_D(T)$$

$$= \frac{Q_s}{\omega_0^2(1-\varsigma^2)}\int_0^T e^{-2\varsigma\omega_0}\begin{bmatrix}\sin\theta \\ \omega_0\left(\sqrt{1-\varsigma^2}\cos\theta-\varsigma\sin\theta\right)\end{bmatrix}\left[\sin\theta \quad \omega_0\left(\sqrt{1-\varsigma^2}\cos\theta-\varsigma\sin\theta\right)\right]d\lambda$$

$$= \frac{Q_s}{\omega_0^2(1-\varsigma^2)}\int_0^T e^{-2\varsigma\omega_0}\begin{bmatrix}\sin^2\theta & \omega_0\sin\theta\left(\sqrt{1-\varsigma^2}\cos\theta-\varsigma\sin\theta\right) \\ \sim & \omega_0^2\left((1-\varsigma^2)\cos^2\theta+\varsigma^2\sin^2\theta-2\varsigma\sqrt{1-\varsigma^2}\sin\theta\cos\theta\right)\end{bmatrix}d\lambda \tag{2.2-64}$$

for $0<\varsigma<1$ with $\theta=\omega_0\lambda\sqrt{1-\varsigma^2}$, or for $\varsigma>0$,

$$\begin{aligned}\mathbf{Q}_D(T) &= \frac{Q_s}{(a+b)^2}\int_0^T\begin{bmatrix}-e^{-a\lambda}+e^{b\lambda} \\ ae^{-a\lambda}+be^{b\lambda}\end{bmatrix}\left[-e^{-a\lambda}+e^{b\lambda} \quad ae^{-a\lambda}+be^{b\lambda}\right]d\lambda \\ &= \frac{Q_s}{(a+b)^2}\int_0^T\begin{bmatrix}e^{-2a\lambda}-2e^{(b-a)\lambda}+e^{2b\lambda} & -ae^{-2a\lambda}+(a-b)e^{(b-a)\lambda}+be^{2b\lambda} \\ \sim & a^2e^{-2a\lambda}+2abe^{(b-a)\lambda}+b^2e^{2b\lambda}\end{bmatrix}d\lambda\end{aligned}. \tag{2.2-65}$$

(Note: The symbol "~" indicates that the term is equal to the transposed element, since the matrix is symmetric.) Carrying out the indicated integrations for the diagonal elements in the underdamped ($0<\varsigma<1$) case gives the result

$$\begin{aligned}Q_D(T)_{11} &= \frac{Q_s}{b^2}\int_0^T e^{-f\lambda}\sin^2(g\lambda)\,d\lambda \\ &= -\frac{Q_s}{g^2}\left[e^{-f\lambda}\left(\frac{1}{4f}+\frac{-2f\cos(2g\lambda)+2g\sin(2g\lambda)}{8(f^2+g^2)}\right)\right]_0^T \\ &= \frac{Q_s}{4}\left[\frac{1}{f(f^2+g^2)}-\frac{e^{-fT}}{g^2}\left(\frac{1}{f}+\frac{-f\cos(2gT)+g\sin(2gT)}{(f^2+g^2)}\right)\right]\end{aligned} \tag{2.2-66}$$

and

$$\begin{aligned}Q_D(T)_{22} &= \frac{Q_s}{g^2}\int_0^T e^{-f\lambda}\left(g^2\cos^2(g\lambda)+f^2\sin^2(g\lambda)-2fg\sin(g\lambda)\cos(g\lambda)\right)d\lambda \\ &= \frac{Q_s}{g^2}\left[e^{-f\lambda}\left(\begin{aligned}&-\frac{f^2+g^2}{4f}+\frac{f^2-g^2}{8(f^2+g^2)}(2f\cos(2g\lambda)-2g\sin(2g\lambda)) \\ &+\frac{2fg}{8(f^2+g^2)}(2f\sin(2g\lambda)+2g\cos(2g\lambda))\end{aligned}\right)\right]_0^T \\ &= \frac{Q_s}{2g^2}\left[e^{-fT}\left(\begin{aligned}&-\frac{f^2+g^2}{2f}+\frac{f^2-g^2}{2(f^2+g^2)}(f\cos(2gT)-g\sin(2gT)) \\ &+\frac{fg}{(f^2+g^2)}(f\sin(2gT)+d\cos(2gT))\end{aligned}\right)+\frac{g^2}{2f}\right]\end{aligned} \tag{2.2-67}$$

where $f=\varsigma\omega_0$, $g=\omega_0\sqrt{1-\varsigma^2}$. Taking the limit as $T\to\infty$, the steady-state variances of the two states are

$$\begin{aligned}\sigma_{x1}^2 &= \frac{Q_s}{4f(f^2+g^2)} = \frac{Q_s}{4\varsigma\omega_0^3} \\ \sigma_{x2}^2 &= \frac{Q_s}{4f} = \frac{Q_s}{4\varsigma\omega_0}\end{aligned} \quad . \tag{2.2-68}$$

for $0 < \varsigma < 1$. $Q_s = 4\varsigma\omega_0\sigma_{x2}^2$ can be used when determining appropriate values of Q_s given σ_{x2}^2. Although not obvious, it can also be shown that $E[x_1x_2] = 0$.

The equivalent relations for the overdamped case are

$$\begin{aligned}Q_D(T)_{11} &= \frac{Q_s}{(a+b)^2}\int_0^T \left(e^{-2a\lambda} - 2e^{-\backslash(b-a)\lambda} + e^{2b\lambda}\right)d\lambda \\ &= \frac{Q_s}{(a+b)^2}\left[-\frac{1}{2a}e^{-2a\lambda} - \frac{2}{b-a}e^{(b-a)\lambda} + \frac{1}{2b}e^{2b\lambda}\right]_0^T \\ &= \frac{Q_s}{(a+b)^2}\left[\frac{1}{2a}(1-e^{-2aT}) + \frac{2}{b-a}(1-e^{(b-a)T}) - \frac{1}{2b}(1-e^{2bT})\right]\end{aligned} \tag{2.2-69}$$

$$\begin{aligned}Q_D(T)_{22} &= \frac{Q_s}{(a+b)^2}\int_0^T \left(a^2e^{-2a\lambda} + 2abe^{(b-a)\lambda} + b^2e^{2b\lambda}\right)d\lambda \\ &= \frac{Q_s}{(a+b)^2}\left[\left(-\frac{a}{2}e^{-2a\lambda} + \frac{2ab}{b-a}e^{(b-a)\lambda} + \frac{b}{2}e^{2b\lambda}\right)\right]_0^T \\ &= \frac{Q_s}{(a+b)^2}\left[\frac{a}{2}(1-e^{-2aT}) - \frac{2ab}{b-a}(1-e^{(a+b)T}) - \frac{b}{2}(1-e^{2bT})\right]\end{aligned} \quad . \tag{2.2-70}$$

In the limit as $T \to \infty$ the steady-state variances of the two states for $\varsigma > 0$ are

$$\begin{aligned}\sigma_{x1}^2 &= \frac{Q_s}{(a+b)^2}\left(\frac{1}{2a} + \frac{2}{b-a} - \frac{1}{2b}\right) = \frac{Q_s}{2ab(b-a)} = \frac{Q_s}{4\varsigma\omega_0^3} \\ \sigma_{x2}^2 &= \frac{Q_s}{(a+b)^2}\left(\frac{a}{2} - \frac{2ab}{b-a} - \frac{b}{2}\right) = -\frac{Q_s}{2(b-a)} = \frac{Q_s}{4\varsigma\omega_0}\end{aligned} \quad . \tag{2.2-71}$$

As expected, these are the same equations as for the underdamped case (eq. 2.2-68). The autocorrelation function of the output state (x_2) can be computed from the steady-state variance and $\mathbf{\Phi}(T)$ using equation (2.2-62) or equation (2.2-63),

$$R_{x2}(T) = \begin{cases} \sigma_{x2}^2\, e^{-\xi\omega_0|T|}\left(\cos\theta - \dfrac{\varsigma}{\sqrt{1-\varsigma^2}}\sin\theta\right) & \text{for } 0 < \varsigma < 1 \\ \dfrac{\sigma_{x2}^2}{2\omega_0\sqrt{\varsigma^2-1}}\left(ae^{-aT} + be^{bT}\right) & \text{for } \varsigma > 1 \end{cases} \quad , \tag{2.2-72}$$

where a, b and θ were previously defined for equations (2.2-62) and (2.2-63).

The PSD of the output can be computed from the Laplace transform of the transfer function corresponding to equation (2.2-60),

$$H(s) = \frac{1}{s^2 + 2\varsigma\omega_0 s + \omega_0^2},$$

which leads to

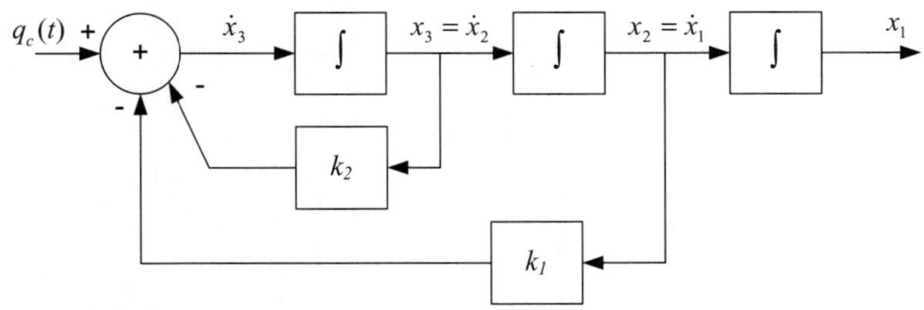

FIGURE 2.5: Integrated second-order Markov process.

$$
\begin{aligned}
S_x(\omega) &= \sigma_x^2 |H(j\omega)|^2 = \sigma_x^2 H(j\omega)H(-j\omega) \\
&= \frac{\sigma_x^2}{(\omega_0^2 - \omega^2)^2 + 4\varsigma^2\omega_0^2\omega^2} .
\end{aligned}
\tag{2.2-73}
$$

As expected, the PSD may exhibit a peak near ω_0 when $0 < \varsigma < 1$. If the output is moved from the second to the first state, the PSD is

$$
S_x(\omega) = \frac{\sigma_x^2\omega^2}{(\omega_0^2 - \omega^2)^2 + 4\varsigma^2\omega_0^2\omega^2},
\tag{2.2-74}
$$

which has a band-pass characteristic.

Low-order Markov process models are mostly used to account for randomness and uncertainty in the highest derivatives of a system model, so it is again of interest to characterize behavior of states that are integrals of the second-order Markov process output. Since the x_1 state in equations (2.2-62) and (2.2-63) is equal to the integral of the model output (x_2), x_1 can be interpreted as proportional to velocity if x_2 represents acceleration. Hence the integral of x_1 can be interpreted as proportional to position. If we imbed the second-order Markov process in the three-state model used previously (for the random walk and first-order Markov process), and retain the state numbering used in equation (2.2-47) or (2.2-56), we obtain the integrated second-order Markov model of Figure 2.5.

For this model the dynamic matrix $\mathbf{F}$ in $\dot{\mathbf{x}} = \mathbf{F}x + \mathbf{q}$ is

$$
\mathbf{F} = \begin{bmatrix} 0 & 1 & 0 \\ 0 & 0 & 1 \\ 0 & \omega_0^2 & -2\varsigma\omega_0 \end{bmatrix}.
$$

To compute the state transition matrix and discrete state noise covariance of this model, we note that adding an extra integrator does not change the dynamics of the second-order Markov process. Hence the lower-right 2×2 block of $\mathbf{\Phi}(t)$ will be equal to the $\mathbf{\Phi}(t)$ given in equation (2.2-62). Further, process noise only drives the third state, so $\Phi(t)_{13}$ is the only element of $\mathbf{\Phi}(t)$ that is required to compute the position (1,1) element of $\mathbf{Q}_D$. We again use $\mathbf{\Phi}(t) = \mathscr{L}^{-1}[s\mathbf{I} - \mathbf{F}]^{-1}$ and obtain

$$
\Phi(T)_{13} = \frac{1}{\omega_0^2}\left[1 - e^{-fT}\left(\cos gT + \frac{f}{g}\sin dT\right)\right]
\tag{2.2-75}
$$

for $0 < \varsigma < 1$ with $f = \varsigma\omega_0$ and $g = \omega_0\sqrt{1-\varsigma^2}$. This is used to compute

$$
\begin{aligned}
Q_D(T)_{11} &= Q_s \int_0^T [\Phi(T)_{13}]^2 \, d\lambda \\
&= \frac{Q_s}{\omega_0^4} \int_0^T \begin{bmatrix} 1 - 2e^{-f\lambda}\left(\cos g\lambda + \dfrac{f}{g}\sin g\lambda\right) \\ + e^{-2f\lambda}[\cos^2 g\lambda + \dfrac{2f}{g}\cos g\lambda \sin g\lambda + \dfrac{f^2}{g^2}\sin^2 g\lambda] \end{bmatrix} d\lambda \\
&= \frac{Q_s}{\omega_0^4} \begin{bmatrix} \lambda + \dfrac{2e^{-f\lambda}}{f^2+g^2}\left(2f\cos g\lambda + \left(-g + \dfrac{f^2}{g}\right)\sin g\lambda\right) \\ + \dfrac{e^{-2f\lambda}}{4(f^2+g^2)}\left(-\dfrac{(f^2+g^2)^2}{fg^2} + f\left(\dfrac{f^2}{g^2} - 3\right)\cos 2g\lambda + \left(g - \dfrac{3f^2}{g}\right)\sin 2g\lambda\right) \end{bmatrix}_0^T \\
&= \frac{Q_s}{\omega_0^4} \begin{bmatrix} T + \dfrac{2e^{-fT}}{f^2+g^2}\left(2f\cos gT + \left(-g + \dfrac{f^2}{g}\right)\sin gT\right) \\ + \dfrac{e^{-2fT}}{4(f^2+g^2)}\left(-\dfrac{(f^2+g^2)^2}{fg^2} + f\left(\dfrac{f^2}{g^2} - 3\right)\cos 2gT + \left(g - \dfrac{3f^2}{g}\right)\sin 2gT\right) \\ + \dfrac{1}{4f} - \dfrac{3f}{f^2+g^2} \end{bmatrix}
\end{aligned}
\tag{2.2-76}
$$

The point of this complicated derivation is to show that $Q_D(T)_{11}$ grows linearly with T for $T \gg c$ $(=\varsigma\omega_0)$.

Figure 2.6 shows the PSD for

1. White noise process (integrated state is random walk)
2. First-order Markov process with $f_0 = 10\,\text{Hz}$
3. Second-order Markov process with $\zeta = 0.1$ and $f_0 = 10\,\text{Hz}$
4. Second-order Markov process with $\zeta = 0.5$ and $f_0 = 10\,\text{Hz}$

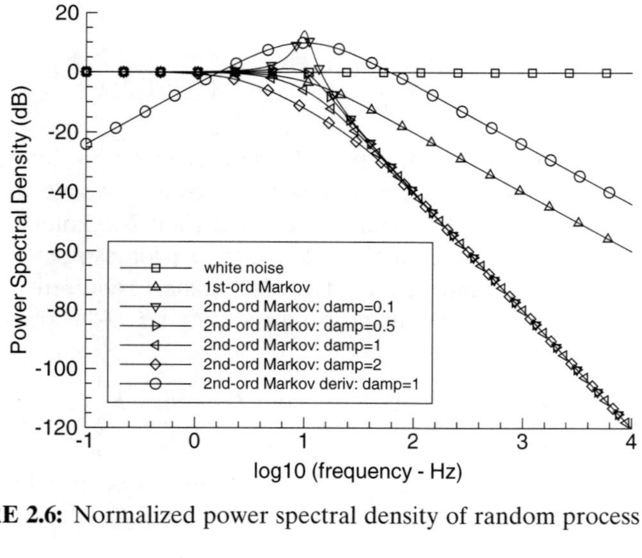

FIGURE 2.6: Normalized power spectral density of random process models.

TABLE 2.1: Random Walk and Markov Process Characteristics

Characteristic	Random Walk	First-Order Markov Process	Second-Order Markov Process
Initial condition (homogenous) response	Remains constant	Decays to 0	Decays to 0 with possible damped oscillation
Growth of variance with time	Linear in T	Constant for $T > \tau$	Constant for $T \gg 1/\omega_0$
Variance growth of first time integral	$\sim T^3$	$\sim T$ for $T \gg \tau$	Constant for $T \gg 1/\omega_0$
Variance growth of second time integral	$\sim T^5$	$\sim T^3$ for $T \gg \tau$	$\sim T$ for $T \gg 1/\omega_0$
PSD versus frequency	Constant	Low-pass: –20 db/decade above frequency $1/\tau$: $S_x(\omega) = \dfrac{Q_s\tau^2}{1+(\omega\tau)^2}$	Low-pass: –40 db/decade above ω_0 for integral state: $S_x(\omega) = \dfrac{\sigma_x^2}{(\omega_0^2-\omega^2)^2 + 4\varsigma^2\omega_0^2\omega^2}$ Band-pass: –20 db/decade above ω_0 for highest derivative state: $S_x(\omega) = \dfrac{\sigma_x^2\omega^2}{(\omega_0^2-\omega^2)^2 + 4\varsigma^2\omega_0^2\omega^2}$ Peak when $\zeta < 1$.
Steady-state variance	None	$\sigma_x^2 = \dfrac{Q_s\tau}{2}$	$\sigma_{x1}^2 = \dfrac{Q_s}{4\varsigma\omega_0^3}$ $\sigma_{x2}^2 = \dfrac{Q_s}{4\varsigma\omega_0}$
Autocorrelation function versus T	$Q_s\delta(T)$	$\sigma_x^2 e^{-\lvert T\rvert/\tau}$	$\sigma_{x2}^2 e^{-\xi\omega_0\lvert T\rvert}\left(\cos\theta - \dfrac{\varsigma}{\sqrt{1-\varsigma^2}}\sin\theta\right)$ for $0<\zeta<1$, $\theta = \omega_0\lambda\sqrt{1-\varsigma^2}$ $\dfrac{\sigma_{x2}^2}{2\omega_0\sqrt{\varsigma^2-1}}\left(ae^{-a\lvert T\rvert} + be^{b\lvert T\rvert}\right)$ for $\zeta > 1$, $a = \omega_0\left(\sqrt{\varsigma^2-1}+\varsigma\right)$, $b = \omega_0\left(\sqrt{\varsigma^2-1}-\varsigma\right)$ $\sigma_{x2}^2 e^{-\omega_0\lvert T\rvert}(1+\omega_0\lvert T\rvert)$ for $\zeta = 1$

5. Second-order Markov process with $\zeta = 1.0$ and $f_0 = 10\,\text{Hz}$
6. Second-order Markov process with $\zeta = 2.0$ and $f_0 = 10\,\text{Hz}$
7. Second-order Markov process with $\zeta = 1.0$, $f_0 = 10\,\text{Hz}$, and output at x_3 in Figure 2.5

This is scaled by −20 dB to keep the PSD on the same plot as other data.

In all cases $Q_s = 1$, but the PSD magnitudes were adjusted to be equal at frequency 0, except for case #7. Notice that both the first and second-order (x_2 output) Markov processes have low-pass characteristics. Power falls off above the break frequency at −20 dB/decade for the first-order model and −40 dB/decade for the second-order model with x_2 as output. The second-order Markov process using x_3 as output has a band-pass characteristic and the PSD falls off at −20 dB/decade above 10 Hz. Also note that the second-order model has a peak near 10 Hz when $0 < \varsigma < 1$.

Table 2.1 compares the characteristics of the three random process models. When selecting a random process model for a particular application, pay particular attention to the initial condition response, the variance growth with time, and the PSD. Random walk models are used more frequently than Markov process models because they do not require additional states in a system model. Furthermore, measurements are often separated from the random process by one or two levels of integration. When measurement noise is significant and measurements are integrals of the driving process noise, it is often difficult to determine whether the process noise is white or colored. In these cases little is gained by using a Markov process model, rather than assuming white process noise for an existing model state.

When used, Markov process noise models are usually first-order. Second-order models are indicated when the PSD has a dominant peak. Otherwise, first-order models will usually perform as well as a second-order models in a Kalman filter. This author has only encountered three applications in which second-order Markov process models were used. One modeled motion of maneuvering tanks where tank drivers used serpentine paths to avoid potential enemy fire. The second application modeled a nearly oscillatory system, and the third modeled spacecraft attitude pointing errors.

2.2.5 Linear Regression Models

Least-squares estimation is sometimes referred to as regression modeling. The term is believed to have been first used by Sir Francis Galton (1822–1911) to describe his discovery that the heights of children were correlated with the parent's deviation from the mean height, but they also tended to "regress" to the mean height. The coefficients describing the observed correlations with the parent's height were less than one, and were called regression coefficients. As the term is used today, regression modeling usually implies fitting a set of observed samples using a linear model that takes into account known input conditions for each sample. In other words, it tries to determine parameters that represent correlations between input and output data. Hence the model is generally based on observed input-output correlations rather than on first-principle concepts. For example, regression modeling is used by medical researchers when trying to determine important parameters for predicting incidence of a particular disease.

To demonstrate the concept, assume that we are researching causative factors for lung cancer. We might create a list of potential explanatory variables that include:

1. Total years of smoking
2. Average number of cigarettes smoked per day
3. Number of years since last smoked
4. Total months of exposure to asbestos above some threshold
5. Urban/suburban/rural location
6. Type of job
7. Age
8. Gender
9. Race
10. Socioeconomic status.

This is not intended to be an exhaustive list but rather a sample of parameters that might be relevant. Then using a large sample of lung cancer incidence and patient histories, we could compute linear correlation coefficients between cancer incidence and the explanatory variables. The model is

$$\begin{bmatrix} y_1 \\ y_2 \\ y_3. \\ \vdots \\ y_{m-1} \\ y_m \end{bmatrix} = \begin{bmatrix} \alpha_{11} & \alpha_{12} & \alpha_{13}. & \cdots & \alpha_{1,n-1} & \alpha_{1n} \\ \alpha_{21} & \alpha_{22} & \alpha_{23} & \cdots & \alpha_{2,n-1} & \alpha_{2n} \\ \alpha_{31} & \alpha_{32} & \alpha_{33} & \cdots & \alpha_{3,n-1} & \alpha_{3n} \\ \vdots & \vdots & \vdots & \ddots & \vdots & \vdots \\ \alpha_{m-1,1} & \alpha_{m-1,2} & \alpha_{m-1,3} & \cdots & \alpha_{m-1,n-1} & \alpha_{m-1,n} \\ \alpha_{m,1} & \alpha_{m,2} & \alpha_{m,3} & \cdots & \alpha_{m,n-1} & \alpha_{m,n} \end{bmatrix} \begin{bmatrix} c_1 \\ c_2 \\ . \\ . \\ . \\ c_n \end{bmatrix}$$

where

y_i variables represent a particular person i's status as whether (+1) or not (0) he/she had lung cancer,

α_{ij} coefficients are person i's response (converted to a numeric value) to a question on explanatory factor j, and

c_j coefficients are to be determined by the method of least squares.

Additional α_{ij} coefficients could also include nonlinear functions of other α_{ij} coefficients—such as products, ratios, logarithms, exponentials, and square roots—with corresponding definitions for the added c_j coefficients. This type of model is very different from the models that we have considered so far. It is not based on a model of a physical system and does not use time as an independent variable (although time is sometimes included). Further, the "measurements" y_i can only have two values (0 or 1) in this particular example. Since measurement noise is not explicitly modeled, the solution is obtained by minimizing a simple (unweighted) sum-of-squares of residuals between the measured and model-computed y_i. It is likely that many of the c_j will be nearly zero, and even for factors that are important, the c_j values will still be small.

This book does not directly address this type of modeling. Many books on the subject are available (e.g., Tukey 1977; Hoaglin et al. 1983; Draper and Smith 1998; Wolberg 2006). However, various techniques used in regression modeling are directly relevant to more general least-squares estimation, and will be discussed in Chapter 6.

2.2.6 Reduced-Order Modeling

In many estimation applications it is not practical to use detailed first-principles models of a system. For example, three-dimensional simulations of mass and energy flow in a system (e.g., atmospheric weather, subsurface groundwater flow) often include hundreds of thousands or even millions of nodes. Although detailed modeling is often of benefit in simulations, use of such a high-order model in an estimator is generally neither necessary nor desirable. Hence estimators frequently use *reduced-order models* (ROMs) of systems. This was essential for several decades after initial development of digital computers because storage and processing power were very limited. Even today it is often necessary because problem sizes have continued to grow as fast (if not faster) than computational capabilities.

The methods used to develop a ROM are varied. Some of the techniques are as follows.

1. Define subsystem elements using lumped-parameter models of overall input-output characteristics (e.g., exponential response model of a heat exchanger) rather than using detailed first-principles models.
2. Ignore high-frequency effects not readily detected at measurement sampling frequencies, or above the dominant dynamic response of the system. Increase the modeled measurement noise variance to compensate.
3. Use first- or second-order Markov processes in a Kalman filter to model time-correlated effects caused by small system discontinuities or other quasi-random behavior. Empirically identify parameters of the Markov processes using moderately large samples of system input-output data (see Chapter 11).
4. Treat some model states as random walk processes (model white noise input to the derivative) rather than modeling the derivative using additional states.
5. Use eigen analysis to define linear combinations of states that behave similarly and can be treated as a single transformed state.
6. Ignore small magnitude states that have little effect on more important states. Compensate with additional process noise.
7. Reduce the number of nodes in a finite-element or finite difference model, and account for the loss of accuracy by either increasing the modeled measurement noise variance, increasing the process noise variance or both.

Some of these techniques are used in examples of the next chapter. It is difficult to define general ROM modeling rules applicable to a variety of systems. Detailed knowledge and first-principles understanding of the system under consideration are still desirable when designing a ROM. Furthermore, extensive testing using real and simulated data is necessary to validate and compare models.

2.3 COMPUTATION OF STATE TRANSITION AND PROCESS NOISE MATRICES

After reviewing the different types of models used to describe continuous systems, we now address computation of the state transition matrix ($\mathbf{\Phi}$) and discrete state noise covariance matrix ($\mathbf{Q}_D$) for a given time interval $T = t_{i+1} - t_i$. Before discussing details, it is helpful to understand how $\mathbf{\Phi}$ and $\mathbf{Q}_D$ are used in the estimation process because usage has an impact on requirements for the methods. Details of least-squares and Kalman estimation will be presented in later chapters, but for the purposes of this section, it is sufficient to know that the propagated state vector $\mathbf{x}$ must be computed at each measurement time (in order to compute measurement residuals), and that $\mathbf{\Phi}$ and $\mathbf{Q}_D$ must be computed for the time intervals between measurements. In fully linear systems, $\mathbf{\Phi}$ may be used to propagate the state $\mathbf{x}$ from one measurement time to the next. In these cases, $\mathbf{\Phi}$ must be very accurate. Alternately, in nonlinear systems the state propagation may be performed by numerical integration, and $\mathbf{\Phi}$ may only be used (directly or indirectly) to compute measurement partial derivatives with respect to the state vector at some epoch time. If $\mathbf{\Phi}$ is only used for purposes of computing partial derivatives, the accuracy requirements are much lower and approximations may be used.

As noted in Section 2.2.1, $\mathbf{\Phi}$ is required in linear or nonlinear least-squares estimation to define the linear relationship between perturbations in the epoch state vector at t_0 and the integrated state vector at the measurement times; that is, $\delta\mathbf{x}(t_i) = \mathbf{\Phi}(t_i,t_0)\delta\mathbf{x}(t_0)$. $\mathbf{\Phi}$ is also used when calculating optimal weighting of the data. Usually $\mathbf{\Phi}$ is only calculated for time intervals between measurements, and $\mathbf{\Phi}$ for the entire interval from epoch t_0 is obtained using the product rule:

$$\mathbf{\Phi}(t_{i+1},t_0) = \mathbf{\Phi}(t_{i+1},t_i)\mathbf{\Phi}(t_i,t_0). \tag{2.3-1}$$

As used in the Kalman filter, $\mathbf{\Phi}$ defines the state perturbations from one measurement time to the next:

$$\delta\mathbf{x}(t_{i+1}) = \mathbf{\Phi}(t_{i+1},t_i)\delta\mathbf{x}(t_i) + \mathbf{q}_D(t_{i+1},t_i). \tag{2.3-2}$$

The process noise covariance $\mathbf{Q}_D = E(\mathbf{q}_D\mathbf{q}_D^T)$ models the integrated effects of random process noise over the given time interval. Both $\mathbf{\Phi}$ and $\mathbf{Q}_D$ are required when calculating optimal weighting of the measurements. The above perturbation definitions apply both when the dynamic model is completely linear, or is nonlinear and has been linearized about some reference state trajectory. Thus we assume in the next section that the system differential equations are linear.

2.3.1 Numeric Computation of $\mathbf{\Phi}$

Section 2.2 briefly discussed three methods for computing the state transition matrix, $\mathbf{\Phi}(T)$, corresponding to the state differential equation

$$\dot{\mathbf{x}}(t) = \mathbf{F}\,\mathbf{x}(t) + \mathbf{q}(t). \tag{2.3-3}$$

To those three we add five other methods for consideration. The options are listed below.

1. Taylor series expansion of $e^{\mathbf{F}T}$
2. Padé series expansion of $e^{\mathbf{F}T}$
3. Inverse Laplace transform of $[s\mathbf{I} - \mathbf{F}]^{-1}$
4. Integration of $\dot{\mathbf{\Phi}} = \mathbf{F}\mathbf{\Phi}$
5. Matrix decomposition methods
6. Scaling and squaring (interval doubling) used with another method
7. Direct numeric partial derivatives using integration
8. Partitioned combinations of methods

We now explore the practicality of these approaches.

2.3.1.1 *Taylor Series Expansion of $e^{\mathbf{F}T}$* When the state dynamics are time-invariant (**F** is constant) or can be treated as time-invariant over the integration interval, $\mathbf{\Phi}(T)$ is equal to the matrix exponential $e^{\mathbf{F}T}$, which can be expanded in a Taylor series as:

$$\mathbf{\Phi}(T) = e^{FT} = \mathbf{I} + \mathbf{F}T + \frac{(\mathbf{F}T)^2}{2!} + \frac{(\mathbf{F}T)^3}{3!} + \ldots \tag{2.3-4}$$

When **F** is time varying, the integral

$$\mathbf{\Phi}(T) = \int_t^{t+T} e^{\mathbf{F}(\lambda)\lambda} d\lambda$$

can sometimes be used, but the conditions under which this applies are rarely satisfied. In the time-invariant case, the Taylor series is useful when the series converges rapidly. For example, the quadratic polynomial model,

$$\mathbf{F} = \begin{bmatrix} 0 & 1 & 0 \\ 0 & 0 & 1 \\ 0 & 0 & 0 \end{bmatrix},$$

converges in two terms to

$$\mathbf{\Phi}(T) = \mathbf{I} + \mathbf{F}T + (\mathbf{F}T)^2/2 = \begin{bmatrix} 1 & T & T^2/2 \\ 0 & 1 & T \\ 0 & 0 & 1 \end{bmatrix}.$$

Full convergence in a few terms is unusual for real problems, but even so, numerical evaluation of the series can be used when "numeric convergence" is achieved in a limited number of terms. That is, the series summation is stopped when the additional term added is numerically insignificant compared with the existing sums. Moler and Van Loan (2003) compared methods for computing $e^{\mathbf{A}}$ using a second-order example in which one eigenvalue of **A** was 17 times larger than the other. They found that 59 terms were required to achieve Taylor series convergence, and the result was only accurate to nine digits when using double precision. They generally regard series approximation as an inefficient method that is sensitive to round-off error and should rarely be used. Despite these reservations, Taylor series approximation for $\mathbf{\Phi}(T)$ is sometimes used in the Kalman filter covariance time update because the covariance propagation can be less accurate than state vector propagation to achieve satisfactory performance *in some cases*. This statement will make more sense after reading Chapter 8 on the Kalman filter. In any case, use of series methods for computing $\mathbf{\Phi}(T)$ is risky and should be considered only after numerically verifying that the approximation is sufficiently accurate for the given problem. Taylor series approximation of $e^{\mathbf{A}}$ is used in MATLAB function EXPM2.

2.3.1.2 Padé Series Expansion of $e^{\mathbf{F}T}$ The (p,q) Padé approximation to $e^{\mathbf{F}T}$ is

$$e^{FT} \cong [\mathbf{D}_{pq}(\mathbf{F}T)]^{-1}\,\mathbf{N}_{pq}(\mathbf{F}T) \tag{2.3-5}$$

where

$$\mathbf{N}_{pq}(\mathbf{F}T) = \sum_{j=0}^{p} \frac{(p+q-j)!\,p!}{(p+q)!\,j!\,(p-q)!}(\mathbf{F}T)^{j} \tag{2.3-6}$$

and

$$\mathbf{D}_{pq}(\mathbf{F}T) = \sum_{j=0}^{q} \frac{(p+q-j)!\,q!}{(p+q)!\,j!\,(q-q)!}(-\mathbf{F}T)^{j}. \tag{2.3-7}$$

Matrix $\mathbf{D}_{pq}(\mathbf{F}T)$ will be nonsingular if p and q are sufficiently large or if the eigenvalues of $\mathbf{F}T$ are negative. Again, round-off error makes Padé approximations unreliable. Cancellation errors can prevent accurate determination of matrices $\mathbf{N}$ and $\mathbf{D}$, or $\mathbf{D}_{pq}(\mathbf{F}T)$ may be poorly conditioned with respect to inversion. In Moler and Van Loan's second-order example described above, the best results were obtained with $p = q = 10$, and the condition number of $\mathbf{D}$ was greater than 10^4. All other values of p, q gave less accurate results. Use of $p = q$ is generally more efficient and accurate than $p \neq q$. However, Padé approximation to $e^{\mathbf{F}T}$ can be used if $\|\mathbf{F}T\|$ is not too large (see Appendix A for definitions of matrix norms). Golub and Van Loan (1996, p 572) present algorithm 11.3.1 for automatic selection of the optimum p, q to achieve a given accuracy. This algorithm combines Padé approximation with the scaling and squaring method described below. Their algorithm 11.3.1 is used in MATLAB function EXPM1.

2.3.1.3 Laplace Tranform The Laplace transform of time-invariant differential equation (2.3-3) is

$$s\,\mathbf{x}(s) = \mathbf{F}\,\mathbf{x}(s) + \mathbf{q}(s) \qquad \text{or} \qquad (s\mathbf{I} - \mathbf{F})\,\mathbf{x}(s) = \mathbf{q}(s).$$

Hence the homogenous (unforced or initial condition) response can be computed using the inverse Laplace transform as

$$\mathbf{x}(t) = \mathscr{L}^{-1}[s\mathbf{I} - \mathbf{F}]^{-1}\,\mathbf{x}(0).$$

This implies that

$$\mathbf{\Phi}(T) = \mathscr{L}^{-1}[s\mathbf{I} - \mathbf{F}]^{-1}. \tag{2.3-8}$$

This method was used in Section 2.2.4 when computing $\mathbf{\Phi}$ for first- and second-order Markov processes. Notice that it involves analytically inverting the matrix $[s\mathbf{I} - \mathbf{F}]$, computing a partial fractions expansion of each term in the matrix, and taking the inverse Laplace transform of each term. Even in the two-state case this involves much work, and the method is really not practical for manual computation when the state dimension exceeds four. However, the approach should be considered for low-order problems when analytic solutions are needed. It is also possible to implement Laplace transform methods recursively in software, or to use other approaches discussed by Moler and Van Loan. However, these methods are computationally expensive $[O(n^4)]$ and may be seriously affected by round-off errors.

2.3.1.4 Integration of $\dot{\boldsymbol{\Phi}} = \mathbf{F}\boldsymbol{\Phi}$ In Section 2.2.1 we showed that

$$\frac{d}{d\tau}\boldsymbol{\Phi}(\tau,t) = \mathbf{F}(\tau)\boldsymbol{\Phi}(\tau,t). \tag{2.3-9}$$

for both time-varying and time-invariant models. Hence it is possible to numerically integrate $\boldsymbol{\Phi}$ over the time interval from t to τ, initialized with $\boldsymbol{\Phi}(t,t) = \mathbf{I}$. Any general-purpose ODE numerical integration method, such as fourth-order Runge-Kutta or Bulirsch-Stoer, employing automatic step size control (see Press et al. 2007, chapter 17) may be used. To use the ODE solver the elements of $\boldsymbol{\Phi}$ must be stored in a vector and treated as a single set of variables to be integrated. Likewise, the corresponding elements of $\dot{\boldsymbol{\Phi}}$ must be stored in a vector and returned to the ODE solver from the derivative function. This approach is easy to implement and is relatively robust. However, there are a total of n^2 elements in $\boldsymbol{\Phi}$ for an n-element state $\mathbf{x}$, and hence the numerical integration involves n^2 variables if matrix structure is ignored. This can be computationally expensive for large problems. At a minimum, the matrix multiplication $\mathbf{F}\boldsymbol{\Phi}$ should be implemented using a general-purpose sparse matrix multiplication algorithm (see Chapter 13) since $\mathbf{F}$ is often very sparse (as demonstrated for the polynomial case). In some real-world problems $\mathbf{F}$ is more than 90% zeroes! If properly implemented, the number of required floating point operations for the multiplication can be reduced from the normal n^3 by the fractional number of nonzero entries in $\mathbf{F}$, for example, nearly 90% reduction for 90% sparseness. If it is also known that $\boldsymbol{\Phi}$ has a sparse structure (such as upper triangular), the number of variables included in the integration can be greatly reduced. This does require, however, additional coding effort. The execution time may be further reduced by using an ODE solver that takes advantage of the linear, constant coefficient nature of the problem. See Moler and Van Loan for a discussion of options. They also note that ODE solvers are increasingly popular and work well for stiff problems.

2.3.1.5 Matrix Decomposition Methods These are based on similarity transformations of the form $\mathbf{F} = \mathbf{S}\mathbf{B}\mathbf{S}^{-1}$, which leads to

$$\boldsymbol{\Phi}(t) = e^{\mathbf{F}t} = \mathbf{S}e^{\mathbf{B}t}\mathbf{S}^{-1}. \tag{2.3-10}$$

A transformation based on eigenvector/eigenvalue decomposition works well when $\mathbf{F}$ is symmetric, but that is rarely the case for dynamic systems of interest. Eigenvector/eigenvalue decomposition also has problems when it is not possible to compute a complete set of linearly independent eigenvectors—possibly because $\mathbf{F}$ is nearly singular. Other decompositions include the Jordan canonical form, Schur decomposition (orthogonal $\mathbf{S}$ and triangular $\mathbf{B}$), and block diagonal $\mathbf{B}$. Unfortunately, some of these approaches do not work when $\mathbf{F}$ has complex eigenvalues, or they fail because of high numerical sensitivity. Moler and Van Loan indicate that modified Schur decomposition methods are still of interest, but applications using this approach are rare. MATLAB function EXPM3 implements an eigenvector decomposition method.

2.3.1.6 Scaling and Squaring The scaling and squaring (interval doubling) method is based on the observation that the homogenous part of **x** can be obtained recursively as

$$\begin{aligned} \mathbf{x}(t_2) &= \mathbf{\Phi}(t_2 - t_1)\mathbf{x}(t_1) \\ &= \mathbf{\Phi}(t_2 - t_1)\mathbf{\Phi}(t_1 - t_0)\mathbf{x}(t_0) \end{aligned} \tag{2.3-11}$$

or

$$\mathbf{\Phi}(t_2 - t_0) = \mathbf{\Phi}(t_2 - t_1)\mathbf{\Phi}(t_1 - t_0). \tag{2.3-12}$$

Hence the transition matrix for the total time interval T can be generated as

$$\mathbf{\Phi}(T) = [\mathbf{\Phi}(T/m)]^m \tag{2.3-13}$$

where m is an integer to be chosen. Since $\mathbf{\Phi}$ is only evaluated for the relatively small time interval T/m, methods such as Taylor series or Padé approximation can be used. If m is chosen so that $\|\mathbf{\Phi}(T/m)\| = \|e^{\mathbf{F}T/m}\| \leq 1$, a Taylor or Padé series can be truncated with relatively few terms. (This criterion is equivalent to selecting T/m to be small compared with the time constants of **F**.) The method is also computationally efficient and is not sensitive to errors caused by a large spread in eigenvalues of **F**. However, it can still be affected by round-off errors.

For efficient implementation, m should be chosen as a power of 2 for which $\mathbf{\Phi}(T/m)$ can be reliably and efficiently computed. Then $\mathbf{\Phi}(T)$ is formed by repeatedly squaring as follows:

$m = \max(\log_2(T/T_{\max}) + 1, 1)$
$T_m = T/2^m$
Compute $\mathbf{\Phi}(T_m)$
For $i = 1$ to m
  $\mathbf{\Phi}(T_m \times 2^i) = \mathbf{\Phi}(T_m \times 2^{(i-1)}) \times \mathbf{\Phi}(T_m \times 2^{(i-1)})$
End loop

where $T_{\max}$ is the maximum allowed step size, which should be less than the shortest time constants of **F**. Notice that the doubling method only involves $\log_2(T/T_{\max}) + 1$ matrix multiplications.

The scaling and squaring method has many advantages and is often used for least-squares and Kalman filtering applications. Since the time interval for evaluation of $\mathbf{\Phi}$ is driven by the time step between measurements, scaling and squaring is only needed when the measurement time step is large compared with system time constants. This may happen because of data gaps or because availability of measurements is determined by sensor geometry, as in orbit determination problems. In systems where the measurement time interval is constant and small compared with system time constants, there is little need for scaling and squaring. Fortunately, many systems use measurement sampling that is well above the Nyquist rate. (The sampling rate of a continuous signal must be equal or greater than two times the highest frequency present in order to uniquely reconstruct the signal: see, for example, Oppenheim and Schafer, 1975.)

2.3.1.7 *Direct Numeric Partial Derivatives Using Integration* Great increases in digital computer speeds during the last decades have allowed another option for calculation of $\mathbf{\Phi}$ that would not have been considered 30 years ago: direct evaluation of numeric partial derivatives via integration. That is, from equation (2.3-1),

$$\mathbf{\Phi}(t+T,t) = \frac{\partial \mathbf{x}(t+T)}{\partial \mathbf{x}(t)} \tag{2.3-14}$$

which works for both linear and nonlinear models. Hence each component j of $\mathbf{x}(t)$ can be individually perturbed by some small value, and $\mathbf{x}(t)$ can be numerically integrated from t to $t + T$. The procedure below shows pseudo-code that is easily converted to Fortran 90/95 or MATLAB:

Allocate arrays $\mathbf{x}(n), \mathbf{x1}(n), \mathbf{x2}(n), \delta\mathbf{x}(n), \mathbf{\Phi}(n,n)$
Set $\delta\mathbf{x}(n)$
For $i = 1$ to n
  Set $\mathbf{x}(:) = \mathbf{x0}(:)$ (input)
  Set $x(i) = x0(i) + \delta x(i)$
  Numerically integrate $\mathbf{x}$ over the interval T to obtain $\mathbf{x1}(:)$
  Set $x(i) = x0(i) - \delta x(i)$
  Numerically integrate $\mathbf{x}$ over the interval T to obtain $\mathbf{x2}(:)$
  $\mathbf{\Phi}$(:, i)=($\mathbf{x1}$(:)-$\mathbf{x2}$(:))/(2 δx(i))
End loop

Notice that central differences are used in the algorithm. You may wonder why they are preferred to one-sided differences since this doubles computations. The primary reason is that one-sided differences cannot be relied upon as accurate, regardless of the selected perturbation $\delta\mathbf{x}$. This can be explained by personal experiences of the author (described below).

In the 1970s and early 1980s, this author routinely checked complicated derivations of analytic partial derivatives using numeric differences. Occasionally it was discovered that analytic and numeric partials did not agree within acceptable tolerances, and it was assumed that a mistake had been made in deriving or coding the analytic partials. Sometimes that was the case, but in several instances no error could be found even after hours of checking. It was eventually realized that errors in the one-sided (forward difference) numeric partials caused the discrepancy. No value of $\delta\mathbf{x}$ could be found for which the agreement between analytic and numeric partials was acceptable. When the numerical partials were changed to central differences, the agreement between the two methods was very good for $\delta\mathbf{x}$ values varying by orders of magnitude. This experience was repeated for several different problems that had little in common. Use of central differences cancels even terms in the Taylor series, and these terms are important down to the levels at which double precision round-off errors are important. Because of this experience, the author has only used central-differences for numerical partial derivatives computed since that time. Also, because of the robustness of central-difference numeric partials, the author has largely abandoned analytic partials for all but the simplest models, or systems in which the extra computational time of numeric partials cannot be tolerated. Use of numeric partials has the added advantage that the system model can be easily modified without having to re-derive analytic partials for the modified system. This makes the code much more flexible.

The primary problem in using numeric partial derivatives is selection of the $\delta\mathbf{x}$ magnitude. A simple rule suggested by Dennis and Schnabel (1983, p. 97) and Press et al. (2007, pp. 229–231) for *simple functions* and *one-sided* numeric derivatives in the absence of further information is that δx_i should be about one-half of the machine accuracy ε_m. That is, the perturbation δx_i should be about $\sqrt{\varepsilon_m}\,|x_i|$. For double precision on a PC, $\varepsilon_m = 2^{-53} = 1.1 \times 10^{-16}$ in IEEE T_floating format, so $\sqrt{\varepsilon_m}\,|x_i| \approx 10^{-8}\,|x_i|$. For central differences, δx_i should be about $\varepsilon_m^{1/3}\,|x_i|$. This step size minimizes the sum of numerical rounding errors and Taylor series truncation errors.

However, the simple rule has several problems. First, x_i could be nominally zero, so the perturbation would then be zero. Hence the rule for central differences should be modified to

$$\delta x_i = \varepsilon_m^{1/3} \max\left(|x_i|, x_{nom\_i}\right),$$

where x_{nom_i} is a user-specified nominal magnitude for x_i. A second problem is that complicated functions can have relative errors much larger than the machine accuracy. Hence the rule is modified to replace ε_m with ε_f, the relative accuracy of the computed function. Unfortunately ε_f is often unknown, so ε_m is used by default. A third problem (related to the second) for computing $\mathbf{\Phi}$ is that differences in magnitude and sensitivity of one state to another may be large. For example, in geosynchronous orbit determination, the earth-centered-inertial (ECI) x and y coordinates of satellite position will behave approximately as sinusoids with a magnitude of about 42,164 km. If the solar radiation pressure coefficient, C_p, with a nominal magnitude of about 1 (dimensionless) is included in the state vector, it is found that a perturbation of 10^{-8} in C_p only results in a change in position of about 2×10^{-10} m over a time interval of 10 min, which is a typical interval between measurements. Hence the relative change in position is $2 \times 10^{-10}/42164000 = 5 \times 10^{-18}$, which is much smaller than the double precision machine precision! In other words, the computed numerical partial derivative will always be zero. To overcome this problem, the perturbation for central differences should be set so that

$$\boxed{\delta x_i = \varepsilon_f^{1/3} \times \max\left\{|x_i|, x_{nom\_i}, \min_j \left|\frac{x_j(t+T)}{\dfrac{\partial x_j(t+T)}{\partial x_i(t)}}\right|\right\}.} \tag{2.3-15}$$

Unfortunately

$$x_j(t+T)\Big/\left(\frac{\partial x_j(t+T)}{\partial x_i(t)}\right)$$

will not be known in advance, so it is generally better to let the user define x_{nom_i} such that it takes into account the sensitivity of other "measured states" to perturbations in x_i.

Another approach defines x_{nom_i} so that $\varepsilon_f^{1/3} x_{nom\_i}$ equals a small fraction, for example, 0.01 to 0.1, of the expected *a-posteriori* 1-σ uncertainty (from the error covariance matrix) for the state. Recall that the purpose of computing $\mathbf{\Phi}$ is to iteratively calculate (for nonlinear least-squares problems) perturbations in the estimate of the state $\mathbf{x}$, where the iterations may continue until those perturbations are small compared with the *a-posteriori* $1 - \sigma$ uncertainty. Hence the numeric partials must be accurate for δx_i perturbations of that magnitude. Unfortunately the *a-posteriori* $1 - \sigma$ uncertainty will not be known until the estimation software is coded and executed on real data. It is usually possible, however, to guess the *a-posteriori* $1 - \sigma$ uncertainty within a factor of 100. This is acceptable because numeric partials computed using central differences are not sensitive to the exact perturbation $\delta\mathbf{x}$, provided that it is reasonable.

Further discussion of step size selection for numerical derivatives may be found in Dennis and Schnabel (1983, pp. 94–99), Vandergraft (1983, pp. 139–147), and Conte (1965, pp. 111–117).

Finally, we note that $\mathbf{\Phi}$ can be obtained using numeric partial derivatives to compute

$$\mathbf{F}(\mathbf{x}) = \frac{\partial \mathbf{f}(\mathbf{x})}{\partial \mathbf{x}}$$

for the nonlinear system $\dot{\mathbf{x}}(t) = \mathbf{f}(\mathbf{x})$. Then $\mathbf{\Phi}$ for the linearized system can be computed using another method, such as Taylor series integration of $\dot{\mathbf{\Phi}} = \mathbf{F}\,\mathbf{\Phi}$.

2.3.1.8 Partitioned Combination of Methods Some systems use both complex and simple dynamic models where the models are independent or loosely coupled. For example, it is not unusual in Kalman filters to have three types of states: core dynamic states, Markov process states, and bias dynamic states, with no coupling between Markov and bias as shown below:

$$\mathbf{\Phi} = \begin{bmatrix} \mathbf{\Phi}_{core} & \mathbf{\Phi}_{core-Markov} & \mathbf{\Phi}_{core-bias} \\ \mathbf{0} & \mathbf{\Phi}_{Markov} & \mathbf{0} \\ \mathbf{0} & \mathbf{0} & \mathbf{\Phi}_{bias} \end{bmatrix}$$

(Least-squares estimators do not include Markov process states since they are not deterministic.) $\mathbf{\Phi}_{core}$ can be computed by one of the listed methods—most likely Taylor series, integration of $\dot{\mathbf{\Phi}} = \mathbf{F}\,\mathbf{\Phi}$, scaling and squaring, or numeric partial derivatives. When dynamics are simple, $\mathbf{\Phi}_{core}$ can sometimes be computed analytically. $\mathbf{\Phi}_{Markov}$ can be computed using the methods listed in Section 2.2.4. $\mathbf{\Phi}_{bias}$ is either the identity matrix or possibly an expansion in basis functions. In many cases $\mathbf{\Phi}_{core-Markov}$ is approximated using low-order Taylor series, because the Markov processes are usually treated as small-driving perturbations in higher derivatives of the core states, moderate accuracy is adequate if $\mathbf{\Phi}$ is not used for propagating the state vector. If higher accuracy is required, $\mathbf{\Phi}_{core}$, $\mathbf{\Phi}_{Markov}$, and $\mathbf{\Phi}_{core-Markov}$ could be computed by integration of $\dot{\mathbf{\Phi}} = \mathbf{F}\,\mathbf{\Phi}$. Likewise $\mathbf{\Phi}_{core-bias}$ could be computed by the same methods used for $\mathbf{\Phi}_{core-Markov}$, or if the bias terms just model measurement errors, then $\mathbf{\Phi}_{core-bias} = \mathbf{0}$.

Example 2.2: GOES I-P Orbit and Attitude Determination (OAD)

A simple example of this partitioning appears in the batch least-squares OAD ground system software that supports the GOES I-P geosynchronous weather imaging satellite (described in Example 2.1). $\mathbf{\Phi}_{core}$ models the six spacecraft orbital dynamic states as described later in Chapter 3. $\mathbf{\Phi}_{bias} = \mathbf{I}$ since the biases are coefficients of the polynomial and Fourier series expansions that model the time-varying instrument "attitude" throughout the day. Since the two models are for different physical systems, the state transition matrix is block diagonal; that is, $\mathbf{\Phi}_{core-bias} = \mathbf{0}$. The two subsystems are coupled through the measurement equations, not through dynamics. Because $\mathbf{\Phi}_{core}$ models complex orbital forces, it is either computed using detailed (but approximate) analytic models or using numerical partial derivatives. The time variation of the five instrument attitude angles is computed as part of the measurement model, and then the five angles—evaluated at the measurement times—are mapped to the instrument scan angles of each observation: $d\mathbf{y}(t) = \mathbf{H}(t)d\mathbf{x}(t_0)$ where

$$\mathbf{H}(t) = \left(\frac{\partial \mathbf{y}(t)}{\partial \mathbf{x}(t)}\right)\left(\frac{\partial \mathbf{x}(t)}{\partial \mathbf{x}(t_0)}\right).$$

Hence the measurement "dynamics" are included in the measurement model, not $\mathbf{\Phi}$. However, there is an option within OAD to include one force model bias—the solar radiation coefficient—as an estimated state. When this bias is estimated by OAD, the 6×1 $\mathbf{\Phi}_{core-bias}$ term is computed using an approximate analytic expression: essentially a first-order Taylor series.

The above methods for computing $\mathbf{\Phi}$ are the principle ones used in estimation software. Moler and Van Loan list several other methods that could be used, but they also have problems. This author is not aware of estimation applications that use these alternate methods.

Notice that methods #4, #7, and possibly #8 are the only listed methods that can be directly used for time-varying systems. These "time-varying" systems include nonlinear systems when the system state $\mathbf{x}(t)$ changes significantly during the interval T so that the linearized

$$\mathbf{F}(\mathbf{x}(t)) = \left.\frac{\partial \dot{\mathbf{x}}}{\partial \mathbf{x}}\right|_t$$

is not a good approximation to $\mathbf{F}(\mathbf{x}(t + T))$. In these cases, $\mathbf{x}(t)$ must be numerically integrated using m time steps of T/m (or non-equal steps), $\mathbf{F}(\mathbf{x})$ must be reevaluated at each intermediate value of $\mathbf{x}(t + jT/m)$, $\mathbf{\Phi}(t + jT/m, t + (j-1)T/m)$ must be recomputed for each value of $\mathbf{F}(\mathbf{x})$, and the total transition matrix must be computed as

$$\mathbf{\Phi}(t+T,t) = \prod_{j=1}^{m} \mathbf{\Phi}\left(t + \frac{jT}{m}, t + \frac{(j-1)T}{m}\right). \tag{2.3-16}$$

All the above methods are subject to a problem called the "hump", but the problem particularly affects the scaling and squaring method. If the dynamics matrix **F** has eigenvalues that are close in magnitude and the off diagonal terms of **F** are large, then the magnitude of $\mathbf{\Phi}(\mathbf{F}T) = e^{\mathbf{F}T}$ may increase with T before it decreases. Moler and Van Loan give one example of this problem, repeated here in a slightly modified form with additional analysis.

Example 2.3: Hump in Computing Φ

For a two-element state, let

$$\mathbf{F} = \begin{bmatrix} -a & c \\ 0 & -b \end{bmatrix}$$

where a, b, and c are all positive. The system is stable since the eigenvalues ($-a$ and $-b$) are negative. This is a realistic example in that **F** matrices for many physical systems have an approximately upper triangular structure and are stable. The problem occurs when c is large, and it can sometimes become worse when a is approximately, but not exactly, equal to b. Again this case is realistic for some real problems. Here $\Phi_{12} = c(e^{-aT} - e^{-bT})/(b - a)$ and this term can become quite large. Figure 2.7 shows the four elements of $\mathbf{\Phi}$ versus T for the case $a = 1.152$, $b = 0.8682$, and $c = 10$. Notice that Φ_{12} reaches a peak of 3.7 before decaying. Hence any round-off errors accumulated in computing $\mathbf{\Phi}$ may grow as time increases to $T = 1$. If the scaling and squaring method is used to compute $\mathbf{\Phi}$ with $T/m < 1$, the relative round-off errors may be magnified when computing $\mathbf{\Phi}(T) = [\mathbf{\Phi}(T/m)]^m$.

Figure 2.8 shows the difference between the $\mathbf{\Phi}$ elements computed using squaring and those computed from the exact solution

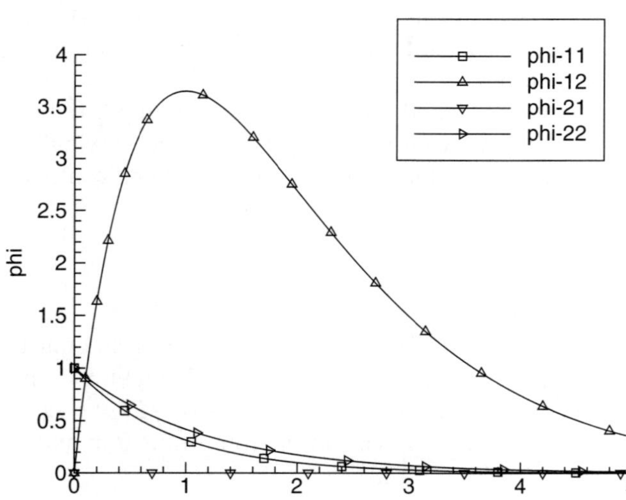

FIGURE 2.7: $\mathbf{\Phi}$ versus T for Example 2.3.

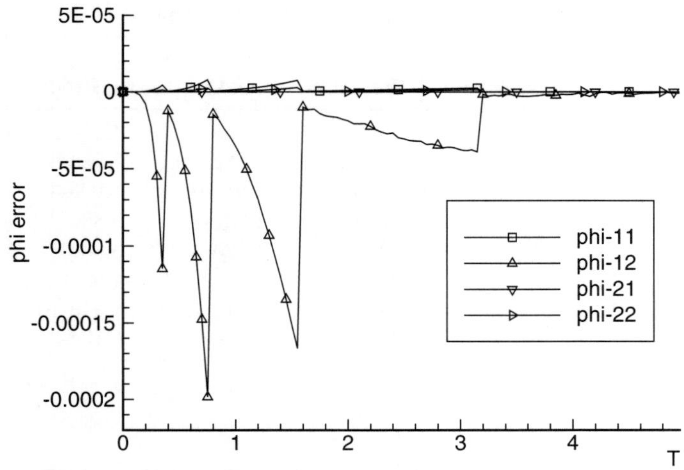

FIGURE 2.8: Error in $\mathbf{\Phi}$: fourth-order Taylor series, single precision, $T_{max} = 0.2$.

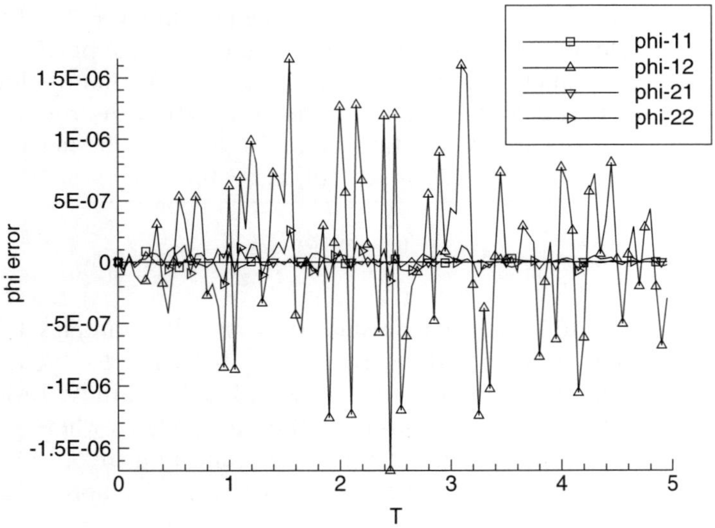

FIGURE 2.9: Error in $\mathbf{\Phi}$: eighth-order Taylor series, single precision, $T_{max} = 0.2$.

$$\mathbf{\Phi}(T) = \begin{bmatrix} e^{-aT} & \dfrac{c(e^{-aT} - e^{-bT})}{b-a} \\ 0 & e^{-bT} \end{bmatrix}$$

for the case where $\mathbf{\Phi}(T/m)$ is computed using a fourth-order Taylor series in single precision (4 bytes) and $T_{max} = 0.2$. Notice that the largest error is in Φ_{12} (as expected), and it exhibits a sawtooth characteristic that corresponds to changes in the number of squaring operations. The error grows with T/m since the dominant error is caused by truncation of the Taylor series to four terms. When an eight-order Taylor series is used, the error drops dramatically and becomes more random, as shown in Figure 2.9. In fact, no hump in error propagation is noticed

TABLE 2.2 Example 2.3 Error in Φ for Scaling and Squaring

Taylor Series Order	T_{max}	# Steps for T = 5 s	Precision	Max $\|\delta\Phi_{11}\|$	Max $\|\delta\Phi_{12}\|$	Max $\|\delta\Phi_{22}\|$
4	0.1	6	Single	5.15e-7	1.22e-5	6.76e-7
4	0.2	5	Single	7.88e-6	1.98e-4	2.82e-6
4	0.2	5	Double	7.91e-6	1.99e-4	2.69e-6
4	0.4	4	Single	1.52e-4	3.88e-3	4.69e-5
4	0.4	4	Double	1.52e-4	3.88e-3	4.96e-5
8	0.2	5	Single	1.44e-7	1.68e-6	2.55e-7
8	0.2	5	Double	6.21e-12	1.94e-10	7.21e-13
8	0.4	4	Double	1.94e-9	6.10e-8	2.15e-10

when using an eight-order series. It appears that time growth of series truncation errors are influenced by the "hump" curve, but that round-off errors are not.

Table 2.2 summarizes the maximum **Φ** errors for various conditions. Notice that an eight-order series can achieve very high accuracy (1.9×10^{-10}) using $T_{max} = 0.2$ s with double-precision calculations, while in single precision calculations are limited to single-precision accuracy (1.7×10^{-6}). There is little difference between single and double precision when using a fourth-order series since series truncation errors are the limiting factor. In this case, the maximum error is 2.0×10^{-4} for $T_{max} = 0.2$ and 1.25×10^{-5} for $T_{max} = 0.1$.

It is always risky to extrapolate results from simple examples to unrelated real-world problems. You should analyze error behavior for your particular problem before deciding on methods to be used. However, these results suggest that when using the scaling and squaring method for systems where growth of **Φ** follows a "hump," it is important to minimize truncation errors by accurately computing $\mathbf{\Phi}(T/m)$. Growth of round-off errors does not appear to follow a "hump" curve.

These results were obtained for a stable system, but many real systems include free integrators where elements of **Φ** grow almost linearly with time. In these cases, round-off error growth will be somewhat linear when using the scaling and squaring method, but growth of truncation errors may still follow sawtooth curves.

To summarize, truncated Taylor series, ODE integration of $\dot{\mathbf{\Phi}} = \mathbf{F}\mathbf{\Phi}$, numeric partials, scaling and squaring, and combinations of these methods (possibly in partitions) are the primary methods used for computing **Φ** in least-squares and Kalman estimators. The series methods should only be used when the time step T is much less than shortest time constants (inverse eigenvalues) of **F**. If the desired time step is larger than the shortest time constants, then scaling and squaring (possibly with series evaluation of the scaled **Φ**), integration of $\dot{\mathbf{\Phi}}$, or numeric partials should be used. For time-varying systems where it cannot be assumed that **F** is constant during the integration interval, integration of $\dot{\mathbf{\Phi}}$ or numeric partials should be used.

2.3.2 Numeric Computation of $\mathbf{Q}_D$

Recall that the discrete state noise matrix $\mathbf{Q}_D$ is used in Kalman filters but not in least-squares estimators. Section 2.2.1 derived the equation for $\mathbf{Q}_D$, repeated here:

$$\mathbf{Q}_D(t_{i+1},t_i)=\int_{t_i}^{t_{i+1}}\mathbf{\Phi}(t_{i+1},\lambda)\,\mathbf{G}(\lambda)\mathbf{Q}_s\,\mathbf{G}^T(\lambda)\mathbf{\Phi}^T(t_{i+1},\lambda)\,d\lambda \tag{2.3-17}$$

where $\mathbf{Q}_s = E[\mathbf{q}(t)\mathbf{q}^T(t)]$ is the PSD of the continuous driving process noise $\mathbf{q}(t)$. When $\mathbf{\Phi}$ is expressed analytically, it is sometimes possible to evaluate equation (2.3-17) analytically—this was demonstrated in Section 2.2.4 for low-order Markov processes. However, analytic evaluation is generally not an option for most applications. Fortunately Kalman filters work quite well using approximations for equation (2.3-17) because $\mathbf{Q}_D$ just accounts for the random error growth of the state $\mathbf{x}$ during time intervals between measurements. Provided that the time step $T = t_{i+1} - t_i$ is small compared with the time constants of $\mathbf{F}$, these approximations usually involve expanding $\mathbf{\Phi}$ in a zeroth, first- or second-order Taylor series, and then evaluating the integral equation (2.3-17). For a first-order expansion of $\mathbf{\Phi}$,

$$\begin{aligned}\mathbf{Q}_D(T)&=\int_0^T[\mathbf{I}+\mathbf{F}\cdot(T-\lambda)]\,\mathbf{Q}'_s\,[\mathbf{I}+\mathbf{F}^T\cdot(T-\lambda)]\,d\lambda\\&=\int_0^T\mathbf{Q}'_s+[\mathbf{F}\mathbf{Q}'_s+\mathbf{Q}'_s\mathbf{F}^T](T-\lambda)+\mathbf{F}\mathbf{Q}'_s\mathbf{F}^T(T-\lambda)^2\,d\lambda\\&=\mathbf{Q}'_sT+[\mathbf{F}\mathbf{Q}'_s+\mathbf{Q}'_s\mathbf{F}^T]T^2/2+\mathbf{F}\mathbf{Q}'_s\mathbf{F}^TT^3/3\end{aligned} \tag{2.3-18}$$

where for simplicity we have set $\mathbf{Q}'_s=\mathbf{G}\,\mathbf{Q}_s\mathbf{G}^T$. This equation is sometimes used to evaluate $\mathbf{Q}_D$, but in many cases it is simplified to

$$\mathbf{Q}_D(T)=\mathbf{Q}'_sT+[\mathbf{F}\mathbf{Q}'_s+\mathbf{Q}'_s\mathbf{F}^T]T^2/2$$

or even just $\mathbf{Q}_D(T)=\mathbf{Q}'_sT$.

Since the accuracy of these approximations depend on the assumption that T is much less than smallest time constants of $\mathbf{F}$, alternate methods must be available when the time step T between measurements is large—possibly because of data gaps. In these cases, $\mathbf{\Phi}$ and $\mathbf{Q}_D$ should be evaluated in steps T/m where m is chosen so that T/m is small compared with shortest $\mathbf{F}$ time constants. This procedure works particularly well using the squaring method of the previous section. This can be derived using a time-invariant form of equation (2.2-13) with the $\mathbf{u}$ term ignored. The state propagation for two time steps is

$$\begin{aligned}\mathbf{x}(t+2T)&=\mathbf{\Phi}(T)\,\mathbf{x}(t+T)+\mathbf{q}_D(t+2T,t+T)\\&=\mathbf{\Phi}(T)\,[\mathbf{\Phi}(T)\,\mathbf{x}(t)+\mathbf{q}_D(t+T,t)]+\mathbf{q}_D(t+2T,t+T)\ .\\&=\mathbf{\Phi}^2(T)\mathbf{x}(t)+[\mathbf{\Phi}(T)\mathbf{q}_D(t+T,t)+\mathbf{q}_D(t+2T,t+T)]\end{aligned} \tag{2.3-19}$$

The last term in brackets is the total effect of process noise over the interval t to $t + 2T$. We define it as $\mathbf{q}_{D2}$. We also define $\mathbf{Q}_D(2T) = E[\mathbf{q}_{D2}\mathbf{q}_{D2}^T]$. Noting that because $\mathbf{q}_s(t)$ is assumed to be stationary white noise,

$$E\left[\mathbf{q}_D(t+T,t)\,\mathbf{q}_D^T(t+2T,t+T)\right]=\mathbf{0}$$

and

$$E\left[\mathbf{q}_D(t+T,t)\mathbf{q}_D^T(t+T,t)\right] = E\left[\mathbf{q}_D(t+2T,t+T)\mathbf{q}_D^T(t+2T,t+T)\right]$$
$$= \mathbf{Q}_D(T)$$

Thus

$$\boxed{\mathbf{Q}_D(2T) = \mathbf{\Phi}(T)\mathbf{Q}_D(T)\mathbf{\Phi}^T(T) + \mathbf{Q}_D(T).} \tag{2.3-20}$$

This equation is valid regardless of the value of T, which suggests that the squaring technique listed previously for $\mathbf{\Phi}$ can be modified to also include $\mathbf{Q}_D$. The modified algorithm is

$m = \max(\log_2(T/T_{\max}) + 1, 1)$
$T_m = T/2^m$
Compute $\mathbf{\Phi}(T_m)$ and $\mathbf{Q}_D(T_m)$
For $i = 1$ to m
$\quad \mathbf{Q}_D(T_m \times 2^i) = \mathbf{\Phi}(T_m \times 2^{(i-1)})\mathbf{Q}_D(T_m \times 2^{(i-1)})\mathbf{\Phi}^T(T_m \times 2^{(i-1)}) + \mathbf{Q}_D(T_m \times 2^{(i-1)})$
$\quad \mathbf{\Phi}(T_m \times 2^i) = \mathbf{\Phi}(T_m \times 2^{(i-1)}) \times \mathbf{\Phi}(T_m \times 2^{(i-1)})$
End loop

The iteration can be written so that arrays $\mathbf{\Phi}$ and $\mathbf{Q}_D$ overwrite existing arrays at each iteration.

Finally we note one other option for computing $\mathbf{Q}_D$, although it is rarely used. If we rewrite equation (2.3-19) as a general relationship for arbitrary times,

$$\mathbf{Q}_D(\lambda) = \mathbf{\Phi}(\lambda - t)\mathbf{Q}_D(t-0)\mathbf{\Phi}^T(\lambda - t) + \mathbf{Q}_D(\lambda - t), \tag{2.3-21}$$

and then differentiate it with respect to time λ evaluated at time t, we obtain

$$\boxed{\dot{\mathbf{Q}}_D(t) = \mathbf{F}\,\mathbf{Q}_D(t) + \mathbf{Q}_D(t)\mathbf{F}^T + \mathbf{Q}_s} \tag{2.3-22}$$

where we have used $\dot{\mathbf{\Phi}}(t) = \mathbf{F}\mathbf{\Phi}(t-t) = \mathbf{F}$ and $\dot{\mathbf{Q}}_D(t) = \mathbf{Q}_s$. Equation (2.3-22) is a *matrix Riccati equation*. It can be integrated using a general-purpose ODE solver, just as for $\dot{\mathbf{\Phi}}$. Note that it is only necessary to integrate the upper triangular elements of the symmetric $\dot{\mathbf{Q}}_D$.

2.4 MEASUREMENT MODELS

Previous sections have concentrated on dynamic models because those are generally more difficult to develop than measurement models. Most measurements represent physical quantities where the functional relationship between measurements and model states is easily defined. For example, measurements in navigation problems (including orbit determination) are usually a nonlinear function of position and velocity. Examples include range, azimuth, elevation, and range rate from a sensor location to the target body-of-interest, or from the body-of-interest to another location or body.

An m-element vector of measurements $\mathbf{y}(t)$ is assumed to be modeled as a linear or nonlinear function of the state vector $\mathbf{x}(t)$ and random measurement noise $\mathbf{r}(t)$. For a nonlinear model,

$$\mathbf{y}(t)=\mathbf{h}(\mathbf{x}(t))+\mathbf{r}(t), \tag{2.4-1}$$

or for a linear model

$$\mathbf{y}(t)=\mathbf{H}(t)\mathbf{x}(t)+\mathbf{r}(t) \tag{2.4-2}$$

where $\mathbf{H}(t)$ is an $m \times n$ matrix that may be time-varying, and $\mathbf{r}(t)$ is an m-vector of zero-mean measurement noise. It is usually (but not always) assumed that $\mathbf{r}(t)$ is uncorrelated white noise with covariance $\mathbf{R}$: $E[\mathbf{r}(t)\mathbf{r}^T(\lambda)] = \mathbf{R}\delta(t-\lambda)$.

When the measurement equations are nonlinear, most least-squares or Kalman filtering methods linearize the model about a reference $\mathbf{x}_{ref}$ (usually the current estimate $\hat{\mathbf{x}}(t)$) to develop perturbation equations of the form

$$\mathbf{y}(t)-\mathbf{h}(\mathbf{x}_{ref}(t))=\mathbf{H}(t)\,(\mathbf{x}(t)-\mathbf{x}_{ref}(t))+\mathbf{r}(t) \tag{2.4-3}$$

where $\mathbf{H}(t)=\left.\dfrac{\partial \mathbf{h}}{\partial \mathbf{x}}\right|_{x\_ref(t)}$.

Example 2.4: Range and Range-Rate Measurements

Measurements of range and range rate from a fixed site to a moving target are an example of a simple nonlinear model. The position vector of the sensor site is defined as $\mathbf{p}_s=[p_{sx}\quad p_{sy}\quad p_{sz}]^T$ and the position of the target is $\mathbf{p}_t=[p_{tx}\quad p_{ty}\quad p_{tz}]^T$ with velocity $\mathbf{v}_t=[v_{tx}\quad v_{ty}\quad v_{tz}]^T$. The range from sensor to target is

$$R=\|\mathbf{p}_s-\mathbf{p}_t\|_2=\sqrt{(\mathbf{p}_s-\mathbf{p}_t)^T(\mathbf{p}_s-\mathbf{p}_t)} \tag{2.4-4}$$

and the range-rate is

$$\dot{R}=\frac{d\|\mathbf{p}_s-\mathbf{p}_t\|_2}{dt}=-\frac{(\mathbf{p}_s-\mathbf{p}_t)^T\mathbf{v}_t}{R}. \tag{2.4-5}$$

If $\mathbf{p}_t$ and $\mathbf{v}_t$ are included in the state vector to be estimated, the measurement partial derivatives are

$$\begin{aligned}\frac{\partial R}{\partial \mathbf{p}_t}&=\frac{(\mathbf{p}_s-\mathbf{p}_t)^T}{R}\\ &=\begin{bmatrix}\dfrac{p_{sx}-p_{tx}}{R} & \dfrac{p_{sy}-p_{ty}}{R} & \dfrac{p_{sz}-p_{tz}}{R}\end{bmatrix}\end{aligned} \tag{2.4-6}$$

and

$$\begin{aligned}\frac{\partial \dot{R}}{\partial \mathbf{p}_t}&=\frac{\mathbf{v}_t^T}{R}+\frac{(\mathbf{p}_s-\mathbf{p}_t)^T\mathbf{v}_t}{R^3}(\mathbf{p}_s-\mathbf{p}_t)^T\\ &=\frac{\mathbf{v}_t^T}{R}-\frac{\dot{R}}{R^2}(\mathbf{p}_s-\mathbf{p}_t)^T\\ \frac{\partial \dot{R}}{\partial \mathbf{v}_t}&=-\frac{(\mathbf{p}_s-\mathbf{p}_t)^T}{R}\end{aligned}. \tag{2.4-7}$$

Most measurement sensors have inherent biases that may or may not be calibrated prior to use. Often the biases change with time, temperature, or other environmental parameters. Uncompensated measurement biases can have a very detrimental effect on the performance of the estimator. This is particularly true when multiple sensors each have their own biases. Hence it is important that either sensor biases be calibrated and biases removed from the measurements, or that the estimator include sensor biases in the state vector of adjusted parameters. Because of the time-varying nature of sensor "biases," some sensors are routinely calibrated using data external to the normal measurements. For example, ground stations that track satellites or aircraft may calibrate range biases by periodically ranging to a ground transponder at a known location.

The measurement noise covariance matrix **R** is also a required input for the Kalman filter and weighted least-squares estimator. If the measurement sampling rate is much higher than the highest frequencies of the dynamic model, it is often possible to approximately determine the measurement noise variance by simply examining plots of the measurement data and calculating the variance about a smooth curve. One can also calculate the correlation between components of the measurement vector—such as range and range rate—using a similar approach. Time correlations (e.g., $E[\mathbf{y}(t)\mathbf{y}^T(t+\lambda)]$) should be computed to determine whether the measurement errors are nearly time-independent, as often assumed. If time-correlation is significant, it will be necessary to account for this in the estimation as discussed later in Chapters 8.

In cases where measurement sampling rates are low or measurements are taken at infrequent intervals, other methods for determining **R** are needed. Sometimes the measurement sensor manufacturer will provide measurement error variances, but actual errors are often a function of the signal-to-noise ratio, so further interpretation may be required. Chapter 6 shows how least-squares measurement residuals can be used to compute the "most likely" value of **R**, and Chapter 11 shows how this is done for Kalman filters.

2.5 SIMULATING STOCHASTIC SYSTEMS

Performance of an estimator should be evaluated using simulation in order to determine whether it meets accuracy, robustness, timing, and other requirements. Simulation can also be used to compare performance for different filter models, state order, parameter values, and algorithms. Ideally this should include use of a detailed simulation that models the real system more accurately than the models used for estimation. The simulation may include effects that are ignored in the estimator, such as high-frequency effects. It should also use random initial conditions, measurement noise, and process noise (if appropriate). Although single-simulation cases allow evaluation of estimator response and behavior, Monte Carlo analysis using many different random number generator seeds is needed to evaluate performance. This allows computation of the mean, variance, and minimum/maximum statistics of the estimate error. One should be aware, however, that many "simple" random number generators produce values that repeat or have structural patterns. Thus they may not produce valid statistics. See Press et al. (2007, chapter 7) for more discussion of this problem and recommended solutions.

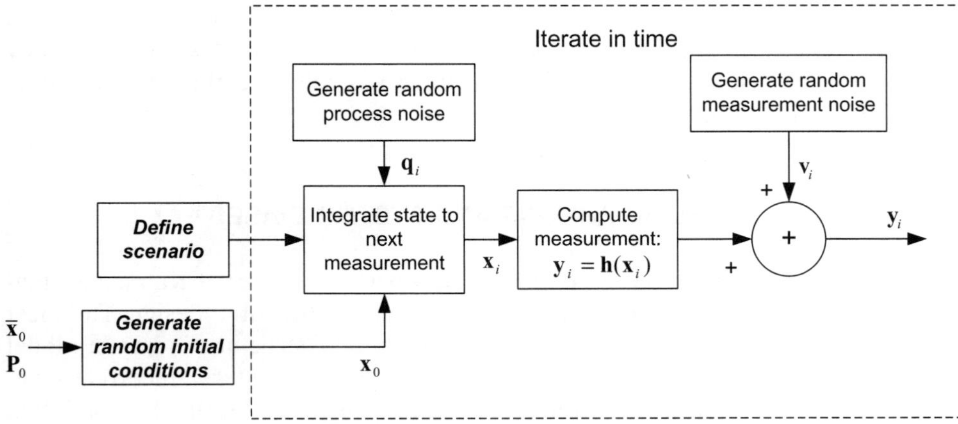

FIGURE 2.10: Simulation structure.

Figure 2.10 shows a general structure for a stochastic simulation. If random initial conditions are to be simulated (generally recommended), the mean ($\bar{\mathbf{x}}_0$) and covariance $\mathbf{P}_0$ of the initial state must be known. Generation of a random vector with a known covariance is easily done by factoring the covariance using Cholesky decomposition. That is, compute

$$\mathbf{P}_0 = \mathbf{L}\mathbf{L}^T \tag{2.5-1}$$

where $\mathbf{L}$ is lower triangular. This factorization is unique provided that $\mathbf{P}_0$ is positive definite, which it should be. Eigen decomposition or other factorizations can be used, but they require far more computation than Cholesky decomposition. Then the random initial state is computed as

$$\mathbf{x}_0 = \bar{\mathbf{x}}_0 + \mathbf{L}\mathbf{w} \tag{2.5-2}$$

where $\mathbf{w}$ is an n-element vector of independent zero-mean random numbers with known distribution (usually Gaussian) and unit variance. It is easily verified that the covariance of $\mathbf{x}_0$ is $\mathbf{P}_0$:

$$\begin{aligned} E\left[(\mathbf{x}_0 - \bar{\mathbf{x}}_0)^T(\mathbf{x}_0 - \bar{\mathbf{x}}_0)\right] &= \mathbf{L}E(\mathbf{w}\mathbf{w}^T)\mathbf{L}^T \\ &= \mathbf{L}\mathbf{L}^T \\ &= \mathbf{P}_0 \end{aligned} \tag{2.5-3}$$

The initial state $\mathbf{x}_0$ is then numerically integrated to the time of the first measurement where the state differential equation is given by equation (2.2-2),

$$\dot{\mathbf{x}}(t) = \mathbf{f}(\mathbf{x}, \mathbf{u}, t) + \mathbf{G}(\mathbf{x}, \mathbf{u}, t)\, \mathbf{q}_c(t),$$

if the system is nonlinear, or equation (2.2-4) if it is linear. In order to properly simulate the effects of process noise $\mathbf{q}_c(t)$, it may be advisable to integrate using smaller steps than the intervals between measurements. Alternately the discrete process noise $\mathbf{q}_D(t_i)$ could be added after integrating $\dot{\mathbf{x}}(t) = \mathbf{f}(\mathbf{x}, \mathbf{u}, t)$, where $\mathbf{q}_D(t_i)$ is computed as $\mathbf{q}_D(t_i) = \mathbf{L}\mathbf{w}$, $\mathbf{L}\mathbf{L}^T = \mathbf{Q}_D$ and $\mathbf{w}$ is a unit random vector. $\mathbf{Q}_D$ can be computed using the methods of Section 2.3.2. Then the measurement vector is computed as $\mathbf{y}(t) = \mathbf{h}(\mathbf{x})$ and random measurement noise of specified covariance $\mathbf{R}$ is added.

If **R** is nondiagonal, Cholesky factorization should be used, as done for the initial state and process noise. Colored noise is generated using the low-order Markov process models discussed previously. This process is repeated through each time step of the simulation.

2.6 COMMON MODELING ERRORS AND SYSTEM BIASES

More effort is often spent trying to determine why least-squares or Kalman estimators are not performing properly than effort spent on the initial design. The most common modeling problems are due to incorrect model structure, incorrect model parameters, incorrect process noise covariance (in Kalman filters), incorrect measurement noise covariance, or mis-modeling of system biases. Of these possibilities, mis-modeling of biases is probably the most common mistake. It is very important that potential process or measurement biases be modeled unless it is known with certainty that they are negligible.

Six examples from the author's experience demonstrate the problem, and show that bias errors in system models are a common cause of estimation errors.

1. In 1973 the National Aeronautics and Space Administration (NASA) launched the IMP-8 spacecraft into a highly elliptical transfer orbit with an apogee altitude of about 233,000 km. At apogee, an injection motor was fired to circularize the orbit. Unfortunately, the eccentricity of the resulting orbit was 0.33 versus the target 0.096. Inclination differed from the target by about 3.4 degrees. A smaller discrepancy was also observed for the IMP-7 spacecraft. Extensive error analysis was conducted to determine the cause of the orbit error. Possibilities included errors in orbit determination, errors in determining spacecraft attitude when pointing the spacecraft for injection motor firing, and errors in the thrust vector generated by the injection motor. Analysis showed that errors in the earth and sun sensors—used for onboard spacecraft attitude determination—were the most likely cause of the problem (Gibbs and Fang 1975; Gibbs et al. 1975). It appeared that the sun sensor calibration transition points were in error by as much as 0.5 degrees and the least significant bit did not register. The earth telescope photodiode sensitivity was greater than expected and detected horizon/terminator crossings when only a small portion of the field of view was lit. This caused an earth central angle-dependent bias, which had the effect of biasing the estimated yaw and pitch. A significant sun-to-earth sensor roll bias was also present. Finally, either the spacecraft spin rate varied or the spacecraft clock drifted, causing a trend in the sensed roll. Analysis was difficult because sensitivity of spacecraft attitude to sensor biases was very high, and geometric observability of sensor biases from the available data was very weak. Also the "time-varying" earth sensor bias caused an attitude-determination Kalman filter to diverge until a correction for the telescope sensitivity error was applied. Batch least-squares attitude estimation was less affected by the earth sensor bias, but the trend in the spacecraft spin rate had to be explicitly modeled.
2. In 1974–1975 NASA and the Maritime Administration (MARAD) conducted two independent ship and aircraft navigation tests using ranging from the

ground to the ATS-5 and ATS-6 geosynchronous satellites, and to ships or aircraft. Unfortunately the computed ship and aircraft positions differed by as much as 55 km from the ground truth positions, both in fixed locations and when moving. Posttest analysis revealed that the ATS-6 ephemeris—used as input to the ship/aircraft navigation algorithm—was in error by up to 11 km in longitude and 3.7 km radially. The satellite ephemeris was a prediction of the orbit determined once every 2 weeks using 1 day of ground range and range rate data. Analysis revealed that the solar radiation coefficient used in orbit determination was 50% smaller than the daily mean value, and that the area normal to the sun line varied by a factor of 3:1 during the day. Fixing these problems greatly reduced prediction errors (Gibbs 1977). Interestingly, the computed ship positions were much more sensitive to ATS-6 ephemeris errors than to ATS-5 ephemeris errors. The sensitivity reduction was due to differences in the algorithms used to preprocess the range measurements. When the MARAD algorithm was modified to process ATS-6 data similarly to ATS-5 data, the ship positions were generally accurate within 4 km even when using the erroneous ATS-6 ephemeris. No such modification was possible with the NASA algorithms used for aircraft tracking. After correcting the ephemeris, ship and aircraft positions were accurate within 4 to 7 km, but patterns in the position errors suggested that time-varying transponder biases were still a problem.

3. In a second part of the NASA/MARAD navigation test, the ATS-6 ephemeris was computed using trilateration data. The measurements available for the navigation were multi-link ranges from the ground transmitting antenna to ATS-6, to one of two ground transponders, and back along the same path. One link just involved the transmitting/receiving antenna and ATS-6. Since the range measurements were computed from phases of the four tones, it was necessary to calibrate the phase delays in the ground transponders. Unfortunately the phase delay calibration procedure used for one ground transponder was flawed, with the result that the calculated range bias was in error by 500 to 1000 m. This resulted in 1500 km crosstrack errors when using trilateration data for orbit determination.

4. During postlaunch testing of the GOES-13 geosynchronous weather satellite, it was discovered that differences between actual ground-spacecraft ranges and 1-day predictions of those ranges exhibited 10–30 m bias-ramp-sinusoidal signatures (Gibbs et al. 2008a). The pattern appeared when the spacecraft was in the "inverted" yaw orientation, but not when in the "upright" orientation. The pattern could be eliminated by adding a bias to observed range measurements, but the required bias changed with spacecraft longitude and yaw attitude, and no physical reason for that behavior was known. Even with this anomaly, GOES-13 met accuracy requirements with margin. Extensive analysis using both real and simulated data was conducted to determine whether various modeling errors could cause the signatures. It was eventually discovered that the satellite transponder delay had not been properly compensated in ground software, but this only partly explained the problem. Only one possibility was found to match observed behavior: an east-west bias between star and landmark angular measurements. Both of these measurement types are

obtained from the *Imager* earth imaging instrument. Landmark observations are obtained when the Imager scans the earth, while star observations are obtained with fixed scan angles as the spacecraft drifts in pitch to maintain earth-nadir pointing. The exact cause of the bias is still unknown. The problem did not appear with GOES-8 to -12 spacecraft that use the same Imager design. Also, the behavior of the bias with GOES-14 (the next satellite in the NOP series) is different. It seems likely that the bias is instrument-dependent. One possibility is uncompensated fixed pattern errors in the Imager.

5. The GOES-13 ground system has the capability to calibrate spacecraft thruster magnitudes and cant angles using maneuver angular momentum (Δh) telemetry, and orbit determination solutions for maneuver velocity change (Δv). The postlaunch calibration worked successfully using data from more than 100 maneuvers, but systematic biases in Δh and Δv measurement residuals were correlated with the spacecraft configuration used for daily momentum-dumping maneuvers (Gibbs et al. May 2008b). Extensive analysis was conducted to determine the source of the biases, and this led to improvements in the thruster and plume impingement models. However, small biases still remained, and one jump in daily momentum-dumping (Δh) measurements occurred without any change in operational conditions. The source of the apparent biases is still not fully understood, but may be partly due to abrupt changes of about 2–3% in thruster performance. Imperfect alignment of the four momentum wheels with respect to spacecraft reference axes may also be a factor: the prelaunch alignment accuracy is sufficient for spacecraft attitude control purposes, but it does affect thruster calibration.
6. The onboard GOES-13 Stellar Inertial Attitude Determination utilizes three star trackers, where only two are used at any given time. The star trackers are mounted on an optical bench and aligned one time shortly after launch, but the alignment is not perfect. Even if it is, diurnal thermal deformation causes additional angular misalignments between the trackers. The onboard Kalman filter does not estimate tracker-dependent alignment biases, so the estimated spacecraft attitude varies somewhat as the filter tries to reconcile differences in measured attitude between sensors. The variation in attitude is small and does not prevent the system from meeting accuracy requirements.

It is difficult to draw general conclusions from the above examples. Most problems were caused by measurement biases, but two cases involved dynamic model biases. Some problems should have been detected before implementation, but others would have been very difficult to identify before operational data became available. Some biases only caused a minor degradation in accuracy, but the system did not meet requirements in others cases. It is concluded that

1. Incorrectly modeled biases are fairly common.
2. Every reasonable attempt should be made to identify potential sources of bias and calibrate them before operations.
3. The estimator should either estimate biases or process measurements in a manner that reduces bias effects.
4. Simulation should be used to determine sensitivity to bias errors.

2.7 SUMMARY

This chapter discussed the different types of models that may be used for least-squares or Kalman estimation. Dynamic models may be either continuous or discrete. Continuous dynamic models may be either basis function expansions in time, first-principle derivations, stochastic processes, linear regression models, or combinations of these. These models may either be parametric or nonparametric, linear or nonlinear, and time-varying or time-invariant. Stochastic models assume that the dynamic system is driven by random process noise. Stochastic models are used in Kalman filters, not least-squares estimation.

Discrete models generally assume that the process is stationary and the sampling interval is constant. Discrete models may be derived as continuous models sampled discretely, but usually the structure is defined in advance and parameters of the model are determined empirically from data. It is often assumed that measurement noise does not appear separately from process noise, but that is not a general restriction. Discrete models are often single-input/single-output, but may be multiple-input/multiple-output. Typical structures for discrete models are

1. AR: a pole-only model ideal for modeling multiple narrowband signals
2. MA: a zero-only model ideal for modeling signals with frequency nulls
3. ARMA: a pole-zero model that can handle more general signals. It is more difficult to determine the model parameters than for AR or MA
4. ARMAX: an ARMA model with exogenous (known) inputs.

Discrete models may be implemented in a variety of equivalent forms, such as the companion form (see Appendix D).

Continuous dynamic models are usually defined as a set of linear or nonlinear first-order differential equations. If the model is nonlinear and stochastic, the effects of process noise are usually treated as linear perturbations of the nonlinear model. Central to both least squares and Kalman filtering is the state transition matrix $\mathbf{\Phi}$. The state transition matrix models the propagation of the state vector from one measurement time to the next in a linear model. It is obtained by integrating the state differential equations $\dot{\mathbf{x}} = \mathbf{F}\mathbf{x}$ with the result $\mathbf{\Phi}(T) = e^{\mathbf{F}T}$. Various methods for numerically computing $\mathbf{\Phi}$ were discussed. Kalman filters also require the state noise covariance matrix $\mathbf{Q}$. $\mathbf{Q}$ is defined as the expected value of a "convolution-like" integral involving the continuous process noise and the state transition matrix. Methods for computing $\mathbf{Q}$ were also discussed.

Various examples showed how continuous dynamic models can be defined using basis functions, first-principle derivations, stochastic processes, linear regression structures, or combinations of these. Additional first-principle examples are presented in Chapter 3 and Appendix C. The properties of random walk and low-order Markov process models were discussed at length since they are central to most Kalman filter applications. Among other properties, it was noted that the variance of a random walk model increases with time while that of stable Markov processes is constant. Markov process models are also called colored noise models since they have a low or band-pass characteristic.

Measurements are often a direct function of dynamic model states, with added measurement noise. Measurement model examples are presented in later chapters.

The importance of simulation in design and evaluation of estimators was emphasized. Methods for simulating random initial conditions, process noise, and measurement noise were discussed. This included use of Cholesky factorization for simulating correlated noise.

Common modeling errors were discussed, and it was emphasized that the mismodeling of system biases is probably the most common modeling error. Several examples of bias problems in operational systems were presented.

Proper design of models to be used in estimation algorithms should include the following steps.

1. Determine the requirements and goals of the estimation because this will impact the level of detail required in the design.
2. Determine if is practical to develop a first-principles model, or combined first-principles/empirical/stochastic model. This will often be the most robust approach. If this is not practical, a basis function or an empirically based model is indicated.
3. Determine which model states are important to estimation: ignore high frequency effects and unobservable linear combinations of states unless those states are important outputs. This will be discussed in greater detail later. If practical, develop several competing models and algorithms and compare performance.
4. Simulate and evaluate performance of the system, and redesign if necessary. Perform covariance or Monte Carlo simulations to verify conformance with requirements.

CHAPTER 3

MODELING EXAMPLES

This chapter presents several examples showing how the concepts of the previous chapter are applied to real problems. The models are also used as the basis of estimation examples in the remaining chapters. The reader may wish to skim this chapter, but refer back to these models when later examples are discussed.

The included examples are

1. Passive angle-only tracking of linear-motion ship or aircraft targets
2. Maneuvering tank tracking using multiple models
3. Aircraft tracking with occasional maneuvers
4. Strapdown inertial navigation
5. Spacecraft orbital dynamics
6. A fossil-fueled power plant

3.1 ANGLE-ONLY TRACKING OF LINEAR TARGET MOTION

Figure 3.1 shows the basic scenario for passive tracking of a target ship moving in a straight line—sometimes called *target motion analysis* (TMA). This is typical of cases in which a ship is using passive sonar to track another ship or submarine. Since sonar is passive to avoid detection, the only measurements are a time series of angle measurements—range is not available. This example is also typical of cases where aircraft emissions are passively tracked from another aircraft. The angle measurements are used in a batch least-squares solution to compute four states: the initial east (e) and north (n) positions and velocities of the target: $\mathbf{x}_t = [p_{te0} \quad p_{tn0} \quad v_{te} \quad v_{tn}]^T$. Notice that the "own-ship" (tracking ship) must make at least one maneuver in order to compute a solution. The 30-degree left maneuver shown may not be operationally realistic since actual maneuvers are a compromise between obtaining a good solution and possibly being in a position to engage the target. While this case may seem trivially simple, it is of great interest for evaluating estimation methods because, without knowledge of range, the nonlinearities of

---

Advanced Kalman Filtering, Least-Squares and Modeling: A Practical Handbook, by Bruce P. Gibbs

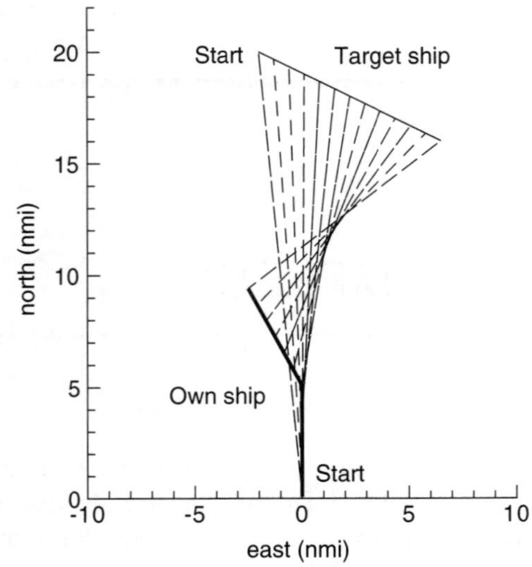

FIGURE 3.1: Passive tracking scenario.

the angle measurements are severe and nonlinear optimization can be difficult. The solution also breaks down when the target makes a maneuver, and that is difficult to detect without another source of information. Changes in received sonar signals may provide that information.

The dynamic model for this problem is

$$\mathbf{x}_t(t) = \begin{bmatrix} 1 & 0 & t & 0 \\ 0 & 1 & 0 & t \\ 0 & 0 & 1 & 0 \\ 0 & 0 & 0 & 1 \end{bmatrix} \mathbf{x}_t(0) \tag{3.1-1}$$

and the measurement model at time t_i is

$$\begin{aligned} y(t_i) &= \tan^{-1}\left(\frac{p_{te}(t_i) - p_{oe}(t_i)}{p_{tn}(t_i) - p_{on}(t_i)}\right) \\ &= \tan^{-1}\left(\frac{p_{te0} + v_{te}t_i - p_{oe}(t_i)}{p_{tn0} + v_{tn}t_i - p_{on}(t_i)}\right) \end{aligned} \tag{3.1-2}$$

where $p_{oe}(t_i)$ and $p_{on}(t_i)$ are the own-vehicle east and north position at t_i. Notice that because the dynamic model is so simple, the dynamics have been incorporated directly into the measurement model. The partial derivatives of the measurement with respect to the epoch target state are

$$\frac{\partial y(t_i)}{\partial \mathbf{x}_{t0}} = \begin{bmatrix} \frac{\Delta n}{R^2} & -\frac{\Delta e}{R^2} & \frac{t_i \Delta n}{R^2} & -\frac{t_i \Delta e}{R^2} \end{bmatrix} \tag{3.1-3}$$

where

$$\Delta e = p_{te0} + v_{te}t_i - p_{oe}(t_i),$$

$$\Delta n = p_{tn0} + v_{tn}t_i - p_{on}(t_i), \text{ and}$$
$$R^2 = (\Delta e)^2 + (\Delta n)^2.$$

For the example shown in Figure 3.1, the target ship is initially at a distance of about 20 nautical miles (nmi), and is moving at a constant velocity of 18.79 knots (kn) at a southeast heading of 115 degrees from north. The own-ship is initially moving directly north at a constant velocity of 20 kn. The own-ship measures bearing from itself to the target every 15 s, with an assumed $1 - \sigma$ accuracy of 1.5 degrees. At 15 min the own-ship turns 30 degrees to the left, and the encounter duration is 30 min. This example is used in Chapters 7 and 9 to demonstrate nonlinear estimation methods.

3.2 MANEUVERING VEHICLE TRACKING

This section discusses tracking models for two different types of maneuvering targets: tanks and aircraft. When in combat situations tanks tend to follow either high-speed straight line (constant velocity) paths, or moderate-speed serpentine paths to minimize the probability of being hit by enemy fire. Hence multiple types of motion are possible, but each motion type is fixed for periods of several seconds. The tank motion may be best tracked using multiple model filters where the best model for a fire control solution is selected using a tracking accuracy metric.

Civilian aircraft primarily fly constant velocity paths for extended periods of time, but occasionally turn, change speed, or climb/descend for periods of tens of seconds. When tracking aircraft for traffic control or collision avoidance purposes, a constant velocity model is optimal for possibly 90% of the tracking time, but occasionally constant rate turns, constant longitudinal acceleration, or constant rate of climb/descent will occur for short periods of time. Hence the tracking can be characterized as a "jump detection" problem, with associated state correction when a jump is detected.

3.2.1 Maneuvering Tank Tracking Using Multiple Models

Tank fire control algorithms are usually based on the assumption that the target vehicle moves in a straight line during the time-of-flight of the projectile. However, military vehicles are capable of turning with accelerations of 0.3 to 0.5 g and their braking capability approaches 0.9 g. At ranges of 1 to 3 km, projectile time-of-flights vary from 0.67 to 2 s. Thus target vehicle maneuvering can significantly reduce the hit probability when tracked by a constant velocity tracker. This explains the interest in higher-order adaptive trackers.

In 1975 the U.S. Army conducted an Anti-Tank Missile Test (ATMT) where accurate positions of moving vehicles were recorded using three vehicles: an M60A1 tank, a Scout armored reconnaissance vehicle, and a Twister prototype reconnaissance vehicle. The M60A1 is a heavy tank and the least maneuverable, while the Twister is the most maneuverable. The drivers were instructed to maneuver tactically at their own discretion. They were also required to make maximum use of terrain and to break the line-of-sight to the gun position. The maneuvers consisted

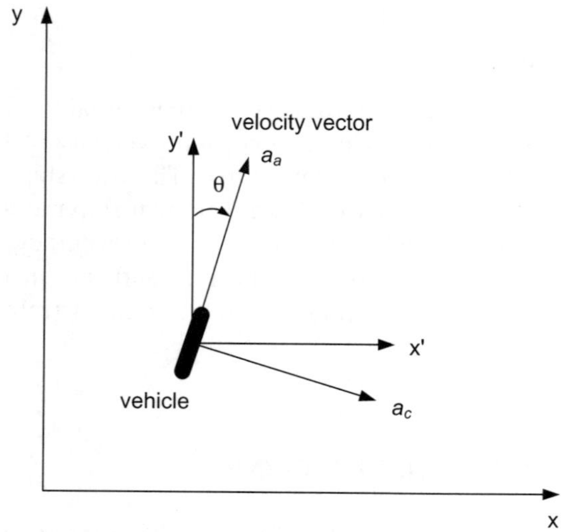

FIGURE 3.2: Tank coordinate system.

of short dashes, random serpentine, and fast turns. Drivers rarely followed the same paths used in previous tests.

The results from those tests showed that typical vehicle motions were very dependent upon the capabilities of the vehicle. Drivers tended to use the full maneuvering capability of the vehicle. For example, the magnitude of the serpentine motion was greater for the Twister than for the M60A1, and the maneuver frequency was higher. This suggested that a multiple-model adaptive tracker might significantly improve the firing hit probability if the fire control prediction was based on the model that best fit observed measurements (Gibbs and Porter 1980).

Referring to Figure 3.2, the accelerations in the Cartesian reference (nonrotating) coordinate system can be written in terms of alongtrack acceleration (a_a) and crosstrack acceleration (a_c)

$$\begin{aligned}\ddot{x} &= a_a \sin\theta + a_c \cos\theta \\ &= a_a(\dot{x}/v) + a_c(\dot{y}/v) \\ \ddot{y} &= a_a \cos\theta - a_c \sin\theta \\ &= a_a(\dot{y}/v) - a_c(\dot{x}/v)\end{aligned} \tag{3.2-1}$$

where $v = \sqrt{\dot{x}^2 + \dot{y}^2}$. Note that the model is nonlinear, but this does not cause problems for a tracking Kalman filter because the measurement data rate is high and accuracy is good.

Then a_a and a_c are defined as first- or second-order Markov (colored noise) processes with parameters chosen to match different characteristics observed in the ATMT. A preliminary evaluation of model structure was based upon power spectral densities computed by double-differencing the ATMT position data and rotating the computed accelerations into alongtrack and crosstrack components. This showed that both a_a and a_c had band-pass characteristics, with a_c having a much narrower passband than a_a. From this observation, the accelerations were modeled as Markov processes with transfer functions of the form

$$\frac{1}{\tau s+1}$$

for first-order or

$$\frac{s+1/\tau}{s^2+2\varsigma\omega_0 s+\omega_0^2}$$

for second-order. Third-order models were also tested but ultimately rejected. In state-space notation the complete dynamic model for second-order Markov processes is

$$\begin{bmatrix} \dot{x}_1 \\ \dot{x}_2 \\ \dot{x}_3 \\ \dot{x}_4 \\ \dot{x}_5 \\ \dot{x}_6 \\ \dot{x}_7 \\ \dot{x}_8 \end{bmatrix} = \begin{bmatrix} x_3 \\ x_4 \\ (a_a x_3 + a_c x_4)/v \\ (a_a x_4 - a_a x_3)/v \\ -\beta_{a1} x_5 - \beta_{a2} x_6 \\ x_5 \\ -\beta_{c1} x_7 - \beta_{c2} x_8 \\ x_7 \end{bmatrix} + \begin{bmatrix} 0 \\ 0 \\ 0 \\ 0 \\ q_a \\ 0 \\ q_c \\ 0 \end{bmatrix} \tag{3.2-2}$$

where

$$\begin{aligned} a_a &= x_5 + x_6/\tau_a \\ a_c &= x_7 + x_8/\tau_c \\ v &= \sqrt{x_3^2 + x_4^2} \end{aligned} \tag{3.2-3}$$

and

$$\begin{aligned} \beta_{a1} &= 2\varsigma_a \omega_{a0}, \quad \beta_{a2} = \omega_{a0}^2 \\ \beta_{c1} &= 2\varsigma_c \omega_{c0}, \quad \beta_{c2} = \omega_{c0}^2 \end{aligned}. \tag{3.2-4}$$

For first-order Markov processes the x_7, x_8 states are eliminated and derivatives of the acceleration states $a_a = x_5$, $a_c = x_6$ are

$$\begin{aligned} \dot{x}_5 &= -\beta_a x_5 + q_a \\ \dot{x}_6 &= -\beta_c x_6 + q_c \end{aligned} \tag{3.2-5}$$

The parameters of these Markov process models were determined using *maximum likelihood* estimation for each ATMT trial. (Chapter 8 shows how the likelihood function can be computed using outputs of a Kalman filter, and Chapter 11 describes how parameters that maximize the likelihood can be computed.) From the multiple models computed using this approach, four maneuver models that capture the dominant characteristics of data were selected. Table 3.1 summarizes the characteristics of these four maneuver models. The listed "vehicle" is a general description and does not necessarily correspond to tracks from that vehicle. Notice that the standard deviation of the "Twister" a_c acceleration is 0.32 g, which is intense maneuvering. In addition to these four models, a "constant velocity" maneuver

TABLE 3.1: Reduced Set of Tank Maneuver Models

	Alongtrack					Crosstrack				
Vehicle	Model Order	τ (s)	Period $2\pi/\omega_0$ (s)	ζ	σ_a (g)	Model Order	τ (s)	Period $2\pi/\omega_0$ (s)	ζ	σ_c (g)
M60	2	2.6	5.0	1.0	0.09	2	4.0	14.3	0.50	0.14
Scout	2	3.2	5.0	1.0	0.23	2	3.6	10.0	0.50	0.20
Twister	2	2.0	5.0	0.7	0.23	2	3.2	7.7	0.60	0.32
Low maneuver	1	0.4	—	—	0.16	1	0.53	—	—	0.11

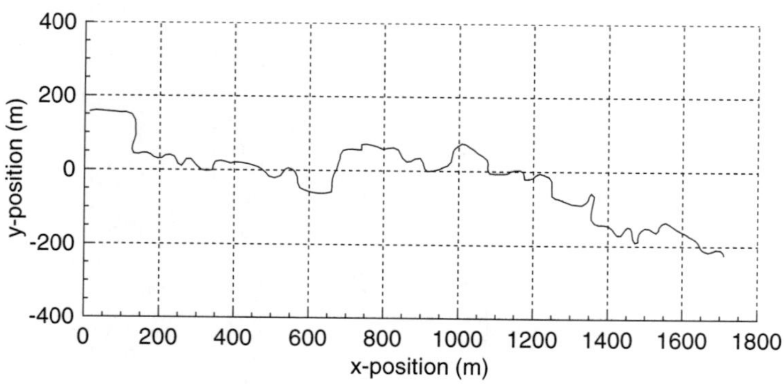

FIGURE 3.3: Scout track.

model (with assumed random walk process noise) was also included for a total of five models used in the adaptive tracking filter.

It should be noted that while these first- and second-order Markov acceleration models work well in tracking Kalman filters, they are not necessarily good long-term simulation models. For example, integration of a first-order Markov process model for acceleration results in infinite velocity. Integration of the above second-order Markov acceleration process has bounded velocity, but does not produce realistic tracks. Hence reduced-order models appropriate for Kalman filters with a short memory may not be accurate models of long-term behavior, particularly for nonlinear systems.

The measurement data available for real-time tracking of the target include observer-to-target azimuth (bearing) angles at a 10-Hz sampling rate, and range measurements every second. The assumed $1-\sigma$ measurement accuracies are 2 m in range and 0.3 milliradians in azimuth. This example is used in Chapter 9 when discussing extended Kalman filters and smoothers, and in Chapter 11 when discussing maximum likelihood parameter identification and multiple-model adaptive trackers. A sample Scout track over a period of about 315 s is shown in Figure 3.3. The average speed when moving is about 7.5 m/s. Unfortunately the position time history data could not be retrieved, so it was necessary to reconstruct the time history from the x-y position data in Figure 3.3. The data were reconstructed using a nonlinear least-squares fitting procedure that included prior knowledge of the average speed. The reconstruction is not perfect because the average speed is closer to 4 m/s, but visually looks very much like Figure 3.3.

3.2.2 Aircraft Tracking

Pilots are taught to maneuver using "coordinated" turns at a fixed rate, to climb/descend at a constant rate, and to use a combination of power and glide path angle changes to alter speed. Coordinated turns are executed so that there is no relative airspeed component in the aircraft lateral direction (parallel to the wings), which minimizes drag and also makes passengers more comfortable because they do not sense any lateral acceleration. Passengers feel that they are being pressed into their seats but, unlike riding in a car when negotiating a curve, are not forced to the outside. Figure 3.4 shows the forces operating in a coordinated turn. To avoid climbing or descending, the vertical component of the lift force must exactly balance the force of gravity, which implies that $L\cos\phi = W = mg$ where m is aircraft mass and g is gravitational acceleration. The lateral (crosstrack) acceleration is $a_c = (L/m)\sin\phi$ or $a_c = g\tan\phi$ where bank angles of 15 to 30 degrees are usually held constant during the turn. The relationship $a_c = g\tan\phi$ is accurate even if moderate climbs or descents are permitted.

Assuming that a_c, alongtrack acceleration (a_a), and vertical velocity ($\dot{z}$) are held constant for periods of time, the following eight-state model of aircraft motion in a fixed Cartesian coordinate frame can be developed.

$$\mathbf{x} = \begin{bmatrix} x \\ y \\ x \\ \dot{x} \\ \dot{y} \\ \dot{z} \\ a_c \\ a_a \end{bmatrix}, \quad \dot{\mathbf{x}} = \begin{bmatrix} \dot{x} \\ \dot{y} \\ \dot{z} \\ a_a(\dot{x}/v) + a_c(\dot{y}/v) \\ a_a(\dot{y}/v) - a_c(\dot{x}/v) \\ 0 \\ 0 \\ 0 \end{bmatrix} + \begin{bmatrix} 0 \\ 0 \\ 0 \\ 0 \\ 0 \\ q_{vz} \\ q_{ac} \\ q_{aa} \end{bmatrix} \tag{3.2-6}$$

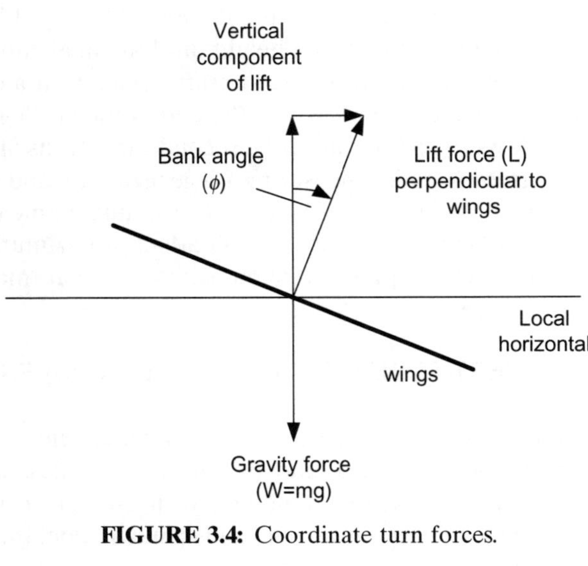

FIGURE 3.4: Coordinate turn forces.

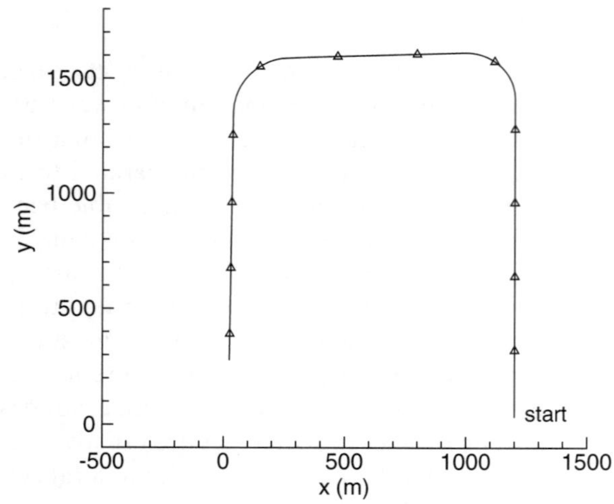

FIGURE 3.5: Simulated ground track for aircraft approach to landing.

where q_{vz}, q_{ac}, q_{aa} represent zero mean random process noise that must be included to allow for changes in a_c, a_a, and $\dot{z}$. As shown, $\dot{z}, a_c, a_a$ are random walk processes if q_{vz}, q_{ac}, q_{aa} are white noise. First-order Markov models (e.g., $\ddot{z} = q_{vz} - \dot{z}/\tau$) may seem preferable since $E[\dot{z}^2]$ is then bounded rather than growing linearly with time, as for a random walk. However, in practice there is usually little difference in tracking performance of the two models if position measurements (e.g., range, azimuth, and elevation from a ground tracker) are available at intervals of a few seconds. As indicated previously, it is often better to assume that process noise is nominally small or zero, and allow jumps in the $\dot{z}, a_a, a_c$ states when changes in acceleration are detected. Methods for implementing this are discussed in Chapter 11.

Figure 3.5 shows a simulated aircraft ground track based on the model of equation (3.2-2). The scenario attempts to model a small aircraft within the traffic pattern for an approach to landing. The aircraft is initially located at $x = 1200\,\text{m}$, $y = 0\,\text{m}$ at an altitude of 300 m, and it remains at that height until the final approach leg of the pattern. During and after the final turn, the aircraft descends at a constant rate of 4 to 6 m/s until reaching an altitude of 38 m at the end of the last leg. Aircraft speed is initially 32 m/s but slows to 28.5 m/s after the second turn. Turns are made at bank angles of 25 to 28 degrees. A radar measuring range, azimuth, and elevation every second is located at the origin (0, 0) on the plot. The simulated measurement noise has a $1-\sigma$ magnitude of 1 m in range and 0.2 milliradians for azimuth and elevation. This example is also used in Chapter 9 to demonstrate optimal smoothing.

3.3 STRAPDOWN INERTIAL NAVIGATION SYSTEM (INS) ERROR MODEL

Inertial navigation has the advantage that it is independent of external data sources. Inertial navigators may be found on aircraft, missiles, submarines, and ground vehicles. They are sometimes combined with Global Positioning System (GPS) receivers to provide more accurate and robust navigation. Inertial navigators are classified as either gimbaled (stable platform) or strapdown. In a gimbaled

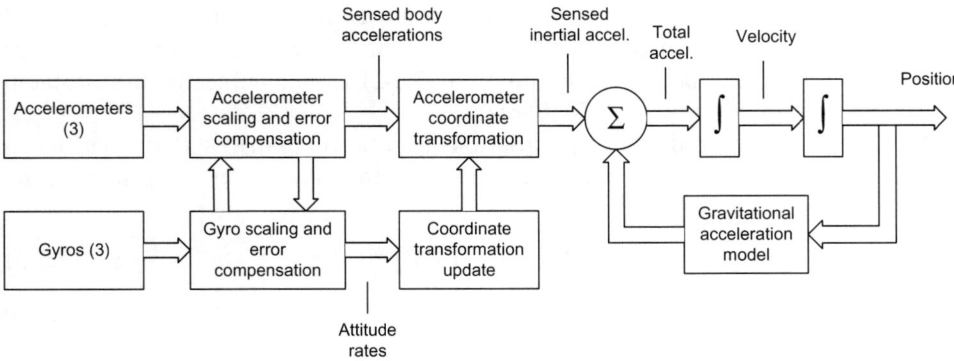

FIGURE 3.6: Strapdown inertial navigator.

system signals from the gyros are used to torque gimbals so that the gyros and accelerometer assemblies remain either fixed in inertial space or locally level. In a strapdown system the gyros and accelerometer assemblies are "strapped" to the vehicle structure and the transformation from the vehicle frame to inertial coordinates is computational. The error characteristics of gimbaled and strapdown INS are very different, but navigation errors in both tend to grow with time (Britting 1971; Grewal and Andrews 2001; Grewal et al. 2001).

INS error behavior is of interest both for system error analysis and because it is used when combining navigation estimates from inertial and GPS systems. Fortunately the error characteristics are well described by linear models. We analyze the strapdown navigator defined in Figure 3.6.

Accelerometers measure the force required to keep a "proof mass" in a reference position within the accelerometer. The proof mass senses the earth's gravitational force, but does not sense linear acceleration experienced by both the accelerometer body and the proof mass, that is, an accelerometer on an earth satellite subject only to gravitational forces will measure no acceleration. We denote the scaled and compensated output of the three-axis accelerometers—the specific force in accelerometer/gyro coordinates—as $\mathbf{f}^G$. The acceleration in computational (inertial) coordinates is calculated by rotating $\mathbf{f}^G$ to inertial coordinates and adding the computed gravitational acceleration, $\mathbf{g}(\mathbf{r}^C)$, at the inertial position $\mathbf{r}^C$:

$$\dot{\mathbf{v}}^C = \mathbf{C}_G^C \mathbf{f}^G + \mathbf{g}(\mathbf{r}^C) \tag{3.3-1}$$

where $\mathbf{C}_G^C$ is the rotation (direction cosine) matrix relating the INS body coordinate frame to a fixed, inertial frame. $\mathbf{C}_G^C$ is computed knowing the initial orientation of the INS body relative to the inertial coordinates, and then adding all subsequent rotations using gyro outputs. The inertial position is computed by numerically integrating $\dot{\mathbf{v}}^C$ to obtain velocity and then integrating again to obtain inertial position $\mathbf{r}^C$.

The error in measured acceleration is defined as the difference between the measured acceleration $\dot{\mathbf{v}}^{Cm}$ and the true acceleration $\dot{\mathbf{v}}^C$: $\delta\dot{\mathbf{v}}^C = \dot{\mathbf{v}}^{Cm} - \dot{\mathbf{v}}^C$. From equation (3.3-1) this is

$$\delta\dot{\mathbf{v}}^C = \delta\mathbf{C}_G^C \mathbf{f}^G + \mathbf{C}_G^C \delta\mathbf{f}^G + \frac{\partial \mathbf{g}(\mathbf{r}^C)}{\partial \mathbf{r}^C} \delta\mathbf{r}^C \tag{3.3-2}$$

where $\delta\mathbf{C}_G^C$ is the error in the computed rotation (due to errors in gyro output) and $\delta\mathbf{f}^G$ is the error in measured acceleration (without gravity). For purposes of this model, we ignore the last term in equation (3.3-2) as negligible for the problem under consideration (short missile flights).

We briefly digress to discuss perturbations in rotation matrices. Perturbations in coordinate transformations from a rotating frame to a fixed (inertial) frame can be represented as

$$\mathbf{C}_{R'}^F = \mathbf{C}_R^F[\mathbf{I}+\boldsymbol{\Psi}] \tag{3.3-3}$$

where

$$\boldsymbol{\Psi} = \begin{bmatrix} 0 & -\psi_z & \psi_y \\ \psi_z & 0 & -\psi_x \\ -\psi_y & \psi_x & 0 \end{bmatrix} \tag{3.3-4}$$

and $\boldsymbol{\psi} = [\psi_x \quad \psi_y \quad \psi_z]^T$ are the three infinitesimal rotation angles (radians) along the x, y, and z axes that define the rotation from the original to the perturbed (primed) coordinate frame. Equation 3.3-3 can be derived by computing the coordinate projections from the primed to the unprimed system for infinitesimal rotations $\boldsymbol{\psi}$ using the small angle assumptions $\cos\psi = 1$ and $\sin\psi = \psi$; that is, $\mathbf{C}_{R'}^F = \mathbf{C}_R^F\mathbf{C}_{R'}^R$ and

$$\mathbf{C}_{R'}^R = \begin{bmatrix} 1 & -\psi_z & \psi_y \\ \psi_z & 1 & -\psi_x \\ -\psi_y & \psi_x & 1 \end{bmatrix} = \mathbf{I}+\boldsymbol{\Psi}$$

for small $\boldsymbol{\psi}$. Hence

$$\begin{aligned} \dot{\mathbf{C}}_R^F &= \frac{d\mathbf{C}_R^F}{dt} = \mathbf{C}_R^F\frac{d\boldsymbol{\Psi}}{dt} \\ &= \mathbf{C}_R^F\boldsymbol{\Omega} \end{aligned} \tag{3.3-5}$$

where we have defined

$$\boldsymbol{\Omega} = \frac{d\boldsymbol{\Psi}}{dt} = \begin{bmatrix} 0 & -\omega_z & \omega_y \\ \omega_z & 0 & -\omega_x \\ -\omega_y & \omega_x & 0 \end{bmatrix} \tag{3.3-6}$$

and $\boldsymbol{\omega} = [\omega_x \quad \omega_y \quad \omega_z]^T$ is the vector of body rotation rates (radians/s).

Notice that the error in the direction cosine matrix of equation (3.3-2) can be represented using equation (3.3-3) as

$$\delta\mathbf{C}_G^C = \mathbf{C}_G^C[\mathbf{I}+\boldsymbol{\Psi}] - \mathbf{C}_G^C = \mathbf{C}_G^C\boldsymbol{\Psi} \tag{3.3-7}$$

where $\boldsymbol{\Psi}$ is the skew-symmetric matrix defining the "tilt error" in rotation angles. Hence

$$\begin{aligned} \delta\dot{\mathbf{v}}^C &= \mathbf{C}_G^C\boldsymbol{\Psi}^G\mathbf{f}^G + \mathbf{C}_G^C\delta\mathbf{f}^G \\ &= \mathbf{C}_G^C(\boldsymbol{\psi}^G \times \mathbf{f}^G) + \mathbf{C}_G^C\delta\mathbf{f}^G \\ &= (\mathbf{C}_G^C\boldsymbol{\psi}^G)\times(\mathbf{C}_G^C\mathbf{f}^G) + \mathbf{C}_G^C\delta\mathbf{f}^G \\ &= \boldsymbol{\psi}^C \times (\mathbf{C}_G^C\mathbf{f}^G) + \mathbf{C}_G^C\delta\mathbf{f}^G \end{aligned} \tag{3.3-8}$$

where

$\boldsymbol{\Psi}^G \mathbf{f}^G = \boldsymbol{\psi}^G \times \mathbf{f}^G$ (from definitions above),

$\boldsymbol{\psi}^G$ is the vector of gyro tilt errors as measured in the gyro coordinate system,

$\boldsymbol{\psi}^C = \mathbf{C}_G^C \boldsymbol{\psi}^G$ is the vector of gyro tilt errors in the computational coordinate system, and

$\times$ denotes vector cross-product.

Gyroscopes used for strapdown navigation systems measure instantaneous body angular rates $\boldsymbol{\omega}^G$ rather than angles. Thus the INS must integrate the angular rates to compute the $\mathbf{C}_G^C$ rotation. From equation (3.3-5),

$$\dot{\mathbf{C}}_G^C = \mathbf{C}_G^C \boldsymbol{\Omega} \tag{3.3-9}$$

with $\boldsymbol{\Omega}$ defined in equation (3.3-6). Since $\delta\dot{\mathbf{v}}^C$ in equation (3.3-8) is defined using $\boldsymbol{\psi}^C = \mathbf{C}_G^C \boldsymbol{\psi}^G$, the tilt error growth in the computational frame can be obtained by integrating

$$\dot{\boldsymbol{\psi}}^C = \mathbf{C}_G^C \dot{\boldsymbol{\psi}}^G = \mathbf{C}_G^C \boldsymbol{\omega}^G \tag{3.3-10}$$

where $\boldsymbol{\omega}^G$ is the gyro output.

Gyroscope designs are varied (e.g., momentum wheels, Coriolis effect, torsion resonance, ring laser, fiber-optic, etc.), but they all tend to have similar error mechanisms. Specific error sources include biases, scale factor errors (SFEs), drift, input-axis misalignments, input/output nonlinearities, acceleration sensitivity, and g-squared sensitivity. Accelerometer error sources include biases, SFEs, drift, g-squared sensitivity, cross-axis coupling, centrifugal acceleration effects, and angular acceleration sensitivity. Detailed INS error models may include hundreds of error parameters, but we define a 27-state reduced-order model as

$$\begin{aligned}
\delta\dot{\mathbf{r}} &= \delta\mathbf{v} \\
\delta\dot{\mathbf{v}} &= \boldsymbol{\psi}^C \times \mathbf{C}_G^C \mathbf{f}^G + \mathbf{C}_G^C \frac{\partial \mathbf{f}^G}{\partial \boldsymbol{\alpha}} \delta\boldsymbol{\alpha} \\
\dot{\boldsymbol{\psi}}^C &= \mathbf{C}_G^C \frac{\partial \boldsymbol{\omega}^G}{\partial \boldsymbol{\beta}} \delta\boldsymbol{\beta}
\end{aligned} \tag{3.3-11}$$

where

$\delta\mathbf{r}$ is the error in computed inertial position,

$\delta\mathbf{v}$ is the error in computed inertial velocity,

$\boldsymbol{\psi}^C$ is the tilt error of the gyro/accelerometer computational frame,

$\mathbf{C}_G^C$ is the transformation from the gyro to guidance computational (inertial) frame,

$\mathbf{f}^G$ is the acceleration expressed in the gyro frame,

$\delta\boldsymbol{\alpha}$ is the total accelerometer error,

$\boldsymbol{\omega}^G$ is the angular rate of the gyro-to-computational frame expressed in the gyro frame,

$\delta\boldsymbol{\beta}$ is the total gyro error.

TABLE 3.2: Inertial Navigator States and Initial Uncertainty

Symbol	State Description (Number)	Initial Uncertainty	Process Noise PSD
$\delta\mathbf{r}$	Error in computed inertial position (3)	4 m	0
$\delta\mathbf{v}$	Error in computed inertial velocity (3)	0.61 mm/s	$5.02 \times 10^{-10}\,(\text{m/s}^2)^2/\text{s}$
$\boldsymbol{\psi}^C$	Tilt error of the gyro/accelerometer computational frame (3)	1440 arc-sec	$0.004\,(\text{arc-s})^2/\text{s}$
$\delta\boldsymbol{\alpha}_b$	Accelerometer bias (3)	60 micro-g	0
$\delta\boldsymbol{\alpha}_{SFE}$	Accelerometer Scale Factor Error (3)	60 PPM	0
$\delta\boldsymbol{\beta}_b$	Gyro rate bias drift (3)	0.2 arc-sec/s	0
$\delta\boldsymbol{\beta}_{SFE}$	Gyro Scale Factor Error (3)	200 PPM	0
$\delta\boldsymbol{\beta}_G$	Gyro G-sensitive error (3)	0.2 arc-sec/s/g	0
$\delta\boldsymbol{\beta}_{G2}$	Gyro G^2-sensitive error (3)	0.012 arc-sec/s/g^2	0

$\delta\boldsymbol{\alpha}$ and $\delta\boldsymbol{\beta}$ are the sum of several other terms. Table 3.2 lists the 27 error model states, assumed state initial uncertainty and process noise power spectral densities (PSD) for a specific missile test system. Random process noise is assumed to drive only velocity and tilt, and the assumed values integrate to 0.5 mm/s (velocity) and 1.55 arc-sec (tilt) $1-\sigma$ in 10 min. The measurements consist of position error in each channel $(\delta x, \delta y, \delta z)$ with a measurement error of 0.61 mm $1-\sigma$. (This extreme high accuracy is only obtained with special test instrumentation, and is not typical for standard measurement systems such as GPS.) Measurements were taken every 10 s during a missile flight of 790 s to give a total of 237 scalar measurements. Note that the dynamic error equations require input of specific force and body rates versus time. A two-stage missile booster with burn times of 70 and 110 s, respectively, flown with a 5 degrees offset between the thrust vector and velocity vector, was simulated. The peak trajectory altitude was approximately 2900 km at 1200 s of flight. After the boost phase, the simulated payload maneuvered every 100 s where the specific force was one cycle of a sinusoid (0.07 g maximum) with a period of 15 s. Figure 3.7 shows the simulated Earth-Centered Inertial (ECI) specific force (acceleration less gravity) for the simulated flight. The acceleration due to booster staging is apparent, but acceleration due to bus maneuvers (0.7 m/s^2 peak) is not evident at this scale.

The dynamics model has the form

$$\begin{bmatrix} \delta\dot{\mathbf{r}} \\ \delta\dot{\mathbf{v}} \\ \dot{\boldsymbol{\psi}}^C \\ \delta\dot{\boldsymbol{\alpha}}_b \\ \delta\dot{\boldsymbol{\alpha}}_{SFE} \\ \delta\dot{\boldsymbol{\beta}}_b \\ \delta\dot{\boldsymbol{\beta}}_{SFE} \\ \delta\dot{\boldsymbol{\beta}}_G \\ \delta\dot{\boldsymbol{\beta}}_{G2} \end{bmatrix} = \begin{bmatrix} \mathbf{0} & \mathbf{I} & \mathbf{0} & \mathbf{0} & \mathbf{0} & \mathbf{0} & \mathbf{0} & \mathbf{0} & \mathbf{0} \\ \mathbf{0} & \mathbf{0} & \mathbf{F}_{23} & \mathbf{F}_{24} & \mathbf{F}_{25} & \mathbf{0} & \mathbf{0} & \mathbf{0} & \mathbf{0} \\ \mathbf{0} & \mathbf{0} & \mathbf{0} & \mathbf{0} & \mathbf{0} & \mathbf{F}_{36} & \mathbf{F}_{37} & \mathbf{F}_{38} & \mathbf{F}_{39} \\ \mathbf{0} & \mathbf{0} & \mathbf{0} & \mathbf{0} & \mathbf{0} & \mathbf{0} & \mathbf{0} & \mathbf{0} & \mathbf{0} \\ \mathbf{0} & \mathbf{0} & \mathbf{0} & \mathbf{0} & \mathbf{0} & \mathbf{0} & \mathbf{0} & \mathbf{0} & \mathbf{0} \\ \mathbf{0} & \mathbf{0} & \mathbf{0} & \mathbf{0} & \mathbf{0} & \mathbf{0} & \mathbf{0} & \mathbf{0} & \mathbf{0} \\ \mathbf{0} & \mathbf{0} & \mathbf{0} & \mathbf{0} & \mathbf{0} & \mathbf{0} & \mathbf{0} & \mathbf{0} & \mathbf{0} \\ \mathbf{0} & \mathbf{0} & \mathbf{0} & \mathbf{0} & \mathbf{0} & \mathbf{0} & \mathbf{0} & \mathbf{0} & \mathbf{0} \\ \mathbf{0} & \mathbf{0} & \mathbf{0} & \mathbf{0} & \mathbf{0} & \mathbf{0} & \mathbf{0} & \mathbf{0} & \mathbf{0} \end{bmatrix} \begin{bmatrix} \delta\mathbf{r} \\ \delta\mathbf{v} \\ \boldsymbol{\psi}^C \\ \delta\boldsymbol{\alpha}_b \\ \delta\boldsymbol{\alpha}_{SFE} \\ \delta\boldsymbol{\beta}_b \\ \delta\boldsymbol{\beta}_{SFE} \\ \delta\boldsymbol{\beta}_G \\ \delta\boldsymbol{\beta}_{G2} \end{bmatrix} \qquad (3.3\text{-}12)$$

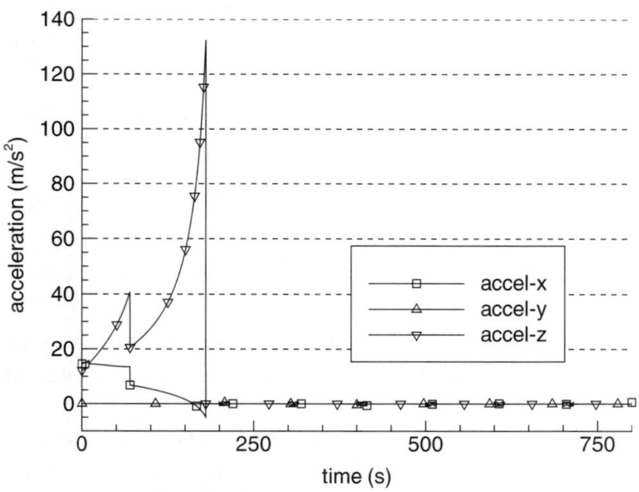

FIGURE 3.7: Simulated ECI missile acceleration.

where all partitions are 3 × 3 and are computed according to equation (3.3-11) using $\mathbf{C}_G^C$, $\mathbf{f}^G$, and $\boldsymbol{\omega}^G$ evaluated along the reference trajectory:

$$\mathbf{F}_{23} = \begin{bmatrix} 0 & f_z^C & -f_y^C \\ -f_z^C & 0 & f_x^C \\ f_y^C & -f_x^C & 0 \end{bmatrix}, \quad (\mathbf{f}^C = \mathbf{C}_G^C \mathbf{f}^G)$$

$$\mathbf{F}_{24} = \mathbf{C}_G^C, \qquad \mathbf{F}_{25} = \mathbf{C}_G^C \begin{bmatrix} f_x^C & 0 & 0 \\ 0 & f_y^C & 0 \\ 0 & 0 & f_z^C \end{bmatrix}$$

$$\mathbf{F}_{36} = \mathbf{C}_G^C, \qquad \mathbf{F}_{37} = \mathbf{C}_G^C \begin{bmatrix} \omega_x^G & 0 & 0 \\ 0 & \omega_y^G & 0 \\ 0 & 0 & \omega_z^G \end{bmatrix}$$

$$\mathbf{F}_{38} = \mathbf{C}_G^C \begin{bmatrix} f_y^G & 0 & 0 \\ 0 & f_x^G & 0 \\ 0 & 0 & f_x^G \end{bmatrix}, \quad \mathbf{F}_{39} = \mathbf{C}_G^C \begin{bmatrix} (f_y^G)^2 & 0 & 0 \\ 0 & (f_x^G)^2 & 0 \\ 0 & 0 & (f_x^G)^2 \end{bmatrix}$$

The somewhat strange use of $\mathbf{f}^G$ components for $\mathbf{F}_{38}$ and $\mathbf{F}_{39}$ is due to the definitions of the gyro frame ($\mathbf{x}$ = along the velocity vector, $\mathbf{y} = \mathbf{r} \times \mathbf{x}$, and $\mathbf{z} = \mathbf{x} \times \mathbf{y}$) and the gyro alignment. Not shown in these equations are unit conversions that may be necessary if accelerometer and gyro errors have units different from those of velocity and tilt.

In this particular problem the differential equations are integrated at a 0.2 s interval with 10 s between measurements. Accelerometer specific force and gyro angular rates are updated at 0.2 s intervals and the state equations (eq. 3.3-11) are integrated using a second-order Taylor series. Because of the very high integration rate and slow dynamics, the state transition matrix is computed as $\boldsymbol{\Phi} = \mathbf{I} + \mathbf{F}T$, and

TABLE 3.3: Simulated INS Jumps

Simulation Jump Time (s)	Jump Parameter	Jump Magnitude
25	*y*-gyro SFE	−60 PPM
50	*y*-gyro G-sensitive error	+0.10 arc-sec/G
100	z-gyro rate bias drift	−0.10 arc-sec/s
120	x-accelerometer SFE	+1.0 PPM
400	y-accelerometer bias	−4 micro-G

$\mathbf{Q}_D$ is assumed to be diagonal over a 0.2-s interval. Accuracy is more important when computing measurement residuals than when propagating the error covariance using $\mathbf{\Phi}$ and $\mathbf{Q}_D$.

By including additional terms of the Taylor series, it can be shown that $\mathbf{\Phi}$ has the form of equation (3.3-12). Notice that $\mathbf{\Phi}$ retains much of the sparse structure of $\mathbf{F}$, but fills in additional terms in the upper triangle. The type of sparse structure depends upon the ordering of states. Filter code execution can often be made more efficient if biases are placed at the end of the state vector and integral states, such as position, at the beginning. In this case $\mathbf{Q}_D$ will only have nonzero elements in the upper left 9 by 9 block.

$$\mathbf{\Phi}=\begin{bmatrix} \mathbf{I} & \mathbf{\Phi}_{12} & \mathbf{\Phi}_{13} & \mathbf{\Phi}_{14} & \mathbf{\Phi}_{15} & \mathbf{\Phi}_{16} & \mathbf{\Phi}_{17} & \mathbf{\Phi}_{18} & \mathbf{\Phi}_{19} \\ \mathbf{0} & \mathbf{I} & \mathbf{\Phi}_{23} & \mathbf{\Phi}_{24} & \mathbf{\Phi}_{25} & \mathbf{\Phi}_{26} & \mathbf{\Phi}_{27} & \mathbf{\Phi}_{28} & \mathbf{\Phi}_{29} \\ \mathbf{0} & \mathbf{0} & \mathbf{I} & \mathbf{0} & \mathbf{0} & \mathbf{\Phi}_{36} & \mathbf{\Phi}_{37} & \mathbf{\Phi}_{38} & \mathbf{\Phi}_{39} \\ \mathbf{0} & \mathbf{0} & \mathbf{0} & \mathbf{I} & \mathbf{0} & \mathbf{0} & \mathbf{0} & \mathbf{0} & \mathbf{0} \\ \mathbf{0} & \mathbf{0} & \mathbf{0} & \mathbf{0} & \mathbf{I} & \mathbf{0} & \mathbf{0} & \mathbf{0} & \mathbf{0} \\ \mathbf{0} & \mathbf{0} & \mathbf{0} & \mathbf{0} & \mathbf{0} & \mathbf{I} & \mathbf{0} & \mathbf{0} & \mathbf{0} \\ \mathbf{0} & \mathbf{0} & \mathbf{0} & \mathbf{0} & \mathbf{0} & \mathbf{0} & \mathbf{I} & \mathbf{0} & \mathbf{0} \\ \mathbf{0} & \mathbf{0} & \mathbf{0} & \mathbf{0} & \mathbf{0} & \mathbf{0} & \mathbf{0} & \mathbf{I} & \mathbf{0} \\ \mathbf{0} & \mathbf{0} & \mathbf{0} & \mathbf{0} & \mathbf{0} & \mathbf{0} & \mathbf{0} & \mathbf{0} & \mathbf{I} \end{bmatrix} \tag{3.3-12}$$

This INS example is used in Chapter 11 to demonstrate how a jump detection algorithm is added to a Kalman filter. The simulated jumps are listed in Table 3.3. The example is also used in Chapter 9 to demonstrate Kalman filter error analysis, and in Chapter 10 to compare the accuracy of different Kalman filter implementations.

3.4 SPACECRAFT ORBIT DETERMINATION (OD)

Dynamics of an earth satellite are of interest (1) for historical reasons, (2) because it is still an active application area, and (3) because the OD problem demonstrates a number of interesting estimation issues involving nonlinear models and poor observability.

Satellite orbital dynamics are usually modeled in an *Earth-Centered Inertial* (ECI) coordinate frame. The ECI *x-y* plane is the earth's equatorial plane, and the *z*-axis is the earth's spin vector. The *x*-axis is defined so that it passes through the intersection (*vernal equinox*) of the earth's equatorial plane and the *ecliptic plane* (the plane containing the mean orbit of the earth around the sun). Due to precession and nutation of the earth's spin axis, ECI frames must be specified at some epoch time, and assumptions about precession and nutation must also be specified. The two most commonly used ECI frames are the *J2000* and *true-of-date* (TOD) frames. The J2000 frame uses the mean (ignoring nutation) equator and equinox of Universal Time Coordinated (UTC) noon on January 1, 2000, called *J2000.0*. The TOD frame uses both earth precession and nutation at the specified epoch when calculating the ECI orientation. Most orbital integration is performed in the J2000 frame because TOD frames are slowly rotating and thus not truly inertial. See Seidelmann (2006), Tapley et al. (2004), Vallado and McClain (2001), Wertz (1978), Escobal (1976), and GTDS Mathematical Theory (1989) for more information on orbital coordinate systems and orbit definitions.

Spacecraft orbits are defined using six orbital parameters. Orbits are often specified using *Kepler elements* (Fig. 3.8) that define the orientation of an orbital ellipse and the position in the ellipse:

1. *Semimajor axis* of ellipse
2. *Eccentricity* of ellipse
3. *Inclination* of ellipse with respect to reference plane (earth equatorial plane for a geocentric orbit)
4. *Right ascension of ascending node*: the central angle in the reference plane, measured from the x-axis, at which the ellipse intersects the reference plane with the satellite moving in a northerly direction

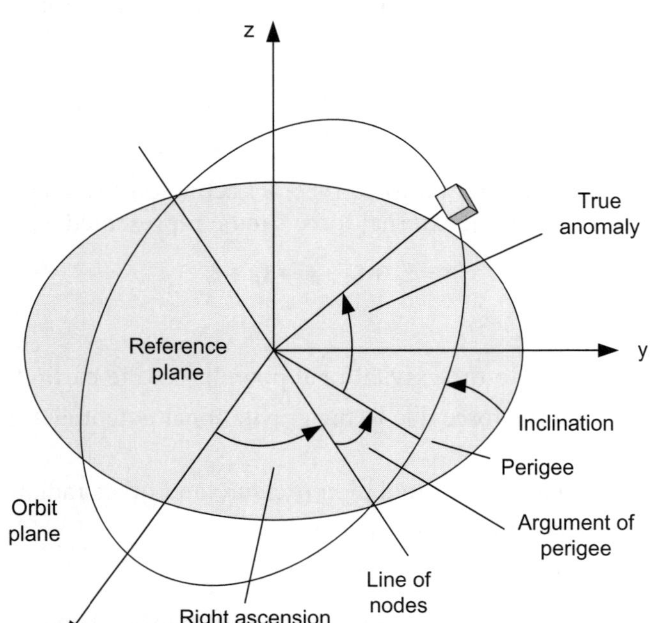

FIGURE 3.8: Kepler orbital elements.

5. *Argument of perigee*: the central angle from the ascending node to the perigee of the ellipse measured in the direction of motion
6. *Mean anomaly*: the central angle from perigee to the current orbital position at the given time, but defined as if the orbit is circular. Mean anomaly is not a physical angle. It is measured assuming that the mean angular rate is constant rather than varying with radial distance. True anomaly is the central angle from perigee to ellipse focus to the current orbital position, but because the angular rate changes throughout the orbit, it is not directly useful for modeling orbit propagation.

Kepler elements are convenient for understanding elliptical orbits. Because the first five parameters are constant for a two-body orbit (one in which the gravitational attraction between the two bodies is the only force acting on the satellite), calculation of the inertial position as a function of time is relatively easy. Since the mean anomaly changes at a constant *mean orbital motion* rate, the change in mean anomaly at any given time is simply mean motion multiplied by the time interval from epoch. Unfortunately, real satellite orbits are affected by forces other than just central gravitation, so propagation for a real orbit is more difficult. Furthermore, Kepler elements are indeterminate for circular or zero-inclination orbits. An alternate representation, *equinoctial elements*, avoids this problem by using two-parameter inclination and eccentricity vectors, and replacing mean anomaly with *mean longitude* (Bate et al. 1971; Vallado and McClain 2001). For circular or zero-inclination orbits, the vector elements are simply zero. However, propagation of equinoctial elements for real orbits is also nontrivial. For that reason, six ECI Cartesian position and velocity elements are usually used for numerical integration of orbits.

Using ECI Cartesian elements **r** and **v**, the orbital dynamics for a constant mass spacecraft are

$$\begin{aligned} \dot{\mathbf{r}} &= \mathbf{v} \\ \dot{\mathbf{v}} &= \mathbf{f}/m \end{aligned} \tag{3.4-1}$$

where **f** is the total ECI external force (vector) acting on the spacecraft and m is the spacecraft mass. The total external force can be represented as

$$\mathbf{f} = \mathbf{f}_E + \mathbf{f}_G + \mathbf{f}_S + \mathbf{f}_D + \mathbf{f}_T \tag{3.4-2}$$

where

$\mathbf{f}_E$ is the force due to the gravitational potential of the earth,

$\mathbf{f}_G$ is the differential force due to the gravitational potential of the sun, moon, and planets,

$\mathbf{f}_S$ is the force due to solar radiation pressure and other radiation,

$\mathbf{f}_D$ is the force due to aerodynamic drag,

$\mathbf{f}_T$ is the force due to spacecraft thrusting.

Note that when the spacecraft is generating thrust, mass changes, so the second line of equation (3.4-1) must be modified. We now examine each of the force components, or in the case of gravitational attractions, compute the acceleration directly.

3.4.1 Geopotential Forces

The earth's gravitational potential U_E at a point r, δ, λ outside the earth is defined by the Laplacian $\nabla^2 U = 0$. The solution can be expressed as a spherical harmonic expansion in Earth-Centered-Fixed (ECF) coordinates:

$$U_E = \frac{\mu}{r}\left(1 + \sum_{n=2}^{\infty}\left(\frac{a_e}{r}\right)^n \sum_{m=0}^{n} P_n^m(\sin\delta)[C_{nm}\cos m\lambda + S_{nm}\sin m\lambda]\right) \qquad (3.4\text{-}3)$$

where

μ = gravitational constant
r = geocentric distance
a_e = mean equatorial radius of the earth
P_n^m = associated Legendre function of the first kind of degree n and order m
δ = geocentric declination (geocentric latitude)
λ = geocentric east longitude
C_{nm}, S_{nm} = harmonic coefficients of degree n and order m

In practice the above series for n is truncated at an order sufficient to achieve the desired accuracy. The n = 1 term in equation (3.4-3) does not normally contribute if the coordinate system origin is at the center of mass. Note that equation (3.4-3) uses geocentric latitude (declination) and longitude as defined in Figure 3.9: the x- and y-axes are in the equatorial plane, the x-axis on the Greenwich meridian, and the z-axis is the earth spin axis. To be rigorously correct, the ECF system should

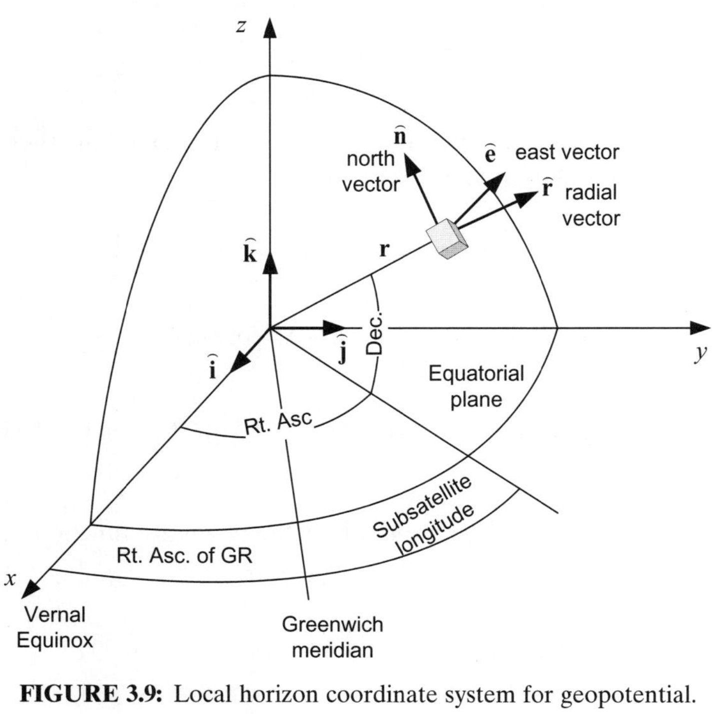

FIGURE 3.9: Local horizon coordinate system for geopotential.

include polar motion, but the approximately two microradian polar motion shift is often ignored as having negligible effect on the computed gravitational field.

Starting with the spacecraft position vector in J2000 coordinates, the position in ECF coordinates is computed using precession/nutation rotations and a rotation for *Greenwich Hour Angle*:

$$\mathbf{r}_{ECF} = \begin{bmatrix} \cos\theta_{Gr} & \sin\theta_{Gr} & 0 \\ -\sin\theta_{Gr} & \cos\theta_{Gr} & 0 \\ 0 & 0 & 1 \end{bmatrix} \mathbf{C}_{J2K}^{TOD} \mathbf{r}_{J2K} \tag{3.4-4}$$

where θ_{Gr} is the right ascension of Greenwich at the given time and $\mathbf{C}_{J2K}^{TOD}$ is the rotation matrix from the J2000 coordinate system (mean-equator and mean-equinox of January 1, 2000 noon) to TOD coordinates (true-equator and true-equinox of current date). $\mathbf{C}_{J2K}^{TOD}$ models precession and nutation of the earth's spin axis (Tapley et al. 2004; Seidelmann 2006). The right ascension of the satellite is the sum of longitude and Greenwich Hour Angle:

$$\alpha = \lambda + \theta_{Gr} \tag{3.4-5}$$

Then the geocentric latitude and longitude can be computed from the ECF coordinates as

$$\begin{aligned} \lambda &= \tan^{-1}(y_{ECF} / x_{ECF}) \\ \delta &= \sin^{-1}(z_{ECF} / r) \end{aligned} \tag{3.4-6}$$

although it is not necessary to work directly with these angles (all computations involve trigonometric functions).

The inertial acceleration vector due to earth gravitation in ECF coordinates is computed using the gradient of the potential function, that is,

$$\ddot{\mathbf{r}}_{E\_ECF} = \frac{\partial U}{\partial r}\left[\frac{\partial r}{\partial \mathbf{r}_{ECF}}\right]^T + \frac{\partial U}{\partial \delta}\left[\frac{\partial \delta}{\partial \mathbf{r}_{ECF}}\right]^T + \frac{\partial U}{\partial \lambda}\left[\frac{\partial \lambda}{\partial \mathbf{r}_{ECF}}\right]^T \tag{3.4-7}$$

The partial derivatives of the potential with respect to radius, longitude, and declination (geocentric latitude) are, respectively,

$$\frac{\partial U_E}{\partial r} = \frac{-\mu}{r^2}\left(\sum_{n=2}^{\infty}(n+1)\left(\frac{a_e}{r}\right)^n \sum_{m=0}^{n} P_n^m(\sin\delta)[C_n^m \cos m\lambda + S_n^m \sin m\lambda]\right) \tag{3.4-8}$$

$$\frac{\partial U_E}{\partial \lambda} = \frac{\mu}{r}\sum_{n=2}^{\infty}\left(\frac{a_e}{r}\right)^n \sum_{m=0}^{n} m P_n^m(\sin\delta)[-C_n^m \sin m\lambda + S_n^m \cos m\lambda] \tag{3.4-9}$$

$$\frac{\partial U_E}{\partial \delta} = \frac{\mu}{r}\sum_{n=2}^{\infty}\left(\frac{a_e}{r}\right)^n \sum_{m=0}^{n} [P_n^{m+1}(\sin\delta) - (m\tan\delta)P_n^m(\sin\delta)][C_n^m \cos m\lambda + S_n^m \sin m\lambda] \tag{3.4-10}$$

The following recursive formulae are used for the Legendre functions:

$$\begin{aligned} P_n^0(\sin\delta) &= \frac{(2n-1)\sin\delta P_{n-1}^0(\sin\delta) - (n-1)P_{n-2}^0(\sin\delta)}{n} \\ P_n^m(\sin\delta) &= (2n-1)\cos\delta P_{n-1}^{m-1}(\sin\delta) + P_{n-2}^m(\sin\delta) \quad (m \neq 0, m < n-1) \\ P_n^m(\sin\delta) &= (2n-1)\cos\delta P_{n-1}^{m-1}(\sin\delta) \quad (m \neq 0, m >= n-1) \end{aligned} \tag{3.4-11}$$

where

$$
\begin{aligned}
P_0^0(\sin\delta) &= 1 \\
P_1^0(\sin\delta) &= \sin\delta \\
P_1^1(\sin\delta) &= \cos\delta
\end{aligned}
\tag{3.4-12}
$$

and

$$
\begin{aligned}
\sin m\lambda &= 2\cos\lambda \sin[(m-1)\lambda] - \sin[(m-2)\lambda] \\
\cos m\lambda &= 2\cos\lambda \cos[(m-1)\lambda] - \cos[(m-2)\lambda] \\
m\tan\delta &= (m-1)\tan\delta + \tan\delta
\end{aligned}
$$

The μ, a_e, C_n^m, and S_n^m coefficients define the gravitational model. Different geopotential fields have been developed using tracking of thousands of satellite arcs, along with other types of measurements. The geopotential field model used by the U.S. Department of Defense and for the GPS is the *World Geodetic System 84* (WGS84; Department of Defense World Geodetic System 2004). The latest revision of WGS84 is called the *Earth Gravitational Model 1996* (EGM96) and includes revisions through 2004. In this system, $\mu = 398600.4418\,\text{km}^3/\text{s}^2$, $a_e = 6378.1370\,\text{km}$, and the spherical harmonic coefficients are computed up to degree and order 360.

Derivatives of the potential from equations (3.4-8) to (3.4-10) are then expressed as gradients in the radial, eastward, and northward directions:

$$
\begin{bmatrix} \dfrac{\partial U}{\partial r} \\ \dfrac{\partial U}{\partial e} \\ \dfrac{\partial U}{\partial n} \end{bmatrix} = \begin{bmatrix} 1 & 0 & 0 \\ 0 & \dfrac{1}{r\cos\delta} & 0 \\ 0 & 0 & \dfrac{1}{r} \end{bmatrix} \begin{bmatrix} \dfrac{\partial U}{\partial r} \\ \dfrac{\partial U}{\partial \lambda} \\ \dfrac{\partial U}{\partial \delta} \end{bmatrix}
\tag{3.4-13}
$$

where $r\cos\delta = \sqrt{x_{ECF}^2 + y_{ECF}^2}$. These local East and North unit vectors are defined from Figure 3.9 in ECI TOD coordinates as:

$$
\hat{\mathbf{u}} = \frac{\mathbf{r}}{r} = \cos\delta(\cos\alpha\hat{\mathbf{i}} + \sin\alpha\hat{\mathbf{j}}) + \sin\delta\hat{\mathbf{k}} \tag{3.4-14}
$$

$$
\hat{\mathbf{e}} = -\sin\alpha\hat{\mathbf{i}} + \cos\alpha\hat{\mathbf{j}} \tag{3.4-15}
$$

$$
\hat{\mathbf{n}} = -\sin\delta(\cos\alpha\hat{\mathbf{i}} + \sin\alpha\hat{\mathbf{j}}) + \cos\delta\hat{\mathbf{k}} \tag{3.4-16}
$$

Then the inertial acceleration in ECI TOD coordinates is obtained using rotations for latitude (declination) and longitude:

$$
\ddot{\mathbf{r}}_{ECF} = \begin{bmatrix} \cos\lambda & -\sin\lambda & 0 \\ \sin\lambda & \cos\lambda & 0 \\ 0 & 0 & 1 \end{bmatrix} \begin{bmatrix} \cos\delta & 0 & -\sin\delta \\ 0 & 1 & 0 \\ \sin\delta & 0 & \cos\delta \end{bmatrix} \begin{bmatrix} \dfrac{\partial U}{\partial r} \\ \dfrac{\partial U}{\partial e} \\ \dfrac{\partial U}{\partial n} \end{bmatrix}
\tag{3.4-17}
$$

where

$$\cos\delta = \sqrt{x_{ECF}^2 + y_{ECF}^2}\,/\,r, \qquad \sin\delta = z_{ECF}\,/\,r$$
$$\cos\lambda = x_{ECF}\,/\sqrt{x_{ECF}^2 + y_{ECF}^2}, \quad \sin\lambda = y_{ECF}\,/\sqrt{x_{ECF}^2 + y_{ECF}^2}$$

Note that equation (3.4-17) is the inertial acceleration due to the geopotential defined in the ECF system. (It is not acceleration with respect to the rotating ECF coordinate system.) It must be rotated into the TOD geocentric inertial coordinate system using the Greenwich Hour Angle:

$$\ddot{\mathbf{r}}_{TOD} = \begin{bmatrix} \cos\theta_G & -\sin\theta_G & 0 \\ \sin\theta_G & \cos\theta_G & 0 \\ 0 & 0 & 1 \end{bmatrix} \ddot{\mathbf{r}}_{ECF} \tag{3.4-18}$$

Alternately the two rotations for longitude and GHA can be combined into a single rotation.

Finally, $\ddot{\mathbf{r}}_{TOD}$ is rotated to the J2000 ECI system as

$$\ddot{\mathbf{r}}_{J2K} = \left[\mathbf{C}_{J2K}^{TOD}\right]^T \ddot{\mathbf{r}}_{TOD} \tag{3.4-19}$$

using the precession/nutation transformation.

Another gravitational effect included in high-accuracy orbit models is due to the deformation of the earth caused by sun and moon gravitation. This *solid earth tide* effect can be 30 cm vertically at some locations, and the bulge lags behind the actual sun and moon position. The tide effect is not only important for modeling ground tracking station locations, but it also has a significant effect on the geopotential field. Solid earth tide models are discussed in Tapley et al. (2004), McCarthy (1996), and McCarthy and Petit 2004). Ocean tides also have a small effect on the geopotential field, and they may be modeled.

Finally, the general relativistic effect on time scales due to earth's gravitational field is sometimes modeled when working with systems that are accurate to centimeter levels, such as the GPS (Parkinson and Spilker 1996; McCarthy and Petit 2004; Tapley et al. 2004).

3.4.2 Other Gravitational Attractions

The total potential acting on a satellite can be written as:

$$U = U_E + U_G \tag{3.4-20}$$

where U_G is the disturbing function due to the gravitational attractions of other bodies. This is written as:

$$U_G = \sum_i \frac{\mu_i}{\|\mathbf{r}_i\|}\left[\frac{\mathbf{r}_i}{\|\mathbf{r}_i - \mathbf{r}\|} - \frac{\mathbf{r}^T\mathbf{r}_i}{\|\mathbf{r}_i\|^2}\right] \tag{3.4-21}$$

where

$\mu_i = \mu(m_i/m_e)$

m_i = mass of the ith body

m_e = mass of the earth

$\mathbf{r}_i$ = geocentric position vector of the ith body

The resulting perturbing acceleration due to U_G is

$$\ddot{\mathbf{r}}_G = \nabla U_G = \sum_i \mu_i \left[\frac{\mathbf{r}_i - \mathbf{r}}{\|\mathbf{r}_i - \mathbf{r}\|^3} - \frac{\mathbf{r}_i}{\|\mathbf{r}_i\|^3} \right] \tag{3.4-22}$$

In words, the perturbing effect on a satellite due to gravitational attractions is the difference of direct attractions of the bodies on the satellite, and the attractions of the bodies on the earth. The sun and moon are the most significant perturbing bodies and are often the only ones modeled.

3.4.3 Solar Radiation Pressure

Solar radiation pressure acting on a spacecraft is usually computed using one of two methods. If the spacecraft is nearly spherical or has large movable solar panels that are continually pointed nearly normal to the sun, it is commonly assumed that the solar radiation force can be computed as if the spacecraft absorbs all photon momentum without regard for the orientation of the spacecraft with respect to the sun. This is sometimes referred to as the *ballistic* or *cannonball* model. The force vector is computed as

$$\mathbf{f}_S = C_r \frac{\nu F_e A}{c} \left(\frac{A_U}{r} \right)^2 \left(\frac{\mathbf{r}_{sc} - \mathbf{r}_{sun}}{r} \right) \tag{3.4-23}$$

where

ν = illumination factor (0 to 1)

A = effective cross-sectional area (m^2)

F_e = mean solar flux at 1AU (~1358 W/m^2)

c = speed of light (m/s)

A_U = 1 Astronomical Unit ($1.49597870 \times 10^{11}$ m)

$\mathbf{r}_{sc} - \mathbf{r}_{sun}$ = sun-to-spacecraft vector (m)

$r = \|\mathbf{r}_{sc} - \mathbf{r}_{sun}\|$ = distance from sun to spacecraft (m)

C_r = radiation pressure coefficient (nominally 1.0).

The illumination factor ν is equal to 1 except during eclipse ($\nu = 0$ when in a full eclipse).

Alternately the total solar radiation pressure force can be computed by modeling each external structure on the spacecraft and summing the force vectors for each structure. For a plate, the force due to direct solar radiation pressure is typically modeled as a combination of diffuse reflection, specular reflection, and absorption:

$$\mathbf{f}_S = \frac{\nu F_e A}{c} \left(\frac{A_U}{r} \right)^2 \cos\alpha \left((1 - C_s) \frac{\mathbf{r}_{pl} - \mathbf{r}_{sun}}{r} - 2\left(C_s \cos\alpha + \frac{C_d}{3} \right) \hat{\mathbf{n}} \right) \tag{3.4-24}$$

where

$\mathbf{r}_{pl} - \mathbf{r}_{sun}$ = sun-to-plate vector (m)

$r = \|\mathbf{r}_{pl} - \mathbf{r}_{sun}\|$ = distance from sun to plate (m)

$$\cos\alpha = \left(\frac{-(\mathbf{r}_{pl} - \mathbf{r}_{sun})^T \hat{\mathbf{n}}}{r} \right)$$

$\hat{\mathbf{n}}$ = unit normal to plate surface

C_d = diffuse reflection coefficient (0 to 1)

C_s = specular reflection coefficient (0 to 1)

The calculation for other structures such as cylinders or cones is similar but requires integration of the differential force over the entire surface. Also, shadowing of one structure by another may be included in the calculation of v.

To determine the illumination factor for eclipses of the sun by the earth, first define

$$r = \|\mathbf{r}_{sc} - \mathbf{r}_{sun}\|$$
$$\beta = \cos^{-1} \frac{\mathbf{r}_{sc} \cdot (\mathbf{r}_{sc} - \mathbf{r}_{sun})}{\|\mathbf{r}_{sc}\| r}. \tag{3.4-25}$$

where $\mathbf{r}_{sc}$ is the geocentric position vector of the spacecraft. If $\beta > \pi/2$, the spacecraft is on the same side of the earth as the sun and no eclipse can occur. Hence $v = 1$. Otherwise an eclipse is possible. In that case, calculate

$$\gamma = \sin^{-1} \frac{a_s}{r}$$
$$d_1 = \|\mathbf{r}_{sc}\| \sin(\beta - \gamma) \tag{3.4-26}$$
$$d_2 = \|\mathbf{r}_{sc}\| \sin(\beta + \gamma)$$

where a_s is the radius of the sun. If $d_2 \le a_e$ where a_e is the radius of the earth, then the spacecraft is inside the umbra and $v = 0$. If $d_1 \ge a_e$ the spacecraft is outside the penumbra and $v = 1$. If $d_1 < a_e$ and $d_2 > a_e$, the eclipse is partial and $v \cong (d_2 - a_e)/(d_2 - d_1)$. Similar equations can be used for sun eclipses by the moon if the vectors and radius are defined accordingly.

The transition from penumbra to umbra only lasts about 2.1 min for geosynchronous orbits and much less for low altitude orbits so it may not be necessary to model the penumbra/umbra transition in the orbit integration. An alternative just tests whether $\|\mathbf{r}_{sc}\| \sin\beta < a_e$ to determine when the spacecraft is in eclipse ($v = 0$).

In some high-accuracy cases, the pressure from solar radiation diffusively reflected from the earth or moon may also be included in the calculation of total radiation force. The average *albedo* of the earth is about 30%, so the effect can be important. Also, the force due to energy radiated from the spacecraft (from antennas or instrument coolers) may sometimes be included.

3.4.4 Aerodynamic Drag

For spacecraft that descend into the earth's atmosphere (below about 2500 km altitude), atmospheric drag may be a significant perturbing force. If it is assumed that all aerodynamic forces are along the relative velocity vector (no transverse aerodynamic forces), the drag force can be modeled as

$$\mathbf{f}_D = -C_D A \left(\frac{\rho \|\mathbf{v}_r\|}{2} \right) \mathbf{v}_r \tag{3.4-27}$$

where

$\mathbf{v}_r = \mathbf{v} - \omega_e \mathbf{r}$ is the relative velocity vector (m/s) between spacecraft and earth (ω_e is the earth rotation rate),

ρ is the atmospheric density (kg/m^3) at the satellite's position

C_D is the drag coefficient (usually between 1 and 2)

As with solar radiation pressure, drag and lift forces can be modeled taking into account the separate shapes of structures attached to the spacecraft, if it is not nearly spherical.

Calculation of ρ at the satellite position is one source of error when computing the drag force. Most atmospheric density models are derived from either the Harris-Priester model or a version of the Jacchia-Roberts models. The Harris-Priester model is a time-dependent diffusion model based on the assumption that the nominal density profile with altitude is fixed, but a diurnal bulge in density lags behind the sun position by 30 degrees. It is a relatively simple model that is easy to implement. The Jacchia-Roberts models are time-invariant, but involve many terms. The *Mass Spectrometer and Incoherent Scatter Radar* (MSIS) Neutral Atmosphere Empirical Model NRLMSISE-00 (Picone et al. 1997) was derived using satellite and rocket data. For further information see GTDS Mathematical Theory (1989), Wertz (1978), Tapley et al. (2004), and the NASA Goddard Space Flight Center web site (ftp://nssdcftp.gsfc.nasa.gov/models/atmospheric/).

3.4.5 Thrust Forces

Orbital maneuvers or momentum-dumping maneuvers executed using low-thrust chemical thrusters generally involve short firing times. Hence they are usually treated as an impulsive velocity change ($\Delta\mathbf{v}$) where orbit propagation proceeds to the time of the maneuver, the $\Delta\mathbf{v}$ is added to velocity, and orbit propagation proceeds. However, orbit injection maneuvers and maneuvers conducted using very low thrust ion propulsion thrusters are conducted over an extended period of time. Hence these maneuvers are modeled by integrating the thrust force

$$\mathbf{f}_T = \mathbf{C}_{body}^{ECI} \mathbf{f}_{t\_body} \tag{3.4-28}$$

where $\mathbf{f}_{t\_body}$ is the thrust force in body coordinates and $\mathbf{C}_{body}^{ECI}$ is the rotation from spacecraft body to ECI coordinates. For moderate to high-thrust orbit injection maneuvers, the change in mass during the maneuver should also be modeled. That is, $\dot{m} = f_T / Isp$ where Isp is the specific impulse of the thruster, and m should be carried as a seventh dynamic state in the orbital model.

In many cases it is also important to model impingement of the chemical thruster plume on solar arrays or other spacecraft structures. Chemical thrusters can have plume half-widths of 20–25 degrees, and the impingement forces and torques can be 1% to 2% of the total maneuver.

3.4.6 Earth Motion

Many internal computations in orbit generation or OD software require the relative position between the spacecraft and fixed points on the earth at specified times. Unfortunately the rotational motion of the earth is not constant: the spin axis changes slightly with time and the rotation rate (and thus earth-referenced time) also changes. The offset between "earth rotation time" and "atomic time" (actually Terrestial Time) and the "polar motion" offset of the spin axis are determined using astronomical and GPS measurements (McCarthy 1996; McCarthy and Petit 2004; Seidelmann 2006). The *International Earth Rotation and Reference Systems Service* (IERS) calculates these changes and issues updates weekly (http://maia.usno.navy.mil/). In addition to these slow changes, occasional 1 s jumps are added or subtracted to UTC time so that the discrepancy between actual earth rotation and UTC time is maintained less than 0.9 s. These leap second jumps, when they occur, are applied at midnight on June 30 or December 31 of the year. IERS Bulletins also list these leap seconds.

The transformation from J2000 coordinates to ECF coordinates involves four separate transformations:

$$\mathbf{r}_{ECF} = \left[\mathbf{C}_{ECFp}^{ECF}(x_p, y_p)\mathbf{C}_{TOD}^{ECFp}(\theta)\mathbf{C}_{TEME}^{TOD}(n)\mathbf{C}_{J2000}^{TEME}(p)\right]\mathbf{r}_{J2000} \tag{3.4-29}$$

where

$\mathbf{C}_{J2000}^{TEME}(p)$ is the direction cosine rotation for earth precession from J2000.0 to the specified date/time,

$\mathbf{C}_{TEME}^{TOD}(n)$ is the rotation for earth nutation from J2000.0 to the specified date/time,

$\mathbf{C}_{TOD}^{ECFp}(\theta)$ is the rotation for Greenwich Apparent Sidereal Time (which is actually an angle θ) at the specified date/time. θ depends on data in IERS Bulletins.

$\mathbf{C}_{ECFp}^{ECF}(x_p, y_p)$ is the rotation for polar motion at the specified date/time (x_p and y_p are defined in IERS bulletins).

The rotation from ECF to J2000 coordinates is the transpose of the above. More information on earth modeling can be found in Seidelmann (2006), Tapley et al. (2004), McCarthy and Petit (2004).

3.4.7 Numerical Integration and Computation of Φ

The nonlinear differential equations defined by equation (3.4-1) must be integrated to obtain the position and velocity at a specified time. Analytic methods have been developed to approximately integrate the equations using different assumptions about applied forces, but these methods are seldom used today in operational software. In the 1960s and for several decades later, predictor-corrector ODE integration algorithms were used for most high-precision satellite orbital dynamic computations. Multistep methods, such as predictor-corrector, are difficult to use because they require single-step methods to start the integration, and they are inflexible when a change in the integration step size is required. Higher order Runge-Kutta integration (a single-step method) was also sometimes used. Today the Bulirsch-Stoer extrapolation and Runge-Kutta methods are the most popular, where Bulirsch-Stoer is generally regarded as more accurate (Press et al. 2007, Chapter 17). When the force model is conservative (does not depend on velocity, which excludes the drag model), computations of the Bulirsch-Stoer method can be reduced by nearly 50% when using Stoermer's rule (Press et al. 2007, Section 17.4).

The differential equations defining orbit propagation are nonlinear, but most OD algorithms require computation of a state transition matrix $\mathbf{\Phi}$ for a linear perturbation model. The earliest OD software computed $\mathbf{\Phi}$ by integrating the *variational equations*—the partial derivatives of equation (3.4-1) with respect to position and velocity and any other parameters of the model that are estimated in the OD; that is, $\mathbf{F} = \partial\dot{\mathbf{x}} / \partial\mathbf{x}$. This required thousands of lines-of-code to compute the partial derivatives, and greatly limited the flexibility of the software. Many new OD programs either compute $\mathbf{F}$ using numerical partial derivatives and then integrate $\dot{\mathbf{\Phi}} = \mathbf{F}\mathbf{\Phi}$, or directly compute $\mathbf{\Phi}$ using numeric partials as described in Section 2.3.1. The generic structure of $\mathbf{\Phi}$ used in OD is

$$\mathbf{\Phi}(t+T,t) = \begin{bmatrix} \dfrac{\partial\mathbf{r}(t+T)}{\partial\mathbf{r}(t)} & \dfrac{\partial\mathbf{r}(t+T)}{\partial\mathbf{v}(t)} & \dfrac{\partial\mathbf{r}(t+T)}{\partial\boldsymbol{\beta}} \\ \dfrac{\partial\mathbf{v}(t+T)}{\partial\mathbf{r}(t)} & \dfrac{\partial\mathbf{r}(t+T)}{\partial\mathbf{v}(t)} & \dfrac{\partial\mathbf{r}(t+T)}{\partial\boldsymbol{\beta}} \\ \mathbf{0} & \mathbf{0} & \mathbf{I} \end{bmatrix} \tag{3.4-30}$$

where all matrix partitions are either 3×3 or $3 \times p$, and $\boldsymbol{\beta}$ is a p-element vector of force model constants—such as C_r, C_D, gravity field coefficients or thrust parameters—included in the OD. The $\partial\mathbf{v}(t + T)/\partial\mathbf{r}(t)$ partition will be zero if the perigee altitude is high enough (e.g., >2500 km) so that drag forces are zero. If no other forces depend on velocity, the force model is conservative.

Another approach to computing $\mathbf{\Phi}$ uses equinoctial elements, rather than Cartesian position/velocity, as states in the OD. The equinoctial elements directly model the dominant force acting on the spacecraft (earth central gravity) so the effect of the other forces (gravity field, sun/moon gravity, solar radiation, drag, thrust) is a small perturbation. Somewhat surprisingly, the upper left 6×6 partition of $\mathbf{\Phi}$ for an equinoctial element state can be computed accurately for some high-altitude satellites by just modeling the spherical harmonic gravity field to degree two; that is, the C_2^0 (also called J_2 when normalized) coefficient with $S_2^0 = 0$ are the only coefficients included. Derivation of the 6×6 $\mathbf{\Phi}$ partition is nontrivial (see Koskela 1967; Escobal 1976; or GOES N-Q OATS SMM 2007, Section 6.3.2.2) but not difficult to compute. A similar approach can also be applied to compute the partitions of $\mathbf{\Phi}$ that model sensitivity of equinoctial elements with respect to C_r, C_D, and thrust parameters.

Most OD is performed using batch least-squares methods since deterministic orbital force models are quite accurate for many satellites. However, it is not possible to model all forces perfectly (particularly for satellites within the earth's atmosphere), and in some high-accuracy cases it is preferable to use a Kalman filter that treats errors in the dynamic model as process noise. Kalman filters are used in GPS satellite OD (Parkinson and Spilker 1996; Misra and Enge 2001; Kaplan and Hegarty 2006,) and for some LEO spacecraft (Gibbs 1978, 1979). Filters often model process acceleration errors in alongtrack, crosstrack, and radial directions as independent white noise (random walk in velocity) or first-order Markov processes. For the first-order Markov model:

$$\begin{aligned} \dot{\mathbf{v}}(t) &= \mathbf{f}(\mathbf{r},\mathbf{v}) + \mathbf{C}_{ACR}^{ECI}\mathbf{a}_p(t) \\ \dot{\mathbf{a}}_p(t) &= \begin{bmatrix} -1/\tau_a & 0 & 0 \\ 0 & -1/\tau_c & 0 \\ 0 & 0 & -1/\tau_r \end{bmatrix} \mathbf{a}_p(t) + \begin{bmatrix} q_a(t) \\ q_c(t) \\ q_r(t) \end{bmatrix} \end{aligned} \tag{3.4-31}$$

where

$\dot{\mathbf{v}}(t)$ is the ECI acceleration at time t,

$\mathbf{f}(\mathbf{r},\mathbf{v})$ is the force model for the gravity field, sun/moon/planet gravity, solar radiation, drag, and thrust,

$\mathbf{C}_{ACR}^{ECI}$ is the rotation from alongtrack, crosstrack, and radial coordinates to ECI (a different transformation could also be used if model errors are more likely to appear in a different direction),

$\mathbf{a}_p(t)$ is the vector of alongtrack, crosstrack, and radial accelerations at time t,

$q_a(t)$, $q_c(t)$, $q_r(t)$ are the alongtrack, crosstrack, and radial random process noise at time t,

τ_a, τ_c, τ_r are the alongtrack, crosstrack, and radial time constants.

In this model the three components of $\mathbf{a}_p(t)$ are included as states in the Kalman filter, and the discrete process noise covariance $\mathbf{Q}_D$ is computed as described in Chapter 2. For a random walk velocity model, white noise is rotated to ECI coordinates and directly added to the velocity rate ($\dot{\mathbf{v}}(t) = \mathbf{f}(\mathbf{r},\mathbf{v}) + \mathbf{C}_{ACR}^{ECI}\mathbf{q}_{ACR}(t)$) so no acceleration states are included in the filter. Process noise may also be used to model thrust acceleration errors during long maneuvers.

3.4.8 Measurements

The measurements typically used in OD include ground-based range, range rate and tracking angles, satellite-to-satellite range and range rate (usually multi-link to ground), satellite measured angles to ground objects, and GPS pseudo-ranges. Other variants on these basic types are also possible. Recall from Section 1.2 that least-squares and Kalman estimation compare actual measurements with model-computed measurements, although the model-computed measurements may be zero in linear least squares.

We now derive the equations for these measurement types given satellite positions and velocity, and ground location.

3.4.8.1 Ground-Based Range, Range Rate, Integrated Range Rate, and Tracking Angles The basic equations for range and range rate measurements were presented in Example 2.4. Ground tracking of a satellite usually involves measuring either the two-way (uplink and downlink) time delay or the two-way phase changes if continuous tones are used for the measurement. Because the distances and velocities are great, and transmitters and receivers are both moving in an inertial frame, it cannot be assumed that the average two-way transit time or phase change can be treated as a measure of the geometric distance between ground observer and satellite at the received measurement time. For example, the distance to a geosynchronous satellite from a ground station may be 37,000 km, and the satellite velocity is about 3 km/s. Using the speed of light in a vacuum (c = 299792.458 km/s), the signal transit time is calculated as 0.123 s, and the satellite will have moved about 369 m in that time. A crude procedure for calculating the downlink range is as follows:

1. Calculate ground location and satellite position/velocity in ECI coordinates at the ground receive time t.

2. Calculate an approximate ground-satellite range ($r' = \|\mathbf{r}_{gnd}(t) - \mathbf{r}_{sc}(t)\|$) and the transit time, $\Delta t = r'/c$.
3. Recalculate satellite position and velocity at $t - \Delta t$.
4. Recalculate ground-satellite range using the satellite position at $t - \Delta t$.

Unfortunately the final range will still not be accurate, and it is necessary to repeat this procedure several times until the computed range does not change. A better method expands the equation for range-squared at different ground/satellite times,

$$r^2(t, t-\Delta t) = \|\mathbf{r}_{gnd}(t) - \mathbf{r}_{sc}(t-\Delta t)\|^2$$

in a second-order Taylor series,

$$r^2(t, t-\tau) = r^2(t,t) + \left.\frac{d(r^2)}{d\tau}\right|_{t=t} \Delta t + \frac{1}{2}\left.\frac{d^2(r^2)}{d\tau^2}\right|_{t=t} \Delta t^2$$

and then solves the quadratic equation for Δt assuming that the satellite acceleration is only due to earth central-body gravity ($\ddot{\mathbf{r}}_{sc} = -\mu \mathbf{r}_{sc} / \|\mathbf{r}_{sc}\|^3$). This method is accurate to better than 3×10^{-6} m for geosynchronous satellites. The same procedure is repeated for the uplink from the ground to the satellite. Range rate measurements can be included as part of the same calculation.

Topocentric elevation and azimuth angles are computed from the position difference vector $\mathbf{r}_{gnd}(t) - \mathbf{r}_{sc}(t - \Delta t)$ rotated into the local level system ($\mathbf{z}$ = local vertical, $\mathbf{y}$ in a northerly direction in the plane normal to $\mathbf{z}$, and $\mathbf{x} = \mathbf{z} \times \mathbf{y}$). These angles are often available from tracking antennas, but they should be used with caution. Antennas used for transmitting/receiving telemetry data may use an angle-tracking control loop that attempts to model the orbit of the satellite. The error characteristics of this angle data are not ideal for OD because the angles may be highly correlated in time and may change rapidly as the tracking algorithm updates the model. The accuracy may also be inadequate for good OD. Check the data before using.

Range, range rate, and angle measurements from the ground to a satellite are influenced by tropospheric and ionospheric effects. Since the index of refraction in the troposphere is greater than 1.0, the effective group propagation speed is less than c and range measurements computed as $\Delta t/c$ will have a positive bias. That bias is a minimum at zenith (because of the shorter path through the troposphere) and it increases approximately as $1/(\sin E + 0.026)$ with decreasing elevation E. Because of this error, tracking is usually terminated below an elevation cutoff in the range of 5 to 15 degrees. Likewise the propagation speed through the ionosphere is also less than c and the range error is a minimum at zenith. Because the ionosphere is a dispersive medium, the delay varies approximately as the inverse squared power of carrier frequency; that is, it is much greater at L-band than C-band frequencies. This dependence on frequency is used in the two-frequency P-code GPS system to calculate the ionospheric effect and remove it. In addition to group delay effects, the change in index of refraction along the signal path bends the ray, and thus angles measurements are biased. Again the effect varies with elevation. Other delays that alter range measurements include satellite transponder and ground link delays (possibly time-varying).

Angular measurements of satellites can be also be obtained by optically comparing the observed satellite position with that of background stars. Measurement processing is complicated by the need to account for *proper motion, parallax, annual aberration, and planetary aberration* of the stars (Seidelmann 2006). Angular measurements are also affected by equipment biases and planetary aberration.

All these effects should be modeled when processing the measurements. There are two ways in which this can be done: model the errors when generating a "calculated measurement" to be compared with the actual measurement in the OD, or approximately calculate the error and correct the measurement before processing in OD. Both approaches are used, but the "preprocessing" method has the advantage that the computations are not included when calculating OD partial derivatives. Measurement preprocessing is routinely used in many systems because it also allows for editing of anomalous measurements prior to processing in the estimation algorithm. It is sometimes difficult to detect anomalous measurements prior to updating the state estimate (particularly when the initial state estimate is poorly known), so editing done during preprocessing can make the system significantly more robust. Sometimes this is done by fitting a linear or quadratic model to a short time series of measurements and editing measurements that deviate more than n-sigma from the curve fit. This works best when the measurement sampling rate is much higher than the highest frequencies at which the dynamic system responds. For these high sampling rate cases, it can be advantageous to compute an "average" measurement over a time window from the curve fit, and use that measurement in the OD. This reduces OD computations and allows applying greater weight to "averaged" measurements than raw measurements because the random errors are smaller. This approach is discussed again in later chapters.

3.4.8.2 Satellite-to-Satellite Range and Range Rate Satellite-to-satellite range and range rate measurements are modeled similarly to ground-based measurements, except that the motion of both satellites must be modeled. Atmospheric effects can be ignored if both satellites are above the atmosphere and the ray path does not pass through the atmosphere.

3.4.8.3 Satellite Measured Angles to Ground Points or Stars Angles from ground points (including the horizon) to satellites are modeled using the inertial vector from the ground to the spacecraft, taking into account the transit time of the ray. Then the vector is rotated from inertial to spacecraft coordinate system and angles are computed as projections on the spacecraft axes. The angle measurements are also affected by planetary aberration, tropospheric refraction, and errors in computing the attitude of the spacecraft. Measured angles to stars (e.g., from a star tracker) are modeled similarly using the inertial coordinates of the stars. Additional error effects include parallax, proper motion of the stars, annual aberration, and planetary aberration. Star measurements provide more information on spacecraft attitude than orbit.

3.4.8.4 GPS Pseudo-Range and Carrier Phase *Pseudo-range* (PR) and *accumulated delta-range* (ADR) or carrier phase measurements obtained using GPS signals are similar to range and integrated range rate measurements, but include a large bias and drift that are due to the unknown time and frequency errors of the receiver. The ADR measurements are less noisy than PR, but both contain tropospheric and ionospheric delay. The PR-ADR difference is not influenced by tropospheric delay modeling error, but contains a satellite-receiver pass-dependent bias. The navigation message transmitted with the GPS signal provides accurate information on the ephemeris and clock errors of the GPS satellites, and also has information allowing single-frequency users to model the ionospheric delay. A typical terrestrial GPS navigator must simultaneously receive signals from at least four GPS satellites in order to uniquely calculate the three components of its position and its local clock error. For satellite receivers, it is not necessary to simultaneously receive four GPS signals and calculate position, since this can be done in the OD from GPS signals received over a period of time. This is convenient because the transmitting antenna of the GPS satellites is designed to only provide earth-coverage, so satellites at a higher altitude (>20.2 km) will receive signals when the signal path is near the earth's edge. Most receivers in geosynchronous orbits can also receive signals transmitted from GPS antenna sidelobes. Sidelobe signals from six or seven GPS satellites can usually be received, but future GPS antenna designs may not retain signal strength in the sidelobes.

Test Case #1: Low Earth Orbit (LEO) Satellite Two cases are used for examples in Chapters 6 and 7. The model for the first case is a *CHAllenging Mini-satellite Payload* (CHAMP) three-axis stabilized earth-pointing spacecraft. It is designed to precisely measure the earth's gravity field using GPS and accelerometer sensors, and to measure characteristics of the upper atmosphere. The nearly circular orbit has an altitude of 350 km (92-min period) and 87.2 degrees inclination. At this altitude CHAMP is well within the earth's atmosphere and is significantly affected by atmospheric drag. Either the Harris-Priester or MSIS atmospheric models are used in the example, but this is the largest source of orbit modeling error. The J2000 parameters of the orbit are defined in Table 3.4, and approximate locations of three ground tracking stations are listed in Table 3.5. Figure 3.10 shows the ground track for a 2-day period (2005 days 141 and 142) and the locations of the ground tracking stations.

For this example the orbit is determined using range, azimuth, elevation, and range rate measurements from each station, where the assumed $1-\sigma$ measurement errors are 15 m, 0.5 degrees, 0.4 degrees, and 2.0 m/s, respectively. Simulated range and range rate data is generated every 120 s when the satellite is in view. Measurement bias is ignored.

TABLE 3.4: Case #1 (CHAMP) Simulated Orbit Parameters

Parameter	Value (J2000 Coordinate Reference)
Epoch time	00:00:00 2008 day 141(UTC)
Semimajor axis	6730.71 km
Eccentricity	0.0012975
Inclination	87.244 degrees
Right ascension of ascending node	189.425 degrees
Argument of perigee	56.556 degrees
Mean anomaly	75.376 degrees
Solar radiation pressure $K_p = C_pA/m$	0.05
Drag coefficient $K_d = C_dA/m$	0.00389

TABLE 3.5: Case #1 (CHAMP) Simulated Ground Stations

Station Designation	Latitude (degrees)	Longitude (degrees East)	Elevation (m)
#1 (Marshall Islands)	9.4	167.5	100
#2 (Hawaii)	20.7	203.7	3000
#3 (southern California)	35.0	240.0	200

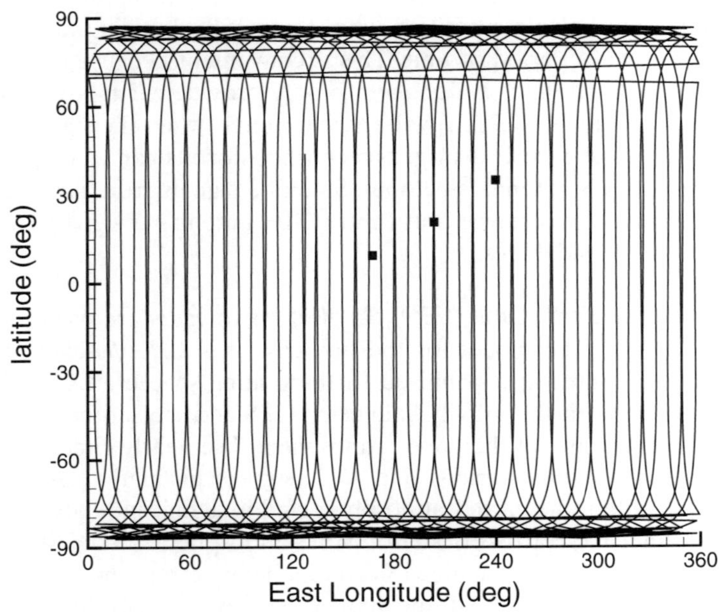

FIGURE 3.10: CHAMP ground track and ground stations.

Test Case #2: Geosynchronous Earth Orbit (GEO) Satellite This case uses orbit elements appropriate for a geosynchronous spacecraft located at approximately 89.5 degrees west longitude. The orbit parameters are listed in Table 3.6. A ground station located at 37.95 degrees latitude, 75.46 degrees west longitude, and −17 m with respect to the reference ellipsoid measures range to the spacecraft. This scenario is typical for the GOES-13 spacecraft during post launch testing.

3.4.9 GOES I-P Satellites

Chapters 2 and 6 include references to Orbit and Attitude Determination (OAD) for the GOES I-P series of spacecraft operated by the National Oceanic and Atmospheric Administration (NOAA). This is a special case of OD where the satellite orbit and misalignment "attitude" of the optical imaging instruments (Imager and Sounder) are simultaneously estimated using measurements of ground-to-satellite ranges, observed angles to ground landmarks, and observed angles to known stars. A coupled orbit and attitude solution is required to meet pointing requirements, as can be understood from a description of the system.

TABLE 3.6: Case #2 GEO Parameters

Parameter	Value
Epoch time	2006 day 276, 12:00.000 UTC
Semimajor axis	42164.844 km
Eccentricity	0.00039150
Inclination	0.3058 degrees
Right ascension of ascending node	275.653 degrees
Argument of perigee	307.625 degrees
Mean anomaly	239.466 degrees
Solar radiation pressure $K_p = C_p A/m$	0.0136

The GOES I-P spacecraft are geosynchronous weather satellites located approximately on the equator at a radius of about 42164 km. The resulting orbit period, 23.934 h, allows the spacecraft to remain nearly at a fixed longitude with respect to the earth. The measurements available for OD include range measurements from a fixed site located in Virginia, and north-south and east-west angles from the spacecraft to fixed landmarks (prominent shorelines) on the earth. The landmark angles are obtained from the earth-imaging instruments as they scan the earth to obtain information on weather patterns. Although both visible (reflected sunlight) and infrared (IR) landmark observations are available, the visible landmarks obtained during daylight are given much more weight in OAD because of large IR random measurement errors and unmodeled visible-to-IR detector co-registration errors.

Pointing of the imaging instruments is affected by thermal deformation caused by non-uniform heating of the instruments. Since the thermal deformation is driven by the angle of the sun with respect to the spacecraft and instruments, the thermal deformation repeats with a 24-hour period, allowing the misalignment of five pointing angles (roll, pitch, yaw, and two internal misalignments) to be modeled as a Fourier series expansion with a 24-hour fundamental period. Provision is also made for a trend. Approximately 100 Fourier coefficients for each instrument must be included to accurately model the daily thermal deformation. Because landmark observations are a function of both orbital position and instrument "attitude" misalignments, orbit and attitude parameters must be determined simultaneously using range and landmark measurements. Optical angle measurements to stars are also available for attitude determination.

For the NOP series of GOES spacecraft, 26 h of observations are normally used in batch weighted least-squares OAD. The orbit and attitude parameters from that solution are used operationally onboard the spacecraft to predict orbit and attitude for the next 24 h. That prediction is used onboard the spacecraft to dynamically adjust the scan angles of the optical imaging instruments while scanning so that the received image appears to have been viewed from a fixed position in space and at fixed attitude.

3.4.10 Global Positioning System (GPS)

A sophisticated form of satellite OD is used operationally in the GPS ground system. The GPS ground segment must accurately calculate the orbits of each GPS satellite and provide that orbital information to GPS receivers so that users may calculate their position. That satellite navigation function is performed using a Kalman filter that estimates hundreds of states. A high-accuracy version of the navigation filter that duplicates most functions of the operational system demonstrates how a Kalman filter for a large complex system can

1. Be further extended to handle disposable pass-dependent bias parameters
2. Detect abrupt changes in clock behavior and immediately correct the filter state by implicitly reprocessing past data (without explicitly reprocessing)
3. Compute smoothed state estimates
4. Use maximum likelihood estimation to calculate optimal values of parameters used in the filter model.

Examples of GPS use appear in Chapters 9 and 11.

The design and operation of the GPS have been well documented. Parkinson and Spilker (1996), Misra and Enge (2001), and Kaplan and Hegarty (2006) provide good descriptions of the system. The GPS space segment consists of up to 32 operational satellites distributed in six orbital planes where the orbital period is 12 h (orbital radius is 26,560 km). The satellites broadcast pseudo-random codes on multiple carrier frequencies where each satellite is assigned one of 32 unique codes. The start of the broadcast code is synchronized with the satellite's onboard clock, where the time delay between each clock and a composite clock reference is calculated by the *master control station* (MCS) of the GPS ground segment. The clock error model and other information are periodically uploaded to each spacecraft in a navigation message that is broadcast to users as part of the GPS signal.

User GPS receivers have their own clocks. By shifting their local copy of each satellite code to match the received signal, they can determine the relative time shift between satellite and receiver codes. That time delay is a function of the signal distance from satellite to receiver, atmospheric path delays, the satellite clock time error, and the receiver clock time error. The broadcast navigation message includes a quadratic model for satellite clock errors, a high-accuracy satellite ephemeris model, and coefficients that allow civilian C/A-code receivers to approximately correct for ionospheric path delays. Military P-code receivers operating at two carrier frequencies can remove most of the ionospheric path error because the effect is a known function of frequency. Using the relative time delay information (equivalent to PR) from at least four GPS satellites plus the ephemeris/clock information in the navigation messages, the GPS receiver can calculate four parameters: the three coordinates of its position and its clock error. Most receivers include a Kalman filter that allows estimation of receiver velocity in addition to position. This greatly simplified overview of the system ignores many details that are required to make GPS work. For example, the system compensates for special and general relativistic effects and deformation of the solid earth due to sun and moon tidal effects.

One function of the MCS is to compute the satellite ephemeris/clock model coefficients that are included in the navigation messages and uploaded to each satellite once or twice per day. The parameters of those models are obtained from the output of a linearized Kalman filter that simultaneously computes six orbital states, two solar radiation parameters, and two to three clock states for each satellite. (Three states are used for rubidium clocks and two for cesium.) For a 30-satellite constellation where 15 clocks are rubidium and 15 are cesium, the filter uses 315 satellite-dependent states. The measurements input to the Kalman filter are the PR and PR minus ADR data from typically 19 ground receivers. The location of each receiver is well known, but the clock errors and local atmospheric errors are imperfectly known. Hence the MCS filter must also estimate three states for each ground receiver for a total of at least 372 states when 19 receivers are used.

3.5 FOSSIL-FUELED POWER PLANT

See FTP site ftp://ftp.wiley.com/public/sci_tech_med/least_squares.

3.6 SUMMARY

This chapter has shown how several different systems are modeled using first-principles concepts. The first two applications involved simple models, but were nonetheless real-world problems. They will be used in later chapters to demonstrate specific problems with estimation performance. The last two applications required extensive knowledge of the system modeled. The last application—power plant modeling—demonstrated the many "tricks" and assumptions are required to develop a good reduced-order model of a complex system.

The examples are summarized as follows:

1. *Angle-only tracking of linear target motion:* This very simple linear deterministic dynamic model and nonlinear measurement model is used later in Chapters 7 and 9 to demonstrate the problems associated with poor observability and nonlinear optimization.
2. *Maneuvering vehicle (tank) tracking using multiple models:* This uses a body-referenced (alongtrack-crosstrack) acceleration model where three different second-order colored noise (stochastic) acceleration models are used to capture the behavior of three different types of maneuvering vehicles. The measurements consist of range and angle data. This example is used in Chapter 9 to demonstrate nonlinear filtering, and in Chapter 11 to demonstrate how multiple-model Kalman filters and the likelihood function are used to compute the optimal prediction of a projectile during the time-of-flight.
3. *Maneuvering vehicle (aircraft) tracking using a nonlinear model:* This also uses a body-referenced (alongtrack-crosstrack) acceleration model, but the acceleration occasionally jumps from zero to a nonzero constant for short periods. The measurements consist of range and angle data. This example is used in Chapter 9 when discussing optimal smoothing, and in Chapter 11 to demonstrate jump detection.
4. *Strapdown INS error model:* This error model is linearized about a nominal missile trajectory consisting of acceleration and body angular rate time history. The 27-state stochastic error model is linear but time-varying. Eighteen states represent constant parameters that affect the error dynamics. Random process noise is assumed to drive the velocity and gyro tilt states. Because of the high

processing rate, the state transition matrix is computed as a first-order Taylor series. Measurements consist of position errors at 10 s intervals. This example is used in Chapter 9 to demonstrate Kalman filter error analysis, in Chapter 10 to compare numerical accuracy of filtering algorithms, and in Chapter 11 to demonstrate jump detection algorithms.

5. *Spacecraft orbital dynamics:* Orbital dynamics are nonlinear, time-varying, and oscillatory. To achieve high accuracy many different forces must be modeled using hundreds of parameters. Depending on the types and frequency of measurements, the state may be well or poorly observable. OD is used as an example in several chapters.
6. *Mass and energy flow in a fossil-fueled power plant:* This example showed that it is possible to develop a 13-state reduced-order "first-principles" model of a very complex system with multiple interconnecting flow paths of two different fluids. Properly calibrated, the model was able to accurately capture the important dynamics of the plant. Because the air/gas path dynamics are very much faster than water/steam thermodynamics, the air/gas path was modeled as steady state, and the resulting set of algebraic equations was solved using Newton iteration. This greatly reduces computations and eliminates dynamics that are much higher than the frequency response of the Kalman filter or of the *nonlinear model-predictive control.* The methods used to simplify the model, such as modeling distributed subsystems as lumped parameter, can be applied to many systems. Although modeling of this particular plant required much more effort than would be practical for many industrial processes, the results demonstrate the improved performance. The success in accurately modeling this complex system using a first-principles approach shows that modeling of simpler systems is certainly practical. The control actuator modeling approach used here is mentioned in Chapter 8. Power plant model parameters were determined using the maximum likelihood parameter estimation method of Chapter 11.

CHAPTER 4

LINEAR LEAST-SQUARES ESTIMATION: FUNDAMENTALS

After laying the groundwork on model types and approaches to modeling, we now address the first estimation topic: the method of least squares. It will be seen that there are several different approaches to least-squares estimation and that the answer depends somewhat on the assumptions used. The simplest approach only assumes that the measurement data represents the output of a model corrupted with "measurement errors" of unknown characteristics, and the goal is to find model parameters that give the "best" fit to the noisy data. Other approaches are similar, but assume that either the variance or the probability distribution of the measurement noise is known. Finally, the "Bayesian" approach treats the state as a random variable and assumes that statistics of both the measurement noise and errors in the *a priori* state estimate are known. These different approaches are described by names such as *weighted least squares, minimum variance, minimum mean-squared error, best linear unbiased estimate, maximum likelihood,* and *maximum a posteriori.* The assumptions and properties of each method are discussed in later sections. Since maximum likelihood and maximum *a posteriori* solutions are based on assumed probability distributions, they are not really "least-squares" methods. However, maximum likelihood and maximum *a posteriori* are included in this chapter because the estimates are identical to other least-squares solutions when the system is linear and distributions are Gaussian.

4.1 LEAST-SQUARES DATA FITTING

Curve fitting and linear regression are the simplest forms of the least-squares method. Often the goal of curve fitting is to fit a low-order polynomial to a time series of noisy data points, $y(t_i)$, $i = 1$ to m. If only two points are available, the solution is obvious: fit a straight line through the two points using one point as a constraint, and compute the slope from the difference between the points. Then the solution is essentially linear interpolation/extrapolation of the form

$$y(t) = y(t_1) + \left(\frac{y(t_2) - y(t_1)}{t_2 - t_1} \right)(t - t_1).$$

Advanced Kalman Filtering, Least-Squares and Modeling: A Practical Handbook, by Bruce P. Gibbs

If more data points are available than the number of coefficients in the polynomial model, then it is generally not possible to fit a curve exactly through all the data points. Thus the question becomes: "What criterion should be used to determine the 'best' fit to the data for the given model?" Two obvious possibilities are (1) the sum of absolute values of the differences (*residuals*) between the data and the polynomial curve; or (2) the sum-of-squared residuals. Gauss and Legendre selected the sum-of-squares as the cost function to be minimized, and that has been the preferred approach ever since. There are two arguments for preferring the residual sum-of-squares. First, this puts the greatest weight on large deviations from the curve, and this is generally a desirable characteristic. (It is intuitively what we tend to do when trying to graphically fit a straight line or curve to a set of points using drafting aids.) Second, the squared cost function has continuous derivatives, and that is a great advantage when attempting to compute mathematical solutions to the problem.

We start by defining the model. Discrete measurements $y_i = y(t_i)$ for $i = 1$ to m are assumed to be a linear function of a set of parameters with random measurement noise r_i added:

$$y_i = \mathbf{h}_i \mathbf{x} + r_i \tag{4.1-1}$$

where

$\mathbf{x}$ is the state vector of n model parameters,

$\mathbf{h}_i$ is a $1 \times n$ **row** vector that defines how the parameters in $\mathbf{x}$ map to the measurement y_i,

r_i is assumed to be zero mean ($E[r_i] = 0$) with a fixed variance ($E[r_i^2] = \sigma^2$), but no other assumptions on statistics are used.

For convenience, all the measurements are often collected into a column vector to give the vector/matrix equation

$$\begin{bmatrix} y_1 \\ y_2 \\ \vdots \\ y_m \end{bmatrix} = \begin{bmatrix} \mathbf{h}_1 \\ \mathbf{h}_2 \\ \vdots \\ \mathbf{h}_m \end{bmatrix} \mathbf{x} + \begin{bmatrix} r_1 \\ r_2 \\ \vdots \\ r_m \end{bmatrix}$$

or

$$\boxed{\mathbf{y} = \mathbf{H}\mathbf{x} + \mathbf{r}} \tag{4.1-2}$$

where $\mathbf{y} = \begin{bmatrix} y_1 \\ y_2 \\ \vdots \\ y_m \end{bmatrix}$, $\mathbf{r} = \begin{bmatrix} r_1 \\ r_2 \\ \vdots \\ r_m \end{bmatrix}$ and $\mathbf{H} = \begin{bmatrix} \mathbf{h}_1 \\ \mathbf{h}_2 \\ \vdots \\ \mathbf{h}_m \end{bmatrix}$ is an $m \times n$ matrix.

In previous chapters we implied that measurement vectors at different times t_i are frequently modeled as

$$\mathbf{y}(t_i) = \mathbf{M}(t_i)\mathbf{\Phi}(t_i, t_0)\mathbf{x}(t_0) + \mathbf{r}(t_i)$$

where

$\mathbf{x}(t_0)$ is the n-element state vector at an epoch time t_0,

$\mathbf{y}(t_i)$ is the l-element measurement vector at time t_i (l may be variable for different times),

$\mathbf{\Phi}(t_i,t_0)$ is the $n \times n$ state transition matrix from time t_0 to t_i, and

$\mathbf{M}(t_i)$ is the $l \times n$ matrix that maps the state vector $\mathbf{x}$ evaluated at time t_i $[\mathbf{x}(t_i) = \mathbf{\Phi}(t_i,t_0)\mathbf{x}(t_0)]$ to the measurement.

However, in this chapter we treat $\mathbf{\Phi}(t_i,t_0)$ as part of the measurement mapping matrix

$$\mathbf{H} = \begin{bmatrix} \mathbf{M}(t_1)\mathbf{\Phi}(t_1,t_0) \\ \mathbf{M}(t_2)\mathbf{\Phi}(t_2,t_0) \\ \vdots \\ \mathbf{M}(t_p)\mathbf{\Phi}(t_p,t_0) \end{bmatrix}$$

where all measurements are stacked in a single vector $\mathbf{y} = \mathbf{Hx} + \mathbf{r}$:

$$\begin{bmatrix} \mathbf{y}_1(t_1) \\ \mathbf{y}_2(t_2) \\ \vdots \\ \mathbf{y}_p(t_p) \end{bmatrix} = \begin{bmatrix} \mathbf{M}(t_1)\mathbf{\Phi}(t_1,t_0) \\ \mathbf{M}(t_2)\mathbf{\Phi}(t_2,t_0) \\ \vdots \\ \mathbf{M}(t_p)\mathbf{\Phi}(t_p,t_0) \end{bmatrix} \mathbf{x} + \begin{bmatrix} \mathbf{r}_1(t_1) \\ \mathbf{r}_2(t_2) \\ \vdots \\ \mathbf{r}_p(t_p) \end{bmatrix}.$$

The total number of scalar measurements over all p measurement times is m. Hence $\mathbf{\Phi}(t_i,t_0)$ does not appear in the equations of this chapter. This change simplifies the notation but otherwise does not imply anything about model structure. In fact, most least-square implementations process the measurements in p "time batches" with information accumulated as m scalar measurements.

The least-squares cost function is defined as

$$\boxed{\begin{aligned} J &= \frac{1}{2}\sum_{i=1}^{m}(y_i - \mathbf{h}_i\hat{\mathbf{x}})^2 \\ &= \frac{1}{2}(\mathbf{y} - \mathbf{H}\hat{\mathbf{x}})^T(\mathbf{y} - \mathbf{H}\hat{\mathbf{x}}) \end{aligned}} \tag{4.1-3}$$

where $\hat{\mathbf{x}}$ represents a "guess" for $\mathbf{x}$. The factor of ½ is not necessary to define the cost function, but it eliminates a factor of two in equations to follow. The gradient of J with respect to $\hat{\mathbf{x}}$ must be zero at the $\hat{\mathbf{x}}$ minimizing J; that is,

$$\begin{aligned} \frac{\partial J}{\partial \hat{\mathbf{x}}} &= -\sum_{i=1}^{m}(y_i - \mathbf{h}_i\hat{\mathbf{x}})\mathbf{h}_i \\ &= -(\mathbf{y} - \mathbf{H}\hat{\mathbf{x}})^T\mathbf{H} \\ &= \mathbf{0} \end{aligned} \tag{4.1-4}$$

where

$$\frac{\partial J}{\partial \hat{\mathbf{x}}} = \begin{bmatrix} \frac{\partial J}{\partial \hat{x}_1} & \frac{\partial J}{\partial \hat{x}_2} & \cdots & \frac{\partial J}{\partial \hat{x}_n} \end{bmatrix}$$

is the transposed gradient of J (abbreviated as $\nabla_x J = [\partial J / \partial \hat{\mathbf{x}}]^T$). There is no universal convention on whether $\partial J / \partial \mathbf{x}$ is a row or column vector, but we treat it as a row vector in this book; for example, $\partial(\mathbf{y}^T\mathbf{x}) / \partial \mathbf{x} = \mathbf{y}^T$. Transposing equation (4.1-4) gives $\mathbf{H}^T(\mathbf{y} - \mathbf{H}\hat{\mathbf{x}}) = \mathbf{0}$ or

$$\boxed{(\mathbf{H}^T\mathbf{H})\hat{\mathbf{x}} = \mathbf{H}^T\mathbf{y}.} \tag{4.1-5}$$

Equation (4.2-4) is called the "normal equation" for least-squares problems. If $\mathbf{H}^T\mathbf{H}$ is rank n, it can be solved by explicitly inverting $(\mathbf{H}^T\mathbf{H})$ to obtain

$$\boxed{\hat{\mathbf{x}} = (\mathbf{H}^T\mathbf{H})^{-1}\mathbf{H}^T\mathbf{y},} \tag{4.1-6}$$

although other solution techniques are possible and will be discussed later. As noted above, formation of $\mathbf{H}^T\mathbf{H}$ and $\mathbf{H}^T\mathbf{y}$ is usually implemented (for efficiency) as sums of vectors for individual measurements, that is,

$$\mathbf{H}^T\mathbf{H} = \sum_{i=1}^{m} \mathbf{h}_i^T\mathbf{h}_i \quad \text{and} \quad \mathbf{H}^T\mathbf{y} = \sum_{i=1}^{m} \mathbf{h}_i^T y_i.$$

A scalar variable minimizes a scalar function when the first derivative with respect to the variable is equal to zero and the second derivative is positive. The equivalent "second-derivative" requirement for a multidimensional problem is that the $n \times n$ symmetric *Hessian matrix*

$$\frac{\partial^2 J}{\partial \hat{\mathbf{x}} \partial \hat{\mathbf{x}}^T} = \frac{\partial}{\partial \hat{\mathbf{x}}}\left(\frac{\partial J}{\partial \hat{\mathbf{x}}}\right)^T = \begin{bmatrix} \frac{\partial^2 J}{\partial \hat{x}_1 \partial \hat{x}_1} & \frac{\partial^2 J}{\partial \hat{x}_1 \partial \hat{x}_2} & \cdots & \frac{\partial^2 J}{\partial \hat{x}_1 \partial \hat{x}_n} \\ \frac{\partial^2 J}{\partial \hat{x}_2 \partial \hat{x}_1} & \frac{\partial^2 J}{\partial \hat{x}_2 \partial \hat{x}_2} & \cdots & \frac{\partial^2 J}{\partial \hat{x}_2 \partial \hat{x}_n} \\ \vdots & \vdots & \ddots & \vdots \\ \frac{\partial^2 J}{\partial \hat{x}_n \partial \hat{x}_1} & \frac{\partial^2 J}{\partial \hat{x}_n \partial \hat{x}_2} & \cdots & \frac{\partial^2 J}{\partial \hat{x}_n \partial \hat{x}_n} \end{bmatrix}$$

be positive definite. That is,

$$\mathbf{z}^T \left[\frac{\partial^2 J}{\partial \hat{\mathbf{x}} \partial \hat{\mathbf{x}}^T}\right] \mathbf{z} > 0 \tag{4.1-7}$$

for all $\|\mathbf{z}\| > 0$ (i.e., for any nonzero norm of $\mathbf{z}$). From equation (4.1-4),

$$\frac{\partial}{\partial \hat{\mathbf{x}}}\left(\frac{\partial J}{\partial \hat{\mathbf{x}}}\right)^T = \mathbf{H}^T\mathbf{H}. \tag{4.1-8}$$

Hence

$$\begin{aligned} \mathbf{z}^T(\mathbf{H}^T\mathbf{H})\mathbf{z} &= (\mathbf{H}\mathbf{z})^T(\mathbf{H}\mathbf{z}) \\ &= \sum_{i=1}^{n}\left(\sum_{j=1}^{n}(\mathbf{H}_{ij}\mathbf{z}_j)\right)^2 \end{aligned} \tag{4.1-9}$$

This is the sum-of-squares of the elements of vector $\mathbf{Hz}$: the square of the *Euclidian* or l_2-norm $\|\mathbf{Hz}\|_2$ of the vector. Obviously $\|\mathbf{Hz}\|_2^2$ must be greater than or equal to zero, and $\|\mathbf{Hz}\|_2^2$ can only be zero for all $\|\mathbf{z}\| > 0$ when the matrix $\mathbf{H}$ has rank less than n. Hence the solution $\hat{\mathbf{x}}$ given in equation (4.1-6) is a least-squares (minimum J) solution when the rank of $\mathbf{H}$ is n. This is in fact the *observability* condition for an estimation problem: the rows $\mathbf{h}_i$ of the measurement matrix $\mathbf{H}$ must span the n-dimensional vector space. In other words, it must be possible to represent any arbitrary n-element vector $\mathbf{v}$ as a weighted sum of $\mathbf{h}_i$, that is,

$$\mathbf{v}^T = \sum_{i=1}^{m} c_i \mathbf{h}_i .$$

For example, it is possible to represent $\mathbf{v}^T = [1 \quad 0]$ as the linear combination of $\mathbf{h}_1 = [1 \quad 1]$ and $\mathbf{h}_2 = [1 \quad -1]$ (i.e., $\mathbf{v}^T = 0.5[1 \quad 1] + 0.5[1 \quad -1]$), but is not possible to represent it as a linear sum of $\mathbf{h}_1 = [1 \quad 1]$ and $\mathbf{h}_2 = [-1 \quad -1]$. In the latter case $\mathbf{h}_1$ and $\mathbf{h}_2$ do not span the vector space.

In addition to knowing that $\hat{\mathbf{x}}$ minimizes J, it is also of interest to know whether $\hat{\mathbf{x}}$ is unbiased; that is, is $E[\hat{\mathbf{x}}] = \mathbf{x}$? Using equations (4.1-6) and (4.1-2), we obtain

$$\begin{aligned} E[\hat{\mathbf{x}}] &= E\left[(\mathbf{H}^T\mathbf{H})^{-1}\mathbf{H}^T(\mathbf{Hx}+\mathbf{r})\right] \\ &= \mathbf{x} + (\mathbf{H}^T\mathbf{H})^{-1}\mathbf{H}^T E(\mathbf{r}) \quad , \\ &= \mathbf{x} \end{aligned} \tag{4.1-10}$$

since $E[\mathbf{r}] = \mathbf{0}$ was assumed in equation (4.1-1). Notice that $\mathbf{x}$ is not assumed to be a random variable with a known distribution. Rather it is treated as a *nonrandom* or *fixed-but-unknown* parameter.

Also recall that we did not use assumptions about the variance or the distribution of $\mathbf{r}$. An obvious question is "Can we infer statistics of $\mathbf{r}$ from the fit residuals $\tilde{\mathbf{y}} = \mathbf{y} - \mathbf{H}\hat{\mathbf{x}}$?" The answer is *yes* under some conditions. Substituting for $\hat{\mathbf{x}}$,

$$\boxed{\begin{aligned} \tilde{\mathbf{y}} &= \mathbf{y} - \mathbf{H}\hat{\mathbf{x}} \\ &= \left(\mathbf{I} - \mathbf{H}(\mathbf{H}^T\mathbf{H})^{-1}\mathbf{H}^T\right)\mathbf{y} \\ &= \left(\mathbf{I} - \mathbf{H}(\mathbf{H}^T\mathbf{H})^{-1}\mathbf{H}^T\right)(\mathbf{Hx}+\mathbf{r}) \\ &= \left(\mathbf{I} - \mathbf{H}(\mathbf{H}^T\mathbf{H})^{-1}\mathbf{H}^T\right)\mathbf{r} \end{aligned}} \tag{4.1-11}$$

and the residual sum-of-squares is

$$\begin{aligned} 2J &= \tilde{\mathbf{y}}^T\tilde{\mathbf{y}} \\ &= \mathbf{r}^T\left(\mathbf{I} - \mathbf{H}(\mathbf{H}^T\mathbf{H})^{-1}\mathbf{H}^T\right)\left(\mathbf{I} - \mathbf{H}(\mathbf{H}^T\mathbf{H})^{-1}\mathbf{H}^T\right)\mathbf{r} \\ &= \mathbf{r}^T\left(\mathbf{I} - \mathbf{H}(\mathbf{H}^T\mathbf{H})^{-1}\mathbf{H}^T\right)\mathbf{r} \end{aligned} \tag{4.1-12}$$

Notice that $E[\tilde{\mathbf{y}}] = E[\mathbf{r}] = \mathbf{0}$. Unfortunately equation (4.1-12) is not directly helpful, but it can be rewritten using the *trace* (sum-of-diagonal elements) operator as

$$\begin{aligned} \tilde{\mathbf{y}}^T\tilde{\mathbf{y}} &= tr(\tilde{\mathbf{y}}\tilde{\mathbf{y}}^T) \\ &= tr\left[\left(\mathbf{I} - \mathbf{H}(\mathbf{H}^T\mathbf{H})^{-1}\mathbf{H}^T\right)(\mathbf{r}\mathbf{r}^T)\left(\mathbf{I} - \mathbf{H}(\mathbf{H}^T\mathbf{H})^{-1}\mathbf{H}^T\right)\right]. \end{aligned} \tag{4.1-13}$$

We further assume that the elements of $\mathbf{r}$ are uncorrelated and that the variance is constant so $E[\mathbf{r}\mathbf{r}^T] = \sigma_r^2\mathbf{I}$. Now taking the expected value of equation (4.1-13),

$$E[\tilde{\mathbf{y}}^T\tilde{\mathbf{y}}] = \sigma_r^2 tr\left[(\mathbf{I} - \mathbf{H}(\mathbf{H}^T\mathbf{H})^{-1}\mathbf{H}^T)\right]. \tag{4.1-14}$$

To proceed, we express $\mathbf{H}$ using a Singular Value Decomposition (SVD):

$$\mathbf{H} = \mathbf{U}\mathbf{S}\mathbf{V}^T \tag{4.1-15}$$

where $\mathbf{U}$ is an $m \times m$ orthogonal matrix ($\mathbf{U}^T\mathbf{U} = \mathbf{I}$), $\mathbf{S}$ is an $m \times n$ diagonal matrix of singular values, and $\mathbf{V}$ is an $n \times n$ orthogonal matrix ($\mathbf{V}^T\mathbf{V} = \mathbf{I}$). Assuming that the system is overdetermined ($m > n$) and observable, there will be n singular values in $\mathbf{S}$ as

$$\mathbf{S} = \begin{bmatrix} s_1 & 0 & \cdots & 0 \\ 0 & s_2 & \cdots & 0 \\ \vdots & \vdots & \ddots & \vdots \\ 0 & 0 & \cdots & s_n \\ 0 & 0 & \cdots & 0 \\ \vdots & \vdots & \ddots & \vdots \\ 0 & 0 & \cdots & 0 \end{bmatrix}.$$

Then

$$\begin{aligned} \mathbf{I} - \mathbf{H}(\mathbf{H}^T\mathbf{H})^{-1}\mathbf{H}^T &= \mathbf{I} - \mathbf{U}\mathbf{S}\mathbf{V}^T(\mathbf{V}\mathbf{S}^T\mathbf{U}^T\mathbf{U}\mathbf{S}\mathbf{V}^T)^{-1}\mathbf{V}\mathbf{S}^T\mathbf{U}^T \\ &= \mathbf{I} - \mathbf{U}\mathbf{S}\mathbf{V}^T\mathbf{V}(\mathbf{S}^T\mathbf{S})^{-1}\mathbf{V}^T\mathbf{V}\mathbf{S}^T\mathbf{U}^T \\ &= \mathbf{I} - \mathbf{U}\begin{bmatrix} 1 & & & & & \\ & \ddots & & & 0 & \\ & & 1 & & & \\ & & & 0 & & \\ & 0 & & & \ddots & \\ & & & & & 0 \end{bmatrix}\mathbf{U}^T \end{aligned} \tag{4.1-16}$$

where there are n ones on the diagonal of $\mathbf{S}(\mathbf{S}^T\mathbf{S})^{-1}\mathbf{S}^T$. Then

$$tr\left[(\mathbf{I} - \mathbf{H}(\mathbf{H}^T\mathbf{H})^{-1}\mathbf{H}^T)\right] = m - n \tag{4.1-17}$$

since $tr(\mathbf{I} - \mathbf{U}\mathbf{A}\mathbf{U}^T) = m - tr(\mathbf{A})$ when $\mathbf{U}$ is orthogonal. Hence

$$\boxed{E[\tilde{\mathbf{y}}^T\tilde{\mathbf{y}}] = \sigma_r^2(m - n)} \tag{4.1-18}$$

In other words, the *expected* fit residual sum-of-squares is equal to the variance of the measurement noise multiplied by the number of measurements minus the number of unknowns in $\mathbf{x}$. This is an important relationship that is frequently used in linear regression problems. Recall, however, that all measurement errors were assumed to be uncorrelated and have a fixed variance. This may not apply when a variety of measurement types with different noise variances are used in the least-squares fit. Also notice that equation (4.1-18) implies that for a system in which the number of equations equals the number of unknown ($m = n$), the fit will exactly match every data point and $E[\tilde{\mathbf{y}}^T\tilde{\mathbf{y}}] = 0$.

Example 4.1: Linear Motion

To demonstrate the technique, consider a simple problem where the position of an object in one direction is measured as a function of time. Although the example is trivial, it demonstrates important properties of least-squares estimation without unnecessary complicated matrix algebra. Assume that the initial position (p_0) of the object is 1.0 m, it is moving at a constant velocity (v) of 1.0 m/s, and that three noisy measurements are recorded at 1.0 s intervals. Table 4.1 lists the true position, the measurement noise (r_i), and the noisy measurements. Figure 4.1 is a plot of the data.

The measurements are written in matrix form as

$$\begin{array}{ccccc} \mathbf{y} & = & \mathbf{H} & \mathbf{x} + & \mathbf{r} \end{array}$$

$$\begin{bmatrix} 0.5 \\ 2 \\ 3.5 \end{bmatrix} = \begin{bmatrix} 1 & 0 \\ 1 & 1 \\ 1 & 2 \end{bmatrix} \begin{bmatrix} p_0 \\ v \end{bmatrix} + \begin{bmatrix} -0.5 \\ 0 \\ +0.5 \end{bmatrix}.$$

Substituting the appropriate matrices into the normal equations using $\hat{\mathbf{x}} = \begin{bmatrix} \hat{p}_0 \\ \hat{v} \end{bmatrix}$, we obtain:

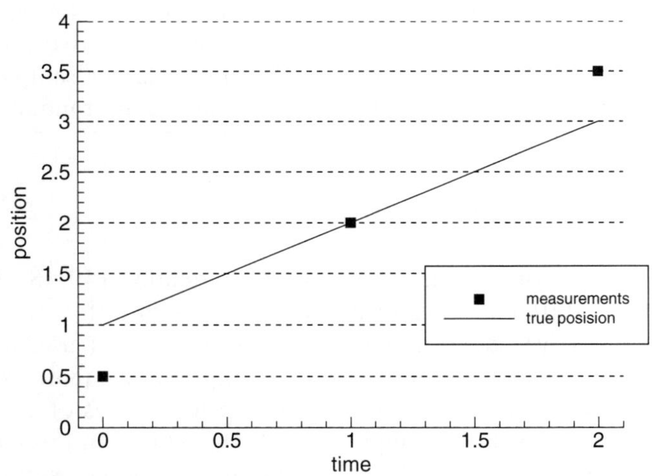

FIGURE 4.1: Linear motion—true position and measurements.

TABLE 4.1: Linear Motion Position and Measurements

Time (t_i)	Position: $p_i = p_0 + vt_i$	Noise: r_i	Measurement
0	1	−0.5	0.5
1	2	0	2.0
2	3	+0.5	3.5

$$(\mathbf{H}^T\mathbf{H})\ \hat{\mathbf{x}} = (\mathbf{H}^T\mathbf{y})$$

$$\begin{bmatrix} 1 & 1 & 1 \\ 0 & 1 & 2 \end{bmatrix}\begin{bmatrix} 1 & 0 \\ 1 & 1 \\ 1 & 2 \end{bmatrix}\begin{bmatrix} \hat{p}_0 \\ \hat{v} \end{bmatrix} = \begin{bmatrix} 1 & 1 & 1 \\ 0 & 1 & 2 \end{bmatrix}\begin{bmatrix} 0.5 \\ 2 \\ 3.5 \end{bmatrix}$$

$$\begin{bmatrix} 3 & 3 \\ 3 & 5 \end{bmatrix}\begin{bmatrix} \hat{p}_0 \\ \hat{v} \end{bmatrix} = \begin{bmatrix} 6 \\ 9 \end{bmatrix}$$

The solution can be obtained by several methods (Gaussian elimination, Crout reduction, Cholesky factoring, or even without forming the normal equations). The answer is

$$\begin{bmatrix} \hat{p}_0 \\ \hat{v} \end{bmatrix} = \begin{bmatrix} 0.5 \\ 1.5 \end{bmatrix}$$

versus the true answer

$$\begin{bmatrix} p_0 \\ v \end{bmatrix} = \begin{bmatrix} 1 \\ 1 \end{bmatrix}.$$

The residual sum-of-squares $2J$ for $\hat{p}_0 = 0.5$, $\hat{v} = 1.5$ is equal to 0 in this case. This example demonstrates a problem with small samples. Although the r_i values in Table 4.1 are unbiased and may be random over large samples, the particular three values in the table have an increasing trend. Hence the solution technique cannot differentiate between the trend in true position and the trend in the measurement errors. Thus $\hat{\mathbf{x}}$ is incorrect. With much larger samples the error in $\hat{\mathbf{x}}$ should be much smaller if the measurement errors are random.

4.2 WEIGHTED LEAST SQUARES

We now continue the discussion appearing after equation (4.1-18). Consider the case involving two different types of measurements with different units. Suppose that in Example 4.1 we added a velocity measurement that used units of km/s. Hence it would not be sensible to compute a simple least-squares solution combining the two measurement types where position had values of approximately 1 to 3 m and velocity had values of approximately 0.001 km/s. In this case the velocity would have so little weight in the solution that it would be essentially ignored. One fix to the problem multiplies the velocity measurements by 1000 so that the values of position and velocity are approximately equal in magnitude. However, this approach becomes less practical as additional measurement types are added, or when the measurement error variances vary with time, or when measurement errors are time-correlated. A more general approach uses variable weighting for different measurements (as done by Gauss). That is, the cost function is modified to

$$\boxed{J = \frac{1}{2}(\mathbf{y} - \mathbf{H}\hat{\mathbf{x}})^T \mathbf{W} (\mathbf{y} - \mathbf{H}\hat{\mathbf{x}})} \qquad (4.2\text{-}1)$$

where **W** is a positive definite symmetric matrix. The most sensible choice for **W** is the inverse covariance matrix of the measurement noise; that is, $\mathbf{W} = \mathbf{R}^{-1}$ where $\mathbf{R} = E[\mathbf{r}\mathbf{r}^T]$. Then the measurements are weighted inversely as the expected errors in the measurements. This method also works when the measurement errors are correlated with each other (**R** is nondiagonal), provided that **R** is full rank and can be inverted. In practice **R** is commonly assumed to be diagonal but this is not a general restriction. However, use of nondiagonal **R** greatly increases solution computations, and it is rarely used.

This weighted least-squares approach is one form of a *generalized least-squares* problem where **H** and/or **W** may be singular. However, in this chapter we assume that they are both full rank.

Using the cost function equation (4.2-1) and $\mathbf{W} = \mathbf{R}^{-1}$, the least-squares solution is computed by setting $\partial J / \partial \hat{\mathbf{x}} = \mathbf{0}^T$ to obtain

$$\boxed{\hat{\mathbf{x}} = (\mathbf{H}^T\mathbf{R}^{-1}\mathbf{H})^{-1}\mathbf{H}^T\mathbf{R}^{-1}\mathbf{y}.} \tag{4.2-2}$$

As before, the solution is unbiased: $E[\hat{\mathbf{x}}] = \mathbf{x}$. The $n \times m$ matrix $(\mathbf{H}^T\mathbf{R}^{-1}\mathbf{H})^{-1}\mathbf{H}^T\mathbf{R}^{-1}$ is one form of a pseudo-inverse, as it "inverts" the **y** measurements to compute **x** values that match **y** in a "minimum norm" sense. Estimates that are unbiased, obtained as a linear function of noisy observables $\mathbf{y} = \mathbf{Hx} + \mathbf{r}$, and minimize the covariance of any linear function of the estimates $\hat{\mathbf{x}}$ are sometimes called *Best Linear Unbiased Estimate* (BLUE). Gauss and Markov showed that the least-squares estimates, equation (4.1-6) or (4.2-2), are BLUE (see Björck 1996, p. 4).

The state estimate error is defined as

$$\begin{aligned}\tilde{\mathbf{x}} &= \mathbf{x} - \hat{\mathbf{x}} \\ &= \mathbf{x} - (\mathbf{H}^T\mathbf{R}^{-1}\mathbf{H})^{-1}\mathbf{H}^T\mathbf{R}^{-1}(\mathbf{Hx} + \mathbf{r}), \\ &= -(\mathbf{H}^T\mathbf{R}^{-1}\mathbf{H})^{-1}\mathbf{H}^T\mathbf{R}^{-1}\mathbf{r}\end{aligned} \tag{4.2-3}$$

and the *a posteriori* state error covariance is

$$\boxed{\begin{aligned}\mathbf{P} &= E[\tilde{\mathbf{x}}\tilde{\mathbf{x}}^T] \\ &= (\mathbf{H}^T\mathbf{R}^{-1}\mathbf{H})^{-1}\mathbf{H}^T\mathbf{R}^{-1}E[\mathbf{r}\mathbf{r}^T]\mathbf{R}^{-1}\mathbf{H}(\mathbf{H}^T\mathbf{R}^{-1}\mathbf{H})^{-1}, \\ &= (\mathbf{H}^T\mathbf{R}^{-1}\mathbf{H})^{-1}\end{aligned}} \tag{4.2-4}$$

where **P** is a positive definite symmetric matrix. Notice that for **P** to equal $(\mathbf{H}^T\mathbf{R}^{-1}\mathbf{H})^{-1}$, the weighting matrix **W** must equal $\mathbf{R}^{-1} = [E(\mathbf{r}\mathbf{r}^T)]^{-1}$. The matrix $\mathbf{H}^T\mathbf{R}^{-1}\mathbf{H} = \mathbf{P}^{-1}$ is called the information matrix (see Fisher 1925, 1934; Van Trees 1968, pp. 79–80; or Schweppe 1973, p. 140) since it represents the theoretical "information" available in the measurements. More will be said about this later.

Notice that the same matrix inverse required to compute the state estimate is also useful for interpreting the accuracy of the solution. In particular the square roots of the **P** diagonals are the *a posteriori* estimate error standard deviations: $\sigma_i = \sqrt{P_{ii}}$ for $i = 1$ to n. The off-diagonal elements of **P**, when divided by the corresponding standard deviations for those states, represent correlation coefficients that must be between −1.0 and 1.0. That is, a correlation matrix **C** can be computed from a covariance matrix as

$$\boxed{\mathbf{C} = \Sigma^{-1}\mathbf{P}\Sigma^{-1},} \tag{4.2-5}$$

where

$$\mathbf{\Sigma} = \begin{bmatrix} \sqrt{P_{11}} & 0 & \cdots & 0 \\ 0 & \sqrt{P_{22}} & \cdots & 0 \\ \vdots & \vdots & \ddots & \vdots \\ 0 & 0 & \cdots & \sqrt{P_{nn}} \end{bmatrix}.$$

C will have the form

$$\mathbf{C} = \begin{bmatrix} 1 & \rho_{12} & \cdots & \rho_{1n} \\ \rho_{21} & 1 & \cdots & \rho_{2n} \\ \vdots & \vdots & \ddots & \vdots \\ \rho_{n1} & \rho_{n2} & \cdots & 1 \end{bmatrix} \tag{4.2-6}$$

where the diagonal elements of **C** are 1 and off-diagonal elements $\rho_{ij} = P_{ji} / \sqrt{P_{ii}P_{jj}}$ are symmetric ($\rho_{ij} = \rho_{ji}$) since **P** is symmetric. If **P** exists, $-1 < \rho_{ij} < 1$. The closer the off-diagonal elements of **C** are to 1 in magnitude, the greater the tendency of errors in one state to be correlated with errors in another state. These off-diagonal correlation coefficients can never be exactly equal to ±1.0 because the matrix **P** would then be singular and it would not have been possible to invert $\mathbf{H}^T\mathbf{R}^{-1}\mathbf{H}$ to obtain **P**.

The weighted residual sum-of-squares for weighted least squares is computed similarly as for the un-weighted case. The measurement residual $\tilde{\mathbf{y}} = \mathbf{y} - \mathbf{H}\hat{\mathbf{x}}$ is

$$\boxed{\tilde{\mathbf{y}} = \left(\mathbf{I} - \mathbf{H}(\mathbf{H}^T\mathbf{R}^{-1}\mathbf{H})^{-1}\mathbf{H}^T\mathbf{R}^{-1}\right)\mathbf{r}} \tag{4.2-7}$$

where again $E[\tilde{\mathbf{y}}] = \mathbf{0}$ since $E[\mathbf{r}] = \mathbf{0}$. We assume that **R** can be factored as $\mathbf{R} = \mathbf{L}\mathbf{L}^T$ where **L** is nonsingular; otherwise the type of factorization is unimportant at this point. (Later we generally assume that **L** is the lower triangular Cholesky factor.) Then the weighted residual sum-of-squares can be written as

$$\begin{aligned} E[\tilde{\mathbf{y}}^T\mathbf{R}^{-1}\tilde{\mathbf{y}}] &= E\left[(\tilde{\mathbf{y}}^T\mathbf{L}^{-T})(\mathbf{L}^{-1}\tilde{\mathbf{y}})\right] \\ &= E\left\{tr\left[(\mathbf{L}^{-1}\tilde{\mathbf{y}})(\tilde{\mathbf{y}}^T\mathbf{L}^{-T})\right]\right\} \\ &= tr\left[\mathbf{L}^{-1}\left(\mathbf{I} - \mathbf{H}(\mathbf{H}^T\mathbf{R}^{-1}\mathbf{H})^{-1}\mathbf{H}^T\mathbf{R}^{-1}\right)\mathbf{R}\left(\mathbf{I} - \mathbf{R}^{-1}\mathbf{H}(\mathbf{H}^T\mathbf{R}^{-1}\mathbf{H})^{-1}\mathbf{H}^T\right)\mathbf{L}^{-T}\right]. \\ &= tr\left[(\mathbf{I} - \mathbf{L}^{-1}\mathbf{H}(\mathbf{H}^T\mathbf{L}^{-T}\mathbf{L}^{-1}\mathbf{H})^{-1}\mathbf{H}^T\mathbf{L}^{-T})\right] \end{aligned} \tag{4.2-8}$$

Using the SVD, we write

$$\mathbf{L}^{-1}\mathbf{H} = \mathbf{U}\mathbf{S}\mathbf{V}^T = [\mathbf{U}_1 \quad \mathbf{U}_2]\begin{bmatrix}\mathbf{S}_1 \\ \mathbf{0}\end{bmatrix}\mathbf{V}^T$$

where $\mathbf{S}_1$ is an $n \times n$ diagonal matrix, $\mathbf{U}_1$ is a $m \times n$ column-orthogonal matrix, and $\mathbf{U}_2$ is an $m \times (m - n)$ column-orthogonal (nullspace) matrix. Substituting $\mathbf{L}^{-1}\mathbf{H} = \mathbf{U}\mathbf{S}\mathbf{V}^T$ in equation (4.2-8) yields

$$\begin{aligned} E[\tilde{\mathbf{y}}^T\mathbf{R}^{-1}\tilde{\mathbf{y}}] &= tr\left[\left(\mathbf{I}-\mathbf{USV}^T(\mathbf{VS}^T\mathbf{U}^T\mathbf{USV}^T)^{-1}\mathbf{VS}^T\mathbf{U}^T\right)\right] \\ &= tr\left[\left(\mathbf{I}-\mathbf{US}(\mathbf{S}^T\mathbf{S})^{-1}\mathbf{S}^T\mathbf{U}^T\right)\right] \\ &= tr\left[\mathbf{I}-\mathbf{U}\begin{bmatrix}\mathbf{S}_1 \\ \mathbf{0}\end{bmatrix}\mathbf{S}_1^{-1}\mathbf{S}_1^{-1}[\mathbf{S}_1 \quad \mathbf{0}]\mathbf{U}^T\right] \\ &= tr\left[\mathbf{I}-\mathbf{U}_1\mathbf{U}_1^T\right] \\ &= tr\left[\mathbf{U}_2\mathbf{U}_2^T\right] \\ &= m-n \end{aligned} \tag{4.2-9}$$

The last step ($tr\left[\mathbf{U}_2\mathbf{U}_2^T\right]=m-n$) is obtained by observing that

$$tr\left[\mathbf{U}_2\mathbf{U}_2^T\right]=\sum_{i=1}^{m}\sum_{j=n+1}^{m}U_{ij}^2.$$

That is, it is the sum-of-squares of all elements in the orthogonal matrix **U** starting with column $n + 1$. Since this is the sum-of-squares of elements in $m - n$ orthonormal columns where the l_2 norm of each column is 1.0, the sum is $m - n$.

Hence the expected value of the weighted sum-of-residuals-squared should be equal to $m - n$ if the system is modeled properly and the measurement noise variances have been properly set. This is one of the first tests that should be used when attempting to validate a solution.

It is also possible to determine the variance of $\tilde{\mathbf{y}}^T\mathbf{R}^{-1}\tilde{\mathbf{y}}$. In the case where **R** is diagonal and the i-th diagonal is denoted as σ_i^2,

$$2J=\tilde{\mathbf{y}}^T\mathbf{R}^{-1}\tilde{\mathbf{y}}=\sum_{i=1}^{m}(\tilde{y}_i/\sigma_i)^2$$

which is a sum of m-squared random variables. If $\tilde{\mathbf{y}}$ is computed using the *true values* of the state $\mathbf{x}$ (that is, $\tilde{\mathbf{y}}_T=\mathbf{y}-\mathbf{Hx}=\mathbf{r}$), then $\tilde{y}_{T\_i}/\sigma_i=r_i/\sigma_i$ are unbiased *independent, identically distributed* (*i.i.d*) random variables with a variance of 1.0. If further **r** is Gaussian-distributed, then

$$2J_T=\sum_{i=1}^{m}(\tilde{y}_{T\_i}/\sigma_i)^2$$

is a Chi-squared (χ^2) distributed variable with m degrees-of-freedom. (See Fisz 1963, p. 339; Papoulis and Pillai 2002, p. 259). For this reason optimally weighted least squares is sometimes called χ^2 least squares. Notice in Figure 4.2 that the mean and variance of a χ^2 distribution increase with the number of degrees-of-freedom: for m degrees-of-freedom, the mean is m, and the variance is $2m$. The χ^2 probability density is

$$p_Y(y)=\frac{y^{m/2-1}}{2^{m/2}\Gamma(m/2)}\exp(-y/2) \qquad y\geq 0$$

where Γ is the gamma function, and the distribution function is

$$F_Y(y)\triangleq \Pr\{Y\leq y\}=P\left(\frac{m}{2},\frac{y}{2}\right)$$

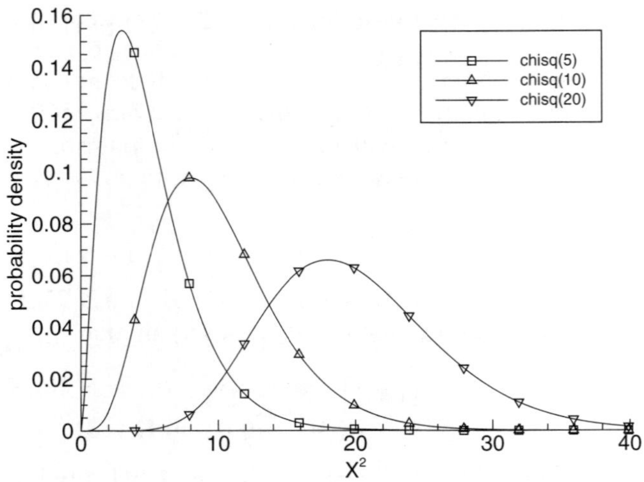

FIGURE 4.2: Chi-squared probability density for 5, 10, and 20 degrees-of-freedom.

where P is the incomplete gamma function (Press et al. 2007, section 6.2).

However, the true value of $\mathbf{x}$ is unknown, so it is not possible to compute $2J_T = \tilde{\mathbf{y}}_T^T \mathbf{R}^{-1} \tilde{\mathbf{y}}_T$ as described above. When $\tilde{\mathbf{y}}$ is computed using the estimated $\mathbf{x}$ (that is, $\tilde{\mathbf{y}} = \mathbf{y} - \mathbf{H}\hat{\mathbf{x}}$), then $\tilde{y}_i / \sigma_i$ are still identically distributed but are no longer independent because $\hat{\mathbf{x}}$ is computed from $\mathbf{y}$. Since there are n parameters in $\hat{\mathbf{x}}$, $\tilde{\mathbf{y}}^T \mathbf{R}^{-1} \tilde{\mathbf{y}}$ has $m - n$ degrees-of-freedom, or equivalently, the variance of $\tilde{y}_i / \sigma_i$ is $(m - n)/m$ rather than 1.0. If $\mathbf{r}$ is Gaussian-distributed, $\tilde{\mathbf{y}}^T \mathbf{R}^{-1} \tilde{\mathbf{y}}$ will be χ^2-distributed with $m - n$ degrees-of-freedom; that is, the mean is $m - n$ and the variance is $2(m - n)$. Notice in Figure 4.2 that the distribution is not symmetric, particularly for small degrees-of-freedom. As the number of degrees-of-freedom increases, the distribution becomes more symmetric and is very close to Gaussian for $m - n > 30$. When $m - n$ is small (e.g., <20), the level of significance should be determined using the cumulative distribution function. When $m - n > 20$ it is usually acceptable to assume symmetry and apply the test

$$\left| \tilde{\mathbf{y}}^T \mathbf{R}^{-1} \tilde{\mathbf{y}} - (m-n) \right| > k\sqrt{2(m-n)} \tag{4.2-10}$$

or equivalently

$$\boxed{\left| \frac{\tilde{\mathbf{y}}^T \mathbf{R}^{-1} \tilde{\mathbf{y}}}{m-n} - 1 \right| > k\sqrt{2/(m-n)},} \tag{4.2-11}$$

where k is a constant to be defined by the user for a specific probability. When the term on the left is greater than the term on the right, the system may not properly be modeled or measurement outliers may be present. Detection of measurement outliers will be discussed in a later chapter. The k constant is often set to 4 to denote a $4 - \sigma$ threshold.

This derivation assumed that the individual r_i / σ_i samples were *i.i.d* and Gaussian $N(0,1)$; that is, mean zero and variance one. If $\mathbf{r}$ is not Gaussian but unbiased *i.i.d* with a variance of 1.0, $\tilde{\mathbf{y}}^T \mathbf{R}^{-1} \tilde{\mathbf{y}}$ will have an expected value of $m - n$ (from eq. 4.2-9) and tend to be somewhat χ^2-distributed for large samples. If the $\mathbf{R}$ matrix is nondiagonal and is factored as before,

$$\tilde{\mathbf{y}}^T\mathbf{R}^{-1}\tilde{\mathbf{y}} = (\mathbf{L}^{-1}\tilde{\mathbf{y}})^T(\mathbf{L}^{-1}\tilde{\mathbf{y}}) = \sum_{i=1}^{m} z_i^2$$

where $\mathbf{z} = \mathbf{L}^{-1}\tilde{\mathbf{y}}$, the expected value is still $m - n$, as shown in equation (4.2-9).

The above derivation was based upon minimizing the weighted sum of measurement residuals squared; that is, weighted least squares. Notice that this approach does not depend upon assumed distributions for the measurement errors or for the state variables to be estimated. The only requirements are that the measurement noise $\mathbf{r}$ is unbiased, $E[\mathbf{r}] = \mathbf{0}$, that $E[\mathbf{r}\mathbf{r}^T] = \mathbf{R}$ is known, and that the measurement residual weighting matrix $\mathbf{W} = \mathbf{R}^{-1}$. In a later section we will examine variations of the least-squares method that depend upon the distribution of $\mathbf{r}$.

Finally we note that the error in the state estimate does not depend on the true value of the state. Since $\mathbf{x}$ is not a random variable and $\hat{\mathbf{x}}$ is an unbiased estimate of $\mathbf{x}$,

$$\begin{aligned} E\left[(\mathbf{x}-\hat{\mathbf{x}})\mathbf{x}^T\right] &= E[(\mathbf{x}-\hat{\mathbf{x}})]\mathbf{x}^T \\ &= \mathbf{0} \end{aligned}.$$

The result will be compared with properties of alternate methods in the next section.

Example 4.2: Linear Motion with Weighted Least Squares

To demonstrate weighted least squares, the problem of Example 4.1 is modified slightly so that different weighting is given to the three measurements. We assume that the noise standard deviation (σ) of the first two measurements is equal to 0.5, and σ of the third measurement is equal to 0.8. We also modify the third measurement slightly so that it does not fit the same line as the first two measurements (which makes it redundant): the measurement noise for the third measurement is set to +0.75 (rather than to +0.5) so that the third measurement is 3.75. Notice that the mean measurement error for these three samples is no longer zero. The normal equation for this problem is

$$(\mathbf{H}^T\mathbf{R}^{-1}\mathbf{H}) \quad \hat{\mathbf{x}} \quad = \quad (\mathbf{H}^T\mathbf{R}^{-1}\mathbf{y})$$

$$\begin{bmatrix} 1 & 1 & 1 \\ 0 & 1 & 2 \end{bmatrix}\begin{bmatrix} 1/0.25 & 0 & 0 \\ 0 & 1/0.25 & 0 \\ 0 & 0 & 1/0.64 \end{bmatrix}\begin{bmatrix} 1 & 0 \\ 1 & 1 \\ 1 & 2 \end{bmatrix}\begin{bmatrix} \hat{p}_0 \\ \hat{v} \end{bmatrix}$$

$$= \begin{bmatrix} 1 & 1 & 1 \\ 0 & 1 & 2 \end{bmatrix}\begin{bmatrix} 1/0.25 & 0 & 0 \\ 0 & 1/0.25 & 0 \\ 0 & 0 & 1/0.64 \end{bmatrix}\begin{bmatrix} 0.5 \\ 2 \\ 3.75 \end{bmatrix}$$

$$\begin{bmatrix} 9.5625 & 7.1250 \\ 7.1250 & 10.250 \end{bmatrix}\begin{bmatrix} \hat{p}_0 \\ \hat{v} \end{bmatrix} = \begin{bmatrix} 15.859375 \\ 19.718750 \end{bmatrix}$$

The solution is obtained by inverting $\mathbf{H}^T\mathbf{R}^{-1}\mathbf{H}$ to compute $\mathbf{P}$.

$$\mathbf{P} = \begin{bmatrix} 9.5625 & 7.1250 \\ 7.1250 & 10.250 \end{bmatrix}^{-1} = \begin{bmatrix} 0.2169312 & -0.1507937 \\ -0.1507937 & 0.2023810 \end{bmatrix}$$

The *a posteriori* standard deviations are $\sigma_{p0} = \sqrt{P_{11}} = 0.4658$, $\sigma_v = \sqrt{P_{22}} = 0.4499$ and the correlation coefficient is $\rho_{pv} = P_{12}/(\sigma_{p0}\sigma_v) = -0.7197$. Because only three measurements are available, the *a posteriori* state σ's are only slightly smaller than the measurement noise σ for the first two measurements. Further, the state estimate are negatively correlated because any positive change in $\hat{p}_0$ due to a larger measurement at $t = 0$ implies a negative change in $\hat{v}$ in order to match measurements at $t = 1$ and 2.

The state estimate is computed as $\hat{\mathbf{x}} = \mathbf{P}(\mathbf{H}^T\mathbf{R}^{-1}\mathbf{y})$, or

$$\begin{bmatrix} \hat{p}_0 \\ \hat{v} \end{bmatrix} = \begin{bmatrix} 0.466931 \\ 1.599206 \end{bmatrix},$$

which is a slightly smaller value for $\hat{p}_0$ and slightly larger value for $\hat{v}$ than obtained in Example 4.1. Notice that the estimate errors, −0.5331 and 0.5992, respectively, are slightly larger than the corresponding σ's. Using $\hat{p}_0 = 0.466931$ and $\hat{v} = 1.599206$, the *a posteriori* measurement residuals are computed as

$$\mathbf{y} - \mathbf{H}\hat{\mathbf{x}} = \begin{bmatrix} 0.5 - 0.466931 \\ 2 - 2.066138 \\ 3.75 - 3.665344 \end{bmatrix} = \begin{bmatrix} 0.033069 \\ -0.066138 \\ -0.084656 \end{bmatrix}$$

and the weighted residual sum-of-squares, $2J$, is 0.0331.

If many more measurements uniformly distributed over the same time interval are available, the *a posteriori* σ's will decrease approximately as $1/\sqrt{m}$ where m is the number of measurements. Figure 4.3 shows σ_{p0} and σ_v as a function of m for measurements uniformly distributed between $t = 0$ to 2 with the measurement noise $\sigma = 0.5$ for $t \le 1.33$ and $\sigma = 0.8$ for $t > 1.33$. For $m > 5$, the standard deviations closely follow the $1/\sqrt{m}$ curve.

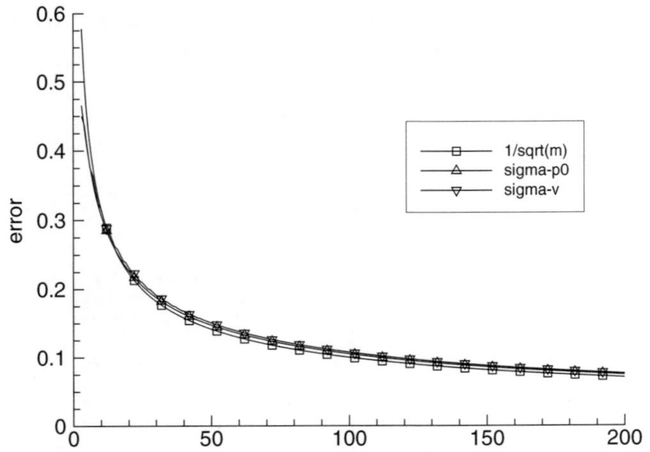

FIGURE 4.3: State estimate standard deviation versus number of measurements.

4.3 BAYESIAN ESTIMATION

The preceding sections assumed that the true state vector $\mathbf{x}$ was a fixed-but-unknown parameter, not a random variable. We now consider the case where $\mathbf{x}$ is a random variable. This change of viewpoint could either mean that $\mathbf{x}$ actually has a known probability density function (or second-order statistics—mean and covariance), or could imply that we have an *a priori* estimate of $\mathbf{x}$ with associated distribution function or covariance. For example, suppose than an instrument (e.g., gyroscope, optical sensing, electronic test equipment) manufacturer produces instruments having certain "biases" that can be calibrated but not completely removed. Examples include scale factor errors, gyro/accelerometer/gimbal misalignments, and thermal deformation. The manufacturer has three possibilities for providing information about the biases to instrument users:

1. The manufacturer may conduct limited accuracy testing on each instrument to verify that it is performing as expected, but no information on biases is provided to the users (or it was provided but the report was lost). Thus the user must treat the biases as fixed-but-unknown if using operational measurements in the field to calibrate the biases. The weighted least-squares approach of the previous section applies in this case.
2. The manufacturer may carefully calibrate the biases of each instrument from early production runs and computes statistics on the bias mean and covariance over all these instruments. In later production, accuracy of each instrument is tested to verify performance, but the bias mean and covariance of the early production is the only information provided to the user. The user can then treat the biases as random variables with known statistics when attempting to calibrate biases using operationally collected measurement data.
3. The manufacturer may carefully calibrate the biases of each instrument and include the calibration test report with the instrument when shipped. The calibration report also includes information on the accuracy of the calibration. That initial calibration may then be used as an *a priori* estimate for an estimation algorithm attempting to improve the initial bias calibration using operationally collected measurement data. Since errors in the initial calibration are independent of errors in the operationally collected measurements, and the calibration mean and covariance are known, the initial calibration estimates can be treated as "measurements" to be included in the weighted least-squares cost function.

Notice that the bias variables are the same regardless of the approach taken. Whether or not a variable is treated as nonrandom or random depends upon the type of information available to the user and how that information is used. We now examine methods for estimating variables that are treated as random.

4.3.1 Bayesian Least Squares

The characteristic of Bayesian estimation that makes it different from least-squares problems previously discussed is availability of prior information on the state mean and covariance (or possibly even a probability density function). This prior information may be the mean and covariance of the state $\mathbf{x}$ across many realizations, or it may be the mean and covariance of an initial measurement of $\mathbf{x}$. For the moment we assume the latter; that is, an unbiased *a priori* state estimate $\mathbf{x}_a$ and associated error covariance $\mathbf{P}_a = E[(\mathbf{x}_a - \mathbf{x})(\mathbf{x}_a - \mathbf{x})^T]$ are to be included in the least-squares cost function to be minimized. The augmented cost function for our linear measurement model problem is

$$J = \frac{1}{2}\left[(\mathbf{y}-\mathbf{H}\hat{\mathbf{x}})^T\mathbf{R}^{-1}(\mathbf{y}-\mathbf{H}\hat{\mathbf{x}}) + (\mathbf{x}_a-\hat{\mathbf{x}})^T\mathbf{P}_a^{-1}(\mathbf{x}_a-\hat{\mathbf{x}})\right]. \tag{4.3-1}$$

The state $\hat{\mathbf{x}}$ that minimizes this J can be computed, as before, by setting the gradient to zero. However, an alternate approach to this problem treats the *a priori* estimate as another measurement type where the error in the prior estimate is uncorrelated with errors in the "conventional" measurements. This viewpoint is taken when the instrument user treats the initial bias calibration as an *a priori* estimate for estimation using other measurements. By augmenting the prior to the measurement vector, the modified measurement model becomes

$$\begin{bmatrix}\mathbf{y}\\ \mathbf{x}_a\end{bmatrix} = \begin{bmatrix}\mathbf{H}\\ \mathbf{I}\end{bmatrix}\mathbf{x} + \begin{bmatrix}\mathbf{r}\\ \mathbf{x}_a-\mathbf{x}\end{bmatrix} \tag{4.3-2}$$

with

$$E\left[\begin{bmatrix}\mathbf{r}\\ \mathbf{x}_a-\mathbf{x}\end{bmatrix}\begin{bmatrix}\mathbf{r} & \mathbf{x}_a-\mathbf{x}\end{bmatrix}\right] = \begin{bmatrix}\mathbf{R} & \mathbf{0}\\ \mathbf{0} & \mathbf{P}_a\end{bmatrix}. \tag{4.3-3}$$

Substituting the augmented measurement model in the weighted least-squares normal equation gives the *a posteriori* state estimate:

$$\begin{aligned}\hat{\mathbf{x}} &= \left\{\begin{bmatrix}\mathbf{H}^T & \mathbf{I}\end{bmatrix}\begin{bmatrix}\mathbf{R}^{-1} & 0\\ 0 & \mathbf{P}_a^{-1}\end{bmatrix}\begin{bmatrix}\mathbf{H}\\ \mathbf{I}\end{bmatrix}\right\}^{-1}\begin{bmatrix}\mathbf{H}^T & \mathbf{I}\end{bmatrix}\begin{bmatrix}\mathbf{R}^{-1} & 0\\ 0 & \mathbf{P}_a^{-1}\end{bmatrix}\begin{bmatrix}\mathbf{y}\\ \mathbf{x}_a\end{bmatrix}.\\ &= (\mathbf{H}^T\mathbf{R}^{-1}\mathbf{H}+\mathbf{P}_a^{-1})^{-1}(\mathbf{H}^T\mathbf{R}^{-1}\mathbf{y}+\mathbf{P}_a^{-1}\mathbf{x}_a)\end{aligned} \tag{4.3-4}$$

This is the least-squares normal equation when using observation weighting and an *a priori* estimate of $\mathbf{x}$. The *a posteriori* information matrix is $\mathbf{H}^T\mathbf{R}^{-1}\mathbf{H}+\mathbf{P}_a^{-1}$ and the state error covariance matrix $\mathbf{P}$ is

$$\mathbf{P} = E\left[(\hat{\mathbf{x}}-\mathbf{x})(\hat{\mathbf{x}}-\mathbf{x})^T\right] = (\mathbf{H}^T\mathbf{R}^{-1}\mathbf{H}+\mathbf{P}_a^{-1})^{-1}. \tag{4.3-5}$$

As before, this is a positive definite symmetric matrix provided that the system is observable and the information matrix can be inverted. As noted above, equations (4.3-4) and (4.3-5) can also be obtained by setting $\partial J/\partial\hat{\mathbf{x}} = \mathbf{0}^T$ using J defined in equation (4.3-1).

It turns out that equation (4.3-4) also applies when the prior information consists of the mean and covariance of $\mathbf{x}$, rather than a measurement of $\mathbf{x}$. That is, equation (4.3-4) becomes

$$\hat{\mathbf{x}} = (\mathbf{H}^T\mathbf{R}^{-1}\mathbf{H}+\mathbf{P}_{xx}^{-1})^{-1}(\mathbf{H}^T\mathbf{R}^{-1}\mathbf{y}+\mathbf{P}_{xx}^{-1}\bar{\mathbf{x}}) \tag{4.3-6}$$

where $\bar{\mathbf{x}} = E[\mathbf{x}]$ and $\mathbf{P}_{xx} = E\left[(\mathbf{x}-\bar{\mathbf{x}})(\mathbf{x}-\bar{\mathbf{x}})^T\right]$ is the covariance of $\mathbf{x}$. From a practical viewpoint there is little difference between viewing the prior information as a noisy measurement versus treating it as a mean and covariance of the random variable **x**.

The assumptions used in deriving equation (4.3-4) or (4.3-6) were

1. The measurement model is linear ($\mathbf{y} = \mathbf{Hx} + \mathbf{r}$)
2. The cost function to be minimized (eq. 4.3-1) is the sum of squared weighted measurement residuals and prior state residuals
3. **r** is a zero-mean random variable vector with covariance **R**
4. **x** is a random variable vector with mean $\bar{\mathbf{x}}$ and covariance $\mathbf{P}_{xx}$ [or $\mathbf{x}_a$ and $\mathbf{P}_a$ for equation (4.3-3)]
5. **r** and $\mathbf{x}-\bar{\mathbf{x}}$ are linearly independent: $E[(\mathbf{x}-\bar{\mathbf{x}})\mathbf{r}^T] = E[(\mathbf{x}-\bar{\mathbf{x}})]E[\mathbf{r}^T] = \mathbf{0}$ for equation (4.3-6), or $E[(\mathbf{x}-\mathbf{x}_a)\mathbf{r}^T] = E[(\mathbf{x}-\mathbf{x}_a)]E[\mathbf{r}^T] = \mathbf{0}$ for (4.3-4).

This is a special case of a *Bayesian estimate*. Notice that the derivation did not use assumptions about probability density functions—only means and covariances. The more general case will be discussed in the next section. It should also be noted that Bayesian estimates can be constructed using nonquadratic cost functions (Van Trees 1968, pp. 54–61).

When using the Bayesian estimates, equation (4.3-4) or (4.3-6), the combined measurement/prior residual sum-of-squares ($2J$ using equation (4.3-1)) will still be χ^2-distributed, but will have an expected value of $m + n - n = m$.

Notice that equation (4.3-6) is equal to the weighted least-squares estimate (eq. 4.2-3) in the limiting case when $\mathbf{P}_{xx}^{-1} = \mathbf{0}$. This fact is sometimes used operationally to obtain a prior-free estimate from a Bayesian least-squares algorithm: the diagonal elements of $\mathbf{P}_{xx}$ are set to very large numbers (e.g., 10^6 times larger than normal) so that the prior has a negligible effect on the solution. One question that often arises is "When should prior information be included and if so, how heavily should it be weighted?" This sometimes leads to emotional discussions between "nonrandom" and "random variable" proponents. The argument against using prior information is that it is sometimes wrong and will bias the solution. However, that argument ignores the fact that measurement data is often wrong or mismodeled, or that weak observability can make computation of an accurate solution difficult. Thus strictly "measurement-based" solutions can sometimes produce unrealistic estimates. This author's opinion, based on extensive experience, is that reasonably weighted prior information should be included in the solution provided that it is consistent with the measurement data. That is, solutions with and without (or weakly weighted) prior information should be compared, and if they are very different, try to determine why. Inclusion of prior information often has the desirable effect of stabilizing a poorly observable solution. It should be included with some caution, and the prior $\mathbf{P}_{xx}$ should not be smaller than can be conservatively justified.

4.3.2 Bayes' Theorem

The term Bayesian estimation refers to the use of Bayes' theorem for computing the *a posteriori* probability of jointly distributed random variables. The formula is derived by expressing the joint probability using conditional probabilities. Consider the probability density of two jointly distributed scalar random variables

X and Y. We use uppercase letters to denote the random variables and lowercase letters to denote realizations of the variables. The joint probability density function of X and Y is

$$p_{X,Y}(x,y) = p_{X|Y}(x \mid y)p_Y(y) = p_{Y|X}(y \mid x)p_X(x) \tag{4.3-7}$$

where

$p_{X|Y}(x \mid y)$ is the conditional probability density function of X given $Y = y$,

$p_{Y|X}(y \mid x)$ is the conditional probability density function of Y given $X = x$, and

$p_X(x)$ and $p_Y(y)$ are unconditional (marginal) probability densities.

This equation follows from the definition of conditional probability for events A and B,

$$\Pr\{A \mid B\} \triangleq \Pr\{AB\}/\Pr\{B\},$$

provided that $\Pr\{B\} > 0$. Equation (4.3-7) can be rearranged to obtain the conditional probability of X given y using the conditional probability of Y given x:

$$\boxed{p_{X|Y}(x \mid y) = \frac{p_{Y|X}(y \mid x)p_X(x)}{p_Y(y)}} \tag{4.3-8}$$

provided that $p_Y(y) > 0$. This gives the *a posteriori* probability density of X given the *a priori* probability density $p_X(x)$. In the estimation problem X represents the state to be estimated, Y represents the measurements, and $p_Y(y)$ is the probability density of Y. The denominator of equation (4.3-8) can be written as

$$p_Y(y) = \int_{-\infty}^{\infty} p_{Y|X}(y \mid x)p_X(x)\,dx \tag{4.3-9}$$

to obtain

$$\boxed{p_{X|Y}(x \mid y) = \frac{p_{Y|X}(y \mid x)p_X(x)}{\int_{-\infty}^{\infty} p_{Y|X}(y \mid x)p_X(x)\,dx}.} \tag{4.3-10}$$

This is Bayes' theorem. Since the integrand is equal to the numerator, $p_Y(y)$ can be viewed as a normalization factor that makes

$$\int_{-\infty}^{\infty} p_{X|Y}(x \mid y)\,dx = 1.$$

For vector random variables, the vector form of equation (4.3-8) is

$$\boxed{p_{\mathbf{X}|\mathbf{Y}}(\mathbf{x} \mid \mathbf{y}) = \frac{p_{\mathbf{Y}|\mathbf{X}}(\mathbf{y} \mid \mathbf{x})p_{\mathbf{X}}(\mathbf{x})}{p_{\mathbf{Y}}(\mathbf{y})}} \tag{4.3-11}$$

where $p_{\mathbf{X}|\mathbf{Y}}(\mathbf{x} \mid \mathbf{y}) = p_{X_1,X_2,\ldots X_n|Y_1,Y_2,\ldots Y_n}(x_1,x_2,\ldots,x_n \mid y_1,y_2,\ldots,y_m)$ and other vector probability densities are defined accordingly.

It is generally more convenient to work with first- and second-order moments (mean and covariances) of random variables rather than probability densities. The mean (expected value) of a random variable is defined as a "center-of-mass" integral, that is,

$$\boxed{\bar{\mathbf{x}} \triangleq E[\mathbf{x}] = \int_{-\infty}^{\infty} \mathbf{x} p_{\mathbf{X}}(\mathbf{x})\, d\mathbf{x}} \tag{4.3-12}$$

where the vector integral is defined as

$$\int_{-\infty}^{\infty} \mathbf{x} p_{\mathbf{X}}(\mathbf{x})\, d\mathbf{x} = \int_{-\infty}^{\infty} \cdots \int_{-\infty}^{\infty} \begin{bmatrix} x_1 \\ x_2 \\ \vdots \\ x_n \end{bmatrix} p_{\mathbf{X}}(x_1, x_2, \ldots, x_n)\, dx_1, \ldots, dx_n.$$

The *conditional mean* is

$$\boxed{E[\mathbf{x} \mid \mathbf{y}] = \int_{-\infty}^{\infty} \mathbf{x} p_{\mathbf{X}|\mathbf{Y}}(\mathbf{x} \mid \mathbf{y})\, d\mathbf{x}.} \tag{4.3-13}$$

The covariance between $\mathbf{x}$ and $\mathbf{y}$ is defined similarly:

$$\begin{aligned} E\left[(\mathbf{x}-\bar{\mathbf{x}})(\mathbf{y}-\bar{\mathbf{y}})^T\right] &= \int_{-\infty}^{\infty}\int_{-\infty}^{\infty} (\mathbf{x}-\bar{\mathbf{x}})(\mathbf{y}-\bar{\mathbf{y}})^T p_{\mathbf{X},\mathbf{Y}}(\mathbf{x},\mathbf{y})\, d\mathbf{x}\, d\mathbf{y} \\ &= \int_{-\infty}^{\infty}\int_{-\infty}^{\infty} (\mathbf{x}-\bar{\mathbf{x}})(\mathbf{y}-\bar{\mathbf{y}})^T p_{\mathbf{X}|\mathbf{Y}}(\mathbf{x} \mid \mathbf{y}) p_Y(\mathbf{y})\, d\mathbf{y}\, d\mathbf{x} \end{aligned} . \tag{4.3-14}$$

If the density functions are jointly Gaussian, it can be shown (Anderson and Moore 1979, p. 25) using equation (4.3-13) that the conditional mean of $\mathbf{x}$ given $\mathbf{y}$ (i.e., mean of the *a posteriori* density) is

$$\boxed{E[\mathbf{x} \mid \mathbf{y}] = \bar{\mathbf{x}} + \mathbf{P}_{yx}^T \mathbf{P}_{yy}^{-1}(\mathbf{y}-\bar{\mathbf{y}})} \tag{4.3-15}$$

and the conditional covariance is

$$\boxed{E\left[(\mathbf{x}-\bar{\mathbf{x}})(\mathbf{x}-\bar{\mathbf{x}})^T \mid \mathbf{y}\right] = \mathbf{P}_{xx} - \mathbf{P}_{yx}^T \mathbf{P}_{yy}^{-1} \mathbf{P}_{yx}} \tag{4.3-16}$$

where $\mathbf{P}_{yx} = E\left[(\mathbf{y}-\bar{\mathbf{y}})(\mathbf{x}-\bar{\mathbf{x}})^T\right]$, $\mathbf{P}_{yy} = E\left[(\mathbf{y}-\bar{\mathbf{y}})(\mathbf{y}-\bar{\mathbf{y}})^T\right]$ and $\mathbf{P}_{xx} = E\left[(\mathbf{x}-\bar{\mathbf{x}})(\mathbf{x}-\bar{\mathbf{x}})^T\right]$. It is generally true that $E[\mathbf{x} \mid \mathbf{y}]$ is Gaussian if $p_{\mathbf{XY}}(\mathbf{x},\mathbf{y})$, $p_{\mathbf{X}}(\mathbf{x})$, and $p_{\mathbf{Y}}(\mathbf{y})$ are Gaussian. Equation (4.3-15) also applies when the density functions are unimodal and symmetric—but not Gaussian—and measurement perturbations are linear in $\mathbf{x}$ perturbations: $\mathbf{y}-\bar{\mathbf{y}} = \mathbf{H}(\mathbf{x}-\bar{\mathbf{x}})$.

Now using our forward linear model $\mathbf{y} = \mathbf{H}\mathbf{x} + \mathbf{r}$ (eq. 4.1-2) with assumption $E\left[(\mathbf{x}-\bar{\mathbf{x}})\mathbf{r}^T\right] = \mathbf{0}$ it is seen that

$$\begin{aligned} \mathbf{P}_{yx} &= E\left[(\mathbf{H}(\mathbf{x}-\bar{\mathbf{x}})+\mathbf{r})(\mathbf{x}-\bar{\mathbf{x}})^T\right] \\ &= \mathbf{H}\mathbf{P}_{xx} \end{aligned} \tag{4.3-17}$$

and

$$\begin{aligned} \mathbf{P}_{yy} &= E\left[(\mathbf{H}(\mathbf{x}-\bar{\mathbf{x}})+\mathbf{r})(\mathbf{H}(\mathbf{x}-\bar{\mathbf{x}})+\mathbf{r})^T\right] \\ &= \mathbf{H}\mathbf{P}_{xx}\mathbf{H}^T + \mathbf{R} \end{aligned} . \tag{4.3-18}$$

Hence the conditional mean $\hat{\mathbf{x}} = E[\mathbf{x} \mid \mathbf{y}]$ from equation (4.3-15) is

$$\boxed{\begin{aligned} \hat{\mathbf{x}} &= \bar{\mathbf{x}} + \mathbf{P}_{xx}\mathbf{H}^T(\mathbf{H}\mathbf{P}_{xx}\mathbf{H}^T + \mathbf{R})^{-1}(\mathbf{y}-\mathbf{H}\bar{\mathbf{x}}) \\ &= \mathbf{P}_{xx}\mathbf{H}^T(\mathbf{H}\mathbf{P}_{xx}\mathbf{H}^T + \mathbf{R})^{-1}\mathbf{y} + [\mathbf{I} - \mathbf{P}_{xx}\mathbf{H}^T(\mathbf{H}\mathbf{P}_{xx}\mathbf{H}^T + \mathbf{R})^{-1}\mathbf{H}]\bar{\mathbf{x}} \end{aligned}} \tag{4.3-19}$$

At first glance this appears to be a very different formula than equation (4.3-6), but it is the same equation. This can be shown using the matrix inversion identity (see Appendix A),

$$(\mathbf{A}+\mathbf{BCB}^T)^{-1} = \mathbf{A}^{-1} - \mathbf{A}^{-1}\mathbf{B}(\mathbf{C}^{-1}+\mathbf{B}^T\mathbf{A}^{-1}\mathbf{B})^{-1}\mathbf{B}^T\mathbf{A}^{-1} \tag{4.3-20}$$

where **A** and **C** are nonsingular square matrices of (usually) different dimensions and **B** is a rectangular or square matrix. Letting $\mathbf{A} = \mathbf{R}$, $\mathbf{B} = \mathbf{H}$, $\mathbf{C} = \mathbf{P}_{xx}$, it follows that

$$\begin{aligned}
\mathbf{P}_{xx}\mathbf{H}^T(\mathbf{HP}_{xx}\mathbf{H}^T+\mathbf{R})^{-1} &= \mathbf{P}_{xx}\mathbf{H}^T\left(\mathbf{R}^{-1} - \mathbf{R}^{-1}\mathbf{H}(\mathbf{P}_{xx}^{-1}+\mathbf{H}^T\mathbf{R}^{-1}\mathbf{H})^{-1}\mathbf{H}^T\mathbf{R}^{-1}\right)\\
&= \mathbf{P}_{xx}\left(\mathbf{H}^T\mathbf{R}^{-1} - \mathbf{H}^T\mathbf{R}^{-1}\mathbf{H}(\mathbf{P}_{xx}^{-1}+\mathbf{H}^T\mathbf{R}^{-1}\mathbf{H})^{-1}\mathbf{H}^T\mathbf{R}^{-1}\right)\\
&= \mathbf{P}_{xx}\left(\mathbf{I} - \mathbf{H}^T\mathbf{R}^{-1}\mathbf{H}(\mathbf{P}_{xx}^{-1}+\mathbf{H}^T\mathbf{R}^{-1}\mathbf{H})^{-1}\right)\mathbf{H}^T\mathbf{R}^{-1}\\
&= \mathbf{P}_{xx}\left(\mathbf{P}_{xx}^{-1}+\mathbf{H}^T\mathbf{R}^{-1}\mathbf{H} - \mathbf{H}^T\mathbf{R}^{-1}\mathbf{H}\right)(\mathbf{P}_{xx}^{-1}+\mathbf{H}^T\mathbf{R}^{-1}\mathbf{H})^{-1}\mathbf{H}^T\mathbf{R}^{-1}\\
&= (\mathbf{H}^T\mathbf{R}^{-1}\mathbf{H}+\mathbf{P}_{xx}^{-1})^{-1}\mathbf{H}^T\mathbf{R}^{-1}
\end{aligned} \tag{4.3-21}$$

Thus equation (4.3-19) becomes

$$\begin{aligned}
\hat{\mathbf{x}} &= (\mathbf{H}^T\mathbf{R}^{-1}\mathbf{H}+\mathbf{P}_{xx}^{-1})^{-1}\mathbf{H}^T\mathbf{R}^{-1}\mathbf{y} + \left(\mathbf{I} - (\mathbf{H}^T\mathbf{R}^{-1}\mathbf{H}+\mathbf{P}_{xx}^{-1})^{-1}\mathbf{H}^T\mathbf{R}^{-1}\mathbf{H}\right)\bar{\mathbf{x}}\\
&= (\mathbf{H}^T\mathbf{R}^{-1}\mathbf{H}+\mathbf{P}_{xx}^{-1})^{-1}(\mathbf{H}^T\mathbf{R}^{-1}\mathbf{y}+\mathbf{P}_{xx}^{-1}\bar{\mathbf{x}})
\end{aligned} \tag{4.3-22}$$

This demonstrates that equation (4.3-19) is equivalent to equation (4.3-6). It will be shown in Chapter 8 that the first line in equation (4.3-19) is the Kalman filter measurement update to $\bar{\mathbf{x}}$ for a measurement **y**. Thus the Kalman filter measurement update is equivalent to the Bayesian least-squares (conditional mean) estimate given the same initial $\bar{\mathbf{x}}$ and measurements **y**. Since the Kalman filter recursively updates the state estimate, equation (4.3-19) can also be written using the *a priori* estimate $\hat{\mathbf{x}}_a$ as

$$\boxed{\hat{\mathbf{x}} = \hat{\mathbf{x}}_a + \mathbf{P}_a\mathbf{H}^T(\mathbf{HP}_a\mathbf{H}^T+\mathbf{R})^{-1}(\mathbf{y}-\mathbf{H}\hat{\mathbf{x}}_a).} \tag{4.3-23}$$

Example 4.3: Linear Motion with Bayesian Least Squares

A modified version of Example 4.2 is used to demonstrate Bayesian weighted least-squares estimation. We assume that

$$\bar{\mathbf{x}} = \begin{bmatrix}\bar{p}_0\\ \bar{v}\end{bmatrix} = \begin{bmatrix}0\\ 0.5\end{bmatrix} \quad\text{and}\quad \mathbf{P}_{xx} = \begin{bmatrix}0.2^2 & 0\\ 0 & 0.2^2\end{bmatrix}.$$

As before, the three measurements are 0.5, 2.0, and 3.75 with measurement noise standard deviations equal to 0.5, 0.5, and 0.8, respectively. Then from equation (4.3-22),

$$(\mathbf{H}^T\mathbf{R}^{-1}\mathbf{H}+\mathbf{P}_{xx}^{-1})\hat{\mathbf{x}} = (\mathbf{H}^T\mathbf{R}^{-1}\mathbf{y}+\mathbf{P}_{xx}^{-1}\bar{\mathbf{x}})$$

is equal to

$$\left\{\begin{bmatrix}1 & 1 & 1\\0 & 1 & 2\end{bmatrix}\begin{bmatrix}1/0.25 & 0 & 0\\0 & 1/0.25 & 0\\0 & 0 & 1/0.64\end{bmatrix}\begin{bmatrix}1 & 0\\1 & 1\\1 & 2\end{bmatrix}+\begin{bmatrix}1/0.04 & 0\\0 & 1/0.04\end{bmatrix}\right\}\begin{bmatrix}\hat{p}_0\\\hat{v}\end{bmatrix}$$

$$=\begin{bmatrix}1 & 1 & 1\\0 & 1 & 2\end{bmatrix}\begin{bmatrix}1/0.25 & 0 & 0\\0 & 1/0.25 & 0\\0 & 0 & 1/0.64\end{bmatrix}\begin{bmatrix}0.5\\2\\3.75\end{bmatrix}+\begin{bmatrix}1/0.04 & 0\\0 & 1/0.04\end{bmatrix}\begin{bmatrix}0\\0.5\end{bmatrix}$$

or

$$\begin{bmatrix}34.562 & 7.1250\\7.1250 & 35.250\end{bmatrix}\begin{bmatrix}p_0\\v\end{bmatrix}=\begin{bmatrix}15.859\\32.219\end{bmatrix}.$$

The resulting estimates are

$$\begin{bmatrix}\hat{p}_0\\\hat{v}\end{bmatrix}=\begin{bmatrix}0.282199\\0.856967\end{bmatrix}$$

with *a posteriori* standard deviations of

$$\begin{bmatrix}\sigma_{\hat{p}0}\\\sigma_{\hat{v}}\end{bmatrix}=\begin{bmatrix}0.17376\\0.17205\end{bmatrix}.$$

Notice that the inclusion of the prior information changed the estimates significantly from those of Example 4.2 ($\hat{p}_0 = 0.466931$, $\hat{v} = 1.599206$). The measurement residual sum-of-squares for this Bayesian case has increased to 7.9602 versus 0.03307 for Example 4.2.

4.3.3 Minimum Variance or Minimum Mean-Squared Error (MMSE)

Kalman and Weiner worked on problems similar to least-squares estimation, but from a different viewpoint. Rather than trying to minimize a measurement/prior estimate residual sum-of-squares, they minimized the variance of the error in an *a posteriori* state (or signal in Wiener's problem) estimate constructed as a linear or affine transformation on the measurements $\mathbf{y}$. That is, they found $\mathbf{K}$ and $\mathbf{b}$ in the equation $\hat{\mathbf{x}} = \mathbf{K}\mathbf{y} + \mathbf{b}$ to minimize the cost function

$$J = \frac{1}{2}E\left[(\mathbf{x}-\hat{\mathbf{x}})^T\mathbf{W}(\mathbf{x}-\hat{\mathbf{x}})\right] \tag{4.3-24}$$

where $\mathbf{W}$ is an arbitrary positive definite symmetric weighting matrix. It turns out that the $\hat{\mathbf{x}}$ that minimizes $E\left[(\mathbf{x}-\hat{\mathbf{x}})^T\mathbf{W}(\mathbf{x}-\hat{\mathbf{x}})\right]$ for a specific $\mathbf{W}$ also minimizes it for all $\mathbf{W}$. This is demonstrated by assuming that $\hat{\mathbf{x}}$ is a function of a set of parameters $\alpha_1, \alpha_2, \ldots$ to be determined. Since the partial derivatives of J with respect to α_j must be zero at the minimum,

$$\frac{\partial}{\partial \alpha_j} E\left[(\mathbf{x}-\hat{\mathbf{x}})^T \mathbf{W}(\mathbf{x}-\hat{\mathbf{x}})/2\right] = -E\left[(\mathbf{x}-\hat{\mathbf{x}})^T \mathbf{W} \frac{\partial \hat{\mathbf{x}}}{\partial \alpha_j}\right]$$
$$= -E\left[tr\left(\mathbf{W}\left(\frac{\partial \hat{\mathbf{x}}}{\partial \alpha_j}\right)(\mathbf{x}-\hat{\mathbf{x}})^T\right)\right].$$

Hence the presence of **W** does not change the condition for a minimum:

$$E\left[\left(\frac{\partial \hat{\mathbf{x}}}{\partial \alpha_j}\right)(\mathbf{x}-\hat{\mathbf{x}})^T\right] = \mathbf{0},$$

so we conveniently make $\mathbf{W} = \mathbf{I}$ and obtain

$$\begin{aligned} J &= \frac{1}{2} E\left[(\mathbf{x}-\hat{\mathbf{x}})^T(\mathbf{x}-\hat{\mathbf{x}})\right] \\ &= \frac{1}{2} E\|\mathbf{x}-\hat{\mathbf{x}}\|^2 \end{aligned} \quad . \tag{4.3-25}$$

This approach is called a *minimum variance* or MMSE method. It minimizes the mean of the *a posteriori* density. To apply the method, hypothesize that the state estimate can be obtained as an affine transformation on the measurements **y**:

$$\hat{\mathbf{x}} = \mathbf{K}\mathbf{y} + \mathbf{b} \tag{4.3-26}$$

where **K** and **b** are to be determined. (The minimum variance method can also be used for $\hat{\mathbf{x}} = \mathbf{g}(\mathbf{y})$ where **g** is nonlinear—and that may possibly yield a smaller J—but the discussion here is limited to affine models.) We want the estimate to be unbiased so

$$\begin{aligned} E[\mathbf{x}-\hat{\mathbf{x}}] &= E[\mathbf{x}-(\mathbf{K}\mathbf{y}+\mathbf{b})] \\ &= E[\mathbf{x}-(\mathbf{K}\mathbf{H}\mathbf{x}+\mathbf{K}\mathbf{r}+\mathbf{b})] \\ &= (\mathbf{I}-\mathbf{K}\mathbf{H})\bar{\mathbf{x}}-\mathbf{b} \\ &= \mathbf{0} \end{aligned}$$

or

$$\mathbf{b} = (\mathbf{I}-\mathbf{K}\mathbf{H})\bar{\mathbf{x}} \tag{4.3-27}$$

Hence

$$\begin{aligned} \mathbf{x}-\hat{\mathbf{x}} &= \mathbf{x}-\mathbf{K}(\mathbf{H}\mathbf{x}+\mathbf{r})-(\mathbf{I}-\mathbf{K}\mathbf{H})\bar{\mathbf{x}} \\ &= (\mathbf{I}-\mathbf{K}\mathbf{H})\tilde{\mathbf{x}}-\mathbf{K}\mathbf{r} \end{aligned} \tag{4.3-28}$$

where $\tilde{\mathbf{x}} \triangleq \mathbf{x}-\bar{\mathbf{x}}$. Thus the cost function becomes

$$\begin{aligned} J &= \frac{1}{2} E[(\mathbf{x}-\hat{\mathbf{x}})^T(\mathbf{x}-\hat{\mathbf{x}})] \\ &= \frac{1}{2}\left[tr\left((\mathbf{I}-\mathbf{K}\mathbf{H})E[\tilde{\mathbf{x}}\tilde{\mathbf{x}}^T](\mathbf{I}-\mathbf{H}^T\mathbf{K}^T)+\mathbf{K}E[\mathbf{r}\mathbf{r}^T]\mathbf{K}^T\right)\right] \\ &= \frac{1}{2} tr\left((\mathbf{I}-\mathbf{K}\mathbf{H})\mathbf{P}_{xx}(\mathbf{I}-\mathbf{H}^T\mathbf{K}^T)+\mathbf{K}\mathbf{R}\mathbf{K}^T\right) \end{aligned} \tag{4.3-29}$$

since $E[\tilde{\mathbf{x}}\mathbf{r}^T] = \mathbf{0}$. Setting the gradient with respect to **K** to zero gives

$$\frac{\partial J}{\partial \mathbf{K}} = -(\mathbf{I} - \mathbf{KH})\mathbf{P}_{xx}\mathbf{H}^T + \mathbf{KR} = \mathbf{0}$$

or

$$\mathbf{K} = \mathbf{P}_{xx}\mathbf{H}^T(\mathbf{HP}_{xx}\mathbf{H}^T + \mathbf{R})^{-1}. \tag{4.3-30}$$

Thus the minimum variance estimate from equation (4.3-26) is

$$\boxed{\hat{\mathbf{x}} = \mathbf{P}_{xx}\mathbf{H}^T(\mathbf{HP}_{xx}\mathbf{H}^T + \mathbf{R})^{-1}\mathbf{y} + [\mathbf{I} - \mathbf{P}_{xx}\mathbf{H}^T(\mathbf{HP}_{xx}\mathbf{H}^T + \mathbf{R})^{-1}\mathbf{H}]\bar{\mathbf{x}}} \tag{4.3-31}$$

which is the same equation as equation (4.3-19) for the conditional mean. This shows that the minimum variance or MMSE estimate is the conditional mean for affine measurement/estimate models. It can be shown (Van Trees 1968, section 2.4.1) that the minimum variance/MMSE solution is always the conditional mean regardless of whether the models are linear or nonlinear.

A minimum variance estimate can also be computed using an *a priori* measurement of the state rather than the mean value. In this case we hypothesize that

$$\hat{\mathbf{x}} = \mathbf{K}_y\mathbf{y} + \mathbf{K}_x\mathbf{x}_a. \tag{4.3-32}$$

A bias term is not necessary in this case because it must be zero. We also require that the estimate be unbiased ($E[\mathbf{x} - \hat{\mathbf{x}}] = \mathbf{0}$) and that the *a priori* estimate $\mathbf{x}_a$ is also unbiased ($E[\mathbf{x} - \mathbf{x}_a] = \mathbf{0}$). Thus

$$\begin{aligned} E[\mathbf{x} - \hat{\mathbf{x}}] &= \mathbf{x} - E[\mathbf{K}_y\mathbf{y} + \mathbf{K}_x\mathbf{x}_a] \\ &= [\mathbf{I} - \mathbf{K}_y\mathbf{H} - \mathbf{K}_x]\bar{\mathbf{x}} \\ &= \mathbf{0} \end{aligned}$$

or

$$\mathbf{K}_x = \mathbf{I} - \mathbf{K}_y\mathbf{H} \tag{4.3-33}$$

for nonzero $\bar{\mathbf{x}}$. Thus

$$\mathbf{x} - \hat{\mathbf{x}} = (\mathbf{I} - \mathbf{K}_y\mathbf{H})(\mathbf{x} - \mathbf{x}_a) - \mathbf{K}_y\mathbf{r} \tag{4.3-34}$$

Substituting equation (4.3-34) in equation (4.3-25) using $\tilde{\mathbf{x}}_a = \mathbf{x} - \mathbf{x}_a$ we obtain

$$\begin{aligned} J &= \frac{1}{2}E[(\mathbf{x} - \hat{\mathbf{x}})^T(\mathbf{x} - \hat{\mathbf{x}})] \\ &= \frac{1}{2}E\left[tr\left((\mathbf{I} - \mathbf{K}_y\mathbf{H})(\tilde{\mathbf{x}}_a\tilde{\mathbf{x}}_a^T)(\mathbf{I} - \mathbf{H}^T\mathbf{K}_y^T) + \mathbf{K}_y(\mathbf{rr}^T)\mathbf{K}_y^T\right)\right]. \\ &= \frac{1}{2}tr\left((\mathbf{I} - \mathbf{K}_y\mathbf{H})\mathbf{P}_a(\mathbf{I} - \mathbf{H}^T\mathbf{K}_y^T) + \mathbf{K}_y\mathbf{RK}_y^T\right) \end{aligned} \tag{4.3-35}$$

Setting

$$\frac{\partial J}{\partial \mathbf{K}_y} = (\mathbf{K}_y\mathbf{H} - \mathbf{I})\mathbf{P}_a\mathbf{H}^T + \mathbf{K}_y\mathbf{R} = \mathbf{0}$$

gives

$$\boxed{\mathbf{K}_y = \mathbf{P}_a\mathbf{H}^T(\mathbf{HP}_a\mathbf{H}^T + \mathbf{R})^{-1}.} \tag{4.3-36}$$

Using equations (4.3-32) and (4.3-33), the minimum variance estimate is

$$\boxed{\begin{aligned}\hat{\mathbf{x}} &= \mathbf{P}_a\mathbf{H}^T(\mathbf{H}\mathbf{P}_a\mathbf{H}^T+\mathbf{R})^{-1}\mathbf{y}+\left[\mathbf{I}-\mathbf{P}_a\mathbf{H}^T(\mathbf{H}\mathbf{P}_a\mathbf{H}^T+\mathbf{R})^{-1}\mathbf{H}\right]\mathbf{x}_a\\ &= \mathbf{x}_a+\mathbf{P}_a\mathbf{H}^T(\mathbf{H}\mathbf{P}_a\mathbf{H}^T+\mathbf{R})^{-1}(\mathbf{y}-\mathbf{H}\mathbf{x}_a)\end{aligned}}. \tag{4.3-37}$$

This is equivalent to equation (4.3-4) and similar to equation (4.3-31), except that $\bar{\mathbf{x}}$ is replaced with $\mathbf{x}_a$ and $\mathbf{P}_{xx}$ is replaced with $\mathbf{P}_a$. For practical purposes there is no difference, other than input numerical values, between assuming that the *a priori* state information is the actual mean and covariance of the state, or is a noisy measurement of the state with associated error covariance.

4.3.4 Orthogonal Projections

Weiner and Kalman obtained the solution to the minimum variance problem using the orthogonality principle. Two random variable (vectors) are considered orthogonal if $E[\mathbf{x}\mathbf{y}^T] = \mathbf{0}$, and uncorrelated if $E\left[(\mathbf{x}-\bar{\mathbf{x}})(\mathbf{y}-\bar{\mathbf{y}})^T\right]=\mathbf{0}$ where $\bar{\mathbf{x}}$ and $\bar{\mathbf{y}}$ are the unconditional means of $\mathbf{x}$ and $\mathbf{y}$, respectively. The value $\hat{\mathbf{x}}$ that minimizes $E\|\mathbf{x}-\hat{\mathbf{x}}\|^2$ when $\hat{\mathbf{x}}$ is constructed as an affine or nonlinear transformation on measurements $\mathbf{y}$ must satisfy the orthogonality condition

$$E\left[(\mathbf{x}-\hat{\mathbf{x}})\mathbf{y}^T\right]=\mathbf{0}. \tag{4.3-38}$$

For an affine estimator of the form $\hat{\mathbf{x}}=\mathbf{K}\mathbf{y}+\mathbf{b}$, the requirement that the estimate be unbiased ($E[\hat{\mathbf{x}}]=E[\mathbf{x}]=\bar{\mathbf{x}}$) implies that $\mathbf{b}=(\mathbf{I}-\mathbf{K}\mathbf{H})\bar{\mathbf{x}}$. Hence

$$\begin{aligned}E\left[(\mathbf{x}-\hat{\mathbf{x}})\mathbf{y}^T\right] &= E\left[(\mathbf{x}-\mathbf{K}\mathbf{y}-\mathbf{b})\mathbf{y}^T\right]\\ &= E\left[(\mathbf{I}-\mathbf{K}\mathbf{H})(\mathbf{x}-\bar{\mathbf{x}})-\mathbf{K}\mathbf{r})(\mathbf{x}^T\mathbf{H}^T+\mathbf{r}^T)\right].\\ &= (\mathbf{I}-\mathbf{K}\mathbf{H})\mathbf{P}_{xx}\mathbf{H}^T-\mathbf{K}\mathbf{R}\end{aligned} \tag{4.3-39}$$

As before, $\mathbf{y}=\mathbf{H}\mathbf{x}+\mathbf{r}$, $\mathbf{P}_{xx}=E\left[(\mathbf{x}-\bar{\mathbf{x}})(\mathbf{x}-\bar{\mathbf{x}})^T\right]$, $\mathbf{R}=E[\mathbf{r}\mathbf{r}^T]$, and $E\left[(\mathbf{x}-\bar{\mathbf{x}})\mathbf{r}^T\right]=\mathbf{0}$. Setting the expectation to $\mathbf{0}$ gives

$$\mathbf{K}=\mathbf{P}_{xx}\mathbf{H}^T\left(\mathbf{H}\mathbf{P}_{xx}\mathbf{H}^T+\mathbf{R}\right)^{-1}$$

which is the same result (eq. 4.3-19) obtained using Bayes' rule or by setting $\partial J/\partial\mathbf{K}=\mathbf{0}$ in equation (4.3-30). We used the specific models $\mathbf{y}=\mathbf{H}\mathbf{x}+\mathbf{r}$ and $\hat{\mathbf{x}}=\mathbf{K}\mathbf{y}+(\mathbf{I}-\mathbf{H}\mathbf{K})\bar{\mathbf{x}}$ to demonstrate the concept, but the orthogonality principle can be applied to find the conditional mean minimizing the mean-squared error for a more general class of models. In fact, the estimation errors $\mathbf{x}-\hat{\mathbf{x}}$ for the nonlinear model $\hat{\mathbf{x}}=\mathbf{g}(\mathbf{y},\bar{\mathbf{x}})$ must be orthogonal to any linear or nonlinear function of $\mathbf{y}$ for $\|\mathbf{x}-\hat{\mathbf{x}}\|^2$ to be a minimum (Papoulis and Pillai 2002, pp. 268–269). It is possible that $\hat{\mathbf{x}}=\mathbf{g}(\mathbf{y},\bar{\mathbf{x}})$ may produce a smaller mean-squared error than an affine model. However, if the random variables $\mathbf{x}$ and $\mathbf{y}$ are jointly Gaussian and zero mean, the affine and nonlinear estimators are equal. See Anderson and Moore (1979, p. 95) or Sorenson (1966) for further information on the orthogonality principle.

A geometric interpretation of the orthogonality principle is shown in Figure 4.4. Assume that the scalar conditional estimate $\hat{x}=E[x\,|\,\mathbf{y}]$ is to be computed as a

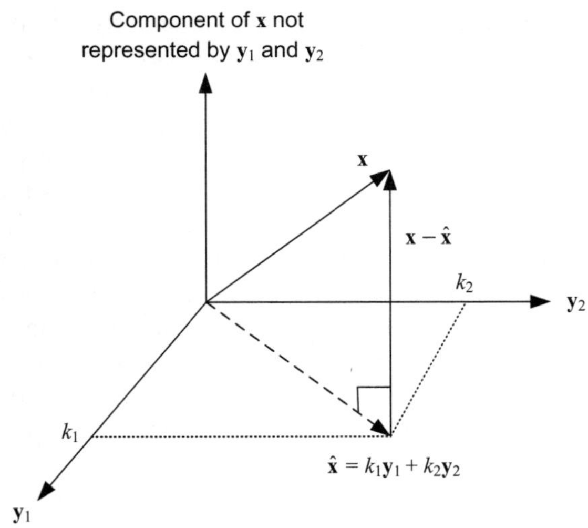

FIGURE 4.4: Orthogonal projections of state and measurements.

linear function of two noisy measurements $\mathbf{y} = [y_1 \quad y_2]^T$, where for convenience we assume that the means of x and $\mathbf{y}$ are zero. Hence $\hat{x}$ can be represented as a vector in the $\mathbf{y}_1 - \mathbf{y}_2$ plane: $\hat{\mathbf{x}} = k_1\mathbf{y}_1 + k_2\mathbf{y}_2$. The coefficients k_1, k_2 are not determined from actual values of y_1, y_2, but rather from the assumed statistical distributions of y and the true state x. Thus x cannot be exactly represented as $k_1y_1 + k_2y_2$ because y_1 and y_2 contain measurement noise. In Figure 4.4 the true state is represented as vector **x**, which has a vertical component representing the part of true state that cannot be represented—in an expected value sense—as a linear transformation on the measurements. That out-of-plane component is entirely due to the presence of measurement noise. If measurements are error-free, $\hat{\mathbf{x}} = \mathbf{x}$ lies in the $\mathbf{y}_1 - \mathbf{y}_2$ plane. It is seen that the expected value of the vector from the point in the $\mathbf{y}_1 - \mathbf{y}_2$ plane closest to **x** must be orthogonal to the $\mathbf{y}_1 - \mathbf{y}_2$ plane. In other words,

$$\begin{aligned} E\left[(\mathbf{x}-\hat{\mathbf{x}})\hat{\mathbf{x}}^T\right] &= E\left[(\mathbf{x}-\hat{\mathbf{x}})\left(\mathbf{y}^T\mathbf{K}^T + \bar{\mathbf{x}}^T(\mathbf{I}-\mathbf{KH})^T\right)\right] \\ &= E\left[(\mathbf{x}-\hat{\mathbf{x}})\mathbf{y}^T\right]\mathbf{K}^T + E\left[(\mathbf{x}-\hat{\mathbf{x}})\right]\bar{\mathbf{x}}^T(\mathbf{I}-\mathbf{KH})^T \\ &= \mathbf{0} \end{aligned} \tag{4.3-40}$$

Thus the minimum variance estimate error is orthogonal to the estimate. When **x** is treated as a fixed-but-unknown (nonrandom) variable, the weighted least-squares estimate error is orthogonal to the true state ($E\left[(\mathbf{x}-\hat{\mathbf{x}})\mathbf{x}^T\right] = \mathbf{0}$) because the estimate is unbiased and **x** is not random.

4.4 PROBABILISTIC APPROACHES—MAXIMUM LIKELIHOOD AND MAXIMUM *A POSTERIORI*

The estimation techniques of the previous sections obtained an "optimal" state estimate using measurement weighting determined from means and covariances of random variables. The underlying probability distributions were only referenced when discussing algorithm assumptions and conditions under which the approaches are valid. We now examine two algorithms that directly use probability density functions when computing the state estimate. The first method, *maximum likelihood*, treats the estimated state as nonrandom (fixed-but-unknown) parameters and finds the state value that maximizes the mean probability density of obtaining the measurements actually received for an assumed measurement probability density. The second method, *maximum a posteriori*, is similar but treats the state as random variables with a known density function, and the state value that maximizes the *a posteriori* probability density (the mode of the *a posteriori* density) is selected as "best." For Gaussian distributions and linear models, the maximum likelihood estimate is equal to the weighted least-squares estimate and the maximum *a posteriori* estimate is equal to the minimum variance estimate.

Both methods can be applied for arbitrary probability density functions, but in practice they usually assume Gaussian densities. Hence we start with a discussion of the Gaussian distribution.

4.4.1 Gaussian Random Variables

For a single continuous random variable Y with mean $\bar{y}$ and standard deviation σ, the *Gaussian* or *normal probability density function* is

$$p_Y(y) = \frac{1}{\sqrt{2\pi}\sigma}\exp\left(-(y-\bar{y})^2/(2\sigma^2)\right). \tag{4.4-1}$$

The integral of the density function is the *probability distribution function* $F_Y(y) \triangleq \Pr\{Y \le y\}$, which is the probability that the random variable Y will be found at or below value y. Hence the probability density

$$p_Y(y) = \lim_{\Delta y \to 0} \frac{F_Y(y+\Delta y) - F_Y(y)}{\Delta y}$$

has units of probability per unit-of-y (provided that the limit exists). For a Gaussian density the distribution function is

$$\begin{aligned} F_Y(y) &= \frac{1}{\sqrt{2\pi}\sigma}\int_{-\infty}^{y} \exp\left(-(\lambda-\bar{y})^2/(2\sigma^2)\right) d\lambda \\ &= \begin{cases} \dfrac{1}{2} + \dfrac{1}{2}\operatorname{erf}\left(\dfrac{y-\bar{y}}{\sqrt{2}\sigma}\right) & \text{for } y \ge \bar{y} \\ \dfrac{1}{2} - \dfrac{1}{2}\operatorname{erf}\left(\dfrac{\bar{y}-y}{\sqrt{2}\sigma}\right) & \text{for } y < \bar{y} \end{cases} \end{aligned} \tag{4.4-2}$$

where $\operatorname{erf}\left((y-\bar{y})/(\sqrt{2}\sigma)\right)$ is the error function. Alternately equation (4.4-2) can be evaluated using

$$\operatorname{erf}\left(\frac{y-\bar{y}}{\sqrt{2}\sigma}\right) = P\left(\frac{1}{2}, \frac{(y-\bar{y})^2}{2\sigma^2}\right) \quad \text{for } y \ge \bar{y}$$

where P is the incomplete gamma function. The denominator $\sqrt{2\pi}\sigma$ is required in equation (4.4-1) so that $F_Y(\infty) = 1$. Figure 4.5 shows the Gaussian density and distribution functions for $\bar{y} = 0$ and $\sigma = 1$, which is often denoted as a $N(0,1^2)$ distribution.

The joint probability density for multiple *independent* Gaussian random variables with a common mean $\bar{y}$ is equal to the products of the individual densities; that is,

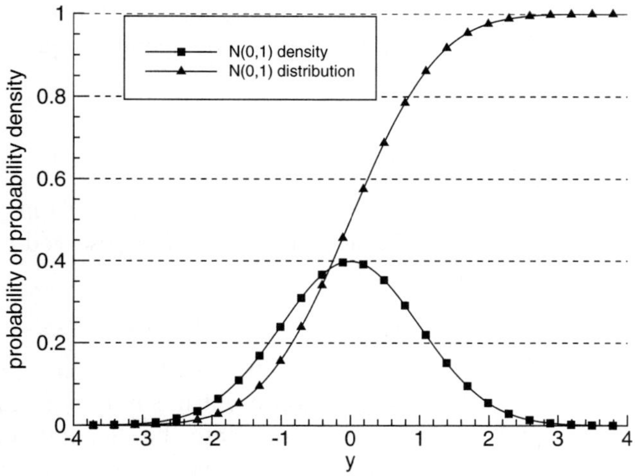

FIGURE 4.5: N(0,1) Gaussian probability density and cumulative distribution functions.

$$p_{Y_1,Y_2,\ldots,Y_m}(y_1, y_2, \ldots, y_m) = \frac{1}{(2\pi)^{m/2} \prod_{i=1}^{m} \sigma_i} \exp\left(-\frac{1}{2} \sum_{i=1}^{m} (y_i - \bar{y})^2 / \sigma_i^2 \right) \tag{4.4-3}$$

or in vector notation

$$p_{\mathbf{Y}}(\mathbf{y}) = \frac{1}{(2\pi)^{m/2} |\mathbf{P}_{yy}|^{1/2}} \exp\left(-\frac{1}{2} (\mathbf{y} - \bar{\mathbf{y}})^T \mathbf{P}_{yy}^{-1} (\mathbf{y} - \bar{\mathbf{y}}) \right) \tag{4.4-4}$$

where

$$\mathbf{P}_{yy} = \begin{bmatrix} \sigma_1^2 & 0 & \cdots & 0 \\ 0 & \sigma_2^2 & \cdots & 0 \\ \vdots & \vdots & \ddots & \vdots \\ 0 & 0 & \cdots & \sigma_m^2 \end{bmatrix}$$

and $|\mathbf{P}_{yy}|$, the determinant of $\mathbf{P}_{yy}$, is equal to the product of the diagonal variances. $p_{\mathbf{Y}}(\mathbf{y})$ has units of probability per (units-of-y)^m; for example, if y_i has units of meters, $p_{\mathbf{Y}}(\mathbf{y})$ has units of probability per (meter)^m. When $p_{\mathbf{Y}}(\mathbf{y})$ is integrated from $-\infty$ to ∞ for each variable in $\mathbf{y}$, the resultant probability is equal to one.

If the individual y_i variables are correlated (off-diagonal elements of $\mathbf{P}_{yy}$ are nonzero), the positive definite symmetric $\mathbf{P}_{yy}$ can be factored as $\mathbf{P}_{yy} = \mathbf{U}\Lambda\mathbf{U}^T$ using eigen decomposition, where $\mathbf{U}$ is an orthogonal $m \times m$ matrix and $\mathbf{\Lambda}$ is a diagonal $m \times m$ matrix of eigenvalues λ_i. Then the transformation $\mathbf{y} = \mathbf{U}\mathbf{z}$ can be substituted in equation (4.4-4) to obtain

$$\begin{aligned} p_{\mathbf{Y}}(\mathbf{z}) &= \frac{1}{(2\pi)^{m/2} |\mathbf{U}\,\mathbf{\Lambda}\,\mathbf{U}^T|^{1/2}} \exp\left(-\frac{1}{2} (\mathbf{z} - \bar{\mathbf{z}})^T \mathbf{\Lambda}^{-1} (\mathbf{z} - \bar{\mathbf{z}}) \right) \\ &= \frac{1}{(2\pi)^{m/2} |\mathbf{\Lambda}|^{1/2}} \exp\left(-\frac{1}{2} (\mathbf{z} - \bar{\mathbf{z}})^T \mathbf{\Lambda}^{-1} (\mathbf{z} - \bar{\mathbf{z}}) \right) \end{aligned} . \tag{4.4-5}$$

Notice that this is exactly the same form as equation (4.4-3) but σ_i^2 is replaced with λ_i and y_i is replaced with z_i. Hence equation (4.4-4) is a general equation that applies whether or not $\mathbf{P}_{yy}$ is diagonal.

4.4.2 Maximum Likelihood Estimation

For jointly distributed continuous random variables y_i, $i = 1$ to m, the *likelihood function* is defined as the joint probability density for the observed y_i given a set of parameters x_i, $i = 1$ to n. That is,

$$L(\mathbf{y} \mid \mathbf{x}) = p_{\mathbf{Y}|\mathbf{X}}(\mathbf{y} \mid \mathbf{x}) = p_{\mathbf{Y}|\mathbf{X}}(y_1, y_2, \ldots, y_m \mid x_1, x_2, \ldots, x_n) \tag{4.4-6}$$

where $\mathbf{y} = [y_1, y_2, \ldots, y_m]^T$ and $\mathbf{x} = [x_1, x_2, \ldots, x_n]^T$. It should be noted that the notation $p_{\mathbf{Y}|\mathbf{X}}(y_1, y_2, \ldots, y_m \mid x_1, x_2, \ldots, x_n)$ has a slightly different meaning here than the conditional density of the previous sections. In this case the elements of vector $\mathbf{x}$ are fixed-but-unknown parameters, not random variables. Also, $\mathbf{x}$ may include parameters that were not considered in previous sections, such as the elements of $\mathbf{R}$ (the measurement noise covariance).

Notice that the likelihood function is a *probability density, not a probability*, and has units of $1 / \prod_i (\text{units-of-}y_i)$.

For the Gaussian density, the log of the likelihood function is

$$\ln L(\mathbf{y} \mid \mathbf{x}) = -\frac{1}{2}\left[m \ln(2\pi) + \ln\left|\mathbf{P}_{yy}\right| + (\mathbf{y} - \bar{\mathbf{y}})^T \mathbf{P}_{yy}^{-1} (\mathbf{y} - \bar{\mathbf{y}})\right] \tag{4.4-7}$$

where $\mathbf{P}_{yy}$ and $\bar{\mathbf{y}}$ are computed using $\mathbf{x}$. The *maximum likelihood* (ML) estimate of $\mathbf{x}$ is computed by setting

$$\frac{\partial \ln L(\mathbf{y} \mid \mathbf{x})}{\partial \mathbf{x}} = \mathbf{0} \tag{4.4-8}$$

which is equivalent to minimizing

$$\boxed{J = -\ln L(\mathbf{y} \mid \mathbf{x}) = \frac{1}{2}\left[m \ln(2\pi) + \ln\left|\mathbf{P}_{yy}\right| + (\mathbf{y} - \bar{\mathbf{y}})^T \mathbf{P}_{yy}^{-1} (\mathbf{y} - \bar{\mathbf{y}})\right]} \tag{4.4-9}$$

Now consider how the log likelihood function can be used for estimation purposes. For the case in which $\mathbf{P}_{yy}$ is not a function of $\mathbf{x}$,

$$\boxed{\frac{\partial J}{\partial \mathbf{x}} = -(\mathbf{y} - \bar{\mathbf{y}})^T \mathbf{P}_{yy}^{-1} \frac{\partial \bar{\mathbf{y}}}{\partial \mathbf{x}}} \tag{4.4-10}$$

which must be $\mathbf{0}$ at the ML estimate $\hat{\mathbf{x}}$. The *Hessian matrix* of the log likelihood,

$$\boxed{\frac{\partial^2 J}{\partial \mathbf{x} \partial \mathbf{x}^T} = \left(\frac{\partial \bar{\mathbf{y}}}{\partial \mathbf{x}}\right)^T \mathbf{P}_{yy}^{-1} \frac{\partial \bar{\mathbf{y}}}{\partial \mathbf{x}} - (\mathbf{y} - \bar{\mathbf{y}})^T \mathbf{P}_{yy}^{-1} \frac{\partial^2 \bar{\mathbf{y}}}{\partial \mathbf{x} \partial \mathbf{x}^T},} \tag{4.4-11}$$

must be positive definite for J to be a minimum at $\hat{\mathbf{x}}$. Notice the second term in equation (4.4-11). The tensor

$$\boldsymbol{\Omega} = \frac{\partial^2 \bar{\mathbf{y}}}{\partial \mathbf{x} \partial \mathbf{x}^T}$$

will be zero for a linear model of the form $\bar{\mathbf{y}} = \mathbf{Hx}$, but for a nonlinear model $\bar{\mathbf{y}} = \mathbf{h}(\mathbf{x})$, the individual terms of $\mathbf{\Omega}$, $\partial^2 \bar{y}_i / (\partial \mathbf{x} \partial \mathbf{x}^T)$, can have positive and negative eigenvalues. When $\partial^2 \bar{\mathbf{y}} / (\partial \mathbf{x} \partial \mathbf{x}^T)$ is multiplied by the vector $(\mathbf{y} - \bar{\mathbf{y}})^T \mathbf{P}_{yy}^{-1}$ (which includes both positive and negative values for different elements of $\mathbf{y} - \bar{\mathbf{y}}$) and subtracted from the first term in equation (4.4-11), the resulting Hessian can have negative eigenvalues. This is more likely to happen when $\hat{\mathbf{x}}$ is far from truth or when measurement errors are large. If the Hessian has a mix of positive and negative eigenvalues when evaluated near the true value of $\mathbf{x}$, the surface of J is a saddle point rather than a minimum, and numerical optimization algorithms will not converge (Dennis and Schnabel 1983, p. 81). This can also make optimization difficult when saddle points occur far from the optimum. Fortunately these problems are rare.

Taking the expectation of equation (4.4-11) gives

$$\boxed{\mathbf{F} = E\left[\frac{\partial^2 J}{\partial \mathbf{x} \partial \mathbf{x}^T}\right] = \left(\frac{\partial \bar{\mathbf{y}}}{\partial \mathbf{x}}\right)^T \mathbf{P}_{yy}^{-1} \frac{\partial \bar{\mathbf{y}}}{\partial \mathbf{x}}} \tag{4.4-12}$$

or element $F_{ij} = E\left[\dfrac{\partial^2 J}{\partial x_i \partial x_j}\right]$.

This $n \times n$ matrix is called the *Fisher information matrix* after the work of Fisher (1922, 1925, 1935). Ideally the partial derivatives should be evaluated at the true value of $\mathbf{x}$, but since this is unknown, the estimate $\hat{\mathbf{x}}$ is often used. Cramér (1946) and Rao (1945) independently showed that for *any* estimate $\hat{\mathbf{x}}$ based on measurements $\mathbf{y}$, the inverse of

$$E\left[\frac{\partial^2 J}{\partial \mathbf{x} \partial \mathbf{x}^T}\right] \text{ or equivalently } E\left[\left(\frac{\partial J}{\partial \mathbf{x}}\right)^T \left(\frac{\partial J}{\partial \mathbf{x}}\right)\right]$$

(with J defined as the *negative* log probability density using any density function), is a lower bound on the estimate error covariance. That is,

$$\boxed{E\left[(\mathbf{x} - \hat{\mathbf{x}})(\mathbf{x} - \hat{\mathbf{x}})^T\right] \geq \left[E\left(\frac{\partial^2 J}{\partial \mathbf{x} \partial \mathbf{x}^T}\right)\right]^{-1}.} \tag{4.4-13}$$

This is called the Cramér-Rao lower bound. An estimator is called *efficient* when the equality is satisfied. The ML estimate (MLE) is asymptotically (as the number of measurements $\rightarrow \infty$) efficient. Further they are *consistent* in that solution of $\partial J / \partial \mathbf{x} = \mathbf{0}^T$ asymptotically converges in probability to $\hat{\mathbf{x}} = \mathbf{x}$, and the MLE is asymptotically Gaussian. See Van Trees (1968, section 2.4.2), Papoulis and Pillai (2002, section 8.3), Fisz (1963, sections 12.4, 13.7), Schweppe (1973, section 11.2.1), or Fisher (1922, 1925, 1934) for further information on properties of ML and other estimators.

For the linear measurement model $\mathbf{y} = \mathbf{Hx} + \mathbf{r}$ of (4.1-2), $\partial \bar{\mathbf{y}} / \partial \mathbf{x} = \mathbf{H}$ and $\mathbf{P}_{yy} = \mathbf{R}$, so equation (4.4-10) becomes $(\mathbf{y} - \mathbf{Hx})^T \mathbf{R}^{-1} \mathbf{H} = \mathbf{0}$ or

$$\boxed{\hat{\mathbf{x}} = (\mathbf{H}^T \mathbf{R}^{-1} \mathbf{H})^{-1} \mathbf{H}^T \mathbf{R}^{-1} \mathbf{y}.} \tag{4.4-14}$$

This is the normal equation (4.2-2) for the weighted least squares method with nonrandom $\mathbf{x}$. Furthermore the Fisher information matrix

$$\mathbf{F} = \left(\frac{\partial \bar{\mathbf{y}}}{\partial \mathbf{x}}\right)^T \mathbf{P}_{yy}^{-1} \frac{\partial \bar{\mathbf{y}}}{\partial \mathbf{x}} = \mathbf{H}^T \mathbf{R}^{-1} \mathbf{H} \tag{4.4-15}$$

is the same information matrix as for weighted least squares. Hence for the linear model $\mathbf{y} = \mathbf{Hx} + \mathbf{r}$ with zero-mean Gaussian noise $\mathbf{r}$, the ML solution is identical to the weighted least-squares solution. However, the ML method is much more powerful as it can be applied to nonlinear problems, problems with non-Gaussian errors, and problems where $\mathbf{P}_{yy}$ is a function of $\mathbf{x}$. ML estimation will be discussed again in Chapter 11.

Example 4.4: MLE of Sample Mean and Variance

This example demonstrates use of the ML method to estimate the mean and variance of a scalar random variable. A Gaussian random number generator is used to produce 50 samples from an $N(0.9,1.2^2)$ population. That is, the population has a mean of 0.9 and variance of $1.2^2 = 1.44$. The measurement model for this problem is simply

$$\mathbf{y} = \begin{bmatrix} 1 \\ 1 \\ \vdots \\ 1 \end{bmatrix} x + \mathbf{r}$$

where $\mathbf{y}$ and $\mathbf{r}$ are 50-element vectors, $E[\mathbf{r}] = \mathbf{0}$ and $\mathbf{P}_{yy} = E[\mathbf{rr}^T] = \sigma^2\mathbf{I}$. The simulated process mean is $x = 0.90$ and the noise standard deviation $\sigma = 1.2$.

When using one particular random number generator seed, the 50-sample mean is 0.931 and the variance is $1.0692^2 = 1.143$. Figure 4.6 shows the 50

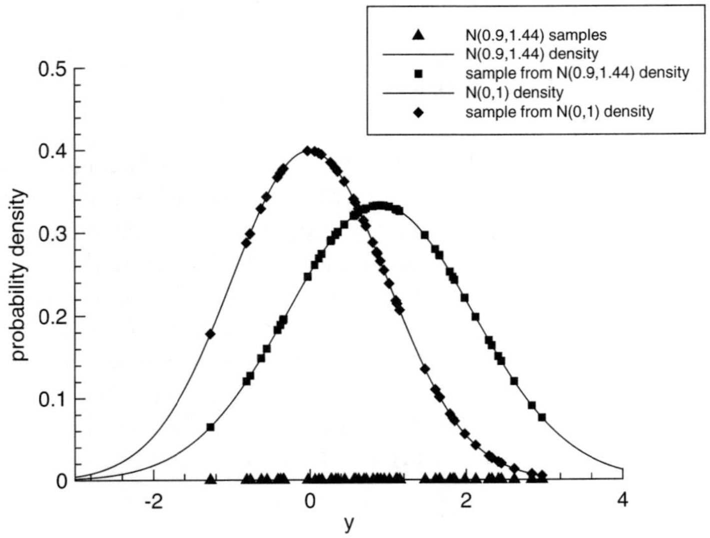

FIGURE 4.6: 50 $N(0.9,1.44)$ samples and Gaussian density functions for true and alternate mean/variance.

individual samples plotted as triangles on the y-axis. Also shown are two Gaussian density curves: one using the true process mean and variance, $N(0.9,1.44)$, and the other using hypothetical parameters, $N(0,1)$. Since the measurement noise samples are uncorrelated, the likelihood function of equation (4.4-6) for a particular value of x is computed as the product of the Gaussian probability densities corresponding to the hypothesized values of x and σ. Equivalently the log likelihood function is computed as the sum of the natural logarithms of these Gaussian probability densities, or from equation (4.4-7),

$$\ln L(\mathbf{y} \mid x,\sigma^2) = \frac{1}{2}\left(m\ln(2\pi) + m\ln(\sigma^2) + \sum_{i=1}^{m} (y_i - x)^2 / \sigma^2 \right).$$

The $N(0.9,1.44)$ probability densities corresponding to the random y-samples are shown in Figure 4.6 as squares. Also shown (as diamonds) are the densities corresponding to the y-samples for the alternate mean/variance hypothesis, $N(0,1)$. It is apparent that the sample densities for the $N(0,1)$ curve have more small values than the samples for the $N(0.9,1.44)$ curve, so it can be concluded that the $N(0,1)$ hypothesis is not optimum. In fact, the log likelihood for the $N(0.9,1.44)$ hypothesis is found to be –74.530, and for the $N(0,1)$ hypothesis it is –95.632.

The "optimal" estimates of x and σ are found by adjusting these values until a maximum of the likelihood function is found. Because the likelihood function is a product of probability densities, samples in the "tails" of the distribution have more impact on the solution than those near the peak. Least-squares estimation also has this property.

The optimum parameter values can also be found by setting the gradient in equation (4.4-10) to zero. In this case the parameter vector is $\mathbf{x} = \begin{bmatrix} x & \sigma^2 \end{bmatrix}^T$. Using equation (4.4-10) and the above equation for $\ln L(\mathbf{y} \mid x,\sigma^2)$, we find that

$$\frac{\partial J}{\partial x} = \frac{1}{\sigma^2}\sum_{i=1}^{m}(y_i - x)$$
$$\frac{\partial J}{\partial \sigma^2} = -\frac{m}{2\sigma^2} + \frac{1}{2\sigma^4}\sum_{i=1}^{m}(y_i - x)^2 .$$

Setting these gradients to zero gives

$$\hat{x} = \frac{1}{m}\sum_{i=1}^{m} y_i$$
$$\hat{\sigma}^2 = \frac{1}{m}\sum_{i=1}^{m}(y_i - x)^2$$

Notice that the MLE of x is the mean of the samples, which does not depend on σ^2. However, the MLE for σ^2 depends on the *true* value of x, which is generally unknown. We can substitute the estimated $\hat{x}$ in the equation for $\hat{\sigma}^2$, but since $\hat{x}$ is computed from the y-samples, $\hat{\sigma}^2$ will be smaller than the true σ^2. This problem is corrected by modifying the divisor to account for the loss of one degree-of-freedom in computing $\hat{x}$ from the measurements. The result is the well-known equation for the sample variance:

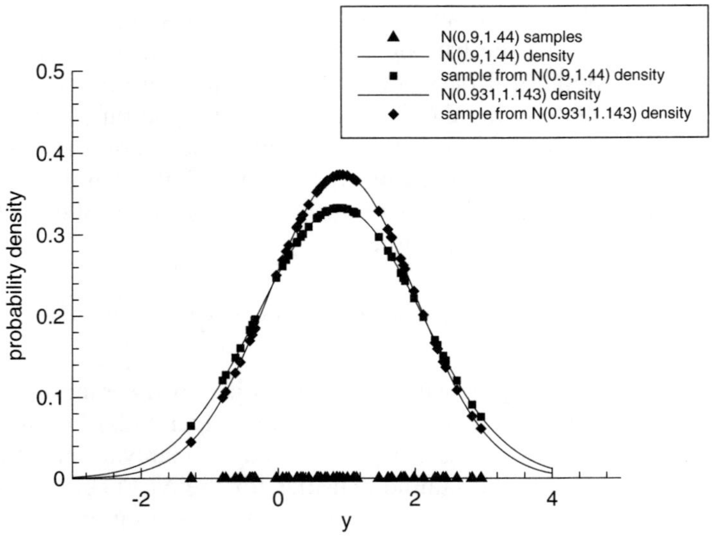

FIGURE 4.7: 50 $N(0.9,1.44)$ samples and Gaussian density functions for true and MLE mean/variance.

$$\hat{\sigma}^2 = \frac{1}{(m-1)}\sum_{i=1}^{m}(y_i - \hat{x})^2$$

For our 50-sample problem, the ML estimates are $\hat{x} = 0.931$ and $\hat{\sigma}^2 = 1.143$. As expected, $\ln L(\mathbf{y} \mid \hat{x}, \hat{\sigma}^2) = -73.792$ is slightly greater than the log likelihood evaluated at the process "truth," $\ln L(\mathbf{y} \mid 0.9, 1.44) = -74.530$. This shows that estimates (of any kind) will generally not match truth because of the effects of measurement noise.

Figure 4.7 shows the true and MLE model probability densities and the samples. Notice that because the $N(0.931,1.143)$ model has a slightly smaller width, the peak density is slightly greater. If σ^2 was assumed to be known—possibly because it was obtained from a much larger set of samples—$\hat{x}$ would not change (because it does not depend on σ^2) and the shape of the $N(0.931,1.143)$ density would match that of the true $N(0.9,1.143)$ density.

Bounds on the $\hat{x}$ error variance can be computed using equation (4.4-15) for the Fisher information matrix:

$$\mathbf{F} = \left(\frac{\partial \bar{\mathbf{y}}}{\partial \mathbf{x}}\right)^T \mathbf{P}_{yy}^{-1} \frac{\partial \bar{\mathbf{y}}}{\partial \mathbf{x}} = \frac{m}{\sigma^2}.$$

Hence

$$E[(\hat{x} - x)^2] = \frac{\sigma^2}{m}$$

which for this 50-sample problem gives $\sqrt{E[(\hat{x} - x)^2]} = 0.170$. The actual error of $0.931 - 0.900 = 0.031$ is considerably smaller than the expected standard deviation for this particular set of samples.

We leave computation of the error variance for $\hat{\sigma}^2$ as an exercise for the reader.

4.4.3 Maximum *A Posteriori*

The *maximum a posteriori* (MAP) estimator is similar to the ML estimator in that it uses conditional probability densities, but it also uses the prior probability density for the estimated parameters **x**. MAP is a form of Bayesian estimation because it starts with the Bayes conditional density

$$p_{\mathbf{X}|\mathbf{Y}}(\mathbf{x}\,|\,\mathbf{y}) = \frac{p_{\mathbf{Y}|\mathbf{X}}(\mathbf{y}\,|\,\mathbf{x})\,p_{\mathbf{X}}(\mathbf{x})}{p_{\mathbf{Y}}(\mathbf{y})}. \tag{4.4-16}$$

As with ML, the solution involves logarithms. The MAP estimate is the value of **x** maximizing

$$\ln p_{\mathbf{X}|\mathbf{Y}}(\mathbf{x}\,|\,\mathbf{y}) = \ln p_{\mathbf{Y}|\mathbf{X}}(\mathbf{y}\,|\,\mathbf{x}) + \ln p_{\mathbf{X}}(\mathbf{x}) - \ln p_{\mathbf{Y}}(\mathbf{y})$$

or minimizing the cost function

$$J = -\ln p_{\mathbf{X}|\mathbf{Y}}(\mathbf{x}\,|\,\mathbf{y}). \tag{4.4-17}$$

The solution is defined by $\partial J/\partial \mathbf{x} = \mathbf{0}^T$. Since $\ln p_{\mathbf{Y}}(\mathbf{y})$ is not a function of **x**, the solution is

$$\frac{\partial J}{\partial \mathbf{x}} = -\frac{\partial \ln p_{\mathbf{Y}|\mathbf{X}}(\mathbf{y}\,|\,\mathbf{x})}{\partial \mathbf{x}} - \frac{\partial \ln p_{\mathbf{X}}(\mathbf{x})}{\partial \mathbf{x}} = \mathbf{0}^T. \tag{4.4-18}$$

For Gaussian densities with

$$p_{\mathbf{X}}(\mathbf{x}) = \frac{1}{(2\pi)^{n/2}|\mathbf{P}_{xx}|^{1/2}} \exp\left[-\frac{1}{2}(\mathbf{x}-\bar{\mathbf{x}})^T \mathbf{P}_{xx}^{-1}(\mathbf{x}-\bar{\mathbf{x}})\right]$$

and

$$p_{\mathbf{Y}|\mathbf{X}}(\mathbf{y}\,|\,\mathbf{x}) = \frac{1}{(2\pi)^{m/2}|\mathbf{P}_{yy}|^{1/2}} \exp\left[-\frac{1}{2}(\mathbf{y}-\bar{\mathbf{y}}(\mathbf{x}))^T \mathbf{P}_{yy}^{-1}(\mathbf{y}-\bar{\mathbf{y}}(\mathbf{x}))\right],$$

then

$$\boxed{J = \frac{1}{2}\begin{pmatrix} (m+n)\ln(2\pi) + \ln|\mathbf{P}_{yy}| + \ln|\mathbf{P}_{xx}| \\ +(\mathbf{y}-\bar{\mathbf{y}}(\mathbf{x}))^T \mathbf{P}_{yy}^{-1}(\mathbf{y}-\bar{\mathbf{y}}(\mathbf{x})) + (\mathbf{x}-\bar{\mathbf{x}})^T \mathbf{P}_{xx}^{-1}(\mathbf{x}-\bar{\mathbf{x}}) \\ +2\ln p_{\mathbf{Y}}(\mathbf{y}) \end{pmatrix}} \tag{4.4-19}$$

so

$$\boxed{\frac{\partial J}{\partial \mathbf{x}} = \mathbf{0}^T = -\left((\mathbf{y}-\bar{\mathbf{y}}(\mathbf{x}))^T \mathbf{P}_{yy}^{-1}\frac{\partial \bar{\mathbf{y}}}{\partial \mathbf{x}} + (\mathbf{x}-\bar{\mathbf{x}})^T \mathbf{P}_{xx}^{-1}\right),} \tag{4.4-20}$$

or using $\bar{\mathbf{y}}(\mathbf{x}) = \mathbf{H}\mathbf{x}$ (not $\bar{\mathbf{y}} = \mathbf{H}\bar{\mathbf{x}}$),

$$(\mathbf{H}^T\mathbf{P}_{yy}^{-1}\mathbf{H} + \mathbf{P}_{xx}^{-1})\hat{\mathbf{x}} = (\mathbf{H}^T\mathbf{P}_{yy}^{-1}\mathbf{y} + \mathbf{P}_{xx}^{-1}\bar{\mathbf{x}}). \tag{4.4-21}$$

This result using Gaussian densities is again the same as for the minimum variance estimate equation (4.3-31) or (4.3-6). Computation of the MAP estimate is somewhat more difficult for non-Gaussian density functions and nonlinear measurement models. J can also have multiple minima. The MAP estimate is the mode of the *a posteriori* probability density function while the minimum variance estimate is the mean of the *a posteriori* density.

A lower bound on the MAP estimate error covariance is similar to the bound for MLE, but involves the joint probability of both **y** and **x**:

$$E\left[(\mathbf{x}-\hat{\mathbf{x}})(\mathbf{x}-\hat{\mathbf{x}})^T\right] \geq \left[E\left[\left(\frac{\partial \ln p_{\mathbf{Y},\mathbf{X}}(\mathbf{y},\mathbf{x})}{\partial \mathbf{x}}\right)\left(\frac{\partial \ln p_{\mathbf{Y},\mathbf{X}}(\mathbf{y},\mathbf{x})}{\partial \mathbf{x}}\right)^T\right]\right]^{-1}$$
$$= \left[-E\left[\frac{\partial^2 \ln p_{\mathbf{Y},\mathbf{X}}(\mathbf{y},\mathbf{x}))}{\partial \mathbf{x}\partial \mathbf{x}^T}\right]\right]^{-1}.$$

Since $\ln p_{\mathbf{Y},\mathbf{X}}(\mathbf{y},\mathbf{x}) = \ln p_{\mathbf{Y}|\mathbf{X}}(\mathbf{y} \mid \mathbf{x}) + \ln p_{\mathbf{X}}(\mathbf{x})$, the covariance bound can be written as

$$E\left[(\mathbf{x}-\hat{\mathbf{x}})(\mathbf{x}-\hat{\mathbf{x}})^T\right] \geq \left[E\left[\left(\frac{\partial \ln p_{\mathbf{Y}|\mathbf{X}}(\mathbf{y}\mid\mathbf{x})}{\partial \mathbf{x}}\right)\left(\frac{\partial \ln p_{\mathbf{Y}|\mathbf{X}}(\mathbf{y}\mid\mathbf{x})}{\partial \mathbf{x}}\right)^T + \left(\frac{\partial \ln p_{\mathbf{X}}(\mathbf{x})}{\partial \mathbf{x}}\right)\left(\frac{\partial \ln p_{\mathbf{X}}(\mathbf{x})}{\partial \mathbf{x}}\right)^T\right]\right]^{-1}$$
$$= -\left[E\left[\frac{\partial^2 \ln p_{\mathbf{Y}|\mathbf{X}}(\mathbf{y}\mid\mathbf{x})}{\partial \mathbf{x}\partial \mathbf{x}^T} + \frac{\partial^2 \ln p_{\mathbf{X}}(\mathbf{x})}{\partial \mathbf{x}\partial \mathbf{x}^T}\right]\right]^{-1} \quad (4.4\text{-}22)$$

For the linear problem of equation (4.4-21), the error covariance is the same as for the minimum variance solution: $E\left[(\mathbf{x}-\hat{\mathbf{x}})(\mathbf{x}-\hat{\mathbf{x}})^T\right] \geq (\mathbf{H}^T\mathbf{P}_{yy}^{-1}\mathbf{H}+\mathbf{P}_{xx}^{-1})^{-1}$.

Example 4.5: MAP Two-Parameter Model with Rayleigh *A Priori* Distribution

This problem uses a two-state model where the *a priori* distribution is Rayleigh, not Gaussian. The time-dependent measurement model uses initial position and velocity states as in previous examples:

$$\begin{bmatrix} y_1 \\ y_2 \\ \vdots \\ y_m \end{bmatrix} = \begin{bmatrix} 1 & t_1 \\ 1 & t_2 \\ \vdots & \vdots \\ 1 & t_m \end{bmatrix} \begin{bmatrix} x_1 \\ x_2 \end{bmatrix} + \begin{bmatrix} r_1 \\ r_2 \\ \vdots \\ r_m \end{bmatrix}$$

or

$$\mathbf{y} = \mathbf{H}\mathbf{x} + \mathbf{r}.$$

The measurement noise **r** is assumed to be Gaussian with $E[\mathbf{r}] = \mathbf{0}$ and $\mathbf{P}_{yy} = E[\mathbf{r}\mathbf{r}^T] = \sigma^2\mathbf{I}$ where we set $\sigma = 1.0$. The two **x** states are assumed to have independent Rayleigh distributions with density

$$p_X(x) = \begin{cases} \dfrac{x}{\sigma_x^2}\exp(-x^2/2\sigma_x^2) & x \geq 0 \\ 0 & \text{otherwise} \end{cases}$$

Figure 4.8 shows the probability density functions for $\sigma_{x1} = 1.0$ and $\sigma_{x2} = 0.5$ used in this example. Rayleigh distributions are often used to model the amplitude of

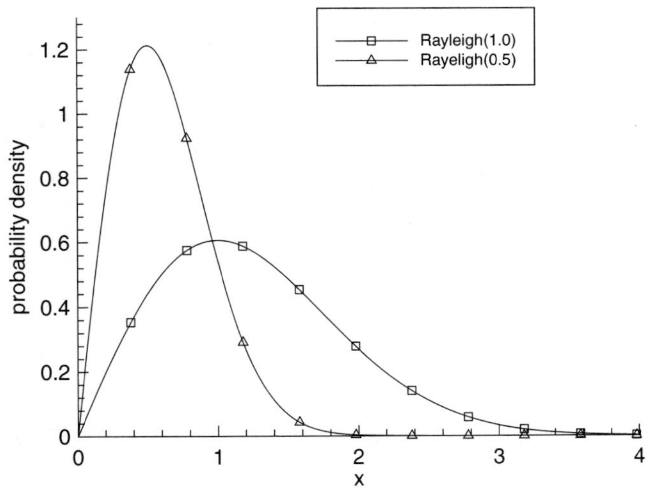

FIGURE 4.8: Rayleigh distributions.

a random communication signal, although there is no physical basis for its use in this example. It was simply selected as one type of nonsymmetric distribution to make the problem more interesting.

The numerator of the *a posteriori* probability density (eq. 4.4-16) is

$$p_{\mathbf{Y|X}}(\mathbf{y}\,|\,\mathbf{x})p_{\mathbf{X}}(\mathbf{x})=\frac{x_1x_2}{(2\pi)^{m/2}\sigma^m\sigma_{x1}^2\sigma_{x2}^2}\exp\left(-\frac{1}{2\sigma^2}(\mathbf{y}-\mathbf{Hx})^T(\mathbf{y}-\mathbf{Hx})-\frac{x_1^2}{2\sigma_{x1}^2}-\frac{x_2^2}{2\sigma_{x2}^2}\right).\qquad x_1\ge 0,\ x_2\ge 0$$

The denominator of equation (4.4-16),

$$p_{\mathbf{Y}}(\mathbf{y})=\int_{-\infty}^{\infty}\int_{-\infty}^{\infty}p_{\mathbf{Y|X}}(\mathbf{y}\,|\,x_1,x_2)p_{\mathbf{X}}(x_1,x_2)\,dx_1dx_2,$$

normalizes the fraction so that $\int_{-\infty}^{\infty}\int_{-\infty}^{\infty}p_{\mathbf{X|Y}}(x_1,x_2\,|\,\mathbf{y})\,dx_1dx_2=1$. As noted, $p_{\mathbf{Y}}(\mathbf{y})$ is not a function of $\mathbf{x}$ and is not required to obtain a MAP solution for $\hat{\mathbf{x}}$.

The cost function for this problem is

$$\begin{aligned}J&=-\ln p_{\mathbf{X|Y}}(\mathbf{x}\,|\,\mathbf{y})\\&=-\ln x_1-\ln x_2+m\ln\sqrt{2\pi}+m\ln\sigma+\ln(\sigma_{x1}^2)+\ln(\sigma_{x2}^2)\\&\quad+\left(\frac{1}{2\sigma^2}\sum_{i=1}^{m}(y_i-x_1-x_2t_i)^2+\frac{x_1^2}{2\sigma_{x1}^2}+\frac{x_2^2}{2\sigma_{x2}^2}\right)+\ln p_{\mathbf{Y}}(\mathbf{y})\end{aligned}$$

for $x_1\ge 0, x_2\ge 0$. In this example we assume that σ_{x1}^2, σ_{x1}^2, and σ^2 are known, so x_1 and x_2 are the only variables to be estimated. The partial derivatives of the cost function with respect to $\mathbf{x}$ are:

$$\begin{aligned}\frac{\partial J}{\partial x_1}&=-\frac{1}{x_1}+\frac{x_1}{\sigma_{x1}^2}-\frac{1}{\sigma^2}\sum_{i=1}^{m}(y_i-x_1-t_ix_2)\\\frac{\partial J}{\partial x_2}&=-\frac{1}{x_2}+\frac{x_2}{\sigma_{x2}^2}-\frac{1}{\sigma^2}\sum_{i=1}^{m}t_i(y_i-x_1-t_ix_2)\end{aligned}.$$

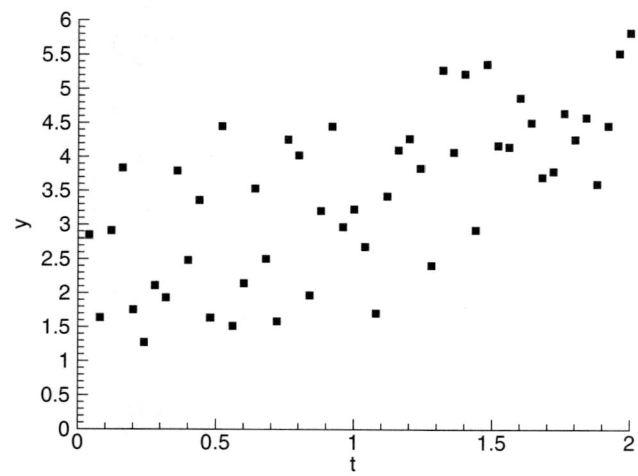

FIGURE 4.9: Measurements for Example 4.5.

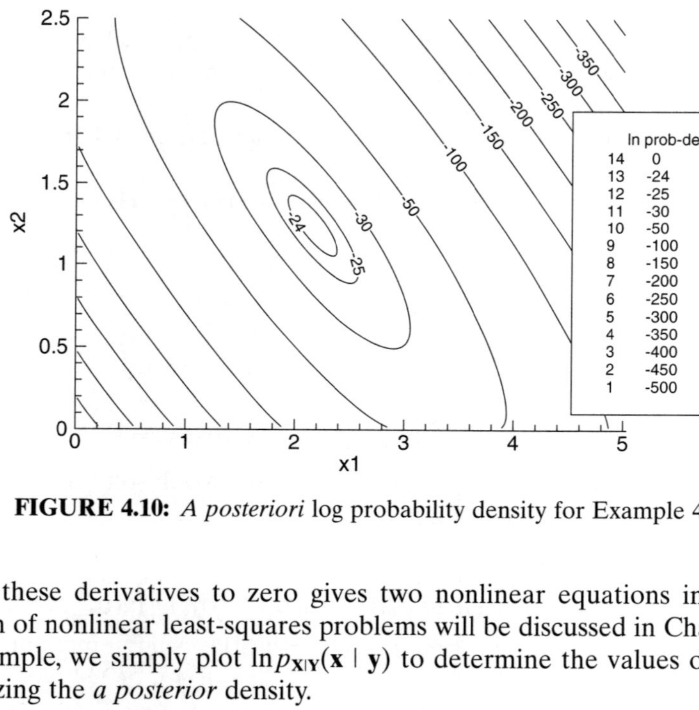

FIGURE 4.10: *A posteriori* log probability density for Example 4.5.

Setting these derivatives to zero gives two nonlinear equations in x_1 and x_2. Solution of nonlinear least-squares problems will be discussed in Chapter 7. For this example, we simply plot $\ln p_{\mathbf{X|Y}}(\mathbf{x} \mid \mathbf{y})$ to determine the values of x_1 and x_2, maximizing the *a posterior* density.

The simulated data for this example used the previously listed Rayleigh density parameters ($\sigma_{x1} = 1.0, \sigma_{x2} = 0.5$) and noise standard deviation ($\sigma = 1$) with $x_1 = 2.0, x_2 = 1.4$. These values of x_1 and x_2 were selected to be in the tails of the Rayleigh distributions so that $p_{\mathbf{x}}(\mathbf{x})$ influenced the solution. Fifty random samples of $\mathbf{y}$ were generated over the interval $0 \le t_i < 2$ as shown in Figure 4.9.

Figure 4.10 shows contours of $\ln p_{\mathbf{X|Y}}(\mathbf{x} \mid \mathbf{y})$ versus x_1 and x_2. As expected, the peak of the $\ln p_{\mathbf{X|Y}}(\mathbf{x} \mid \mathbf{y})$ contours is broad because the magnitude of random measurement noise is comparable to the trend in the data (Fig. 4.9). The maximum of $\ln p_{\mathbf{X|Y}}(\mathbf{x} \mid \mathbf{y})$ occurs at $x_1 = 2.157$ and $x_2 = 1.239$, which are close to the true values of 2.0 and 1.4, respectively. Even with the relatively large measurement noise, the influence of the measurements is much greater than that of $\ln p_{\mathbf{X}}(\mathbf{x})$. Also notice the negative slope of the ellipse semimajor axes in Figure 4.10. This indicates that the errors in $\hat{x}_1$ and $\hat{x}_2$ are somewhat negatively correlated.

The bound on estimate uncertainty is obtained from the inverse of the Fisher information matrix

$$-E\left[\frac{\partial^2 \ln p_{\mathbf{X}|\mathbf{Y}}(\mathbf{x}\,|\,\mathbf{y})}{\partial\mathbf{x}\partial\mathbf{x}^T}\right]=\begin{bmatrix} 1/x_1+1/\sigma_{x1}^2+m/\sigma^2 & \sum_{i=1}^{m} t_i/\sigma^2 \\ \sum_{i=1}^{m} t_i/\sigma^2 & 1/x_2+1/\sigma_{x2}^2+\sum_{i=1}^{m} t_i^2/\sigma^2 \end{bmatrix}.$$

Bounds on the estimate error standard deviations are computed as the square roots of diagonals of

$$\left(-E\left[\frac{\partial^2 \ln p_{\mathbf{X}|\mathbf{Y}}(\mathbf{x}\,|\,\mathbf{y})}{\partial\mathbf{x}\partial\mathbf{x}^T}\right]\right)^{-1},$$

which are 0.252 and 0.211, respectively. Hence the estimate errors of $\hat{x}_1 - x_1 = 0.157$ and $\hat{x}_2 - x_2 = -0.161$ are well within the error standard deviations. The errors are negatively correlated, as expected. The influence of the prior density $\ln p_{\mathbf{X}}(\mathbf{x})$ is more evident in $\hat{x}_2$ than in $\hat{x}_1$ since x_2 is further in the tails of the Rayleigh distribution; that is,

$$p_{x2}(1.40)=\frac{1.40}{0.5^2}\exp(-1.40^2/(2\cdot 0.5^2)=0.111$$

is only about 9% of the peak value at 0.5 (see Fig. 4.8).

4.5 SUMMARY OF LINEAR ESTIMATION APPROACHES

We have derived and analyzed six different approaches to "least squares" and related estimation problems. Some methods were shown to be equivalent for linear models and Gaussian densities. Table 4.2 summarizes the characteristics. In trying to decide which technique to use for a given problem, we offer a few suggestions:

1. For linear models and Gaussian densities the Bayesian, minimum variance/MMSE and MAP algorithms produce the same result. A Bayesian solution (eq. 4.3-4 or 4.3-6) should work for most problems. If prior information is not included, simply set the diagonals of $\mathbf{P}_{xx}$ or $\mathbf{P}_a$ to a large number, although this can cause numerical problems for some solution techniques.
2. The Bayesian algorithm can also be used for nonlinear measurement models if the $\mathbf{H}$ matrix is defined as the Jacobian evaluated at the state $\hat{\mathbf{x}}_i$ for iteration i of a nonlinear optimization algorithm. (This will be discussed in Chapter 7.)
3. For simple curve fitting, use an unweighted least-squares method or use a weighted least-squares algorithm with equal weighting.

TABLE 4.2: Summary of "Least-Squares" Methods

Approach	Measurement Weighting	Inclusion of *A Priori* Statistics	Assumes Gaussian Densities	Metric Minimized/ Maximized	Cost Function Is Quadratic?
Unweighted least squares (BLUE)	No	No	No	Minimize measurement residual sum-of-squares	Yes
Weighted least squares (BLUE)	Yes	No	No	Minimize weighted measurement residual sum-of-squares	Yes
Bayes (conditional mean)	Yes	Yes	No but some solutions do	Estimate is mean of *a posteriori* probability density	Not necessarily
Minimum variance/ MMSE (same as conditional mean)	Yes	Yes	No but some solutions do	Minimize posterior state error variance (mean of *a posteriori* probability density)	Yes
Maximum likelihood (ML) estimation	Yes	No	Usually	Finds state that maximizes conditional probability of observed data	Not necessarily
Maximum *A Posteriori* (MAP)	Yes	Yes	Often	Finds mode of *a posteriori* probability density	Not necessarily

4. For problems where the measurement probability density (and thus the covariance $\mathbf{P}_{yy}$) is a function of parameters that are to be estimated as part of the state $\mathbf{x}$, use ML or MAP algorithms.
5. For problems where the probability density of the measurement or state is not Gaussian, use ML or MAP algorithms.
6. For linear or nonlinear problems where data are processed per test for the purposes of computing the mean and covariance across tests, use ML (or weighted least squares) because this will not reuse prior information.

CHAPTER 5

LINEAR LEAST-SQUARES ESTIMATION: SOLUTION TECHNIQUES

This chapter discusses numerical methods for solving least-squares problems, sensitivity to numerical errors, and practical implementation issues. These topics necessarily involve discussions of matrix inversion methods (Gauss-Jordan elimination, Cholesky factorization), orthogonal transformations, and iterative refinement of solutions. The orthogonalization methods (Givens rotations, Householder transformations, and Gram-Schmidt orthogonalization) are used in the QR and Singular Value Decomposition (SVD) methods for computing least-squares solutions. The concepts of observability, numerical conditioning, and pseudo-inverses are also discussed. Examples demonstrate numerical accuracy and computational speed issues.

5.1 MATRIX NORMS, CONDITION NUMBER, OBSERVABILITY, AND THE PSEUDO-INVERSE

Prior to addressing least-squares solution methods, we review four matrix-vector properties that are useful when evaluating and comparing solutions.

5.1.1 Vector-Matrix Norms

The first is property of interest is the norm of a vector or matrix. Norms such as root-sum-squared, maximum value, and average absolute value are often used when discussing scalar variables. Similar concepts apply to vectors. The Hölder p-norms for vectors are defined as

$$\|\mathbf{x}\|_p = \left(\sum_{i=1}^{n} |x_i|^p \right)^{1/p}. \tag{5.1-1}$$

The most important of these p-norms are

Advanced Kalman Filtering, Least-Squares and Modeling: A Practical Handbook, by Bruce P. Gibbs

$$\|\mathbf{x}\|_1 = \sum_{i=1}^{n} |x_i| \tag{5.1-2}$$

$$\|\mathbf{x}\|_2 = \left(\sum_{i=1}^{n} x_i^2 \right)^{1/2} \tag{5.1-3}$$

$$\|\mathbf{x}\|_\infty = \max_i |x_i|. \tag{5.1-4}$$

The l_2-norm $\|\mathbf{x}\|_2$, also called the Euclidian or root-sum-squared norm, is generally the most often used. To apply these concepts to matrices, the elements of matrix $\mathbf{A}$ could be treated as a vector. For an l_2-like norm, this leads to the Frobenius norm

$$\|\mathbf{A}\|_F = \left(\sum_{i=1}^{m} \sum_{j=1}^{n} A_{ij}^2 \right)^{1/2}. \tag{5.1-5}$$

One can also compute *induced matrix norms* based on the ability of matrix $\mathbf{A}$ to modify the magnitude of a vector, that is,

$$\|\mathbf{A}\| = \max_{\|\mathbf{x}\|=1} \left(\frac{\|\mathbf{Ax}\|}{\|\mathbf{x}\|} \right) \tag{5.1-6}$$

where the norms for $\mathbf{A}$, $\mathbf{Ax}$, and $\mathbf{x}$ need not be the same, but usually are. The l_1-norm is

$$\|\mathbf{A}\|_1 = \max_j \|\mathbf{a}_{:j}\|_1$$

where $\mathbf{a}_{:j}$ is the j-th column of $\mathbf{A}$, and the l_∞-norm is

$$\|\mathbf{A}\|_\infty = \max_i \|\mathbf{a}_{i:}\|_1$$

where $\mathbf{a}_{i:}$ is the i-th row of $\mathbf{A}$. Unfortunately it is more difficult to compute an l_2-norm based on this definition than a Frobenius norm. It turns out that $\|\mathbf{A}\|_2$ is equal to the square root of the maximum eigenvalue of $\mathbf{A}^T\mathbf{A}$—or equivalently the largest singular value of $\mathbf{A}$—but it is not always convenient to compute the eigen decomposition or SVD.

Two properties of matrix norms will be of use in the next sections. From definition (eq. 5.1-6) it follows that

$$\|\mathbf{Ax}\| \le \|\mathbf{A}\| \|\mathbf{x}\|, \tag{5.1-7}$$

and this can be extended to products of matrices:

$$\|\mathbf{AB}\| \le \|\mathbf{A}\| \|\mathbf{B}\|. \tag{5.1-8}$$

However, the l_2 and Frobenius matrix norms are unchanged by orthogonal transformations, that is,

$$\left. \begin{aligned} \|\mathbf{AB}\|_2 &= \|\mathbf{A}\|_2 \\ \|\mathbf{BA}\|_2 &= \|\mathbf{A}\|_2 \end{aligned} \right\} \quad \text{if} \quad \mathbf{B}^T\mathbf{B} = \mathbf{I}. \tag{5.1-9}$$

More information on matrix-vector norms may be found in Dennis and Schnabel (1983), Björck (1996), Lawson and Hanson (1974), Golub and van Loan (1996), and Stewart (1998).

5.1.2 Matrix Pseudo-Inverse

When matrix $\mathbf{A}$ is square and nonsingular, the equation $\mathbf{Ax} = \mathbf{b}$ can be solved for $\mathbf{x}$ using matrix inversion ($\mathbf{x} = \mathbf{A}^{-1}\mathbf{b}$). When $\mathbf{A}$ is rectangular or singular, $\mathbf{A}$ does not have an inverse, but Penrose (1955) defined a *pseudo-inverse* $\mathbf{A}^{\#}$ uniquely determined by four properties:

$$\begin{aligned} \mathbf{A}\mathbf{A}^{\#}\mathbf{A} &= \mathbf{A} \\ \mathbf{A}^{\#}\mathbf{A}\mathbf{A}^{\#} &= \mathbf{A}^{\#} \\ (\mathbf{A}\mathbf{A}^{\#})^T &= \mathbf{A}\mathbf{A}^{\#} \\ (\mathbf{A}^{\#}\mathbf{A})^T &= \mathbf{A}^{\#}\mathbf{A} \end{aligned} \quad (5.1\text{-}10)$$

This Moore-Penrose pseudo-inverse is of interest in solving least-squares problems because the *normal equation* solution $\hat{\mathbf{x}} = (\mathbf{H}^T\mathbf{H})^{-1}\mathbf{H}^T\mathbf{y}$ for $\mathbf{y} = \mathbf{Hx} + \mathbf{r}$ uses a pseudo-inverse; that is, $\mathbf{H}^{\#} = (\mathbf{H}^T\mathbf{H})^{-1}\mathbf{H}^T$ is the pseudo-inverse of $\mathbf{H}$. Notice that when $\mathbf{H}$ has more rows than columns (the normal overdetermined least-squares case), the pseudo-inverse $(\mathbf{H}^T\mathbf{H})^{-1}\mathbf{H}^T$ does not provide an exact solution—only the closest solution in a least-squares sense. This pseudo-inverse can be also written using the SVD of $\mathbf{H}$

$$\mathbf{H} = \mathbf{U}\mathbf{S}\mathbf{V}^T$$

where $\mathbf{U}$ is an $m \times m$ orthogonal matrix, $\mathbf{V}$ is an $n \times n$ orthogonal matrix, and $\mathbf{S} = [\mathbf{S}_1 \quad \mathbf{0}]^T$ is an $m \times n$ upper diagonal matrix of singular values ($\mathbf{S}_1$ is an $n \times n$ diagonal matrix). The pseudo-inverse is

$$\begin{aligned} \mathbf{H}^{\#} &= (\mathbf{V}\mathbf{S}^T\mathbf{U}^T\mathbf{U}\mathbf{S}\mathbf{V}^T)^{-1}\mathbf{V}\mathbf{S}^T\mathbf{U}^T \\ &= \mathbf{V}(\mathbf{S}^T\mathbf{S})^{-1}\mathbf{V}^T\mathbf{V}\mathbf{S}^T\mathbf{U}^T \\ &= \mathbf{V}(\mathbf{S}^T\mathbf{S})^{-1}\mathbf{S}^T\mathbf{U}^T \\ &= \mathbf{V}[\mathbf{S}_1^{\#} \quad \mathbf{0}]\mathbf{U}^T \end{aligned} \quad (5.1\text{-}11)$$

where $\mathbf{S}_1^{\#}$ is the pseudo-inverse of $\mathbf{S}_1$. This equation shows that a pseudo-inverse can be computed even when $\mathbf{H}^T\mathbf{H}$ is singular (the underdetermined least-squares case). Singular values that are exactly zero are replaced with zeroes in the same locations when forming $\mathbf{S}_1^{\#}$. The term "pseudo-inverse" often implies an inverse computed for a rank deficient (rank < min(m,n)) $\mathbf{H}$ matrix rather than a full-rank inverse such as $(\mathbf{H}^T\mathbf{H})^{-1}\mathbf{H}^T$. In this case the pseudo-inverse provides the minimal norm solution, as will be discussed in Section 5.8.

5.1.3 Condition Number

Another useful matrix property describes the sensitivity of $\mathbf{x}$ in

$$\mathbf{Ax} = \mathbf{b} \quad (5.1\text{-}12)$$

to perturbations in $\mathbf{A}$ and $\mathbf{b}$; that is, the magnitude of $\delta\mathbf{x}$ due to perturbations $\delta\mathbf{A}$ and $\delta\mathbf{b}$ in

$$(\mathbf{A}+\delta\mathbf{A})(\mathbf{x}+\delta\mathbf{x})=\mathbf{b}+\delta\mathbf{b}. \tag{5.1-13}$$

Note that $\mathbf{A}$ can be a square matrix, but this is not a requirement. The following analysis of the problem is from Dennis and Schnabel (1983, p. 52), which is based on Wilkinson's (1965) backward error analysis. First consider the effects of perturbation $\delta\mathbf{b}$ such that $\mathbf{x}+\delta\mathbf{x}=\mathbf{A}^{-1}(\mathbf{b}+\delta\mathbf{b})$ or

$$\delta\mathbf{x}=\mathbf{A}^{-1}\delta\mathbf{b}. \tag{5.1-14}$$

This equation applies for a square nonsingular $\mathbf{A}$, but it also applies if $\mathbf{A}^{-1}$ is interpreted as a pseudo-inverse of a full-rank rectangular $\mathbf{A}$. Using equation (5.1-7),

$$\|\delta\mathbf{x}\|\le\|\mathbf{A}^{-1}\|\|\delta\mathbf{b}\|. \tag{5.1-15}$$

Equation (5.1-7) can also be used on equation (5.1-12) to obtain

$$\frac{1}{\|\mathbf{x}\|}\le\frac{\|\mathbf{A}\|}{\|\mathbf{b}\|} \tag{5.1-16}$$

Merging the two equations yields

$$\frac{\|\delta\mathbf{x}\|}{\|\mathbf{x}\|}\le\|\mathbf{A}\|\|\mathbf{A}^{-1}\|\frac{\|\delta\mathbf{b}\|}{\|\mathbf{b}\|}. \tag{5.1-17}$$

Similarly for perturbation $\delta\mathbf{A}$, $(\mathbf{A}+\delta\mathbf{A})(\mathbf{x}+\delta\mathbf{x})=\mathbf{A}\mathbf{b}$, or ignoring the $\delta\mathbf{A}\delta\mathbf{x}$ term,

$$\mathbf{A}\delta\mathbf{x}=-\delta\mathbf{A}\mathbf{x}. \tag{5.1-18}$$

Using (5.1-8),

$$\|\delta\mathbf{x}\|\le\|\mathbf{A}^{-1}\|\|\delta\mathbf{A}\|\|\mathbf{x}\| \tag{5.1-19}$$

or

$$\frac{\|\delta\mathbf{x}\|}{\|\mathbf{x}\|}\le\|\mathbf{A}\|\|\mathbf{A}^{-1}\|\frac{\|\delta\mathbf{A}\|}{\|\mathbf{A}\|}. \tag{5.1-20}$$

In both cases the relative sensitivity is bounded by $\|\mathbf{A}\|\|\mathbf{A}\|^{-1}$, which is called a *condition number* of $\mathbf{A}$ and is denoted as

$$\boxed{\kappa_p(\mathbf{A})=\|\mathbf{A}\|_p\|\mathbf{A}^{-1}\|_p} \tag{5.1-21}$$

when using the corresponding l_p induced matrix norm for $\mathbf{A}$. Again these equations still apply if $\mathbf{A}^{-1}$ is replaced with $\mathbf{A}^{\#}$ when $\mathbf{A}$ is rectangular and full rank (rank = min(m,n)). Notice that $\kappa_p(\mathbf{A})$ is unaffected by scaling of $\mathbf{A}$ that multiplies the entire matrix; that is, $\kappa_p(\alpha\mathbf{A})=\kappa_p(\mathbf{A})$. However, $\kappa_p(\mathbf{A})$ is affected by scaling that affects only parts of $\mathbf{A}$.

The effects of numerical errors (due to computational round-off) on minimal norm solutions for $\mathbf{x}$ in $\mathbf{A}\mathbf{x}=\mathbf{b}$ are analyzed using $\kappa_p(\mathbf{A})$. It can be shown that most numerical solution techniques limit $\|\delta\mathbf{A}\|/\|\mathbf{A}\|$ to a constant (c) times the computer numerical precision (ε), so that

$$\frac{\|\delta\mathbf{x}\|}{\|\mathbf{x}\|} \leq c\varepsilon\kappa(A).$$

Notice that $\kappa_p(\mathbf{A}) = 1$ when $\mathbf{A} = \alpha\mathbf{I}$, and $\kappa_p(\mathbf{A})$ increases as $\mathbf{A}$ approaches singularity.

Example 5.1: Matrix Condition Number for Rank-2 Matrix

First consider

$$\mathbf{A} = \alpha\begin{bmatrix} 1 & 0 \\ 0 & 1 \end{bmatrix}$$

for which $\kappa_1(\mathbf{A}) = \alpha/\alpha = 1$. We use the l_1-norm for convenience. This is an example of a well-conditioned matrix. Now consider

$$\mathbf{A} = \begin{bmatrix} 1 & \beta \\ \beta & 1 \end{bmatrix}$$

for which

$$\mathbf{A}^{-1} = \frac{1}{1-\beta^2}\begin{bmatrix} 1 & -\beta \\ -\beta & 1 \end{bmatrix}.$$

Assuming that $|\beta| < 1$ (which is a condition for $\mathbf{A}$ to be nonsingular), $\|\mathbf{A}\|_1 = 1$, and $\|\mathbf{A}^{-1}\|_1 = 1/(1 - \beta^2)$, so that

$$\kappa_1(\mathbf{A}) = 1/(1-\beta^2).$$

Hence as $|\beta| \to 1$, $\kappa_1(\mathbf{A}) \to \infty$. For example, if $\beta = 0.999$, $\kappa_1(\mathbf{A}) = 500.25$. It is obvious that solution of $\mathbf{x}$ in equation $\mathbf{x} = \mathbf{A}^{-1}\mathbf{b}$ will become increasingly sensitive to errors in either $\mathbf{A}$ or $\mathbf{b}$ as $|\beta| \to 1$.

As a third example, consider

$$\mathbf{A} = \begin{bmatrix} 1 & 0 \\ 0 & \beta \end{bmatrix}$$

for which $\kappa_1(\mathbf{A}) = \beta$ if $|\beta| > 1$. $\mathbf{A}$ is clearly nonsingular, but can still have a large condition number when β is large. This demonstrates a problem in treating condition numbers as a measure of "closeness to singularity" or "non-uniqueness." For $\kappa_p(\mathbf{A})$ to be used as a test for *observability* (a term defined in the next section), matrix $\mathbf{A}$ should be scaled so that the diagonal elements are equal. The effects of scaling are often overlooked when condition numbers are used for tests of observability. This issue will be explored further in Section 5.7.

It is generally not convenient or accurate to compute condition numbers by inverting a matrix. Condition numbers can also be computed from the singular values of matrix $\mathbf{A}$. Since $\mathbf{A}$ can be decomposed as $\mathbf{A} = \mathbf{USV}^T$, where $\mathbf{U}$ and $\mathbf{V}$ are orthogonal, and $\mathbf{A}^{-1} = \mathbf{VS}^{-1}\mathbf{U}^T$ when $\mathbf{A}$ is square, then $\|\mathbf{A}\|_2 = \max_i(s_i)$ and $\|\mathbf{A}^{-1}\|_2 = 1/\min_i(s_i)$ where s_i are the singular values in $\mathbf{S}$. Hence

$$\boxed{\kappa_2(\mathbf{A}) = \frac{\max\limits_i(s_i)}{\min\limits_i(s_i)}.} \tag{5.1-22}$$

However, it is computationally expensive to evaluate the SVD, so simpler methods are often used. One such method assumes that the **A** matrix has been transformed to an upper triangular form using an orthogonal transformation; that is, $\mathbf{T}^T\mathbf{A} = \mathbf{U}$ where $\mathbf{T}^T\mathbf{T} = \mathbf{I}$ and **U** is upper triangular. (This *QR* transformation is discussed in Section 5.3.) Since **T** is orthogonal, $\|\mathbf{A}_2\| = \|\mathbf{U}\|_2$ and $\|\mathbf{A}\|_F = \|\mathbf{U}\|_F$. Bierman's Estimation Subroutine Library (ESL) subroutine RINCON.F—used for examples in this chapter—computes the condition number

$$\kappa_F(\mathbf{U}) = \|\mathbf{U}\|_F \|\mathbf{U}^{-1}\|_F \tag{5.1-23}$$

using the Frobenius norm because RINCON explicitly inverts **U**. $\kappa_F(\mathbf{U})$ may be larger than $\kappa_2(\mathbf{U})$, although it is usually quite close.

Computation of $\kappa_2(\mathbf{U})$ without using the SVD depends upon computation of the matrix l_2-norm, which is difficult to compute. Thus alternatives involving the l_1-norm are often used. For the l_1-norm of $n \times n$ matrix **A**,

$$\frac{1}{n}\kappa_1(\mathbf{A}) \le \kappa_1(\mathbf{U}) \le n\kappa_1(\mathbf{A}). \tag{5.1-24}$$

In practice it is normally found that $\kappa_1(\mathbf{U}) \approx \kappa_1(\mathbf{A})$, as will be shown in a later example. An estimate of $\kappa_1(\mathbf{U})$ can be computed using a method developed by Cline, Moler, Stewart, and Wilkinson (1979), and is included in the LINPACK linear algebra library (Dongarra et al. 1979). First, $\|\mathbf{U}\|_1 = \max\limits_j \|\mathbf{u}_{:j}\|_1$ (where $\mathbf{u}_{:j}$ is column j of **U**) is computed. Then a bound on $\|\mathbf{U}^{-1}\|_1$ is computed from the inequality

$$\|\mathbf{U}^{-1}\|_1 \ge \frac{\|\mathbf{U}^{-1}\mathbf{z}\|_1}{\|\mathbf{z}\|_1}$$

for any nonzero **z**. The appropriate **z** is chosen by solving $\mathbf{U}^T\mathbf{z} = \mathbf{e}$, where **e** is a vector of ±1's with the signs chosen to maximize $|z_i|$ for $i = 1, 2, \ldots, n$ in sequence. Then $\mathbf{U}\mathbf{y} = \mathbf{z}$ is solved for **y** (i.e., $\mathbf{y} = \mathbf{U}^{-1}\mathbf{U}^{-T}\mathbf{e}$), and the estimated $\kappa_1(\mathbf{U})$ is evaluated as

$$\hat{\kappa}_1(\mathbf{U}) = \|\mathbf{U}\|_1 \frac{\|\mathbf{y}\|_1}{\|\mathbf{z}\|_1}. \tag{5.1-25}$$

As noted by Dennis and Schnabel (1983, p.55) and Björck (1996, p. 116), even though $\|\mathbf{y}\|_1/\|\mathbf{z}\|_1$ is only a lower bound on $\|\mathbf{U}^{-1}\|_1$, it is usually close. Another $\kappa_1(\mathbf{U})$ estimator developed by Hagar and Higham is found in LAPACK (Anderson et al. 1999). This algorithm does not require transforming **A** to upper triangular form. It only requires computation of products $\mathbf{Ax}$ and $\mathbf{A}^T\mathbf{x}$ where **x** is a vector. Björck (1996, p. 117) claims that the estimates produced by this algorithm are sharper than those of the LINPACK algorithm.

5.1.4 Observability

Previous chapters have generally assumed that the system $\mathbf{y} = \mathbf{Hx} + \mathbf{r}$ is *observable*, but this term has not been clearly defined. Theoretical observability is defined in several ways that are equivalent:

1. The rank of the $m \times n$ matrix $\mathbf{H}$ must be n.
2. $\mathbf{H}$ must have n nonzero singular values.
3. The rows of $\mathbf{H}$ must span the n-dimensional vector space such that any n-vector can be represented as
$$\mathbf{x}^T = \sum_{i=1}^{m} k_i \mathbf{h}_{i:}$$
where $\mathbf{h}_{i:}$ is row i of $\mathbf{H}$.
4. The eigenvalues of $\mathbf{H}^T\mathbf{H}$ must be greater than zero.
5. The determinant of $\mathbf{H}^T\mathbf{H}$ must be greater than zero.
6. $\mathbf{H}^T\mathbf{H}$ must be positive definite; that is, $\mathbf{x}^T\mathbf{H}^T\mathbf{Hx} > 0$ for all nonzero $\mathbf{x}$.
7. It must be possible to uniquely factor $\mathbf{H}^T\mathbf{H}$ as $\mathbf{LL}^T$ (lower triangular) or $\mathbf{UU}^T$ (upper triangular) where all diagonals of $\mathbf{L}$ or $\mathbf{U}$ are nonzero.
8. $|\mathbf{A}_{ij}| < \sqrt{\mathbf{A}_{ii}\mathbf{A}_{jj}}$ (where $\mathbf{A} = \mathbf{H}^T\mathbf{H}$) for $i \neq j$ is a necessary but not sufficient condition for observability.

When the rank of $\mathbf{H}$ is less than n (or any of the equivalent conditions), the system is unobservable; that is, $\mathbf{H}^T\mathbf{H}$ of the normal equations is singular and a unique solution cannot be obtained. However, even when $\mathbf{x}$ is theoretically observable from a given set of measurements, numerical errors may cause tests for the conditions listed above to fail. Conversely, numerical errors may also allow tests for many of these conditions to pass even when the system is theoretically unobservable. Hence numerical observability may be different from theoretical observability. The matrix condition number is one metric used to determine numerical observability. Practical tests for observability are addressed in Section 5.7.

5.2 NORMAL EQUATION FORMATION AND SOLUTION

5.2.1 Computation of the Normal Equations

The normal equation for the weighted least-squares case of equation (4.2-3) is

$$(\mathbf{H}^T\mathbf{R}^{-1}\mathbf{H})\hat{\mathbf{x}} = \mathbf{H}^T\mathbf{R}^{-1}\mathbf{y} \tag{5.2-1}$$

When the number of scalar measurements (m) in $\mathbf{y}$ is much larger than the number of unknowns (n) in $\mathbf{x}$—which is usually the case—and explicit or implicit integration is required to compute $\mathbf{H}$, it is generally most efficient to accumulate information in the normal equations individually for each measurement or set of measurements with a common time. That is, the measurement loop is the outer loop in forming $\mathbf{b} = \mathbf{H}^T\mathbf{R}^{-1}\mathbf{y}$ and $\mathbf{A} = \mathbf{H}^T\mathbf{R}^{-1}\mathbf{H}$. For generality we assume that the state vector $\mathbf{x}$ is referenced to an epoch time t_0, and that measurements $\mathbf{y}(t_i)$ at other times t_i

for $i = 1, 2, 3, \ldots, p$ are modeled as the product of a measurement sensitivity matrix $\mathbf{M}(t_i)$ and the state transition matrix from time t_0 to t_i; that is,

$$\mathbf{y}(t_i) = \mathbf{M}(t_i)\mathbf{\Phi}(t_i - t_0)\mathbf{x}(t_0) + \mathbf{r}(t_i). \tag{5.2-2}$$

$\mathbf{y}(t_i)$ may be a scalar or a vector, but the sum of the number of measurements for all p times is assumed to be m. If the independent variable in the model is not time but simply a measurement index, equation (5.2-2) still applies, but $\mathbf{\Phi}(t_i - t_0) = \mathbf{I}$. In either case the total set of all measurements is represented as

$$\begin{bmatrix} \mathbf{y}(t_1) \\ \mathbf{y}(t_2) \\ \mathbf{y}(t_3) \\ \vdots \\ \mathbf{y}(t_p) \end{bmatrix} = \begin{bmatrix} \mathbf{M}(t_1)\mathbf{\Phi}(t_1 - t_0) \\ \mathbf{M}(t_2)\mathbf{\Phi}(t_2 - t_0) \\ \mathbf{M}(t_3)\mathbf{\Phi}(t_3 - t_0) \\ \vdots \\ \mathbf{M}(t_p)\mathbf{\Phi}(t_p - t_0) \end{bmatrix} \mathbf{x}(t_0) + \begin{bmatrix} \mathbf{r}(t_1) \\ \mathbf{r}(t_2) \\ \mathbf{r}(t_3) \\ \vdots \\ \mathbf{r}(t_p) \end{bmatrix}. \tag{5.2-3}$$

The most commonly used software structure for solving the least-squares normal equation is:

$$\begin{aligned}
&\mathbf{b} = \mathbf{0} \\
&\mathbf{A} = \mathbf{0} \\
&\mathbf{\Phi}(t_0 - t_0) = \mathbf{I} \\
&\text{for } i = 1 \text{ to } p \\
&\quad \text{Compute } \mathbf{\Phi}(t_i - t_{i-1}) \\
&\quad \mathbf{\Phi}(t_i - t_0) = \mathbf{\Phi}(t_i - t_{i-1})\mathbf{\Phi}(t_{i-1} - t_0) \\
&\quad \mathbf{H}_i(t_i) = \mathbf{M}(t_i)\mathbf{\Phi}(t_i - t_0)\mathbf{R}(t_i)^{-1/2} \\
&\quad \mathbf{b} \Leftarrow \mathbf{b} + \mathbf{H}_i^T(t_i)\mathbf{y}(t_i) \\
&\quad \mathbf{A} \Leftarrow \mathbf{A} + \mathbf{H}_i^T(t_i)\mathbf{H}_i(t_i) \\
&\text{end loop}
\end{aligned} \tag{5.2-4}$$

where $\mathbf{R}(t_i)^{-1/2}$ is an inverse "square-root" factor of the measurement noise covariance matrix $\mathbf{R}(t_i)$ at time t_i; that is, $\mathbf{R}(t_i) = \mathbf{R}(t_i)^{1/2}\mathbf{R}(t_i)^{T/2}$ and $\mathbf{R}(t_i)^{-1/2} = (\mathbf{R}(t_i)^{1/2})^{-1}$. The notation $\mathbf{b} \Leftarrow \mathbf{b} + \mathbf{c}$ means that $\mathbf{b} + \mathbf{c}$ replaces $\mathbf{b}$ in memory. In most cases $\mathbf{R}(t_i)$ is assumed to be a diagonal matrix with diagonal elements in row i equal to $\sigma_i^2(t_i)$. Thus formation of $\mathbf{M}(t_i)\mathbf{\Phi}(t_i - t_0)\mathbf{R}(t_i)^{-1/2}$ just involves dividing each row of $\mathbf{M}(t_i)\mathbf{\Phi}(t_i - t_0)$ by $\sigma_i(t_i)$. If $\mathbf{R}(t_i)$ is not diagonal, $\mathbf{R}(t_i)^{1/2}$ is usually a lower triangular Cholesky factor of $\mathbf{R}(t_i)$.

Notice that the total number of *floating point operations* (flops) required to accumulate the $n \times n$ matrix $\mathbf{A}$ and n-element vector $\mathbf{b}$ in equation (5.2-4) is $mn(n + 3)/2$ if only the upper triangular partition of $\mathbf{A}$ is formed (as recommended). It will be seen later that alternate algorithms require more than $mn(n + 3)/2$ flops. It will also be seen that the general structure of equation (5.2-4) can be applied even when using these alternate algorithms.

An example demonstrates how the special structure of certain partitioned problems can suggest alternate solution approaches.

Example 5.2: Partitioned Orbit Determination

In some least-squares problems a subset of model states contribute to all measurements while other states influence only selected subsets of the measurements. For example, in orbit determination of a low-earth-orbiting satellite, ground station tracking biases only affect measurements from that ground station. For a case in which three ground stations track a satellite, the measurement sensitivity matrix $\mathbf{H}$ could have the structure

$$\begin{bmatrix} \mathbf{y}_1 \\ \mathbf{y}_2 \\ \mathbf{y}_3 \end{bmatrix} = \begin{bmatrix} \mathbf{H}_{11} & \mathbf{0} & \mathbf{0} & \mathbf{H}_{14} \\ \mathbf{0} & \mathbf{H}_{22} & \mathbf{0} & \mathbf{H}_{24} \\ \mathbf{0} & \mathbf{0} & \mathbf{H}_{33} & \mathbf{H}_{34} \end{bmatrix} \begin{bmatrix} \mathbf{x}_1 \\ \mathbf{x}_2 \\ \mathbf{x}_3 \\ \mathbf{x}_4 \end{bmatrix} + \begin{bmatrix} \mathbf{r}_1 \\ \mathbf{r}_2 \\ \mathbf{r}_3 \end{bmatrix}$$

where

$\mathbf{y}_1$ includes all measurements from station #1 (for all passes),

$\mathbf{y}_2$ includes all measurements from station #2,

$\mathbf{y}_3$ includes all measurements from station #3,

$\mathbf{x}_1$, $\mathbf{x}_2$, $\mathbf{x}_3$ are the measurement biases for ground stations 1, 2, and 3, respectively,

$\mathbf{x}_4$ are the common states such as the satellite epoch orbit states and any other force model parameters,

$\mathbf{H}_{ij}$ is the measurement sensitivity matrix for measurement set i with respect to states j,

$\mathbf{r}_1$, $\mathbf{r}_2$, $\mathbf{r}_3$ are the measurement noises for the three stations.

Notice that this numbering scheme does not make any assumptions about measurement time order.

The information matrix for this problem is

$$\mathbf{H}^T\mathbf{R}^{-1}\mathbf{H} = \begin{bmatrix} \mathbf{H}_{11}^T\mathbf{R}_1^{-1}\mathbf{H}_{11} & \mathbf{0} & \mathbf{0} & \mathbf{H}_{11}^T\mathbf{R}_1^{-1}\mathbf{H}_{14} \\ \mathbf{0} & \mathbf{H}_{22}^T\mathbf{R}_2^{-1}\mathbf{H}_{22} & \mathbf{0} & \mathbf{H}_{22}^T\mathbf{R}_2^{-1}\mathbf{H}_{24} \\ \mathbf{0} & \mathbf{0} & \mathbf{H}_{33}^T\mathbf{R}_3^{-1}\mathbf{H}_{33} & \mathbf{H}_{33}^T\mathbf{R}_3^{-1}\mathbf{H}_{34} \\ \mathbf{H}_{14}^T\mathbf{R}_1^{-1}\mathbf{H}_{11} & \mathbf{H}_{24}^T\mathbf{R}_2^{-1}\mathbf{H}_{22} & \mathbf{H}_{34}^T\mathbf{R}_3^{-1}\mathbf{H}_{33} & \mathbf{H}_{14}^T\mathbf{R}_1^{-1}\mathbf{H}_{14} + \mathbf{H}_{24}^T\mathbf{R}_2^{-1}\mathbf{H}_{24} + \mathbf{H}_{34}^T\mathbf{R}_3^{-1}\mathbf{H}_{34} \end{bmatrix}$$

and the information vector is

$$\mathbf{H}^T\mathbf{R}^{-1}\mathbf{y} = \begin{bmatrix} \mathbf{H}_{11}^T\mathbf{R}_1^{-1}\mathbf{y}_1 \\ \mathbf{H}_{21}^T\mathbf{R}_2^{-1}\mathbf{y}_2 \\ \mathbf{H}_{31}^T\mathbf{R}_3^{-1}\mathbf{y}_3 \\ \mathbf{H}_{14}^T\mathbf{R}_1^{-1}\mathbf{y}_1 + \mathbf{H}_{24}^T\mathbf{R}_2^{-1}\mathbf{y}_2 + \mathbf{H}_{34}^T\mathbf{R}_3^{-1}\mathbf{y}_3 \end{bmatrix}.$$

The zero partitions in $\mathbf{H}^T\mathbf{R}^{-1}\mathbf{H}$ allow the normal equation solution $(\mathbf{H}^T\mathbf{R}^{-1}\mathbf{H})\hat{\mathbf{x}} = \mathbf{H}^T\mathbf{R}^{-1}\mathbf{y}$ to be obtained in partitions without inverting the entire matrix. Specifically, for the case

$$\mathbf{H}^T\mathbf{R}^{-1}\mathbf{H} = \begin{bmatrix} \mathbf{A}_{11} & \mathbf{0} & \mathbf{0} & \mathbf{A}_{14} \\ \mathbf{0} & \mathbf{A}_{22} & \mathbf{0} & \mathbf{A}_{24} \\ \mathbf{0} & \mathbf{0} & \mathbf{A}_{33} & \mathbf{A}_{34} \\ \mathbf{A}_{14}^T & \mathbf{A}_{24}^T & \mathbf{A}_{34}^T & \mathbf{A}_{44} \end{bmatrix} \quad \text{and} \quad \mathbf{H}^T\mathbf{R}^{-1}\mathbf{y} = \begin{bmatrix} \mathbf{b}_1 \\ \mathbf{b}_2 \\ \mathbf{b}_3 \\ \mathbf{b}_4 \end{bmatrix},$$

$(\mathbf{H}^T\mathbf{R}^{-1}\mathbf{H})\hat{\mathbf{x}} = \mathbf{H}^T\mathbf{R}^{-1}\mathbf{y}$ can be solved for $\hat{\mathbf{x}}$ in partitions, starting with the common parameters. The partitioned solution is easily determined to be

$$\begin{aligned}
\hat{\mathbf{x}}_4 &= \mathbf{P}_{44}(\mathbf{b}_4 - \mathbf{A}_{14}^T\mathbf{A}_{11}^{-1}\mathbf{b}_1 - \mathbf{A}_{24}^T\mathbf{A}_{22}^{-1}\mathbf{b}_2 - \mathbf{A}_{34}^T\mathbf{A}_{33}^{-1}\mathbf{b}_3) \\
\hat{\mathbf{x}}_1 &= \mathbf{A}_{11}^{-1}(\mathbf{b}_1 - \mathbf{A}_{14}^T\hat{\mathbf{x}}_4) \\
\hat{\mathbf{x}}_2 &= \mathbf{A}_{22}^{-1}(\mathbf{b}_2 - \mathbf{A}_{24}^T\hat{\mathbf{x}}_4) \\
\hat{\mathbf{x}}_3 &= \mathbf{A}_{33}^{-1}(\mathbf{b}_3 - \mathbf{A}_{34}^T\hat{\mathbf{x}}_4)
\end{aligned}$$

where

$$\mathbf{P}_{44} = \left(\mathbf{A}_{44} - \mathbf{A}_{14}^T\mathbf{A}_{11}^{-1}\mathbf{A}_{14} - \mathbf{A}_{24}^T\mathbf{A}_{22}^{-1}\mathbf{A}_{24} - \mathbf{A}_{34}^T\mathbf{A}_{33}^{-1}\mathbf{A}_{34}\right)^{-1}$$

is the error covariance for $\hat{\mathbf{x}}_4$. Likewise

$$\begin{aligned}
\mathbf{P}_{11} &= \mathbf{A}_{11}^{-1} - \mathbf{A}_{11}^{-1}\mathbf{A}_{14}\mathbf{P}_{44}\mathbf{A}_{14}^T\mathbf{A}_{11}^{-1} \\
\mathbf{P}_{22} &= \mathbf{A}_{22}^{-1} - \mathbf{A}_{21}^{-1}\mathbf{A}_{24}\mathbf{P}_{44}\mathbf{A}_{24}^T\mathbf{A}_{22}^{-1} \\
\mathbf{P}_{33} &= \mathbf{A}_{33}^{-1} - \mathbf{A}_{33}^{-1}\mathbf{A}_{34}\mathbf{P}_{44}\mathbf{A}_{34}^T\mathbf{A}_{33}^{-1}
\end{aligned}$$

are the error covariances for $\hat{\mathbf{x}}_1, \hat{\mathbf{x}}_2, \hat{\mathbf{x}}_3$, respectively. Computation of the covariances between $\hat{\mathbf{x}}_1, \hat{\mathbf{x}}_2, \hat{\mathbf{x}}_3$, and $\hat{\mathbf{x}}_4$ are left as an exercise for the reader. Hint: see the partitioned matrix solution in Appendix A.

This example shows how it is possible to solve separately for subsets of states when some subsets only influence subsets of the measurements. This is a great advantage when the number of states and subsets is very large because it is only necessary to compute and store information arrays for the nonzero partitions. For example, when computing earth gravity field models, thousands of satellite arcs from multiple satellites are processed, and the orbit elements from each arc are solved for as separate states. This would not be practical without partitioning. For problems with a large number of states and ground stations (or equivalent "pass-dependent" parameters), the savings in computations and memory usage may be orders of magnitude.

Block partitioning of the measurement matrix also appears in other systems when the states are spatially distributed parameters and either the force model or measurement models have a limited spatial extent. Finite difference and finite element models of physical systems are another example of this structure.

Although the partitioned solution works best when using the normal equations, it is sometimes possible to exploit block partitioned structure when using some of the other least-squares solution techniques described below. An example appears in Section 5.3.

5.2.2 Cholesky Decomposition of the Normal Equations

In this section we address solution of the least-squares normal equations or equations that are equivalent to the normal equations. Consider the weighted least-squares normal equations of equation (4.2-3)

$$(\mathbf{H}^T\mathbf{R}^{-1}\mathbf{H})\hat{\mathbf{x}} = \mathbf{H}^T\mathbf{R}^{-1}\mathbf{y}, \tag{5.2-5}$$

or for the Bayesian estimate,

$$(\mathbf{H}^T\mathbf{R}^{-1}\mathbf{H} + \mathbf{P}_a^{-1})\hat{\mathbf{x}} = \mathbf{H}^T\mathbf{R}^{-1}\mathbf{y} + \mathbf{P}_a^{-1}\mathbf{x}_a. \tag{5.2-6}$$

Solution techniques for the two equations are identical because equation (5.2-6) can be viewed as equation (5.2-5) with the *a priori* estimate $\mathbf{x}_a$ treated as a measurement. Hence we concentrate on equation (5.2-5). This can be solved by a number of methods. For example, Gauss-Jordan elimination can be used to transform $(\mathbf{H}^T\mathbf{R}^{-1}\mathbf{H})$ to the identity matrix. This is done by computing both $(\mathbf{H}^T\mathbf{R}^{-1}\mathbf{H})^{-1}$ and $(\mathbf{H}^T\mathbf{R}^{-1}\mathbf{H})^{-1}\mathbf{H}^T\mathbf{R}^{-1}\mathbf{y}$ in one operation, or $(\mathbf{H}^T\mathbf{R}^{-1}\mathbf{H})^{-1}\mathbf{H}^T\mathbf{R}^{-1}\mathbf{y}$ can be computed without inverting the matrix (Press et al. 2007, Section 2.1). It is important that pivoting be used in Gauss-Jordan elimination to maintain accuracy—particularly when $\mathbf{H}^T\mathbf{R}^{-1}\mathbf{H}$ is poorly conditioned. Another approach uses the *LU* decomposition to factor $\mathbf{H}^T\mathbf{R}^{-1}\mathbf{H} = \mathbf{LU}$ where $\mathbf{L}$ is a lower triangular matrix and $\mathbf{U}$ is an upper triangular matrix. Then the $\mathbf{L}$ and $\mathbf{U}$ factors can be easily inverted to form $\mathbf{P} = (\mathbf{H}^T\mathbf{R}^{-1}\mathbf{H})^{-1} = \mathbf{U}^{-1}\mathbf{L}^{-1}$. With partial pivoting, this has approximately the same accuracy as Gauss-Jordan elimination with full pivoting, but does not take advantage of the fact that $\mathbf{H}^T\mathbf{R}^{-1}\mathbf{H}$ is a positive definite symmetric matrix.

Cholesky decomposition of a positive definite symmetric matrix is similar to the LU decomposition but with the constraint that $\mathbf{L} = \mathbf{U}^T$. That is,

$$\mathbf{H}^T\mathbf{R}^{-1}\mathbf{H} = \mathbf{U}^T\mathbf{U}. \tag{5.2-7}$$

Similar factorizations can also be defined using $\mathbf{UU}^T$ where $\mathbf{U}$ is upper triangular, or as $\mathbf{UDU}^T$ where $\mathbf{U}$ is *unit* upper triangular (1's on the diagonal) and $\mathbf{D}$ is a diagonal matrix of positive values. The $\mathbf{UDU}^T$ factorization has the advantage that square roots are not required in the computation. However, we work with $\mathbf{U}^T\mathbf{U}$ because that is the usual definition of Cholesky factors. The factorization can be understood using a three-state example.

Example 5.3: Cholesky Factorization

Consider the Cholesky factorization of a 3×3 positive definite symmetric matrix $\mathbf{A} = \mathbf{U}^T\mathbf{U}$:

$$\begin{aligned}\begin{bmatrix} A_{11} & A_{12} & A_{13} \\ A_{12} & A_{22} & A_{23} \\ A_{13} & A_{23} & A_{33} \end{bmatrix} &= \begin{bmatrix} U_{11} & 0 & 0 \\ U_{12} & U_{22} & 0 \\ U_{13} & U_{23} & U_{33} \end{bmatrix} \begin{bmatrix} U_{11} & U_{12} & U_{13} \\ 0 & U_{22} & U_{23} \\ 0 & 0 & U_{33} \end{bmatrix} \\ &= \begin{bmatrix} U_{11}^2 & U_{11}U_{12} & U_{11}U_{13} \\ U_{11}U_{12} & U_{12}^2 + U_{22}^2 & U_{12}U_{13} + U_{22}U_{23} \\ U_{11}U_{13} & U_{12}U_{13} + U_{22}U_{23} & U_{13}^2 + U_{23}^2 + U_{33}^2 \end{bmatrix}\end{aligned} \tag{5.2-8}$$

It is evident that the U elements can be computed starting from the upper left element:

$$\begin{aligned}
&U_{11}=\sqrt{a_{11}}, \quad U_{12}=A_{12}/U_{11}, \quad U_{13}=A_{13}/U_{11} \\
&U_{22}=\sqrt{A_{22}-U_{12}^2}, \quad U_{23}=(A_{23}-U_{12}U_{13})/U_{22}. \\
&U_{33}=\sqrt{A_{33}-U_{13}^2-U_{23}^2}
\end{aligned} \tag{5.2-9}$$

The general algorithm for a $n \times m$ matrix is

$$\begin{aligned}
&\text{for } i=1 \text{ to } n \\
&\quad U_{ii}=\sqrt{A_{ii}-\sum_{k=1}^{i-1}U_{ki}^2} \\
&\quad \text{for } j=i+1 \text{ to } n \\
&\quad\quad U_{ij}=\left(A_{ij}-\sum_{k=1}^{i-1}U_{ki}U_{kj}\right)\Big/U_{ii} \\
&\quad \text{end loop} \\
&\text{end loop}
\end{aligned} \tag{5.2-10}$$

Note that the elements of **U** and the upper triangular portion of **A** can occupy the same storage locations because A_{ij} is never referenced after setting U_{ij}. The factorization can also be implemented in column order or as vector outer products rather than the row order indicated here (Björck 1996, Section 2.2.2).

The stability of the Cholesky factorization is due to the fact that the elements of **U** must be less than or equal to the diagonals of **A**. Note that

$$U_{ji}^2 \le \sum_{k=1}^{i} U_{ki}^2 = A_{ii} \quad \text{for } j \le i.$$

Hence numerical errors do not grow and the pivoting used with other matrix inversion methods is not necessary. However, Björck (1996, Section 2.7.4) shows that pivoting can be used to solve unobservable (singular) least-squares problems when $\mathbf{H}^T\mathbf{R}^{-1}\mathbf{H}$ is only positive semi-definite.

In coding Cholesky and other algorithms that work with symmetric or upper triangular matrices, it is customary to store these arrays as upper-triangular-by-column vectors such that

$$\mathbf{A}=\begin{bmatrix} a_1 & a_2 & a_4 \\ a_2 & a_3 & a_5 \\ a_4 & a_5 & a_6 \end{bmatrix}, \quad \mathbf{U}=\begin{bmatrix} u_1 & u_2 & u_4 \\ 0 & u_3 & u_5 \\ 0 & 0 & u_6 \end{bmatrix}.$$

Use of vectors for array storage cuts the required memory approximately in half, which is beneficial when the dimension is large. For this nomenclature, the Cholesky algorithm becomes

$$
\begin{aligned}
&\text{for } i=1 \text{ to } n\\
&\quad p=\text{floor}(i(i-1)/2)\\
&\quad u_{p+i}=\sqrt{a_{p+i}-\sum_{k=1}^{i-1}u_{p+k}^2}\\
&\quad \text{for } j=i+1 \text{ to } n\\
&\quad\quad q=\text{floor}(j(j-1)/2)\\
&\quad\quad u_{q+i}=\left(a_{q+i}-\sum_{k=1}^{i-1}u_{p+k}u_{q+k}\right)\Big/u_{p+i}\\
&\quad \text{end loop}\\
&\text{end loop}
\end{aligned}
\tag{5.2-11}
$$

where floor(x) is the largest integer less than or equal to x. Again the elements of **U** and **A** can occupy the same storage locations.

To invert **U**, the equation **UM** = **I** is solved for **M** where $\mathbf{M}=\mathbf{U}^{-1}$ is an upper triangular matrix; that is,

$$
\begin{bmatrix} u_1 & u_2 & u_4 \\ 0 & u_3 & u_5 \\ 0 & 0 & u_6 \end{bmatrix}
\begin{bmatrix} m_1 & m_2 & m_4 \\ 0 & m_3 & m_5 \\ 0 & 0 & m_6 \end{bmatrix}
=
\begin{bmatrix} 1 & 0 & 0 \\ 0 & 1 & 0 \\ 0 & 0 & 1 \end{bmatrix}
$$

which leads to

$$
\begin{aligned}
&m_6=1/u_6\\
&m_3=1/u_3,\quad m_5=-u_5m_6/u_3\\
&m_1=1/u_1,\quad m_2=-u_2m_3/u_1,\quad m_4=-(u_2m_5+u_4m_6)/u_1
\end{aligned}
$$

Again **M** and **U** may occupy the same storage. We leave it as an exercise for the reader to develop a general algorithm for computing $\mathbf{M}=\mathbf{U}^{-1}$ and then forming $\mathbf{A}^{-1}=\mathbf{U}^{-1}\mathbf{U}^{-T}$. This can also be implemented using the same storage as **A**. Hint: Subroutine SINV in software associated with this book and LAPACK routine DPOTF2 implement this algorithm, although the indexing can be somewhat confusing.

The number of flops required to Cholesky factor $\mathbf{A}=\mathbf{U}^T\mathbf{U}$ is approximately $n^3/6$, where a multiplication or division plus an addition or subtraction are counted as a single flop. $n^3/6$ flops are also required to invert **U**, and the same to form $\mathbf{A}^{-1}=\mathbf{U}^{-1}\mathbf{U}^{-T}$. Hence the total number of flops to invert **A** is about $n^3/2$. This is half of the $\sim n^3$ operations required to implement inversion based on LU decomposition or Gauss-Jordan elimination. However, since m is generally much greater than n, the flops required to invert $\mathbf{H}^T\mathbf{R}^{-1}\mathbf{H}$ are small compared with flops for computing $\mathbf{H}^T\mathbf{R}^{-1}\mathbf{H}$ and $\mathbf{H}^T\mathbf{R}^{-1}\mathbf{y}$.

Use of the Cholesky factorization to invert $\mathbf{H}^T\mathbf{R}^{-1}\mathbf{H}$ has the potential advantage of greater accuracy compared with the LU or Gauss-Jordan elimination algorithms.

However, the accuracy of the LU and Gauss-Jordan elimination can equal that of Cholesky factorization when full pivoting is used. Cholesky factorization is insensitive to scaling issues even without pivoting. That is, the states estimated using the least-squares method can be scaled arbitrarily with little effect on the numerical accuracy of the matrix inversion. This is demonstrated shortly in Example 5.4.

Another advantage of the Cholesky factorization is that it automatically provides information about numerical conditioning of $\mathbf{H}^T\mathbf{R}^{-1}\mathbf{H}$. Notice the second line in equation (5.2-10):

$$U_{ii} = \sqrt{A_{ii} - \sum_{k=1}^{i-1} U_{ki}^2}.$$

When $\sum_{k=1}^{i-1} U_{ki}^2$ is very close in magnitude to A_{ii}, the subtraction will result in a loss of precision. This happens when $\mathbf{H}^T\mathbf{R}^{-1}\mathbf{H}$ is nearly singular. The metric

$$\eta_i = \log_{10}\left(A_{ii} \bigg/ \left(A_{ii} - \sum_{k=1}^{i-1} U_{ki}^2 \right) \right) \tag{5.2-12}$$

indicates the number of numerical digits lost when processing column i. The maximum η_i for $i = 1$ to n is somewhat correlated with $\log_{10}$("*condition number*") for matrix $\mathbf{H}^T\mathbf{R}^{-1}\mathbf{H}$. Hence when the number of digits lost approaches half of the total floating point precision digits, the accuracy of the solution should be suspect. The following example demonstrates these concepts.

Example 5.4: Least Squares Using a Polynomial Model

Use of the polynomial model

$$\begin{aligned} y(t) &= x_1 + x_2 t + x_3 t^2 \ldots + x_n t^{n-1} \\ &= \sum_{i=1}^{n} x_i t^{i-1} \end{aligned}$$

for least-squares data fitting results in a poorly conditioned solution when n values are not "small." We evaluate the loss of accuracy for this polynomial model using 101 measurement samples separated in time by 0.01 s over the interval 0 to 1.0 s. For simplicity, the measurement noise variance is assumed to be 1.0, so $\mathbf{R} = \mathbf{I}$. The measurement matrix $\mathbf{H}$ for this case is

$$\mathbf{H} = \begin{bmatrix} 1 & 0 & 0 & \cdots & 0 \\ 1 & 0.01 & 0.01^2 & \cdots & 0.01^{n-1} \\ 1 & 0.02 & 0.02^2 & \cdots & 0.02^{n-1} \\ 1 & 0.03 & 0.03^2 & \cdots & 0.03^{n-1} \\ \vdots & \vdots & \vdots & \ddots & \vdots \\ 1 & 1 & 1 & \cdots & 1 \end{bmatrix}.$$

This structure, consisting of integer powers of variables, is called a *Vandermonde matrix*. Conditioning of Vandermonde matrices is very poor for even moderate values of n, and it is critical that calculations be performed in double precision. If possible, the problem should be restructured to use orthogonal basis functions such as Chebyshev polynomials. However, the polynomial problem offers a good test case for least-squares algorithms.

Notice that for the **H** matrix defined above,

$$\mathbf{H}^T\mathbf{H} = \begin{bmatrix} m & \sum_{i=1}^{m} t_i & \cdots & \sum_{i=1}^{m} t_i^{n-1} \\ \sum_{i=1}^{m} t_i & \sum_{i=1}^{m} t_i^2 & \cdots & \sum_{i=1}^{m} t_i^n \\ \vdots & \vdots & \ddots & \vdots \\ \sum_{i=1}^{m} t_i^{n-1} & \sum_{i=1}^{m} t_i^n & \cdots & \sum_{i=1}^{m} t_i^{2(n-1)} \end{bmatrix} \quad \text{and} \quad \mathbf{H}^T\mathbf{y} = \begin{bmatrix} \sum_{i=1}^{m} y_i \\ \sum_{i=1}^{m} y_i t_i \\ \vdots \\ \sum_{i=1}^{m} y_i t_i^{n-1} \end{bmatrix}.$$

$\mathbf{H}^T\mathbf{H}$ and $\mathbf{H}^T\mathbf{y}$ for this particular problem can be efficiently calculated as shown using sums of powers of t_i. However, we use the method described in Section 5.2.1.

When evaluating accuracy of the solution, the quantity of most interest is the error in the estimated state:

$$\begin{aligned}\hat{\mathbf{x}} - \mathbf{x} &= (\mathbf{H}^T\mathbf{H})^{-1}\mathbf{H}^T\mathbf{y} - \mathbf{x} \\ &= \left((\mathbf{H}^T\mathbf{H})^{-1}(\mathbf{H}^T\mathbf{H}) - \mathbf{I}\right)\mathbf{x} + (\mathbf{H}^T\mathbf{H})^{-1}\mathbf{H}^T\mathbf{r}\end{aligned}$$

The first term should be zero and the second is the error due to measurement noise. The extent to which $(\mathbf{H}^T\mathbf{H})^{-1}(\mathbf{H}^T\mathbf{H}) - \mathbf{I} \neq \mathbf{0}$ is a measure of numerical errors in forming and inverting the normal equations. We use two metrics to evaluate accuracy of the polynomial solution. The first is the Frobenius norm,

$$\left\|\mathbf{A}^{-1}\mathbf{A} - \mathbf{I}\right\|_F = \sqrt{\sum_{i=1}^{n}\sum_{j=1}^{n}(\mathbf{A}^{-1}\mathbf{A} - \mathbf{I})_{ij}^2},$$

where $\mathbf{A} = \mathbf{H}^T\mathbf{H}$. The second metric is the induced l_1-norm

$$\left[\mathbf{A}^{-1}\mathbf{A} - \mathbf{I}\right]_1 = \max_j \sum_{i=1}^{n}\left|(\mathbf{A}^{-1}\mathbf{A} - \mathbf{I})_{ij}\right|.$$

The base 10 logarithm of these metrics is computed as function of n when using double precision (53 bits mantissa) in IEEE T_floating format on a PC. We also tabulate expected precision, calculated as $\log_{10}(1.1\times 10^{-16}) + \max_i \eta_i$, computed during the Cholesky factorization of $\mathbf{H}^T\mathbf{H}$.

Figure 5.1 shows the results when using Cholesky factorization and Gauss-Jordan elimination with full pivoting to invert $\mathbf{H}^T\mathbf{H}$. Figure 5.2 is a similar plot obtained when the first four columns of **H** are scaled by 100. Several conclusions can be drawn from these results:

1. Even with 101 measurements, no normal equation method will work for $n > 13$.

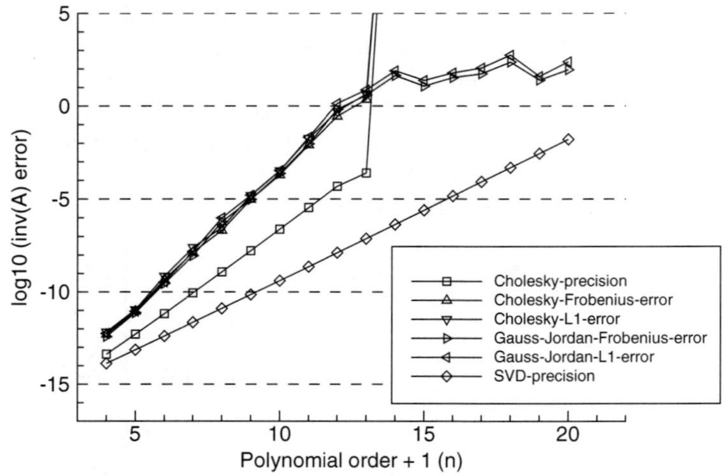

FIGURE 5.1: Least-squares normal equations inversion accuracy using polynomial model.

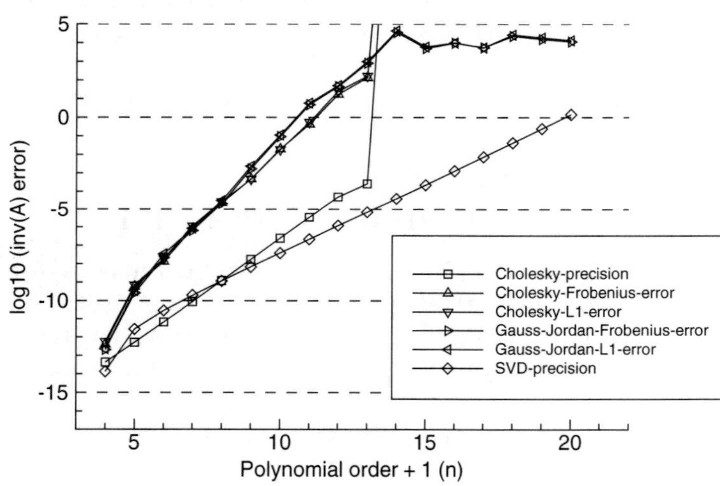

FIGURE 5.2: Least-squares normal equations inversion accuracy using polynomial model with H(:,1:4) multiplied by 100.

2. The Cholesky inversion is slightly more accurate than Gauss-Jordan elimination with full pivoting, but the difference is negligible. Gauss-Jordan elimination will continue to compute solutions for $n > 13$ even though all precision is lost.
3. The Cholesky precision calculation for number of digits lost,

$$\eta_i = \log_{10}\left(A_{ii} \Big/ \left(A_{ii} - \sum_{j=1}^{i-1} U_{ji}^2 \right) \right),$$

tends to underestimate the actual loss of precision by a factor of about 10 for $\eta_i < 8$, and a factor of 10^4 as the matrix approaches singularity. In fact, the actual loss of precision seems to follow

$$\eta_i = 1.3\log_{10}\left(A_{ii} \bigg/ \left(A_{ii} - \sum_{j=1}^{i-1} U_{ji}^2 \right) \right),$$

although the reason for the factor of 1.3 is unknown. Nonetheless, $\max_i \eta_i$ is still a useful indicator of observability problems as it loosely tracks the loss of precision. Any solution where $\max_i \eta_i$ is greater than eight should be highly suspect.

4. The "SVD precision" line in the figures is the expected precision, calculated as $\log_{10}[1.1 \times 10^{-16} \times \kappa_2(\mathbf{H})]$ from the singular values of **H**. Since formation of $\mathbf{H}^T\mathbf{H}$ doubles the condition number, the expected precision of the normal equations is $\log_{10}(1.1 \times 10^{-16}) + 2\log_{10}(\kappa_2(\mathbf{H}))$, which has twice the slope of the "SVD precision" line. This roughly matches the actual "Cholesky" or "Gauss-Jordan" inversion accuracy up to the point where they lose all precision. Hence the precision predicted from the condition number of **H** is appropriate for the normal equation solution.
5. Scaling of the first four columns of the **H** matrix by 100 degrades the accuracy of the matrix inversion by the same factor. However, it has negligible effect on calculation of $\max_i \eta_i$. This result is somewhat misleading because multiplying the columns of **H** by 100 implies that the corresponding elements of **x** must be divided by 100 if **y** is unchanged. Hence the fact that the errors increased by 100 in Figure 5.2 implies that the Cholesky factorization is relatively insensitive to scale changes.

Monte Carlo evaluation is also used as an alternate check on numeric accuracy. One hundred independent samples of state vector **x** were randomly generated where each individual element of **x** was obtained from a Gaussian random number generator using a standard deviation of 1.0. Then $\mathbf{y} = \mathbf{H}\mathbf{x}$ (with no measurement noise) was used to calculate $\hat{\mathbf{x}} - \mathbf{x} = (\mathbf{H}^T\mathbf{H})^{-1}\mathbf{H}^T\mathbf{y} - \mathbf{x}$. The two metrics used for the evaluation are averaged values of the l_2- and l_∞-norms:

$$\text{RMS error} = \sqrt{\frac{1}{100}\sum_{i=1}^{100} \|\hat{\mathbf{x}}_i - \mathbf{x}_i\|_2^2}$$

and

$$\text{max error} = \max_i \|\hat{\mathbf{x}}_i - \mathbf{x}_i\|_\infty$$

where $\max_i \|\hat{\mathbf{x}}_i - \mathbf{x}_i\|_\infty$ is the maximum absolute value of the elements in $\hat{\mathbf{x}}_i - \mathbf{x}_i$ over all samples of $\hat{\mathbf{x}}_i - \mathbf{x}_i$. Figures 5.3–5.4 show the results plotted on a log scale. The root-mean-squared (RMS) error is approximately equal to the Frobenius norm of $\mathbf{A}^{-1}\mathbf{A} - \mathbf{I}$ in Figures 5.1 and 5.2 for the same conditions, and the maximum error is approximately equal to the l_1-norm of $\mathbf{A}^{-1}\mathbf{A} - \mathbf{I}$. This suggests that numerical errors in $(\mathbf{H}^T\mathbf{H})^{-1}(\mathbf{H}^T\mathbf{H})$ are positively correlated rather than independent. Also notice that the RMS errors in $\hat{\mathbf{x}}$ are about 20 to 4000

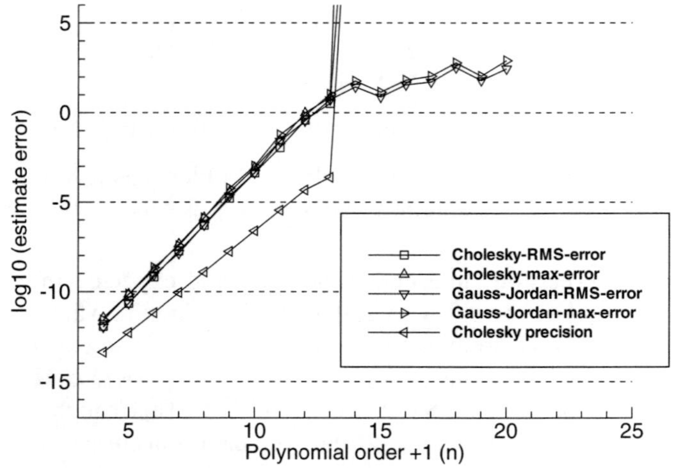

FIGURE 5.3: Monte Carlo evaluation of LS normal equations solution accuracy using polynomial model.

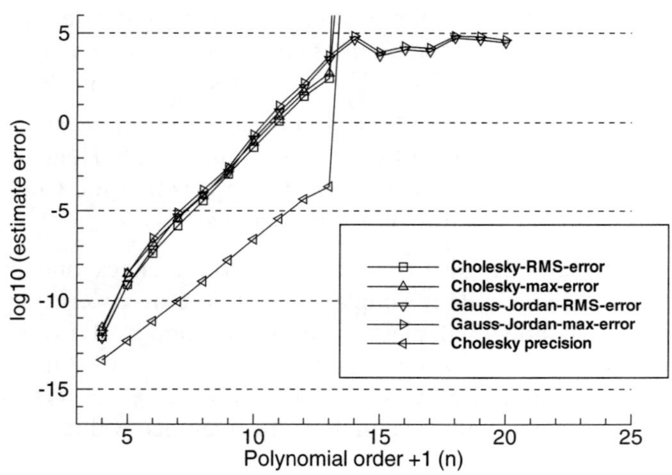

FIGURE 5.4: Monte Carlo evaluation of LS normal equations solution accuracy using polynomial model with H(:,1:4) multiplied by 100.

times larger than $1.1\times10^{-16}\times10^{\max_i \eta_i}$ (labeled "Cholesky precision"). Again when columns of **H** are scaled by 100, the errors are about 100 times larger than when unscaled.

5.3 ORTHOGONAL TRANSFORMATIONS AND THE QR METHOD

As noted previously, solution of the normal equations is not the only method that can be used to compute least-squares estimates. Previous sections used the normal equations because they are easily understood and often used in practice. However, the normal equations are more sensitive to the effects of numerical round-off errors than alternative algorithms. This can be understood by expanding the information matrix using the SVD of matrix $\mathbf{L}^{-1}\mathbf{H}$ where $\mathbf{R} = \mathbf{L}\mathbf{L}^T$:

$$\mathbf{H}^T\mathbf{R}^{-1}\mathbf{H} = (\mathbf{H}^T\mathbf{L}^{-T})(\mathbf{L}^{-1}\mathbf{H}) = \mathbf{V}\mathbf{S}^T\mathbf{U}^T\mathbf{U}\mathbf{S}\mathbf{V}^T = \mathbf{V}(\mathbf{S}^T\mathbf{S})\mathbf{V}^T. \qquad (5.3\text{-}1)$$

Notice that equation (5.3-1) is an eigen decomposition of $\mathbf{H}^T\mathbf{R}^{-1}\mathbf{H}$ since $\mathbf{V}$ is an orthogonal matrix and $\mathbf{S}^T\mathbf{S}$ is the diagonal matrix of eigenvalues. Hence the condition number for $\mathbf{H}^T\mathbf{R}^{-1}\mathbf{H}$ is twice as large as that for $\mathbf{L}^{-1}\mathbf{H}$ since the singular values are squared in $\mathbf{S}^T\mathbf{S}$. Thus a least-squares solution technique that does not form $\mathbf{H}^T\mathbf{R}^{-1}\mathbf{H}$ would be expected to have approximately twice the accuracy of a solution that inverts (directly or indirectly) $\mathbf{H}^T\mathbf{R}^{-1}\mathbf{H}$.

One solution method that avoids formation of the normal equations is based on the properties of orthogonal transformations. To use this *QR method* (originally developed by Francis [1961] for eigen decomposition), it is necessary that the measurement equation be expressed as $\mathbf{y} = \mathbf{H}\mathbf{x} + \mathbf{r}$ where $E[\mathbf{r}\mathbf{r}^T] = \mathbf{I}$. If $E[\mathbf{r}\mathbf{r}^T] = \mathbf{R} \neq \mathbf{I}$, first factor $\mathbf{R}$ as $\mathbf{R} = \mathbf{L}\mathbf{L}^T$ and then multiply the measurement equation by $\mathbf{L}^{-1}$ as

$$\mathbf{L}^{-1}\mathbf{y} = (\mathbf{L}^{-1}\mathbf{H})\hat{\mathbf{x}} + \mathbf{L}^{-1}\mathbf{r}$$

so that $E[\mathbf{L}^{-1}\mathbf{r}\mathbf{r}^T\mathbf{L}^{-T}] = \mathbf{I}$. In most cases $\mathbf{R}$ is diagonal so this normalization step just involves dividing the rows of $\mathbf{y}$ and $\mathbf{H}$ by the measurement standard deviation for each measurement. Then the QR algorithm proceeds using $\mathbf{L}^{-1}\mathbf{y}$ as the measurement vector and $\mathbf{L}^{-1}\mathbf{H}$ as the measurement matrix. For simplicity in notation, we assume that the normalization has been applied before using the QR algorithm, and we drop the use of $\mathbf{L}^{-1}$ in the equations to follow. It should be noted that when measurement weighting results in a "stiff system" (one in which certain linear combinations of states are weighted much more heavily than others), additional pivoting operations are necessary in the QR algorithm to prevent possible numerical problems (Björck 1996, Section 2.7.3). Other approaches for including the weights directly in the QR algorithm are also possible (Björck 1996, p. 164). This topic is addressed again in Chapter 7 when discussing constraints.

In the QR algorithm the $m \times n$ measurement matrix $\mathbf{H}$ is factored as $\mathbf{H} = \mathbf{Q}^T\mathbf{R}$, or equivalently $\mathbf{Q}\mathbf{H} = \mathbf{R}$, where $\mathbf{Q}$ **is** an $m \times n$ *orthogonal* matrix and $\mathbf{R}$ is an upper triangular $m \times n$ matrix. Although $\mathbf{Q}$ and $\mathbf{R}$ are commonly used as the array symbols in the QR algorithm (for obvious reasons), this notation will be confusing in this book because we have already used $\mathbf{Q}$ for the discrete state noise covariance matrix and $\mathbf{R}$ for the measurement noise covariance. Hence we substitute array $\mathbf{T}$ for $\mathbf{Q}$ and $\mathbf{U}$ for $\mathbf{R}$ in the QR algorithm so that $\mathbf{H}$ is factored as $\mathbf{H} = \mathbf{T}^T\mathbf{U}$. (The $\mathbf{U}$ matrix represents an upper triangular matrix, not the orthogonal matrix of the SVD algorithm.) This implies that the measurement equation is multiplied by $\mathbf{T} = (\mathbf{T}^T)^{-1}$ (since $\mathbf{T}$ is orthogonal) such that $\mathbf{T}\mathbf{H} = \mathbf{U}$; that is,

$$\mathbf{T}\mathbf{y} = (\mathbf{T}\mathbf{H})\mathbf{x} + \mathbf{T}\mathbf{r}$$

or

$$\begin{bmatrix} \mathbf{T}_1\mathbf{y} \\ \mathbf{T}_2\mathbf{y} \end{bmatrix} = \begin{bmatrix} \mathbf{U}_1 \\ \mathbf{0} \end{bmatrix} \mathbf{x} + \begin{bmatrix} \mathbf{T}_1\mathbf{r} \\ \mathbf{T}_2\mathbf{r} \end{bmatrix} \qquad (5.3\text{-}2)$$

where $\mathbf{T} = \begin{bmatrix} \mathbf{T}_1 \\ \mathbf{T}_2 \end{bmatrix}$, $\mathbf{T}_1$ is $n \times m$, $\mathbf{T}_2$ is $(m-n) \times m$, and $\mathbf{U}_1$ is $n \times n$. The least-squares cost function is unchanged by this transformation because

$$\begin{aligned} J &= \frac{1}{2}(\mathbf{y} - \mathbf{H}\hat{\mathbf{x}})^T \mathbf{T}^T \mathbf{T}(\mathbf{y} - \mathbf{H}\hat{\mathbf{x}}) \\ &= \frac{1}{2}(\mathbf{y} - \mathbf{H}\hat{\mathbf{x}})^T (\mathbf{y} - \mathbf{H}\hat{\mathbf{x}}) \end{aligned} . \tag{5.3-3}$$

This can also be written as

$$\begin{aligned} J &= \frac{1}{2}(\mathbf{T}\mathbf{y} - \mathbf{T}\mathbf{H}\hat{\mathbf{x}})^T (\mathbf{T}\mathbf{y} - \mathbf{T}\mathbf{H}\hat{\mathbf{x}}) \\ &= \frac{1}{2}\begin{bmatrix} \mathbf{T}_1\mathbf{y} - \mathbf{U}_1\hat{\mathbf{x}} \\ \mathbf{T}_2\mathbf{y} \end{bmatrix}^T \begin{bmatrix} \mathbf{T}_1\mathbf{y} - \mathbf{U}_1\hat{\mathbf{x}} \\ \mathbf{T}_2\mathbf{y} \end{bmatrix} \\ &= \frac{1}{2}\left((\mathbf{T}_1\mathbf{y} - \mathbf{U}_1\hat{\mathbf{x}})^T (\mathbf{T}_1\mathbf{y} - \mathbf{U}_1\hat{\mathbf{x}}) + (\mathbf{T}_2\mathbf{y})^T (\mathbf{T}_2\mathbf{y})\right) \end{aligned} . \tag{5.3-4}$$

Since the second term in equation (5.3-4) is not a function of $\hat{\mathbf{x}}$, the value of $\hat{\mathbf{x}}$ that minimizes J is

$$\boxed{\hat{\mathbf{x}} = \mathbf{U}_1^{-1}(\mathbf{T}_1\mathbf{y})} \tag{5.3-5}$$

where $\mathbf{T}_1\mathbf{y}$ is an n-vector. If $\mathbf{H}$ is of rank n, the upper triangular $n \times n$ matrix $\mathbf{U}_1^{-1}$ is easily computed by back-solving $\mathbf{U}_1\mathbf{U}_1^{-1} = \mathbf{I}$. Notice that the cost function is $J = (\mathbf{T}_2\mathbf{y})^T(\mathbf{T}_2\mathbf{y})/2$ for the $\hat{\mathbf{x}}$ of equation (5.3-5).

The error covariance matrix for $\hat{\mathbf{x}}$ is

$$\boxed{\begin{aligned} \mathbf{P} &= (\mathbf{H}^T\mathbf{H})^{-1} \\ &= (\mathbf{U}_1^T\mathbf{U}_1)^{-1} = \mathbf{U}_1^{-1}\mathbf{U}_1^{-T} \end{aligned}} . \tag{5.3-6}$$

Since $\mathbf{T}$ is orthogonal, it does not amplify numerical errors, and the condition number of $\mathbf{U}_1$ is the same as that of $\mathbf{H}$. Furthermore, $\mathbf{U}_1$ is never squared in the solution, so the condition number does not increase as it does when forming the normal equations. Hence sensitivity to numerical errors is approximately one-half that of the normal equations.

The $\mathbf{T}$ matrix is computed using elementary Givens rotations, orthogonal Householder transformations, or Modified Gram-Schmidt (MGS) orthogonalization (see Lawson and Hanson 1974; Bierman 1977b; Dennis and Schnabel 1983; Björck 1996; Press et al. 2007). It will be seen in the next section that it is not necessary to explicitly calculate $\mathbf{T}$.

5.3.1 Givens Rotations

Givens rotations are defined as

$$\mathbf{T} = \begin{bmatrix} c & s \\ -s & c \end{bmatrix}$$

where $c^2 + s^2 = 1$. The coefficients can be interpreted as cosine and sine functions of an angle, but trigonometric functions are not required. A Givens rotation can be applied to zero an element of a vector. For example

$$\begin{bmatrix} c & s & 0 \\ -s & c & 0 \\ 0 & 0 & 1 \end{bmatrix}\begin{bmatrix} x_1 \\ x_2 \\ x_3 \end{bmatrix} = \begin{bmatrix} \sqrt{x_1^2 + x_2^2} \\ 0 \\ x_3 \end{bmatrix}$$

when $c = x_1 / \sqrt{x_1^2 + x_2^2}$ and $s = x_2 / \sqrt{x_1^2 + x_2^2}$. Individual Givens rotations can be sequentially applied to zero all elements in a column of the **H** matrix below the diagonal where only the diagonal element is modified at each step. Then this sequence of operations is applied for remaining columns until **TH** = **U** where **T** represents the product of all individual rotations. Since the individual rotations do not modify the magnitude of the **H** columns, the transformation is numerically stable. Givens rotations are typically used to modify a few elements of a matrix or vector. While Givens rotations can be used to triangularize matrices, Householder transformations require one-half or fewer flops compared with "fast" versions of Givens rotations (Björck 1996, p. 56; Golub and Van Loan 1996, p. 218). However, Givens rotations offer more flexibility in zeroing specific sections within a matrix, and can be less sensitive to scaling problems (Björck 1996).

5.3.2 Householder Transformations

Householder transformations are called reflection operations because they behave like specular reflection of a light ray from a mirror (Fig. 5.5)

An elementary Householder reflection is defined as

$$\mathbf{T} = \mathbf{I} - \frac{\mathbf{v}\mathbf{v}^T}{(\mathbf{v}^T\mathbf{v})/2} \tag{5.3-7}$$

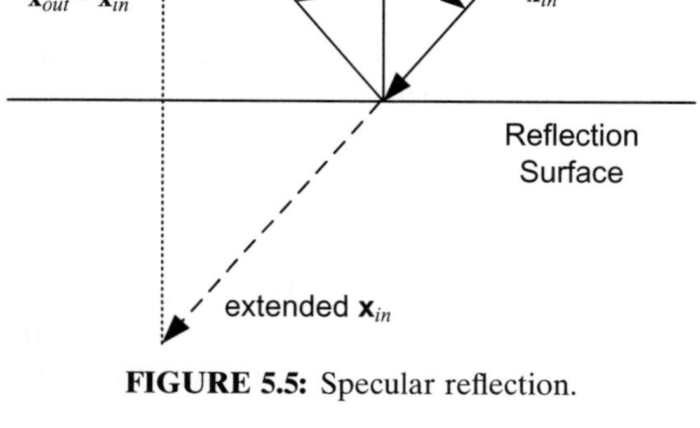

FIGURE 5.5: Specular reflection.

where **v** is the vector normal to the hyperplane of reflection. It is easily shown that **T** is orthogonal:

$$\begin{aligned}\mathbf{T}^T\mathbf{T} &= \left(\mathbf{I} - \frac{\mathbf{v}\mathbf{v}^T}{(\mathbf{v}^T\mathbf{v})/2}\right)\left(\mathbf{I} - \frac{\mathbf{v}\mathbf{v}^T}{(\mathbf{v}^T\mathbf{v})/2}\right) \\ &= \mathbf{I} - \frac{2\mathbf{v}\mathbf{v}^T}{(\mathbf{v}^T\mathbf{v})/2} + \frac{\mathbf{v}(\mathbf{v}^T\mathbf{v})\mathbf{v}^T}{(\mathbf{v}^T\mathbf{v})^2/4} \\ &= \mathbf{I}\end{aligned}.$$

The reflection property can be seen by differencing the output and input vectors,

$$\begin{aligned}\mathbf{T}\mathbf{x} - \mathbf{x} &= \mathbf{x} - \frac{\mathbf{v}\mathbf{v}^T\mathbf{x}}{(\mathbf{v}^T\mathbf{v})/2} - \mathbf{x} \\ &= -\left(\frac{2\mathbf{v}^T\mathbf{x}}{\mathbf{v}^T\mathbf{v}}\right)\mathbf{v}\end{aligned}. \tag{5.3-8}$$

The sign of all components of **x** in direction **v** are reversed by the transformation. A sequence of Householder transformations can be used to zero the lower triangle of the measurement H matrix as shown in the example below.

Example 5.5: Use of Householder Transformations to Triangularize an Overdetermined System of Equations

Assume that four measurements are used to solve for three states where the measurement equation is $\mathbf{y} = \mathbf{H}\mathbf{x} + \mathbf{r}$. We wish to multiply the left- and right-hand side of that equation by an orthogonal transformation **T** that makes **TH** an upper triangular matrix. We start by appending the measurement vector **y** to **H** to form an augmented measurement matrix:

$$[\mathbf{H} \quad \mathbf{y}] = \begin{bmatrix} H_{11} & H_{12} & H_{13} & y_1 \\ H_{21} & H_{22} & H_{23} & y_2 \\ H_{31} & H_{32} & H_{33} & y_3 \\ H_{41} & H_{42} & H_{43} & y_4 \end{bmatrix}.$$

The first transformation should eliminate H_{21}, H_{31}, and H_{41}. Since the Householder transformation produces output vector

$$\mathbf{x}_1 - \left(\frac{2\mathbf{v}_1^T\mathbf{x}_1}{\mathbf{v}_1^T\mathbf{v}_1}\right)\mathbf{v}_1$$

(where $\mathbf{x}_1$ is defined as the first column of **H**), it is apparent that the last three elements of $\mathbf{v}_1$ should equal the last three elements of $\mathbf{x}_1$, with $2\mathbf{v}_1^T\mathbf{x} = \mathbf{v}_1^T\mathbf{v}_1$. Let $\mathbf{v}_1 = [a_1 \quad H_{21} \quad H_{31} \quad H_{41}]^T$ with a_1 to be defined such that

$$2\mathbf{v}_1^T \mathbf{x}_1 = \mathbf{v}_1^T \mathbf{v}_1$$
$$2(a_1 H_{11} + H_{21}^2 + H_{31}^2 + H_{41}^2) = a_1^2 + H_{21}^2 + H_{31}^2 + H_{41}^2$$

or

$$a_1^2 - 2aH_{11} - (H_{21}^2 + H_{31}^2 + H_{41}^2) = 0.$$

Thus

$$a_1 = H_{11} \pm \sqrt{H_{11}^2 + H_{21}^2 + H_{31}^2 + H_{41}^2}.$$

To minimize numerical cancellation, it is better to use the same sign (above) as the sign of H_{11}; that is,

$$a_1 = H_{11} + \sqrt{H_{11}^2 + H_{21}^2 + H_{31}^2 + H_{41}^2}\ \mathrm{sgn}(H_{11}).$$

Thus the transformed first column of **H** equals

$$\left[\sqrt{H_{11}^2 + H_{21}^2 + H_{31}^2 + H_{41}^2} \quad 0 \quad 0 \quad 0\right]^T.$$

Notice that this transformation preserves the magnitude of the vector, which is a desirable property for numerical stability. The remaining columns of [**H** **y**] are also modified by the transformation using

$$\mathbf{x}_j - \left(\frac{2\mathbf{v}_1^T \mathbf{x}_j}{\mathbf{v}_1^T \mathbf{v}_1}\right)\mathbf{v}_1$$

where $\mathbf{x}_j$ is sequentially defined as each remaining column of [**H** **y**]. Then the process is repeated to zero the two (modified) elements of the second column below the diagonal using

$$\mathbf{v}_2 = [0 \quad a_2 \quad H_{32} \quad H_{42}]^T.$$

Finally the modified H_{43} is zeroed using

$$\mathbf{v}_3 = [0 \quad 0 \quad a_3 \quad H_{43}]^T.$$

Notice that each subsequent transformation works with one fewer elements in $\mathbf{v}_j$ than the preceding transformation ($\mathbf{v}_{j-1}$) since it only operates on **H** columns from the diagonal down. The three transformations (corresponding to the three columns) are applied as

$$\left(\mathbf{I} - \frac{2\mathbf{v}_3 \mathbf{v}_3^T}{\mathbf{v}_3^T \mathbf{v}_3}\right)\left(\mathbf{I} - \frac{2\mathbf{v}_2 \mathbf{v}_2^T}{\mathbf{v}_2^T \mathbf{v}_2}\right)\left(\mathbf{I} - \frac{2\mathbf{v}_1 \mathbf{v}_1^T}{\mathbf{v}_1^T \mathbf{v}_1}\right)[\mathbf{H} \quad \mathbf{y}] = [\mathbf{U} \quad \mathbf{Ty}]$$

to produce the upper triangular matrix **U** and transformed vector **Ty**. In implementing this algorithm it is only necessary to store the current $\mathbf{v}_i$ vector at step i, not the transformation matrix **T**.

The total number of flops required to triangularize the $m \times (n + 1)$ matrix [**H** **y**] is approximately $n^2(m - n/3)$, which is about twice the flops required to form the normal equation. Hence there is a penalty in using the QR algorithm, but the benefit is greater accuracy.

Example 5.6: Upper Triangular Factors of Sparse Normal Equations

In Example 5.2, the information matrix from the normal equations was assumed to have the structure

$$\mathbf{H}^T\mathbf{R}^{-1}\mathbf{H} = \begin{bmatrix} \mathbf{A}_{11} & \mathbf{0} & \mathbf{0} & \mathbf{A}_{14} \\ \mathbf{0} & \mathbf{A}_{22} & \mathbf{0} & \mathbf{A}_{24} \\ \mathbf{0} & \mathbf{0} & \mathbf{A}_{33} & \mathbf{A}_{34} \\ \mathbf{A}_{14}^T & \mathbf{A}_{24}^T & \mathbf{A}_{34}^T & \mathbf{A}_{44} \end{bmatrix}.$$

If one attempts to find an upper triangular matrix $\mathbf{U}$ such that $\mathbf{U}\mathbf{U}^T = \mathbf{H}^T\mathbf{R}^{-1}\mathbf{H}$, it is found that $\mathbf{U}$ will generally not contain zero partitions. However, if the state ordering is changed so that common parameters represented by $\mathbf{x}_4$ are moved to the top of the state vector, then the structure is changed to

$$\mathbf{H}^T\mathbf{R}^{-1}\mathbf{H} = \begin{bmatrix} \mathbf{A}_{44} & \mathbf{A}_{14}^T & \mathbf{A}_{24}^T & \mathbf{A}_{34}^T \\ \mathbf{A}_{14} & \mathbf{A}_{11} & \mathbf{0} & \mathbf{0} \\ \mathbf{A}_{24} & \mathbf{0} & \mathbf{A}_{22} & \mathbf{0} \\ \mathbf{A}_{34} & \mathbf{0} & \mathbf{0} & \mathbf{A}_{33} \end{bmatrix}$$

which can be factored as

$$\begin{bmatrix} \mathbf{A}_{44} & \mathbf{A}_{14}^T & \mathbf{A}_{24}^T & \mathbf{A}_{34}^T \\ \mathbf{A}_{14} & \mathbf{A}_{11} & \mathbf{0} & \mathbf{0} \\ \mathbf{A}_{24} & \mathbf{0} & \mathbf{A}_{22} & \mathbf{0} \\ \mathbf{A}_{34} & \mathbf{0} & \mathbf{0} & \mathbf{A}_{33} \end{bmatrix} = \begin{bmatrix} \mathbf{U}_1 & \mathbf{U}_2 & \mathbf{U}_3 & \mathbf{U}_4 \\ \mathbf{0} & \mathbf{U}_5 & \mathbf{0} & \mathbf{0} \\ \mathbf{0} & \mathbf{0} & \mathbf{U}_6 & \mathbf{0} \\ \mathbf{0} & \mathbf{0} & \mathbf{0} & \mathbf{U}_7 \end{bmatrix} \begin{bmatrix} \mathbf{U}_1^T & \mathbf{0} & \mathbf{0} & \mathbf{0} \\ \mathbf{U}_2^T & \mathbf{U}_5^T & \mathbf{0} & \mathbf{0} \\ \mathbf{U}_3^T & \mathbf{0} & \mathbf{U}_6^T & \mathbf{0} \\ \mathbf{U}_4^T & \mathbf{0} & \mathbf{0} & \mathbf{U}_7^T \end{bmatrix}.$$

Hence it is possible to develop a version of the QR algorithm that only operates with the nonzero partitions of $\mathbf{U}$ when processing measurements from a given ground station; for example, one need not include $\mathbf{U}_6$ and $\mathbf{U}_7$ when processing measurements from ground station #1 (corresponding to $\mathbf{U}_5$ on the diagonal). However, $\mathbf{U}_1$, $\mathbf{U}_2$, $\mathbf{U}_3$, $\mathbf{U}_4$ must be included when processing all measurements that are affected by the common parameters. The unused matrix partitions may be swapped in and out from disk to memory as needed. Implementation of this block partitioning using the QR method is more difficult than for the normal equations, but the savings can still be substantial.

5.3.3 Modified Gram-Schmidt (MGS) Orthogonalization

Another approach that can be used to triangularize a matrix is the *Modified Gram-Schmidt* (MGS) orthogonalization. In this method, the columns of $\mathbf{H}$ are sequentially transformed to a set of orthogonal columns. In the original Gram-Schmidt algorithm, the columns could gradually lose orthogonality due to numerical errors, with consequent loss of accuracy. The modification used in MGS first normalizes the columns at each step, with the result that the numerical errors are bounded. A row-oriented version of MGS for computing the QR decomposition

$$\mathbf{TH} = \begin{bmatrix} \mathbf{U} \\ \mathbf{0} \end{bmatrix}, \quad \mathbf{Ty} = \begin{bmatrix} \mathbf{z} \\ \mathbf{0} \end{bmatrix}$$

where **H** is $m \times n$, **U** is $n \times n$, **y** is $m \times 1$, and **z** is $n \times 1$ is implemented as follows:

for $k = 1$ to n
  $\mathbf{v} = \mathbf{H}(:,k)$
  $U(k,k) = \sqrt{\mathbf{v}^T\mathbf{v}}$
  $\mathbf{w}_k = \mathbf{v}/\sqrt{\mathbf{v}^T\mathbf{v}}$
  for $j = k+1$ to n
    $U(k,j) = \mathbf{w}_k^T\mathbf{H}(:,j)$
    $\mathbf{H}(:,j) = \mathbf{H}(:,j) - U(k,j)\mathbf{w}_k$
  end loop
  $z(\mathrm{k}) = \mathbf{w}_k^T\mathbf{y}$
  $\mathbf{y} = \mathbf{y} - z(k)\mathbf{w}_k$
end loop

Notice that **T** is not explicitly formed, but the $\mathbf{w}_k$ vectors $(k = 1 : n)$ contain the information required to generate **T**. Björck (1996, p. 65) shows that the MGS QR method applied to least-squares problems is equivalent to the Householder QR method for a slightly modified problem. Rather than use the Householder QR method to directly triangularize the [**H** **y**] matrix, it is applied to an augmented matrix with n rows of zeroes added at the top:

$$\begin{bmatrix} \mathbf{0}_{n\times n} & \mathbf{0}_{n\times 1} \\ \mathbf{H}_{m\times n} & \mathbf{y}_{m\times 1} \end{bmatrix}.$$

This corresponds to the modified measurement equation

$$\begin{bmatrix} \mathbf{0} \\ \mathbf{y} \end{bmatrix} = \begin{bmatrix} \mathbf{0} \\ \mathbf{H} \end{bmatrix}\mathbf{x} + \begin{bmatrix} \mathbf{0} \\ \mathbf{r} \end{bmatrix}.$$

Then a sequence of Householder transformations, denoted as **T**, is applied to obtain:

$$\mathbf{T}\begin{bmatrix} \mathbf{0} \\ \mathbf{y} \end{bmatrix} = \mathbf{T}\begin{bmatrix} \mathbf{0} \\ \mathbf{H} \end{bmatrix}\mathbf{x} + \mathbf{T}\begin{bmatrix} \mathbf{0} \\ \mathbf{r} \end{bmatrix}$$
$$\text{or} \qquad (5.3\text{-}9)$$
$$\begin{bmatrix} \mathbf{z} \\ \mathbf{e} \end{bmatrix} = \begin{bmatrix} \mathbf{U} \\ \mathbf{0} \end{bmatrix}\mathbf{x} + \mathbf{T}\begin{bmatrix} \mathbf{0} \\ \mathbf{r} \end{bmatrix}$$

where **T** is an orthogonal $(n + m) \times (n + m)$ matrix,

$$\begin{bmatrix} \mathbf{z} \\ \mathbf{e} \end{bmatrix} = \mathbf{T} \begin{bmatrix} \mathbf{0} \\ \mathbf{y} \end{bmatrix},$$

and $\mathbf{U} = \mathbf{TH}$ is an upper triangular $n \times n$ matrix. A recursive version of this algorithm (where the upper rows may already contain **U** and **z** arrays) is presented in Chapter 7 when discussing recursive least squares.

The least squares cost function is not modified by these changes, as shown by

$$\begin{aligned} J &= \frac{1}{2}(\mathbf{y} - \mathbf{H}\hat{\mathbf{x}})^T (\mathbf{y} - \mathbf{H}\hat{\mathbf{x}}) \\ &= \frac{1}{2} \begin{bmatrix} \mathbf{0} \\ \mathbf{y} - \mathbf{H}\hat{\mathbf{x}} \end{bmatrix}^T \begin{bmatrix} \mathbf{0} \\ \mathbf{y} - \mathbf{H}\hat{\mathbf{x}} \end{bmatrix} \\ &= \frac{1}{2} \begin{bmatrix} \mathbf{0} \\ \mathbf{y} - \mathbf{H}\hat{\mathbf{x}} \end{bmatrix}^T \mathbf{T}^T \mathbf{T} \begin{bmatrix} \mathbf{0} \\ \mathbf{y} - \mathbf{H}\hat{\mathbf{x}} \end{bmatrix} \\ &= \frac{1}{2}\left((\mathbf{z} - \mathbf{U}\hat{\mathbf{x}})^T (\mathbf{z} - \mathbf{U}\hat{\mathbf{x}}) + \mathbf{e}^T \mathbf{e}\right) \end{aligned} \quad (5.3\text{-}10)$$

Provided that the upper triangular $n \times n$ matrix **U** is of rank n, $\hat{\mathbf{x}} = \mathbf{U}^{-1}\mathbf{z}$ minimizes J with the result that $J = \mathbf{e}^T\mathbf{e}/2$.

Throughout this book we use "Householder QR" to mean the standard Householder QR method, "MGS QR" to mean the MGS/augmented Householder method (since they are equivalent), and "QR" algorithm to refer to either.

Golub and Van Loan (1996, p. 241) note that the MGS QR method has the same theoretical accuracy bounds as Householder QR, provided that it is implemented as an augmented system so that transformations also operate on **y** (as described above). However, several investigators have found that the MGS QR method is slightly more accurate than alternative orthogonalization methods (Björck 1996, p. 66), although the reason is not clear. Björck notes that the Householder QR method is sensitive to row permutations of **H**, whereas MGS is almost insensitive because it zeroes entire columns of **H** rather than just from the diagonal down. Björck suggests that when using the Householder QR method for weighted least-squares problems, the rows should be sorted in decreasing row norm before factorization, or column pivoting should be used. Unfortunately sorting in decreasing row norm order does not always solve the problem, as will be shown later.

Another benefit of the MGS QR approach is that measurements need not be processed all at once in a single batch, which is desirable (to avoid storing **H**) when n and m are large. That is, measurements may be processed in small batches (e.g., 100) where the nonzero **U** and **z** arrays from one batch are retained when processing the next batch. However, to implement this approach, additional terms must be included in the "row-oriented version of MGS" algorithm listed above. This recursive procedure is, in fact, the measurement update portion of Bierman's (1977b) Square Root Information Filter (SRIF) that will be discussed in Chapter 10. Recursive updating of the Householder QR method can also be accomplished using Givens rank-1 updates (Björck 1996, p. 132). For more information on orthogonal transformations and the QR algorithm see Lawson and Hanson (1974), Björck (1996), Dennis and Schnabel (1983), Golub and Van Loan (1996), Bierman (1977b), and Press et al. (2007).

5.3.4 QR Numerical Accuracy

To evaluate the numerical accuracy of the QR algorithm, we again examine the error in the state estimate due to numerical errors only. Using equation (5.3-5),

$$\begin{aligned}\hat{\mathbf{x}} - \mathbf{x} &= \mathbf{U}_1^{-1}(\mathbf{T}_1\mathbf{y}) - \mathbf{x} \\ &= \mathbf{U}_1^{-1}\mathbf{T}_1(\mathbf{H}\mathbf{x} + \mathbf{r}) - \mathbf{x} \\ &= (\mathbf{U}_1^{-1}\mathbf{U}_1 - \mathbf{I})\mathbf{x} + \mathbf{U}_1^{-1}\mathbf{T}_1\mathbf{r}\end{aligned} \quad .$$

However, $(\mathbf{U}_1^{-1}\mathbf{U}_1 - \mathbf{I})$ is likely to underestimate the actual errors since it ignores errors in forming $\mathbf{U}_1$. Therefore, Monte Carlo evaluation of $\mathbf{U}_1^{-1}(\mathbf{T}_1\mathbf{y}) - \mathbf{x}$, where $\mathbf{y} = \mathbf{H}\mathbf{x}$, is deemed a more reliable indicator of numerical errors. Numerical results using the QR algorithm for our example polynomial problem are given later in Section 5.6.

5.4 LEAST-SQUARES SOLUTION USING THE SVD

Another technique based on orthogonal transformations, the SVD, can also be used to solve the least-squares problem. As before, we start by computing the SVD of the measurement matrix: $\mathbf{H} = \mathbf{U}\mathbf{S}\mathbf{V}^T$ where $\mathbf{U}$ is an $m \times m$ orthogonal (not upper triangular) matrix, $\mathbf{V}$ is an $n \times n$ orthogonal matrix, and $\mathbf{S}$ is an $m \times n$ upper diagonal matrix. When $m > n$ (the overdetermined case), the SVD can be written as

$$\mathbf{H} = [\mathbf{U}_1 \quad \mathbf{U}_2]\begin{bmatrix}\mathbf{S}_1 \\ \mathbf{0}\end{bmatrix}\mathbf{V}^T$$

where $\mathbf{U}_1$ is $n \times n$, $\mathbf{U}_2$ is $(m - n) \times n$, and $\mathbf{S}_1$ is $n \times n$. $\mathbf{U}_2$ is the nullspace of $\mathbf{H}$ as it multiplies $\mathbf{0}$. Hence it is not needed to construct a least-squares solution and is often not computed by SVD algorithms. Golub and Van Loan (1996, p. 72) call SVD solutions without the $\mathbf{U}_2$ term a "thin SVD."

The least-squares solution can be obtained by substituting this replacement for $\mathbf{H}$ in the cost function and then setting the gradient to zero. However, it is easier to make the substitution in the normal equations $\hat{\mathbf{x}} = (\mathbf{H}^T\mathbf{H})^{-1}\mathbf{H}^T\mathbf{y}$ to obtain:

$$\boxed{\begin{aligned}\hat{\mathbf{x}} &= (\mathbf{V}\mathbf{S}^T\mathbf{U}^T\mathbf{U}\mathbf{S}\mathbf{V}^T)^{-1}\mathbf{V}\mathbf{S}^T\mathbf{U}^T\mathbf{y} \\ &= \mathbf{V}(\mathbf{S}^T\mathbf{S})^{-1}\mathbf{V}^T\mathbf{V}\mathbf{S}^T\mathbf{U}^T\mathbf{y} \\ &= \mathbf{V}[\mathbf{S}_1^{-1} \quad \mathbf{0}]\mathbf{U}^T\mathbf{y} \\ &= \mathbf{V}\mathbf{S}_1^{-1}\mathbf{U}_1^T\mathbf{y}\end{aligned}} \tag{5.4-1}$$

where $\mathbf{S}_1^{-1}$ is the $n \times n$ diagonal matrix of inverse singular values and $\mathbf{U}_1^T$ is the transpose of the first n columns of $\mathbf{U}$. The error in the least-squares estimate is

$$\begin{aligned}\hat{\mathbf{x}} - \mathbf{x} &= \mathbf{V}\mathbf{S}_1^{-1}\mathbf{U}_1^T(\mathbf{H}\mathbf{x} + \mathbf{r}) - \mathbf{x} \\ &= (\mathbf{V}\mathbf{S}_1^{-1}\mathbf{U}_1^T\mathbf{H} - \mathbf{I})\mathbf{x} + \mathbf{V}\mathbf{S}_1^{-1}\mathbf{U}_1^T\mathbf{r}\end{aligned} \cdot$$

Hence either the magnitude of elements in $\mathbf{V}\mathbf{S}_1^{-1}\mathbf{U}_1^T\mathbf{H} - \mathbf{I}$ or the Monte Carlo statistics on $(\mathbf{V}\mathbf{S}_1^{-1}\mathbf{U}_1^T\mathbf{H} - \mathbf{I})\mathbf{x}$ can be used as a measure of numerical accuracy. Numerical results for our polynomial problem are given in Section 5.6.

The most successful SVD algorithms use Householder transformations to first convert **H** to a bi-diagonal form (nonzero elements are limited to the diagonal, and either immediately above or below the diagonal). Then iterative or divide-and-conquer techniques are used to converge on the singular values in **S**. Many methods are based on a routine written by Forsythe et al. (1977, chapter 9), which was based on the work of Golub and Reinsch (see Wilkinson and Reinsch 1971; Dongarra et al. 1979; Stoer and Bulirsch 1980). Since the solution is iterative, the total number of flops varies with the problem. It can be generally expected that the computational time required to obtain an SVD solution will be five or more times greater than the time required to solve the problem using the QR methods, although the ratio will be problem-dependent.

One popular reference (Press et al. 2007) states that the SVD is the method of choice for solving most least-squares problems. That advice may reflect the experience of the authors, but one suspects that their experience is limited to small problems. The arguments for using the SVD are

1. The SVD is numerically accurate and reliable.
2. The SVD automatically indicates when the system of equations is numerically unobservable, and it provides a mechanism to obtain a minimal norm solution in these cases.
3. The SVD provides information on linear combinations of states that are unobservable or weakly observable.

However, the first two arguments are not by themselves convincing (because QR methods also have these properties), and there are several reasons why use of the SVD should be avoided for problems where the number of measurement is large and much greater than the number of unknowns:

1. Multistep algorithms used to compute the SVD may result in an accumulation of numerical errors, and they can potentially be less accurate than the QR algorithm.
2. In some real problems tens of thousands of measurements are required to obtain an accurate solution with thousands of unknown states. Hence storage of the **H** matrix may require gigabytes of memory, and computational time for the SVD may be an order of magnitude (or more) greater than for the QR algorithms. If observability is to be analyzed via the SVD, the QR algorithms should first be used to transform the measurement equations to the compact n-element form $\mathbf{z} = \mathbf{U}\mathbf{x} + \mathbf{e}$. Then the SVD can be used on $\mathbf{z}$ and $\mathbf{U}$ with no loss of accuracy. This is the method recommended by Lawson and Hanson (1974, p. 290) when mn is large and $m \gg n$. Results later in this chapter show that accuracy is enhanced and execution time is greatly reduced when using this procedure.
3. Most other solution techniques also provide information on conditioning of the solution. Minimal norm solutions in unobservable cases can be computed using other techniques.
4. Use of the SVD to obtain a minimal-norm solution in cases where the system of equations is unobservable implicitly applies a linear constraint on the estimated parameters. That is, the state estimate from the pseudo-inverse is a

linear combination of the model states. If the only use of the least-squares model is to obtain smoothed or predicted values of the same quantities that are measured, this is usually acceptable. However, in many problems the states represent physical parameters, and estimation of those states is the goal. Other estimation applications attempt to predict quantities not directly measured, or predict to untested conditions. In these cases it is often not acceptable to compute a minimal-norm (pseudo-inverse) solution when the measurements alone are not sufficient to obtain a solution—this is demonstrated in a later example. Better approaches either obtain more measurements or use prior information.

5.5 ITERATIVE TECHNIQUES

Linear least-squares problems can also be solved using iterative techniques that generally fall into two categories: linear iteration and Krylov space. Iterative methods for solving nonlinear least-squares problems are discussed in Chapter 7. Before addressing iterative approaches, we first review sparse matrix storage methods because iterative methods are mostly used for large sparse systems.

5.5.1 Sparse Array Storage

One commonly used sparse array storage method places all nonzero elements of sparse $m \times n$ matrix $\mathbf{A}$ in a real vector $\mathbf{a}$ of dimension p. The column index of elements in $\mathbf{a}$ are stored in integer vector $\mathbf{ja}$, and another integer vector $\mathbf{ia}$ of dimension $m + 1$ stores the $\mathbf{ja}$ and $\mathbf{a}$ indices of the first element belonging to each $\mathbf{A}$ row. That is, elements in row i of matrix $\mathbf{A}$ will be found in elements $ia(i)$ to $ia(i + 1) - 1$ of vectors $\mathbf{ja}$ and $\mathbf{a}$, where $ja(k)$ specifies the column index for $a(k)$. Hence a matrix-vector product $\mathbf{c} = \mathbf{Ab}$ can be implemented using this storage scheme as follows:

$$
\begin{aligned}
&\text{for } i = 1 \text{ to } m \\
&\quad c(i) = 0. \\
&\quad \text{for } k = ia(i) \text{ to } ia(\mathrm{i}+1)-1 \\
&\qquad c(i) = c(i) + a(k)*b(ja(k)) \\
&\quad \text{end loop} \\
&\text{end loop}
\end{aligned}
$$

The transpose product $\mathbf{c} = \mathbf{A}^T\mathbf{b}$ is implemented in a slightly different order:

$$
\begin{aligned}
&\mathbf{c}(1:n) = 0. \\
&\text{for } i = 1 \text{ to } m \\
&\quad \text{for } k = ia(i) \text{ to } ia(\mathrm{i}+1)-1 \\
&\qquad jc = ja(k) \\
&\qquad c(jc) = c(jc) + a(k) * b(i) \\
&\quad \text{end loop} \\
&\text{end loop}
\end{aligned}
$$

where we have used the MATLAB/Fortran 90 convention for specifying partitions of arrays and vectors; that is, $\mathbf{c}(1\!:\!n) = c(i),\ 1 \le i \le n$. Although both of the above matrix-vector multiplications involve some "hopping" in computer memory, compilers effectively optimize the above loops and execution can be efficient. Other sparse storage methods are possible (Björck 1996, p. 270; Anderson et al. 1999; Press et al. 2007, p. 71), and they may be more efficient for operations involving specific types of sparse arrays. Finally it should be noted that matrix-matrix products can be efficient even when sparsity is evaluated "on-the-fly." For example, a general-purpose version of $\mathbf{C} = \mathbf{AB}$, where $\mathbf{A}$ is sparse, can be implemented without using sparse storage methods as follows:

```
for i = 1 to m
  k = 0
  for j = 1 to n
    if |A(i,j)| > ε, then
      k = k + 1
      ja(k) = j
      d(k) = A(i,j)
    end if
  end loop
  C(i,:) = 0.
  for j = 1 to k
    C(i,:) = C(i,:) + d(j) * B(ja(j),:)
  end loop
end loop
```

where ε is a threshold for nonzero elements (e.g., 10^{-60} or even 0). Temporary n-element vectors **d** and **ja** are used to store the nonzero element values and column indices of each **A** row. This method becomes somewhat less efficient when **B** is a vector, but even when **A** is a full matrix the execution time may only increase slightly compared with a normal matrix product.

5.5.2 Linear Iteration

Iterative techniques can be used to refine an approximate solution of the equation $\mathbf{Ax} = \mathbf{b}$. The simplest method is linear iteration. Let $\hat{\mathbf{x}}$ be the exact solution to the normal equations $(\mathbf{H}^T\mathbf{H}\hat{\mathbf{x}} = \mathbf{H}^T\mathbf{y})$ and let $\hat{\mathbf{x}}'$ be an approximate solution. Then assuming linearity,

$$\mathbf{H}^T\mathbf{H}\hat{\mathbf{x}}' = \mathbf{H}^T\mathbf{y} + \mathbf{H}^T\mathbf{H}(\hat{\mathbf{x}}' - \hat{\mathbf{x}}) \tag{5.5-1}$$

which can be arranged to obtain

$$\mathbf{H}^T\mathbf{H}(\hat{\mathbf{x}} - \hat{\mathbf{x}}') = \mathbf{H}^T\mathbf{y} - \mathbf{H}^T\mathbf{H}\hat{\mathbf{x}}'. \tag{5.5-2}$$

The quantity on the right should be evaluated before inverting $\mathbf{H}^T\mathbf{H}$ to obtain $\hat{\mathbf{x}} - \hat{\mathbf{x}}'$. Iterative refinement of this type is rarely used because accuracy is usually not much better than that of direct normal equation solutions. This statement also applies when $\mathbf{H}^T\mathbf{H}$ is Cholesky factored and the factors are back-solved to compute the solution to equation (5.5-2). The loss of precision in forming $\mathbf{H}^T\mathbf{H}$ is the limiting factor.

Other classical iteration techniques include *Jacobi*, *Gauss-Seidel*, *successive overrelaxation* (SOR), and *Richardson's first-order method* (Kelley 1995; Björck 1996). These methods split the $\mathbf{A}$ matrix in $\mathbf{Ax} = \mathbf{b}$ as $\mathbf{A} = \mathbf{A}_1 + \mathbf{A}_2$ where $\mathbf{A}_1$ is a nonsingular matrix constructed so that $\mathbf{A}_1\mathbf{x} = (\mathbf{b} - \mathbf{A}_2\mathbf{x})$ can be easily solved for $\mathbf{x}$.

5.5.3 Least-Squares Solution for Large Sparse Problems Using Krylov Space Methods

Another category of iterative techniques is used to solve large sparse linear least-squares problems when it is not practical to store the entire information matrix $\mathbf{H}^T\mathbf{H}$ in computer memory. $\mathbf{H}^T\mathbf{H}$ generally fills in sparsity present in the original $\mathbf{H}$ matrix, and thus the storage requirements can be prohibitive. However, several Krylov space methods can solve the least-squares problem using $\mathbf{H}$ without forming $\mathbf{H}^T\mathbf{H}$.

A Krylov subspace is defined as $\kappa_k(\mathbf{A},\mathbf{x}) = \text{span}\{\mathbf{x},\mathbf{Ax},\mathbf{A}^2\mathbf{x},\ldots,\mathbf{A}^{k-1}\mathbf{x}\}$ where $\mathbf{A}$ is an $n \times n$ matrix and $\mathbf{x}$ is an n-element vector. Krylov space methods are iterative techniques used to either find the zeros of an affine function $\mathbf{Ax} - \mathbf{b} = \mathbf{0}$, or to locate the minimum of a multidimensional cost function $J = (\mathbf{y} - \mathbf{Hx})^T(\mathbf{y} - \mathbf{Hx})$. Krylov space algorithms fall into three classes. The first, *conjugate gradient* (CG) methods, assume that $\mathbf{A}$ is square and positive definite symmetric (Kelley 1995, Section 2.4; Björck 1996, Section 7.4). CG solvers work well with sparse array storage and are often used to solve finite-difference and finite-element problems. Unlike steepest descent methods for minimization problems, they do not take negative gradient steps. Rather, they attempt to take a sequence of steps that are orthogonal to previous steps within the Krylov subspace and in directions that minimize the cost function. As a worst case, CG algorithms should converge exactly when the number of iterations equals the number of unknown states, but if the conditioning is poor, this may not happen because of numerical errors. Furthermore, the convergence rate is unacceptably slow when the number of unknowns is thousands or hundreds of thousands. Conversely, CG algorithms can converge in tens of iterations if the conditioning is good. The *bi-CG stabilized* (Bi-CGSTAB) algorithm is one of the most successful of the CG methods applicable to general square matrices (see Van der Vorst 1992; Barrett et al. 1993, p.27; Kelley 1995, p. 50). Bi-CGSTAB can also handle square positive definite nonsymmetric $\mathbf{A}$ matrices, but convergence may be slower than for symmetric $\mathbf{A}$.

The second class of Krylov space algorithms also takes a sequence of steps that are orthogonal to previous steps, but rather than recursively accumulating information "summarizing" past search directions—as done in CG solvers—these methods store past basis vectors. Because of this, storage requirements increase as iterations proceed. Implementations usually limit the storage to a fixed number of past vectors.

Generalized minimum residual (GMRES) and ORTHOMIN are two of the most successful algorithms in this class (see Vinsome 1976; Kelley 1995; Barrett et al. 1993). ORTHOMIN and GMRES also work directly with overdetermined (non-square) **H** matrices to minimize a quadratic cost function.

This author found that the Bi-CGSTAB method converged faster than ORTHOMIN or GMRES when solving steady-state groundwater flow problems of the form $\mathbf{Ax} - \mathbf{b} = \mathbf{0}$. However, Bi-CGSTAB requires that matrix **A** be square, so it can only be used for least-squares problems when starting with the normal equations. This is undesirable because the condition number of $\mathbf{H}^T\mathbf{H}$ is twice that of **H**, and convergence of CG solvers is a function of the conditioning. Also the sparsity of $\mathbf{H}^T\mathbf{H}$ is generally much less than that of **H**, so storage and computations may greatly increase.

The third class of Krylov space methods was designed to work directly with least-squares problems. The most popular methods include *CG on the normal equations to minimize the residual* (CGNR; Barrett et al. 1993, p. 18; Kelley 1995, p. 25; Björck 1996, p. 288) and *LSQR* (Paige and Saunders 1982). CGNR is also called CGLS by Paige and Saunders. Even though these methods do not form $\mathbf{H}^T\mathbf{H}$, they are still affected by the doubling of the condition number implied by forming $\mathbf{H}^T\mathbf{H}$. CGNR and LSQR both use the measurement equations $\mathbf{v} = \mathbf{y} - \mathbf{H}\hat{\mathbf{x}}$ and $\mathbf{H}^T\mathbf{v} = \mathbf{0}$, and iterate many times until the residuals are acceptably small. The LSQR algorithm attempts to minimize the effect of conditioning problems by solving the linearized least-squares problem using a Lanczos bi-diagonalization approach. LSQR is claimed to be more numerically stable than CGNR. Storage and computations are generally low, but for many flow-type problems, the performance of the LSQR algorithm *without preconditioning* appears to be almost identical to that of CG solutions on the normal equations. That is, the iterations tend to stall (do not reduce the residual error) after a few thousand iterations, and they can produce a solution that is unacceptable (Gibbs 1998).

The convergence rate of Krylov space methods can be greatly improved if *preconditioning* is used to transform **H** or $\mathbf{H}^T\mathbf{H}$ to a matrix in which the eigenvalues are clustered close to 1. This can be implemented as either left or right preconditioning; for example, the root-finding problem $\mathbf{Hx} = \mathbf{y}$ can be preconditioned as

$$(\mathbf{S}_L\mathbf{H}\mathbf{S}_R)\mathbf{S}_R^{-1}\mathbf{x} = \mathbf{S}_L\mathbf{y} \tag{5.5-3}$$

where $\mathbf{S}_L$ and $\mathbf{S}_R$ are the left and right preconditioning matrices. Left preconditioning modifies scaling of the measurement residuals—which may be desirable in poorly scaled problems—while right preconditioning modifies scaling of the state **x**. Generally either left or right preconditioning is used, but not both. For the least-squares problem, direct preconditioning of $\mathbf{H}^T\mathbf{H}$ is not desirable because this destroys matrix symmetry. Rather, preconditioning is of the form

$$(\mathbf{S}_R^T\mathbf{H}^T\mathbf{S}_L^T)(\mathbf{S}_L\mathbf{H}\mathbf{S}_R)\mathbf{S}_R^{-1}\mathbf{x} = \mathbf{S}_R^T\mathbf{H}^T\mathbf{S}_L^T\mathbf{S}_L\mathbf{y}. \tag{5.5-4}$$

The goal is to find matrix $\mathbf{S}_L$ or $\mathbf{S}_R$ such that $(\mathbf{S}_R^T\mathbf{H}^T\mathbf{S}_L^T)(\mathbf{S}_L\mathbf{H}\mathbf{S}_R)$ is close to the identity matrix. Preconditioning can either be applied as a first step where $\mathbf{S}_L\mathbf{H}\mathbf{S}_R$ is explicitly computed, or the multiplication can be included as part of the CG algorithm. Commonly used preconditioners include scaling by the inverse diagonals of $\mathbf{H}^T\mathbf{H}$ (the sum-of-squares of the columns of $\mathbf{H}$), incomplete sparse Cholesky decomposition of the form $\mathbf{L}\mathbf{L}^T + \mathbf{E} = \mathbf{H}^T\mathbf{H}$ where $\mathbf{E}$ is small, incomplete $\mathbf{LU}$ decomposition for square $\mathbf{H}$, or polynomial preconditioning (see Simon 1988; Barrett et al. 1993; Kelley 1995; Björck 1996). Diagonal scaling is generally not very effective and it can sometimes be difficult to compute incomplete Cholesky factors or polynomial preconditioners.

Another alternative to direct application of the CGNR method changes variables and inverts $(\mathbf{H}^T\mathbf{H})\hat{\mathbf{x}} = \mathbf{H}^T\mathbf{y}$ in partitions, where the partitions of $\mathbf{H}$ are implicitly inverted using other methods. This approach is somewhat similar to the red-black partitioning (Dongarra et al. 1991; Barrett et al. 1993, section 5.3) often used in iterative solution of sparse systems, and it can also be interpreted as a form of preconditioning. Numerical conditioning for the CG solver is improved because the dimension of the system to be solved is reduced. For flow and similar models resulting from finite-difference equations, those portions of the $\mathbf{H}$ matrix can be (implicitly) inverted efficiently using nonsymmetric *preconditioned CG* (PCG) solvers such as the Bi-CGSTAB method.

The CGNR algorithm applied to equation $(\mathbf{H}^T\mathbf{H})\hat{\mathbf{x}} = \mathbf{H}^T\mathbf{y}$ is given below. Initial values of the residual and scaled state are computed as

$$\begin{aligned}
\hat{\mathbf{x}}_0 &= \mathbf{S}^{-1}\hat{\mathbf{x}}_{in} \\
\mathbf{s}_0 &= \mathbf{S}^T\mathbf{H}^T(\mathbf{y} - \mathbf{H}\hat{\mathbf{x}}_{in}) \\
\mathbf{p}_1 &= \mathbf{s}_0
\end{aligned} \tag{5.5-5}$$

where $\mathbf{S}$ is a right preconditioning matrix. Then the CGNR iterations for $i = 1$, 2, ... are

$$\begin{aligned}
&\tau_i = \mathbf{s}_{i-1}^T\mathbf{s}_{i-1} \\
&\text{if } i > 1, \mathbf{p}_i = \mathbf{s}_{i-1} + (\tau_i / \tau_{i-1})\mathbf{p}_{i-1} \\
&\mathbf{q}_i = \mathbf{H}\mathbf{S}\mathbf{p}_i \\
&\alpha = \tau_i / \mathbf{q}_i^T\mathbf{q}_i \\
&\hat{\mathbf{x}}_i = \hat{\mathbf{x}}_{i-1} + \alpha\mathbf{p}_i \\
&\mathbf{s}_i = \mathbf{s}_{i-1} - \alpha\mathbf{S}^T\mathbf{H}^T\mathbf{q}_i \\
&\text{if } \|\mathbf{s}_i\| > \varepsilon\|\mathbf{s}_0\|, \text{set } i = i + 1 \text{ and return to top} \\
&\mathbf{x}_{out} = \mathbf{S}\mathbf{x}_i
\end{aligned} \quad . \tag{5.5-6}$$

The LSQR algorithm (Paige and Saunders 1982) without preconditioning is initialized as

$$\begin{aligned}
&\beta_1 = \|\mathbf{y}\|, \quad \mathbf{u}_1 = \mathbf{y}/\beta_1 \\
&\alpha_1 = \|\mathbf{A}^T\mathbf{u}_1\|, \quad \mathbf{v}_1 = \mathbf{A}^T\mathbf{u}_1/\alpha_1 \\
&\mathbf{w}_1 = \mathbf{v}_1, \quad \mathbf{x}_0 = \mathbf{0} \\
&\bar{\phi}_1 = \beta_1, \quad \bar{\rho}_1 = \alpha_1
\end{aligned} \quad . \tag{5.5-7}$$

The LSQR iterations for $i = 1, 2, \ldots$ are

$$
\begin{aligned}
\beta_{i+1} &= \|\mathbf{A}\mathbf{v}_i - \alpha_i \mathbf{u}_i\|_2, \quad \mathbf{u}_{i+1} = (\mathbf{A}\mathbf{v}_i - \alpha_i \mathbf{u}_i)/\beta_{i+1} \\
\alpha_{i+1} &= \|\mathbf{A}^T \mathbf{u}_{i+1} - \beta_{i+1}\mathbf{v}_i\|_2, \quad \mathbf{v}_{i+1} = (\mathbf{A}^T \mathbf{u}_{i+1} - \beta_{i+1}\mathbf{v}_i)/\alpha_{i+1} \\
\rho_i &= (\bar{\rho}_i^2 + \beta_{i+1}^2)^{1/2} \\
c_i &= \bar{\rho}_i / \rho_i \\
s_i &= \beta_{i+1} / \rho_i \\
\theta_{i+1} &= s_i \alpha_{i+1} \\
\bar{\rho}_{i+1} &= -c_i \alpha_{i+1} \\
\phi_i &= c_i \bar{\phi}_i \\
\bar{\varphi}_{i+1} &= s_i \bar{\varphi}_i \\
\mathbf{x}_i &= \mathbf{x}_{i-1} + (\phi_i / \rho_i)\mathbf{w}_i \\
\mathbf{w}_{i+1} &= \mathbf{v}_{i+1} - (\theta_{i+1} / \rho_i)\mathbf{w}_i
\end{aligned}
\tag{5.5-8}
$$

where $\bar{\varphi}_{i+1}$ is the residual vector norm $\|\mathbf{r}_k\|_2 = \bar{\varphi}_{i+1}$. Iterations stop if any of these tests are satisfied:

$$
\begin{aligned}
&\|\mathbf{r}_k\|_2 \le (\mathrm{BTOL}) \cdot \|\mathbf{b}\|_2 + (\mathrm{ATOL}) \cdot \|\mathbf{x}_k\|_2 \\
&\|\mathbf{A}^T \mathbf{r}_k\|_2 \le (\mathrm{ATOL}) \cdot \|\mathbf{A}\|_2 \|\mathbf{r}_k\|_2 \\
&cond(\mathbf{A}) \ge \mathrm{CONLIM}
\end{aligned}
\tag{5.5-9}
$$

ATOL, BTOL, and CONLIM are user-specified tolerances and $cond(\mathbf{A})$ is an estimate of the condition number of **A** as explained by Paige and Saunders (1982). To obtain a solution accurate to six digits, suggested values for ATOL and BTOL are 10^{-6}, and the suggested CONLIM is $1/(10\sqrt{\mathrm{RELPR}})$ where RELPR is the relative precision of the computer (e.g., 1.1×10^{-16} for double precision in IEEE T_floating). Preconditioning can be applied to LSQR or CGNR by redefining the **A** matrix.

The primary function of an iterative least-squares solver is to obtain a solution faster and with less storage than a direct method such as QR. If the estimate error covariance matrix or log likelihood function is required, one final iteration of a direct method must be used.

Example 5.7: Groundwater Flow Modeling Calibration Problem

Consider a flow modeling least-squares problem (Porter et al. 1997) consisting of three linearized matrix equations:

$$
\begin{bmatrix} \mathbf{B}_h & \mathbf{B}_k & \mathbf{B}_b \\ \mathbf{0} & \mathbf{C} & \mathbf{0} \\ \mathbf{H}_h & \mathbf{H}_k & \mathbf{H}_b \end{bmatrix}
\begin{bmatrix} \delta\mathbf{h} \\ \delta\mathbf{k} \\ \delta\mathbf{b} \end{bmatrix}
=
\begin{bmatrix} \mathbf{0} \\ \mathbf{v} \\ \mathbf{r} \end{bmatrix}
$$

where

$\delta\mathbf{h}$ are perturbations in hydraulic "head" at each node (n total) of a three-dimensional (3-D) finite difference grid,

$\delta\mathbf{k}$ are perturbations in hydraulic conductivity at each node of a 3-D finite difference grid,

$\delta\mathbf{b}$ are perturbations in p flow common parameters such as boundary conditions,

$\mathbf{B}_h = \partial\mathbf{f}/\partial\mathbf{h}$ and $\mathbf{B}_k = \partial\mathbf{f}/\partial\mathbf{k}$ are $n \times n$ linearizations of the steady-state flow equation $\mathbf{f}(\mathbf{h},\mathbf{k},\mathbf{b}) = \mathbf{0}$ where $\mathbf{f}$ is the net flow to the nodes,

$\mathbf{B}_b = \partial\mathbf{f}/\partial\mathbf{b}$ is a $n \times p$ linearization of the steady-state flow equation $\mathbf{f}(\mathbf{h},\mathbf{k},\mathbf{b}) = \mathbf{0}$,

$\mathbf{C}$ is a "soft" $n \times n$ spatial continuity conductivity constraint implemented as a 3-D autoregressive model,

$\mathbf{v}$ is the conductivity constraint error at each node,

$\mathbf{H}_h$ is the $m \times n$ measurement partial matrix (Jacobian) for hydraulic head measurements (each measurement is a function of head at four surrounding nodes),

$\mathbf{H}_k$ is the $m \times n$ measurement partial matrix for hydraulic conductivity measurements (each measurement is a function of conductivity at four surrounding nodes),

$\mathbf{H}_b$ is the $m \times p$ measurement partial matrix for hydraulic conductivity measurements with respect to common parameters (usually $\mathbf{H}_b = \mathbf{0}$), and

$\mathbf{r}$ is random measurement noise.

All the above arrays are very sparse because each node connects to six (or less) other nodes. Typically only a few hundred head and conductivity measurements are available for site models consisting of 10^4 to 10^6 nodes. Hence the spatial continuity "pseudo-measurement" conductivity constraint is required to obtain a solution.

Since the first (flow) equation is a hard constraint and the square $\mathbf{B}_h$ matrix should be nonsingular, $\delta\mathbf{h} = -\mathbf{B}_h^{-1}(\mathbf{B}_k\delta\mathbf{k} + \mathbf{B}_b\delta\mathbf{b})$ can be substituted to obtain the reduced equations

$$\begin{bmatrix} \mathbf{C} & \mathbf{0} \\ \mathbf{H}_k - \mathbf{H}_h\mathbf{B}_h^{-1}\mathbf{B}_k & \mathbf{H}_b - \mathbf{H}_h\mathbf{B}_h^{-1}\mathbf{B}_b \end{bmatrix} \begin{bmatrix} \delta\mathbf{k} \\ \delta\mathbf{b} \end{bmatrix} = \begin{bmatrix} \mathbf{v} \\ \mathbf{r} \end{bmatrix}.$$

Since square matrix $\mathbf{C}$ is formed using an autoregressive model applied to nearest neighbors, it should also be nonsingular. This suggests that $\mathbf{C}$ can be used as a right preconditioner to obtain the modified measurement equation

$$\begin{bmatrix} \mathbf{I} & \mathbf{0} \\ (\mathbf{H}_k - \mathbf{H}_h\mathbf{B}_h^{-1}\mathbf{B}_k)\mathbf{C}^{-1} & (\mathbf{H}_b - \mathbf{H}_h\mathbf{B}_h^{-1}\mathbf{B}_b) \end{bmatrix} \begin{bmatrix} \mathbf{C}\delta\mathbf{k} \\ \delta\mathbf{b} \end{bmatrix} = \begin{bmatrix} \mathbf{v} \\ \mathbf{r} \end{bmatrix}.$$

Notice that the CGNR operations in equation (5.5-6) involving $\mathbf{H}$ appear as either $(\mathbf{HS})\mathbf{p}_i$ or $(\mathbf{HS})^T\mathbf{q}_i$—a matrix times a vector. Hence $(\mathbf{H}_k - \mathbf{H}_h\mathbf{B}_h^{-1}\mathbf{B}_k)\mathbf{C}^{-1}\mathbf{p}_i$ can be evaluated as follows:

1. $\mathbf{z} = \mathbf{C}^{-1}\mathbf{p}_i$ is computed using the Bi-CGSTAB solver (use incomplete LU decomposition preconditioning with the same fill as the original matrix, and red-black partitioning),

2. $\mathbf{s} = \mathbf{B}_h^{-1}(\mathbf{B}_k \mathbf{z})$ is computed using the Bi-CGSTAB solver (again use incomplete LU decomposition preconditioning with the same fill as the original matrix, and red-black partitioning),
3. $\mathbf{a} = \mathbf{H}_k \mathbf{z} - \mathbf{H}_h \mathbf{s}$.

Likewise $\mathbf{C}^{-T}(\mathbf{H}_k^T - \mathbf{B}_k^T \mathbf{H}_h^T \mathbf{B}_h^{-T})\mathbf{q}_i$ can be evaluated using a similar sequence of operations. The substitution procedure can also be used for terms involving the boundary condition parameters.

When using this procedure for the groundwater flow modeling calibration problem, CGNR convergence is usually achieved in less than 100 CGNR iterations even when using 200000 node models (Gibbs 1998). Although this problem is unusual in that the number of actual measurements (not pseudo-measurements) is much smaller than the number of unknowns, the solution techniques can be applied to other 2-D or 3-D problems.

Example 5.8: Polynomial Problem Using CGNR and LSQR Solvers

The CGNR solver was also tested on the polynomial problem of Example 5.4. Right diagonal preconditioning was based on the inverse l_2-norms of the columns of $\mathbf{H}$, so the columns of $\mathbf{HS}$ were normalized to one. However, the results did not change significantly without preconditioning. Table 5.1 lists the results for CGNR and LSQR when using $\varepsilon = 10^{-13}$ for CGNR and ATOL = BTOL = 10^{-6} for LSQR. For $n < 8$, the CGNR solver achieved accuracy comparable to Cholesky solutions of the normal equations, and the LSQR accuracy was almost as accurate as the QR algorithm for $n < 9$. This is not an entirely fair comparison because coding of the LSQR algorithm (LSQR 1982) was far more sophisticated than the elementary algorithm used for CGNR. For $n \geq 9$, both the CGNR and LSQR solutions were unacceptable. (In interpreting the values of Table 5.1, recall that the $\mathbf{x}$ samples were generated with a standard deviation of 1.0) These results are compared with the other solution methods in the next section.

Both algorithms may perform better when using alternate preconditioners. We again caution that results for simple problems should not be interpreted as general characteristics applicable to all large-scale problems.

TABLE 5.1: CGNR and LSQR Accuracy on Least-Squares Polynomial Problem

n	Average CGNR Iterations	$\sqrt{\frac{1}{100}\sum_{i=1}^{100}\lVert\hat{\mathbf{x}}_i - \mathbf{x}\rVert_2^2}$ Using CGNR	Average LSQR Iterations	$\sqrt{\frac{1}{100}\sum_{i=1}^{100}\lVert\hat{\mathbf{x}}_i - \mathbf{x}\rVert_2^2}$ Using LSQR
4	5	2.2e-12	7	1.4e-15
5	7	1.5e-11	10	7.8e-15
6	10	3.2e-10	12	4.8e-14
7	13	5.1e-9	17	2.6e-13
8	17	3.5e-4	23	1.5e-12
9	24	2.6e-2	18	0.30
10	22	0.29	23	0.30
11	22	0.36	22	0.43
12	30	0.41	22	0.49

5.6 COMPARISON OF METHODS

We now compare numerical results when using different least-squares methods on the polynomial problem of Example 5.4 and on another problem. Execution times are also discussed.

5.6.1 Solution Accuracy for Polynomial Problem

The following examples discuss solution accuracy for the basic polynomial model of Example 5.4 and minor variations.

Example 5.9: Polynomial Problem

Figure 5.6 shows the 100-sample Monte Carlo RMS solution errors

$$\sqrt{\frac{1}{100}\sum_{i=1}^{100}\|\hat{\mathbf{x}}_i - \mathbf{x}_i\|_2^2}$$

for six different least-squares solution techniques versus polynomial order. As before, results are plotted on a log scale. Also shown are the accuracies obtained using condition numbers computed as the maximum ratio of singular values, and as $\kappa_F(\mathbf{U}) = \|\mathbf{U}\|_F\|\mathbf{U}^{-1}\|_F$, which uses the Frobenius norm of the upper triangular $\mathbf{U}$ matrix produced by the QR method and its inverse (as discussed in Section 5.1.3). The plotted variables are labeled as follows:

1. Cholesky-RMS-error: the RMS error obtained when using the normal equations solved using Cholesky factorization.
2. Gauss-Jordan-RMS-error: the RMS error obtained when using the normal equations solved using Gauss-Jordan elimination.

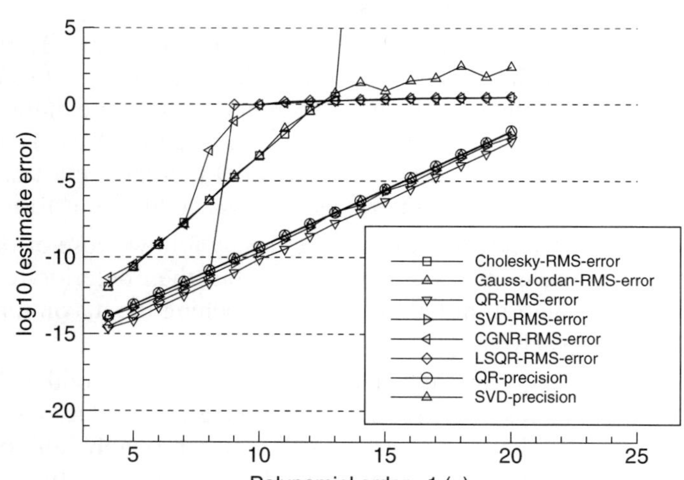

FIGURE 5.6: Monte Carlo evaluation of LS solution accuracy using polynomial model.

3. QR-RMS-error: the RMS error obtained when using the QR method. Unless stated otherwise, the QR method is implemented using the MGS algorithm.
4. SVD-RMS-error: the RMS error obtained when using SVD. Unless stated otherwise, the SVD results are computed using the LAPACK DGESDD divide-and-conquer routine.
5. CGNR-RMS-error: the RMS error obtained when using CGNR method.
6. LSQR-RMS-error: the RMS error obtained when using LSQR method.
7. QR precision#: the accuracy computed as 1.1e-16 × condition number $\kappa_F(\mathbf{U}) = \|\mathbf{U}\|_F\|\mathbf{U}^{-1}\|_F$.
8. SVD precision: the accuracy computed as $1.1\text{e-}16 \times \kappa_2(\mathbf{H}) = \max_i(s_i)/\min_i(s_i)$

The conclusions from Figure 5.6 are:

1. Errors of the normal equation method solved using Cholesky factorization and Gauss-Jordan method are approximately equal in magnitude up to the point at which all precision is lost ($n = 13$). These errors grow at twice the rate at which the condition numbers $\kappa_2(\mathbf{H})$ or $\kappa_F(\mathbf{U})$ grow with polynomial order.
2. The condition number estimates $\kappa_2(\mathbf{H})$ and $\kappa_F(\mathbf{U})$ closely agree for this problem. Although not shown in the plot, the LINPACK condition number estimate $\kappa_1(\mathbf{U}) \approx \kappa_1(\mathbf{A})$ closely follows the other two estimates.
3. Errors of the SVD solution closely follow the accuracy computed using the condition numbers $\kappa_2(\mathbf{H})$ and $\kappa_F(\mathbf{U})$. When using the LAPACK DGESVD (SVD using QR iteration) subroutine, the errors are similar to those of DGESDD, but those of the numerical recipes (Press et al. 2007) SVDCMP routine are approximately 5.5 (= $\log_{10}$ 0.74) times *larger*.
4. Errors of the MGS QR method are a factor of approximately 3.5 (= $\log_{10}$ 0.54) *smaller* than those of the LAPACK SVD methods. When the QR method is changed to use the Householder QR method, the errors are a factor of 12.6 (= $\log_{10}$ 1.10) greater, or about 3.6 times greater than those of the LAPACK SVD methods.
5. When the MGS QR method is first used to transform $[\mathbf{H} \;\; \mathbf{y}]$ to $[\mathbf{U} \;\; \mathbf{z}]$, and then either the LAPACK or Numerical Recipes SVD routine is used on $[\mathbf{U} \;\; \mathbf{z}]$, the estimation errors are comparable to those of the direct MGS QR method; that is, 3.5 times smaller than when the SVD is used directly on **H**. This is unexpected because the SVD methods first transform **H** to a nearly upper triangular matrix using Householder transformations.
6. Errors of the CGNR method at first closely follow those of the Cholesky and Gauss-Jordan methods up to $n = 7$, but then deteriorate rapidly. No indication of algorithm failure (such as reaching the maximum number of iterations) was given.
7. Errors of the LSQR method at first closely follow those of the SVD method, but LSQR then loses all precision at $n = 9$. LSQR produced a "termination flag" indicating that "The system A*X = B is probably not compatible. A least-squares solution has been obtained which is sufficiently accurate, given the value of ATOL." However, the flag did not indicate complete failure.

Based on comments by Björck regarding the sensitivity of the QR method to row-norm ordering, the row order of **H** was changed from smallest norm to largest (as in Example 5.4), to largest to smallest. Figure 5.7 shows the result: errors of the normal equation and SVD methods increased very slightly, but the changes are barely noticeable in the plot. The lack of sensitivity to row ordering was also observed when the Householder QR method replaced the MGS QR method. (This result is counter to Björck's suggestion that sorting by decreasing row norm will minimize numerical errors.) However, errors when using the Householder QR method are more than an order of magnitude greater than those obtained using the MGS QR method, as shown in Figure 5.8.

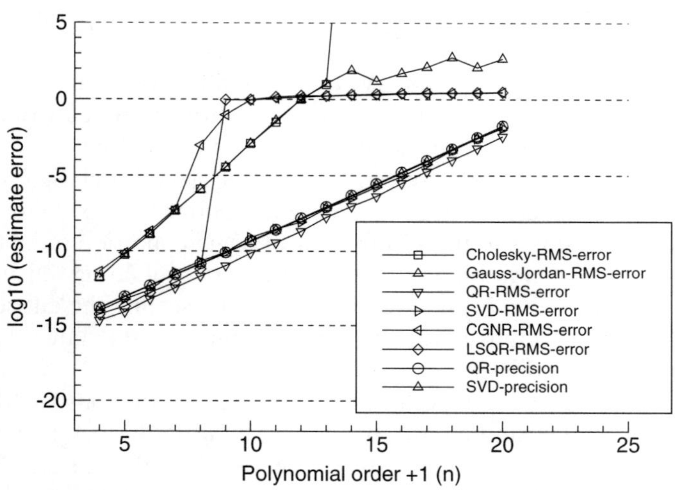

FIGURE 5.7: Monte Carlo evaluation of LS solution accuracy using polynomial model with H rows reordered by decreasing norm.

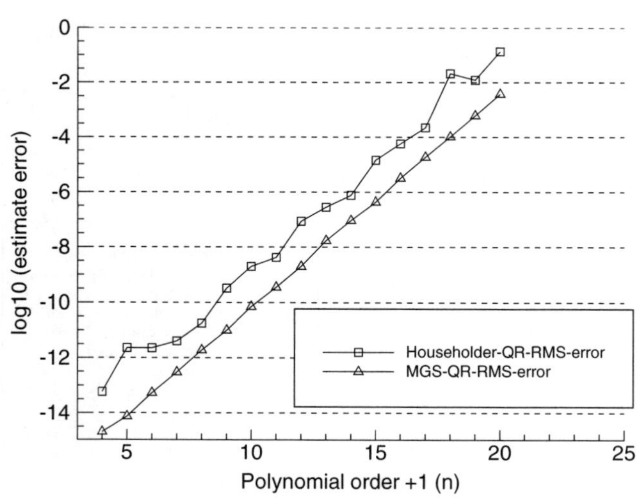

FIGURE 5.8: Monte Carlo evaluation of LS solution accuracy using QR methods on polynomial model with H rows reordered by decreasing norm.

Example 5.10: Polynomial Problem with H(:,1:4) Multiplied by 100

As before, we examine the effects of scaling the first four columns of **H** by 100. Figure 5.9 shows the result. Compared with Figure 5.6, all the curves for $n > 4$ increase by a factor of 100 (= $\log_{10}$ 2.0), except that the LSQR method breaks down for $n \geq 7$ rather than 9. However, this example is not entirely realistic because standard deviations of the first four states in the Monte Carlo simulation are the same (1.0) as the remaining states, so these four states contribute 100 times more to the measurements than the other states. In Figure 5.10 the standard deviations of the first four states are set to 0.01, so all states contribute equally to the measurements. Notice that compared with Figure 5.9, the RMS errors drop by a factor of about 100 and are again comparable to those of Figure 5.6.

Notice, however, that QR and SVD condition numbers are still 100 times larger in Figure 5.10 than in Figure 5.6 since they are computed from **H** and are not dependent on actual measurements **y** or the states **x**. Recall that the condition number is defined as a relative error sensitivity, so this behavior is expected. This emphasizes a deficiency of using condition numbers to evaluate the importance of individual states to the solution, or of using condition numbers to infer observability. While condition numbers are useful in evaluating sensitivity of the solution to numerical errors and are not affected by scaling the entire H matrix, they are affected by relative scaling of measurements and states. This topic is discussed again in Section 5.7.

Finally, we note that the loss of numerical accuracy demonstrated above for the polynomial model can be entirely avoided by using basis functions that are orthogonal (or nearly orthogonal) over the fit interval $0 \leq t \leq 1$. Figure 5.11 shows the 100-sample RMS estimate error, maximum error, and Cholesky precision versus model order when using a Chebyshev polynomial model and the

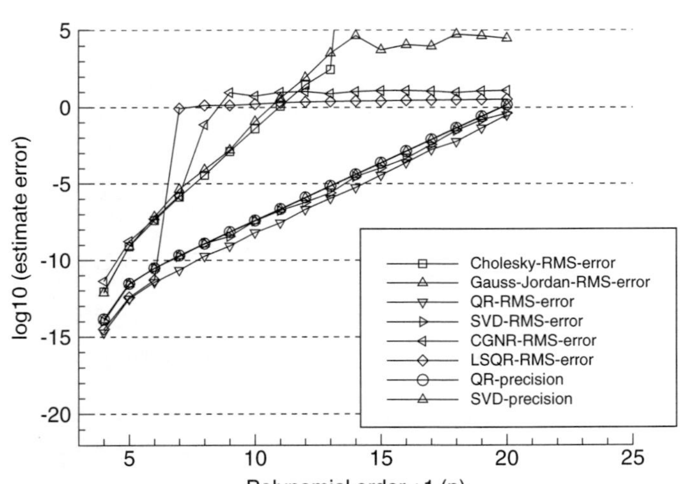

FIGURE 5.9: Monte Carlo evaluation of LS solution accuracy using polynomial models: H(:,1:4) multiplied by 100.

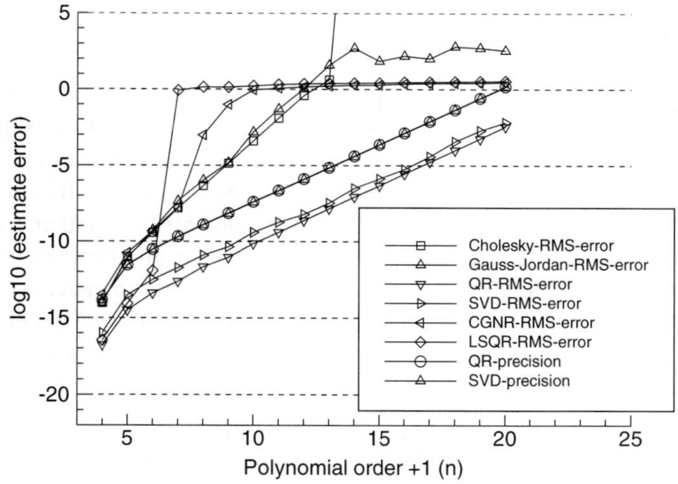

FIGURE 5.10: Monte Carlo evaluation of LS solution accuracy using polynomial models: H(:,1:4) multiplied by 100 and x(1:4) scaled by 0.01.

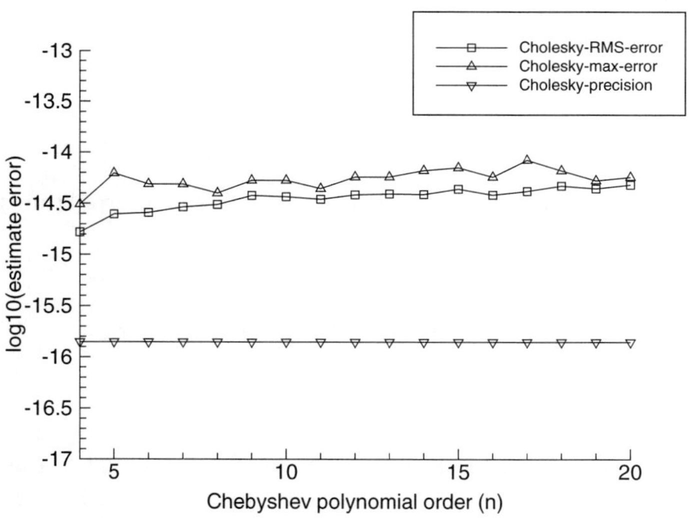

FIGURE 5.11: Monte Carlo evaluation of LS solution accuracy using Chebyshev polynomial models.

normal equations solved with Cholesky factorization. Notice that the estimate errors are always less than 10^{-14} regardless of model order. The measurement residual magnitudes are also less than 10^{-14} over the fit interval, and they remain less than 10^{-14} for predictions over the next second ($1 \le t \le 2$). This accuracy is remarkable considering the results obtained using ordinary polynomials. Use of an orthogonal polynomial results in a least-squares information matrix ($\mathbf{H}^T\mathbf{R}^{-1}\mathbf{H}$) that is diagonal, and therefore inversion is very accurate. Use of other basis functions that are orthogonal over the fit interval, such as Fourier series, should also have comparable accuracy.

Example 5.11: Quasi "Bi-Diagonal" H

While the polynomial example is a good test case for evaluating the effects of poor conditioning, it may not be representative of many real-world problems. For example, results on the polynomial problem indicate that normal equation methods quickly lose all accuracy and should be avoided. However, normal equation methods (particularly Cholesky factorization) are used successfully on most least-squares problems. Therefore, an alternate evaluation test case similar to some real-world problems (but still simple) should be instructive. As before, we use a model with 101 (= m) measurements and 4 to 20 (= n) states. However, each row of the **H** matrix has three nonzero elements: two approximately on the diagonal and the third in the last column. For example, when $m = 8$ and $n = 6$, **H** has the structure

$$\mathbf{H} = \begin{bmatrix} H_{11} & H_{12} & 0 & 0 & 0 & 1 \\ H_{21} & H_{22} & 0 & 0 & 0 & 1 \\ 0 & H_{32} & H_{33} & 0 & 0 & 1 \\ 0 & H_{42} & H_{43} & 0 & 0 & 1 \\ 0 & 0 & H_{53} & H_{54} & 0 & 1 \\ 0 & 0 & H_{63} & H_{64} & 0 & 1 \\ 0 & 0 & 0 & H_{74} & H_{75} & 1 \\ 0 & 0 & 0 & H_{84} & H_{85} & 1 \end{bmatrix}$$

where the first two elements in each row i are linear interpolations from the diagonal, computed as

$$\begin{aligned} t &= \text{real}(i)/\text{real}(m) \\ j &= \min\{\max[\text{floor}[(n-1)t],1],n-2\} \\ H_{i,j} &= \text{real}(j)/\text{real}(m) - t \\ H_{i,j+1} &= [\text{real}(j)+1]/\text{real}(m) - t \end{aligned}$$

The *floor* function is the largest integer smaller than or equal to the real argument, and *real* converts an integer variable to floating point. Hence each measurement can be viewed as a linear interpolation on a local "position," plus a common bias parameter. Using $m = 101$, the numerical results for the least-squares solution methods are shown in Figure 5.12. We notice the following:

1. The largest condition number for $n = 40$ is about 1160. Compared with the polynomial problem, this is a very well-conditioned case.
2. All methods except CGNR produce results accurate to more than 12.5 digits. The CGNR solution method is much less accurate than the others.
3. The LSQR solution is the most accurate of all up to $n = 22$, but then breaks down.
4. The accuracy of all other solution methods is similar, with the MGS QR solution somewhat better, and the SVD solution at times the worst.

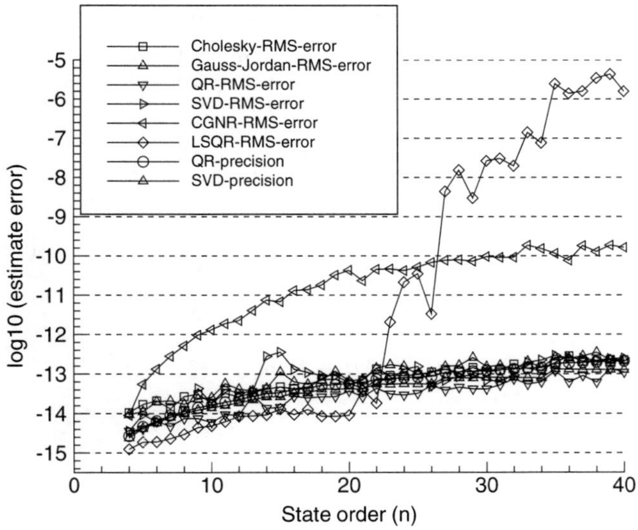

FIGURE 5.12: Monte Carlo evaluation of LS solution accuracy using "bi-diagonal" H.

Interestingly, the normal equation Cholesky method is more accurate than the SVD for all $n > 8$.

5. The QR and SVD condition number accuracy bounds are again close, and roughly characterize the estimate errors. However, the SVD estimate errors are almost always larger than the condition number accuracy bound, and the QR estimate errors are almost always smaller.

These results show that the relative solution accuracy can be problem-dependent, but the MGS QR method can usually be relied upon as accurate. Even the Cholesky normal equation method works very well for well-conditioned problems. The major problem in using normal equation methods is in recognizing when conditioning is a problem. When using Cholesky factorization, the loss-of-precision metric $\max_i \eta_i$ from equation (5.2-12) should always be monitored, and a more accurate algorithm should be used when $\max_i \eta_i$ is greater than half of the machine computational precision digits; for example, 8 digits lost when precision is 16 digits.

5.6.2 Algorithm Timing

Execution time is also an important criteria when evaluating least-squares solution methods because least-squares methods are sometimes used on problems with thousands (occasionally tens of thousands) of states. Table 5.2 lists the approximate flop count and execution times for solution methods used on the polynomial problem with $n = 13$. This polynomial order is large enough so that $n^3 \gg n^2$ and the n^2 terms can be ignored. It is also the order at which the normal equation methods have lost accuracy, but this is irrelevant when computing execution times. In all cases the

TABLE 5.2: FLOPS and Timing of Least-Squares Solution Methods: Polynomial (n = 13)

Method	Approximate Flops	Time (Microsec)[a]
Normal equations with Cholesky inversion	$mn^2/2 + n^3/2$	41
Normal equations with Gauss-Jordan inversion	$mn^2/2 + n^3$	61
Householder QR	$mn^2 - n^3/3$	75 (79 for LAPACK)
MGS QR	mn^2	50
MGS QR plus SVD on **R** (numerical recipes SVDCMP)[b]	$mn^2 + (14/3 + 6k)n^3$	58
SVD on **H** computing full **V** and n columns of **U** (numerical recipes SVDCMP)[b]	$(6 + 4k)mn^2 + (2k - 1)n^3$	285
SVD on **H** computing full **U**, **V** (LAPACK divide and conquer DGESDD)[c]	unknown	830
SVD on **H** computing full **V** and n columns of **U** (LAPACK divide and conquer DGESDD)[c]	unknown	368
SVD on **H** computing full **V** and n columns of **U** (LAPACK bi-diagonal QR iteration DGESVD)[c]	unknown	362
SVD on **H** computing full **V** and no **U** (LAPACK bi-diagonal QR iteration DGESVD)[c]	unknown	273

[a]All code used double precision, compiled with default optimization (O4) and executed on 3.0 GHz P4 PC.
[b]k in the flop count is the variable number of internal iterations.
[c]LAPACK compiled with generic BLAS using default (O4) optimization.

algorithms were executed 100,000 times so that timing was not significantly affected by clock precision. All LAPACK *Basic Linear Algebra Subprograms* (BLAS) were compiled with default optimization rather than using special object code optimized for the PC. Although the execution times in Table 5.2 may seem negligibly small, execution times tend to grow as either n^3 or mn^2. Since the state size of least-squares problems continues to grow with the capabilities of available computers, it is not unusual to have execution times measured in hours or even days. Hence algorithm execution time can be important.

Flop counts indicate that the MGS QR method should take twice the execution time as the normal equations with Cholesky factorization, but in practice it only takes 22% more time for this problem and is 33% faster than the Householder QR method. MGS QR is also generally the most accurate algorithm. Note that the MGS QR method followed by the SVD only takes 16% more time than the MGS QR method alone, while direct use of the thin SVD requires at least 724% as much time. Computation of the full **U** matrix in the SVD requires much more time and is of little benefit for general least-squares solutions. The relative performance of

TABLE 5.3: FLOPS and Timing of Least-Square Solution Methods: "Bi-Diagonal" H, $n = 200$, $m = 2000$

Method	Time (millisecond)[a]
Normal equations with Cholesky inversion	378
Normal equations with Gauss-Jordan inversion	472
Householder QR	120
MGS QR	126
MGS QR plus SVD on **R** (numerical recipes SVDCMP)	126 + 249 = 375
SVD on **H** computing full **V** and n columns of **U** (numerical recipes SVDCMP)	2234
SVD on **H** computing full **U**, **V** (LAPACK divide and conquer DGESDD)[b]	5258
SVD on **H** computing full **V** and n columns of **U** (LAPACK divide and conquer DGESDD)[b]	683
SVD on **H** computing full **V** and n columns of **U** (LAPACK bi-diagonal QR iteration DGESVD)[b]	858
SVD on **H** computing full **V** and no **U** (LAPACK bi-diagonal QR iteration DGESVD)[b]	336

[a]All codes used double precision, compiled with default optimization (O4) and executed on 3.0 GHz P4 PC.
[b]LAPACK compiled with generic BLAS using default (O4) optimization.

LAPACK may be better when used for large problems with a machine-specific BLAS.

Table 5.3 shows the algorithm timing for the "bidiagonal H" case when $m = 2000$ and $n = 200$. In this case the Householder QR algorithm is the fastest with MGS QR only slightly slower. It is surprising that both QR methods are much faster than the normal equations with Cholesky factorization. QR methods are generally slower than normal equations with Cholesky factorization, but not by a factor of two as would be expected from the flop count. Also notice that the "full **U**" SVD methods (computing all of **U**) are at least 5.4 times slower than the MGS QR method.

Relative execution times of the algorithms may be quite different when used on large-scale problems or on different computers, but these results suggest trends in expected performance.

5.7 SOLUTION UNIQUENESS, OBSERVABILITY, AND CONDITION NUMBER

Theoretical observability of states $\mathbf{x}$ in measurement equation $\mathbf{y} = \mathbf{Hx} + \mathbf{r}$ was defined in Section 5.1.4 by various conditions directly or indirectly implying that all singular values of $\mathbf{H}$ must be greater than zero. In Section 5.1.3 the matrix condition number $\kappa_p(\mathbf{H})$ was defined as a "scale-free" measure of the relative sensitivity of $\mathbf{x}$ to relative numerical errors in $\mathbf{H}$ or $\mathbf{y}$. These two concepts are related but not necessarily equivalent. Example 5.4 demonstrated that both the condition number and relative errors in the least-squares solution $\hat{\mathbf{x}}$ are affected by scaling of the $\mathbf{H}$

columns, but when the elements of $\mathbf{x}$ are also adjusted so that the magnitude of $\mathbf{y}$ is unchanged, then the errors in $\hat{\mathbf{x}}$ return to the unscaled values.

Condition numbers $\kappa_p(\mathbf{H})$, or more specifically the relative magnitudes of the smallest to the largest singular values, are often used to determine whether specific states or linear combinations of states are observable. The fundamental problem in this approach is that singular values are properties of $\mathbf{H}$, and they do not take into account the relative magnitudes of states, measurements and measurement noise.

Consider a reordered version of the two-state measurement model $\mathbf{y} = \mathbf{Hx} + \mathbf{r}$ used in Example 5.1:

$$\begin{bmatrix} y_1 \\ y_2 \end{bmatrix} = \begin{bmatrix} \beta & 0 \\ 0 & 1 \end{bmatrix} \begin{bmatrix} x_1 \\ x_2 \end{bmatrix} + \begin{bmatrix} r_1 \\ r_2 \end{bmatrix}.$$

The singular values of $\mathbf{H}$ are β and 1, so $\kappa_2(\mathbf{H}) = \beta$. If $\beta = 10^{18}$ and operations are conducted in 16-digit precision, the wide range in singular values suggests that the second state is unobservable and should be eliminated from the system. However, the system is fully observable, so the ratio of singular values is not a sufficient measure of observability. Now consider three possibilities:

1. $|x_1| \cong 1$ and $|x_2| \cong 1$, but β is large so y_1 is extremely sensitive (possibly because of different units) to x_1. Hence if $E(\mathbf{rr}^T) = \mathbf{I}$, the weighted least-squares estimate $\hat{x}_1$ will be extremely accurate because the *a posteriori* error variance is $E[(\hat{x}_1 - x_1)^2] = 1/\beta^2$. Both x_1 and x_2 are highly observable in this case.
2. $|x_1| \cong 1/\beta$ and $|x_2| \cong 1$. Hence y_1 will be slightly larger than one in magnitude if $E(\mathbf{rr}^T) = \mathbf{I}$. Again both x_1 and x_2 are highly observable but the *a posteriori* variance $E[(\hat{x}_1 - x_1)^2] \cong E[x_1^2]$, so the measurements have not reduced the prior uncertainty.
3. $|x_1| \cong 1$, $|x_2| \cong 1$, and $E(r_1 r_1^T) = 1$, but the random measurement noise on y_2 is very large; for example, $E(r_2 r_2^T) = 100\beta^2$ with $\beta = 10^{18}$. In this case y_2 is entirely measurement noise because the contribution from x_2 is much smaller than the machine precision. Here the relative magnitudes of the measurement noise make x_2 unobservable—the problem is not caused by a large range in singular values of $\mathbf{H}$.

These examples show that before using singular values to infer observability, both rows and columns of $\mathbf{H}$ should be appropriately scaled. First, the rows of $\mathbf{y} = \mathbf{Hx} + \mathbf{r}$ should be normalized so that all measurements have the same error variance due to measurement noise. That is, for systems in which all measurement noise variances are not equal, a weighted least-squares solution should be used with weights defined as the inverse measurement noise variances. Second, the states should be scaled so that the state magnitudes are approximately equal. More accurately, the contribution of each state x_i to the measurements $\mathbf{y}$ should be approximately equal. This can be implemented by scaling the states by the norms of the columns of $\mathbf{H}$. Thus the original measurement model $\mathbf{y} = \mathbf{Hx} + \mathbf{r}$ is scaled as:

$$(\mathbf{R}^{-1/2}\mathbf{y}) = (\mathbf{R}^{-1/2}\mathbf{HD}^{-1})(\mathbf{Dx}) + (\mathbf{R}^{-1/2}\mathbf{r})$$

where

$\mathbf{R} = E[\mathbf{r}\mathbf{r}^T]$ (usually diagonal),

$\mathbf{R}^{1/2}$ is a square-root factor of $\mathbf{R}$ such that $\mathbf{R} = \mathbf{R}^{1/2}\mathbf{R}^{T/2}$ and $\mathbf{R}^{-1} = \mathbf{R}^{-T/2}\mathbf{R}^{-1/2}$,

$\mathbf{D}$ is a diagonal $n \times n$ matrix with elements equal to the l_2-norm of the corresponding column of $\mathbf{H}$, that is, $D_i = \|\mathbf{H}_{:i}\|_2$,

$\mathbf{H}' = \mathbf{R}^{-1/2}\mathbf{H}\mathbf{D}^{-1}$ is the scaled $m \times n$ measurement matrix, and

$\mathbf{Dx}$ is the scaled n-element state vector.

The scaled measurement matrix to be used for SVD analysis of observability is $\mathbf{H}' = \mathbf{R}^{-1/2}\mathbf{H}\mathbf{D}^{-1}$. Notice that the singular values and basis vectors of $\mathbf{H}'$ will be different from those of $\mathbf{H}$. If it is found that the smallest singular values of $\mathbf{H}'$ are numerically insignificant compared with the largest singular values, then it may be concluded that the system is unobservable.

For real-world problems the difference in calculated condition numbers between scaled and unscaled $\mathbf{H}$ can be orders of magnitude. For example, consider geosynchronous orbit determination where the epoch orbit states have units of km for position and km/s for velocity, and 1 day of range-only measurements from a single-ground station are available for orbit determination. The unscaled $\mathbf{H}$ condition number is 1.2×10^{11} for a poor tracking geometry, but when $\mathbf{H}$ is scaled as described, the condition number is reduced by four orders of magnitude to 1.2×10^7. One might conclude from the unscaled condition number that the solution is nearly indeterminate. However, the scaled condition number indicates that observability is poor, but not hopeless.

Although the SVD is generally regarded as the best method for evaluating observability and numerical conditioning, a *rank-revealing QR decomposition* can also be used for this purpose. This is implemented using a state permutation matrix $\mathbf{\Pi}$ in the QR decomposition such that $\mathbf{H\Pi} = \mathbf{T}^T\mathbf{U}$ where, as before, $\mathbf{T}$ is an $m \times m$ orthogonal matrix and $\mathbf{U}$ is an $m \times n$ upper triangular matrix. $\mathbf{\Pi}$ is an $n \times n$ matrix consisting of only 1's and 0's that permute the order of the states $\mathbf{x}$ such that $\mathbf{U}$ is partitioned with the l_2-norm of the lower right partition smaller than a specified constant times a singular-value threshold of $\mathbf{H}$. More about the rank-revealing QR decomposition can be found in Björck (1996, p. 108) and Golub and Van Loan (1996, p. 248).

5.8 PSEUDO-INVERSES AND THE SINGULAR VALUE TRANSFORMATION (SVD)

If SVD analysis shows that linear combinations of states in the system $\mathbf{y} = \mathbf{Hx} + \mathbf{r}$ are unobservable, the obvious question is "Can anything be done to fix the problem?" If possible, additional measurements that provide information on the unobservable components should be collected. Unfortunately that is often not an option. Alternately, prior information on some or all states can often be obtained, thus allowing computation of a least squares solution. Approximate magnitudes of many physical parameters can often be defined from knowledge of physics, from past history, or from previous testing. For example, solar radiation pressure and atmospheric drag coefficients used for satellite orbital dynamic models are constrained by shapes and areas of individual surfaces, angles of those surfaces with respect to the sun or relative velocity vector, and other factors. Hence it is often possible to specify *a priori* estimates for many parameters with *a priori* uncertainties that are reasonably small. Inclusion of prior information in the solution is often sufficient to eliminate unreasonable state estimates. The *a posteriori* estimate uncertainty of some states may still be quite large, but that is a fundamental problem when working with a limited set of measurements.

Another approach uses the pseudo-inverse to compute a solution that minimizes both the norm of the measurement residuals and the norm of the estimated state. That is, the pseudo-inverse applies an additional constraint on the system that eliminates the unobservable components. This is implemented using the procedure described in Section 5.1.2. Specifically, the SVD is used on the scaled measurement matrix to obtain $\mathbf{H}' = \mathbf{USV}^T$, where $\mathbf{U}$ is an $m \times m$ orthogonal matrix, $\mathbf{V}$ is an $n \times n$ orthogonal matrix, and $\mathbf{S}$ is an $m \times n$ upper diagonal matrix of singular values. Then the pseudo-inverse of $\mathbf{H}'$ is computed as $\mathbf{H}^{\#} = \mathbf{V}\begin{bmatrix}\mathbf{S}_1^{\#} & \mathbf{0}\end{bmatrix}\mathbf{U}^T$ where $\mathbf{S}_1^{\#}$ is the pseudo-inverse of $\mathbf{S}_1$. Singular values that are numerically negligible relative to the given machine precision are replaced with zero in the same location when forming $\mathbf{S}_1^{\#}$. If the "thin SVD" is used and only the first n columns of $\mathbf{U}$ are computed, $\mathbf{H}^{\#} = \mathbf{VS}_1^{\#}\mathbf{U}_1^T$ and the minimal norm least-squares solution for the scaled problem is

$$\hat{\mathbf{x}} = \mathbf{D}^{-1}(\mathbf{VS}_1^{\#}\mathbf{U}_1^T)(\mathbf{R}^{-1/2}\mathbf{y}). \tag{5.8-1}$$

It is easily shown that zeroing the elements of $\mathbf{S}_1^{\#}$ corresponding to numerically negligible elements of $\mathbf{S}_1$ also minimizes $\|\mathbf{D}\hat{x}\|_2$:

$$\|\mathbf{D}\hat{x}\|_2 \le \|\mathbf{S}_1^{\#}\|_2 \|\mathbf{U}_1^T\mathbf{R}^{-1/2}\mathbf{y}\|_2 = \left(\sum_{i=1}^{n}(S_1^{\#})_i^2\right)^{1/2} \|\mathbf{U}_1^T\mathbf{R}^{-1/2}\mathbf{y}\|_2. \tag{5.8-2}$$

Hence replacing a nonzero element $(S_1^{\#})_i = 1/(S_1)_i$ with zero reduces $\|\mathbf{D}\hat{x}\|_2$. The fact that certain diagonal elements of $\mathbf{S}_1$ are numerically negligible implies that a unique solution to $\hat{\mathbf{x}}$ does not exist, so an infinite number of solutions are possible.

The pseudo-inverse should be used with great caution because it fundamentally changes the nature of the least-squares solution. Rather than accepting that we have no information on a linear combination of states, we are stating that knowledge of that combination of states is perfect. Hence the definition of states included in that linear combination is altered. If the only purpose of the least-squares solution is to obtain a smooth interpolation of noisy data, then use of the pseudo-inverse is justified. However, if the purpose is to predict system variables outside the range of measured conditions (such as prediction past the last measurement time), then use of the pseudo-inverse can produce bad predictions. Likewise if the purpose of least-squares modeling is to determine the values of physical parameters (such as system biases and force model coefficients), then the redefinition of state variables implied by use of the pseudo-inverse is potentially disastrous. Satellite orbit determination is an example where pseudo-inverses should be avoided because typical measurements—such as occasional range and range rates from ground tracking—are not by themselves sufficient to estimate orbit elements. It is only through tracking over a period of time, coupled with a very strong orbital dynamic constraint, that it is possible to obtain a solution. If the model orbit states are redefined by use of a pseudo-inverse, orbit predictions may be poor. Fortunately this problem is seldom encountered because most reasonable orbit determination tracking scenarios result in condition numbers less than 10^7 (if properly scaled). When larger condition numbers are encountered, iterative methods used to obtain a solution to the nonlinear problem usually do not converge.

When unobservable systems are encountered and the goal of the estimation is other than simply interpolating measurements, it is almost always preferable to either obtain additional measurements or to include prior information on the states (use a Bayesian solution).

The following examples demonstrate problems in using the pseudo-inverse.

Example 5.12: Pseudo-Inverse for Two-State Singular Model

Consider the two-measurement, two-state singular model

$$\begin{bmatrix} y_1 \\ y_2 \end{bmatrix} = \begin{bmatrix} 1 & 1 \\ 1 & 1 \end{bmatrix} \begin{bmatrix} x_1 \\ x_2 \end{bmatrix} + \begin{bmatrix} r_1 \\ r_2 \end{bmatrix}$$

with $E[\mathbf{r}\mathbf{r}^T] = \mathbf{I}$ for simplicity. Obviously it is not possible to separately estimate the two states because the mapping to both measurements is identical. The SVD of the measurement matrix $\mathbf{H}$ is found to be

$$\mathbf{H} = \begin{bmatrix} 1 & 1 \\ 1 & 1 \end{bmatrix} = \begin{bmatrix} 1/\sqrt{2} & -1/\sqrt{2} \\ 1/\sqrt{2} & 1/\sqrt{2} \end{bmatrix} \begin{bmatrix} 2 & 0 \\ 0 & 0 \end{bmatrix} \begin{bmatrix} 1/\sqrt{2} & 1/\sqrt{2} \\ -1/\sqrt{2} & 1/\sqrt{2} \end{bmatrix}.$$

Again for simplicity, we assume that the measurement noise for both measurements is zero, so the measurements are

$$\begin{bmatrix} y_1 \\ y_1 \end{bmatrix} = \begin{bmatrix} x_1 + x_2 \\ x_1 + x_2 \end{bmatrix}.$$

Using the pseudo-inverse,

$$\begin{aligned} \begin{bmatrix} \hat{x}_1 \\ \hat{x}_2 \end{bmatrix} &= \begin{bmatrix} 1/\sqrt{2} & -1/\sqrt{2} \\ 1/\sqrt{2} & 1/\sqrt{2} \end{bmatrix} \begin{bmatrix} 1/2 & 0 \\ 0 & 0 \end{bmatrix} \begin{bmatrix} 1/\sqrt{2} & 1/\sqrt{2} \\ -1/\sqrt{2} & 1/\sqrt{2} \end{bmatrix} \begin{bmatrix} x_1 + x_2 \\ x_1 + x_2 \end{bmatrix} \\ &= \begin{bmatrix} (x_1 + x_2)/2 \\ (x_1 + x_2)/2 \end{bmatrix} \end{aligned}.$$

As expected, the estimated states $\hat{x}_1$ and $\hat{x}_2$ are both equal to the average of the measurements. This solution fits the measurements and minimizes the norm $\sqrt{\hat{x}_1^2 + \hat{x}_2^2}$, but it may be a very poor solution. Suppose that the true state values are $\mathbf{x} = [10^{10} \; -10^{10}]$, which give $\hat{\mathbf{x}} = \mathbf{0}$. In this case the state estimate error is very large, which demonstrates that a minimal norm assumption may produce bad results.

As another example, assume that we are given a single noisy measurement, $y = h_1 - h_2 + r$, of the difference in height $(h_1 - h_2)$ for two people. Assume that the actual heights are 69 and 71 inches, so the difference is +2 inches. It can be shown that the pseudo-inverse solution for the two heights are −1 and +1 inches—a physically unrealistic answer. However, if we know that the two people are males living in the United States, then statistical surveys indicate that the mean height should be about 70 inches, with a standard deviation of approximately 3 inches. This can be used as prior information in a Bayesian solution. If the measurement noise standard deviation ($\sqrt{E[r^2]}$) is 0.1 inch, then the estimated heights of the two people are found to be 69.0006 and 70.9992 inches—a much more accurate answer than when using the pseudo-inverse. Of course the answer would not have been as accurate if the mean height of the two people was not exactly 70 inches, but results would still have been more accurate than the pseudo-inverse solution. Although this example is trivially simple, somewhat similar behavior may be observed for more realistic "unobservable" problems.

Example 5.13: Pseudo-Inverse Solution for Polynomial Problem

We return to our polynomial problem of Example 5.4 in which the data span was $0 \le t \le 1$. Let the threshold for computing a pseudo-inverse be ε, where ε ranges from 10^{-11} to 10^{-13}. That is, for all scaled singular values ε times smaller than the largest singular value, the corresponding diagonal elements of $\mathbf{S}_1^{\#}$ are set to zero. Figure 5.13 shows the QR and SVD 100-sample RMS solution accuracy,

$$\sqrt{\frac{1}{100}\sum_{i=1}^{100}\|\hat{\mathbf{x}}_i - \mathbf{x}\|_2^2},$$

versus polynomial order. The lines labeled

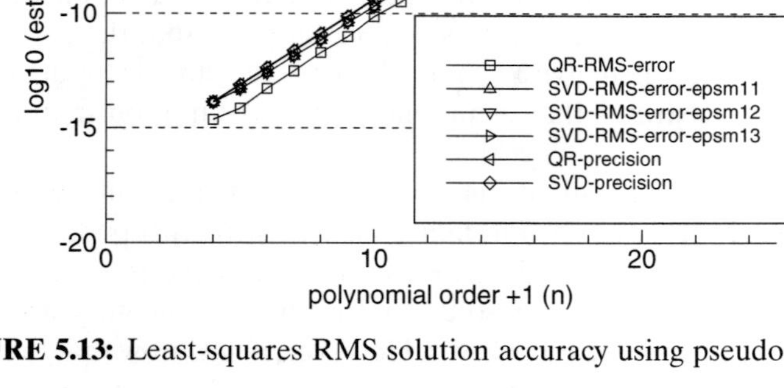

FIGURE 5.13: Least-squares RMS solution accuracy using pseudo-inverse.

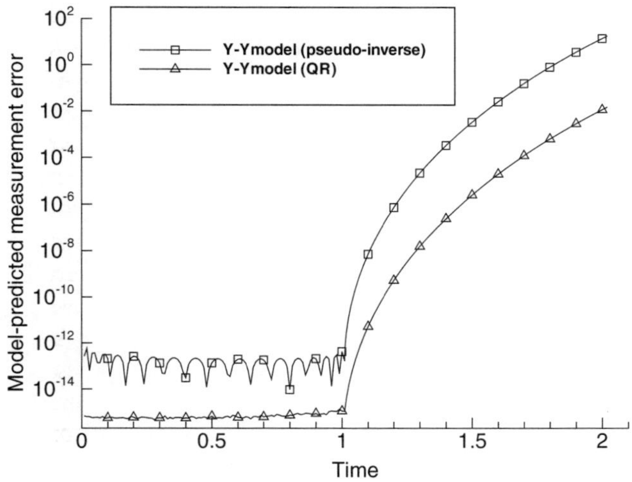

FIGURE 5.14: RMS QR and pseudo-inverse (ε = 1.e-12) $n = 20$ polynomial model prediction accuracy versus time.

"SVD-RMS-error_epsm11" used $\varepsilon = 10^{-11}$,

"SVD-RMS-error_epsm12" used $\varepsilon = 10^{-12}$, and

"SVD-RMS-error_epsm13" used $\varepsilon = 10^{-13}$

for the SVD solution. Notice that $\hat{\mathbf{x}}$ has lost all accuracy whenever a pseudo-inverse solution is computed because the unobservable linear combinations of $\hat{\mathbf{x}}$ have been set to zero.

Figure 5.14 shows the 100-sample RMS fit and 1 s measurement ($\mathbf{y}$) prediction accuracy of the QR and pseudo-inverse solutions for $n = 20$ and $\varepsilon = 10^{-12}$. The QR RMS error is about 0.5×10^{-15} over the first 1 s (the "fit" interval"), and rises to a maximum error of 0.015 at 2.0 s. However, the pseudo-inverse RMS error is about 0.2×10^{-12} over the first second and rises to a maximum of 18.1 at 2.0 s. Although the pseudo-inverse error at 2.0 s is only 2.8×10^{-5} times the RMS magnitude of the measurement, it is 1200 times larger than the error of the full-rank QR solution. The pseudo-inverse prediction error grows much faster with time than the full-rank prediction, which is a consequence of the constraint imposed by use of the pseudo-inverse. As shown in Figure 5.15, the pseudo-inverse RMS error rises to 78.2 at 2 s when using $\varepsilon = 10^{-11}$. For some problems, changes in ε can result in dramatic changes in errors for predictions outside tested conditions. This is another problem when using pseudo-inverses because the appropriate level of ε is not always obvious.

Although not directly relevant to the pseudo-inverse, notice that the estimate errors $\|\hat{\mathbf{x}}_i - \mathbf{x}\|_2^2$ from Figure 5.13 are about 0.002 for $n = 20$ when using the QR algorithm, but the computed model can still fit the measurements to an accuracy of 0.5×10^{-15} over the fit interval (Fig. 5.14).This shows that numerical errors may cause the state estimate to be erroneous, but this does not affect the ability of the model to fit the data.

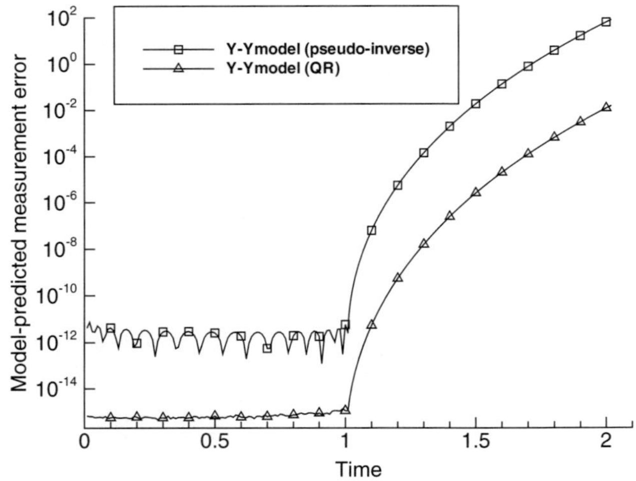

FIGURE 5.15: RMS QR and pseudo-inverse (ε = 1.e-11) n = 20 polynomial model prediction accuracy versus time.

5.9 SUMMARY

The important lessons from this chapter are

1. The matrix condition number $\kappa_p(\mathbf{A}) = \|\mathbf{A}\|_p\|\mathbf{A}^{-1}\|_p$ is a measure of the l_p-norm *relative* sensitivity of $\mathbf{x}$ to numerical perturbations in $\mathbf{A}$ or $\mathbf{b}$ for the equation $\mathbf{Ax} = \mathbf{b}$, that is,

$$\left(\frac{\|\delta\mathbf{x}\|_p}{\|\mathbf{x}\|_p}\right) \le \kappa_p(\mathbf{A})\left(\frac{\|\delta\mathbf{b}\|_p}{\|\mathbf{b}\|_p}\right) \quad \text{and} \quad \left(\frac{\|\delta\mathbf{x}\|_p}{\|\mathbf{x}\|_p}\right) \le \kappa_p(\mathbf{A})\left(\frac{\|\delta\mathbf{A}\|_p}{\|\mathbf{A}\|_p}\right).$$

   For the l_2-norm,

$$\kappa_2(\mathbf{A}) = \frac{\max_i(s_i)}{\min_i(s_i)},$$

   where s_i are the singular values of $\mathbf{A}$.

2. The Moore-Penrose matrix pseudo-inverse, defined by four properties, is used to compute minimum norm matrix inversions for rectangular and/or singular matrices.
3. Theoretical observability of the measurement equation $\mathbf{y} = \mathbf{Hx} + \mathbf{r}$ is defined by various conditions that all equivalent to the requirement that the rank of the $m \times n$ matrix $\mathbf{H}$ must be n. This implies that all singular values of $\mathbf{H}$ must be greater than zero.
4. The normal equation $(\mathbf{H}^T\mathbf{R}^{-1}\mathbf{H})\hat{\mathbf{x}} = \mathbf{H}^T\mathbf{R}^{-1}\mathbf{y}$ is the most frequently used technique for solving the weighted least-squares problem. The most commonly used solution technique uses Cholesky factorization of $\mathbf{H}^T\mathbf{R}^{-1}\mathbf{H} = \mathbf{LL}^T$, where $\mathbf{L}$ is lower triangular. The solution for $\hat{\mathbf{x}}$ is computed by explicitly or implicitly inverting $\mathbf{L}$. Cholesky factorization is numerically stable, generally does

not require pivoting, and provides a metric for determining loss of numeric precision. It can also be easily adapted for partitioned solutions of sparse systems. However, formation of the normal equation doubles the condition number of matrix **H**, and that can cause a loss of accuracy for poorly conditioned problems.

5. The QR factorization $(\mathbf{R}^{-1/2}\mathbf{H}) = \mathbf{T}^T\mathbf{U}$, where **T** is an $m \times m$ orthogonal matrix formed using elementary rotations and **U** is a $m \times n$ upper triangular matrix, is a more numerically stable method for solving the least-squares problem. The solution $\hat{\mathbf{x}}$ is obtained by back-solving $\mathbf{U}\hat{\mathbf{x}} = \mathbf{T}^T\mathbf{R}^{-1/2}\mathbf{y}$. The condition number of **U** is the same as that of $\mathbf{R}^{-1/2}\mathbf{H}$, so numerical precision is unaltered. The QR factorization is most often computed using Householder transformations or MGS orthogonalization. The MGS method requires slightly more computation, but some results suggest that it is more accurate than Householder transformations. Both methods require approximately twice the flops of normal equation methods, but execution times indicate that it can actually be faster than normal equation methods for some problems. (Note: the QR factorization is usually defined as $\mathbf{H} = \mathbf{Q}^T\mathbf{R}$ but we have changed the notation [using **T** and **U**] to avoid confusion with our previous definitions of **Q** and **R**.)
6. The SVD can also be used to solve the least-squares problem. It has the advantages of automatically computing the solution condition number and indicating which linear combinations of states are unobservable, but it can be less accurate than the MGS QR method and requires far more execution time than the QR or normal equation methods. Its primary strength is in determining which states should be included in the system model. If it is to be used operationally for large problems, the QR method should first be used as a "data compression" technique to reduce the system size from $m \times n$ to $n \times n$.
7. The normal equations, or equations equivalent to the normal equations, can also be solved using iterative methods. Krylov space methods are particularly useful for solution of large sparse systems. The CGNR and LSQR methods are two iterative methods that solve the least-squares problem without forming the normal equations, but they are both affected by the doubling of the condition number implied by the normal equations. This author has used the CGNR method with success on flow-type problems when preconditioning was used, but on our sample problems without preconditioning, both methods failed completely (without indication of failure) for moderate condition numbers. The LSQR method was generally more accurate than the CGNR method, but also failed more abruptly than CGNR.
8. To use condition numbers as a test for state observability, or equivalently to use the ratio of singular values to determine which linear combinations of variables are unobservable, both row and column scaling of **H** should be applied before using the SVD. In particular, $\mathbf{H}' = \mathbf{R}^{-1/2}\mathbf{H}\mathbf{D}^{-1}$ should be used where $\mathbf{R}^{-1/2}$ scales the rows to obtain unit measurement noise variance, and $\mathbf{D}^{-1}$ scales the columns for unit column norms.
9. When unobservable systems are encountered, it is preferable to either obtain additional measurements or include prior information so that all states are

observable. Alternately the pseudo-inverse solution $\hat{\mathbf{x}} = \mathbf{D}^{-1}(\mathbf{V}\mathbf{S}_1^{\#}\mathbf{U}_1^T)(\mathbf{R}^{-1/2}\mathbf{y})$ can be used by replacing diagonals of $\mathbf{S}_1^{\#}$ corresponding to negligible-small singular values with zero. However, use of a pseudo-inverse fundamentally alters the definition of estimated states, and can lead to poor predictions of unmeasured conditions, or can result in physically unreasonable state estimates.

10. Use of model basis functions that are orthogonal over the measurement data span will eliminate most numerical problems in the least-squares solution because the information matrix $\mathbf{H}^T\mathbf{R}^{-1}\mathbf{H}$ will be diagonal. Even when using physically based models, selection of "approximately orthogonal" states can improve the conditioning.

CHAPTER 6

LEAST-SQUARES ESTIMATION: MODEL ERRORS AND MODEL ORDER

The year of this writing, 2009, is the 200th anniversary of Gauss's (1809) *Theory of the Motion of Heavenly Bodies Moving About the Sun in Conic Sections*. One might suspect that after 200 years there is little to be learned about least squares, and that applications of the method are so routine as to be boring. While literature on new theory may be sparse, applications of the theory can be challenging. This author still spends a considerable fraction of his professional time trying to determine why least-squares and Kalman filter applications do not perform as expected. In fact, it is not unusual to spend more time trying to "fix" implementations than designing and developing them. This may suggest that insufficient time was spent in the design phase, but in most cases the data required to develop a better design were simply not available until the system was built and tested. Usually the problems are due to incorrect modeling assumptions, severe nonlinearities, poor system observability, or combinations of these.

This chapter discusses practical usage of least-squares techniques. Of particular interest are methods that allow analysis of solution validity, model errors, estimate confidence bounds, and selection of states to be estimated. Specific questions that are addressed (at least partially) in this chapter include:

1. Is the solution valid?
2. How unique is the solution?
3. What is the expected total error in the state estimate taking into account not only errors in measurements, but also errors in model structure and in unestimated parameters?
4. What set of estimated states produces results with the smallest total state error?
5. What subset of estimated states (reduced-order model [ROM]) meets accuracy and execution time requirements?
6. What set of estimated states produces a result that is most robust to errors in modeling assumptions?

Advanced Kalman Filtering, Least-Squares and Modeling: A Practical Handbook, by Bruce P. Gibbs

7. Can the search for the "best" set of estimated states be automated?
8. What fit data span produces the best predictions?
9. Can the variance of the measurement noise be determined from fit residuals?

6.1 ASSESSING THE VALIDITY OF THE SOLUTION

When a least squares, Minimum Mean-Squared Error (MMSE), Maximum Likelihood (ML), or Maximum *A Posteriori* (MAP) solution has been obtained by some method described in the previous chapters, it is important to know whether the solution is valid and accurate. This is a particularly important step in the estimation process because anomalous measurements and modeling errors are always potential problems.

6.1.1 Residual Sum-of-Squares (SOS)

For weighted least squares, MMSE/minimum variance, ML, and MAP estimates, the first step in the validation process should involve checking the cost function. As shown in Chapter 4, the expected sum of weighted measurement residuals $\tilde{\mathbf{y}}^T\mathbf{R}^{-1}\tilde{\mathbf{y}}$ (where $\tilde{\mathbf{y}}=\mathbf{y}-\mathbf{H}\hat{\mathbf{x}}$) for an optimally weighted, full-rank, non-Bayesian least-squares solution should be equal to the number of scalar measurements (m) minus the number of estimated states (n): $m - n$. If the measurement errors $\mathbf{r}$ are Gaussian-distributed, $\tilde{\mathbf{y}}^T\mathbf{R}^{-1}\tilde{\mathbf{y}}$ should be χ^2-distributed with degrees-of-freedom equal to $m - n$. Hence the expected value is $m - n$ with variance $2(m - n)$. When $m - n > 20$, the distribution is approximately $N(m - n, 2m - 2n)$. Thus when

$$\left|\tilde{\mathbf{y}}^T\mathbf{R}^{-1}\tilde{\mathbf{y}}-(m-n)\right|>k\sqrt{2(m-n)}, \tag{6.1-1}$$

where k is the desired level of significance (e.g., $4 - \sigma$), the residuals are statistically "large," and either the fitted model is invalid, the measurement noise level is incorrectly specified, or some outlying measurements have not been edited. Since the χ^2 distribution is not symmetric when $m - n$ is small (e.g., <10), significance tests should use

$$\Pr\left\{\tilde{\mathbf{y}}^T\mathbf{R}^{-1}\tilde{\mathbf{y}}<x\right\}=F_X(x)=P\left(\frac{m-n}{2},\frac{x}{2}\right)$$

where P is the incomplete gamma function. Even when $\mathbf{r}$ is not Gaussian-distributed, $\tilde{\mathbf{y}}^T\mathbf{R}^{-1}\tilde{\mathbf{y}}$ is often close to $m - n$ provided that $E[\mathbf{r}] = \mathbf{0}$ and $E[\mathbf{r}\mathbf{r}^T] = \mathbf{R}$. Hence equation (6.1-1) can still be used as a rough check provided that anomalous measurements (outliers) are eliminated from the solution. More will be said about this topic later.

Even when $\left|\tilde{\mathbf{y}}^T\mathbf{R}^{-1}\tilde{\mathbf{y}}/(m-n)-1\right|$ is larger than $k\sqrt{2/(m-n)}$, this does not necessarily mean that the state solution $\hat{\mathbf{x}}$ is invalid. For example, consider a case with $m = 1000$, $n = 10$, and $k = 4$, which gives $k\sqrt{2/(m-n)}=0.18$. If the measurement noise standard deviations (square roots of **R** diagonals) were incorrectly specified by 8.6%, test equation (6.1-1) would fail. It is often difficult to define the measurement noise σ's accurate within 10%, so in many cases failure of equation (6.1-1) simply implies that the noise σ's are incorrectly scaled. Notice that if all diagonals of **R** are scaled by the same multiplier, the weighted least-squares solution $\hat{\mathbf{x}}$ will not change, although the computed state error covariance $(\mathbf{H}^T\mathbf{R}^{-1}\mathbf{H})^{-1}$ will be incorrectly scaled by the same factor.

As discussed in Chapter 4, when Bayesian least-squares (MMSE) solutions are used with *a priori* state estimate and covariance specified realistically, the weighted sum of measurement and prior state estimate residuals, $\tilde{\mathbf{y}}^T\mathbf{R}^{-1}\tilde{\mathbf{y}}+(\mathbf{x}_a-\hat{\mathbf{x}})^T\mathbf{P}_a^{-1}(\mathbf{x}_a-\hat{\mathbf{x}})$, should have an expected value of m, and equation (6.1-1) should be modified accordingly.

6.1.2 Residual Patterns

The next step in the evaluation process should include plotting of measurement residuals ($\tilde{\mathbf{y}}$) and/or normalized residuals ($\tilde{\mathbf{y}}^T\mathbf{R}^{-1/2}$) versus time (or whatever independent variable is used in the model). If the model structure is correct and the number of measurements is large compared with the number of unknowns, the residuals should not have significant patterns. The state estimate $\hat{\mathbf{x}}$ that minimizes the least-squares cost function tends to eliminate systematic patterns from the measurement residuals. That happens because patterns within the observable subspace of the model can be eliminated by adjustment of $\hat{\mathbf{x}}$, thus decreasing the least-squares cost function. However, $\tilde{\mathbf{y}}$ will still be somewhat correlated (nonwhite) even for an optimal model. This is explained using equation (4.2-8) and recalling that $E[\tilde{\mathbf{y}}]=\mathbf{0}$ if $E[\mathbf{r}]=\mathbf{0}$:

$$\begin{aligned}E[\tilde{\mathbf{y}}\tilde{\mathbf{y}}^T]&=(\mathbf{I}-\mathbf{H}(\mathbf{H}^T\mathbf{R}^{-1}\mathbf{H})^{-1}\mathbf{H}^T\mathbf{R}^{-1})E[\mathbf{r}\mathbf{r}^T](\mathbf{I}-\mathbf{R}^{-1}\mathbf{H}(\mathbf{H}^T\mathbf{R}^{-1}\mathbf{H})^{-1}\mathbf{H}^T)\\&=\mathbf{R}-\mathbf{H}(\mathbf{H}^T\mathbf{R}^{-1}\mathbf{H})^{-1}\mathbf{H}^T\end{aligned}. \tag{6.1-2}$$

Thus $E[\tilde{\mathbf{y}}\tilde{\mathbf{y}}^T]$ will generally not be diagonal. To explore this further, we compute the covariance of $\mathbf{L}^{-1}\tilde{\mathbf{y}}$ where $\mathbf{R}=\mathbf{L}\mathbf{L}^T$. Matrix $\mathbf{L}=\mathbf{R}^{1/2}$ may be the Cholesky factor of **R**, but this is not a requirement. (Usually **R** is diagonal so **L** is also diagonal. In many least-squares implementations measurements are normalized as $\mathbf{L}^{-1}\mathbf{y}$ to simplify processing, but this is not required for the following steps.) Thus

$$\mathbf{L}^{-1}E[\tilde{\mathbf{y}}\tilde{\mathbf{y}}^T]\mathbf{L}^{-T}=\mathbf{I}-\mathbf{L}^{-1}\mathbf{H}(\mathbf{H}^T\mathbf{L}^{-T}\mathbf{L}^{-1}\mathbf{H})^{-1}\mathbf{H}^T\mathbf{L}^{-T}.$$

We now use the Singular Value Decomposition (SVD) to express

$$\mathbf{L}^{-1}\mathbf{H}=\mathbf{U}\mathbf{S}\mathbf{V}^T=[\mathbf{U}_1\quad\mathbf{U}_2]\begin{bmatrix}\mathbf{S}_1\\\mathbf{0}\end{bmatrix}\mathbf{V}^T$$

where $\mathbf{S}_1$ is an $n\times n$ diagonal matrix, $\mathbf{U}_1$ is an $m\times n$ column-orthogonal matrix, $\mathbf{U}_2$ is an $m\times(m-n)$ column-orthogonal nullspace matrix, and **V** is an $n\times n$ orthogonal matrix. Substituting this in equation (6.1-2) yields

$$
\begin{aligned}
\mathbf{L}^{-1}E[\tilde{\mathbf{y}}\tilde{\mathbf{y}}^T]\mathbf{L}^{-T} &= \mathbf{I}-\mathbf{USV}^T(\mathbf{VS}^T\mathbf{U}^T\mathbf{USV}^T)^{-1}\mathbf{VS}^T\mathbf{U}^T \\
&= \mathbf{I}-\mathbf{USV}^T\mathbf{V}(\mathbf{S}^T\mathbf{S})^{-1}\mathbf{V}^T\mathbf{VS}^T\mathbf{U}^T \\
&= \mathbf{I}-\mathbf{U}\begin{bmatrix}\mathbf{S}_1\\ \mathbf{0}\end{bmatrix}\mathbf{S}_1^{-1}\mathbf{S}_1^{-T}\begin{bmatrix}\mathbf{S}_1^T & \mathbf{0}\end{bmatrix}\mathbf{U}^T \\
&= \mathbf{I}-\mathbf{U}_1\mathbf{U}_1^T \\
&= \mathbf{U}_2\mathbf{U}_2^T
\end{aligned}
\quad .
$$

Notice that $\mathbf{U}_1^T\mathbf{U}_1=\mathbf{I}$, $\mathbf{U}_2^T\mathbf{U}_2=\mathbf{I}$, $\mathbf{U}_1^T\mathbf{U}_2=\mathbf{0}$, $\mathbf{U}_2^T\mathbf{U}_1=\mathbf{0}$, and $\mathbf{U}_1\mathbf{U}_1^T+\mathbf{U}_2\mathbf{U}_2^T=\mathbf{I}$. However, $\mathbf{U}_1\mathbf{U}_1^T$ and $\mathbf{U}_2\mathbf{U}_2^T$ will generally not be diagonal, so $\mathbf{L}^{-1}E[\tilde{\mathbf{y}}\tilde{\mathbf{y}}^T]\mathbf{L}^{-T}$ will have nonzero off-diagonal elements even when $\mathbf{R}$ is diagonal. In other words, the residuals will be correlated. Evidence of correlation may appear as systematic patterns in the plotted residuals. Notice that the correlation is defined by the nullspace of the measurement model. When the number of measurements is equal to the number of unknowns, the nullspace does not exist and the residuals will be zero.

It should be noted that for Bayesian estimation (MMSE or MAP) with prior information weighted in the solution, correlations between measurement residuals can be somewhat larger than for the prior-free (nonrandom state) case discussed above.

To summarize, a valid weighted least-squares solution may still produce measurement residuals that are somewhat correlated, but signatures appearing in the residuals are not characteristic of any linear combination of estimated states. In practice, when model structure is correct and the number of measurements is significantly larger than the number of unknowns, it is usually difficult to detect patterns in measurement residuals. If systematic patterns are evident, one should consider the strong possibility that the model is not correct.

6.1.3 Subsets of Residuals

If the solution consists of several different types of measurements, or measurements from different sensors, it is good practice to compute residuals statistics (mean, variance, root-mean-squared [RMS], and min/max) for each measurement subset. This can sometimes reveal problems with particular types of measurements or sensors. However, beware: problems with one measurement subset can sometimes be caused by problems with modeling a different measurement subset. Furthermore, while the mean of all residuals must be zero for the solution to minimize the least-squares cost function, the residual means for individual subsets need not be zero, particularly if the sample size is small. Hence a large residual mean for a specific measurement subset does not necessarily imply that there is a modeling problem. Experience with similar problems is very helpful when validating a least-squares (or Kalman filter) solution.

6.1.4 Measurement Prediction

Another useful test for validating least-squares models involves predicting the measurements outside the data span used for fitting the model, or predicting to untested conditions if parameters other than time are more important in defining model limits. Prediction residuals often show model limitations more readily than fit residuals. Solutions for time span t_0 to t_1 can be used to compute residuals for time span t_1 to t_2, and vice versa. This concept can be extended to different types of measurements, or measurements from different sensors. For example, if a navigation problem uses both range and angular measurements, and it is possible to compute a solution using only range data, one should use that solution to compute angle residuals. If the residuals appear larger than expected or have significant patterns, then model errors may be suspected. Likewise a solution based on only angle data can be used to compute range residuals. Unfortunately it has been this author's experience that redundancy of measurement types or sets is rare. If you have such a problem, take advantage of the redundancy for model validation.

6.1.5 Estimate Comparison

Finally, the state estimates should be checked for reasonableness. If the states values are physically impossible or very unlikely, modeling errors should be considered. When Bayesian estimation is used, compare the difference between the *a priori* and *a posteriori* state estimates normalized by the *a priori* standard deviation: $|\hat{x}_j - \hat{x}_{a\_j}| / \sqrt{P_{a\_jj}}$ for state j. If the state change is many times larger than the prior uncertainty, either the prior estimate, prior uncertainty, or the model is incorrect. Also notice the ratio of the *a priori* and *a posteriori* standard deviations: $\sqrt{P_{a\_jj}} / \sqrt{P_{jj}}$. If the *a posteriori* standard deviation is much smaller than the *a priori*, the estimator believes that the information in the measurements is much greater than that in the prior estimate. Problems with the measurements or model are likely when the *a posteriori* state estimate is suspicious under these conditions. The state correlations computed from the *a posteriori* error covariance matrix may provide insight when several state estimates appear suspect. Additional insight on expected performance may be obtained by testing the estimator using simulated measurements generated using the same conditions and parameters as for the real data.

We now demonstrate some of these concepts using two examples. The first is a variant on the low-order polynomial model previously used, and the second is an actual orbit and attitude solution for the GOES-13 satellite.

Example 6.1: Fourth-Order Polynomial Model

The polynomial problem of Example 5.9 used 101 measurements modeled as

$$y_i = \sum_{j=1}^{n} x_j t_i^{j-1} + r_i$$

with time samples (t_i) uniformly spaced from 0 to 1. The samples of random measurement noise r_i were generated as $N(0,1)$, and the random state x_i values were also $N(0,1)$. However, in this example the standard deviation of the x_i samples was increased to 50 so that the signature of the polynomial is much larger than that of the noise, and easily recognized in plots. Figure 6.1 shows the least-squares fit measurement residuals for a case in which the measurements were simulated using a third-order polynomial ($n_t = 4$), and the weighted least-squares

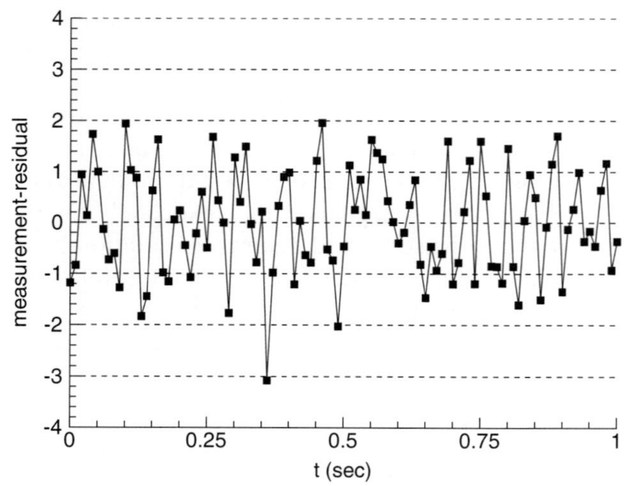

FIGURE 6.1: Fit measurement residuals for $n_t = n = 4$ polynomial example.

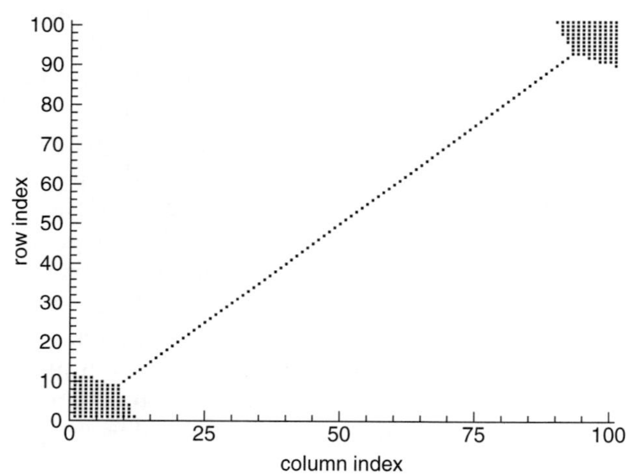

FIGURE 6.2: Measurement residual covariances >0.05 for $n_t = n = 4$ polynomial.

estimator used the same polynomial order ($n = 4$). Hence measurement noise is the only error source in the estimator. The weighted residual SOS for this case is 110.30 with an expected value of $101 - 4 = 97$. The difference of 13.3 is within the standard deviation ($\sqrt{2 \cdot 97} = 13.9$) for the χ^2 distribution. Notice that the residuals have a slight periodicity of about four samples (0.04 s), but otherwise appear zero-mean and random.

Figure 6.2 shows the location in the measurement residual covariance $E(\tilde{\mathbf{y}}\tilde{\mathbf{y}}^T)$ of all absolute value covariances greater than 0.05 for the $n_t = n = 4$ case. Notice that except at the beginning and end of the data span, the residuals are nearly uncorrelated with adjacent samples: the computed (diagonal) variances are about 0.98 at the middle of the data span (column index 50) and adjacent sample covariances are 0.02. The variances only drop to 0.96 at 10 samples from the span ends. Near the beginning and end of the data span, the adjacent sample covariances are more significant (0.10 to 0.13) and are spread over more samples. This result suggests that patterns in the residuals will be minimal, except near the ends of the data span.

Figure 6.3 shows measurement residuals for a higher order case in which $n_t = n = 8$. The weighted residual SOS for this case is 105.15 with an expected value of $101 - 8 = 93$. Again the difference of 12.15 is within the standard deviation ($\sqrt{2 \cdot 93} = 13.6$) for the χ^2 distribution. The same measurement noise sequence was used for this case as for the $n = 4$ case, and the residuals look similar to those in Figure 6.1. Figure 6.4 shows the location in the measurement residual covariance matrix of all absolute value covariances greater than 0.05.

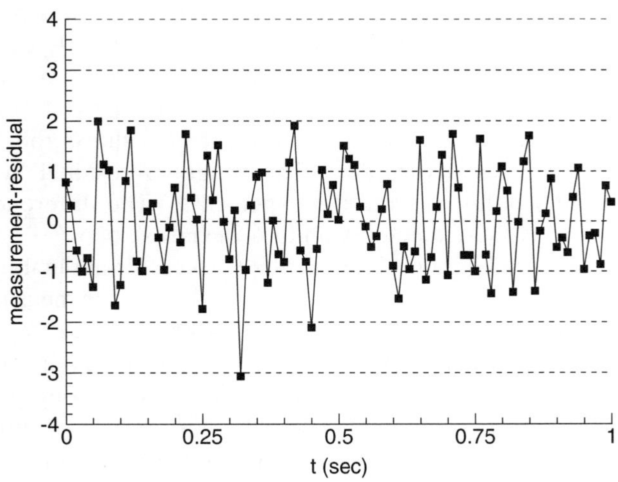

FIGURE 6.3: Fit measurement residuals for $n_t = n = 8$ polynomial example.

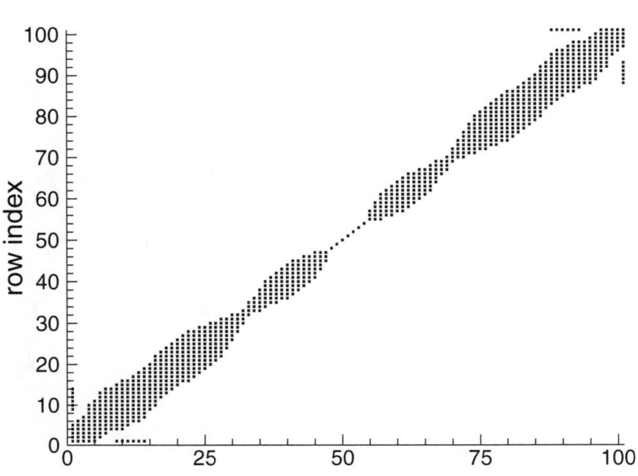

FIGURE 6.4: Measurement residual covariances >0.05 for $n_t = n = 8$ polynomial.

Now the variances are slightly smaller (about 0.95 at the span middle) but covariances are slightly above 0.05 for several adjacent samples. The residuals for the first and last measurements are somewhat more correlated with residuals of nearby measurements.

Extrapolating for larger polynomial orders, we would expect the variances to continue dropping and the adjacent sample covariance magnitudes to increase as polynomial order is increased for a fixed number of measurements. That is the actual behavior: with $n = 15$, the maximum variances are about 0.91 and adjacent covariances are about −0.09, −0.08, −0.07, −0.07 for measurements offsets of 1 to 4, respectively. However, as the number of model states (n) continues to increase, eventually all covariances tend toward zero as n approaches m.

These two cases demonstrate residual behavior when measurement noise is the only model error. However, the primary reasons for plotting residuals are to verify that the measurement noise variance has been correctly defined (so that measurement weighting is appropriate), and to determine if the estimation model is capturing all important behavior of the actual data. To demonstrate a mis-modeling case, we set $n_t = 8$ and $n = 4$. Figure 6.5 shows the residuals: the nonrandom pattern is clearly evident. The weighted residual SOS is 166.43, which is significantly different from the expected 97. The difference of 69.43 is 4.99 times as large as the expected standard deviation of $\sqrt{2 \cdot 97} = 13.9$, which also indicates that a modeling problem is likely. To better understand the effects of the modeling error, Figure 6.6 shows the residuals with the measurement noise removed. The pattern will of course vary with different state sample values, but it is often true for a variety of problems that low-frequency residual patterns appear when systems are mis-modeled. One should look for long-period residual patterns when attempting to determine whether or not a model is valid. Alternately if the residuals appear mostly random with no significant systematic patterns but

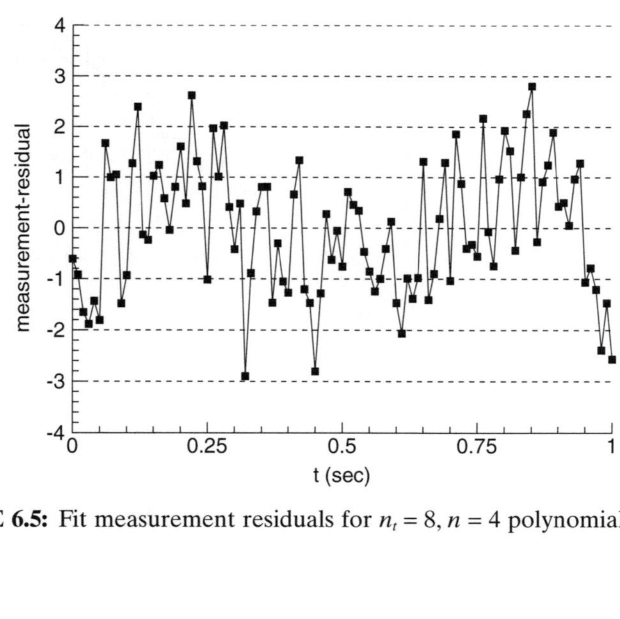

FIGURE 6.5: Fit measurement residuals for $n_t = 8$, $n = 4$ polynomial example.

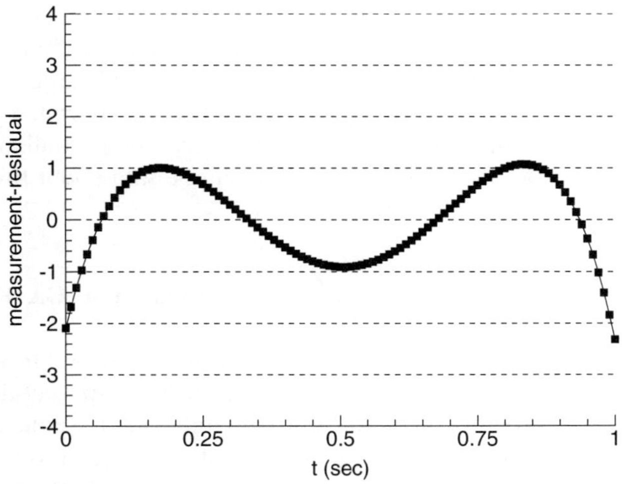

FIGURE 6.6: Fit measurement residuals for $n_t = 8$, $n = 4$ polynomial example: no measurement noise.

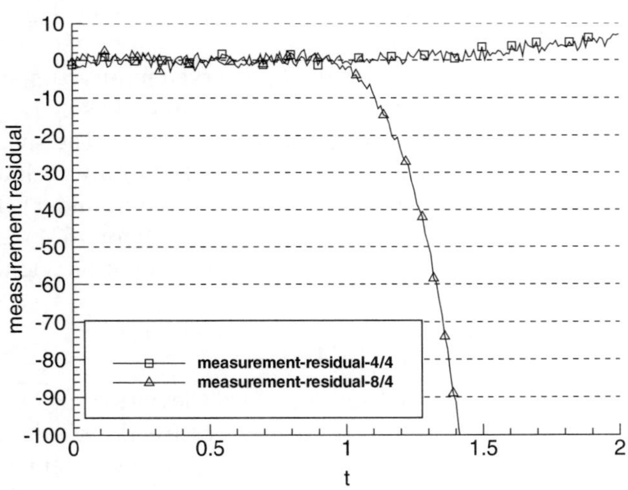

FIGURE 6.7: Measurement fit and prediction residuals for $n_t = 4$, $n = 4$ and $n_t = 8$, $n = 4$ polynomials.

$$\left| \frac{\tilde{\mathbf{y}}^T \mathbf{R}^{-1} \tilde{\mathbf{y}}}{m-n} - 1 \right| \gg k\sqrt{2/(m-n)},$$

it may be suspected that the measurement noise variance $\mathbf{R}$ has been specified incorrectly, and scaling is necessary.

Finally we examine the ability of the models to predict outside the data span. Figure 6.7 shows the fit ($0 \leq t \leq 1$) and prediction ($1 < t \leq 2$) residuals for the $n_t = n = 4$, and $n_t = 8$, $n = 4$ models. Notice that the residuals for both models are approximately the same magnitude during the fit interval, but the prediction residuals are more than two orders of magnitude larger when the model has the wrong state order. This clearly demonstrates the impact of model errors and the importance of calculating prediction residuals when evaluating estimation models. Unfortunately the behavior is not usually as dramatic for many real-world problems because the effective signal-to-noise ratio for many states is not nearly as high. Hence many mis-modeling effects can be partially modeled by a linear combination of estimated states. More will be said about this topic later.

Example 6.2: GOES-13 Orbit and Attitude Determination (OAD)

We now demonstrate model validation issues using actual GOES-13 measurement data. As explained in Section 3.4.9, the GOES I-P spacecraft are geosynchronous weather satellites located approximately on the equator at a fixed longitude. Measurements of imaging instrument angles to ground landmarks, angles to known stars, and ground-to-spacecraft ranges are used in 26-h OAD solutions. The estimated orbit and instrument misalignment attitude coefficients (Fourier series) are used to predict the orbit and instrument attitude of the spacecraft for the next 24 h. That prediction is used onboard the spacecraft to dynamically adjust the scan angles of the optical imaging instruments while scanning.

Table 6.1 lists the measurement residual (observed minus computed) statistics for a 26-h GOES-13 (GOES N) OAD solution computed using range, landmark, and star observations from 2006 day 268.90 to day 269.97. The spacecraft was in the "upright" yaw orientation at this time and OAD only included stars and landmarks from the Imager instrument (not from the Sounder). The weighted measurement residual SOS for this solution was 1687.9 for 1624 scalar measurements, with 114 states estimated in the least-squares solution. Hence

$$\frac{\tilde{\mathbf{y}}^T \mathbf{R}^{-1} \tilde{\mathbf{y}}}{m-n} - 1 = \frac{1687.9}{1624-114} - 1 = 0.1178,$$

which is 3.24 times the normalized standard deviation ($\sqrt{2/1510} = 0.0364$) expected for a χ^2 distribution with 1510 degrees-of-freedom. This is only slightly higher than the expected range for a properly modeled system, so there is no reason to believe that modeling problems exist. In examining Table 6.1, it is helpful to compare the residual statistics with the measurement noise standard deviations ($1-\sigma$) assumed for the OAD fit. These $1-\sigma$ values were set to 3 m

TABLE 6.1: OAD Fit Residual Statistics for Upright GOES-13 on 2006 Day 269: Range Bias = 0

	Range or EW Angle Residuals			NS Angle Residuals		
Measurement Type	#	Mean	RMS	#	Mean	RMS
Ranges (m)	102	0.49	4.41	—	—	—
Imager star observations (μrad)	308	1.19	8.46	308	0.46	9.13
Imager visible landmark observations (μrad)	453	−0.79	7.16	453	−0.18	6.64

for ranges, 7.48 east-west (EW) and 9.71 north-south (NS) microradian (μrad) for Imager star observations, and 7.30 (EW) and 7.40 (NS) μrad for Imager landmark observations. Hence only the EW star residual RMS is larger than the specified $1-\sigma$, and only by 13%. Furthermore, the mean residuals for all observation types except ranges are less than 16% of the specified noise $1-\sigma$'s. The mean range residual is 16% of the specified noise $1-\sigma$, but this is not particularly suspicious. The least-squares solution condition number (scaled) was only 4284, which indicates that solution observability is acceptable.

Figure 6.8 shows the fit and prediction range and visible landmark residuals for this OAD fit. Notice that the range residuals exhibit a strong downward trend with an added sinusoidal pattern. The visible landmark fit residuals appear to

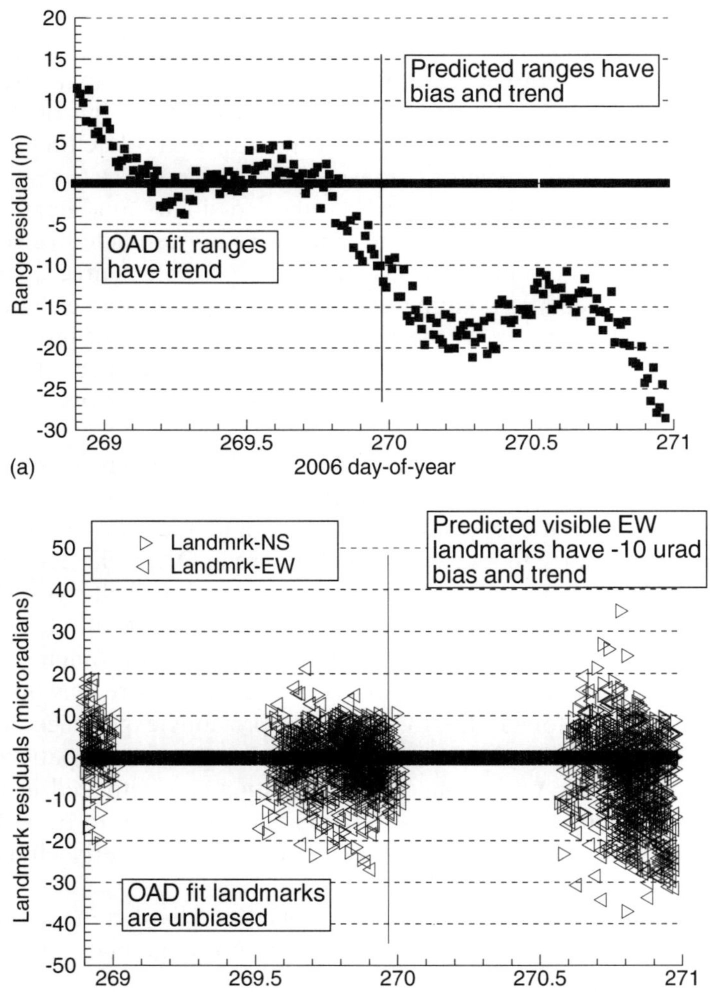

FIGURE 6.8: (a) 26-h OAD fit/prediction range residuals for upright GOES-13: range bias = 0. (b) 26-h OAD fit/prediction landmark residuals for upright GOES-13: range bias = 0.

have little bias, but the EW prediction residuals have a −10 μrad bias and slight trend. Similar patterns appeared in every daily OAD solution when the spacecraft was upright. Even though the residual statistics do not suggest modeling problems, the residual patterns strongly indicate that some characteristics of the GOES-13 system are not modeled properly. This result demonstrates the importance of using multiple evaluation criteria.

Subsequent investigations attempted to determine types of modeling errors that could cause the observed range residual signature. Because the spacecraft is so far from earth, the only significant forces acting on the spacecraft are gravitational (earth, sun, moon), solar radiation pressure, and spacecraft thrusting during maneuvers. At that distance the gravitational forces are accurately modeled and can be eliminated as a possible modeling error. Solar radiation pressure was a possibility, but an error in the assumed spacecraft reflection coefficient would not cause the observed range residual pattern. It was further eliminated as an error source by estimating the coefficient in the OAD solution. Errors in modeling the thrust vector for momentum-dumping maneuvers during the daily housekeeping (HK) period were also a possibility, but this was eliminated using special 2-day OAD solutions centered on the HK time, where the maneuver velocity change vector ($\Delta\mathbf{v}$) was estimated in OAD.

Systematic measurement errors were another possibility. Spacecraft or Imager angular biases affecting both stars and landmarks are estimated as coefficients of the Fourier attitude model, so this was not a possible source of bias error. Other types of systematic pointing errors that have repeatable patterns from day-to-day are also handled by the Fourier series. One type of instrument misalignment is not modeled by the five-parameter attitude model, but the modeling error caused by this omission is too small to produce the errors of Figure 6.8. Hence investigation switched focus to range biases.

Range biases are not adjusted in OAD for the previous series of GOES spacecraft (designated I-M prior to launch and 8–12 after launch) because the ground system calibrates ground loop delays, and the calibrated biases are removed from the range measurements before use in OAD. The calibrations are routinely performed whenever equipment in the signal path changes. This works well and it has not been necessary to estimate GOES-8, 9, 10, 11, or 12 range biases in OAD. Since the same ground equipment is used for ranging to GOES-13, and the spacecraft range delays are well calibrated, there was no reason to suspect a range bias for GOES-13. However, previous experience had shown that uncompensated range biases could produce range residual patterns similar to those of Figure 6.8. When range bias estimation was activated in GOES-13 OAD for the day 269 solution, the estimated bias was found to be +54.0 m, and the trends in both range and landmark residuals were nearly eliminated, as shown in Figure 6.9.

Unfortunately this is not the end of the story. Even though range bias adjustment eliminated the fit and prediction problem, no range bias adjustment was necessary when the spacecraft was operated in the "inverted" yaw orientation (Fig. 6.10), which occurred twice during GOES-13 Post Launch Testing (PLT). Since there is no known reason for the range bias to change with spacecraft orientation, this was suspicious. Furthermore, the spacecraft was intentionally drifted 15 degrees west in longitude during PLT, and the OAD-estimated range

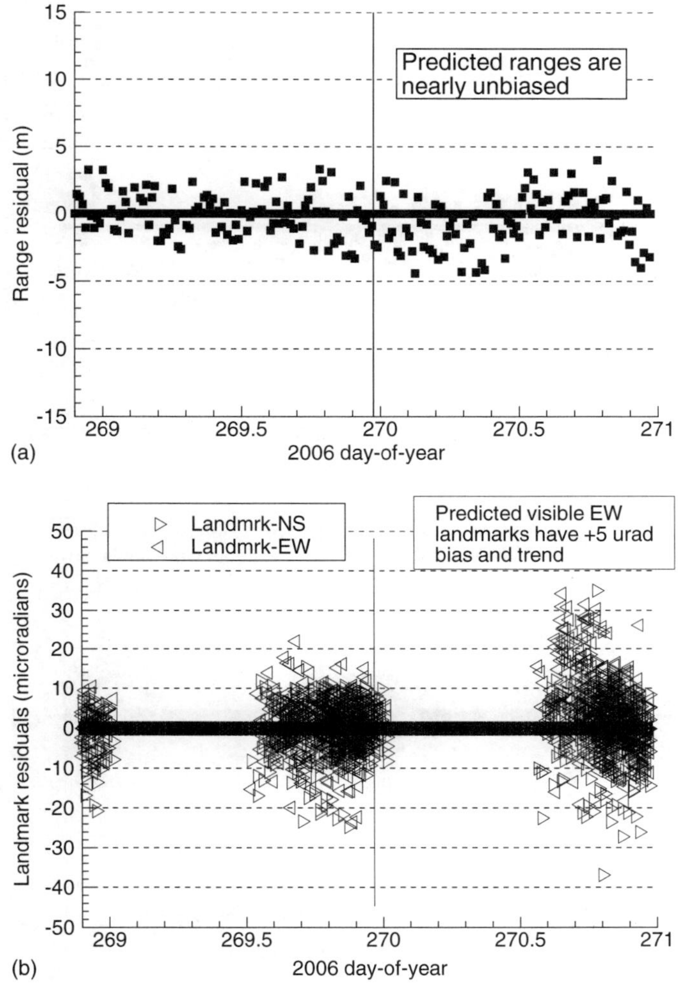

FIGURE 6.9: (a) OAD fit and prediction range residuals for upright GOES-13 with range bias estimated (+54.0 m). (b) OAD fit and prediction landmark residuals for upright GOES-13 with range bias estimated (+54.0 m).

bias was found to increase linearly with longitude offset from the ground station longitude. Again there was no known physical reason for this change in range bias. An extensive investigation over many months was initiated to determine the source of the problem. It was eventually found that an EW bias offset between star and landmark observations of about 32 μrad could produce the same behavior as a range bias, and was invariant with spacecraft longitude. Each EW image pixel is 16 μrad, so the bias appeared to be 2 pixels. Ground and spacecraft processing of star and landmark observations was investigated to determine whether an implementation error could cause the EW bias to appear when GOES-13 was upright but not inverted. No processing error could be found. More information about these GOES-13 investigations is provided in Gibbs et al. (2008a) and Carr et al. (2008).

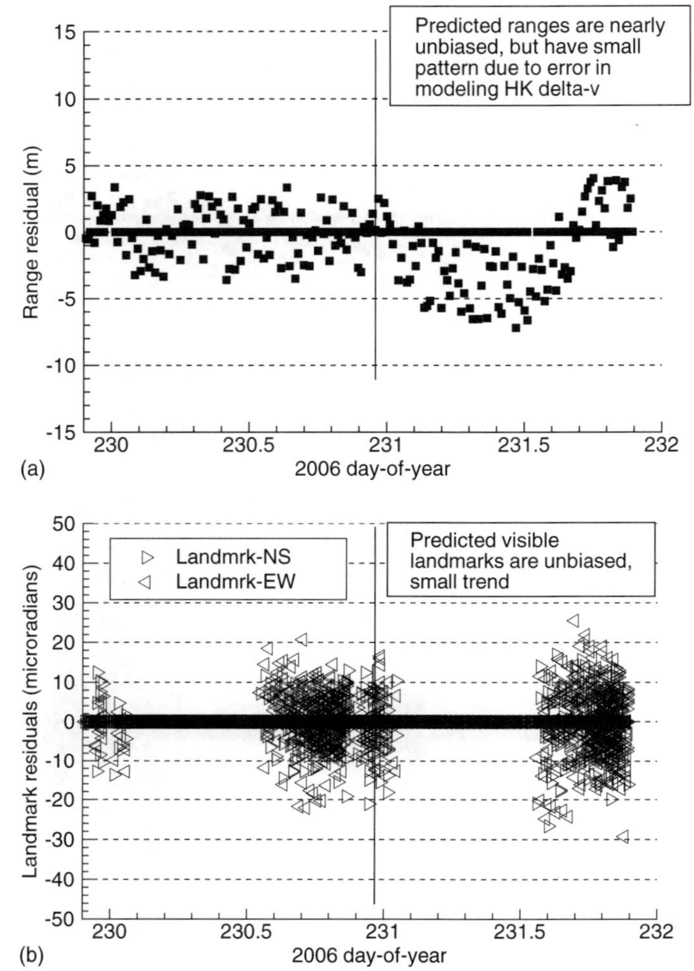

FIGURE 6.10: (a) OAD fit and prediction range residuals for inverted GOES-13 with range bias = 0. (b) OAD fit and prediction landmark residuals for inverted GOES-13 with range bias = 0.

The next spacecraft in the GOES NOP series, GOES-14, was launched in June 2009. Large trends and prediction biases were also observed in GOES-14 range residuals for both spacecraft yaw orientations. Since it is difficult to separate effects of range and landmark biases using just GOES observations, it was requested that the National Aeronautics and Space Administration (NASA) track GOES-14 using the Deep Space Network (DSN). This provided an orbit solution that did not depend on operational GOES range, landmark, and star observations. Comparison of GOES-14 range measurements with ranges computed using the DSN orbit solutions verified that there was indeed a bias of about +50 m in the GOES-14 range measurements. Further investigation revealed that the measured transponder delays for GOES-13 and 14 had never been set in the ground equipment component that computes the two-way range measurements:

the values used were defaults appropriate for GOES I-M spacecraft. Hence the error in one-way range measurements was +42.2 m for GOES-13 and +48.6 m for GOES-14.

When the proper transponder correction was applied, GOES-14 OAD solutions produced nearly unbiased range predictions for both spacecraft orientations. There was still evidence of a small landmark bias that probably changed with yaw orientation, but the error was small enough to be ignored. With the transponder correction properly set for GOES-13, the trend in the prediction range residuals is reduced, producing a mean bias of about −15 m when in the upright orientation at 105 degrees west longitude. (Fig. 6.8 was obtained at 89.5 degrees west longitude, where the range bias effect is smaller.) GOES-13 has not been operated inverted since the transponder correction was applied, but analysis using 2006 data shows that the range adjustment will not eliminate the difference in behavior between upright and inverted orientations. One possibility for a yaw-dependent landmark bias is uncorrected Imager fixed-pattern scan angle errors. This would produce a different bias in the two orientations because ground landmarks appear in a small set of geographic areas. However, this theory has not been verified and the exact cause of the bias is still unknown at the time of this writing.

Although the GOES system is more complex than most systems for which least-squares estimation or Kalman filtering is applied, the procedures used to investigate possible modeling problems are fairly routine and apply to most estimation problems. Hence this case study demonstrates a general approach for isolating modeling problems. When model problems are suspected, it is very helpful to know the residual signatures of various modeling errors: this topic will be discussed in Section 6.2.

In addition to the various residual tests discussed above, other tests provide insight on the expected performance of the estimation. One should always compute the condition number of the least-squares solution—appropriately scaled as discussed in Chapter 5—and be aware that the state estimate will be sensitive to both measurement noise and model errors when the condition number is large. The *a posteriori* state error standard deviations ($\sigma_i = \sqrt{P_{ii}}$ for state $\hat{x}_i$, where P is the inverse of the Fisher information matrix) provide guidance on the expected accuracy of the state solution. Correlation coefficients ($\rho_{ij} = P_{ij}/(\sigma_i\sigma_j)$) close to 1.0 indicate that linear combinations of states are difficult to separately estimate. More information on weakly observable combinations of states is provided by the eigenvectors of the information matrix (or equivalently the right singular vectors of the measurement matrix) corresponding to the smallest eigenvalues. Finally, when using Bayesian MMSE or MAP solutions, the ratio of *a priori* to *a posteriori* state standard deviations ($\sqrt{P_{a\_ii} / P_{ii}}$) allows assessment of how much information is provided by the measurements compared with the prior information.

6.2 SOLUTION ERROR ANALYSIS

6.2.1 State Error Covariance and Confidence Bounds

In Chapter 4 we noted that the inverse of the Fisher information matrix is a lower bound on the state estimate error covariance. Specifically,

$$E[\tilde{\mathbf{x}}\tilde{\mathbf{x}}^T] \geq \left[E\left[\frac{\partial^2 J}{\partial \mathbf{x}\partial \mathbf{x}^T}\right]\right]^{-1} = \left[E\left[\left(\frac{\partial J}{\partial \mathbf{x}}\right)^T\left(\frac{\partial J}{\partial \mathbf{x}}\right)\right]\right]^{-1} \tag{6.2-1}$$

with J defined as the negative log probability density and $\tilde{\mathbf{x}} = \mathbf{x} - \hat{\mathbf{x}}$. For the maximum likelihood problem using the linear model $\mathbf{y} = \mathbf{H}\mathbf{x} + \mathbf{r}$, where $\mathbf{r}$ is $N(\mathbf{0},\mathbf{R})$ random noise, the negative log probability density is

$$J = -\ln L(\mathbf{y}\,|\,\mathbf{x}) = \frac{1}{2}\left[m\ln(2\pi) + \ln|\mathbf{R}| + (\mathbf{y}-\overline{\mathbf{y}})^T\mathbf{R}^{-1}(\mathbf{y}-\overline{\mathbf{y}})\right]$$

so the error covariance bound is

$$\mathbf{P} = \left[E\left[\frac{\partial^2 J}{\partial \mathbf{x}\partial \mathbf{x}^T}\right]\right]^{-1} = (\mathbf{H}^T\mathbf{R}^{-1}\mathbf{H})^{-1}. \tag{6.2-2}$$

The same equation is also obtained when computing second-order error statistics of the weighted least-squares method (without considering the underlying probability density of $\mathbf{r}$). For the Bayesian least-squares or MAP problem, the error covariance bound is

$$\mathbf{P} = \left[E\left[\frac{\partial^2 J}{\partial \mathbf{x}\partial \mathbf{x}^T}\right]\right]^{-1} = (\mathbf{H}^T\mathbf{R}^{-1}\mathbf{H} + \mathbf{P}_a^{-1})^{-1} \tag{6.2-3}$$

where $\mathbf{P}_a$ is the error covariance of the *a priori* state estimate.

We now consider the meaning of $\mathbf{P}$, often called the *a posteriori* or *formal* error covariance matrix. When $\mathbf{r}$ is zero-mean Gaussian-distributed, the state estimate errors will also be Gaussian, that is,

$$p_{\tilde{X}}(\tilde{\mathbf{x}}\,|\,\mathbf{y}) = \frac{1}{(2\pi)^{n/2}|\mathbf{P}|^{1/2}}\exp\left(-\frac{1}{2}\tilde{\mathbf{x}}^T\mathbf{P}^{-1}\tilde{\mathbf{x}}\right). \tag{6.2-4}$$

When $\mathbf{x}$ is a scalar ($n = 1$), the probability that $|\tilde{x}|/\sigma < a$ for $a > 0$ is $1-2[1-F_{\tilde{X}}(a)]$, where $F_{\tilde{X}}$ is the Gaussian distribution function. For example, the probability that $|x-\hat{x}|/\sigma < a$ is $0.683, 0.955, 0.997$ for $a = 1, 2, 3$ respectively. When $n > 1$, the quantity in the exponent, $\tilde{\mathbf{x}}^T\mathbf{P}^{-1}\tilde{\mathbf{x}}$, is a χ^2-distributed variable with n degrees-of-freedom. Hence values of $\tilde{\mathbf{x}}$ for which $\tilde{\mathbf{x}}^T\mathbf{P}^{-1}\tilde{\mathbf{x}}$ has a fixed value represent contours in multi-dimensional space of constant probability. It is easily shown using eigen decomposition that $\tilde{\mathbf{x}}^T\mathbf{P}^{-1}\tilde{\mathbf{x}} = c$ is the equation of an n-dimensional ellipsoid. The state covariance is first factored as

$$\mathbf{P} = \mathbf{M}\mathbf{\Lambda}\mathbf{M}^T, \tag{6.2-5}$$

where $\mathbf{M}$ is the modal matrix of eigenvectors and $\mathbf{\Lambda}$ is a diagonal matrix of eigenvalues $\lambda_1, \lambda_2, \ldots, \lambda_n$. Then $\tilde{\mathbf{x}}^T\mathbf{P}^{-1}\tilde{\mathbf{x}} = \mathbf{z}^T\mathbf{\Lambda}^{-1}\mathbf{z}$, where $\mathbf{z} = \mathbf{M}^T\tilde{\mathbf{x}}$. The constant χ^2 contours are defined by

$$\mathbf{z}^T \boldsymbol{\Lambda}^{-1} \mathbf{z} = \frac{z_1^2}{\lambda_1} + \frac{z_2^2}{\lambda_2} + \ldots + \frac{z_n^2}{\lambda_n} = c, \tag{6.2-6}$$

which is the equation of an n-dimensional ellipsoid. Notice that the square roots of the eigenvalues are the lengths of the ellipsoid semimajor axes, and the eigenvectors (columns of $\mathbf{M}$) define the directions of the principal axes. The mean value of $\tilde{\mathbf{x}}^T \mathbf{P}^{-1} \tilde{\mathbf{x}}$ should be n if the model is correct, so approximately 50% of the random samples of $\tilde{\mathbf{x}}$ should lie within the ellipsoid defined by $c = n$. Exact probabilities or values of c for other cumulative probabilities can be computed from the χ^2 distribution using $\Pr\{\mathbf{z}^T \boldsymbol{\Lambda}^{-1} \mathbf{z} < c\} = P(n/2, c/2)$ where P is the incomplete gamma function.

Chi-squared analysis is frequently used to characterize the uncertainty in a least-squares or Kalman filter state estimate. Often the distribution for a given confidence level is summarized by *confidence bounds* or *regions* for each state variable, or for linear combinations of the variables. Risk analysis and system accuracy specification are two examples sometimes based on this approach. This type of analysis should be used with caution because it depends on knowledge of the error distributions. Recall that the inverse of the Fisher information matrix is a lower bound on the error covariance. For linear systems with Gaussian measurement errors where the model structure has been validated, the inverse information matrix accurately characterizes the solution accuracy. However, for highly nonlinear systems, systems with non-Gaussian errors, or models that are only approximations to reality, the computed covariance may significantly underestimate actual errors. More will be said about these issues later.

Example 6.3: Two-Dimensional (2-D) Constant Probability Contours

Use of the χ^2 distribution for computing confidence limits can be demonstrated using a two-dimensional problem. Let the error covariance be

$$\mathbf{P} = \begin{bmatrix} 2.5 & 1.5 \\ 1.5 & 2.5 \end{bmatrix} = \begin{bmatrix} 1/\sqrt{2} & 1/\sqrt{2} \\ 1/\sqrt{2} & -1/\sqrt{2} \end{bmatrix} \begin{bmatrix} 4 & 0 \\ 0 & 1 \end{bmatrix} \begin{bmatrix} 1/\sqrt{2} & 1/\sqrt{2} \\ 1/\sqrt{2} & -1/\sqrt{2} \end{bmatrix}$$

where the eigenvalues are 4 and 1, and $\begin{bmatrix} 1/\sqrt{2} & 1/\sqrt{2} \end{bmatrix}^T$, $\begin{bmatrix} 1/\sqrt{2} & -1/\sqrt{2} \end{bmatrix}^T$ are the eigenvectors. Hence the ellipsoid semimajor and semiminor axes are 2 and 1, respectively, with major and minor axes oriented at +45 degrees and +135 degrees with respect to the x-axis. Using the incomplete gamma function for 2 degrees-of-freedom, we find that $c = 9.21$ defines the ellipse corresponding to 99% cumulative probability, and $c = 2.30$ defines the ellipse for 68.3% cumulative probability. (Note that 68.3% corresponds to the Gaussian $1 - \sigma$ limit for a single variable.) Figure 6.11 shows the 99%ile and 68.3%ile ellipses, with 200 random samples generated to have the above covariance. (A Gaussian random number generator produced $N(0,4)$ samples of z_1 and $N(0,1)$ samples of z_2. Then the $\mathbf{x}$ samples were computed as $\tilde{\mathbf{x}} = \mathbf{Mz}$.) Notice that all but one of the 200 samples lie within the outer ellipse, which is consistent with the 99%ile bound. Also notice that the 68.3% ellipse crosses the x_1 and x_2 axes at about ±1.93, which

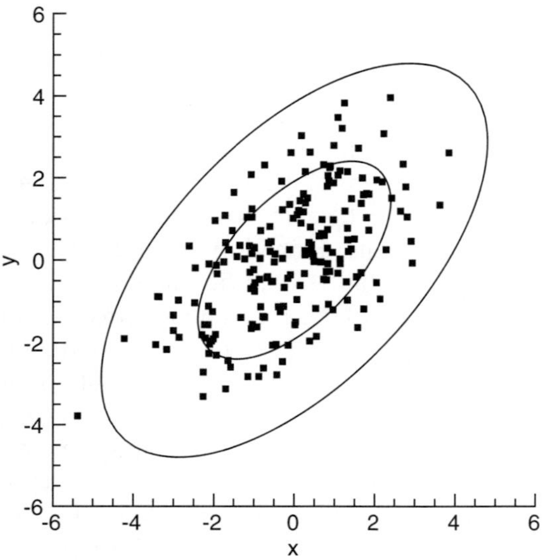

FIGURE 6.11: 2-D random samples, 68.3%ile, and 99%ile ellipses.

is 22% larger than the standard deviation computed from the square roots of the covariance matrix diagonals ($\sqrt{2.50} = 1.581$). Hence when samples are correlated, the confidence limits for each variable necessary to meet a given confidence level must be larger than the limits required for independent x_1 and x_2 samples. Further discussion of confidence limits may be found in Press et al. (2007, section 15.6.3) and many probability texts. Press et al. also discuss use of the χ^2 distribution in non-normal cases, and the relationship between measurement fit error and the state estimate error confidence limits.

Example 6.4: Satellite Collision Avoidance

Suppose that we want to know whether the probability of impact between two earth-orbiting satellites is greater than 1%. We are given the earth-centered-inertial position-velocity orbit elements of the two satellites at a fixed epoch time t_0. The distance between the two satellites is then tabulated as a function of time by integrating the orbit elements. Let t_1 be the time at which the distance is a minimum, although this is often not the point of maximum collision probability. To determine whether the probability of impact is significant, the three-dimensional (3-D) error covariance of the position difference between the two satellites and the sizes of each satellite must be known. To make the problem somewhat simpler, assume that the satellites have zero physical size and that the orbit error covariance of one satellite is much smaller than the other. In fact, assume that the orbit of satellite #2 is known perfectly, and thus it is only necessary to integrate the error covariance of satellite #1. The position error covariance of satellite #1 can be computed as

$$\mathbf{P}_{p1}(t_1) = \mathbf{T}\mathbf{\Phi}(t_1 - t_0)\mathbf{P}_1(t_0)\mathbf{\Phi}^T(t_1 - t_0)\mathbf{T}^T$$

where $\mathbf{\Phi}(t_1 - t_0)$ is the 6×6 position-velocity state transition matrix from time t_0 to t_1 and $\mathbf{T} = [\mathbf{I}_{3\times3} \quad \mathbf{0}_{3\times3}]$. The position difference at time t_1 is designated as $\tilde{\mathbf{x}}_p(t_1)$.

Since the problem is 3-D, the mean value of the χ^2-distributed variable $\tilde{\mathbf{x}}_p^T \mathbf{P}_{p1}^{-1} \tilde{\mathbf{x}}_p$ should be 3. Tables of χ^2 distributions for $n = 3$ show that only 1% of samples have $\tilde{\mathbf{x}}_p^T \mathbf{P}_{p1}^{-1} \tilde{\mathbf{x}}_p > 11.3$. Hence the relative positions of the two satellites at times near the closest approach can be used to compute $\tilde{\mathbf{x}}_p^T \mathbf{P}_{p1}^{-1} \tilde{\mathbf{x}}_p$, and if that value exceeds 11.3, we can conclude that the probability of impact for the assumed conditions is less than 1%. A more accurate computation should also include uncertainty in the satellite #2 orbit, the size of the spacecraft, and the fact that large alongtrack errors may allow collisions at times slightly offset from predicted closest approach. Campbell and Udrea (2002) provide more information on the satellite collision avoidance problem.

For visualization purposes, 3-D ellipsoid contours for 1% probability can be computed by finding the $\tilde{\mathbf{x}}_p$ values that yield $\tilde{\mathbf{x}}_p^T \mathbf{P}_{p1}^{-1} \tilde{\mathbf{x}}_p = 11.3$. One contour drawing method that produces visually appealing plots defines the plotted coordinates as

$$\tilde{\mathbf{x}} = \sqrt{c}\, \mathbf{M} \mathbf{\Lambda}^{1/2} \hat{\mathbf{y}}$$

where $\hat{\mathbf{y}}$ is a unit vector in ellipsoid principal axes and $c = 11.3$ in our example. Since eigenvalue/eigenvector functions typically order the eigenvectors from largest to smallest, the first eigenvector is the ellipsoid major axis. Thus $\hat{\mathbf{y}}$ can be specified using fixed "slices" of $\hat{y}_1$ (e.g., $\hat{y}_1 = -1.0, -0.95, \ldots, 0.95, 1.0$) with the other two coordinates specified as $\hat{y}_2 = \sqrt{1 - \hat{y}_1^2} \cos\theta$, $\hat{y}_3 = \sqrt{1 - \hat{y}_1^2} \sin\theta$, and θ stepped in fixed increments (e.g., 0.2 radians). This approach works well even for high eccentricity ellipsoids. Woodburn and Tanygin (2002) provide more information on visualization of position error ellipsoids.

For illustrative purposes, we use the 3×3 position partition of the 6×6 position-velocity error covariance obtained from orbit determination of a low-earth-orbit satellite,

$$\mathbf{P}_{p1}(t_1) = \begin{bmatrix} 30.28910 & 17.86375 & -35.32536 \\ 17.86375 & 19.14262 & -23.78177 \\ -35.32536 & -23.78177 & 68.79261 \end{bmatrix},$$

where position is specified in meters in true-of-date earth-centered-inertial coordinates. The modal matrix of eigenvectors for this covariance is

$$\mathbf{M} = \begin{bmatrix} -0.4901768 & 0.5327286 & -0.6898746 \\ -0.3418741 & 0.6105519 & 0.7143867 \\ 0.8017785 & 0.5860261 & -0.1171523 \end{bmatrix}$$

and the eigenvalues are

$$\mathbf{\Lambda} = \begin{bmatrix} 100.5296 & 0 & 0 \\ 0 & 11.90293 & 0 \\ 0 & 0 & 5.791791 \end{bmatrix}.$$

Figure 6.12 shows the 99%ile ellipsoid of this example, plotted as described. The major axis roughly corresponds to the alongtrack (velocity vector) of the satellite.

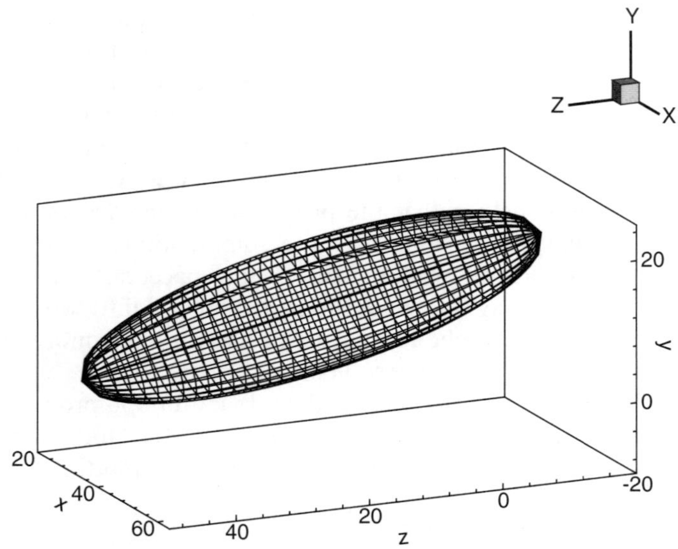

FIGURE 6.12: Satellite position error (m) 99% probability ellipsoid.

The collision probability can also be visualized in two dimensions by projecting the 3-D ellipsoid into the plane normal to the relative velocity of the two spacecraft. If the distance between spacecraft is outside the 2-D ellipse, the probability of collision at that point is less that the threshold. The 2-D projections should also be computed for small time offsets from the time of closest approach because the alongtrack error is usually far larger than crosstrack or radial errors. Hence the closest approach is often not the point of maximum collision probability.

6.2.2 Model Error Analysis

We now consider modeling error effects that are not included in the formal covariance matrix of the previous section. There are three categories of modeling errors:

1. Model states included in the estimator are a subset of the true model states.
2. The model structure used in the estimator differs from the true model structure. This includes differences in both **H** and measurement noise characteristics (modeled in the estimator as fixed covariance **R**).
3. The estimator model is linear but the true model is nonlinear and linearization of the nonlinear model results in significant model errors.

The tools available for analysis of modeling errors include covariance analysis and Monte Carlo simulation. We address the error categories in reverse order because options for the latter two are limited.

6.2.2.1 System Nonlinearities Most real-world systems are nonlinear at some level. The least-squares method is applied to these systems by linearizing the model about the current state estimate and then iterating (with new linearization at each step) using a Newton-like method. Nonlinear least-squares solution techniques are discussed in Chapter 7. System nonlinearities tend to have more impact on convergence of the iterations than on accuracy of the final solution because the iterations are stopped when the steps of $\Delta\hat{\mathbf{x}}$ are "small." When measurement noise is large enough that the estimated state $\hat{\mathbf{x}}$ is far from the true $\mathbf{x}$, nonlinearities of the model may cause $\mathbf{x}-\hat{\mathbf{x}}$ to be larger than would be predicted from the formal error covariance.

Although the effects of nonlinearities could (in theory) be analyzed in a covariance sense using Taylor series expansions given approximate magnitudes of higher order terms, this is generally not practical. Hence nonlinear error effects are usually analyzed via Monte Carlo simulation where the simulated measurements are based on the nonlinear model and the estimate is obtained using a linearized model.

6.2.2.2 *Incorrect Model Structure* When the estimator model has a structure different from that of the true model, either Monte Carlo simulation or covariance analysis can be used to analyze error behavior. However, the analysis is difficult because of fundamental inconsistencies in the models, as we now demonstrate. Let the true model for measurements used in the least-squares fit be

$$\mathbf{y}_f = \mathbf{H}_{tf}\mathbf{x}_t + \mathbf{r}_{tf}, \tag{6.2-7}$$

where $\mathbf{x}_t$ has n_t elements. Subscript "f" denotes "fit" measurements because we will later use "p" to indicate prediction measurements. As before, the estimator model is

$$\mathbf{y}_f = \mathbf{H}_f\mathbf{x} + \mathbf{r}_f$$

where $\mathbf{x}$ has n elements. The Bayesian least-squares estimate is

$$\begin{aligned}\hat{\mathbf{x}} &= (\mathbf{H}_f^T\mathbf{R}_f^{-1}\mathbf{H}_f + \mathbf{P}_a^{-1})^{-1}[\mathbf{H}_f^T\mathbf{R}_f^{-1}\mathbf{y}_f + \mathbf{P}_a^{-1}\mathbf{x}_a] \\ &= \mathbf{P}^{-1}[\mathbf{H}_f^T\mathbf{R}_f^{-1}(\mathbf{H}_{tf}\mathbf{x}_t + \mathbf{r}_{tf}) + \mathbf{P}_a^{-1}\mathbf{x}_a]\end{aligned} \tag{6.2-8}$$

where $\mathbf{x}_a$ and $\mathbf{P}_a$ are the *a priori* estimate and error covariance, respectively, $\mathbf{R}_f = E[\mathbf{r}_f\mathbf{r}_f^T]$, and

$$\mathbf{P} = (\mathbf{H}_f^T\mathbf{R}_f^{-1}\mathbf{H}_f + \mathbf{P}_a^{-1})^{-1}$$

is the calculated *a posteriori* state error covariance. Note that $\mathbf{P}$ does not characterize the actual estimate errors when $\mathbf{H}_f$, $\mathbf{P}_a$, or $\mathbf{R}_f$ do not match reality.

When the model structure is correct, a "true" version ($\mathbf{x}$) of the estimator state $\hat{\mathbf{x}}$ will exist and it is possible to compute the covariance of the estimate error $\tilde{\mathbf{x}} = \mathbf{x} - \hat{\mathbf{x}}$. However, when the model structure of $\mathbf{H}_f$ and $\hat{\mathbf{x}}$ are incorrect, a "true" version of the model state $\hat{\mathbf{x}}$ does not exist. Rather than computing the covariance of $\tilde{\mathbf{x}}$, we must be content with characterizing the error in either the fit measurements or measurements not used in the fit (predicted measurements). The fit measurement residuals are

$$\begin{aligned}\tilde{\mathbf{y}}_f &= \mathbf{y}_f - \mathbf{H}_f\hat{\mathbf{x}} \\ &= \mathbf{H}_{tf}\mathbf{x}_t + \mathbf{r}_{tf} - \mathbf{H}_f\mathbf{P}[\mathbf{H}_f^T\mathbf{R}_f^{-1}(\mathbf{H}_{tf}\mathbf{x}_t + \mathbf{r}_{tf}) + \mathbf{P}_a^{-1}\mathbf{x}_a] \\ &= (\mathbf{I} - \mathbf{H}_f\mathbf{P}\mathbf{H}_f^T\mathbf{R}_f^{-1})(\mathbf{H}_{tf}\mathbf{x}_t + \mathbf{r}_{tf}) - \mathbf{H}_f\mathbf{P}\mathbf{P}_a^{-1}\mathbf{x}_a\end{aligned} \tag{6.2-9}$$

and the residual covariance matrix is

$$\begin{aligned}E[\tilde{\mathbf{y}}_f\tilde{\mathbf{y}}_f^T] &= (\mathbf{I} - \mathbf{H}_f\mathbf{P}\mathbf{H}_f^T\mathbf{R}_f^{-1})E[\mathbf{r}_{tf}\mathbf{r}_{tf}^T](\mathbf{I} - \mathbf{H}_f\mathbf{P}\mathbf{H}_f^T\mathbf{R}_f^{-1})^T \\ &+ \left[(\mathbf{I} - \mathbf{H}_f\mathbf{P}\mathbf{H}_f^T\mathbf{R}_f^{-1})\mathbf{H}_{tf} \quad -\mathbf{H}_f\mathbf{P}\mathbf{P}_a^{-1}\right] E\begin{bmatrix}\mathbf{x}_t\mathbf{x}_t^T & \mathbf{x}_t\mathbf{x}_a^T \\ \mathbf{x}_a\mathbf{x}_t^T & \mathbf{x}_a\mathbf{x}_a^T\end{bmatrix}\begin{bmatrix}\mathbf{H}_{tf}^T(\mathbf{I} - \mathbf{H}_f\mathbf{P}\mathbf{H}_f^T\mathbf{R}_f^{-1})^T \\ -\mathbf{P}_a^{-1}\mathbf{P}\mathbf{H}_f^T\end{bmatrix}.\end{aligned} \tag{6.2-10}$$

Although $\mathbf{x}_t$ and $\mathbf{x}_a$ will generally have different dimensions, it is quite possible that $\mathbf{x}_t$ and $\mathbf{x}_a$ will have subsets of states in common. Hence it may be possible to specify portions of

$$E\begin{bmatrix}\mathbf{x}_t\mathbf{x}_t^T & \mathbf{x}_t\mathbf{x}_a^T \\ \mathbf{x}_a\mathbf{x}_t^T & \mathbf{x}_a\mathbf{x}_a^T\end{bmatrix},$$

and unknown off-diagonal partitions could be reasonably set to zero. Notice that there is potential for significant cancellation in the last term of equation (6.2-10), so the fit residual covariance may be only slightly larger than the measurement noise (first) term.

For measurements not included in the fit (e.g., predicted), defined as

$$\mathbf{y}_p = \mathbf{H}_{tp}\mathbf{x}_t + \mathbf{r}_{tp}, \tag{6.2-11}$$

the residuals are

$$\begin{aligned}\tilde{\mathbf{y}}_p &= \mathbf{y}_p - \mathbf{H}_p\hat{\mathbf{x}} \\ &= \mathbf{H}_{tp}\mathbf{x}_t + \mathbf{r}_{tp} - \mathbf{H}_p\mathbf{P}[\mathbf{H}_f^T\mathbf{R}_f^{-1}(\mathbf{H}_{tf}\mathbf{x}_t + \mathbf{r}_{tf}) + \mathbf{P}_a^{-1}\mathbf{x}_a] \\ &= \mathbf{H}_{tp}\mathbf{x}_t + \mathbf{r}_{tp} - \mathbf{H}_p\mathbf{P}\mathbf{H}_f^T\mathbf{R}_f^{-1}(\mathbf{H}_{tf}\mathbf{x}_t + \mathbf{r}_{tf}) - \mathbf{H}_p\mathbf{P}\mathbf{P}_a^{-1}\mathbf{x}_a\end{aligned}, \tag{6.2-12}$$

where $\mathbf{H}_p$ is the estimator model of predicted measurement sensitivity to estimator states. Assuming that $E[\mathbf{r}_{tp}\mathbf{r}_t^T] = \mathbf{0}$, the covariance is

$$\begin{aligned}E[\tilde{\mathbf{y}}_p\tilde{\mathbf{y}}_p^T] &= E[\mathbf{r}_{tp}\mathbf{r}_{tp}^T] + \mathbf{H}_p\mathbf{P}\mathbf{H}_f^T\mathbf{R}_f^{-1}E[\mathbf{r}_f\mathbf{r}_f^T]\mathbf{R}_f^{-1}\mathbf{H}_f\mathbf{P}\mathbf{H}_p^T \\ &+ \left[\mathbf{H}_{tp} - \mathbf{H}_p\mathbf{P}\mathbf{H}_f^T\mathbf{R}_f^{-1}\mathbf{H}_{tf} \quad -\mathbf{H}_p\mathbf{P}\mathbf{P}_a^{-1}\right] E\begin{bmatrix}\mathbf{x}_t\mathbf{x}_t^T & \mathbf{x}_t\mathbf{x}_a^T \\ \mathbf{x}_a\mathbf{x}_t^T & \mathbf{x}_a\mathbf{x}_a^T\end{bmatrix}\begin{bmatrix}\mathbf{H}_{tp}^T - \mathbf{H}_{tf}\mathbf{R}_f^{-1}\mathbf{H}_f\mathbf{P}\mathbf{H}_p^T \\ -\mathbf{P}_a^{-1}\mathbf{P}\mathbf{H}_p^T\end{bmatrix}.\end{aligned} \tag{6.2-13}$$

There is much less potential for error cancellation of predicted measurements in equation (6.2-13), so the prediction residuals may be much larger than fit residuals when model errors are present. The predicted measurement residuals are usually of more interest than fit residuals because the goal of estimation problems is often either to use the model for insight on parameter values, or to predict linear combinations of states under conditions different from those of the fit.

Equations (6.2-9) to (6.2-13) are based on the assumption that the Bayesian estimator equation (6.2-8) is used. If a nonrandom model is assumed and the prior information of equation (6.2-8) is removed, equations (6.2-9) and (6.2-12) simplify to

$$\boxed{\tilde{\mathbf{y}}_f = (\mathbf{I} - \mathbf{H}_f \mathbf{P} \mathbf{H}_f^T \mathbf{R}_f^{-1})(\mathbf{H}_{tf}\mathbf{x}_t + \mathbf{r}_{tf})} \tag{6.2-14}$$

for the fit residuals and

$$\boxed{\tilde{\mathbf{y}}_p = \mathbf{H}_{tp}\mathbf{x}_t + \mathbf{r}_{tp} - \mathbf{H}_p \mathbf{P} \mathbf{H}_f^T \mathbf{R}_f^{-1}(\mathbf{H}_{tf}\mathbf{x}_t + \mathbf{r}_{tf})} \tag{6.2-15}$$

for the prediction residuals where $\mathbf{P} = (\mathbf{H}_f^T \mathbf{R}_f^{-1} \mathbf{H}_f)^{-1}$. The measurement residual covariance equations are modified accordingly.

While the covariance equations (6.2-10) and (6.2-13) show how differences in models impact measurement residuals, they are of limited usefulness because many of the inputs such as $\mathbf{H}_{tf}$, $\mathbf{H}_{tp}$, $E[\mathbf{r}_{tf}\mathbf{r}_{tf}]$, $E[\mathbf{r}_{tp}\mathbf{r}_{tp}]$, and

$$E\begin{bmatrix} \mathbf{x}_t\mathbf{x}_t^T & \mathbf{x}_t\mathbf{x}_a^T \\ \mathbf{x}_a\mathbf{x}_t^T & \mathbf{x}_a\mathbf{x}_a^T \end{bmatrix}$$

may not be known. $\mathbf{H}_{tf}$ is usually only an approximation to reality: if it is well known the states in the estimator can often be modified so that $\mathbf{H}_f = \mathbf{H}_{tf}$ and thus modeling errors due to this term are eliminated. Hence mis-modeling analysis is generally of most interest when the true models are poorly known. A further problem is that equations (6.2-10) and (6.2-13) do not provide much guidance on the magnitudes or general characteristics of the modeling error because the behavior depends on differences in quantities that are poorly known.

So how can the above equations be used? One approach hypothesizes a specific type of "true" model and then plots the residual signatures using equations (6.2-9) and (6.2-12) without the measurement noise terms. This is similar to a "residual derivative" approach. If similar signatures appear when processing actual measurements, one can speculate that the true model is similar to the hypothesized model and the estimator model can be improved. Whether or not this is less work than simply testing different models in the estimator is debatable, but it can provide insight about model structure.

Even in cases when sufficient information is available to compute the covariance equations (6.2-10) and (6.2-13), it is often less work to use Monte Carlo simulation than covariance analysis.

6.2.2.3 Unadjusted Analyze (Consider) States In some cases, states included in the estimator are a subset of states available in a high-accuracy model. That is, a high-accuracy state model may have p states but only $n < p$ states are included in the estimator state vector. This is a *Reduced-Order Model* (ROM). Use of a ROM may be desirable either to achieve optimum accuracy (when some states are negligibly small), to make the estimator more robust, or to reduce the execution time of the estimator. In some cases a subset of states is determined using multiple large sets of measurements, and then those estimated states are treated as known when an estimator is used on a small set of new data. This may be done because the unadjusted (fixed) states are not observable solely from the limited number of new measurements. In any case, errors in the values of unadjusted states will cause errors in the adjusted states.

There are two reasons for analyzing the effects of the errors in unadjusted states on the estimate of the adjusted states: to compute a total error budget and to optimize selection of states to be included in the estimator. We assume that the states can be partitioned such that the fit measurement model is

$$\mathbf{y}_f = [\mathbf{H}_{f1} \quad \mathbf{H}_{f2}]\begin{bmatrix}\mathbf{x}_1 \\ \mathbf{x}_2\end{bmatrix} + \mathbf{r}_f \tag{6.2-16}$$

where the $\mathbf{x}_1$ states are adjusted in the estimator. The prior estimate $\mathbf{x}_{2a}$ of the $\mathbf{x}_2$ states is used to compute the residual $\tilde{\mathbf{y}}_f = \mathbf{y}_f - (\mathbf{H}_{f1}\hat{\mathbf{x}}_1 + \mathbf{H}_{f2}\mathbf{x}_{2a})$, but $\mathbf{x}_{2a}$ is not adjusted. The $\mathbf{x}_2$ states are often called *consider parameters* because they are assumed to be constants, and uncertainty in their values is "considered" when computing the error covariance for $\hat{\mathbf{x}}_1$. Using equation (6.2-8) for a Bayesian solution, the estimate $\hat{\mathbf{x}}_1$ is computed as

$$\begin{aligned}\hat{\mathbf{x}}_1 &= (\mathbf{H}_{f1}^T\mathbf{R}_f^{-1}\mathbf{H}_{f1} + \mathbf{P}_{a1}^{-1})^{-1}[\mathbf{H}_{f1}^T\mathbf{R}_f^{-1}(\mathbf{y}_f - \mathbf{H}_{f2}\mathbf{x}_{a2}) + \mathbf{P}_{a1}^{-1}\mathbf{x}_{a1}] \\ &= \mathbf{P}_1[\mathbf{H}_{f1}^T\mathbf{R}_f^{-1}(\mathbf{H}_{f1}\mathbf{x}_1 + \mathbf{H}_{f2}(\mathbf{x}_2 - \mathbf{x}_{a2}) + \mathbf{r}_f) + \mathbf{P}_{a1}^{-1}\mathbf{x}_{a1}] \\ &= \mathbf{P}_1[\mathbf{H}_{f1}^T\mathbf{R}_f^{-1}\mathbf{H}_{f1}\mathbf{x}_1 + \mathbf{H}_{f1}^T\mathbf{R}_f^{-1}\mathbf{H}_{f2}(\mathbf{x}_2 - \mathbf{x}_{a2}) + \mathbf{H}_{f1}^T\mathbf{R}_f^{-1}\mathbf{r}_f + \mathbf{P}_{a1}^{-1}\mathbf{x}_{a1}]\end{aligned} \tag{6.2-17}$$

where $\tilde{\mathbf{x}}_{a1} \triangleq \mathbf{x}_1 - \mathbf{x}_{a1}$, $\mathbf{P}_{a1} \triangleq E[\tilde{\mathbf{x}}_{a1}\tilde{\mathbf{x}}_{a1}^T]$, and

$$\mathbf{P}_1 = (\mathbf{H}_{f1}^T\mathbf{R}_f^{-1}\mathbf{H}_{f1} + \mathbf{P}_{a1}^{-1})^{-1}. \tag{6.2-18}$$

A similar equation for $\hat{\mathbf{x}}_1$ can be computed using a non-Bayesian weighted least-squares solution by eliminating the $\mathbf{P}_{a1}^{-1}\mathbf{x}_{a1}$ and $\mathbf{P}_{a1}^{-1}$ terms, but the $\mathbf{x}_{a2}$ term is still needed to compute the effects of errors in $\mathbf{x}_{a2}$ on $\hat{\mathbf{x}}_1$. The reason for assuming a Bayesian solution will be seen shortly when using equation (6.2-17) and related equations to optimize selection of adjusted states.

Defining $\tilde{\mathbf{x}}_{a2} \triangleq \mathbf{x}_2 - \mathbf{x}_{a2}$ and $\mathbf{P}_{a2} \triangleq E[\tilde{\mathbf{x}}_{a2}\tilde{\mathbf{x}}_{a2}^T]$, then

$$\boxed{\begin{aligned}\tilde{\mathbf{x}}_1 &\triangleq \mathbf{x}_1 - \hat{\mathbf{x}}_1 \\ &= (\mathbf{I} - \mathbf{P}_1\mathbf{H}_{f1}^T\mathbf{R}_f^{-1}\mathbf{H}_{f1})\mathbf{x}_1 - \mathbf{P}_1\mathbf{H}_{f1}^T\mathbf{R}_f^{-1}\mathbf{H}_{f2}\tilde{\mathbf{x}}_{a2} - \mathbf{P}_1\mathbf{H}_{f1}^T\mathbf{R}_f^{-1}\mathbf{r}_f - \mathbf{P}_1\mathbf{P}_{a1}^{-1}\mathbf{x}_{a1} \\ &= \mathbf{P}_1(\mathbf{P}_{a1}^{-1}\tilde{\mathbf{x}}_{a1} - \mathbf{H}_{f1}^T\mathbf{R}_f^{-1}\mathbf{H}_{f2}\tilde{\mathbf{x}}_{a2} - \mathbf{H}_{f1}^T\mathbf{R}_f^{-1}\mathbf{r}_f)\end{aligned}} \tag{6.2-19}$$

and

$$\begin{aligned}E[\tilde{\mathbf{x}}_1\tilde{\mathbf{x}}_1^T] &= \mathbf{P}_1(\mathbf{H}_{f1}^T\mathbf{R}_f^{-1}\mathbf{H}_{f1} + \mathbf{P}_{a1}^{-1})\mathbf{P}_1 + \mathbf{P}_1\mathbf{H}_{f1}^T\mathbf{R}_f^{-1}\mathbf{H}_{f2}\mathbf{P}_{a2}\mathbf{H}_{f2}^T\mathbf{R}_f^{-1}\mathbf{H}_{f1}\mathbf{P}_1 \\ &= \mathbf{P}_1 + \mathbf{P}_1\mathbf{H}_{f1}^T\mathbf{R}_f^{-1}\mathbf{H}_{f2}\mathbf{P}_{a2}\mathbf{H}_{f2}^T\mathbf{R}_f^{-1}\mathbf{H}_{f1}\mathbf{P}_1\end{aligned}. \tag{6.2-20}$$

Similarly using equation (6.2-19),

$$E[\tilde{\mathbf{x}}_1\tilde{\mathbf{x}}_{a2}^T] = -\mathbf{P}_1\mathbf{H}_{f1}^T\mathbf{R}_f^{-1}\mathbf{H}_{f2}\mathbf{P}_{a2},$$

and the joint covariance for both terms is:

$$\boxed{E\begin{bmatrix}\tilde{\mathbf{x}}_1\tilde{\mathbf{x}}_1^T & \tilde{\mathbf{x}}_1\tilde{\mathbf{x}}_{a2}^T \\ \tilde{\mathbf{x}}_{a2}\tilde{\mathbf{x}}_1^T & \tilde{\mathbf{x}}_{a2}\tilde{\mathbf{x}}_{a2}^T\end{bmatrix} = \begin{bmatrix}\mathbf{P}_1 + \mathbf{P}_1\mathbf{H}_{f1}^T\mathbf{R}_f^{-1}\mathbf{H}_{f2}\mathbf{P}_{a2}\mathbf{H}_{f2}^T\mathbf{R}_f^{-1}\mathbf{H}_{f1}\mathbf{P}_1 & -\mathbf{P}_1\mathbf{H}_{f1}^T\mathbf{R}_f^{-1}\mathbf{H}_{f2}\mathbf{P}_{a2} \\ -\mathbf{P}_{a2}\mathbf{H}_{f2}^T\mathbf{R}_f^{-1}\mathbf{H}_{f1}\mathbf{P}_1 & \mathbf{P}_{a2}\end{bmatrix}.}$$

(6.2-21)

Note that $\mathbf{P}_1$ will overestimate the error when the filter order is greater than the true order ($n > p$) because $\mathbf{H}_{f1}$ then includes terms that are actually zero. This property will be useful when attempting to determine the true order of the system.

Equations (6.2-19) and (6.2-21) show the impact of errors in unadjusted parameters on errors in the adjusted parameters. The equations are fundamental for both error analysis and optimal selection of adjusted states. The latter topic is addressed in Section 6.3. Equation (6.2-20) or the upper left partition of equation (6.2-21) is often called the *consider* or *unadjusted analyze* covariance of the adjusted states. This type of covariance analysis has been used for orbit determination since the early days of the space program. Typical consider parameters used in orbit determination are tracking station position errors, atmospheric refraction coefficients, and gravity field coefficients. These parameters are usually determined using measurements from hundreds or even thousands of satellite data arcs. Parameters that may be typically either adjusted or treated as consider parameters include solar radiation pressure coefficients, atmospheric drag coefficients, and tracking station biases.

Since accumulation of the normal equation information arrays is usually the most computationally expensive step in the least-squares error computations, consider parameter analysis is often implemented by initially treating all parameters as adjusted when forming the information arrays. That is, an **A** matrix and **b** vector are formed by processing all measurements and prior as

$$\begin{bmatrix} \mathbf{A}_{11} & \mathbf{A}_{12} \\ \mathbf{A}_{21} & \mathbf{A}_{22} \end{bmatrix} = \mathbf{H}_f^T \mathbf{R}_f^{-1} \mathbf{H}_f + \mathbf{P}_a^{-1} = \begin{bmatrix} \mathbf{H}_{f1}^T \mathbf{R}_f^{-1} \mathbf{H}_{f1} + \mathbf{P}_{a1}^{-1} & \mathbf{H}_{f1}^T \mathbf{R}_f^{-1} \mathbf{H}_{f2} \\ \mathbf{H}_{f2}^T \mathbf{R}_f^{-1} \mathbf{H}_{f1} & \mathbf{H}_{f2}^T \mathbf{R}_f^{-1} \mathbf{H}_{f2} + \mathbf{P}_{a2}^{-1} \end{bmatrix} \tag{6.2-22}$$

$$\begin{bmatrix} \mathbf{b}_1 \\ \mathbf{b}_2 \end{bmatrix} = \mathbf{H}_f^T \mathbf{R}_f^{-1} \mathbf{y}_f + \mathbf{P}_a^{-1} \mathbf{x}_a = \begin{bmatrix} \mathbf{H}_{f1}^T \mathbf{R}_f^{-1} \mathbf{y}_f + \mathbf{P}_a^{-1} \mathbf{x}_{a1} \\ \mathbf{H}_{f2}^T \mathbf{R}_f^{-1} \mathbf{y}_f + \mathbf{P}_a^{-1} \mathbf{x}_{a2} \end{bmatrix}. \tag{6.2-23}$$

The consider error covariance is obtained by first forming

$$\mathbf{C} = \mathbf{P}_1 \mathbf{A}_{12} \mathbf{P}_{a2}^{1/2} \tag{6.2-24}$$

where $\mathbf{P}_1 = \mathbf{A}_{11}^{-1}$ and $\mathbf{P}_{a2} = \mathbf{P}_{a2}^{1/2} \mathbf{P}_{a2}^{T/2}$. Then

$$\begin{aligned} E[\tilde{\mathbf{x}}_1 \tilde{\mathbf{x}}_1^T] &= \mathbf{P}_1 + \mathbf{P}_1 \mathbf{A}_{12} \mathbf{P}_{a2} \mathbf{A}_{12}^T \mathbf{P}_1 \\ &= \mathbf{P}_1 + \mathbf{C}\mathbf{C}^T \end{aligned} \tag{6.2-25}$$

and

$$\begin{aligned} E[\tilde{\mathbf{x}}_1 \tilde{\mathbf{x}}_{a2}^T] &= -\mathbf{P}_1 \mathbf{A}_{12} \mathbf{P}_{a2} \\ &= -\mathbf{C}\mathbf{P}_{a2}^{T/2} \end{aligned}. \tag{6.2-26}$$

Although the **A** and **b** arrays are shown partitioned into adjusted and consider parameters, that partitioning is not necessary when forming the arrays. Specific states may be selected for treatment as either adjusted or consider by extracting the appropriate rows and columns from the **A** and **b** arrays during the analysis step, reordering the array elements, and forming $\mathbf{A}_{11}$ and $\mathbf{A}_{12}$ as needed. Thus parameters may be easily moved from adjusted to consider or vice versa and the solution accuracy can be evaluated as a function of the selection.

Consider Analysis Using the QR Algorithm Consider covariance analysis can also be implemented using the arrays produced by the QR algorithm. Recall from Chapter 5 that the QR algorithm uses orthogonal transformations **T** to zero lower rows of $\mathbf{H}_f$ and create an upper triangular matrix **U**:

$$\begin{bmatrix} \mathbf{T}_1 \\ \mathbf{T}_2 \end{bmatrix} \mathbf{H}_f \hat{\mathbf{x}} = \begin{bmatrix} \mathbf{T}_1 \mathbf{y} \\ \mathbf{T}_2 \mathbf{y} \end{bmatrix}$$

or

$$\begin{bmatrix} \mathbf{U} \\ \mathbf{0} \end{bmatrix} \hat{\mathbf{x}} = \begin{bmatrix} \mathbf{T}_1 \mathbf{y} \\ \mathbf{T}_2 \mathbf{y} \end{bmatrix}.$$

where $\mathbf{T}$ is implicitly defined from elementary Householder transformations—it is never explicitly formed. Then the least-squares solution $\hat{\mathbf{x}}$ is obtained by back-solving $\mathbf{U}\hat{\mathbf{x}} = (\mathbf{T}_1\mathbf{y})$. As when using the normal equations, measurement and prior information processing proceeds as if all states are to be adjusted. Then when specific states are selected as consider parameters, they are moved to the end of the state vector and the corresponding columns of the $\mathbf{U}$ matrix are moved to the right partition of $\mathbf{U}$. Unfortunately the resulting $\mathbf{U}$ is no longer upper triangular, so it is necessary to again apply orthogonal transformations to zero the lower partition. This second orthogonalization step does not change any properties of the solution—it simply reorders the states. The $\mathbf{U}$, $\mathbf{x}$, and $\mathbf{z}$ arrays are thus partitioned as

$$\begin{bmatrix} \mathbf{U}_{11} & \mathbf{U}_{12} \\ \mathbf{0} & \mathbf{U}_{22} \end{bmatrix} \begin{bmatrix} \hat{\mathbf{x}}_1 \\ \hat{\mathbf{x}}_2 \end{bmatrix} = \begin{bmatrix} \mathbf{z}_1 \\ \mathbf{z}_1 \end{bmatrix}$$

where $\mathbf{x}_1$ includes n adjusted states and $\mathbf{x}_2$ includes p consider parameters. To understand the error propagation it is necessary to examine partitioning of the orthogonal transforms. We partition the orthogonal $\mathbf{T}$ matrix into row blocks corresponding to partitioning of the states, where $\mathbf{T}_1$ is $n \times m$, $\mathbf{T}_2$ is $p \times m$, and $\mathbf{T}_3$ is $(m - n - p) \times m$. Then the transformation on $\mathbf{H}_f \hat{\mathbf{x}} = \mathbf{y}_f$ may be written as

$$\begin{aligned} \begin{bmatrix} \mathbf{T}_1 \\ \mathbf{T}_2 \\ \mathbf{T}_3 \end{bmatrix} [\mathbf{H}_{f1} \quad \mathbf{H}_{f2}] \begin{bmatrix} \hat{\mathbf{x}}_1 \\ \hat{\mathbf{x}}_2 \end{bmatrix} &= \begin{bmatrix} \mathbf{T}_1 \mathbf{y}_f \\ \mathbf{T}_2 \mathbf{y}_f \\ \mathbf{T}_3 \mathbf{y}_f \end{bmatrix} \\ &= \begin{bmatrix} \mathbf{T}_1 \\ \mathbf{T}_2 \\ \mathbf{T}_3 \end{bmatrix} [\mathbf{H}_{f1} \quad \mathbf{H}_{f2}] \begin{bmatrix} \mathbf{x}_1 \\ \mathbf{x}_2 \end{bmatrix} + \begin{bmatrix} \mathbf{T}_1 \mathbf{r}_f \\ \mathbf{T}_2 \mathbf{r}_f \\ \mathbf{T}_3 \mathbf{r}_f \end{bmatrix} \end{aligned}.$$

Subtracting to form

$$\begin{bmatrix} \tilde{\mathbf{x}}_1 \\ \tilde{\mathbf{x}}_2 \end{bmatrix} = \begin{bmatrix} \mathbf{x}_1 - \hat{\mathbf{x}}_1 \\ \mathbf{x}_2 - \hat{\mathbf{x}}_2 \end{bmatrix}$$

and using

$$\begin{bmatrix} \mathbf{T}_1 \\ \mathbf{T}_2 \\ \mathbf{T}_3 \end{bmatrix} [\mathbf{H}_{f1} \quad \mathbf{H}_{f2}] = \begin{bmatrix} \mathbf{U}_{11} & \mathbf{U}_{12} \\ \mathbf{0} & \mathbf{U}_{22} \\ \mathbf{0} & \mathbf{0} \end{bmatrix},$$

we find that

$$\mathbf{U}_{11}\tilde{\mathbf{x}}_1 + \mathbf{U}_{12}\tilde{\mathbf{x}}_2 = -\mathbf{T}_1 \mathbf{r}_f$$

or

$$\boxed{\tilde{\mathbf{x}}_1 = -\mathbf{U}_{11}^{-1}(\mathbf{T}_1\mathbf{r} + \mathbf{U}_{12}\tilde{\mathbf{x}}_2).} \tag{6.2-27}$$

This equation applies for any definition of $\tilde{\mathbf{x}}_2$, not necessarily for the optimal solution $\tilde{\mathbf{x}}_2 = \mathbf{x}_2 - \mathbf{U}_{22}^{-1}(\mathbf{T}_2\mathbf{y}_f)$, so we use $\tilde{\mathbf{x}}_{a2} = \mathbf{x}_2 - \mathbf{x}_{a2}$ where $\mathbf{P}_{a2} = E[\tilde{\mathbf{x}}_{a2}\tilde{\mathbf{x}}_{a2}^T]$. Assuming that $E[\mathbf{r}\tilde{\mathbf{x}}_{a2}^T] = \mathbf{0}$ and $E[\mathbf{r}\mathbf{r}^T] = \mathbf{I}$ (as required in the QR algorithm), the error covariance for the adjusted states is found to be

$$\boxed{\begin{aligned} E(\tilde{\mathbf{x}}_1\tilde{\mathbf{x}}_1^T) &= \mathbf{U}_{11}^{-1}(\mathbf{T}_1\mathbf{I}\mathbf{T}_1^T + \mathbf{U}_{12}\mathbf{P}_{a2}\mathbf{U}_{12}^T)\mathbf{U}_{11}^{-T} \\ &= \mathbf{U}_{11}^{-1}\mathbf{U}_{11}^{-T} + \mathbf{U}_{11}^{-1}\mathbf{U}_{12}\mathbf{P}_{a2}\mathbf{U}_{12}^T\mathbf{U}_{11}^{-T} \\ &= \mathbf{P}_1 + \mathbf{C}\mathbf{C}^T \end{aligned}} \tag{6.2-28}$$

where $\mathbf{P}_1 = \mathbf{U}_{11}^{-1}\mathbf{U}_{11}^{-T}$ and $\mathbf{C} = \mathbf{U}_{11}^{-1}\mathbf{U}_{12}\mathbf{P}_{a2}^{1/2}$. Consider parameter analysis can be efficiently performed when using the QR method, but it is slightly more difficult to implement than when using the normal equations.

Residual Derivatives and Covariance Systematic patterns in measurement residuals are often the only evidence of modeling problems. Hence knowledge of measurement residual sensitivity to errors in unadjusted states can be very helpful when analyzing these problems. Other useful information includes the expected total residual variance as a function of measurement index, time, and type. As before, we are interested in both measurement fit and prediction residuals. Using $\mathbf{y}_f$ to designate fit measurements and $\mathbf{y}_p$ to designate prediction measurements, the residuals for fit and prediction data spans are

$$\boxed{\begin{aligned} \begin{bmatrix} \tilde{\mathbf{y}}_f \\ \tilde{\mathbf{y}}_p \end{bmatrix} &= \begin{bmatrix} \mathbf{y}_f \\ \mathbf{y}_p \end{bmatrix} - \begin{bmatrix} \mathbf{H}_{f1} & \mathbf{H}_{f2} \\ \mathbf{H}_{p1} & \mathbf{H}_{p2} \end{bmatrix} \begin{bmatrix} \hat{\mathbf{x}}_1 \\ \mathbf{x}_{a2} \end{bmatrix} \\ &= \begin{bmatrix} \mathbf{r}_f \\ \mathbf{r}_p \end{bmatrix} + \begin{bmatrix} \mathbf{H}_{f1} & \mathbf{H}_{f2} \\ \mathbf{H}_{p1} & \mathbf{H}_{p2} \end{bmatrix} \begin{bmatrix} \tilde{\mathbf{x}}_1 \\ \tilde{\mathbf{x}}_{a2} \end{bmatrix} \\ &= \begin{bmatrix} \mathbf{r}_f \\ \mathbf{r}_p \end{bmatrix} + \begin{bmatrix} \mathbf{H}_{f1} & \mathbf{H}_{f2} \\ \mathbf{H}_{p1} & \mathbf{H}_{p2} \end{bmatrix} \begin{bmatrix} \mathbf{P}_1\mathbf{P}_{a1}^{-1}\tilde{\mathbf{x}}_{a1} - \mathbf{P}_1\mathbf{H}_{f1}^T\mathbf{R}_f^{-1}\mathbf{H}_{f2}\tilde{\mathbf{x}}_{a2} - \mathbf{P}_1\mathbf{H}_{f1}^T\mathbf{R}_f^{-1}\mathbf{r}_f \\ \tilde{\mathbf{x}}_{a2} \end{bmatrix} \\ &= \begin{bmatrix} (\mathbf{I} - \mathbf{H}_{f1}\mathbf{P}_1\mathbf{H}_{f1}^T\mathbf{R}_f^{-1})\mathbf{r}_f \\ \mathbf{r}_p - \mathbf{H}_{p1}\mathbf{P}_1\mathbf{H}_{f1}^T\mathbf{R}_f^{-1}\mathbf{r}_f \end{bmatrix} + \begin{bmatrix} \mathbf{H}_{f1}\mathbf{P}_1\mathbf{P}_{a1}^{-1} & (\mathbf{I} - \mathbf{H}_{f1}\mathbf{P}_1\mathbf{H}_{f1}^T\mathbf{R}_f^{-1})\mathbf{H}_{f2} \\ \mathbf{H}_{p1}\mathbf{P}_1\mathbf{P}_{a1}^{-1} & \mathbf{H}_{p2} - \mathbf{H}_{p1}\mathbf{P}_1\mathbf{H}_{f1}^T\mathbf{R}_f^{-1}\mathbf{H}_{f2} \end{bmatrix} \begin{bmatrix} \tilde{\mathbf{x}}_{a1} \\ \tilde{\mathbf{x}}_{a2} \end{bmatrix} \end{aligned}} \tag{6.2-29}$$

Hence the fit residual derivative is

$$\boxed{\frac{\partial \tilde{\mathbf{y}}_f}{\partial \mathbf{x}_2} = (\mathbf{I} - \mathbf{H}_{f1}\mathbf{P}_1\mathbf{H}_{f1}^T\mathbf{R}_f^{-1})\mathbf{H}_{f2}.} \tag{6.2-30}$$

This can be useful when attempting to identify unmodeled parameters that can generate observed signatures in the residuals.

Since $\mathbf{r}_f$, $\tilde{\mathbf{x}}_{a1}$ and $\tilde{\mathbf{x}}_{a2}$ are assumed to be uncorrelated, equation (6.2-29) can be used to compute the measurement fit residual covariance as

$$\begin{aligned} E[\tilde{\mathbf{y}}_f\tilde{\mathbf{y}}_f^T] &= \mathbf{R}_f - \mathbf{H}_{f1}\mathbf{P}_1\mathbf{H}_{f1}^T \\ &\quad + (\mathbf{I} - \mathbf{H}_{f1}\mathbf{P}_1\mathbf{H}_{f1}^T\mathbf{R}_f^{-1})\mathbf{H}_{f2}\mathbf{P}_{a2}\mathbf{H}_{f2}^T(\mathbf{I} - \mathbf{R}_f^{-1}\mathbf{H}_{f1}\mathbf{P}_1\mathbf{H}_{f1}^T) \end{aligned} \quad (6.2\text{-}31)$$

The scaled residual covariance (used in computing the least-squares cost function) is

$$\begin{aligned} \mathbf{R}_f^{-1/2}E[\tilde{\mathbf{y}}_f\tilde{\mathbf{y}}_f^T]\mathbf{R}_f^{-T/2} &= \mathbf{I} - \mathbf{R}_f^{-1/2}\mathbf{H}_{f1}\mathbf{P}_1\mathbf{H}_{f1}^T\mathbf{R}_f^{-T/2} \\ &\quad + \mathbf{R}_f^{-1/2}(\mathbf{I} - \mathbf{H}_{f1}\mathbf{P}_1\mathbf{H}_{f1}^T\mathbf{R}_f^{-1})\mathbf{H}_{f2}\mathbf{P}_{a2}\mathbf{H}_{f2}^T(\mathbf{I} - \mathbf{R}_f^{-1}\mathbf{H}_{f1}\mathbf{P}_1\mathbf{H}_{f1}^T)\mathbf{R}_f^{-T/2} \end{aligned} \quad (6.2\text{-}32)$$

From equation (6.2-29) the measurement prediction residuals are

$$\begin{aligned} \tilde{\mathbf{y}}_p &= \mathbf{r}_p + \mathbf{H}_{p1}\tilde{\mathbf{x}}_1 + \mathbf{H}_{p2}\tilde{\mathbf{x}}_{a2} \\ &= \mathbf{r}_p + \mathbf{H}_{p1}\mathbf{P}_1(\mathbf{P}_{a1}^{-1}\tilde{\mathbf{x}}_{a1} - \mathbf{H}_{f1}^T\mathbf{R}_f^{-1}\mathbf{r}_f) + (\mathbf{H}_{p2} - \mathbf{H}_{p1}\mathbf{P}_1\mathbf{H}_{f1}^T\mathbf{R}_f^{-1}\mathbf{H}_{f2})\tilde{\mathbf{x}}_{a2} \end{aligned} \quad (6.2\text{-}33)$$

and the prediction residual covariance is

$$\begin{aligned} E(\tilde{\mathbf{y}}_p\tilde{\mathbf{y}}_p^T) &= \mathbf{R}_p + \mathbf{H}_{p1}\mathbf{P}_1\mathbf{H}_{p1}^T \\ &\quad + (\mathbf{H}_{p2} - \mathbf{H}_{p1}\mathbf{P}_1\mathbf{H}_{f1}^T\mathbf{R}_f^{-1}\mathbf{H}_{f2})\mathbf{P}_{a2}(\mathbf{H}_{p2} - \mathbf{H}_{p1}\mathbf{P}_1\mathbf{H}_{f1}^T\mathbf{R}_f^{-1}\mathbf{H}_{f2})^T \end{aligned} \quad (6.2\text{-}34)$$

Notice that $\mathbf{H}_{f1}\mathbf{P}_1\mathbf{H}_{f1}^T$ in equation (6.2-31) reduces the residual covariance, but $\mathbf{H}_{p1}\mathbf{P}_1\mathbf{H}_{p1}^T$ in equation (6.2-34) increases the covariance. As expected, prediction residuals tend to be greater than fit residuals even when all states are adjusted. Also notice that the unadjusted state error ($\tilde{\mathbf{x}}_{a2}$) contribution to the fit and prediction residual covariance is positive-semi-definite. That is, errors in unadjusted states will always increase the residual variances—another unsurprising result.

The residual between the *a posteriori* state estimate $\hat{\mathbf{x}}_1$ and the *a priori* value $\mathbf{x}_{a1}$ is another component of the Bayesian cost function. That residual is:

$$\begin{aligned} \mathbf{x}_{a1} - \hat{\mathbf{x}}_1 &= -\mathbf{P}_1(\mathbf{H}_{f1}^T\mathbf{R}_f^{-1}\mathbf{H}_{f1})\mathbf{x}_1 - \mathbf{P}_1\mathbf{H}_{f1}^T\mathbf{R}_f^{-1}\mathbf{H}_{f2}\tilde{\mathbf{x}}_{a2} - \mathbf{P}_1\mathbf{H}_{f1}^T\mathbf{R}_f^{-1}\mathbf{r} + (\mathbf{I} - \mathbf{P}_1\mathbf{P}_{a1}^{-1})\mathbf{x}_{a1} \\ &= -\mathbf{P}_1\mathbf{H}_{f1}^T\mathbf{R}_f^{-1}(\mathbf{H}_{f1}\tilde{\mathbf{x}}_{a1} + \mathbf{H}_{f2}\tilde{\mathbf{x}}_{a2} + \mathbf{r}) \end{aligned} \quad (6.2\text{-}35)$$

The covariance of that residual is

$$E[(\mathbf{x}_{a1} - \hat{\mathbf{x}}_1)(\mathbf{x}_{a1} - \hat{\mathbf{x}}_1)^T] = \mathbf{P}_1\mathbf{H}_{f1}^T\mathbf{R}_f^{-1}(\mathbf{H}_{f1}\mathbf{P}_{a1}\mathbf{H}_{f1}^T + \mathbf{H}_{f2}\mathbf{P}_{a2}\mathbf{H}_{f2}^T + \mathbf{R}_f)\mathbf{R}_f^{-1}\mathbf{H}_{f1}\mathbf{P}_1 \quad (6.2\text{-}36)$$

and the contribution to the cost function is

$$\begin{aligned} &E[(\mathbf{x}_{a1} - \hat{\mathbf{x}}_1)^T\mathbf{P}_{a1}^{-1}(\mathbf{x}_{a1} - \hat{\mathbf{x}}_1)] \\ &= E\left[tr[\mathbf{P}_{a1}^{-1}(\mathbf{x}_{a1} - \hat{\mathbf{x}}_1)(\mathbf{x}_{a1} - \hat{\mathbf{x}}_1)^T]\right] \\ &= tr\left[\mathbf{P}_{a1}^{-1}\mathbf{P}_1\mathbf{H}_{f1}^T\mathbf{R}_f^{-1}(\mathbf{H}_{f1}\mathbf{P}_{a1}\mathbf{H}_{f1}^T + \mathbf{H}_{f2}\mathbf{P}_{a2}\mathbf{H}_{f2}^T + \mathbf{R}_f)\mathbf{R}_f^{-1}\mathbf{H}_{f1}\mathbf{P}_1\right] \end{aligned} \quad (6.2\text{-}37)$$

The above sensitivity and covariance equations are useful when analyzing the impact of unadjusted parameters on fit and prediction measurement residuals and prior estimate residuals. The residual sensitivity equation (6.2-30) and the last term in equation (6.2-33) are easily calculated using partitions of the total information matrix computed with all states adjusted. The residual covariances may be obtained either as shown, or using Monte Carlo simulation with repeated samples of both measurement noise and state "truth" values. Monte Carlo analysis involves less coding effort, but the covariance equations can be more easily used to analyze different modeling possibilities. These methods will be demonstrated in an example appearing after the next topic.

6.2.2.4 Estimating Model Order The previous error analysis was based on the assumption that the estimator model order is smaller than the order of the "truth" model. "Truth models" may include states (parameters) that are negligibly small or—more generally—are nonzero but accuracy of the prior estimate is much better than modeled by the *a priori* covariance. Hence inclusion of these states in the estimator may make it overly sensitive to measurement noise. That is, the extra states may be adjusted by the estimator to fit measurement noise rather than true model signatures. This reduces measurement residuals but increases state estimate errors. Although there is little penalty for including extra states if the estimator is only used to "smooth" measured quantities, accuracy can be significantly degraded if the state estimates are used to predict measurements or other quantities derived from the measurements. For these reasons it is important that the effective model order be accurately determined.

There are several methods that can be used to evaluate whether individual states should be included in the "adjusted" set. Perhaps the most obvious metric is the least-squares cost function. The fit measurement residuals (and thus the cost function) will decrease as unadjusted states are moved to the adjusted category. If the cost reduction when adding a given state is smaller than a yet-to-be-specified threshold, it may be concluded that the state should not be adjusted. Another useful metric is the log likelihood function, which for Gaussian models includes the least-squares cost function as the data-dependent component. A third metric is the variance of prediction residuals, which (as noted previously) is more sensitive to modeling errors than is the variance of fit residuals. Other metrics, such as the *Akaike information criteria* (Akaike 1974b), are also used and will be discussed in Chapter 12.

Least-Squares Cost Function We start by analyzing the reduction in the least-squares cost function when a parameter is added to the state vector of adjusted states. It is assumed that prior estimates (possibly zero) and a nominal error covariance for all unadjusted states are available. The change in the least-squares cost function is used as a metric to determine whether the error in the prior estimate of a given state is sufficiently large to justify including it as an adjusted state, or whether the prior estimate is sufficiently accurate and the state should not be adjusted.

For notational simplicity, the cost function is expressed using a scaled and augmented $m+n$ element measurement vector consisting of actual measurements and prior estimates of states:

$$\mathbf{z} = \begin{bmatrix} \mathbf{R}^{-1/2}\mathbf{y} \\ \mathbf{P}_a^{-1/2}\mathbf{x}_a \end{bmatrix}_{m+n} = \mathbf{M}\mathbf{x} + \mathbf{e}$$

where $\mathbf{R} = \mathbf{R}^{1/2}\mathbf{R}^{T/2}$, $\mathbf{P}_a = \mathbf{P}_a^{1/2}\mathbf{P}_a^{T/2}$,

$$\mathbf{M} = \begin{bmatrix} \mathbf{R}^{-1/2}\mathbf{H} \\ \mathbf{P}_a^{-1/2} \end{bmatrix}_{(m+n)\times n}$$

is an $(m + n) \times n$ matrix, and

$$\mathbf{e} = \begin{bmatrix} \mathbf{R}^{-1/2}\mathbf{r} \\ \mathbf{P}_a^{-1/2}(\mathbf{x}_a - \mathbf{x}) \end{bmatrix}_{m+n}$$

is an $m + n$ measurement error vector. Since the only measurements used in this section are fit measurements, the subscript "f" is dropped from $\mathbf{y}$ to simplify the notation.

Twice the Bayesian least-squares cost function is now written as

$$\begin{aligned} 2J &= \tilde{\mathbf{y}}^T\mathbf{R}^{-1}\tilde{\mathbf{y}} + (\mathbf{x}_a - \hat{\mathbf{x}})^T\mathbf{P}_a^{-1}(\mathbf{x}_a - \hat{\mathbf{x}}) \\ &= (\mathbf{z} - \mathbf{M}\hat{\mathbf{x}})^T(\mathbf{z} - \mathbf{M}\hat{\mathbf{x}}) \end{aligned}$$

where the Bayesian estimate is

$$\hat{\mathbf{x}} = (\mathbf{M}^T\mathbf{M})^{-1}\mathbf{M}^T\mathbf{z}$$

and the measurement/prior residual vector is

$$\begin{aligned} \mathbf{z} - \mathbf{M}\hat{\mathbf{x}} &= [\mathbf{I} - \mathbf{M}(\mathbf{M}^T\mathbf{M})^{-1}\mathbf{M}^T](\mathbf{M}\mathbf{x} + \mathbf{e}) \\ &= [\mathbf{I} - \mathbf{M}(\mathbf{M}^T\mathbf{M})^{-1}\mathbf{M}^T]\,\mathbf{e} \end{aligned}.$$

We now divide the adjusted states adjusted into two sets. States currently in the estimator are represented as $\mathbf{x}_1$, while x_2 represents a single state that was initially unadjusted but is now added as an adjusted state located at the bottom of the state vector. This analysis only considers adding one state at a time to the adjusted states. For consistency, prior estimates of all unadjusted states must be used when computing measurement residuals. That is, measurements used for estimation of $\hat{\mathbf{x}}_1$ are computed as $\mathbf{y} - \mathbf{H}_u\mathbf{x}_{au}$ where $\mathbf{x}_{au}$ is the prior estimate of all unadjusted parameters ($\mathbf{x}_u$), and $\mathbf{H}_u$ is the sensitivity matrix of the measurements with respect to $\mathbf{x}_u$. If the same corrected measurements are used when testing whether x_2 should be adjusted, then the Bayesian solution will estimate the correction δx_2 to the prior estimate, where the prior estimate of $\delta\hat{x}_{a2}$ is zero since the measurements have already been correction for the prior value x_{a2}.

The corrected measurement vector is defined in terms of adjusted states as

$$\mathbf{y} - \mathbf{H}_u\mathbf{x}_{au} - \mathbf{h}_2 x_{a2} = [\mathbf{H}_1 \quad \mathbf{h}_2]\begin{bmatrix} \mathbf{x}_1 \\ \delta x_2 \end{bmatrix} + \mathbf{r}. \qquad (6.2\text{-}38)$$

The augmented $\mathbf{z}$ vector consists of corrected measurements, prior estimates for the $\mathbf{x}_1$ states, and the prior estimate for the δx_2 state as

$$\mathbf{z} = \begin{bmatrix} \mathbf{R}^{-1/2}(\mathbf{y} - \mathbf{H}_u\mathbf{x}_{au}) \\ \mathbf{P}_{a1}^{-1/2}\mathbf{x}_{a1} \\ P_{a2}^{-1/2}\cdot 0 \end{bmatrix}.$$

We now combine the first two sets of $\mathbf{z}$ elements—since they are the measurements used when only $\mathbf{x}_1$ states are adjusted in the estimator—and denote those measurements as $\mathbf{z}_1$. The remaining prior measurement for δx_2 is denoted as scalar $z_2 = 0$, and the model for these measurements is

$$\mathbf{z} = \begin{bmatrix} \mathbf{z}_1 \\ z_2 \end{bmatrix} = \begin{bmatrix} \mathbf{M}_1 & \mathbf{m}_2 \\ \mathbf{0} & m_3 \end{bmatrix} \begin{bmatrix} \mathbf{x}_1 \\ x_2 \end{bmatrix} + \begin{bmatrix} \mathbf{e}_1 \\ e_2 \end{bmatrix},$$

where

$$\mathbf{M}_1 = \begin{bmatrix} \mathbf{R}^{-1/2}\mathbf{H}_1 \\ \mathbf{P}_{a1}^{-1/2} \end{bmatrix}, \quad \mathbf{m}_2 = \begin{bmatrix} \mathbf{R}^{-1/2}\mathbf{h}_2 \\ \mathbf{0} \end{bmatrix}, \quad m_3 = P_{a2}^{-1/2}$$

and

$$\mathbf{e}_1 = \begin{bmatrix} \mathbf{R}^{-1/2}\mathbf{r} \\ \mathbf{P}_{a1}^{-1/2}(\mathbf{x}_{a1} - \mathbf{x}_1) \end{bmatrix}, \quad e_2 = P_{a2}^{-1/2}(0 - x_2).$$

The least-squares information matrix and vector (denoted as $\mathbf{A}$ and $\mathbf{b}$) are defined as

$$\begin{aligned} \mathbf{A} &= \begin{bmatrix} \mathbf{A}_1 & \mathbf{a}_2 \\ \mathbf{a}_2^T & a_3 \end{bmatrix} = \mathbf{M}^T\mathbf{M} \\ &= \begin{bmatrix} \mathbf{M}_1^T\mathbf{M}_1 & \mathbf{M}_1^T\mathbf{m}_2 \\ \mathbf{m}_2^T\mathbf{M}_1 & \mathbf{m}_2^T\mathbf{m}_2 + m_3^2 \end{bmatrix} \\ &= \begin{bmatrix} \mathbf{H}_1^T\mathbf{R}^{-1}\mathbf{H}_1 + \mathbf{P}_{a1}^{-1} & \mathbf{H}_1^T\mathbf{R}^{-1}\mathbf{h}_2 \\ \mathbf{h}_2^T\mathbf{R}^{-1}\mathbf{H}_1 & \mathbf{h}_2^T\mathbf{R}^{-1}\mathbf{h}_2 + P_{a2}^{-1} \end{bmatrix} \end{aligned} \tag{6.2-39}$$

and

$$\begin{aligned} \mathbf{b} &= \begin{bmatrix} \mathbf{b}_1 \\ b_2 \end{bmatrix} = \mathbf{M}^T\mathbf{z} \\ &= \begin{bmatrix} \mathbf{M}_1^T\mathbf{z}_1 \\ \mathbf{m}_2^T\mathbf{z}_1 + m_3 z_2 \end{bmatrix} \\ &= \begin{bmatrix} \mathbf{H}_1^T\mathbf{R}^{-1}\mathbf{y} + \mathbf{P}_{a1}^{-1}\mathbf{x}_{a1} \\ \mathbf{h}_2^T\mathbf{R}^{-1}\mathbf{y} \end{bmatrix} \end{aligned} \tag{6.2-40}$$

Hence the normal equations can be written as:

$$\begin{bmatrix} \mathbf{A}_1 & \mathbf{a}_2 \\ \mathbf{a}_2^T & a_3 \end{bmatrix} \begin{bmatrix} \hat{\mathbf{x}}_1 \\ \hat{x}_2 \end{bmatrix} = \begin{bmatrix} \mathbf{b}_1 \\ \mathbf{b}_2 \end{bmatrix} \tag{6.2-41}$$

where $\mathbf{A}_1 = \mathbf{H}_1^T\mathbf{R}^{-1}\mathbf{H}_1 + \mathbf{P}_{a1}^{-1}$, etc. Solving the lower equation and substituting it in the upper yields:

$$\hat{x}_2 = (b_2 - \mathbf{a}_2^T\hat{\mathbf{x}}_1)/a_3$$

$$\mathbf{A}_1\hat{\mathbf{x}}_1 + \mathbf{a}_2(b_2 - \mathbf{a}_2^T\hat{\mathbf{x}}_1)/a_3 = \mathbf{b}_1$$

or

$$
\begin{aligned}
\hat{\mathbf{x}}_1 &= (\mathbf{A}_1 - \mathbf{a}_2\mathbf{a}_2^T / a_3)^{-1}(\mathbf{b}_1 - \mathbf{a}_2 b_2 / a_3) \\
&= \left(\mathbf{A}_1^{-1} + \frac{\mathbf{A}_1^{-1}\mathbf{a}_2\mathbf{a}_2^T\mathbf{A}_1^{-1}}{a_3 - \mathbf{a}_2^T\mathbf{A}_1^{-1}\mathbf{a}_2}\right)(\mathbf{b}_1 - \mathbf{a}_2 b_2 / a_3) \\
&= \hat{\mathbf{x}}_{1\_0} - \mathbf{A}_1^{-1}\mathbf{a}_2 b_2 / a_3 + \frac{\mathbf{A}_1^{-1}\mathbf{a}_2\mathbf{a}_2^T\mathbf{A}_1^{-1}(\mathbf{b}_1 - \mathbf{a}_2 b_2 / a_3)}{a_3 - \mathbf{a}_2^T\mathbf{A}_1^{-1}\mathbf{a}_2} \\
&= \hat{\mathbf{x}}_{1\_0} + \frac{-\mathbf{A}_1^{-1}\mathbf{a}_2(b_2 - \mathbf{a}_2^T\mathbf{A}_1^{-1}\mathbf{a}_2 b_2 / a_3) + \mathbf{A}_1^{-1}\mathbf{a}_2\mathbf{a}_2^T\mathbf{A}_1^{-1}(\mathbf{b}_1 - \mathbf{a}_2 b_2 / a_3)}{a_3 - \mathbf{a}_2^T\mathbf{A}_1^{-1}\mathbf{a}_2} \\
&= \hat{\mathbf{x}}_{1\_0} + \frac{\mathbf{A}_1^{-1}\mathbf{a}_2(-b_2 + \mathbf{a}_2^T\mathbf{A}_1^{-1}\mathbf{b}_1)}{a_3 - \mathbf{a}_2^T\mathbf{A}_1^{-1}\mathbf{a}_2}
\end{aligned}
\tag{6.2-42}
$$

where $\hat{\mathbf{x}}_{1\_0} = \mathbf{A}_1^{-1}\mathbf{b}_1$ is the estimate of $\hat{\mathbf{x}}_1$ before state $\hat{x}_2$ was added to the adjusted states. Then

$$
\begin{aligned}
\hat{x}_2 &= (b_2 - \mathbf{a}_2^T\hat{\mathbf{x}}_1) / a_3 \\
&= b_2 / a_3 - \mathbf{a}_2^T\left(\mathbf{A}_1^{-1} + \frac{\mathbf{A}_1^{-1}\mathbf{a}_2\mathbf{a}_2^T\mathbf{A}_1^{-1}}{a_3 - \mathbf{a}_2^T\mathbf{A}_1^{-1}\mathbf{a}_2}\right)(\mathbf{b}_1 - \mathbf{a}_2 b_2 / a_3) / a_3 \\
&= \left(\frac{b_2(a_3 - \mathbf{a}_2^T\mathbf{A}_1^{-1}\mathbf{a}_2) - \mathbf{a}_2^T\mathbf{A}_1^{-1}[(a_3 - \mathbf{a}_2^T\mathbf{A}_1^{-1}\mathbf{a}_2) + \mathbf{a}_2\mathbf{a}_2^T\mathbf{A}_1^{-1}](\mathbf{b}_1 - \mathbf{a}_2 b_2 / a_3)}{a_3(a_3 - \mathbf{a}_2^T\mathbf{A}_1^{-1}\mathbf{a}_2)}\right) \\
&= \frac{b_2 - \mathbf{a}_2^T\mathbf{A}_1^{-1}\mathbf{b}_1}{a_3 - \mathbf{a}_2^T\mathbf{A}_1^{-1}\mathbf{a}_2}
\end{aligned}
\tag{6.2-43}
$$

and the Bayesian least-squares cost function is

$$
\begin{aligned}
(\mathbf{z} - \mathbf{M}\hat{\mathbf{x}})^T(\mathbf{z} - \mathbf{M}\hat{\mathbf{x}}) &= \mathbf{z}^T\mathbf{z} - \mathbf{z}^T\mathbf{M}\hat{\mathbf{x}} \\
&= \mathbf{z}_1^T\mathbf{z}_1 + z_2^2 - \mathbf{z}_1^T\mathbf{M}_1\hat{\mathbf{x}}_1 - (\mathbf{z}_1^T\mathbf{m}_2 + z_2 m_3)\hat{x}_2 \\
&= \mathbf{z}_1^T\mathbf{z}_1 + z_2^2 - \mathbf{b}_1^T\hat{\mathbf{x}}_1 - b_2\hat{x}_2 \\
&= \mathbf{z}_1^T\mathbf{z}_1 + z_2^2 - \mathbf{b}_1^T\left[\hat{\mathbf{x}}_{1\_0} + \frac{\mathbf{A}_1^{-1}\mathbf{a}_2(-b_2 + \mathbf{a}_2^T\mathbf{A}_1^{-1}\mathbf{b}_1)}{a_3 - \mathbf{a}_2^T\mathbf{A}_1^{-1}\mathbf{a}_2}\right] - b_2\left[\frac{b_2 - \mathbf{a}_2^T\mathbf{A}_1^{-1}\mathbf{b}_1}{a_3(a_3 - \mathbf{a}_2^T\mathbf{A}_1^{-1}\mathbf{a}_2)}\right] \\
&= (\mathbf{z}_1^T\mathbf{z}_1 - \mathbf{b}_1^T\hat{\mathbf{x}}_{1\_0}) + z_2^2 + \left[\frac{2(\mathbf{b}_1^T\mathbf{A}_1^{-1}\mathbf{a}_2)b_2 - (\mathbf{b}_1^T\mathbf{A}_1^{-1}\mathbf{a}_2)(\mathbf{a}_2^T\mathbf{A}_1^{-1}\mathbf{b}_1) - b_2^2}{a_3 - \mathbf{a}_2^T\mathbf{A}_1^{-1}\mathbf{a}_2}\right] \\
&= (\mathbf{z}_1^T\mathbf{z}_1 - \mathbf{b}_1^T\hat{\mathbf{x}}_{1\_0}) + z_2^2 - \frac{(b_2 - \mathbf{a}_2^T\mathbf{A}_1^{-1}\mathbf{b}_1)^2}{a_3 - \mathbf{a}_2^T\mathbf{A}_1^{-1}\mathbf{a}_2} \\
&= (\mathbf{z}_1^T\mathbf{z}_1 - \mathbf{b}_1^T\hat{\mathbf{x}}_{1\_0}) + z_2^2 - \frac{\delta\hat{x}_2^2}{\sigma_2^2} \\
&= (\mathbf{z}_1^T\mathbf{z}_1 - \mathbf{b}_1^T\hat{\mathbf{x}}_{1\_0}) - \frac{\delta\hat{x}_2^2}{\sigma_2^2}
\end{aligned}
\tag{6.2-44}
$$

where

$\sigma_{a2}^2 = P_{a2}$ is the error variance of the *a priori* estimate $\hat{x}_{a2}$, and

$\sigma_2^2 = P_2 = \dfrac{1}{a_3 - \mathbf{a}_2^T \mathbf{A}_1^{-1} \mathbf{a}_2}$ is the error variance of the *a posteriori* estimate $\hat{x}_2$.

Hence the cost function will change by

$$-\frac{\delta \hat{x}_2^2}{\sigma_2^2}$$

when δx_2 is moved from unadjusted to adjusted. That is, the cost function will drop by the square of the estimate magnitude divided by the estimate uncertainty. If that ratio is small, it can be assumed that δx_2 is unimportant for modeling the measurements and can be eliminated from the estimator model; that is, $x_2 = x_{a2}$.

It is also of interest to determine the expected change in the cost function under the assumption that P_{a2} accurately models the variance of $\delta \hat{x}_2$. The expected value of the cost function after adding adjusted state $\delta \hat{x}_2$ is

$$E\left[\mathbf{z}_1^T \mathbf{z}_1 - \mathbf{b}_1^T \hat{\mathbf{x}}_{1\_0} - \frac{\hat{x}_2^2}{\sigma_2^2}\right] = E[\mathbf{z}_1^T \mathbf{z}_1 - \mathbf{b}_1^T \hat{\mathbf{x}}_{1\_0}] - \frac{E[(\delta \hat{x}_2)^2]}{\sigma_2^2}.$$

$E[\delta \hat{x}_2^2]$ can be computed indirectly using

$$\begin{aligned} E[\delta x_2^2] &= E[(\delta \hat{x}_2 - (\delta \hat{x}_2 - \delta x_2))^2] \\ &= E[(\delta \hat{x}_2)^2] + E[(\delta x_2 - \delta \hat{x}_2)^2] \\ &= E[(\delta \hat{x}_2)^2] + \sigma_2^2 \end{aligned}$$

since $E[\delta \hat{x}_2(\delta \hat{x}_2 - \delta x_2)] = 0$ from the orthogonality principle if $\mathbf{x}_1$ and x_2 completely define the system. Since $E[\delta x_2] = 0$ and $E[(\delta x_2)^2] = P_{a2} = \sigma_{a2}^2$, then $E[(\delta \hat{x}_2)^2] = \sigma_{a2}^2 - \sigma_2^2$ and the expected value of the cost function is

$$E[(\mathbf{z} - \mathbf{M}\hat{\mathbf{x}})^T (\mathbf{z} - \mathbf{M}\hat{\mathbf{x}})] = E[\mathbf{z}_1^T \mathbf{z}_1 - \mathbf{b}_1^T \hat{\mathbf{x}}_{1\_0}] + 1 - \frac{\sigma_{a2}^2}{\sigma_2^2}. \tag{6.2-45}$$

Thus the expected change in the cost function is

$$1 - \frac{\sigma_{a2}^2}{\sigma_2^2} = 1 - \frac{P_{a2}}{P_2} \tag{6.2-46}$$

when state δx_2 is moved to the adjusted states. The ratio $\sqrt{P_{ai} / P_i} = \sigma_{ai} / \sigma_i$ is a measure of the information in the measurements relative to the prior information for state x_i. When that ratio for the last state added is much greater than 1.0 (indicating that the *a posteriori* estimate of $\hat{x}_i$ is much more accurate than the *a priori* estimate), the reduction in the cost function will be great when $|\delta x_2| > \sigma_2$.

The actual change from equation (6.2-44) for a specific set of measurements will be different. When

$$-\frac{\delta \hat{x}_2^2}{\sigma_2^2} < 1 - \frac{\sigma_{a2}^2}{\sigma_2^2} \qquad \text{or equivalently} \qquad \delta \hat{x}_2^2 > \sigma_{a2}^2 - \sigma_2^2,$$

it may be concluded that the error in the prior estimate x_{a2} is sufficiently large to justify including δx_2 as adjusted.

The Log Likelihood Function and Model Order The negative log likelihood for a MAP estimator using Gaussian models was defined in Chapter 4 as:

$$-\ln p_{\mathbf{X|Y}}(\mathbf{x}\,|\,\mathbf{y}) = \frac{1}{2}\begin{pmatrix} (m+n)\ln(2\pi) + \ln|\mathbf{P}_{yy}| + \ln|\mathbf{P}_{xx}| \\ +[\mathbf{y}-\bar{\mathbf{y}}(\mathbf{x})]^T \mathbf{P}_{yy}^{-1}[\mathbf{y}-\bar{\mathbf{y}}(\mathbf{x})] + (\mathbf{x}-\bar{\mathbf{x}})^T \mathbf{P}_{xx}^{-1}(\mathbf{x}-\bar{\mathbf{x}}) \\ +2\ln p_{\mathbf{Y}}(\mathbf{y}) \end{pmatrix}. \tag{6.2-47}$$

Notice that it includes a data-dependent part

$$[\mathbf{y}-\bar{\mathbf{y}}(\mathbf{x})]^T \mathbf{P}_{yy}^{-1}[\mathbf{y}-\bar{\mathbf{y}}(\mathbf{x})] + (\mathbf{x}-\bar{\mathbf{x}})^T \mathbf{P}_{xx}^{-1}(\mathbf{x}-\bar{\mathbf{x}})$$

that is equal to twice the Bayesian cost function for the fit,

$$2J = \tilde{\mathbf{y}}^T \mathbf{R}^{-1}\tilde{\mathbf{y}} + (\hat{\mathbf{x}}_1 - \mathbf{x}_{a1})^T \mathbf{P}_{a1}^{-1}(\hat{\mathbf{x}}_1 - \mathbf{x}_{a1}), \tag{6.2-48}$$

at the MAP estimate $\hat{\mathbf{x}}_1$ when $\tilde{\mathbf{y}} = \mathbf{y} - \bar{\mathbf{y}}(\hat{\mathbf{x}}_1)$ and $\mathbf{x}_{a1} = \bar{\mathbf{x}}$. The $\ln p_{\mathbf{Y}}(\mathbf{y})$ and $\ln|\mathbf{P}_{yy}|$ terms are not a function of the data or the model order, so they can be ignored. The $(m + n)\ln(2\pi)$ and $\ln|\mathbf{P}_{xx}|$ terms are a function of model order, but they are present in the Gaussian density function as normalization factors so that the integral of the density over the probability space is equal to 1.0. Both terms increase with model order, so one might suspect that the model order minimizing the negative log likelihood is optimal. However, inclusion of these terms results in selection of a model order that is lower than the actual order. Hence the negative log likelihood cannot be directly used to determine the correct model order.

The optimal model order can be determined using the data-dependent part of the likelihood function ($2J$) plus a "penalty" correction for the expected change in cost as parameters are changed from unadjusted to adjusted. That correction is obtained by summing the expected changes from equation (6.2-46) for all previously unadjusted states that are currently adjusted:

$$f = -\sum_{i=n1}^{n}\left(1 - \frac{\sigma_{a2}^2}{\sigma_2^2}\right)_i \tag{6.2-49}$$

where $n1 - 1$ is the "base" model order corresponding to the smallest model to be considered. Since the terms in the summation are negative, f will be positive and will tend to offset the reduction in $2J$ as unadjusted states are made adjusted. The minimum of the metric $2J + f$ defines the optimal model order. Use of this approach is demonstrated in Example 6.5.

Prediction Residuals Since prediction residuals are more sensitive to errors in model structure than fit residuals, the weighted variance of prediction residuals (or more generally measurement residuals not included in the fit) can be a better test of model order than the least-squares cost function. Of course this can only be done when the available data span is much larger than the span required for individual fits. Assuming that time is the independent variable of the system, the measurement data are broken into a number of time spans (possibly overlapping) where a least-squares fit is computed for each span. Each fit model is used to predict measurements for the next time span, and the sample variance of the prediction residuals is tabulated versus prediction interval and model order. The optimal order is the one yielding the smallest prediction variances. This technique is of most use when the purpose of the modeling is to either compute linear combinations of the states under conditions that were not directly measured, or when the state estimates are the direct goal of the estimation.

One practical issue associated with model order selection involves grouping of states to be adjusted in the estimator. Often several states must be grouped together for a meaningful model. For example, error sources often appear in three axes and there is no reason to believe that one or two axes dominate. In these cases the states should be tested as a group. Another issue is ordering of states to be adjusted if testing is performed sequentially. Different results can be obtained when sequential testing is performed in different orders. This topic will be discussed again in Section 6.3.

We now show how methods described in previous sections can be used to analyze errors and optimize selection of adjusted states.

Example 6.5: Consider Parameter Analysis and Selection of Polynomial Order

The polynomial problem of the previous chapter is used to demonstrate analysis of unmodeled parameters and optimal selection of model order. It may seem inconsistent to continue using polynomial examples when previous chapters have emphasized first-principles modeling and real-world examples, but there are good reasons for using basis functions to demonstrate selection of model order. If the inertial navigation system (INS) model of Section 3.3 had been used for this example, it would have been found that the SOS of fit and prediction measurement residuals jumped in groups and that the behavior was not monotonic. Recall that the INS model includes states in groups of three: tilt errors, accelerometer biases, accelerometer scale factors, gyro biases, gyro scale factors, gyro g-sensitive errors, and gyro g^2-sensitive errors. While this grouping of states if often typical in physical systems, it would tend to obscure the relationship between estimator behavior and number of adjusted states. For that reason a basis function expansion is preferable for this example. Although the polynomial model is used here, the behavior of Chebyshev polynomials and Fourier expansions is somewhat similar.

This example again uses 101 simulated measurements of the form

$$y_i = \sum_{j=1}^{n_t} x_j t_i^{j-1} + r_i$$

with $n_t = 9$ and time samples (t_i) uniformly spaced from 0 to 1. The random measurement noise samples r_i are $N(0,1)$ and the simulated random state x_i values are $N(0,10^2)$. For the Bayesian least-squares estimator, the total model order is assumed to be 13 and the number of adjusted parameters (n) in $\hat{\mathbf{x}}_1$ is varied from 1 to 13; for example, when estimator $n = 6$, the number of $\mathbf{x}_2$ consider states is 7. The prior covariance is set as $\mathbf{P}_a = diag(10^2)$ to give the consider states the proper uncertainty for the simulated data. Both Monte Carlo and covariance analysis are used in the evaluation.

Figure 6.13 shows the covariance-based noise-only fit measurement residual standard deviations

$$\sigma_{yres\_i} \triangleq \sqrt{E[\tilde{y}_i^2]} = \sqrt{(\mathbf{R}_f - \mathbf{H}_{f1}\mathbf{P}_1\mathbf{H}_{f1}^T)_{ii}}$$

computed from the first two terms in equation (6.2-31) as a function of estimator model order. Only three measurement time points are plotted: #1 at $t = 0$, #51 at $t = 0.5$, and #101 at $t = 1$. Since the plotted residual $1-\sigma$ does not include the effects of unadjusted parameters, σ_{yres_i} should nearly equal $E[r_i^2] = 1.0$ when the number of adjusted parameters is small. For measurement #51 at the middle of the data span, σ_{yres_i} does not drop significantly as the model order is increased, but for the first and last measurements, σ_{yres_i} is smaller when more degrees-of-freedom are available for the fit model. It is interesting that σ_{yres_i} for the last measurement continues to drop with model order while σ_{yres_i} for the first point remains almost flat above $n = 4$. This behavior is due to the inclusion of prior information ($\mathbf{P}_a = diag(10^2)$) on the states. When the prior information of the Bayesian solution is removed to produce a weighted least-squares solution, the behavior of the first and last points is similar. By weakly constraining the estimated polynomial coefficients using prior information, the modeled first measurement is more constrained than the last measurement because higher polynomial coefficients have less influence on the first point.

Figure 6.14 shows the total fit residual standard deviations computed from all terms in equation (6.2-31) as a function of estimator model order. (Because $\mathbf{R}_f = \mathbf{I}$, the results are unchanged when using equation [6.2-32].) Notice that the fit residuals are quite large when the model order is low, but they drop to approximately 1.0 at order 7 or 8. Since the simulated data were generated using order 9, the residuals should match those of Figure 6.13 when estimator $n \geq 9$. Also plotted in Figure 6.14 is "$2J + f - m$," computed as

$$\begin{aligned} 2J + f - m &= f - m + E[\tilde{\mathbf{y}}_f^T \mathbf{R}_f^{-1} \tilde{\mathbf{y}}_f] + E[\tilde{\mathbf{x}}_1^T \mathbf{P}_{a1}^{-1} \tilde{\mathbf{x}}_1] \\ &= f - m + tr[\mathbf{R}_f^{-1/2} E[\tilde{\mathbf{y}}_f \tilde{\mathbf{y}}_f^T] \mathbf{R}_f^{-T/2}] + tr[\mathbf{P}_{a1}^{-1/2} E[\tilde{\mathbf{x}}_1 \tilde{\mathbf{x}}_1^T] \mathbf{P}_{a1}^{-T/2}] \end{aligned}$$

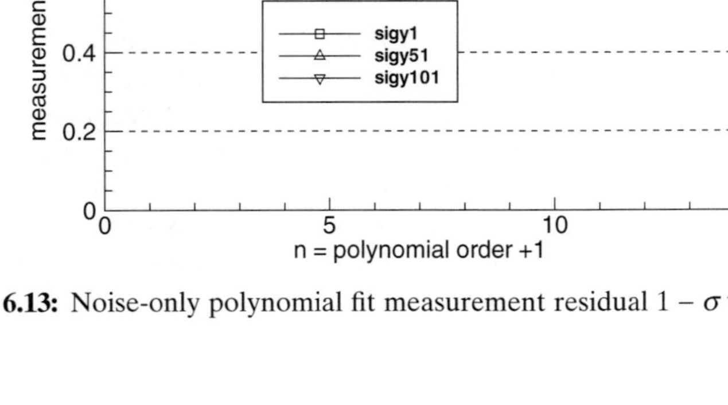

FIGURE 6.13: Noise-only polynomial fit measurement residual $1-\sigma$ versus order.

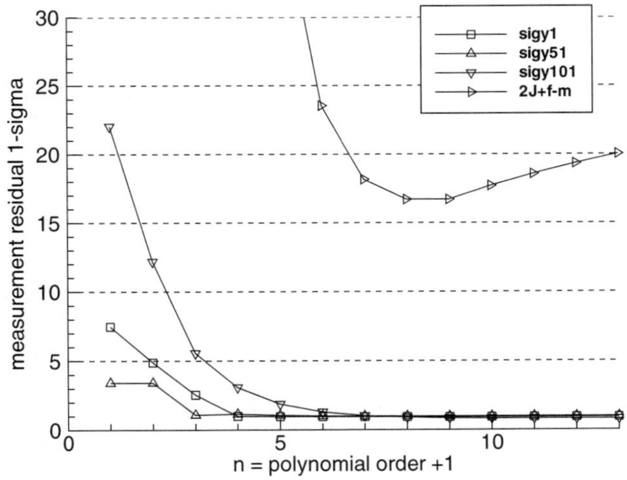

FIGURE 6.14: Covariance-computed fit measurement residual $1-\sigma$ versus order.

where $\tilde{\mathbf{y}}_f = \mathbf{y}_f - (\mathbf{H}_{f1}\hat{\mathbf{x}}_1 + \mathbf{H}_{f2}\mathbf{x}_{2a})$, $\tilde{\mathbf{x}}_1 = \mathbf{x}_{a1} - \hat{\mathbf{x}}_1$, and

$$f = -\sum_{i=4}^{n}\left(1-\frac{\sigma_{a2}^2}{\sigma_2^2}\right)_i$$

penalizes the cost function for the expected change due to adding a state. The summation does not include the first three terms because the difference is not defined for $i = 1$, and the next two terms are dropped to allow plotting on the same scale as for the residuals.

The expected values above were computed using the covariance equations (6.2-32) and (6.2-37). The number of measurements (m) was subtracted from the plotted values to allow plotting on the same scale as the residual $1-\sigma$. Notice that "$2J+f-m$" initially drops as order increases and has a minimum at $n = 9$ (the true n). For $n < 9$, the drop in $2J$ is much greater than the increase in f because the calculated f does not allow for additional unadjusted terms beyond the current order. For $n > 9$,

$$tr\left[\mathbf{R}_f^{-1/2}E[\tilde{\mathbf{y}}_f\tilde{\mathbf{y}}_f^T]\mathbf{R}_f^{-T/2}\right] + tr[\mathbf{P}_{a1}^{-1/2}E[\tilde{\mathbf{x}}_1\tilde{\mathbf{x}}_1^T]\mathbf{P}_{a1}^{-T/2}] - m$$

computed from equations (6.2-32) and (6.2-37) should be equal to zero because equations (6.2-32) and (6.2-37) do not include the reduction in residuals due to over-fitting the data. Hence the plotted change represents the effects of f.

Figure 6.15 is a similar plot based on 1000 samples of a Monte Carlo simulation where $\mathbf{x}_a = \mathbf{0}$; that is, the mean value of all simulated states is zero, and the prior value of all estimated states is also zero. Simulations using randomly selected $\mathbf{x}_a$ elements produce curves almost identical to Figure 6.15. In this plot all variables are computed as sums of actual realizations of the random variables—as done when processing real data—and the plotted values are averages of the sample values. For example, the standard deviation of measurement residuals at time t_i is computed as

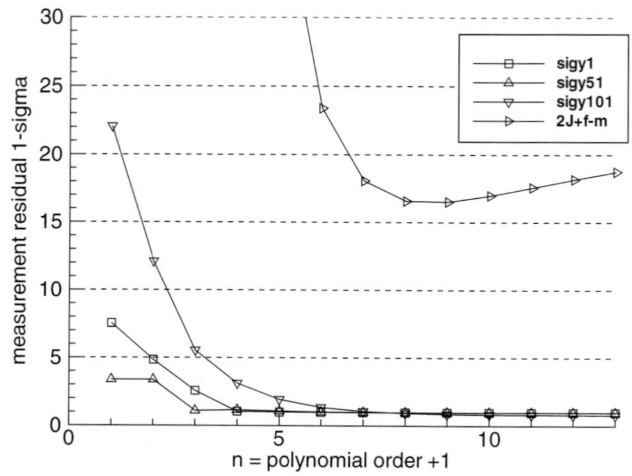

FIGURE 6.15: Monte Carlo fit measurement residual $1-\sigma$ versus order.

$$\sigma_{Ri} = \sqrt{\frac{1}{1000}\sum_{j=1}^{1000}(y_{i\_j} - \mathbf{h}_{1i}\hat{\mathbf{x}}_{1\_j} - \mathbf{h}_{2i}\mathbf{x}_{2a\_j})^2}$$

where $y_{i\_j} = \mathbf{h}_{1i}\mathbf{x}_{1\_j} - \mathbf{h}_{2i}\mathbf{x}_{2\_j} + r_{i\_j}$ is the simulated noisy measurement at t_i for Monte Carlo case j. Also $2J + f - m$ is computed using equations (6.2-48) and (6.2-49).

The behavior in Figure 6.15 is very close to that of Figure 6.14, except that "$2J + f - m$" increases less rapidly with order when $n > 9$ due to data "over-fitting" that reduces the residuals. Again the minimum occurs at $n = 9$, which is correct. When the prior covariance $\mathbf{P}_a$ is reduced to $diag(1^2)$ rather than $diag(10^2)$, the curves shift downward, but the minimum in "$2J + f - m$" still occurs at $n = 9$.

To summarize, use of the cost metric $2J + f - m$ appears to be a reliable indicator of true model order for this example.

Figure 6.16 shows the total *a posteriori* $1 - \sigma$ uncertainty for the first seven states, computed as the square roots of the diagonals of equation (6.2-21). Notice that the $1 - \sigma$ uncertainty for unadjusted states is equal to 10, but the uncertainty increases significantly when a state is first made adjusted because of the aliasing effect of other unadjusted states. The uncertainty then drops as other states are moved from unadjusted to adjusted.

Figure 6.17 shows the covariance-computed total measurement residual $1 - \sigma$ for predicted measurements at times 1.0, 1.5, and 2.0. The plot for $t = 1.0$ is identical to that of Figure 6.14, but on a different scale. However, for $t = 1.5$ the residual $1 - \sigma$ has a definite minimum at $n = 9$, and the minimum is quite pronounced for $t = 2.0$. Figure 6.18 is the comparable plot obtained from the Monte Carlo simulation. Again the two plots are quite similar, except that the covariance-computed errors are larger than actual errors for $n > 9$ because the covariance does not account for data over-fitting.

As noted before, errors in model structure generally have more effect on an estimator's prediction accuracy than on its ability to match data used in the fit. This behavior is somewhat general for different types of systems. Hence

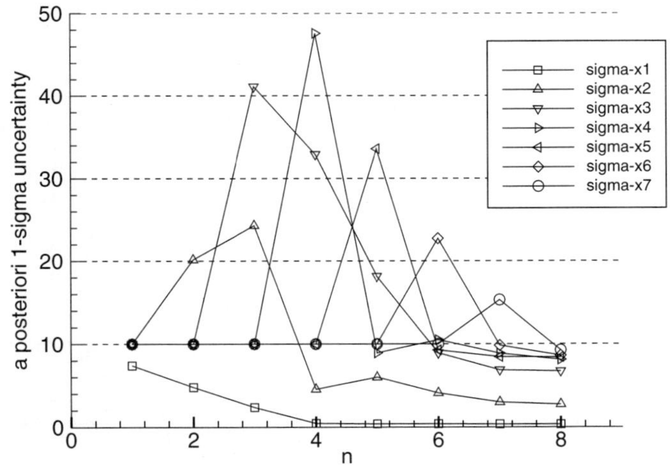

FIGURE 6.16: Total *a posteriori* state $1-\sigma$ uncertainty.

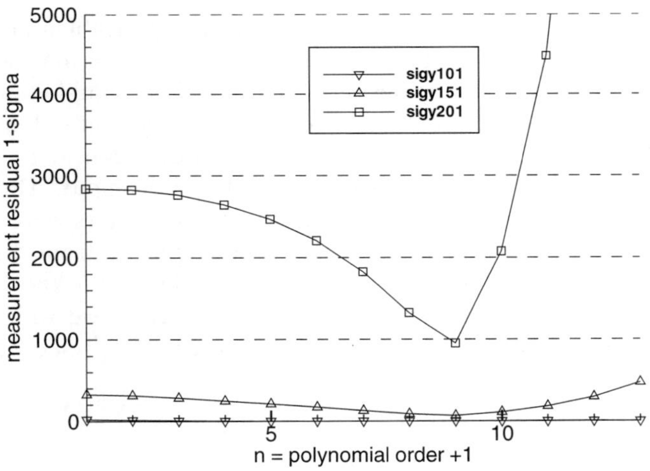

FIGURE 6.17: Covariance-computed prediction measurement residual $1-\sigma$ versus order.

prediction (or nonfit) measurement residuals can often be a more sensitive indicator of model order than the least-squares cost function.

A similar analysis was conducted using Chebyshev orthogonal polynomial and Fourier series basis functions rather than ordinary polynomials. Since Chebyshev polynomials are only defined over a finite time interval, it was necessary to use $0 \le t \le 2$ s as the interval so that the polynomials were reasonably bounded in the prediction interval. The Fourier series repeats rather than "blowing up" outside the fundamental period, so a fundamental period of 2 s was used to avoid repeating the function over the prediction interval $1 \le t \le 2$. Although the "measurement residual versus order" plots were less smooth than those of Figures 6.15 and 6.18, and the prediction residuals were smaller, the general trends and behavior were almost the same as shown for ordinary polynomials. It may be

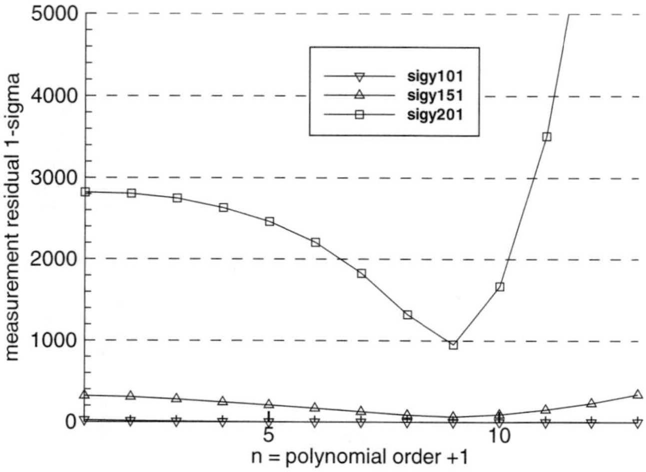

FIGURE 6.18: Monte-Carlo prediction measurement residual $1 - \sigma$ versus order.

expected that the results of this example are generally applicable to other problems, even though the residual plots may be less well behaved than shown.

Finally, we explain why it was desirable to use a Bayesian solution for the consider analysis. If all state parameters were treated as nonrandom, it would not be possible to compute the fit and prediction residual consider covariances equations (6.2-31) and (6.2-34) because $\mathbf{P}_{a2}$ is a required input. Alternately, if the $\mathbf{x}_1$ states were treated as nonrandom but the $\mathbf{x}_2$ parameters were treated as random, results would be inconsistent as states were moved from the adjusted to unadjusted category. By treating all states as random with known *a priori* covariance, the treatment is consistent. In the next section we will show how stepwise regression allows the optimum model order to be selected when all parameters are assumed to be nonrandom.

Example 6.6: ROM of Optical Imaging Instrument Misalignments

This example demonstrates a slightly different approach to selecting states and optimal model order for a ROM. Rather than using a single metric to determine whether individual parameters should be adjusted in the estimator, a preliminary analysis is conducted using the SVD to determine the approximate model order necessary to accurately model angular measurements of an optical imaging instrument. Singular vectors from the SVD help identify the most important linear combinations of states. Then Monte Carlo analysis is used to evaluate accuracy in predicting pointing of image pixels on the earth—a required function of the operational system. A trial and error process—constrained by knowledge of approximate model order and important states—is used to determine the optimal choice of states to meet accuracy requirements.

One design for an optical earth-imaging instrument on a geosynchronous satellite uses two separate mirrors that can be independently rotated to vary the line-of-sight for the detector array. Rotation of one mirror allows the line-of-sight to scan in an EW direction, while the other mirror allows scanning in an NS direction. Various misalignments within the instrument cause the actual line-of-sight to deviate from the ideal. Although most misalignments are calibrated before launch, thermal deformation caused by solar heating changes the misalignments, so it is necessary to estimate misalignments from measurements available on-orbit. Spacecraft pointing error is another factor in total pointing error. In designing an estimator for this purpose, it is desirable that unobservable, nearly unobservable, or negligibly small misalignments not be included in the estimator state vector because they will increase the onboard computational load and potentially reduce accuracy. The following analysis demonstrates methods that can be used to determine the appropriate states needed for the system to meet pointing accuracy requirements.

The optical path from an individual detector to space passes through a telescope with reflections on a fixed mirror and two rotating mirrors. By following the path for a detector at the array center, allowing for misalignments, the optical line-of-sight unit vector in the Instrument Coordinate System (ICS) is found to be:

$$
\begin{aligned}
\hat{\mathbf{s}}_{ICS} = &\begin{bmatrix} SE \\ -SN\,CE \\ CN\,CE \end{bmatrix} + \begin{bmatrix} CE \\ SN\,SE \\ -CN\,SE \end{bmatrix} m_{d2} + \begin{bmatrix} 0 \\ -CN \\ -SN \end{bmatrix} m_{d3} + \begin{bmatrix} -2CE \\ -2SN\,SE \\ 2CN\,SE \end{bmatrix} (m_{em3} + m_{e3}) \\
&+ \begin{bmatrix} 0 \\ -CN(1+SE) \\ -SN(1+SE) \end{bmatrix} m_{e2} + \begin{bmatrix} 0 \\ CN\,CE \\ SN\,CE \end{bmatrix} m_{e1} + \sqrt{2} \begin{bmatrix} 0 \\ CN\,(Ce+Se) \\ SN\,(Ce+Se) \end{bmatrix} m_{em1} \\
&+ \begin{bmatrix} 0 \\ -2CE\,CN \\ -2CE\,SN \end{bmatrix} (m_{nm1} + m_{n1}) + \sqrt{2} \begin{bmatrix} CE(Cn-Sn) \\ SE\,(Cn-Sn) \\ -SE\,(Cn+Sn) \end{bmatrix} m_{nm3} \\
&+ \begin{bmatrix} CE\,CN \\ SE\,CN \\ -SE(1-SN) \end{bmatrix} m_{n2} + \begin{bmatrix} (1+SN)CE \\ SE(1+SN) \\ -SE\,CN \end{bmatrix} m_{n3}
\end{aligned}
$$

where

m_{d2}, m_{d3} are angular misalignments of the detector array,

m_{em1}, m_{em3} are mirror misalignments of the EW mirror with respect to the EW rotating shaft,

m_{e1}, m_{e2}, m_{e3} are misalignments of the EW shaft axis with respect to the ICS,

m_{nm1}, m_{nm3} are mirror misalignments of the NS mirror with respect to the NS rotating shaft,

m_{n1}, m_{n2}, m_{n3} are misalignments of the NS shaft axis with respect to the ICS,

$SE = \sin(2e)$, $CE = \cos(2e)$ where e is the rotation of the EW mirror from the reference position,

$SN = \sin(-2n)$, $CN = \cos(2n)$ where n is the rotation of the NS mirror from the reference position.

It is immediately seen that m_{em3} and m_{e3} can be treated as a single parameter because they only appear as a sum. Likewise m_{nm1} and m_{n1} can also be treated as a single parameter. Hence a total of 10 parameters are needed to model internal instrument misalignments. Another three angles (roll, pitch, and yaw angles) are need to model the rotation of the ICS from the ideal spacecraft coordinate system—this rotation also includes spacecraft pointing error. Hence a total of 13 parameters are potentially needed to model total optical pointing. Computational load considerations make it desirable that about half that number of states be used, provided that pointing accuracy requirements can be met.

Angular measurements to known stars are used for computing misalignments. Mirror angles are adjusted to point near specific stars. With the spacecraft rotating in pitch to maintain earth-pointing, optical pointing angles to the star can be computed from the time at which the star is sensed in a particular detector. After preprocessing, the star measurements consist of an EW and NS pointing error for each star.

Seventeen star observations outside the earth disk were simulated for purposes of analyzing pointing errors. One extra star observation was intentionally located in the upper right quadrant so that the star distribution was not symmetric. Only about three stars would typically be observed within a period of a few minutes, but within an hour all quadrants would be covered with multiple observations. Figure 6.19 shows the location of simulated star positions and the earth in the optical field. Requirements specify pointing accuracy on the earth, so 17 "landmarks" uniformly distributed on the earth were used to compute

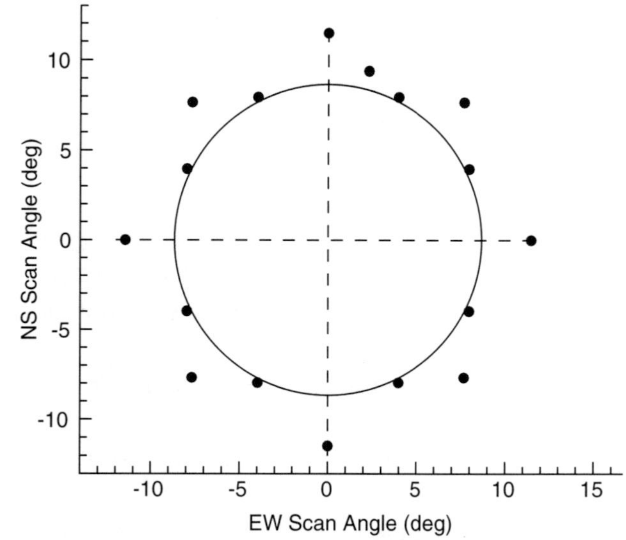

FIGURE 6.19: Simulated star observation angles relative to the earth.

pointing error statistics for evaluation purposes. Notice that because none of the star observation angles were within the earth's radius, use of landmarks on the earth for evaluation is "prediction to untested conditions." However, because the effects of misalignments tend to increase in magnitude with angular distance from nadir, pointing errors on the earth will generally be smaller than for star observations.

The first step in the analysis involves SVD computation of singular values and vectors for the 13 misalignment parameters when using 17 star observations. The SVD provides insight on the most important parameters for least-squares estimation. Table 6.2 lists the seven largest singular values (normalized as described in Section 5.7) and corresponding singular vectors. The m_{em3} and m_{nm1} parameters are not included in the analysis because they are not separable from m_{e3} and m_{n1}. From Table 6.2 we note the following:

1. Singular value magnitudes indicate that a minimum of four or five states will be required to accurately model the system. The maximum required model order is probably seven because singular values beyond that are 2 to 10 orders of magnitude smaller.
2. The third singular vector is the only one in which one parameter (yaw) is dominant. Other parameters with singular vector components greater than 0.5 are pitch, m_{d3}, m_{e2}, m_{nm3}, and m_{n2}. All dominant singular vectors except the third are a nearly equal combination of multiple parameters, with few parameters appearing strongly in multiple singular vectors.

The SVD analysis is helpful but not conclusive either to required model order or to the most important parameters (other than yaw). The physical parameters do not appear to have distinct signatures that allow a unique subset to mostly span the measurement space. However, because the parameters appear in multiple singular vectors, we cannot rule out the possibility that a subset can accurately model the measurements. Another option is to use the first five to seven singular vectors as the mapping between states of the estimation model and the physical parameters; that is, use estimation model states that are a transformation on physical parameters. However, system design issues make it desirable to retain the attitude states (roll, pitch, and yaw) as estimation model states.

There are three limitations of this SVD analysis:

1. It does not directly indicate which physical parameters should be used as ROM states. However, the example of this problem is unusually difficult.
2. It does not take into account the different magnitudes of the parameters. Instrument manufacturers perform extensive analysis of misalignment sources, calibration accuracy, and thermal deformation, and use that information to predict misalignment magnitudes. Mirror shaft angle misalignments can be an order of magnitude larger than mirror-to-shaft misalignments, and also much larger than detector location errors. This information is not taken into account in the SVD analysis. The different expected magnitudes can be used to scale the columns of the **H** matrix before computing the SVD, but this is also not conclusive.

TABLE 6.2: Largest Singular Values and Vectors for Attitude and Misalignments

	Normalized Singular Vector												
Singular Value	Roll	Pitch	Yaw	m_{d2}	m_{d3}	m_{em1}	m_{e1}	m_{e2}	m_{e3}	m_{nm3}	m_{n1}	m_{n2}	m_{n3}
2.5	–0.34	0.23	0.03	0.23	–0.34	0.34	0.34	–0.34	–0.23	0.23	–0.34	0.23	0.23
2.4	–0.23	–0.34	0.01	–0.34	–0.23	0.23	0.23	–0.23	0.34	–0.34	–0.23	–0.34	–0.34
1.02	0.03	–0.07	0.98	–0.07	0.03	0.01	–0.03	–0.05	0.07	0.06	0.03	0.02	0.11
0.115	–0.19	0.18	0.09	0.19	–0.18	–0.19	0.19	0.56	–0.19	–0.18	–0.19	–0.57	0.20
0.013	–0.12	0.66	0.06	0.33	–0.03	–0.11	0.12	0.28	0.33	–0.04	–0.12	0.34	0.31
0.006	0.34	0.33	0.01	–0.01	–0.67	0.17	–0.33	–0.16	0.01	–0.13	0.34	–0.19	–0.00
0.001	–0.00	0.07	0.05	0.16	0.14	0.03	0.00	–0.11	–0.16	–0.88	–0.00	0.32	0.17

3. It only models the error in fitting the star observation angles. It does not identify the combinations of parameters that allow the most accurate pointing for positions on the earth—the goal of the system.

The second analysis phase attempted to address these issues. A Monte Carlo simulation was used to provide more rigorous evaluation of various ROMs. Misalignment angles were randomly selected using values realistic for the class of instruments under consideration, but not representative of any particular instrument. Three sigma values ranged from 30 to 60 μrad for instrument attitude errors, 70 to 150 μrad for detector location errors, 50 to 240 μrad for mirror shaft angle errors, and 10 to 60 μrad for mirror misalignments. Random detector locations were also simulated, but random measurement noise was not added to simulated scan angles since this is irrelevant for the purpose of analyzing modeling errors. The estimated parameters always included roll, pitch, and yaw rotations, but the selected misalignment parameters were varied to determine the set that produced the smallest pointing residuals for "evaluation landmarks" uniformly distributed on the earth.

The maximum and RMS residual statistics for various combinations of estimated attitude parameters were tabulated. The results were

1. A five-parameter misalignment model (roll, pitch, yaw, m_{d2}, m_{e2}) can predict positions on earth with worst-case errors of 3.4 μrad EW and 2.1 μrad NS over the full range of detector locations.
2. Another five-parameter misalignment model (roll, pitch, yaw, m_{n3}, m_{e2}) can predict positions on the earth with worst-case errors of 3.6 μrad EW and 1.6 μrad NS over the full range of detector locations.
3. A seven-parameter misalignment model (roll, pitch, yaw, m_{n2}, m_{n3}, m_{e1}, m_{e2}) can predict positions on the earth with worst-case errors of 0.1 μrad in both EW and NS over the full range of detector locations.
4. Compared with the five-parameter model, a six-parameter model reduces either the maximum EW or NS residuals, but not both.

Based on these results, it is concluded that a five-parameter misalignment model can meet pointing requirements if other error sources, such as random noise, are of nominal magnitudes. Another approach for selecting states of a ROM—stepwise regression—is demonstrated in a later example.

6.3 REGRESSION ANALYSIS FOR WEIGHTED LEAST SQUARES

The consider parameter and model order selection analysis of the previous section was based on Bayesian estimation and random parameters. Standard regression analysis usually assumes that the solved-for states are nonrandom parameters and measurement noise variances are unknown. Hence solution methods are either unweighted least squares or weighted least squares where the assumed measurement noise variance is treated as a guess that can be scaled up or down. In the latter case, the scale factor is computed to make the measurement *residual* SOS equal to the χ^2 distribution mean for the given number of degrees-of-freedom. We now discuss two aspects of regression analysis that provide insight on the modeling problem and allow optimal selection of model order.

6.3.1 Analysis of Variance

Recall that the cost function to be minimized in weighed least squares is $J = \tilde{\mathbf{y}}^T \mathbf{W}\tilde{\mathbf{y}}/2$ where $\tilde{\mathbf{y}} = \mathbf{y} - \mathbf{H}\hat{\mathbf{x}}$ is the vector of *a posteriori* measurement residuals and $\mathbf{W}$ is a weighting matrix that is usually set equal to the inverse of the measurement noise covariance matrix $\mathbf{R}$; that is, $J = \tilde{\mathbf{y}}^T \mathbf{R}^{-1}\tilde{\mathbf{y}}/2$. In unweighted least squares $\mathbf{W} = \mathbf{I}$. For the moment we ignore the weighting because it is not critical to the analysis. Using results from previous sections, the *a posteriori* unweighted measurement residual SOS can be written as

$$\begin{aligned}\tilde{\mathbf{y}}^T\tilde{\mathbf{y}} &= \mathbf{y}^T\mathbf{y} - \mathbf{y}^T\mathbf{H}\hat{\mathbf{x}} \\ &= \mathbf{y}^T\mathbf{y} - \mathbf{y}^T\mathbf{H}(\mathbf{H}^T\mathbf{H})^{-1}\mathbf{H}^T\mathbf{y}\end{aligned} \tag{6.3-1}$$

using the normal equation solution $\hat{\mathbf{x}} = (\mathbf{H}^T\mathbf{H})^{-1}\mathbf{H}^T\mathbf{y}$. In other words, the residual SOS is equal to the total SOS of measurements ($\mathbf{y}^T\mathbf{y}$) minus a term that represents the ability of the estimated state $\hat{\mathbf{x}}$ to model to measurements: $\mathbf{y}^T\mathbf{H}(\mathbf{H}^T\mathbf{H})^{-1}\mathbf{H}^T\mathbf{y}$. $\mathbf{y}^T\mathbf{y}$ is called the *total measurement SOS*. Since it is the sum of m squared random variables, it has m degrees-of-freedom. $\mathbf{y}^T\mathbf{H}(\mathbf{H}^T\mathbf{H})^{-1}\mathbf{H}^T\mathbf{y}$ is called the *SOS due to the regression* and can be expressed (using the SVD) as the sum of n (the number of states in $\mathbf{x}$) squared random variables. Therefore it has n degrees-of-freedom. Finally $\tilde{\mathbf{y}}^T\tilde{\mathbf{y}}$ is called the *SOS about the regression* and has $m - n$ degrees-of-freedom. These sums, divided by the corresponding degrees-of-freedom, give mean-squared errors. For example, $\tilde{\mathbf{y}}^T\tilde{\mathbf{y}}/(m-n)$ is the *mean-square error about the regression*. Table 6.3 summarizes these concepts. Further information on regression variance analysis may be found in Draper and Smith (1998).

The above analysis used whole values of measurements without removing means. However, analysis of variance is most frequently applied to variance about the mean. While a scalar sample mean (e.g., $\bar{y} = \sum_{i=1}^{m} y_i$) is often used in simple cases when one of the states is a measurement bias, it is not generally possible to find a state vector that exactly maps to a bias in the measurements; that is, it is not usually possible to find $\mathbf{x}_b$ such that $\mathbf{H}\mathbf{x}_b = [1 \;\; 1 \;\; \cdots \;\; 1]^T\,\bar{y}$ is exactly satisfied. The measurement mean of most interest is the mapping of the true state $\mathbf{x}$ into measurement space; that is, $\bar{\mathbf{y}} = \mathbf{H}\mathbf{x}$ where as before, $\mathbf{y} = \mathbf{H}\mathbf{x} + \mathbf{r}$ and $\mathbf{r}$ is zero-mean measurement noise. Using this definition for $\bar{\mathbf{y}}$, the residual vector can be expressed as

$$\begin{aligned}\tilde{\mathbf{y}} &= \mathbf{y} - \mathbf{H}\hat{\mathbf{x}} \\ &= (\mathbf{y} - \mathbf{H}\mathbf{x}) + (\mathbf{H}\mathbf{x} - \mathbf{H}\hat{\mathbf{x}}) \\ &= \mathbf{r} + \mathbf{H}\tilde{\mathbf{x}}\end{aligned}$$

TABLE 6.3: Analysis of Variance Summary

Source of Variance	Degrees-of-Freedom	Sum-of-Squares (SOS)	Mean Square (MS)
Due to regression	n	$\mathbf{y}^T\mathbf{H}(\mathbf{H}^T\mathbf{H})^{-1}\mathbf{H}^T\mathbf{y}$	$\mathbf{y}^T\mathbf{H}(\mathbf{H}^T\mathbf{H})^{-1}\mathbf{H}^T\mathbf{y}/n$
About the regression	$m - n$	$\tilde{\mathbf{y}}^T\tilde{\mathbf{y}}$	$\tilde{\mathbf{y}}^T\tilde{\mathbf{y}}/(m-n)$
Total	m	$\mathbf{y}^T\mathbf{y}$	$\mathbf{y}^T\mathbf{y}/m$

where $\tilde{\mathbf{x}} = \mathbf{x} - \hat{\mathbf{x}}$. Hence

$$\begin{aligned}\tilde{\mathbf{y}}^T\tilde{\mathbf{y}} &= \mathbf{r}^T\mathbf{r} - \mathbf{r}^T\mathbf{H}(\mathbf{H}^T\mathbf{H})^{-1}\mathbf{H}^T\mathbf{r} \\ &= \mathbf{r}^T[\mathbf{I} - \mathbf{H}(\mathbf{H}^T\mathbf{H})^{-1}\mathbf{H}^T]\mathbf{r}\end{aligned}. \qquad (6.3\text{-}2)$$

$\mathbf{r}^T\mathbf{r}$ is the measurement *SOS about the mean* and $\mathbf{r}^T\mathbf{H}(\mathbf{H}^T\mathbf{H})^{-1}\mathbf{H}^T\mathbf{r}$ is the *SOS about the mean due to the fit*. Notice that $(\mathbf{H}^T\mathbf{H})^{-1}$ times the measurement noise variance is the error covariance of the state estimate $\hat{\mathbf{x}}$, so $\mathbf{r}^T\mathbf{H}(\mathbf{H}^T\mathbf{H})^{-1}\mathbf{H}^T\mathbf{r}$ is the squared mapping of errors in $\hat{\mathbf{x}}$ to the measurement space. It is the part of $\mathbf{r}^T\mathbf{r}$ that can be removed by erroneous adjustments in $\hat{\mathbf{x}}$.

Since for zero-mean Gaussian $\mathbf{r}$, $\tilde{\mathbf{y}}^T\tilde{\mathbf{y}}$ should be χ^2–distributed with $m - n$ degrees-of-freedom and $\tilde{\mathbf{y}}^T\tilde{\mathbf{y}}/(m-n)$ is an estimate of the variance about the regression, which should equal $\sigma_r^2 = E[r^2]$ when using unweighted least squares. When weighting of $\mathbf{R} = \sigma_r^2\mathbf{I}$ is used, $\tilde{\mathbf{y}}^T\tilde{\mathbf{y}}/[(m-n)\sigma_r^2]$ is an estimate of the scale factor required to correct σ_r^2 to match the true measurement noise variance.

6.3.2 Stepwise Regression

We now show how previously discussed concepts can be used to determine appropriate parameters to be included in the regression model. While many approaches are available, stepwise regression is one of the most commonly used techniques.

Stepwise regression is automatic procedure used to determine a subset of model variables providing a good fit to observed data. It is primarily used when there are a large number of potential explanatory variables (e.g., patient age, sex, environment, and occupation for cancer epidemiology) and there is no underlying theory on which to base the model selection. Although based on statistical tests, the implementation is somewhat ad hoc and is generally used in cases when a physically based model is not available; it is used more often to model data correlations rather than to determine physical model parameters. Stepwise regression is not guaranteed to find the unique set of adjusted parameters minimizing the least-squares cost function. In fact, when given a large set of explanatory variables, it is possible to find multiple "optimal" sets with nearly the same cost function. The variables selected for inclusion in the model depend on the order in which the variables are selected because the sequential F-tests (on which selections are based) are assumed to be independent, which they are not. Neglecting "statistically small" parameters that are actually significant can bias the estimates. However, these problems tend to be less important for physically based models, when engineering judgment is applied to limit the number of variables and eliminate unrealistic solutions.

Alternative regression techniques attempt to search many combinations of parameters to find the set with the minimum norm (see Furnival and Wilson 1971; Hocking 1983; Draper and Smith 1998). When the number of possibilities is large, the computational burden of "search all subsets" techniques can be very high. Also these alternatives are subject to many of the same problems as stepwise regression. Draper and Smith still state a preference for stepwise regression over many alternate techniques.

Stepwise regression tests the statistical significance of adding or removing individual states from the adjusted set. Hence prior to discussing the algorithm, we first compute the change in the least-squares cost function when a state is added. The analysis is similar to that used for the Bayesian least-squares case, but is simpler without the prior estimate. We again partition the state vector, where $\mathbf{x}_1$ represents states that were previously included in the regression and x_2 represents a scalar state that is conditionally added to the adjusted set. The weighted least-squares normal equations for this system are

$$\begin{bmatrix} \mathbf{H}_1^T\mathbf{R}^{-1}\mathbf{H}_1 & \mathbf{H}_1^T\mathbf{R}^{-1}\mathbf{h}_2 \\ \mathbf{h}_2^T\mathbf{R}^{-1}\mathbf{H}_1 & \mathbf{h}_2^T\mathbf{R}^{-1}\mathbf{h}_2 \end{bmatrix}\begin{bmatrix} \hat{\mathbf{x}}_1 \\ \hat{x}_2 \end{bmatrix} = \begin{bmatrix} \mathbf{H}_1^T\mathbf{R}^{-1}\mathbf{y} \\ \mathbf{h}_2^T\mathbf{R}^{-1}\mathbf{y} \end{bmatrix} \tag{6.3-3}$$

or more compactly

$$\begin{bmatrix} \mathbf{A}_1 & \mathbf{a}_2 \\ \mathbf{a}_2^T & a_3 \end{bmatrix}\begin{bmatrix} \hat{\mathbf{x}}_1 \\ \hat{x}_2 \end{bmatrix} = \begin{bmatrix} \mathbf{b}_1 \\ \mathbf{b}_2 \end{bmatrix} \tag{6.3-4}$$

where the substitutions are obvious. Solving the lower equation and substituting it in the upper yields:

$$\begin{aligned} \hat{x}_2 &= (b_2 - \mathbf{a}_2^T\hat{\mathbf{x}}_1)/a_3 \\ \mathbf{A}_1\hat{\mathbf{x}}_1 + \mathbf{a}_2(b_2 - \mathbf{a}_2^T\hat{\mathbf{x}}_1)/a_3 &= \mathbf{b}_1 \end{aligned}.$$

Using these revised definitions of $\mathbf{A}$ and $\mathbf{b}$, the solutions are the same as equations (6.2-42) and (6.2-43):

$$\boxed{\hat{\mathbf{x}}_1 = \hat{\mathbf{x}}_{1\_0} + \frac{\mathbf{A}_1^{-1}\mathbf{a}_2(-b_2 - \mathbf{a}_2\mathbf{A}_1^{-1}\mathbf{b}_1)}{a_3 - \mathbf{a}_2^T\mathbf{A}_1^{-1}\mathbf{a}_2}} \tag{6.3-5}$$

and

$$\boxed{\hat{x}_2 = \frac{b_2 - \mathbf{a}_2^T\mathbf{A}_1^{-1}\mathbf{b}_1}{a_3 - \mathbf{a}_2^T\mathbf{A}_1^{-1}\mathbf{a}_2}} \tag{6.3-6}$$

where

$$\boxed{\hat{\mathbf{x}}_{1\_0} = \mathbf{A}_1^{-1}\mathbf{b}_1 = (\mathbf{H}_1^T\mathbf{R}^{-1}\mathbf{H}_1)^{-1}\mathbf{H}_1^T\mathbf{R}^{-1}\mathbf{y}} \tag{6.3-7}$$

is the estimate of $\hat{\mathbf{x}}_1$ before state $\hat{x}_2$ is added to the adjusted states. Then the weighted residual SOS is

$$\boxed{\begin{aligned} \tilde{\mathbf{y}}^T\mathbf{R}^{-1}\tilde{\mathbf{y}} &= \mathbf{y}^T\mathbf{R}^{-1}\mathbf{y} - \tilde{\mathbf{y}}^T\mathbf{R}^{-1}(\mathbf{H}_1\hat{\mathbf{x}}_1 + \mathbf{h}_2\hat{x}_2) \\ &= \mathbf{y}^T\mathbf{R}^{-1}\mathbf{y} - \mathbf{b}_1^T\hat{\mathbf{x}}_1 - b_2\hat{x}_2 \\ &= (\mathbf{y}^T\mathbf{R}^{-1}\mathbf{y} - \mathbf{b}_1^T\hat{\mathbf{x}}_{1\_0}) - \frac{\hat{x}_2^2}{\sigma_2^2} \end{aligned}} \tag{6.3-8}$$

where $\sigma_2^2 = a_3 - \mathbf{a}_2^T\mathbf{A}_1^{-1}\mathbf{a}_2$. Hence the change in the cost function is $-\hat{x}_2^2/\sigma_2^2$ when state x_2 is adjusted. When $\mathbf{R} = E[\mathbf{r}\mathbf{r}^T]$ and the true $x_2 = 0$, $\hat{x}_2^2/\sigma_2^2$ is a χ^2-distributed variable with one degree-of-freedom. Thus the χ^2 distribution can be used to test the statistical significance of including $\hat{\mathbf{x}}_2$ as an adjusted state. More often stepwise regression assumes that $E[\mathbf{r}\mathbf{r}^T] = \sigma_r^2\mathbf{I}$ where σ_r^2 (the constant measurement noise variance) is unknown. Hence the solution uses unweighted least squares where σ_r^2 is computed from the mean-squared error of the fit. In this case $\hat{x}_2^2 / \sigma_2^2$ has an F-distribution, as explained below after first describing the stepwise regression algorithm.

Most implementations of stepwise regression are based on an algorithm by Efroymson (1960), which is a variation on forward selection. After a new variable is added to the regression, additional tests attempt to determine whether variables already in the regression can be removed without significantly increasing the residual SOS. The basic steps are

1. The sample information matrix and information vector are computed from the measurement data and measurement sensitivity matrix; that is, given the measurement equation $\mathbf{y} = \mathbf{Hx} + \mathbf{r}$, where $E[\mathbf{rr}^T] = \sigma_r^2\mathbf{I}$, compute the unweighted least-squares information arrays $\mathbf{A} = \mathbf{H}^T\mathbf{H}$ and $\mathbf{b} = \mathbf{H}^T\mathbf{y}$. Stepwise regression attempts to estimate the subset of $\mathbf{x}$, $\hat{\mathbf{x}}_x$, of dimension p that minimizes the SOS of the residual $\mathbf{y} - \mathbf{H}\hat{\mathbf{x}}_x$.
2. The $\mathbf{A}$ and $\mathbf{B}$ arrays are partitioned into the independent variables currently in the regression and the variables not in the regression. Let $\mathbf{A}_x$ and $\mathbf{b}_x$ represent the partitions of $\mathbf{A}$ and $\mathbf{B}$ corresponding to variables in the regression. Also, the residual degrees-of-freedom, SOS, and mean square are computed as

$$\begin{aligned} d_f &= m - p \\ S_s &= \mathbf{y}^T\mathbf{y} - \mathbf{b}_x^T\mathbf{A}_x^{-1}\mathbf{b}_x \\ M_s &= S_s / d_f \end{aligned} \tag{6.3-9}$$

3. Estimates of states in the regression are computed as $\hat{\mathbf{x}}_x = \mathbf{A}_x^{-1}\mathbf{b}_x$, and the standard error for variable k, σ_k, is computed as the square root of the k-th diagonal of $\mathbf{A}_x^{-1}M_s$. Then, the F-value for each variable k is computed as if it were the last variable added to the regression:

$$F_k = (\hat{\mathbf{x}}_{xk} / \sigma_k)^2. \tag{6.3-10}$$

If one or more independent variables in the regression have F-values less than the "F-to-remove" threshold, the one with the smallest F value is removed from the regression and control passes to step 2.
4. For each independent variable not in the regression, the tolerance value and F value are computed as if each independent variable is the only one added to the regression. This computation is implemented by means of a partitioned matrix inverse of $\mathbf{A}$ for each independent variable (see Dixon 1975). The tolerance value for a given variable is defined as the variance computed as if it is the only variable in the regression, divided by the variable variance computed as part of the regression with other variables. This is inversely related to the correlation between the variable and variables already in the regression. For variables not in the regression with tolerance greater than the tolerance threshold and with the F value greater than the "F-to-add" threshold, the variable with the highest F value is added and control passes to step 2.
5. The stepwise procedure is terminated if no variables are added or removed in previous steps.

The above procedure is a somewhat simplified description of basic stepwise regression algorithms included in statistical packages. Other computations and logic may be added to fine-tune the process or to use alternate techniques. Draper and Smith suggest that the "F-to-add" and "F-to-remove" thresholds be based on the same significance level α for the current degrees-of-freedom. They use $\alpha = 0.05$ in their example, but note that some workers use a fixed number such as 4.0 for both "F-to-add" and "F-to-remove".

It is instructive to examine the "F-statistic" $F_k = (\hat{\mathbf{x}}_{xk}/\sigma_k)^2$, which is algebraically equivalent to the ratio of two random variables:

$$\frac{[S_s(\hat{\mathbf{x}}_{x\_j-1}) - S_s(\hat{\mathbf{x}}_{x\_j})]/\sigma_r^2}{[S_s(\hat{\mathbf{x}}_{x\_j-1})]/\sigma_r^2} \tag{6.3-11}$$

where σ_r^2 is the measurement noise variance, and $S_s(\hat{\mathbf{x}}_{x\_j})$ denotes the residual SOS computed using the estimated parameters $\hat{\mathbf{x}}_x$ obtained at regression iteration j. At step 4 of iteration j, equation (6.3-10) is executed for all parameters not currently in the regression and the *reduction* in S_s is computed for each parameter *as if it is the only one added to the regression*. Thus, for the null hypothesis of no additional jumps, the numerator of equation (6.3-11) is χ^2-distributed with one degree-of-freedom. The denominator will also be χ^2-distributed, but with d_f degrees-of-freedom at iteration $j - 1$. A ratio of two χ^2 variables with 1 and d_f degrees-of-freedom, respectively, is modeled by the $F(1,d_f)$ distribution. Thus the probability of false alarm *for a single parameter* (assuming the null hypothesis) can be calculated from tables of F-distributions. Unfortunately, this statistic will not be exactly F-distributed when multiple hypotheses are considered (see Draper and Smith 1998, section 15.2, or Dixon 1979). This same problem is also present when using the log likelihood ratio statistic of *Generalized Likelihood Ratio* (GLR) methods (see Kerr 1983 or Basseville and Beneviste 1986).

A hypothesis test based on equation (6.3-10) is often used in stepwise regression because the true measurement noise variance, σ_r^2, is unknown. Since σ_r^2 appears in both the numerator and denominator, it cancels and the test can be simply based on the ratio of residual SOS. In many applications the noise variance is well known, and thus a χ^2 test on the numerator of equation (6.3-10) could be used. Under certain conditions, it can be shown that this is exactly equivalent to the log likelihood ratio statistic used in GLR (Gibbs 1992).

Stepwise regression is used as the basis of a Kalman filter jump detection algorithm presented in Chapter 11.

Example 6.7: Stepwise Regression for Optical Imaging Instrument

Here we use stepwise regression to continue the analysis of Example 6.6, which involved selection of a reduced set of misalignments for an optical imaging instrument. The basic stepwise regression algorithm is applied in each of 500 Monte Carlo trials using the 17 noiseless star observation pairs (north and east angles) to determine the best subset of states. The F-to-add threshold is set to 4, F-to-remove is set to 2, and tolerance is set to 0.001. This low tolerance threshold is necessary because the measurements are noiseless.

TABLE 6.4: Number of Times Parameters Were Included in Stepwise Regression

Parameter	No. Times Selected with Regression Initialized Using Zero States	No. Times Selected with Regression Initialized Using Roll, Pitch, Yaw
Roll	101	493
Pitch	121	479
Yaw	241	473
m_{d2}	128	1
m_{d3}	56	0
m_{em1}	200	137
m_{e1}	178	2
m_{e2}	182	134
m_{e3}	125	2
m_{nm3}	181	136
m_{n1}	89	0
m_{n2}	156	93
m_{n3}	230	22

Table 6.4 shows the number of times each state was selected for two different conditions: with no states initially included in the regression, and with roll, pitch, and yaw initially included. Notice that only three parameters were chosen in more than 40% of samples when started with zero states in the regression. The result is very different when roll, pitch, and yaw are initially included: the regression retains roll, pitch, and yaw in 95% of the samples, and typically adds only one or two other parameters. In both cases m_{em1}, m_{e2}, and m_{nm3} are frequently selected, but m_{n3} is frequently selected only when roll, pitch, and yaw are not initially included. Tests were also conducted with an F-to-remove of 0.5, and with measurement noise added: the results were not significantly different in character.

When the regression was initialized with zero states, a total of 44 different state sets were selected in the regression. Table 6.5 shows states, number of times the set was selected, and the measurement residual RMS for the seven most commonly selected sets. When the regression was initialized with roll, pitch, and yaw, a total of only 28 sets were used: Table 6.6 shows the results. Again notice the differences in selected states. The fit to the noiseless data, as measured by the residual RMS, is slightly better in Table 6.5, but the difference is small.

This difference in behavior with different initialization demonstrates the non-uniqueness problem of stepwise regression for a case in which SVD analysis showed that many parameters are highly correlated. However, the problem of this example is not typical: the similarity of parameter signatures makes the problem unusually difficult. While this example demonstrates many reduced-order modeling issues, it may give a pessimistic impression of the approach. In many problems the dominant parameters can more easily be determined than demonstrated here. However, even with the difficulties of this case, it was possible to find low-order models that met earth-pointing requirements, as shown in Example 6.6.

TABLE 6.5: Most Commonly Selected State Sets When Initialized with Zero States

Parameter	Set 1	Set 2	Set 3	Set 4	Set 5	Set 6	Set 7
Roll	—	—	—	—	—	—	—
Pitch	—	x	—	x	—	—	—
Yaw	—	—	x	x	—	x	—
m_{d2}	x	—	—	—	—	x	—
m_{d3}	—	—	—	—	—	—	—
m_{em1}	—	—	—	—	x	x	x
m_{e1}	x	x	—	—	—	—	—
m_{e2}	x	—	x	—	x	—	x
m_{e3}	—	—	—	—	—	—	x
m_{nm3}	—	—	—	x	—	—	—
m_{n1}	—	—	x	x	—	x	—
m_{n2}	—	x	x	—	x	—	—
m_{n3}	x	x	x	x	x	—	x
# times selected	34	32	29	28	27	26	26
Residual RMS	0.062	0.058	0.068	0.059	0.051	0.057	0.060

TABLE 6.6: Most Commonly Selected State Sets When Initialized with Three States

Parameter	Set 1	Set 2	Set 3	Set 4	Set 5	Set 6	Set 7
Roll	x	x	x	x	x	x	x
Pitch	x	x	x	x	—	x	—
Yaw	x	x	x	x	x	x	x
m_{d2}	—	—	—	—	—	—	—
m_{d3}	—	—	—	—	—	—	—
m_{em1}	x	x	—	—	—	—	—
m_{e1}	—	—	—	—	—	—	—
m_{e2}	—	—	x	—	—	—	—
m_{e3}	—	—	—	—	—	—	—
m_{nm3}	—	—	—	x	x	x	—
m_{n1}	—	—	—	—	—	—	—
m_{n2}	—	—	—	—	—	x	x
m_{n3}	x	—	—	—	—	—	—
# times selected	132	128	128	127	108	97	93
Residual RMS	0.071	0.072	0.067	0.069	0.064	0.069	0.066

6.3.3 Prediction and Optimal Data Span

When the ultimate goal of least-squares modeling is to predict system output for a period in the future, it can be important to determine the fit data span producing the best predictions. You may suspect that more data are better, and it is usually found that adding measurements does improve prediction accuracy when the estimator model accurately matches truth. However, when the estimator model does not exactly match truth—either because some states have been deliberately unadjusted or because the true model is not known—fit data spans much longer than the prediction span can sometimes degrade predictions because the batch estimator tends to weight long-term behavior greater than short term. On the other hand, fit data spans much shorter than the prediction span can also cause poor predictions because the estimator may miss important long-term behavior.

Again it is difficult to define general rules for specifying the optimal fit data span when model errors are present, but a good starting point is to make the fit span equal to the prediction span (if that is an option). Other factors that may be important are the ratio between the longest time constants (or basis function periods) and the fit and prediction spans, the measurement accuracy, and the stability of the system. (Is it really deterministic or is there a small stochastic component?) If the fit model is suboptimal because some parameters were deliberately unadjusted, then the prediction residual equation (6.2-33), covariance equation (6.2-34), or Monte Carlo simulation can be used to analyze the prediction accuracy as a function of the fit span. Otherwise evaluation on different segments of real data (if available) should be used.

6.4 SUMMARY

This chapter discussed practical usage of least-squares techniques. Specific topics included solution validation, model error analysis, confidence bound calculations, and optimal selection of states to be estimated.

There are many reasons why a least-squares solution may not be valid. Some of the more common problems include incorrect model structure, incorrect model assumptions, incorrect selection of model states or order, poor state observability, and anomalous measurements. Hence it is very important that solutions be validated. The methods used to validate the solutions include:

1. Check the least-squares cost function for weighted least squares, MMSE/minimum variance, ML, and MAP estimates. For weighted least squares, the weighted residual sum $\tilde{\mathbf{y}}^T\mathbf{R}^{-1}\tilde{\mathbf{y}}$ should be χ^2-distributed with a mean value of m-n and variance equal to 2(m-n). For Bayesian estimates the sum should include the prior residuals and the degrees-of-freedom should be modified accordingly. When $\tilde{\mathbf{y}}^T\mathbf{R}^{-1}\tilde{\mathbf{y}}$ deviates significantly from m-n, model errors may be present or the assumed measurement noise variance may be wrong.
2. Plot the residuals to verify that systematic patterns are not significant. The residuals may be slightly correlated when the number of measurements is small, but systematic patterns should not be large.
3. Compute residual means and variances separately for each subset (sensor, measurement type, etc.) to verify that no one subset is anomalous.
4. When possible, compute separate solutions using different subsets of measurements and check the prediction accuracy for other measurement subtypes.
5. Use the model to predict measurements either past the time period used for the model fit or to untested conditions (using alternate test data), and check the magnitude and character of prediction residuals. Prediction residuals are generally more sensitive to model problems than fit residuals.

6. Compare the behavior of the estimator using real measurements with behavior obtained using simulated measurements. Is the behavior expected?
7. Check the state estimates for reasonableness. For Bayesian estimates, compare the change in *a priori* and *a posteriori* state values with the *a priori* standard deviation.

Two examples were given to demonstrate the concepts. The first was a fourth-order polynomial model. It was shown that the fit residuals are slightly correlated near the ends of the data span, but essentially uncorrelated near the middle. However, when measurements generated using an eighth-order polynomial are fit using a fourth-order model, systematic patterns in the fit residuals are readily evident. Also the prediction residuals diverge quickly when the model order is incorrect.

The second example was a case study of GOES-13 performance that demonstrated many of the analysis steps described above. It was noticed that 1-day predictions of range and landmark (but not star) measurements deviated significantly from actual measurements when the spacecraft was operated in the "inverted" yaw orientation. Analysis to determine the source of the problem included examination of fit and prediction residuals, comparisons of estimation results for different time spans and under different operating conditions, selection of different adjusted states, and testing using simulated data. The problem was found to be due to incorrect compensation for spacecraft transponder delay, and an EW bias in landmark or star measurements.

Section 6.2 of this chapter first showed how the *a posteriori* or formal state estimate error covariance—computed as the inverse of the Fisher information matrix—can be used to compute confidence limits on errors in the state estimate. Examples demonstrate plotting of error ellipses and contours for a given probability limit. The formal error covariance is a lower bound on the true estimate error covariance because it only includes the effects of measurement noise: nonlinearities and model errors are not included. Effects of nonlinearities are best analyzed via Monte Carlo simulation. The effects of incorrect model structure can be analyzed using either covariance or Monte Carlo analysis, but some of the required statistical inputs are likely to be poorly known. Error equations show that model structure errors are likely to have a larger effect on prediction residuals than fit residuals.

Covariance error analysis is better defined when a subset of the total states is adjusted in the estimator. The effects of the unadjusted analyze (consider) states on estimated states were computed in Section 6.2.2.3, and the total estimate error covariance was derived for both normal equation and QR solution methods. The sensitivity of fit and prediction residuals with respect to unadjusted states was also derived and used to compute the residual covariance. These equations are used to compute the expected total error in estimated states, and are also used when attempting to determine the optimal choice of states for an estimator model. Monte Carlo analysis can also be used for the same purpose.

It is generally undesirable to adjust states that have negligible effect on the measurements or have the same measurement signature as other linear combinations of adjusted states. Estimation of unnecessary states can degrade the accuracy of estimates and predictions, and increase computations. The weighted least-squares cost function is an important metric used when determining the optimal model order. Section 6.2.2.4 derived the expected change in the sum of weighted residuals when an unadjusted state is changed to adjusted in a Bayesian solution. The difference between the actual change and the expected change for nominal *a priori* variances can be used to determine the statistical significance of the state. A polynomial example demonstrated how fit and prediction residual statistics change with model order, and showed that the minimum of a modified log likelihood function occurs at the true model order used to generate the simulated measurements. It was also shown that the prediction residuals are more sensitive to incorrect order selection than fit residuals. Somewhat similar behavior was also observed when using Chebyshev polynomials or Fourier series models. Another example analyzed the effect of misalignments on pointing error of an optical imaging instrument, and demonstrated how the optimal selection of adjusted states is determined.

Unlike the Bayesian approach described above, standard regression analysis usually assumes that the solved-for (adjusted) states are nonrandom parameters, and that the measurement noise variance is unknown. The *analysis of variance* method computes the "residual about the mean" SOS due to the regression and about the regression, and compares those sums with degrees-of-freedom for each. An estimate of the measurement noise variance can be computed from the variance about the regression.

The concepts of the previous sections are used as the basis of stepwise regression methods for determining optimal selection of adjusted states in a non-Bayesian solution. Sequential F-tests and correlation tests are used to determine which unadjusted states are "statistically significant" and should be adjusted. Also at each step, currently adjusted states are tested for significance to determine if they should be removed from the regression. Unfortunately stepwise regression is not guaranteed to find the unique set of adjusted parameters minimizing the least-squares cost function, and different results can be obtained when the order in which variables are tested is changed. Alternate "search all subsets" methods are subject to some of the same problems as stepwise regression. This uniqueness problem is somewhat less of an issue with physical systems because engineering judgment can be used to limit the number of possible solutions. In practice stepwise regression often works well. However, an example based on the imaging instrument misalignment model showed the difficulty in using stepwise regression for a weakly observable problem.

Selection of the optimal fit data span to minimize prediction errors was the final topic. Use of fit data spans equal to the desired prediction data span is often a good starting point, but the optimal span in any particular problem depends on the dynamic model time constants, model errors, measurement accuracy, and inherent randomness (if any) of the system. Simulation should be used to investigate the trade-offs.

CHAPTER 7

LEAST-SQUARES ESTIMATION: CONSTRAINTS, NONLINEAR MODELS, AND ROBUST TECHNIQUES

The previous chapter addressed least-squares topics related to modeling, model errors, and accuracy assessment. This chapter discusses extensions of basic linear least-squares techniques, including constrained least-squares estimation (equality and inequality), recursive least squares, nonlinear least squares, robust estimation (including data editing), and measurement preprocessing.

7.1 CONSTRAINED ESTIMATES

It is sometimes necessary to enforce constraints on the estimated states. These constraints may be either linear equality constraints of the form $\mathbf{C}\hat{\mathbf{x}} = \mathbf{d}$, where $\mathbf{C}$ is a full-rank $p \times n$ matrix with $p < n$, or inequality constraints such as $\mathbf{C}\hat{\mathbf{x}} > \mathbf{d}$, $\mathbf{C}\hat{\mathbf{x}} < \mathbf{d}$, $\|\mathbf{C}\hat{\mathbf{x}} - \mathbf{d}\|_a < \alpha$, or combinations of equality and inequality constraints. This section presents a brief summary of the techniques. More information may be found in Lawson and Hanson (1974), Björck (1996), and Golub and Van Loan (1996). Also see Anderson et al. (1999) for information on the LAPACK routine S/DGGLSE.

7.1.1 Least-Squares with Linear Equality Constraints (Problem LSE)

Many methods are available for implementing equality constraints of the form $\mathbf{C}\hat{\mathbf{x}} = \mathbf{d}$ in least-squares problems. In all cases it is assumed that the $p \times n$ matrix $\mathbf{C}$ is rank $p < n$, the augmented matrix

$$\begin{bmatrix} \mathbf{R}^{-1/2}\mathbf{H} \\ \mathbf{C} \end{bmatrix}$$

is of rank n (column dimension) for minimization of $\|\mathbf{R}^{-1/2}(\mathbf{H}\mathbf{x} - \mathbf{y})\|_2$, and the constraints in $\mathbf{C}$ are consistent. A few of the methods include:

Advanced Kalman Filtering, Least-Squares and Modeling: A Practical Handbook, by Bruce P. Gibbs

1. Algebraically solve the constraint equation for a subset of states, and use the substitution to eliminate states from the estimator. This is the method used in Example 5.7, but unfortunately this is often not feasible for general linear equations involving multiple states. For constraints on individual parameters, the state can be removed from the problem by setting it to the constraint and solving for the remaining states (with residuals adjusted for constrained states). This must either be done before processing measurements, or by adjusting the information arrays after processing.
2. Include the constraint using Lagrange multipliers. This redefines the least-squares problem.
3. Implement the constraint as a measurement with very high weighting. This has the advantage that software for unconstrained least squares can be applied without modification, and for that reason it is often used. However, it can lead to numerical problems.
4. Derive an equivalent unconstrained least-squares problem of lower dimension using either *direct elimination* or the *nullspace method*. Both methods use the QR algorithm for the solution and are numerically stable. However, explanations are nontrivial and we refer to Björck (1996, section 5.1) or Lawson and Hanson (1974, chapters 20 and 21) for further information.
5. Implement the constraint using a generalized Singular Value Decomposition (GSVD) that includes both minimization of $\|\mathbf{R}^{-1/2}(\mathbf{Hx} - \mathbf{y})\|_2$ and the constraint $\mathbf{Cx} = \mathbf{d}$. This is the most general approach as it can be used for other types of constraints.

7.1.1.1 Constraints Using Lagrange Multipliers We first demonstrate how Lagrange multipliers can be used to enforce the constraint. The constraint is added to the least-squares cost function, which for (non-Bayesian) weighted least squares is

$$J = \frac{1}{2}(\mathbf{y} - \mathbf{Hx})^T \mathbf{R}^{-1}(\mathbf{y} - \mathbf{Hx}) + \boldsymbol{\lambda}^T(\mathbf{Cx} - \mathbf{d}) \tag{7.1-1}$$

where $\boldsymbol{\lambda}$ is the p-vector of Lagrange multipliers. The equivalent equation for Bayesian estimates can be inferred by treating the prior estimate as a measurement. The gradients of J with respect to $\mathbf{x}$ and $\boldsymbol{\lambda}$ are

$$\frac{\partial J}{\partial \mathbf{x}} = -(\mathbf{y} - \mathbf{Hx})^T \mathbf{R}^{-1}\mathbf{H} + \boldsymbol{\lambda}^T \mathbf{C} \tag{7.1-2}$$

$$\frac{\partial J}{\partial \boldsymbol{\lambda}} = (\mathbf{Cx} - \mathbf{d})^T. \tag{7.1-3}$$

Setting equation (7.1-2) to zero gives the "normal equation" version of the constrained least-squares estimate

$$\hat{\mathbf{x}}_c = (\mathbf{H}^T \mathbf{R}^{-1}\mathbf{H})^{-1}(\mathbf{H}^T \mathbf{R}^{-1}\mathbf{y} - \mathbf{C}^T \boldsymbol{\lambda}) \tag{7.1-4}$$

which is substituted in $\mathbf{C}\hat{\mathbf{x}}_c = \mathbf{d}$ to obtain

$$\boldsymbol{\lambda} = (\mathbf{CPC}^T)^{-1}(\mathbf{CPH}^T\mathbf{R}^{-1}\mathbf{y} - \mathbf{d})$$

where $\mathbf{P} = (\mathbf{H}^T\mathbf{R}^{-1}\mathbf{H})^{-1}$. This is then substituted in equation (7.1-4) to obtain

$$\begin{aligned}\hat{\mathbf{x}}_c &= \mathbf{P}\left[\mathbf{H}^T\mathbf{R}^{-1}\mathbf{y} - \mathbf{C}^T(\mathbf{CPC}^T)^{-1}(\mathbf{CPH}^T\mathbf{R}^{-1}\mathbf{y} - \mathbf{d})\right] \\ &= \left[\mathbf{P} - \mathbf{PC}^T(\mathbf{CPC}^T)^{-1}\mathbf{CP}\right]\mathbf{H}^T\mathbf{R}^{-1}\mathbf{y} + \mathbf{PC}^T(\mathbf{CPC}^T)^{-1}\mathbf{d}\end{aligned} \quad (7.1\text{-}5)$$

The constrained error covariance $E\left[(\hat{\mathbf{x}}_c - \mathbf{x})(\hat{\mathbf{x}}_c - \mathbf{x})^T\right]$ is

$$\mathbf{P}_c = \mathbf{P} - \mathbf{PC}^T(\mathbf{CPC}^T)^{-1}\mathbf{CP}. \quad (7.1\text{-}6)$$

A similar equation can be derived using the QR approach of Chapter 5, starting with

$$\begin{aligned} J &= \frac{1}{2}[(\mathbf{y} - \mathbf{Hx})^T\mathbf{R}^{-T/2}]\mathbf{T}^T\mathbf{T}[\mathbf{R}^{-1/2}(\mathbf{y} - \mathbf{Hx})] + \boldsymbol{\lambda}^T(\mathbf{Cx} - \mathbf{d}) \\ &= \frac{1}{2}[(\mathbf{z} - \mathbf{Ux})^T(\mathbf{z} - \mathbf{Ux}) + \boldsymbol{\lambda}^T(\mathbf{Cx} - \mathbf{d})\end{aligned} \quad (7.1\text{-}7)$$

where $\mathbf{T}$ is orthogonal, $\mathbf{z} = \mathbf{TR}^{-1/2}\mathbf{y}$ and $\mathbf{U} = \mathbf{TR}^{-1/2}\mathbf{H}$ (upper triangular). The result is

$$\begin{aligned}\hat{\mathbf{x}}_c &= \mathbf{U}^{-1}\left[\mathbf{I} - \mathbf{U}^{-T}\mathbf{C}^T(\mathbf{CU}^{-1}\mathbf{U}^{-T}\mathbf{C}^T)^{-1}\mathbf{CU}^{-1}\right]\mathbf{z} + \mathbf{U}^{-1}\mathbf{U}^{-T}\mathbf{C}^T(\mathbf{CU}^{-1}\mathbf{U}^{-T}\mathbf{C}^T)^{-1}\mathbf{d} \\ &= \mathbf{U}^{-1}\mathbf{z} + \mathbf{U}^{-1}\mathbf{E}^T(\mathbf{EE}^T)^{-1}(\mathbf{d} - \mathbf{Ez})\end{aligned} \quad (7.1\text{-}8)$$

with $\mathbf{E} = \mathbf{CU}^{-1}$. Notice that this is equivalent to equation (7.1-5) since $\mathbf{P} = (\mathbf{H}^T\mathbf{R}^{-1}\mathbf{H})^{-1} = \mathbf{U}^{-1}\mathbf{U}^{-T}$. Björck (1996, Section 5.1.6) calls this method an "updating" technique.

Implementation of equation (7.1-5) or (7.1-8) requires some additional coding above that required for an unconstrained solution. However, the equations simplify for certain types of constraints. For example, a single constraint on a single state (such that $\mathbf{C}$ is a row vector of zeroes with element j equal to 1) implies that $[\mathbf{P} - \mathbf{PC}^T(\mathbf{CPC}^T)^{-1}\mathbf{CP}]$ is equal to $\mathbf{P}$ minus the outer product of the j-th column of $\mathbf{P}$ divided by P_{jj}. Likewise $\mathbf{PC}^T(\mathbf{CPC}^T)^{-1}\mathbf{d}$ is the j-th column of $\mathbf{P}$ multiplied by d/P_{jj}. This can be easily extended to multiple constraints of the same type.

Another problem with equations (7.1-5) and (7.1-8) is the requirement that the original unconstrained least-squares problem be nonsingular ($\mathbf{H}$ is full column rank) and $\mathbf{CPC}$ be nonsingular. Requiring $\mathbf{H}$ to be full rank may be undesirable because constraints are sometimes used to compensate for lack of observability.

7.1.1.2 Constraints as Heavily Weighted Measurements

A simpler approach to handling equality constraints processes them as heavily weighted measurements. This allows use of software designed for unconstrained least squares. With some caution, this approach works quite well when using the Modified Gram-Schmidt (MGS) QR method (Chapter 5). It also works reasonably well when using the normal equations for constraints on individual states, but can result in great loss of precision when the constraints involve multiple states.

To see how the constraint-as-measurement approach works with normal equations, assume that the constraint is processed as

$$J = \frac{1}{2}\left((\mathbf{y}-\mathbf{Hx})^T\mathbf{R}^{-1}(\mathbf{y}-\mathbf{Hx}) + \mu^2(\mathbf{Cx}-\mathbf{d})^T(\mathbf{Cx}-\mathbf{d})\right) \tag{7.1-9}$$

where μ^2 is many orders of magnitude larger than the inverse measurement noise variances ($\mathbf{R}^{-1}$ diagonals). For the case of constraining a single state in the middle of the state vector, $\mathbf{C} = [\mathbf{0}^T \quad 1 \quad \mathbf{0}^T]$, the information matrix and vector of the normal equations will be

$$\mathbf{H}^T\mathbf{R}^{-1}\mathbf{H} + \mu^2\mathbf{C}^T\mathbf{C} = \begin{bmatrix} \mathbf{A}_{11} & \mathbf{A}_{12} & \mathbf{A}_{13} \\ \mathbf{A}_{12}^T & A_{22}+\mu^2 & \mathbf{A}_{23} \\ \mathbf{A}_{13}^T & \mathbf{A}_{23}^T & \mathbf{A}_{33} \end{bmatrix}$$

and

$$\mathbf{H}^T\mathbf{R}^{-1}\mathbf{y} + \mu^2\mathbf{C}^T d = \begin{bmatrix} \mathbf{b}_1 \\ b_2 + \mu^2 d \\ \mathbf{b}_3 \end{bmatrix}.$$

When the information matrix is inverted using Cholesky factorization, the large value of μ^2 makes the Cholesky factor diagonal of the middle row approximately equal to μ. Factor elements to the right of the middle diagonal are multiplied by $1/\mu$ and contribute little to the solution. The middle diagonal element is eventually inverse-squared and multiplied by $\mu^2 d$ in the right-hand side to give the desired result $\hat{x}_2 = d$. Estimates of the other states ($\hat{\mathbf{x}}_1$ and $\hat{\mathbf{x}}_3$) are influenced by the constraint through the Cholesky factor row above the middle diagonal. Unfortunately numerics are not nearly as simple when the constraint involves multiple states, and thus precision is lost.

The constraint-as-measurement approach for the QR algorithm is implemented similarly. The scaled constraint is appended to the measurement equations, and orthogonal transformations $\mathbf{T}$ are applied to the augmented matrix. For the MGS QR algorithm, this is represented as

$$\mathbf{T}\begin{bmatrix} \mathbf{0} \\ \mathbf{R}^{-1/2}\mathbf{H} \\ \mu\mathbf{C} \end{bmatrix}\mathbf{x} = \mathbf{T}\begin{bmatrix} \mathbf{0} \\ \mathbf{R}^{-1/2}\mathbf{y} \\ \mu\mathbf{d} \end{bmatrix} - \mathbf{T}\begin{bmatrix} \mathbf{0} \\ \mathbf{R}^{-1/2}\mathbf{r} \\ \mathbf{0} \end{bmatrix}. \tag{7.1-10}$$

The result is the upper triangular $n \times n$ matrix $\mathbf{U}_c$ and n-vector $\mathbf{z}_c$ with lower elements zeroed:

$$\begin{bmatrix} \mathbf{U}_c \\ \mathbf{0} \\ \mathbf{0} \end{bmatrix}\mathbf{x} = \begin{bmatrix} \mathbf{z}_c \\ \mathbf{e}_y \\ \mathbf{e}_d \end{bmatrix} - \begin{bmatrix} \mathbf{e}_c \\ \mathbf{e}_y \\ \mathbf{e}_d \end{bmatrix},$$

which is solved for $\hat{\mathbf{x}} = \mathbf{U}_c^{-1}\mathbf{z}_c$. Unfortunately the large magnitudes in $\mu\mathbf{C}$ can cause a loss of precision in forming and applying the Householder/MGS vector (Björck 1996, section 4.4.3). For the Householder algorithm, column pivoting and row interchanges are required. Row interchanges are not needed for the MGS algorithm or when using Givens rotations, but precision can still be lost if μ is not chosen with care. Björck (1996, section 5.1.5) discusses column pivoting and other numerical issues of the "weighting" method. Numerical problems of the MGS QR constraint-as-measurement approach without column pivoting do not appear to be serious for most values of μ. However, values of μ close to the inverse of the machine precision can cause problems, as demonstrated later in this chapter.

7.1.1.3 Constraints Using the GSVD The GSVD for $m \times n$ matrix $\mathbf{A}$ and $p \times n$ matrix $\mathbf{B}$ is defined (see Björck 1996, section 4.2; Golub and Van Loan 1996, section 8.7.3) for the $(m+p) \times n$ augmented system

$$\mathbf{M} = \begin{bmatrix} \mathbf{A} \\ \mathbf{B} \end{bmatrix}$$

with result

$$\mathbf{A} = \mathbf{U}_A \begin{bmatrix} \mathbf{S}_A \\ \mathbf{0} \end{bmatrix} \mathbf{Z}, \qquad \mathbf{B} = \mathbf{U}_B \begin{bmatrix} \mathbf{S}_B & \mathbf{0} \\ \mathbf{0} & \mathbf{0} \end{bmatrix} \mathbf{Z} \tag{7.1-11}$$

where

$\mathbf{S}_A = \text{diag}[\alpha_1, \alpha_2, \ldots, \alpha_k]$ is a $k \times k$ matrix, $k \le n$,
$\mathbf{S}_B = \text{diag}[\beta_1, \beta_2, \ldots, \beta_q]$ is a $q \times q$ matrix, $q = \min(p, k)$,
$\mathbf{U}_A$ is an orthogonal $m \times m$ matrix,
$\mathbf{U}_B$ is an orthogonal $p \times p$ matrix, and
$\mathbf{Z}$ is a $k \times n$ matrix of rank k.

For most least-squares problems with a unique solution, $m > n$ and $k = n$, but that is not a requirement of the GSVD definition. Further

$$\begin{aligned} &0 \le \alpha_1 \le \ldots \le \alpha_k \le 1, \qquad 0 \le \beta_q \le \ldots \le \beta_1 \le 1 \\ &\alpha_i^2 + \beta_i^2 = 1, \quad i = 1:q \\ &\alpha_i = 1, \qquad\quad i = q+1:k \end{aligned} \tag{7.1-12}$$

and the singular values of $\mathbf{Z}$ equal the nonzero singular values of $\mathbf{M}$. To use the GSVD for constrained least squares, we define $\mathbf{A} = \mathbf{R}^{-1/2}\mathbf{H}$ and $\mathbf{B} = \mathbf{C}$. Solution to the constrained problem is possible because $\mathbf{Z}$ is common to both GSVD equations. One approach finds the state $\hat{\mathbf{x}}$ that minimizes

$$\left\| \begin{bmatrix} \mathbf{R}^{-1/2}\mathbf{H} \\ \mu\mathbf{C} \end{bmatrix} \mathbf{x} - \begin{bmatrix} \mathbf{R}^{-1/2}\mathbf{y} \\ \mu\mathbf{d} \end{bmatrix} \right\|_2 \tag{7.1-13}$$

where μ is a very large number (Golub and Van Loan 1996, section 12.1.5). For convenience we temporarily use the normal equations

$$[(\mathbf{H}^T\mathbf{R}^{-T/2})(\mathbf{R}^{-1/2}\mathbf{H}) + \mu^2\mathbf{C}^T\mathbf{C}]\hat{\mathbf{x}} = \mathbf{H}^T\mathbf{R}^{-1}\mathbf{y} + \mu^2\mathbf{C}^T\mathbf{d}$$

but substitute the GSVD representation to obtain

$$\mathbf{Z}^T \left(\mathbf{S}_A^2 + \mu^2 \begin{bmatrix} \mathbf{S}_B^2 & \mathbf{0} \\ \mathbf{0} & \mathbf{0} \end{bmatrix} \right) \mathbf{Z}\hat{\mathbf{x}} = \mathbf{Z}^T \left([\mathbf{S}_A \quad \mathbf{0}]\mathbf{U}_A^T\mathbf{R}^{-1/2}\mathbf{y} + \mu^2 \begin{bmatrix} \mathbf{S}_B & \mathbf{0} \\ \mathbf{0} & \mathbf{0} \end{bmatrix} \mathbf{U}_B^T\mathbf{d} \right).$$

Assuming that $\mathbf{Z}$ is square ($k = n$) and nonsingular (required for a unique solution),

$$\left(\mathbf{S}_A^2+\mu^2\begin{bmatrix}\mathbf{S}_B^2 & \mathbf{0}\\ \mathbf{0} & \mathbf{0}\end{bmatrix}\right)\mathbf{w}=[\mathbf{S}_A \quad \mathbf{0}]\mathbf{e}+\mu^2\begin{bmatrix}\mathbf{S}_B & \mathbf{0}\\ \mathbf{0} & \mathbf{0}\end{bmatrix}\mathbf{f} \tag{7.1-14}$$

where

$$\boxed{\begin{aligned}\mathbf{w}&=\mathbf{Z}\hat{\mathbf{x}}\\ \mathbf{e}&=\mathbf{U}_A^T\mathbf{R}^{-1/2}\mathbf{y}.\\ \mathbf{f}&=\mathbf{U}_B^T\mathbf{d}\end{aligned}} \tag{7.1-15}$$

As $\mu \to \infty$, the first q diagonal elements of

$$\mathbf{S}_A^2+\mu^2\begin{bmatrix}\mathbf{S}_B^2 & \mathbf{0}\\ \mathbf{0} & \mathbf{0}\end{bmatrix}$$

will approach μ^2 diag$[\beta_1, \beta_2,\ldots, \beta_q]$. This implies that

$$\boxed{w_i = f_i / \beta_i, \qquad i=1:q} \tag{7.1-16}$$

and

$$\boxed{w_i = e_i / \alpha_i = e_i, \qquad i=q+1:k.} \tag{7.1-17}$$

Finally the solution is computed as $\hat{\mathbf{x}}=\mathbf{Z}^T\mathbf{w}$. See Björck (1996, section 5.1.4) or Golub and Van Loan (1996, section 12.1.1) for more detail. The linear equality-constrained least-squares problem is solved using LAPACK routine S/DGGLSE, and the GSVD is computed using S/DGGSVD—although $\mathbf{Z}$ is factored as $\mathbf{Z} = [\mathbf{0} \quad \mathbf{G}]\mathbf{Q}^T$ where $\mathbf{G}$ is a $(k + q) \times (k + q)$ upper triangular matrix and $\mathbf{Q}$ is an $n \times n$ orthogonal matrix. Notice that both the constraint-as-measurement and GSVD approaches allow solution of problems where neither $\mathbf{R}^{-1/2}\mathbf{H}$ nor $\mathbf{C}$ are individually rank n. The GSVD method also allows solution of the inequality problem $\|\mathbf{Cx} - \mathbf{d}\|_2 < \alpha$.

Example 7.1: Constrained Polynomial Least Squares

We again turn to the polynomial problem of Example 5.9 with 101 measurements uniformly spaced in time from $0 \le t_i \le 1$. The Monte Carlo evaluation included 1000 samples of simulated $N(\mathbf{0},\mathbf{I})$ random state vectors $\mathbf{x}$, but no measurement noise was included so that the only errors were due to numerical round-off. The model order n was set to 11, which was high enough to make numerical sensitivity high (condition number $\cong 10^7$) but not so high that all precision of the normal equations was lost: the unconstrained normal equation solution should have nearly two digits of precision. Two constraints were applied using

$$\mathbf{C}=\begin{bmatrix}0&0&0&0&0&1&0&-1&0&0&0\\0&0&0&0&0&0&0&0&0&1&0\end{bmatrix}$$

so that the constraint equation $\mathbf{Cx} = \mathbf{d}$ forces $x_6 - x_8 = d_1$ and $x_{10} = d_2$. For each set of the simulated random $\mathbf{x}$ samples, the $\mathbf{d}$ vector was computed as $\mathbf{d} = \mathbf{Cx}$ rather than constraining all samples to have a common $\mathbf{d}$.

TABLE 7.1: Constrained Least-Squares RMS Solution Accuracy

Solution Method	RMS Error
Normal equations with constraint as measurement: $\mu^2 = 10^5$	0.061
Lagrange multiplier MGS QR using equation (7.1-8)	0.54e-11
MGS QR with constraint as measurement: $\mu = 10^{20}$	0.24e-9
LAPACK constrained least squares (DGGLSE)	0.56e-10
LAPACK generalized SVD (DGGSVD) using equations (7.1-16) and (7.1-17)	0.24e-10

Table 7.1 shows the resulting accuracy of the solution computed as

$$RMS \text{ error} = \sqrt{\frac{1}{1000n}\sum_{i=1}^{1000}\|\hat{\mathbf{x}}_i - \mathbf{x}_i\|_2^2}\,.$$

First, notice that the constrained normal equation solution using Cholesky factorization with $\mu^2 = 10^5$ has little precision: the root-mean-squared (RMS) state error is 0.061 for states that have a standard deviation of 1.0. The accuracy improves to 0.71×10^{-2} when $\mu^2 = 10^4$, and degrades to 0.38 with $\mu^2 = 10^6$. For $\mu^2 > 10^6$ the Cholesky factorization fails. The accuracy improves greatly to 0.73×10^{-6} when the dual-state constraint $x_6 - x_8 = d_1$ is replaced with $x_8 = d_1$. This suggests that the constraint-as-measurement approach using the normal equations should only be used for single-state constraints. This author has successfully used the normal equations for a number of real problems involving single-state constraints.

The Lagrange multiplier MGS QR method with equation (7.1-8) as the constraint was the most accurate method in all cases, retaining about 11 digits of accuracy. When the MGS QR was used with the constraint-as-measurement approach, the accuracy was about two orders of magnitude worse, *provided that* $\mu \geq 10^{20}$. For $\mu = 10^{15}$ the solution lost all accuracy: the RMS error = 0.80. However, a smaller value of $\mu = 10^{10}$ gives about the same accuracy as $\mu = 10^{20}$. Apparently values of μ approximately equal to the machine precision cause numerical problems. It should be noted that the MGS QR implementation did not column pivot. Accuracy may improve when pivoting.

The LAPACK constrained least-squares method DGGLSE was about an order of magnitude less accurate than the MGS QR method using equation (7.1-8), but still much better than the constraint-as-measurement QR. Interestingly, use of the LAPACK GSVD routine DGGSVD with equations (7.1-16) and (7.1-17) reduced the error by a factor of two compared with DGGLSE.

Most of the error was in states x_5 to x_{11} for all solution methods, with the directly constrained state x_{10} always accurate to about 15 digits. This example problem was also tested with alternate constraints and model orders. The relative results were generally as described above. However, we caution that these conclusions should not be treated as generally applicable to all constrained least-squares problems.

7.1.2 Least-Squares with Linear Inequality Constraints (Problem LSI)

In some least-squares problems the states represent physical parameters that must be positive (e.g., electrical resistance, hydraulic conductivity, variances) or are otherwise constrained by inequality constraints. For parameters that must be greater than zero, a nonlinear change of variables (such as a logarithm) can implement the constraint. Somewhat surprisingly, this nonlinear change of variables not only solves the constraint problem, but sometimes also improves iteration convergence in inherently nonlinear problems. However, that approach does not work for all constraints, so a more general approach is needed.

Variants of the least-squares problem with linear inequality constraints include:

1. Problem LSI: $\min_{\mathbf{x}} \|\mathbf{Ax}-\mathbf{b}\|_2$ subject to $\mathbf{d} \le \mathbf{Cx} \le \mathbf{e}$.
2. Problem BLS (simple bounds): $\min_{\mathbf{x}} \|\mathbf{Ax}-\mathbf{b}\|_2$ subject to $\mathbf{d} \le \mathbf{x} \le \mathbf{e}$.
3. Problem NNLS (nonnegative least squares): $\min_{\mathbf{x}} \|\mathbf{Ax}-\mathbf{b}\|_2$ subject to $\mathbf{x} \ge \mathbf{0}$.
4. Problem LSD (least distance): $\min_{\mathbf{x}} \|\mathbf{x}\|_2$ subject to $\mathbf{d} \le \mathbf{Cx} \le \mathbf{e}$.

These problems and solutions are discussed in Lawson and Hanson (1974) and Björck (1996) among other sources. It is not unusual to encounter the BLS and NNLS problems in real applications. In addition to these linear problems, quadratic inequality constraints (problem LSQI) of the form $\min_{\mathbf{x}} \|\mathbf{Ax}-\mathbf{b}\|_2$ subject to $\|\mathbf{Cx} - \mathbf{d}\|_2 \le \alpha$ are also of interest. However, the LSQI problem is somewhat different from the LSI problems and is not discussed here (see Golub and Van Loan 1996).

Björck (1996, section 5.2.2) notes that the general LSI problem is equivalent to a quadratic programming problem. Although many quadratic programming algorithms are available, they are not numerically safe if used directly with the **C** matrix. However, quadratic programming algorithms can be adapted if the problem is transformed to a least-distance problem.

Other approaches for solving the general LSI problem are called *active set algorithms*. The term active set indicates that during the sequential solution a subset of constraints $\mathbf{d} \le \mathbf{Cx} \le \mathbf{e}$ will be active. The active set changes from one iteration to the next, and iterations continue until a minimum cost solution is found. It can be proven that the number of iterations is finite. Gill et al. (1991, chapter 5.2) describe general active set algorithms, and Björck (1996, section 5.2.3) provides a summary. Lawson and Hanson (1974) describe an iterative algorithm for solving the NNLS problem, and show that it can be used for the LSD problem.

Figure 7.1 graphically shows the problem of inequality constraints for a two-dimensional (2-D) least-squares problem. Contours of constant least-squares "cost" are plotted versus x and y. The unconstrained minimum cost occurs at $x = 0.4$, $y = -1.0$, but we specify the constraints $x \ge 0$, $y \ge 0$. If the constraint $y = 0$ is applied first, the minimum is found to be located at $x = -0.17$ (the "diamond"). This now violates the $x \ge 0$ constraint. If the constraint $x = 0$ is applied first, the minimum is found to be located at $y = -0.75$ (the "square"), which violates the $y \ge 0$ constraint. The global solution is actually at located at $x = 0$, $y = 0$. This is found by first applying one of the two constraints to find the minimum. Since the new point violates a constraint, the problem is then solved with both constraints applied. Of course that does not leave any free variables, so the solution is simply the boundary for both constraints. This generally does not happen, so individual

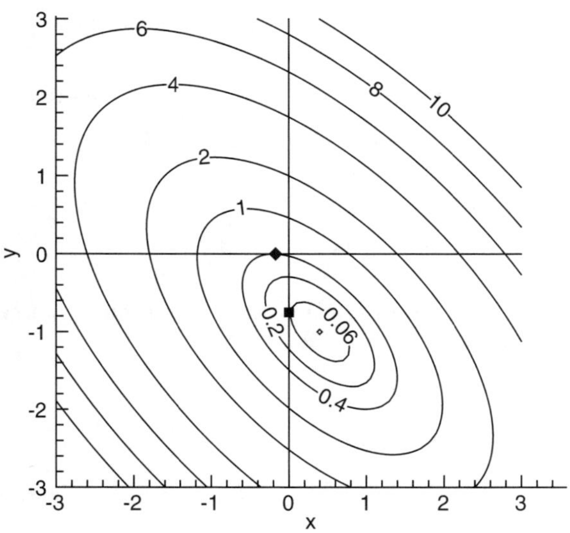

FIGURE 7.1: Constant cost contours for 2-D constrained least-squares problem.

constraints are sequentially applied until all constraints are satisfied. However, as new constraints are applied, old constraints may no longer be necessary, so iterative testing is required until a solution is found. This is the general approach used in active set methods.

7.2 RECURSIVE LEAST SQUARES

Most least-squares problems are solved using batch processing. That is, all measurements are collected and a single solution is computed using all measurements. In some applications it is desirable to process measurements and compute new solutions as measurements (or batches of measurements) become available. One approach to the problem simply accumulates measurements in the information arrays (either normal equations or square-root array of the QR algorithm) and "inverts" the information array when a solution is needed. However, this may not be a particularly efficient method.

Both Householder and MGS QR algorithms may be sequentially updated, but the MGS algorithm can be adapted for sequential updating with only slight modifications. Recall from Chapter 5 that the MGS method is equivalent to

$$\mathbf{T}\begin{bmatrix}\mathbf{0}\\ \mathbf{R}^{-1/2}\mathbf{H}\end{bmatrix}\mathbf{x}=\mathbf{T}\begin{bmatrix}\mathbf{0}\\ \mathbf{R}^{-1/2}\mathbf{y}\end{bmatrix}-\mathbf{T}\begin{bmatrix}\mathbf{0}\\ \mathbf{R}^{-1/2}\mathbf{r}\end{bmatrix}$$
$$\Rightarrow$$
$$\begin{bmatrix}\mathbf{U}\\ \mathbf{0}\end{bmatrix}\mathbf{x}=\begin{bmatrix}\mathbf{z}\\ \mathbf{e}_r\end{bmatrix}-\begin{bmatrix}\mathbf{e}_z\\ \mathbf{e}_r\end{bmatrix}$$

where $\mathbf{T}$ is an orthogonal $(n+m)\times(n+m)$ matrix, and $\mathbf{U}$ is an upper triangular $n \times n$ matrix. However, there is no reason why the upper rows of the arrays must initially be zero. They can be replaced with the information arrays from a previous solution as:

$$\begin{gathered}\mathbf{T}\begin{bmatrix}\mathbf{U}_a \\ \mathbf{R}^{-1/2}\mathbf{H}\end{bmatrix}\mathbf{x}=\mathbf{T}\begin{bmatrix}\mathbf{z}_a \\ \mathbf{R}^{-1/2}\mathbf{y}\end{bmatrix}-\mathbf{T}\begin{bmatrix}\mathbf{e}_a \\ \mathbf{R}^{-1/2}r\end{bmatrix} \\ \Rightarrow \\ \begin{bmatrix}\mathbf{U} \\ \mathbf{0}\end{bmatrix}\mathbf{x}=\begin{bmatrix}\mathbf{z} \\ \mathbf{e}_r\end{bmatrix}-\begin{bmatrix}\mathbf{e}_z \\ \mathbf{e}_r\end{bmatrix}\end{gathered} \quad . \tag{7.2-1}$$

The modified version of the MGS QR algorithm (assuming that **H** and **y** have been previously multiplied by $\mathbf{R}^{-1/2}$) becomes

$$\begin{aligned}
&\text{for } k=1 \text{ to } n \\
&\quad \mathbf{v}=\mathbf{H}(:,k) \\
&\quad \alpha=U_a(k,k)^2+\mathbf{v}^T\mathbf{v} \\
&\quad \lambda=\sqrt{\alpha}\,\operatorname{sgn}(U_a(k,k)) \\
&\quad \varphi=U_a(k,k)+\lambda \\
&\quad U(k,k)=-\lambda \\
&\quad \beta=1/(\alpha+\lambda U_a(k,k)) \\
&\quad \text{for } j=k+1 \text{ to } n \\
&\qquad \gamma=-\beta[\varphi U_a(k,j)+\mathbf{v}^T\mathbf{H}(:,j)] \\
&\qquad U(k,j)=U_a(k,j)+\gamma\,\varphi \\
&\qquad \mathbf{H}(:,j)=\mathbf{H}(:,j)+\gamma\,\mathbf{v} \\
&\quad \text{end loop} \\
&\quad \gamma=-\beta[\varphi\, z_a(k)+\mathbf{v}^T\mathbf{y}] \\
&\quad z(k)=z_a(k)+\gamma\,\varphi \\
&\quad \mathbf{y}=\mathbf{y}+\gamma\,\mathbf{v} \\
&\text{end loop.}
\end{aligned} \tag{7.2-2}$$

Notice in line four that the sign is chosen to maximize the magnitude of φ. Algorithm equation (7.2-2) is in fact the measurement update portion of Bierman's (1977b) Square Root Information Filter (SRIF), which is an "information" version of the Kalman filter and is discussed in Chapter 10.

Alternately, the conditional mean or minimum variance solution to the least-squares problem (presented in Section 4.3) provides a simple algorithm for this sequential processing, repeated here as

$$\hat{\mathbf{x}}=\hat{\mathbf{x}}_a+\mathbf{P}_a\mathbf{H}^T(\mathbf{H}\mathbf{P}_a\mathbf{H}^T+\mathbf{R})^{-1}(\mathbf{y}-\mathbf{H}\hat{\mathbf{x}}_a) \tag{7.2-3}$$

where $\hat{\mathbf{x}}_a$ is the *a priori* solution from a previous set of measurements and $\mathbf{P}_a$ is the associated covariance matrix. It will be shown in Chapter 8 that this is also the measurements update of the Kalman filter. The updated error covariance is computed as

$$\mathbf{P}=\mathbf{P}_a-\mathbf{P}_a\mathbf{H}^T(\mathbf{H}\mathbf{P}_a\mathbf{H}^T+\mathbf{R})^{-1}\mathbf{H}\mathbf{P}_a, \tag{7.2-4}$$

where $\hat{\mathbf{x}}$ and $\mathbf{P}$ become prior information for the next set of measurements. Notice, however, that this solution is in the covariance (not information) domain and can potentially be less accurate than information-based methods because of the subtraction appearing in equation (7.2-4).

The above recursive least-squares methods apply equal weight to measurements collected at different times, so state estimates from a batch solution are identical to those obtained from recursive processing after all measurements have been processed. Eventually the recursive processor will tend to ignore new measurements as the information matrix continues to grow in magnitude. One variant of recursive least squares is designed to have a fading memory or exponential forgetting characteristic. That is, new data are given more weight than old data (see Strejc 1980 or Hayes 1996, section 9.4). This is implemented by introducing a scalar factor $0 < \alpha \le 1$ into equations (7.2-3) and (7.2-4):

$$\boxed{\begin{aligned} &\mathbf{P}_a' = \mathbf{P}_a/\alpha \\ &\mathbf{K} = \mathbf{P}_a'\mathbf{H}^T(\mathbf{H}\mathbf{P}_a'\mathbf{H}^T + \alpha\,\mathbf{R})^{-1} \\ &\hat{\mathbf{x}} = \hat{\mathbf{x}}_a + \mathbf{K}(\mathbf{y} - \mathbf{H}\hat{\mathbf{x}}_a) \\ &\mathbf{P} = (\mathbf{I} - \mathbf{K}\mathbf{H})\mathbf{P}_a' \end{aligned}}. \tag{7.2-5}$$

With $\alpha = 1$ the solution is identical to the equally weighted batch solution, and with $\alpha \ll 1$ only the most recent data are given much weight in the solution. With this modification equation (7.2-3) behaves similarly to the Kalman filter, which also has a fading memory characteristic. (However, the Kalman filter is a far more comprehensive solution to the problem.) It should be noted that α can be dynamically adjusted based on the current measurement residual norm $\|\mathbf{y} - \mathbf{H}\hat{\mathbf{x}}_a\|_2$ so that the recursive processing is adaptive.

7.3 NONLINEAR LEAST SQUARES

Nearly all physical systems are nonlinear at some level, but may appear linear over restricted operating ranges. Hence the linear least-squares methods above are often not directly applicable to physical systems. We now address options for solving nonlinear least-squares problems. The discussion is extensive because it has been this author's experience that nonlinear optimization methods sometimes do not converge well for real problems.

We start by again assuming that Bayesian least squares can be treated as a weighted least-squares problem by including the prior information as a measurement. This avoids the necessity of carrying separate terms for the prior information in equations to follow. The weighted least-squares cost function for nonlinear measurement model $\mathbf{y} = \mathbf{h}(\mathbf{x}) + \mathbf{r}$ can be written as

$$J(\mathbf{x}) = \frac{1}{2}[\mathbf{y} - \mathbf{h}(\mathbf{x})]^T \mathbf{R}^{-1}[\mathbf{y} - \mathbf{h}(\mathbf{x})].$$

This is typically expanded in a Taylor series about $\mathbf{x}_0$ as

$$J(\mathbf{x}) = J(\mathbf{x}_0) + \left.\frac{\partial J}{\partial \mathbf{x}}\right|_{\mathbf{x}_0} (\mathbf{x}-\mathbf{x}_0) + \frac{1}{2}(\mathbf{x}-\mathbf{x}_0)^T \left.\frac{\partial^2 J}{\partial \mathbf{x}\partial \mathbf{x}^T}\right|_{\mathbf{x}_0} (\mathbf{x}-\mathbf{x}_0) + \ldots. \tag{7.3-1}$$

The solution is obtained by truncating the Taylor series at order two and setting the gradient of J to zero as

$$\begin{aligned}\frac{\partial J(\mathbf{x})}{\partial \mathbf{x}} &= \left.\frac{\partial J}{\partial \mathbf{x}}\right|_{\mathbf{x}_0} + (\mathbf{x}-\mathbf{x}_0)^T \left.\frac{\partial^2 J}{\partial \mathbf{x}\partial \mathbf{x}^T}\right|_{\mathbf{x}_0} \\ &= -\tilde{\mathbf{y}}^T(\mathbf{x}_0)\mathbf{R}^{-1}\mathbf{H}(\mathbf{x}_0) + (\mathbf{x}-\mathbf{x}_0)^T\left(\mathbf{H}^T(\mathbf{x}_0)\mathbf{R}^{-1}\mathbf{H}(\mathbf{x}_0) - \tilde{\mathbf{y}}^T(\mathbf{x}_0)\mathbf{R}^{-1}\left.\frac{\partial \mathbf{H}(\mathbf{x})}{\partial \mathbf{x}}\right|_{\mathbf{x}_0}\right) \\ &= \mathbf{0}\end{aligned} \tag{7.3-2}$$

where

$$\tilde{\mathbf{y}}(\mathbf{x}) = \mathbf{y} - \mathbf{h}(\mathbf{x}) \tag{7.3-3}$$

and

$$\mathbf{H}(\mathbf{x}_0) = \left.\frac{\partial \mathbf{h}(\mathbf{x})}{\partial \mathbf{x}}\right|_{\mathbf{x}_0}. \tag{7.3-4}$$

To simplify notation we define $\tilde{\mathbf{y}}_i = \mathbf{y} - \mathbf{h}(\mathbf{x}_i)$ and $\mathbf{H}_i = \left.\frac{\partial \mathbf{h}(\mathbf{x})}{\partial \mathbf{x}}\right|_{\mathbf{x}_i}$ for iteration i to obtain

$$\boxed{\left(\mathbf{H}_i^T\mathbf{R}^{-1}\mathbf{H}_i - \frac{\partial \mathbf{H}_i^T}{\partial \mathbf{x}}\mathbf{R}^{-1}\tilde{\mathbf{y}}_i\right)(\mathbf{x}-\mathbf{x}_i) = \mathbf{H}_i^T\mathbf{R}^{-1}\tilde{\mathbf{y}}_i.} \tag{7.3-5}$$

The *Hessian matrix*

$$\boxed{\frac{\partial^2 J}{\partial \mathbf{x}\partial \mathbf{x}^T} = \mathbf{H}_i^T\mathbf{R}^{-1}\mathbf{H}_i - \frac{\partial \mathbf{H}_i^T}{\partial \mathbf{x}}\mathbf{R}^{-1}\tilde{\mathbf{y}}_i}$$

must be positive definite at $\hat{\mathbf{x}}$ for J to be a minimum. Given an initial estimate $\hat{\mathbf{x}}_0$, *Newton steps* are computed as

$$\boxed{\hat{\mathbf{x}}_{i+1} = \hat{\mathbf{x}}_i + \left(\mathbf{H}_i^T\mathbf{R}^{-1}\mathbf{H}_i - \frac{\partial \mathbf{H}_i^T}{\partial \mathbf{x}}\mathbf{R}^{-1}\tilde{\mathbf{y}}_i\right)^{-1}\mathbf{H}_i^T\mathbf{R}^{-1}\tilde{\mathbf{y}}_i.} \tag{7.3-6}$$

Notice that the second term in the Hessian uses the measurement residual $\tilde{\mathbf{y}}(\hat{\mathbf{x}})$, which has an expected value of zero at the least-squares estimate $\hat{\mathbf{x}}$. Hence the expected value of the Hessian at $\hat{\mathbf{x}}$ is $\mathbf{H}(\hat{\mathbf{x}})^T\mathbf{R}^{-1}\mathbf{H}(\hat{\mathbf{x}})$, which is positive definite if $\mathbf{H}$ is full rank. Since the second term in the Hessian may have negative eigenvalues when far from the optimal $\hat{\mathbf{x}}$, the Hessian many not be positive definite and the stability of iterations equation (7.3-6) cannot be guaranteed. Gauss avoided this problem by assuming that the Hessian term involving the residuals could be ignored. Thus *Gauss-Newton* (GN) iterations are defined using the information matrix $\mathbf{H}_i^T\mathbf{R}^{-1}\mathbf{H}_i$ as:

$$\boxed{\hat{\mathbf{x}}_{i+1} = \hat{\mathbf{x}}_i + (\mathbf{H}_i^T\mathbf{R}^{-1}\mathbf{H}_i)^{-1}\mathbf{H}_i^T\mathbf{R}^{-1}\tilde{\mathbf{y}}_i.} \tag{7.3-7}$$

This iteration or an equivalent form (QR or SVD) is routinely used for nonlinear least-squares problems. The partial derivatives $\mathbf{H}(\mathbf{x})$ and residuals $\tilde{\mathbf{y}}(\hat{\mathbf{x}})$ are computed at the new point $\hat{\mathbf{x}}_{i+1}$ and iterations continue until $\hat{\mathbf{x}}_{i+1} - \hat{\mathbf{x}}_i$ is negligible.

GN iterations also have a significant computational advantage compared with full Newton iterations. Notice that $\mathbf{H}_i^T\mathbf{R}^{-1}\mathbf{H}_i$ of the GN iterations uses the same partial derivatives ($\mathbf{H}_i$) required to compute the cost gradient. If $\mathbf{H}_i$ is computed numerically, each of the n states in $\mathbf{x}$ must be perturbed. However, computation of

$$\frac{\partial \mathbf{H}_i^T}{\partial \mathbf{x}}\mathbf{R}^{-1}\tilde{\mathbf{y}}_i$$

in the full Hessian requires a total of $n(n + 1)/2$ partial derivatives to be numerically computed. This is prohibitively expensive for many problems, and rarely results in a corresponding improvement in Newton iteration convergence.

GN iterations will converge for many mildly nonlinear problems if the initial guess for $\hat{\mathbf{x}}_0$ is reasonably close to truth. However, in many nonlinear applications, either the nonlinearities are great or $\hat{\mathbf{x}}_0$ is far from truth, and initial iterations may increase the least-squares cost function rather than decreasing it. This leads to unnecessary iterations, or in severe cases, the iterations do not converge or converge to a local (rather than global) minimum. Hence modifications of the basic GN iterations are necessary. A number of algorithms for restricting (damping) or modifying the GN steps have been developed. However, before discussing alternate methods for multidimensional problems, we review the problems associated with nonlinear least squares by examining a one-dimensional (1-D) problem.

Example 7.2: 1-D Nonlinear Least Squares

Consider the 1-D nonlinear least-squares problem with a single measurement

$$y = x + ax^2 - 0.5$$

where −0.5 is the measurement noise, $a = -0.04$, and the true value of x is 1.0. The estimation model is $\hat{y} = x + ax^2$ with cost function and derivatives

$$J = [y - (x + ax^2)]^2 / 2$$

$$\frac{dJ}{dx} = -[y - (x + ax^2)](1 + 2ax)$$

$$\frac{d^2J}{dx^2} = (1 + 2ax)^2 - 2ax[y - (x + ax^2)].$$

Figure 7.2 shows the cost function and first derivative as a function of x. Notice the slight downward curvature of the derivative caused by $a = -0.04$. Also shown are two Newton iterations starting with an initial guess $\hat{x} = 2.0$ at point A. The line from point A to the "cost/derivative = 0" line is the derivative tangent line at point A used for the Newton step. Notice that the first Newton iteration computes an excessively large step: $x = 0.052$ at point B. The next step recovers and

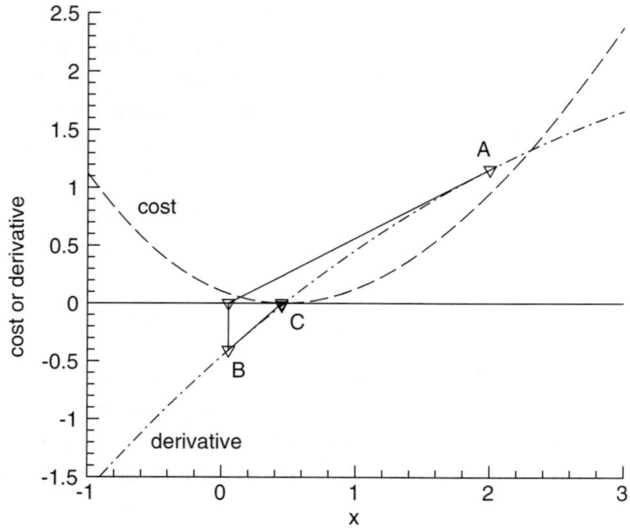

FIGURE 7.2: Least-squares cost, gradient, and Newton iterations for 1-D problem.

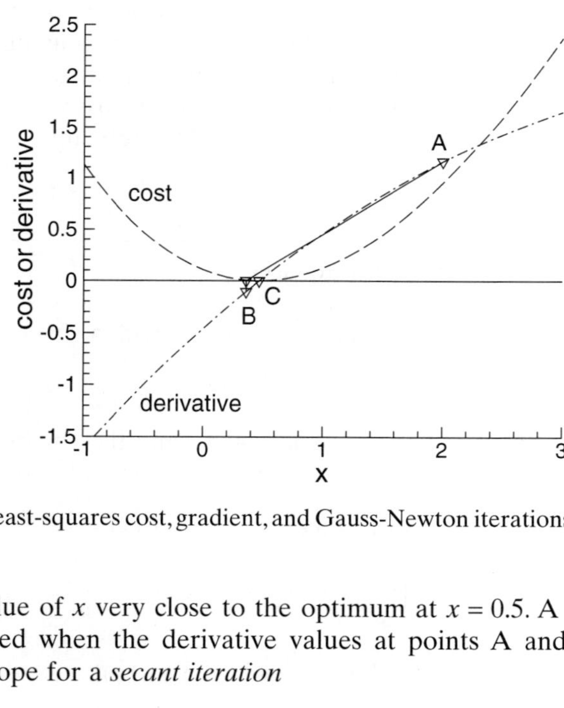

FIGURE 7.3: Least-squares cost, gradient, and Gauss-Newton iterations for 1-D problem.

computes a value of x very close to the optimum at $x = 0.5$. A result nearly as good is obtained when the derivative values at points A and B are used to compute the slope for a *secant iteration*

$$x_{i+1} = x_i - g_i(x_i - x_{i-1})/(g_i - g_{i-1})$$

where $g_i = (dJ/dx)|_{x_i}$. Newton iterations converge q-quadratically (accuracy at least doubles in each iteration) while convergence of secant iterations is less than quadratic but better than linear.

Figure 7.3 is a similar plot showing convergence of GN iterations that use the expected Hessian rather than the true Hessian. The first iteration is much closer to the optimum x than the first Newton iteration because the GN iteration does not model the downward curvature of the derivative. In fact, when $a = -0.1$ Newton iterations are almost unstable: four iterations are required for convergence while one GN iteration nearly converges. When $a = -0.2$, Newton iterations diverge to a cost maximum while GN converges correctly to the minimum. The better convergence of GN iterations cannot be expected as a general result. Dennis and Schnabel (1983, section 10.3) claim that Newton's method is quickly locally convergent on almost all problems, while damped GN methods may be slowly locally convergent when nonlinearities are severe or residuals are large. However, Press et al. (2007) note that inclusion of the second derivative term can be "destabilizing" when residuals are large. This author has not observed faster or more robust convergence of Newton's method compared with GN.

7.3.1 1-D Nonlinear Least-Squares Solutions

We now review the options for finding the minimum of the least-squares cost function for 1-D problems. Some of these methods are applicable to multidimensional problems while others are not. As expected, the options depend on the available information.

1. Only the cost function can be computed: The options are limited and convergence is slow if derivatives are not used.
   a. Find the minimum by bracketing: An initial step is taken in either direction and the cost function is evaluated at the new point. If the new point decreases the cost, continue in that direction until it increases. Then bisect the outer two of the three lowest points to compute the next step, and continue until the improvement or the steps are small.
   b. Use several points to fit the cost to a quadratic or cubic model: Then analytically compute the minimum of the function as the next step.
2. The cost function and an *initial* first derivative (possibly computed numerically) are available: Unfortunately the scale of the model is unknown and thus the magnitude of the step required to minimize the cost must be determined experimentally. This is little more information than case 1. The main benefit is that the first derivative can be used in the quadratic or cubic cost model for method 1b.
3. The cost function and first derivatives at each step (possibly computed numerically) are available:
   a. Take "steepest-descent" steps in the negative derivative direction to find the point where the derivative is zero: Again the magnitude of the step required to zero the derivative must be determined experimentally. A modified version of method 1b can be used at each iteration (using two or three last values) where the quadratic or cubic model fits the derivative rather than cost.
   b. Use the derivative values in secant iterations. This is generally the best method when first derivatives are available.

4. The cost function, first derivatives at each step, and an initial second derivative or "expected second derivative" (possibly computed numerically) are available: This is sufficient information for Newton or GN iterations. Unfortunately full Newton or GN steps can overshoot the minimum and increase the cost function for highly nonlinear problems: The algorithms are not globally convergent. As noted above, it is necessary in these cases to damp the nominal step to find the minimum. Step halving or backtracking using a quadratic or cubic model can be used to find the optimal step.

7.3.2 Optimization for Multidimensional Unconstrained Nonlinear Least Squares

Some of the above methods can be adapted for multidimensional problems, and some of these lead to multiple algorithms. Optimization algorithms for solving multidimensional least-squares problems include:

7.3.2.1 "Cost-Only" Algorithms Cost-only algorithms, such as the simplex method (see Press et al. 2007, section 10.5), are used for general optimization problems, but they are not efficient for unconstrained least-squares problems.

7.3.2.2 "Cost Function and Gradient" Algorithms These tend to be used more for general optimization problems than for nonlinear least squares because the measurement model on which the least-squares method is based allows easy computation of the Fisher information matrix (expected Hessian) using the same partial derivatives used for the gradient. However, these cost plus gradient methods are still of interest because one is the basis of a hybrid algorithm that also uses the Fisher information matrix.

1. Steepest descent: Stepping in the negative gradient direction will always reduce the cost function for small steps, but the cost will eventually increase as step size grows. In addition to not knowing the optimal step length, negative gradient steps can be very inefficient when state variables are correlated. Since the gradient direction changes with the point of evaluation, the steepest descent steps can follow a path that is far from direct (see Press et al. 2007, section 10.8). This is not a problem when variables are uncorrelated, but can result in very slow convergence for real systems.
2. Conjugate gradient: The conjugate gradient method attempts to overcome the inefficiencies of the steepest descent method by taking steps in directions that are conjugate (orthogonal) to previous steps. This concept was discussed in Section 5.5.3 on Krylov space methods. For quadratic models it should theoretically converge to the optimum in a finite number of steps that are less than or equal to the number of unknowns. In practice this is not guaranteed if conditioning is poor. Conjugate gradient methods do not require computation and storage of an approximate Hessian matrix, but instead accumulate information on past search directions using sequences of two vectors. The Fletcher-Reeves-Polak-Ribiere algorithm is recommended by Press et al. (2007, section 10.8).
3. Secant methods (variable metric or quasi-Newton): These are multidimensional versions of the simple secant method. They use past values of the gradient to accumulate an approximate Hessian matrix (Dennis and Schnabel

1983, section 9.2; Press et al. 2007, section 10.9). The algorithm due to Broyden-Fletcher-Goldfarb-Shanno (BFGS) is regarded as the most robust, and has the advantage of automatically taking into account deviations of the true cost from the ideal quadratic model. However, the basic algorithm does not use the approximate Hessian (Fisher information matrix) easily computed in least-squares problems. A variant of the method due to Dennis et al. (1981a) uses the information matrix as the starting point and only computes corrections using the gradients. Dennis and Schnabel (1983, section 10.3) provide a good summary.

7.3.2.3 "Cost, Gradient, and Initial Hessian" Algorithms These are the most successful algorithms for multidimensional least-squares problems. Because robust implementations of the algorithms are quite complex, we only provide summaries and refer to references and supplied code for further details.

1. Newton or GN with step-halving or backtracking: Because the Newton and GN methods are not globally convergent, it is often necessary to constrain or "damp" the steps. In step-halving, the steps are repeatedly reduced by a factor of two until the cost function is reduced. Backtracking uses the most recent two or three cost evaluations in the step direction to compute a quadratic or cubic model that is then solved for the minimum (Dennis and Schnabel 1983, section 6.3; Press et al. 2007, section 9.7.1). Backtracking works well for many, if not most, nonlinear problems, but does not work for all. The backtracking algorithm with other refinements is implemented in the OPTL_BACK.F subroutine of the supplied code on the Wiley FTP site (ftp://ftp.wiley.com/public/sci_tech_med/least_squares).
2. Trust region constraint of Newton or GN steps: Rather than simply damp Newton or GN steps, a more sophisticated approach attempts to compute the optimum step within a region where the quadratic model can be "trusted" (Dennis and Schnabel 1983, section 6.4). Hence the step may deviate from Newton or GN directions. Basically the trust region defines the step length, and the quadratic model is used to determine the optimal step direction. The step length is adaptively modified in each iteration, based on performance of the previous iteration. The solution is obtained by adding a constant to the diagonal elements of the Hessian before solving for the step; that is,

$$\Delta\mathbf{x} = -\left[\frac{\partial^2 J}{\partial\mathbf{x}\partial\mathbf{x}^T} + \lambda\mathbf{I}\right]^{-1}\frac{\partial J}{\partial\mathbf{x}}. \tag{7.3-8}$$

   Unfortunately there is no direct method for computing the optimal λ to meet the specified step length, so two approximate methods (hook or dogleg step) are used. See Dennis and Schnabel (1983) for details.
3. Rectangular trust region: In another version of the trust region method due to Vandergraft (1987), the trust region is defined as a "rectangular" region where the step for each component of $\mathbf{x}$ is constrained within symmetric limits defined from the computed variable uncertainty (inverse of Fisher information matrix). The solution is computed using an iterative constrained least-squares method. This approach has the advantage that it can easily include hard physical constraints on variables in $\mathbf{x}$, but it is relatively slow and sometimes does not perform quite as well as alternatives.

4. Levenberg-Marquardt (LM) method: This algorithm computes steps that vary smoothly between GN and steepest descent steps. It is sometimes considered a trust region algorithm because the solution is obtained using a modified version of equation (7.3-8). However, the expected Hessian (Fisher information matrix) of the GN method is used rather than the true Hessian. Other implementations multiply λ by the diagonals of the Fisher information matrix (not $\mathbf{I}$) to scale the gradient component. Most implementations do not directly constrain the step length. Rather λ is initially set to zero, and the GN step is accepted if the GN step reduces the cost function sufficiently. If the GN step does not reduce the cost, λ is increased until a cost reduction is obtained. A successful factor λ on one iteration is remembered for the next iteration and adjusted accordingly. Good implementations must scale the states appropriately, and use somewhat complex logic to handle all conditions. The LM algorithm is generally regarded as one of the best optimization methods for nonlinear least squares (Dennis and Schnabel 1983, section 10.2; Press et al. 2007, section 15.5.2). This author found that the LM algorithm sometimes computed solutions when backtracking failed. However, it has been difficult to find initial values and limits on λ that work for all problems. Hence these parameters are sometimes treated as problem-dependent variables. The algorithm is implemented in the OPTL_LM.F subroutine of the supplied code.
5. Enhanced secant method: As indicated previously, the secant method of Dennis et al. (1981a) uses the information matrix as the starting point, and approximates the full Hessian using the sequence of gradients. This approach has the potential advantage of better convergence on large residual or very nonlinear problems, and is the basis of the NL2SOL code (Dennis et al. 1981b) that incorporates model trust region concepts to improve global convergence. The code is quite complex, and we refer to references for more information. Although claimed to be quite robust, NL2SOL may not work properly with automated editing logic (described in Section 7.4.2) that can change the number of measurements edited in each iteration.

The optimization methods usually regarded as best for solving nonlinear least-squares problems are GN with backtracking, LM, and N2LSOL. The following sections provide more details on the backtracking and LM methods, and stopping criteria. Three realistic examples follow.

7.3.2.4 Backtracking Method Backtracking algorithms are discussed in Press et al. (2007, section 9.7.1) and Dennis and Schnabel (1983, section 6.3). Optimization based on backtracking always starts with a full GN step, and if that does not reduce the cost function, the step length is reduced to find the minimum cost point along the GN step direction. The basic optimization method is described below, but other logic may be required to handle abnormal cases such as failure of the model function to compute $J(\hat{\mathbf{x}}_k)$.

1. Initialize variables: state $\hat{\mathbf{x}}_0$, convergence threshold ε, maximum GN iterations k_{max}, and maximum backtracking iterations j_{max}
2. Loop k for GN iterations ($k = 1, 2,\ldots, k_{max}$)

a. Evaluate $J(\hat{\mathbf{x}}_k), \mathbf{b}_k = -\partial J(\hat{\mathbf{x}})/\partial \hat{\mathbf{x}}$ and Fisher information matrix $\mathbf{A}_k$ (or equivalent for QR or SVD methods)
b. Compute GN step $\Delta\mathbf{x}_{GN} = \mathbf{A}_k^{-1}\mathbf{b}_k$
c. Set $\lambda_1 = 1$ and $\Delta\mathbf{x}_C = \Delta\mathbf{x}_{GN}$
d. Loop j for backtracking ($j = 1, 2, \ldots, j_{\max}$)
   i. Compute $\hat{\mathbf{x}}_{k,j} = \hat{\mathbf{x}}_k + \lambda_j \Delta\mathbf{x}_C$ (Note: $\Delta\mathbf{x}_C$ may be modified for constraints)
   ii. Compute $g_\lambda = \partial J(\hat{\mathbf{x}}_{k,j})/\partial \lambda_j = -\Delta\mathbf{x}_C^T \mathbf{b}_k$ (used in backtracking calculation)
   iii. Evaluate $J(\hat{\mathbf{x}}_{k,j})$. If function evaluation fails or measurement editing is excessive, set $\lambda_{j+1} = 0.5\lambda_j$ and cycle j loop
   iv. If $J(\hat{\mathbf{x}}_{k,j}) < J(\hat{\mathbf{x}}_k)$, accept $\hat{\mathbf{x}}_{k+1} = \hat{\mathbf{x}}_{k,j}$ and exit loop
   v. If $\sqrt{|\Delta\mathbf{x}_C^T \mathbf{b}_k|/n} < \varepsilon$ and $J(\hat{\mathbf{x}}_{k,j}) < 1.01 J(\hat{\mathbf{x}}_k)$, then accept $\hat{\mathbf{x}}_{k+1} = \hat{\mathbf{x}}_{k,j}$ and exit loop
   vi. If $j = j_{\max}$, declare failure and exit optimization
   vii. Compute λ_{j+1} to minimize $J(\hat{\mathbf{x}}_k + \lambda_{j+1}\Delta\mathbf{x}_C)$, where $0.2\lambda_j < \lambda_{j+1} < 0.7\lambda_j$
e. End backtracking loop j
f. If $\sqrt{|\Delta\mathbf{x}_C^T \mathbf{b}_k|/n} < \varepsilon$, accept step $\hat{\mathbf{x}}_{k+1}$ and exit loop
g. If $k > 3$ and $\sqrt{|\Delta\mathbf{x}_C^T \mathbf{b}_k|/n} < 2\varepsilon$ and $J(\hat{\mathbf{x}}_{k-2}) < 1.01 J(\hat{\mathbf{x}}_k)$, accept $\hat{\mathbf{x}}_{k+1}$ and exit loop
h. If $k = k_{\max}$, declare failure and exit optimization

3. End iteration loop

The convergence test $\sqrt{|\Delta\mathbf{x}_C^T \mathbf{b}_k|/n} < \varepsilon$ will be explained in the next section. Alternate stopping criteria are also discussed.

The backtracking algorithm computes the step length to minimize cost using the initial cost $f_0 = J(\hat{\mathbf{x}}_k)$, derivative $f_0' = g_\lambda$ evaluated at $\hat{\mathbf{x}}_k$, J evaluated at another candidate point $f_1 = J(\hat{\mathbf{x}}_k + \lambda_1 \Delta\mathbf{x}_C)$, and optionally J at a third point $f_2 = J(\hat{\mathbf{x}}_k + \lambda_2 \Delta\mathbf{x}_C)$. If more than two backtracks have been used without success, λ_1 and λ_2 represent the two most recent values.

When only f_0 and f_1 are available, the backtracking algorithm finds the minimum of the quadratic model

$$f(\lambda) = f_0 + f_0'\lambda + a\lambda^2 \tag{7.3-9}$$

where f_0 and f_0' are known, and a is computed from

$$\boxed{a = (f_1 - f_0 - f_0'\lambda_1)/\lambda_1^2.} \tag{7.3-10}$$

Then the minimum is computed as

$$\boxed{\lambda_{\min} = -f_0'/(2a)} \tag{7.3-11}$$

provided that a is positive. If a is negative the quadratic model is invalid and $\lambda_{\min} = 0.5\lambda_1$ is used.

If the first backtracking is not successful, a third point is used in the cubic model

$$f(\lambda) = f_0 + f_0'\lambda + a\lambda^2 + b\lambda^3. \tag{7.3-12}$$

Coefficients are computed from

$$\begin{bmatrix} f_1 - f_0 - f_0'\lambda_1 \\ f_2 - f_0 - f_0'\lambda_2 \end{bmatrix} = \begin{bmatrix} \lambda_1^2 & \lambda_1^3 \\ \lambda_2^2 & \lambda_2^3 \end{bmatrix} \begin{bmatrix} a \\ b \end{bmatrix} \tag{7.3-13}$$

which has the solution

$$\begin{bmatrix} a \\ b \end{bmatrix} = \frac{1}{\lambda_2 - \lambda_1} \begin{bmatrix} \lambda_2 / \lambda_1^2 & -\lambda_1 / \lambda_2^2 \\ -1/\lambda_1^2 & 1/\lambda_2^2 \end{bmatrix} \begin{bmatrix} f_1 - f_0 - f_0'\lambda_1 \\ f_2 - f_0 - f_0'\lambda_2 \end{bmatrix}. \tag{7.3-14}$$

The minimum point is computed as

$$\lambda_{\min} = \frac{-a + \sqrt{a^2 - 3bf_0'}}{3b}. \tag{7.3-15}$$

(The alternate solution $\lambda = -\left(a + \sqrt{a^2 - 3bf_0'}\right)/(3b)$ corresponds to a maximum.) If $a^2 - 3bf_0' < 0$ the cubic model is not valid and the quadratic solution is used with λ_2, the most recent value. In all cases $\lambda_{\min}$ is constrained as $0.2\lambda_2 < \lambda_{\min} < 0.7\lambda_2$ (assuming that $\lambda_2 < \lambda_1$) to prevent excessively large or small steps at any one backtracking iteration.

7.3.2.5 LM Method The LM method is described in Press et al. (2007, section 15.5.2), Dennis and Schnabel (1983, section 10.2), and Moré (1977). As indicated previously, it computes steps that are a combination of GN and negative gradient directions. As with backtracking, it always tests the GN step first, and if that is not successful, it computes the GN version of step equation (7.3-8):

$$\Delta\mathbf{x} = [\mathbf{A} + \lambda\,\mathbf{D}]^{-1}\mathbf{b} \tag{7.3-16}$$

where

$$\mathbf{A} = E\left[\frac{\partial^2 J}{\partial \mathbf{x} \partial \mathbf{x}^T}\right], \quad \mathbf{D} = diag(\mathbf{A}), \quad \text{and} \quad \mathbf{b} = -\frac{\partial J}{\partial \mathbf{x}}.$$

Notice that the diagonal matrix multiplied by λ contains the diagonals of **A**. This automatically adjusts for different state scaling. One advantage of the LM method compared with backtracking GN steps is that $\mathbf{A} + \lambda\mathbf{D}$ will be nonsingular even when **A** is less than full rank. This may potentially allow a solution when GN iterations fail.

When using the QR algorithm (with arrays **U** and **z**), the equivalent to equation (7.3-16) is obtained by first computing $\mathbf{D} = diag(\mathbf{U}^T\mathbf{U})$, which can be done without forming $\mathbf{U}^T\mathbf{U}$. Then orthogonal transformations are used to make the augmented equations upper triangular:

$$\mathbf{T}\begin{bmatrix} \mathbf{U} \\ \mathbf{E} \end{bmatrix} \Delta\mathbf{x} = \mathbf{T}\begin{bmatrix} \mathbf{z} \\ \mathbf{0} \end{bmatrix} \Rightarrow \begin{bmatrix} \mathbf{U}^* \\ \mathbf{0} \end{bmatrix} \Delta\mathbf{x} = \begin{bmatrix} \mathbf{z}^* \\ \mathbf{0} \end{bmatrix} \tag{7.3-17}$$

where **E** is diagonal with diagonal elements $\mathbf{E}_{ii} = \sqrt{\lambda_{k,j}\mathbf{D}_{ii}}$. This can be implemented as a series of rank-1 updates or using the standard MGS version of the QR algorithm. Then $\mathbf{U}^*\Delta\mathbf{x} = \mathbf{z}^*$ is solved for $\Delta\mathbf{x}$ by back-solving.

The initial value of λ should always be small. Unfortunately it is difficult to specify a minimum value that will work for all problems. Usually λ_0 is set to 0.001 or 0.0001, but values from 0.01 to 0.00001 can be used. It is also necessary to specify a maximum value to prevent endless iterations. Typically λ_{max} is set to 10 or 100. If a tested value of λ does not reduce J, λ is increased by a factor of 10 and tested again. When a successful value of λ has been found, it is divided by 100 and used as the initial λ for the next outer (GN) iteration. The basic optimization method is described below. As before, other logic may be required to handle abnormal cases.

1. Initialize variables: $\hat{\mathbf{x}}_0$, ε, λ_0, λ_{max}, j_{max} and k_{max}
2. Loop k for GN iterations ($k = 1, 2,\ldots, k_{max}$)
   a. Evaluate $J(\hat{\mathbf{x}}_k)$, $\mathbf{b}_k = -\partial J(\hat{\mathbf{x}})/\partial \hat{\mathbf{x}}$ and $\mathbf{A}_k$ (or equivalent)
   b. Compute GN step $\hat{\mathbf{x}}_{k,1} = \hat{\mathbf{x}}_k + \mathbf{A}_k^{-1}\mathbf{b}_k$
   c. Set $\lambda_{k,0} = \max(0.01\lambda_{k-1}, \lambda_0)$
   d. Loop j for LM iterations ($j = 1, 2,\ldots, j_{max}$)
      i. Evaluate $J(\hat{\mathbf{x}}_{k,j})$. If function evaluation fails, set $\hat{\mathbf{x}}_{k,j+1} = 0.5(\hat{\mathbf{x}}_k + \hat{\mathbf{x}}_{k,j})$ and cycle loop
      ii. If $J(\hat{\mathbf{x}}_{k,j}) < J(\hat{\mathbf{x}}_k)$, set $\hat{\mathbf{x}}_{k+1} = \hat{\mathbf{x}}_{k,j}$, $j_{max} = 5$, and exit loop
      iii. If $j = j_{max}$ or $\lambda_{k,j} \geq \lambda_{max}$, then
         1. Set $j_{max} \Leftarrow j_{max} - 1$
         2. If $\sqrt{|\Delta\mathbf{x}_C^T\mathbf{b}_k|/n} < \varepsilon$, set $\hat{\mathbf{x}}_{k+1} = \hat{\mathbf{x}}_k$
         3. Exit j and k loops
      iv. Set $\lambda_{k,j} = 10\lambda_{k,j-1}$
      v. Compute $\hat{\mathbf{x}}_{k,j+1} = \hat{\mathbf{x}}_k + [\mathbf{A}_k + \lambda_{k,j}\mathbf{D}]^{-1}\mathbf{b}_k$
   e. End j loop
   f. If $\sqrt{|\Delta\mathbf{x}_C^T\mathbf{b}_k|/n} < \varepsilon$, accept $\hat{\mathbf{x}}_{k+1} = \hat{\mathbf{x}}_{k,j+1}$ and exit k loop
   g. If $k > 3$ and $\sqrt{|\Delta\mathbf{x}_C^T\mathbf{b}_k|/n} < 2\varepsilon$ and $J(\hat{\mathbf{x}}_{k-2}) < 1.01J(\hat{\mathbf{x}}_k)$, accept $\hat{\mathbf{x}}_{k+1} = \hat{\mathbf{x}}_{k,j+1}$ and exit k loop
   h. If $k = k_{max}$, declare failure and exit optimization
3. End k loop

7.3.3 Stopping Criteria and Convergence Tests

There are three criteria that can be used to determine whether nonlinear least-squares iterations have converged:

1. The change in the cost function between iterations is "small."
2. The computed cost gradient is "small."
3. The norm of the computed step change in $\mathbf{x}$ is "small."

As you may suspect, these criteria are related. Consider the GN step equation (7.3-7)

$$\hat{\mathbf{x}}_{i+1} - \hat{\mathbf{x}}_i = \mathbf{A}^{-1}\mathbf{b}$$

where $\mathbf{A} = \mathbf{H}_i^T \mathbf{R}^{-1} \mathbf{H}_i$ and $\mathbf{b} = \mathbf{H}_i^T \mathbf{R}^{-1} \tilde{\mathbf{y}}_i$ is the negative gradient ($\mathbf{b} = -\nabla_x J = -\partial J / \partial \mathbf{x}$). Using the methods of Sections 6.2 and 6.3, it can be shown that the modeled change in the cost function for the GN update is

$$2(J_i - J_{i+1}) = (\hat{\mathbf{x}}_{i+1} - \hat{\mathbf{x}}_i)^T \mathbf{b} = \mathbf{b}^T \mathbf{A}^{-1} \mathbf{b}. \tag{7.3-18}$$

(If using the QR algorithm to transform $\mathbf{R}^{-1/2}\mathbf{y} = \mathbf{R}^{-1/2}(\mathbf{Hx} + \mathbf{r})$ to $\mathbf{z} = \mathbf{Ux} + \mathbf{e}$, then substitute $\mathbf{b}^T\mathbf{A}^{-1}\mathbf{b} = \mathbf{z}^T\mathbf{z}$.) For nonlinear problems the actual cost reduction may not be as great as indicated, but the model becomes more linear near the optimal solution. Notice that the term $\mathbf{b}^T\mathbf{A}^{-1}\mathbf{b}$ is a quadratic form that can be interpreted as the squared-norm of the gradient normalized by the expected Hessian. When divided by the number of states (n), it is a scale-free measure of the squared gradient magnitude. It also has another interpretation. Recall from Chapter 4 that for J defined as the log of the probability density function,

$$\mathbf{A}^{-1} = \left[E\left(\frac{\partial^2 J}{\partial \mathbf{x} \partial \mathbf{x}^T} \right) \right]^{-1}$$

is the Cramér-Rao lower bound on the state error covariance matrix; that is, $E[\tilde{\mathbf{x}}_{i+1}\tilde{\mathbf{x}}_{i+1}^T] \geq \mathbf{A}^{-1}$. Thus

$$\mathbf{b}^T \mathbf{A}^{-1} \mathbf{b} = (\mathbf{x}_{i+1} - \mathbf{x}_i)^T \mathbf{A} (\mathbf{x}_{i+1} - \mathbf{x}_i) \tag{7.3-19}$$

can be interpreted as the squared-norm of the step in **x** normalized by the estimate error covariance. Ideally we want to stop when the step **x** is small compared with the estimate uncertainty. Hence

$$\boxed{\sqrt{(\mathbf{x}_{i+1} - \mathbf{x}_i)^T \mathbf{A} (\mathbf{x}_{i+1} - \mathbf{x}_i) / n} = \sqrt{\mathbf{b}^T \mathbf{A}^{-1} \mathbf{b} / n} < \varepsilon} \tag{7.3-20}$$

is often used where as a test for convergence, where ε is typically set to 0.01 in mildly nonlinear, simple problems with partial derivatives computed analytically. This can be interpreted to mean that the RMS step for the states is less than 1% of the $1 - \sigma$ uncertainty (statistical error). However, in more nonlinear problems, or when the computations are extensive (e.g., orbit determination), or when partial derivatives are computed numerically, or when model errors are significant, it may be necessary to set ε as large as 0.2 to account for the fact that the model equations are not completely accurate.

You may wonder whether this test can be used with damped GN or LM methods. Even though the actual step may not follow the GN model, we still want the idealized step to be small. Hence the $\sqrt{\mathbf{b}^T \mathbf{A}^{-1} \mathbf{b} / n} < \varepsilon$ can still be used provided that **A** does not include the $\lambda\mathbf{I}$ term of LM. For some problems it will still be difficult to satisfy this convergence test. In these cases it is useful to apply an alternate convergence test based on the change in residuals observed for the actual (not GN) steps:

$$J_i - J_{i+1} < \mu J_i \tag{7.3-21}$$

where μ is a small number such as 0.001. Notice that ε is testing a square root while μ is testing a sum-of-squared normalized residuals, so the magnitudes should not be equal. Equation (7.3-21) should only be applied when $\sqrt{\mathbf{b}^T \mathbf{A}^{-1} \mathbf{b} / n}$ has been small for two iterations but not small enough to satisfy equation (7.3-20). Finally, when automated editing logic of Section 7.4 is used, the convergence test should only be applied when comparing metrics computed using the same number of non-edited measurements.

We now present examples demonstrating these nonlinear least-squares methods and issues.

Example 7.3: Falling Body with Range Tracking

The following simple example demonstrates the differences between the true cost function and the modeled quadratic contours for a simple 2-D example. It also demonstrates the difference between Newton and GN performance.

Consider the scenario shown in Figure 7.4. A motionless body at initial location $x = 30$ m and $y = 110$ m is allowed to drop under the influence of gravity. A sensor located at the origin measures range to the body every 0.05 s for 3.0 s (61 measurements total) with a $1 - \sigma$ accuracy of 4 m. Lines between sensor and body are shown in the figure every 0.25 s. The unknowns to be estimated from the range data are the initial positions $x(0)$ and $y(0)$. The model for body coordinates as a function of time is

$$x(t) = x(0) \qquad y(t) = y(0) - 0.5gt^2$$

where $g = 9.80\ \text{m/s}^2$ and time t is measured in seconds. The measurement model is

$$r(t) = \sqrt{x(t)^2 + y(t)^2}$$

with measurement partials

$$\left.\frac{\partial r}{\partial x(0)}\right|_t = \frac{x(t)}{r(t)}, \quad \left.\frac{\partial r}{\partial y(0)}\right|_t = \frac{y(t)}{r(t)}.$$

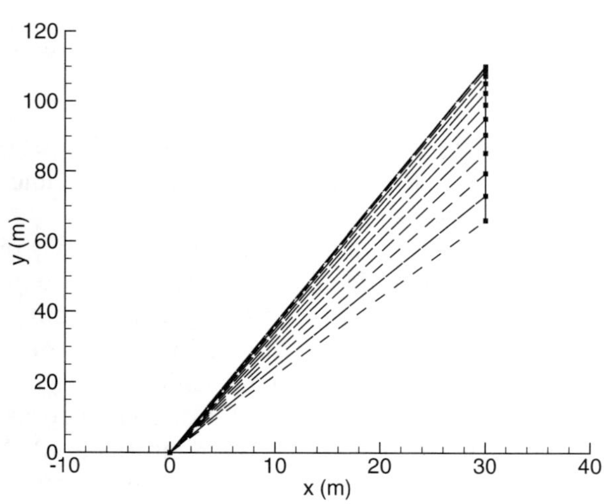

FIGURE 7.4: Falling body scenario.

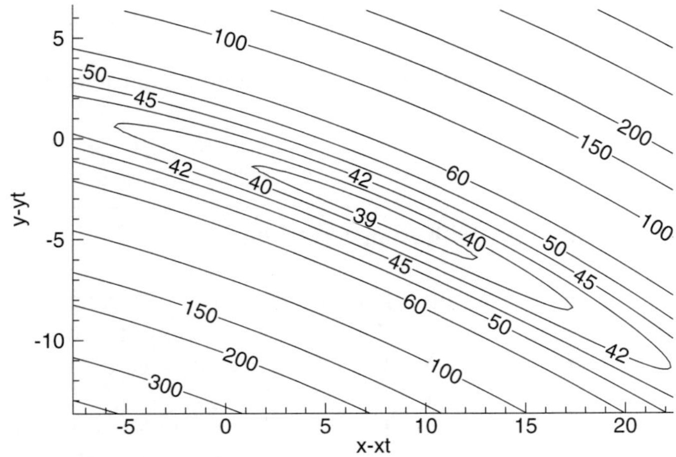

FIGURE 7.5: Falling body true constant-cost contours.

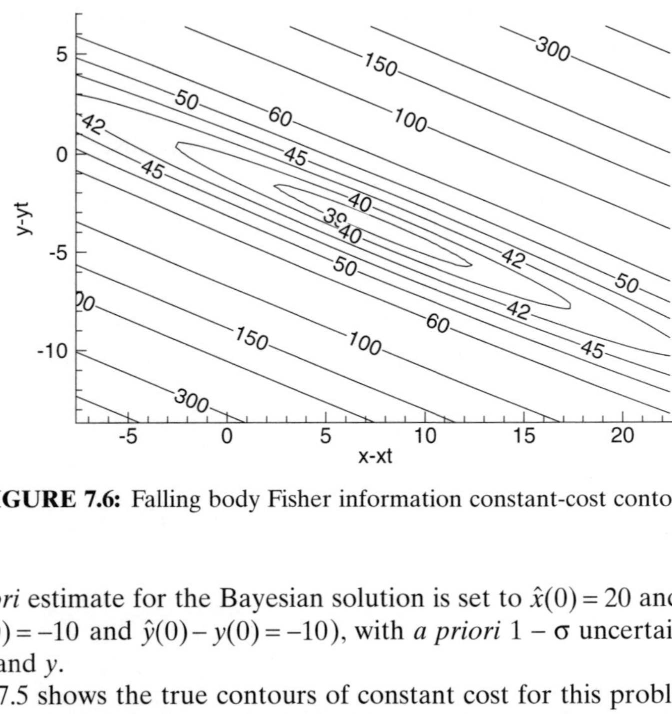

FIGURE 7.6: Falling body Fisher information constant-cost contours.

The *a priori* estimate for the Bayesian solution is set to $\hat{x}(0) = 20$ and $\hat{y}(0) = 100$ ($\hat{x}(0) - x(0) = -10$ and $\hat{y}(0) - y(0) = -10$), with *a priori* $1 - \sigma$ uncertainty of 10 m in both x and y.

Figure 7.5 shows the true contours of constant cost for this problem, plotted as a function of distance in x and y from the true values. Notice that the "ellipses" are highly elongated and curved. Also notice that the minimum cost point is located at about $x - xt = 7.3$ m and $y - yt = -3.7$ m, which deviates from zero because of measurement noise and prior information.

Figure 7.6 shows the contours of constant cost computed using a quadratic model based on the expected Hessian (Fisher information). Notice that the contours are ellipses, as expected. Figure 7.7 shows the contours of constant cost computed using a quadratic model based on the actual Hessian. These are also

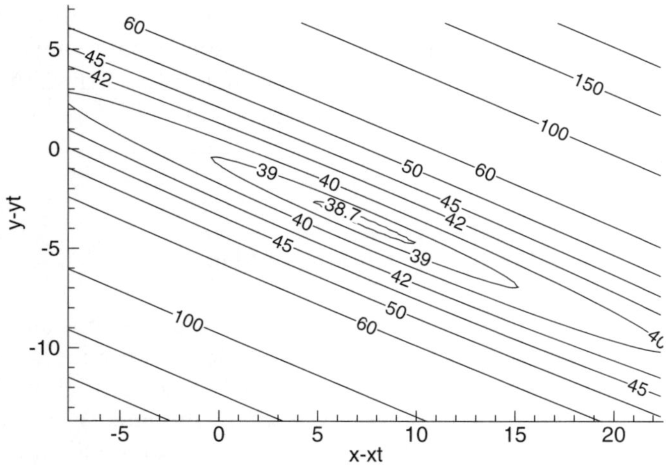

FIGURE 7.7: Falling body Hessian constant-cost contours.

TABLE 7.2: Falling Body Least-Squares GN Iterations

Iteration	Cost Function J	Convergence Metric	Estimated $x(0)$	Estimated $y(0)$
0	40.69	1.124	20.00	100.00
1	38.77	0.272	35.94	107.16
2	38.66	0.025	37.06	106.47
3	38.66	0.005	37.30	106.36
4	—	—	37.35	106.34

TABLE 7.3: Falling Body Least-Squares Newton Iterations

Iteration	Cost Function J	Convergence Metric	Estimated $x(0)$	Estimated $y(0)$
0	40.69	1.093	20.00	100.00
1	38.78	0.264	34.22	107.71
2	38.66	0.064	36.80	106.62
3	38.66	0.008	37.27	106.38
4	—	—	37.35	106.34

ellipses, but the surface is flatter than that of either the Fisher information surface or the true cost surface.

Table 7.2 shows details of the GN iterations. Notice that four iterations are required to achieve convergence because curvature of the true cost ellipses in Figure 7.5 is not handled by the quadratic model. Hence the GN directions are always slightly in error, but the discrepancy is not so large that backtracking was required. When using full Newton iterations (Table 7.3), the flatter cost surface slows convergence slightly, but does not increase the number of required iterations. The results can, of course, be different with other initial conditions or measurement noise sequences, but general behavior is similar. To summarize, the nonlinear nature of the model slows convergence somewhat (compared with the one iteration required for a linear model), but there is no significant difference between Newton and GN iterations for this problem.

Example 7.4: Passive Angle-Only Tracking of Ship

This example is a modified version of a real least-squares application that uses simulated random measurement noise. The example was described in Section 3.1. As shown in Figure 7.8, a target ship at an initial distance of about 20 nautical miles (nmi) is moving at a constant velocity of 18.79 knots (kn) at a southeast heading of 115 degrees from north. The tracking ship is initially moving directly north (+y) at a constant velocity of 20 kn. The tracking ship measures bearing from itself to the target every 15 s, with simulated random measurement noise of 1.5 degrees $1-\sigma$. The four unknowns to be estimated are the target initial position and velocity for assumed constant velocity motion. Since the target position and velocity are not observable without a maneuver by the tracking ship, at 15 min it turns 30 degrees to the left in the direction opposite the target ship velocity. Tactically this maneuver may not be ideal, but turning away from the target motion increases the bearing change and improves observability. The encounter duration is 30 min, resulting in a total of 121 measurements.

For least-squares estimation purposes, the target initial position is "guessed" to be at (east, north) = (−3, 30) nmi with (east velocity, north velocity) = (0, 0) kn. The initial $1-\sigma$ position and velocity uncertainties are set to 15 nmi and 20 kn, respectively. The guesses are in error from the true values by (−1.0, +10.0) nmi

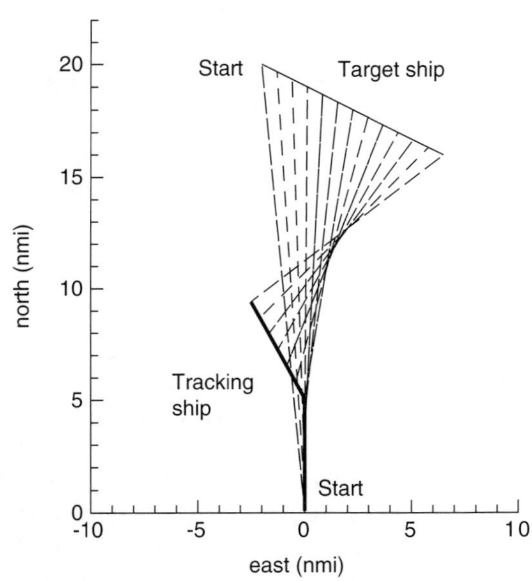

FIGURE 7.8: Ship passive tracking scenario.

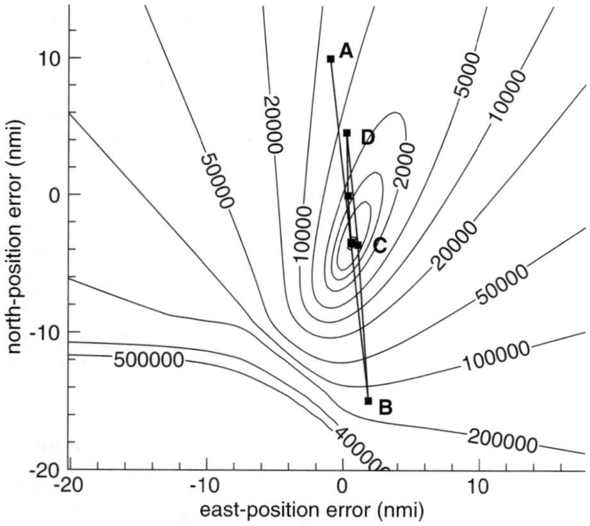

FIGURE 7.9: Ship passive tracking true constant-cost contours and Levenburg-Marquardt (LM) position optimization path.

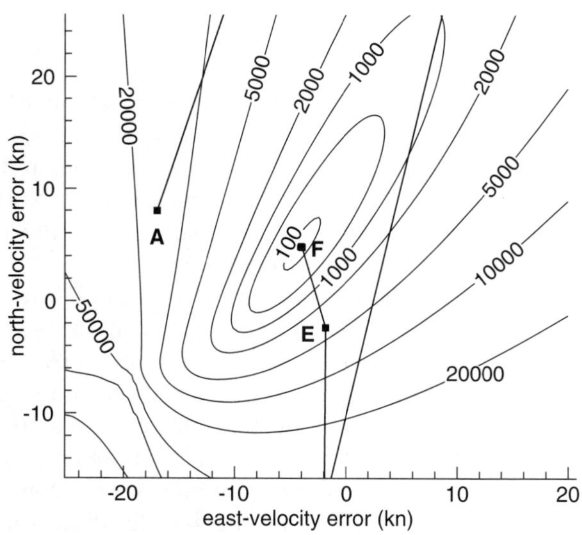

FIGURE 7.10: Ship passive tracking true constant-cost contours and Levenburg-Marquardt (LM) velocity optimization path.

and (–17.0, +8.0) kn, respectively. Since the bearing measurements are accurate to 1.5 degrees $1-\sigma$, the initial east position can be estimated to an accuracy of about 0.5 nmi $1-\sigma$ at the actual range, but there is no prior information on either target range or velocity. The target range and perhaps speed can be guessed from sonar information, but it is not accurate.

Figure 7.9 shows the least-squares cost function as a function of the deviation in east and north position from the true values. Figure 7.10 is the corresponding

TABLE 7.4: Passive Tracking Levenburg-Marquardt (LM) Iterations

Iteration	Residual RMS	Convergence Test	No. LM Steps (λ)	East Error (nmi)	North Error (nmi)	East Velocity Error (kn)	North Velocity Error (kn)
1	12.50	98.61	3 (0.10)	–1.0	10.0	–17.0	8.0
2	6.15	45.86	2 (0.10)	1.86	–14.93	9.13	84.74
3	3.63	27.66	1 (0.01)	1.07	–3.62	11.00	35.41
4	3.39	26.16	0	0.22	4.59	–1.95	–18.29
5	0.86	3.93	0	0.32	–0.03	–1.85	–2.36
6	0.71	0.260	0	0.57	–3.42	–3.96	4.81
7	0.71	0.002	0	0.58	–3.52	–4.09	4.86

TABLE 7.5: Passive Tracking *A Posteriori* State 1-σ and Correlation Coefficients

State	$1-\sigma$	East-Position	North-Position	East-Velocity	North-Velocity
East-position	0.86 nmi	1.0	—	—	—
North-position	7.93 nmi	–0.9859	1.0	—	—
East-velocity	6.53 kn	–0.9966	0.9922	1.0	—
North-velocity	9.82 kn	0.9681	–0.9940	–0.9741	1.0

plot for velocity error. The actual minimum occurs at an offset of (+0.58, –3.52) nmi and (–4.09, +4.86) kn from the true values because of measurement noise. In position plot 7.3-8 the velocity is set to the estimated value, and in velocity plot 7.3-9 the position is set to the estimated value. Notice that model nonlinearities cause the constant-cost contours to deviate significantly from symmetric ellipses—except when very near the minimum.

Also shown on the plots is the path taken by the LM algorithm in the seven iterations required for convergence (Table 7.4) when using $\lambda_0 = 0.0001$ and $\varepsilon = 0.01$ in convergence test equation (7.3-20). In each plot point A is the starting location, B is the first iteration, C is the second iteration, and so on. Notice that the estimated *y* position greatly overshoots the true position on the first iteration (B), recovers at iteration three (C), but then overshoots again (D) before closing on the correct values. The erratic position estimates were accompanied by even more erratic velocity estimates. Notice that the first iteration computes a north-velocity error of +84.7 kn, then +35.4 kn on the second iteration, and -18.2 kn on the third iteration. This unstable behavior occurs even though the LM algorithm restricts the step sizes on the first three iterations.

The condition number for this four-state problem is only 206, which is small compared with the 10^{16} range of double precision. However, the solution correlation coefficient computed from the *a posteriori* error covariance matrix are very close to 1.0, as shown in Table 7.5. Also notice that the $1-\sigma$ uncertainty is much greater in the north-south direction than in east-west This is not surprising because the bearing measurement are sensitive to east-west position but the lack of a range measurement leads to large north-south uncertainty.

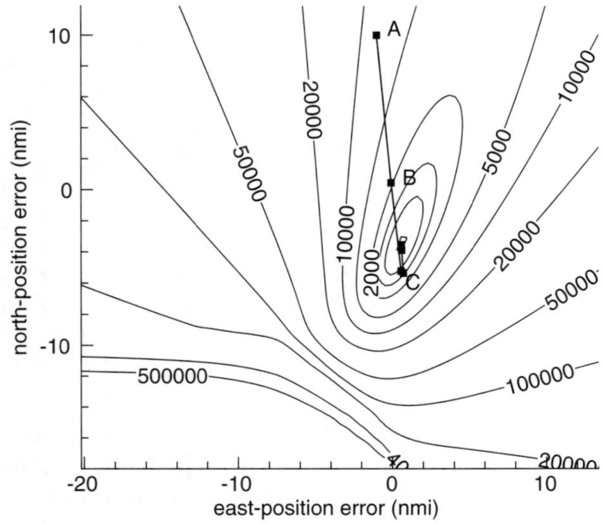

FIGURE 7.11: Ship passive tracking true constant-cost contours and backtracking position optimization path.

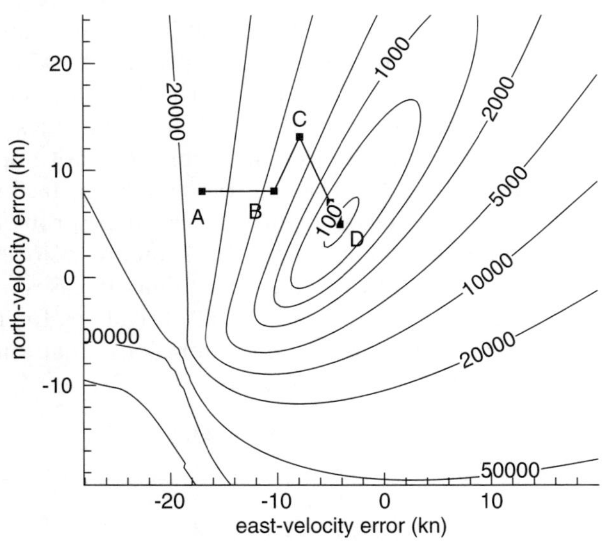

FIGURE 7.12: Ship passive tracking true constant-cost contours and backtracking velocity optimization path.

Figures 7.11 and 7.12 show comparable results for the backtracking algorithm. Notice that even with only one backtrack step for the first two iterations (Table 7.6), the position estimates did not significantly overshoot the minimum, and the velocity estimates were relatively stable. Backtracking converged to the same estimate as LM in one less iteration.

TABLE 7.6: Passive Tracking Backtracking GN Iterations

Iteration	Residual RMS	Convergence Test	No. Backtracks (Step)	East Error (nmi)	North Error (nmi)	East Velocity Error (kn)	North Velocity Error (kn)
1	12.50	98.61	1 (0.20)	−1.0	10.0	−17.0	8.0
2	8.15	64.22	1 (0.38)	−0.07	0.44	−10.27	8.01
3	4.14	32.29	0	0.57	−5.22	−7.87	13.10
4	1.05	6.15	0	0.72	−5.35	−5.03	6.93
5	0.71	0.522	0	0.61	−3.88	−4.30	5.40
6	0.71	0.009	0	0.58	−3.54	−4.10	4.90

The observed difference in behavior between the two algorithms is due to the difference in approaches. Backtracking tries to find the minimum-cost point along the GN direction, while LM increases the weight on the scaled gradient step relative to the GN step until the cost is less than the starting point. Thus LM may accept a step that overshoots the minimum. Also gradient weighting may result in a step that is not in the optimal direction.

The results shown in these figures should not be interpreted as representative for all cases. Small differences in initial conditions, scenario (e.g., tracking ship turns, target heading) or measurement noise sequences can change the results dramatically. For example, moving the initial east position by +2 nmi completely changes the LM optimization path in that position is more unstable and 10 iterations are required for convergence. For the same case backtracking is still well behaved and converges in six iterations. This result does not imply that backtracking is a better algorithm. Backtracking may converge faster than LM in many cases, but occasionally it does not converge. For most problems LM seems to converge reliably to the minimum, even if it is sometimes slower.

Finally we mention results for three other algorithms. In cases when considerable computation is required to evaluate partial derivatives for the **H** matrix, computation can be reduced by taking "inner iterations" that only update the residual vector, not **H**. That is, steps of the form

$$\hat{\mathbf{x}}_{i,j+1} = \hat{\mathbf{x}}_{i,j} + \mathbf{C}_i[\mathbf{y} - \mathbf{h}(\hat{\mathbf{x}}_{i,j})] \tag{7.3-22}$$

where

$$\mathbf{C}_i = (\mathbf{H}_i^T \mathbf{R}^{-1} \mathbf{H}_i)^{-1} \mathbf{H}_i^T \mathbf{R}^{-1}$$

is evaluated using $\mathbf{H}_i = \partial\mathbf{h}(\mathbf{x})/\partial\mathbf{x}$ computed at $\hat{\mathbf{x}}_i$ of the last GN iteration. Then the state is initialized as $\hat{\mathbf{x}}_{i,1} = \hat{\mathbf{x}}_i$ and iterations (7.3-22) for $j = 1, 2, \ldots$ continue until the cost function no longer decreases significantly. If the QR or the SVD algorithms are used to compute GN steps, the equation for $\mathbf{C}_i$ must be modified accordingly. This technique can reduce total computations if nonlinearities are mild, but the ship passive tracking example is not one of those cases. In early iterations equation (7.3-22) took steps that were too large, and in later iterations it did not reduce the cost function. Only in the third iteration did it take a step reducing cost, and even then the total number of backtracking GN iterations was unchanged compared with backtracking GN only.

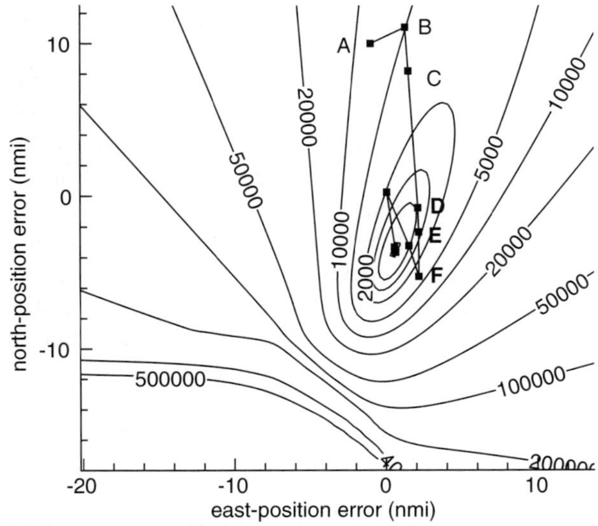

FIGURE 7.13: Ship passive tracking true constant-cost contours and NL2SOL position optimization path.

The NL2SOL code was also tested on this example. It should be noted that NL2SOL expects the "Jacobian" computed by the user subroutine CALCJ to be defined as $-\mathbf{H}_i\mathbf{R}^{-1/2}$ rather than $+\mathbf{H}_i\mathbf{R}^{-1/2}$. Figure 7.13 shows the NL2SOL position estimate path: notice that is less direct than for LM or backtracking GN. The velocity estimates (not shown) were even more erratic than for LM. A total of 14 gradient evaluations were required for convergence, although the accuracy at the 11th evaluation was comparable to the converged LM or backtracking GN iteration. Even so, NL2SOL does not appear to provide faster convergence than LM or backtracking GN methods for this example.

An alternate method for initializing the iterations was also tested, since the first GN iterations (when velocity is poorly known) are the most troublesome. For the first two iterations, the H matrix was a linearization about the measured bearing angles using the estimated range. That is, the partial derivatives

$$\frac{\partial b}{\partial e} = \frac{\Delta n}{\Delta e^2 + \Delta n^2}, \quad \frac{\partial b}{\partial n} = \frac{-\Delta e}{\Delta e^2 + \Delta n^2}$$

where b is bearing angle, e = east position, and n = north position, were replaced with

$$\frac{\partial b}{\partial e} = \frac{\cos b}{\sqrt{\Delta e^2 + \Delta n^2}}, \quad \frac{\partial b}{\partial n} = \frac{\sin b}{\sqrt{\Delta e^2 + \Delta n^2}}$$

for two iterations. Later iterations reverted to the first equation. This approach is somewhat similar to methods used operationally. The state estimate computed using this change was more accurate than the "guess" used to initialize the normal GN iterations, with the result that no backtracking was required. However, it did not reduce the total number of GN iterations—in fact, the number increased by one. In other cases the approach can reduce the number of GN steps.

It should be mentioned that the convergence would be much worse if the target made a maneuver. A model error of this type would either result in convergence to an incorrect solution, or would prevent convergence.

To summarize, this example demonstrated algorithms for solving nonlinear least-squares problems using a highly nonlinear poorly observable case. The GN backtracking method converged in the fewest iterations for the tested initial conditions, and behavior was stable. The LM method was less stable—exhibiting large oscillations about the true state values—but it still converged in only one iteration more than required by backtracking. The NL2SOL code required even more iterations for convergence. These results may or may not be repeated for other problems, but the general behavior for LM and backtracking GN has been observed by the author on other real problems.

Example 7.5: Low Earth Orbit (LEO) Satellite Orbit Determination

The previous example was low order, very nonlinear, and poorly observable. Our final example is slightly higher order (seven) and less nonlinear, but the model is imperfect. The problem was described as Case #1 (CHAMP) in Section 3.4 on orbit determination. At 350 km altitude, the CHAMP spacecraft is significantly affected by atmospheric drag, and models of the atmosphere are only approximations to reality. Because of this modeling problem, it is necessary to estimate the drag "ballistic coefficient" in addition to the six orbit elements.

Table 7.7 summarizes the LM iterations when determining the CHAMP orbit for 2005 days 140.0 to 142.6 using actual range, azimuth, elevation, and range rate measurements from three stations. The column labeled "Weighted Residual RMS" is computed as

$$\sqrt{\frac{1}{m}\sum_{i=1}^{m}(\tilde{y}_i/\sigma_i)^2}$$

TABLE 7.7: 2.6-Day CHAMP OD LM Iterations—Real Data, Harris-Priester Atmosphere

Outer Iteration	Weighted Residual RMS (254 meas.)	No. Meas. Edited	Convergence Metric	No. LM Steps (λ)	LM (λ)
1	4.279	0	24.2	0	—
2	1.663	0	4.50	0	—
3	1.359	2	0.290	0	—
4	1.314	3	0.450	1	1e-6
5	1.314	3	0.191	1	1e-6
6	1.314	3	0.346	5	1e-2
7	1.314	3	0.277	1	1e-3
8	1.314	3	0.291	5	1.0
9	1.314	3	0.217	4	100.0

where $\tilde{y}_i$ is the measurement residual and σ_i is the specified weighting standard deviation. The editing threshold is set to $5 - \sigma$, where the editing algorithm is explained in Section 7.4.2. The initial orbit elements in this case are fairly accurate, so the GN steps behave well for several outer (partial derivative) iterations, and the fourth step is essentially converged. However, the convergence metric equation (7.3-20) is equal to 0.450, which is much larger than desired to stop the iterations. Also the GN step did not reduce the least-squares cost function, so one LM step with $\lambda = 10^{-6}$ was used. On subsequent iterations the convergence metric was still large and additional LM steps using larger values of λ were required. Basically the modeling errors are comparable in magnitude to the measurement noise, and the GN steps do not improve the cost function even though the model believes that they should. Measurement biases were not the problem: they were calibrated using tracking from multiple satellites. Also estimation of range biases did not improve convergence.

The results in Table 7.7 were obtained using the Harris-Priester atmospheric model, but use of another model (MSIS: http://modelweb.gsfc.nasa.gov/atmos/msise.html) with daily values of solar flux measurements did not improve the convergence—in fact it was worse.

The convergence behavior observed in this example is not unusual when modeling errors are present. To study the problem further, 2 days of simulated measurements were generated at 120 s intervals (when CHAMP was visible at ground stations), and Gaussian random noise with $1 - \sigma$ values listed in Section 3.4 was added. The initial estimate of Cartesian orbit elements was perturbed randomly with $1 - \sigma$ perturbations of 1 km in position and 0.050 m/s in velocity. To introduce a modeling error, the simulated orbit was generated using the MSIS atmospheric model but the least-squares estimator used the Harris-Priester model. Table 7.8 lists the resulting LM iterations. The behavior when using backtracking GN steps is very similar to that of Table 7.7, although one additional iteration is required to obtain convergence. However, there is no significant difference between LM and backtracking GN for this example. Because the simulated initial errors are much larger than those of Table 7.7, seven iterations are required for convergence and only one of those required an LM step. Notice that the first step greatly increases the epoch position and velocity errors, but still

TABLE 7.8: Simulated 2-Day CHAMP OD LM Iterations

Iteration	Weighted Residual RMS (168 meas.)	No. Meas. Edited	Convergence Test	No. LM Steps (λ)	RSS Epoch Position Error (m)	RSS Epoch Velocity Error (m/s)
1	10861	1	47466	0	3262	0.050
2	4666	0	22286	1 (1e-5)	142310	132.80
3	2866	0	13938	0	135920	135.91
4	2668	0	12988	0	14296	14.376
5	258	0	1265	0	8868	7.607
6	18.2	0	88.8	0	74	0.076
7	1.04	0	1.584	0	17	0.035
8	0.99	0	0.027	0	18	0.031

reduces the measurement residuals because the first measurements are 4h from the epoch. Four more iterations are required before the errors are reduced to those of the initial estimate. The scaled condition number for this problem is approximately 2000, which is large enough to slow convergence, but not large enough to cause serious observability problems.

Convergence behavior in the last few iterations is the most significant difference between simulated and real results: the slow convergence encountered with the real data is not observed when using simulated data. This suggests significant modeling problems. The atmospheric drag model is the most likely error source, but systematic measurement errors are another possibility. As noted previously, models of physical systems often do not accurately match true behavior.

7.4 ROBUST ESTIMATION

The term *robust estimation* describes an estimator that is insensitive to small errors in assumptions or models used to design the estimator. These errors may be due to an incorrect model structure, incorrect choice of states, failure to estimate all relevant states, or errors in modeling the distribution or characteristics of measurement noise. When applied to least-squares estimation, robust estimation usually refers to an estimator that is insensitive to "small" measurement modeling errors. Here small can mean fractionally small departures from the modeled distribution for all points, or large errors occurring more often than expected by the modeled probability density; that is, outlying measurements. The first interpretation is of less concern. However, outliers can significantly degrade the accuracy of state estimates because the least-squares method inherently applies more weight to large residuals than to small ones. There are generally two approaches for dealing with outliers. The first approach gives less weight to large residuals than done by the least-squares algorithm, and the second method attempts to edit large residuals from the solution.

7.4.1 De-Weighting Large Residuals

It is often observed in real systems that large values of random variables appear more often than expected from the Gaussian distribution, which is explicitly or implicitly the basis for many least-squares solutions. There are various reasons for this phenomenon, and the problem occurs with many different systems. Tukey observed that least-squares estimators lose much of their statistical efficiency for even slightly non-Gaussian errors. That is, the number of observations needed to obtain a desired level of reliability is much larger for non-Gaussian errors. Robust estimation techniques were originally developed to improve statistical efficiency.

M-estimates (maximum likelihood type), L-estimates (linear combinations of order statistics), R-estimates (based on rank transformation), and others are designed to minimize sensitivity to errors in the assumed measurement distribution (see Mosteller and Tukey 1977; Huber 1981; Siegel 1982; Hampel et al. 1986; Cressie 1991; Maronna et al. 2006; Press et al. 2007). M-estimates are the class most relevant for model parameter estimation. This approach attempts to find a maximum likelihood solution for the specified (non-Gaussian) error distribution. Example distributions include two-sided (absolute value) exponential, Lorentzian, or combinations of various distributions. The solution is obtained by applying a weighting function to the measurement residuals. M-estimates are called local if the assumed distribution depends only on the difference between measured and modeled quantities, as in $[y_i - h_i(\hat{\mathbf{x}})]/\sigma_{yi}$. M-estimate problems usually involve nonlinear equations because the weighting is a function of the estimate.

Tukey, Andrews, and Hampel weighting functions increase rapidly with residual magnitude up to a threshold, but then the weights are fixed. Huber (1981) weighting is based on a merger of Gaussian and two-sided exponential distributions. Here residual-squared weighting is used in the cost function up to a certain η-sigma value, and then weighting increases linearly above that threshold; that is, for measurement y_i,

$$\begin{aligned}
&\text{if } |y_i - \mathbf{H}_i\hat{\mathbf{x}}| / \sigma_{yi} < \eta, \text{ then} \\
&\quad J_i = J_{i-1} + \frac{1}{2}\left\|(y_i - \mathbf{H}_i\hat{\mathbf{x}}) / \sigma_{yi}\right\|_2^2 \\
&\text{else} \\
&\quad J_i = J_{i-1} + \frac{1}{2}\left[2\eta|y_i - \mathbf{H}_i\hat{\mathbf{x}}| / \sigma_{yi} - \eta^2\right] \\
&\text{end if}
\end{aligned} \qquad (7.4\text{-}1)$$

Partial derivatives of J must follow equation (7.4-1). Notice that both J and the first derivative are continuous at the transition point η. This algorithm has the desirable feature that it behaves like a least-squares estimator provided that the residuals do not exceed $\eta\sigma_{yi}$ where η is typically set to about 3. Hence all the statistical properties of least-squares estimation—such as using the inverse Fisher information matrix to define the error covariance—are mostly applicable. However, large residuals do not influence the solution as much as in a least-squares approach. Notice that the transition in weighting depends on $\hat{\mathbf{x}}$, so solution methods must be iterative and the weighting will change slightly from one iteration to the next. This can potentially cause convergence problems.

Measurement de-weighting approaches can make estimators more efficient and can provide acceptable solutions when the outlying measurement errors are not "too large." However, state estimates can be erroneous when large residuals appear, and de-weighting does not directly address the fact that erroneous measurements should not be included in the solution. We now describe a simple method that directly edits outlying measurements and has been shown to work remarkably well on a wide set of batch least-squares problems.

7.4.2 Data Editing

An automated measurement editing algorithm has been routinely used with much success in satellite orbit determination since at least the early 1970s (Chin et al. 1972). This author has used the method in a wide variety of nonlinear batch least-squares applications (e.g., Gibbs et al. 1975; Gibbs 1977, 1981, 1992, 1995, 1997, 2008), and has found it to be very effective and to function with few problems.

The algorithm is based on the assumption that all measurements with errors above a specified statistical threshold should be edited. The problem arises in trying to determine the magnitude of the measurement error, since that is not known.

Statistical tests are instead based on the difference between actual measurements and expected measurements as computed from a model. It is assumed that an initial estimate of the state vector $\hat{\mathbf{x}}_0$ is available, as necessary when iteratively solving nonlinear least-squares problems. The normalized measurement residual for each measurement (with independent measurement errors) is then defined as

$$\phi_i = |y_i - \mathbf{h}_i(\hat{\mathbf{x}}_{j-1})| / \sigma_{yi}, \tag{7.4-2}$$

where $\hat{\mathbf{x}}_{j-1}$ is the state estimate from the previous GN iteration. If measurement errors are known to be correlated, all measurements residuals should be transformed to an uncorrelated form using $\mathbf{z} = \mathbf{R}^{-1/2}[\mathbf{y} - \mathbf{h}_i(\hat{\mathbf{x}}_{j-1})]$ before edit testing. This transformation may not be practical in all cases, but when correlation is characteristic of a local sensor, simultaneous measurements from a single sensor can be transformed on a set-by-set basis.

Significance of the residual is tested to determine if

$$\phi_i > \eta \cdot RMS_{j-1} \tag{7.4-3}$$

where η is the desired significance level (typically 4 or 5), and RMS_{j-1} is the root-mean-squared residual from iteration j–1, computed as

$$RMS_{j-1} = \sqrt{\frac{1}{m}\sum_{i=1}^{m}[y_i - \mathbf{h}_i(\hat{\mathbf{x}}_{j-1})]^2 / \sigma_{yi}^2}. \tag{7.4-4}$$

When $\phi_i > \eta \cdot RMS_{j-1}$, the measurement is not included in the solution. Since the residual RMS from the previous iteration is included in the test, large residuals on early iterations (caused by inaccurate initial estimates of the states) will not cause excessive editing. In later iterations the test becomes much tighter as RMS_{j-1} drops. However, RMS_0 does not exist, so it is necessary that the user specify an initial value characterizing an upper bound on errors in $\hat{\mathbf{x}}_0$. For example, when $\hat{\mathbf{x}}_0$ is reasonably well known, $RMS_0 = 200$ might be used, while in cases where $\hat{\mathbf{x}}_0$ is a wild guess, $RMS_0 = 10^4$ might be more appropriate. It is generally not advisable to use extremely large values unless preprocessing has eliminated "1000 – σ" measurements that occasionally appear because of communication errors or other malfunctions: a single undetected 1000-σ measurement can cause complete failure of iterative least-squares methods.

If the nonlinear iterations are restarted from a previous solution, it may be desirable to use the residual RMS from the last iteration of the previous solution, or to use a value slightly larger if the system model has been modified.

When a sensor simultaneously obtains multiple measurements as a set (e.g., range, azimuth, and elevation), the definition of ϕ_i is sometimes modified to edit-test the measurement set:

$$\phi_i = \sqrt{(\mathbf{y}_i - \mathbf{H}_i\hat{\mathbf{x}}_{j-1})^T \mathbf{R}_i^{-1}(\mathbf{y}_i - \mathbf{H}_i\hat{\mathbf{x}}_{j-1}) / m_i}, \tag{7.4-5}$$

where $\mathbf{y}_i$ is the vector of measurements at time t_i, $\mathbf{R}_i$ is the assumed measurement noise covariance for $\mathbf{y}_i$, and m_i is the number of measurements in $\mathbf{y}_i$. Again $\mathbf{R}_i$ is usually assumed to be diagonal unless known otherwise, so equation (7.4-2) can be implemented as a sum of normalized squared quantities.

Notice that editing procedure described above is inherently a nonlinear process. Even with linear problems the measurements selected for editing may change after the first iteration. Thus batch processing should be iterative, regardless of whether or not the measurement models are nonlinear. It should also be noted that editing logic of any type (including manual editing) works best when many measurements are available to provide redundancy in observing the system state. If measurements are few and sparse, it can be difficult to determine whether large residuals in early iterations are outliers or whether the state estimate is inaccurate.

It is critically important that the measurements edited do not change when testing the change in the least-squares cost function for candidate steps of a nonlinear least-squares solver. Automated editing logic should be enabled when computing partial derivatives during optimization, but indices of the edited measurements should be saved and used to edit measurements when testing candidate steps. If this is not done, the cost function may change because of the change in editing rather than because of the improvement in state estimates. This can fool logic designed to determine optimal step size. It is relatively easy to implement this edit tracking logic if measurements are stored in a structure and the edit indices are stored as part of the structure.

One might attempt to improve the basic algorithm by separate testing of measurements using "mean" (and possibly trend) residual models for each sensor or other logical grouping of measurements. That is, the mean and standard deviation of residuals for each sensor are computed separately. Then ϕ_i in test equation (7.4-3) is modified to remove the sensor residual mean of the previous iteration before testing, and RMS_{j-1} is replaced with the sample standard deviation for the sensor as computed on the previous iteration. Unfortunately this approach can lead to "thrashing" of the editing logic from one iteration to the next, with the result that iterations never converge. While well-designed algorithms for iterative solutions of nonlinear least-squares problems can guarantee that the residual RMS of each iteration is smaller than that of the previous iteration, the sample residual standard deviations for subsets of measurements will not necessarily be smaller. This is particularly evident in early iterations. For that reason edit testing based on the overall residual RMS is much more robust. If subset testing is desired, it can be performed after a converged solution is obtained. Then when additional editing is indicated, the nonlinear iterations can be repeated with selected measurements manually removed.

7.5 MEASUREMENT PREPROCESSING

While the editing tests of the previous section are usually quite effecting in eliminating anomalous measurements, it is generally preferable to eliminate outliers before use in least-squares processing. This is so important that preprocessing is routinely used in many operational systems, such as the ground systems for GOES and Global Positioning System (GPS) satellites. Preprocessing often has other functions besides measurement editing, although they are generally complementary with the editing function. Some of these functions include:

1. Subsampling and smoothing of high-rate data: Some sensors collect measurements at a sampling interval much shorter than time constants of the dynamic system being modeled. For example, ground-to-vehicle range measurements may be available at 10 samples per second even though dynamics of many vehicles (particularly satellites) are measured in minutes. Hence it is not desirable to process all these measurements directly in a least-squares estimator or

Kalman filter because computation time would increase greatly without any significant improvement in accuracy. Instead, an affine bias/trend model is used to fit groups of measurements over small intervals, and the mean value at the midpoint is used as the measurement for subsequent estimation. Since many raw measurements are averaged, the effective measurement noise of the averaged measurement is reduced by approximately $1/\sqrt{m}$ where m is the number of measurements in the group. Another advantage of this method is that the editing tests of the previous section can be used to identify and remove isolated outliers from the averaged measurement.

2. Filtering to detect jumps: In systems where measurements are continuously available at a high sampling rate, batch averaging of the previous method can be replaced with a digital filter having a "low-pass" characteristic. This is particularly effective for detecting outlying measurements if the passband is chosen wide enough to follow actual signal variations, but not so wide that output follows measurement noise. The filter can also detect jumps in system characteristics. It is somewhat less desirable to use filter output as a measurement for the estimator because the filter response lags the measurements. If the output is used as a measurement, sampling should be much longer than filter time constants so that measurement errors are uncorrelated.
3. Removal of known errors: In some systems various corrections are applied to measurements before use in the estimator. These corrections may include bias and scale factor calibrations, and removal of known effects where it is awkward to model the effect in the estimator. For example, optical angle measurements of stars include effects such as star proper motion, parallax, and aberration. Measurement time tag errors are also a problem with some measurements. Any corrections applied should be accurately modeled or the correction may introduce additional errors.
4. Measurement transformations: Some sensors produce measurements in coordinate systems or forms that are awkward to use directly in the estimator. For example, multiple measurements from a sensor may have correlated errors or may be correlated in time. Transformations described in previous sections may be used to "uncorrelate" the measurements. Differencing of consecutive measurements may also be used to remove temporal correlations (described in Chapter 8). However, before using transformations, it should be verified that additional errors are not introduced.
5. Comparison of measurements from different sensors: When multiple sensors measure similar properties of a system, the measurements may be compared to determine if all are consistent. This works best when a simple system model can approximately predict measured quantities. GPS ground system preprocessing uses this approach to detect unexpected satellite and ground clock timing errors.

7.6 SUMMARY

This chapter covered five topics that are extensions of basic linear least-squares techniques: constrained least-squares estimation, recursive least squares, nonlinear least squares, robust estimation, and measurement preprocessing.

In some least-squares problems it is necessary to constrain linear combinations of the estimated states with equality or inequality constraints. Equality constraints of the form $\mathbf{C}\hat{\mathbf{x}} = \mathbf{d}$, where there are fewer constraint equations than unknowns in $\hat{\mathbf{x}}$, can be handled using at least six different methods:

1. Algebraically eliminate a subset of states using the constraint equations, and solve for the remaining states using unconstrained least squares. Unfortunately this is not a general approach and only works for a limited number of problems.
2. Use the method of Lagrange multipliers to redefine the least-squares problem. This works when the unconstrained least-squares problem is full rank. The method is fairly general and was shown to work well in examples.
3. Implement the constraint as a measurement with very high weighting using software for unconstrained least squares. This approach is often used but is subject to loss of precision, particularly when using the normal equations on problems where individual constraints involve multiple states. When solved using the QR method, numerical problems tend to occur when the constraint weighting multiplier is approximately equal to the numerical precision of the computer—otherwise it works well.
4. Solve an equivalent unconstrained least-squares problem of lower dimension using either direct elimination or the nullspace method. These approaches are more general versions of method 1, but are somewhat complex. Two references are provided for further information.
5. Implement the constraint using a GSVD. The GSVD method is probably the most general approach for handling equality constraints, and it can also be used for inequality constraints. However, in polynomial examples implemented using LAPACK routines DGGLSE or DGGSVD, results were found to be an order of magnitude less accurate than those of the Lagrange multiplier method using the MGS QR solution.

Simple minimum/maximum constraints on individual states are probably the most common type of inequality constraint. When physical states must be greater than a specified value, a change of variable (e.g., logarithm) sometimes allows the constraint to be handled without difficulty. Problems with more general inequality constraints can be handled using quadratic programming algorithms if the problem is transformed to a least-distance problem. Another approach uses active set algorithms where a subset of constraints are active in individual iterations, and the active set changes from one iteration to the next. The solution for each iteration is obtained using equality-constrained least squares.

Although batch processing of measurements is by far the most common method for solving least-squares problems, the state solution can also be recursively updated as individual measurements or batches of measurements are processed. The recursive algorithm is related to the batch algorithm through the matrix inversion identity (see Section 4.3 and Appendix A.3.3). The conditional mean or minimum variance solution to the Bayesian least-squares problem (Section 4.3) provides the recursive algorithm—this is the measurement update step of the Kalman filter. Recursive least squares can also be implemented using the MGS version of the QR algorithm. Another recursive least-squares variant has a fading memory characteristic because recent measurements are given more weight than older data.

Since physical systems are inherently nonlinear for large parameter excursions, linear least-squares techniques must be adapted to work with models that are used outside the linear region. The GN method linearizes the model about the current state estimate, and uses the expected Hessian (Fisher information matrix) rather than the true Hessian. Then the solution is obtained iteratively as the model is re-linearized at each iteration. However, the GN method is not globally convergent for large parameter deviations or when measurement errors are large. One globally convergent method "damps" the GN step by backtracking along the GN step direction to find the approximate minimum cost point along the GN step direction. Other approaches are called trust region methods because the iteration step is limited to a region of "trust" about the current estimate. The most successful variant of trust region concepts is due to Levenberg and Marquardt. This LM method smoothly varies the step direction between GN and negative gradient steps by adding positive values to the diagonals of the Fisher information matrix before inversion. The multiplier of the diagonals is iteratively increased (effectively shortening the step length and changing its direction slightly) until the resulting step reduces the least-squares cost. References claim that the LM method is more robust than backtracking GN, and this author's experience supports the conclusion. However, LM requires more iterations than backtracking GN on many problems, and it is unusual to encounter problems where LM works and backtracking GN does not. Another approach recursively updates an approximation to the Hessian matrix using a secant method. Since the GN and secant algorithms have different responses to the large residuals of early iterations, the methods are combined in the NL2SOL code developed by Dennis, Gay, and Welsch. However, the NL2SOL method did not converge as quickly as either backtracking GN or LM on example problems.

Robust estimators are designed to be insensitive to small errors in assumptions or models, particularly with respect to non-Gaussian measurement error distributions. Most robust "least-squares" techniques de-weight large residuals relative to the quadratic weighting of least squares. M-estimates are the class of robust algorithms most relevant for model parameter estimation. Huber weighting is based on a merger of Gaussian and two-sided exponential distributions: residual-squared weighting is used in the cost function up to a certain threshold, and then weighting increases linearly above that threshold.

Measurement de-weighting make estimators more efficient, but erroneous measurements can still influence the solution. A very simple data editing algorithm has been used in operational nonlinear least-squares software for decades and has been found to be quite robust on a wide variety of problems.

While measurement de-weighting or editing can be effective, many operational systems preprocess measurements to eliminate outliers before use in least-squares processing. This offers an additional level of protection against anomalous measurements. In cases where measurements are available at a high rate or when measurements are to some extent redundant, editing in the preprocessing stage can be more effective than when done in the estimator. Other reasons for preprocessing include sub-sampling and smoothing of high-rate data, jump detection, removal of known errors, and transformations to reduce correlation.

CHAPTER 8

KALMAN FILTERING

The first seven chapters were an extended introduction to the next topic: Kalman filtering. Recall from Chapter 1 that there are two primary differences between least-squares estimators and the Kalman filter: (1) the Kalman filter models stochastic (possibly nonstationary) systems while least squares is used for deterministic systems, and (2) the Kalman filter processes measurements recursively while most least-squares implementations use batch processing. However, the second difference is not unique as Chapter 7 described recursive implementations of least-squares estimators. Hence the ability of Kalman filters to model nonstationary dynamic systems driven by random process noise is its most important characteristic.

Much of Kalman filter theory has already been presented in previous chapters. Chapter 2 discussed stochastic dynamic models, integrated a continuous stochastic dynamic model over a finite time interval (eq. 2.2-14), defined the state transition matrix (eq. 2.2-7), and derived the state noise covariance matrix (eq. 2.2-20). Chapter 4 discussed the minimum variance estimator and showed how the recursive measurement update (eqs. 4.3-34 and 4.3-35) can be derived by setting the cost gradient to zero or using the orthogonality principle. We only need to connect these concepts to derive the discrete Kalman filter. Many of the solution techniques in Chapter 5 also apply to Kalman filtering.

Most of the material in this chapter is well covered in other texts on Kalman filtering. For that reason some sections are brief, and readers are referred to other sources for more information. We particularly recommend books by Simon (2006), Gelb (1974), Anderson and Moore (1979), Maybeck (1979, 1982), Jazwinski (1970), Kailath et al. (2000), Grewal and Andrews (2001), and Bar-Shalom et al. (2001). Simon compares the emphasis and strengths of the other books. This chapter presents basic Kalman filter theory and standard extensions. Discussions of bias state handling, methods for steady-state filtering, and the Wiener filter are more extensive than that for other topics. Chapter 9 continues with discussions of nonlinear filtering, smoothing, and error analysis; Chapter 10 discusses factorized filter implementations; and Chapter 11 includes more sophisticated extensions of Kalman filtering.

8.1 DISCRETE-TIME KALMAN FILTER

This section discusses the discrete-time Kalman filter for linear systems. Section 8.3 addresses continuous-time Kalman-Bucy filters and Chapter 9 discusses extensions of the Kalman filter for nonlinear systems. Most Kalman filter implementations use the discrete, rather than continuous implementation (even when the system modeled is actually continuous), because measurements are almost always obtained at discrete points in time.

8.1.1 Truth Model

We start by reviewing the "truth" model for the discrete system. The n-element state vector at time t_i is assumed to be a function of the state at time t_{i-1} and the integrated effect of random process noise and "control" inputs over the interval t_{i-1} to t_i as

$$\mathbf{x}_i = \mathbf{\Phi}_{i,i-1}\mathbf{x}_{i-1} + \mathbf{q}_{i,i-1} + \mathbf{u}_{i,i-1} \tag{8.1-1}$$

where

$\mathbf{\Phi}_{i,i-1}$ is an abbreviation for the $n \times n$ state transition matrix $\mathbf{\Phi}(t_i,t_{i-1})$

$\mathbf{q}_{i,i-1}$ is an abbreviation for n-element integrated zero-mean random state noise vector $\mathbf{q}_D(t_i,t_{i-1})$, and

$\mathbf{u}_{i,i-1}$ is an abbreviation for the n-element vector of integrated control or exogenous inputs $\mathbf{u}_D(t_i,t_{i-1})$.

These discrete variables can be obtained by integrating the linear continuous time-invariant system

$$\dot{\mathbf{x}}(t) = \mathbf{F}\,\mathbf{x}(t) + \mathbf{B}\,\mathbf{u}(t) + \mathbf{G}\,\mathbf{q}_c(t) \tag{8.1-2}$$

over the time interval $T = t_i - t_{i-1}$:

$$\mathbf{\Phi}_{i,i-1} = e^{\mathbf{F}T}$$

$$\mathbf{q}_{i,i-1} = \int_{t_{i-1}}^{t_i} e^{\mathbf{F}(t_i-\lambda)}\mathbf{G}\,\mathbf{q}_c(\lambda)\,d\lambda$$

$$\mathbf{u}_{i,i-1} = \int_{t_{i-1}}^{t_i} e^{\mathbf{F}(t_i-\lambda)}\mathbf{B}\,\mathbf{u}(\lambda)\,d\lambda.$$

It is assumed that $E[\mathbf{q}_c(t)] = \mathbf{0}$ and $E[\mathbf{q}_c(t)\mathbf{q}_c^T(\lambda)] = \mathbf{Q}_s\delta(t-\lambda)$ so that

$$E[\mathbf{q}_{i,i-1}] = \mathbf{0}$$

$$E[\mathbf{q}_{i,i-1}\mathbf{q}_{j,j-1}^T] = \mathbf{Q}_{i,i-1}\delta_{ij}$$

where δ_{ij} is the Kronecker delta function ($\delta_{ij} = 0$ for $i \neq j$ and $\delta_{ii} = 1$). Notice that elements of $\mathbf{q}_{i,i-1}$ and $\mathbf{u}_{i,i-1}$ may be zero since process noise and control inputs do not necessarily drive all elements of $\mathbf{x}$. In fact $\mathbf{u}_{i,i-1}$ is not present in most noncontrol applications. Furthermore, $\mathbf{Bu}(t)$ is constant over time intervals T in many control applications because digital control systems only update control commands at discrete time intervals. Of course the discrete model need not be obtained as the integral of a continuous system model: discrete autoregressive moving average (ARMA) models are sometimes used directly in Kalman filters.

The true model of the m-element measurement vector at time t_i is assumed to be

$$\mathbf{y}_i = \mathbf{H}_i \mathbf{x}_i + \mathbf{r}_i \tag{8.1-3}$$

where $E[\mathbf{r}_i] = \mathbf{0}$, $E[\mathbf{r}_i \mathbf{r}_j^T] = \mathbf{R}_i \delta_{ij}$, $E(\mathbf{r}_i \mathbf{q}_{j,j-1}^T) = \mathbf{0}$ for all i, j, and $\mathbf{H}_i$ is an $m \times n$ matrix. Note that the dimension of $\mathbf{y}_i$ and types of measurements included in $\mathbf{y}_i$ may change from one time point to the next.

8.1.2 Discrete-Time Kalman Filter Algorithm

The filter dynamic and measurement models are assumed to follow the true models, except that because $E[\mathbf{q}_{i,i-1}] = \mathbf{0}$ and $E[\mathbf{r}_i] = \mathbf{0}$, the random inputs are ignored in modeling the filter state. However, the effect on the error covariance is modeled. The predicted (*a priori*) filter state at time t_i is based on a state estimate at time t_{i-1}. Letting $\hat{\mathbf{x}}_{i-1/i-1}$ denote the *a posteriori* filter state estimate at time t_{i-1} based on measurements up to and including time t_{i-1}, the *a priori* filter state at time t_i is

$$\boxed{\hat{\mathbf{x}}_{i/i-1} = \mathbf{\Phi}_{i,i-1} \hat{\mathbf{x}}_{i-1/i-1} + \mathbf{u}_{i,i-1}.} \tag{8.1-4}$$

(As before, "boxed" equations are those actually implemented.) Note that $\mathbf{u}_{i,i-1}$ is assumed to be known and used (if it exists) to predict the filter state, but $\mathbf{q}_{i,i-1}$ is ignored because $E[\mathbf{q}_{i,i-1}] = \mathbf{0}$. Weighting of measurements in the Kalman filter is based on the state error covariance matrix, so the equations for the covariance time update are also needed. Letting $\tilde{\mathbf{x}}_{i-1/i-1} = \mathbf{x}_{i-1} - \hat{\mathbf{x}}_{i-1/i-1}$ and $\tilde{\mathbf{x}}_{i/i-1} = \mathbf{x}_i - \hat{\mathbf{x}}_{i/i-1}$, the *a posteriori* error covariance is computed as

$$\boxed{\begin{aligned}
\mathbf{P}_{i/i-1} &\triangleq E[\tilde{\mathbf{x}}_{i/i-1} \tilde{\mathbf{x}}_{i/i-1}^T] \\
&= E\left[(\mathbf{\Phi}_{i,i-1} \tilde{\mathbf{x}}_{i-1/i-1} + \mathbf{q}_{i,i-1})(\tilde{\mathbf{x}}_{i-1/i-1}^T \mathbf{\Phi}_{i,i-1}^T + \mathbf{q}_{i,i-1}^T)\right] \\
&= \mathbf{\Phi}_{i,i-1} E[\tilde{\mathbf{x}}_{i-1/i-1} \tilde{\mathbf{x}}_{i-1/i-1}^T] \mathbf{\Phi}_{i,i-1}^T + E[\mathbf{q}_{i,i-1} \mathbf{q}_{i,i-1}^T] \\
&= \mathbf{\Phi}_{i,i-1} \mathbf{P}_{i-1/i-1} \mathbf{\Phi}_{i,i-1}^T + \mathbf{Q}_{i,i-1}
\end{aligned}} \tag{8.1-5}$$

where the discrete state noise covariance $\mathbf{Q}_{i,i-1} = E[\mathbf{q}_{i,i-1} \mathbf{q}_{i,i-1}^T]$ is computed using equation (2.2-20). Notice that a term corresponding to $\mathbf{u}_{i,i-1}$ is not included since it is assumed to be perfectly known. Any errors in application of control inputs should be included in $\mathbf{q}_{i,i-1}$ if random, or modeled by extra states (as done in the power plant example of Section 3.5) if systematic. Equations (8.1-4) and (8.1-5) represent the time update step of the Kalman filter.

The filter measurement model is assumed to be

$$\boxed{\hat{\mathbf{y}}_i = \mathbf{H}_i \hat{\mathbf{x}}_{i/i-1}} \tag{8.1-6}$$

so the measurement residual is

$$\begin{aligned}\tilde{\mathbf{y}}_i &= \mathbf{y}_i - \mathbf{H}_i\hat{\mathbf{x}}_{i/i-1} \\ &= \mathbf{H}_i\tilde{\mathbf{x}}_{i/i-1} + \mathbf{r}_i\end{aligned}. \tag{8.1-7}$$

Notice that $E[\tilde{\mathbf{y}}_i] = \mathbf{0}$ and the covariance of the measurement residual $\tilde{\mathbf{y}}_i$ is

$$\mathbf{C}_i = E[\tilde{\mathbf{y}}_i\tilde{\mathbf{y}}_i^T] = \mathbf{H}_i\mathbf{P}_{i/i-1}\mathbf{H}_i^T + \mathbf{R}_i\,. \tag{8.1-8}$$

Following the development of the minimum variance estimate in Section 4.3, the *a posteriori* state estimate at time t_i is calculated as

$$\hat{\mathbf{x}}_{i/i} = \hat{\mathbf{x}}_{i/i-1} + \mathbf{K}_i\tilde{\mathbf{y}}_i\,. \tag{8.1-9}$$

The gain matrix $\mathbf{K}_i$ minimizing the covariance of $\tilde{\mathbf{x}}_{i/i} = \mathbf{x}_i - \hat{\mathbf{x}}_{i/i}$ is computed using the orthogonality principle:

$$\begin{aligned}E\left[(\mathbf{x}_i - \hat{\mathbf{x}}_{i/i})\mathbf{y}^T\right] &= E\left[(\tilde{\mathbf{x}}_{i/i-1} - \mathbf{K}_i(\mathbf{H}_i\tilde{\mathbf{x}}_{i/i-1} + \mathbf{r}_i))(\mathbf{H}_i\mathbf{x}_i + \mathbf{r}_i)^T\right] \\ &= E\left[((\mathbf{I} - \mathbf{K}_i\mathbf{H}_i)\tilde{\mathbf{x}}_{i/i-1} - \mathbf{K}_i\mathbf{r}_i)\left((\hat{\mathbf{x}}_{i/i-1} + \tilde{\mathbf{x}}_{i/i-1})^T\mathbf{H}_i^T + \mathbf{r}_i^T\right)\right] \\ &= (\mathbf{I} - \mathbf{K}_i\mathbf{H}_i)\mathbf{P}_{i/i-1}\mathbf{H}_i^T - \mathbf{K}_i\mathbf{R}_i \\ &= \mathbf{0}\end{aligned}$$

where we have assumed that $\hat{\mathbf{x}}_{i-1/i-1}$ at t_{i-1} is also minimum variance so that $E\left[\tilde{\mathbf{x}}_{i/i-1}\hat{\mathbf{x}}_{i/i-1}^T\right] = \mathbf{0}$. Thus the $n \times m$ gain matrix is

$$\begin{aligned}\mathbf{K}_i &= \mathbf{P}_{i/i-1}\mathbf{H}_i^T(\mathbf{H}_i\mathbf{P}_{i/i-1}\mathbf{H}_i^T + \mathbf{R}_i)^{-1} \\ &= \mathbf{P}_{i/i-1}\mathbf{H}_i^T\mathbf{C}_i^{-1}\end{aligned}, \tag{8.1-10}$$

the *a posteriori* state estimate is

$$\hat{\mathbf{x}}_{i/i} = \hat{\mathbf{x}}_{i/i-1} + \mathbf{K}_i(\mathbf{y}_i - \mathbf{H}_i\hat{\mathbf{x}}_{i/i-1}), \tag{8.1-11}$$

and the *a posteriori* state error covariance is

$$\begin{aligned}\mathbf{P}_{i/i} &\triangleq E(\tilde{\mathbf{x}}_{i/i}\tilde{\mathbf{x}}_{i/i}^T) \\ &= E\left[((\mathbf{I} - \mathbf{K}_i\mathbf{H}_i)\tilde{\mathbf{x}}_{i/i-1} - \mathbf{K}_i\mathbf{r}_i)((\mathbf{I} - \mathbf{K}_i\mathbf{H}_i)\tilde{\mathbf{x}}_{i/i-1} - \mathbf{K}_i\mathbf{r}_i)^T\right]. \\ &= (\mathbf{I} - \mathbf{K}_i\mathbf{H}_i)\mathbf{P}_{i/i-1}(\mathbf{I} - \mathbf{K}_i\mathbf{H}_i)^T + \mathbf{K}_i\mathbf{R}_i\mathbf{K}_i^T\end{aligned} \tag{8.1-12}$$

Equation (8.1-12) is applicable whether or not $\mathbf{K}_i$ is computed using equation (8.1-10). (Readers skeptical of the orthogonality principle may wish to verify that the above $\mathbf{K}_i$ minimizes trace($\mathbf{P}_{i/i}$) by differentiation of equation (8.1-12) with respect to $\mathbf{K}_i$.) However, when equation (8.1-10) is substituted in equation (8.1-12), we obtain

$$\begin{aligned}\mathbf{P}_{i/i} &= \mathbf{P}_{i/i-1} - \mathbf{K}_i\mathbf{H}_i\mathbf{P}_{i/i-1} - \mathbf{P}_{i/i-1}\mathbf{H}_i^T\mathbf{K}_i^T + \mathbf{K}_i(\mathbf{H}_i\mathbf{P}_{i/i-1}\mathbf{H}_i^T + \mathbf{R}_i)\mathbf{K}_i^T \\ &= (\mathbf{I} - \mathbf{K}_i\mathbf{H}_i)\mathbf{P}_{i/i-1}\end{aligned}. \tag{8.1-13}$$

Equations (8.1-10), (8.1-11), and (8.1-13) are the measurement update equations of the Kalman filter. (Because Kalman used the orthogonality principle to derive the optimal gain, equation (8.1-12) did not appear in his original 1960 paper.) The time

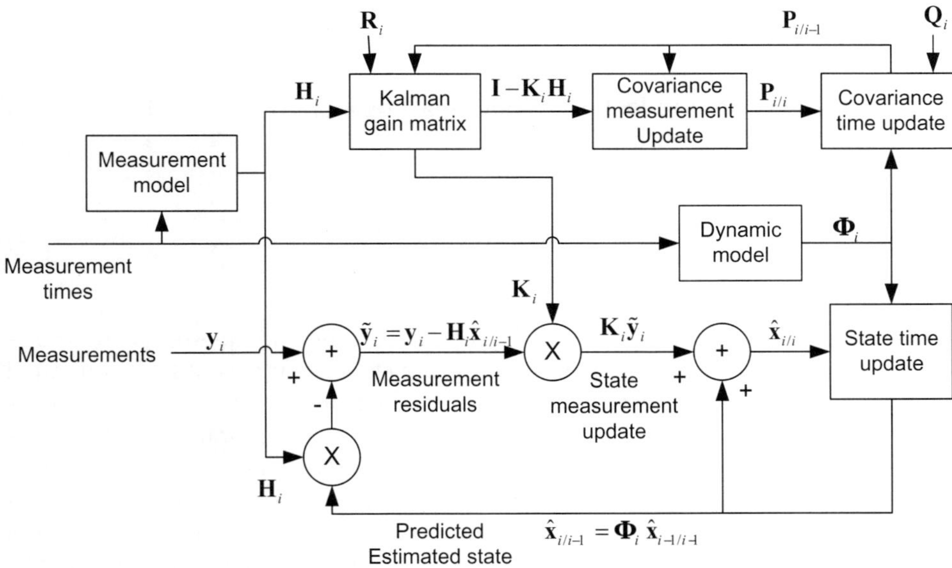

FIGURE 8.1: Discrete-time Kalman filter data flow.

update and measurement update equations are computed at every new measurement time. Figure 8.1 shows the overall data flow of the Kalman filter. Notice that the gain matrix $\mathbf{K}_i$ depends only on the measurement sensitivity matrix $\mathbf{H}_i$ and the predicted state error covariance $\mathbf{P}_{i/i-1}$, not on the actual measurements $\mathbf{y}_i$. This implies that actual measurements are not necessary to compute the error covariance $\mathbf{P}_{i/i}$, so state estimate errors can be analyzed (within the limitations of the model) before actual measurements are available.

8.1.2.1 Summary of Assumptions The assumptions used in deriving the discrete Kalman filter are:

1. The system is linear.
2. The model matrices $\mathbf{\Phi}$, $\mathbf{Q}$, $\mathbf{H}$, and $\mathbf{R}$ exactly match truth (no model errors).
3. The process and measurement noises are unbiased: $E[\mathbf{q}_{i,i-1}] = \mathbf{0}$ and $E[\mathbf{r}_i] = \mathbf{0}$.
4. The process and measurement noises are white: $E[\mathbf{q}_{i,i-1}\mathbf{q}_{j,j-1}^T] = \mathbf{Q}_{i,i-1}\delta_{ij}$ and $E[\mathbf{r}_i\mathbf{r}_j^T] = \mathbf{R}_i\delta_{ij}$.
5. Process and measurement noises are uncorrelated: $E[\mathbf{q}_{j,j-1}\mathbf{r}_i] = \mathbf{0}$ for any i, j.
6. The solution is Bayesian, as the recursion must be initialized with $\hat{\mathbf{x}}_{0/0}$ and $\mathbf{P}_{0/0}$. $\mathbf{P}_{0/0}$ is sometimes initialized with "very large" values to approximate a non-Bayesian solution, but this can lead to numerical problems.

Although not an assumption, the state estimate at each time step is computed to minimum the variance; that is, $\hat{\mathbf{x}}_{i/i}$ at t_i is computed to minimize $E(\tilde{\mathbf{x}}_{i/i}\tilde{\mathbf{x}}_{i/i}^T)$, which implies that $E(\tilde{\mathbf{x}}_{i/i}\hat{\mathbf{x}}_{i/i}^T) = \mathbf{0}$.

Note that Gaussian noise is *not* assumed for this linear system, and unlike a Wiener filter, the system may be time-varying and statistically nonstationary (time variable $\mathbf{\Phi}$, $\mathbf{H}$, $\mathbf{Q}$, $\mathbf{R}$). Unfortunately one or more of the above assumptions are violated for many real-system applications. Later sections discuss methods that handle non-ideal cases.

8.1.2.2 Implementation Issues Notice that equations (8.1-10) and (8.1-13) can be implemented efficiently as

$$\boxed{\begin{aligned}\mathbf{D} &= \mathbf{P}_{i/i-1}\mathbf{H}_i^T \\ \mathbf{K}_i &= \mathbf{D}(\mathbf{H}_i\mathbf{D}+\mathbf{R}_i)^{-1}, \\ \mathbf{P}_{i/i} &= \mathbf{P}_{i/i-1} - \mathbf{K}_i\mathbf{D}^T\end{aligned}} \tag{8.1-14}$$

requiring approximately $mn(1.5n + 2m + 0.5n) + 0.5m^3$ floating-point multiply-accumulate operations if only the upper (or lower) triangular part of $\mathbf{P}_{i/i}$ is computed and no attempt is made to take advantage of sparseness in $\mathbf{H}$. (Flexible filter implementations should use general-purpose sparse matrix multiplication functions because $\mathbf{H}$ is often very sparse.) Furthermore, when $\mathbf{R}_i$ is diagonal (measurement errors in vector $\mathbf{y}_i$ are uncorrelated), measurements in $\mathbf{y}_i$ may be processed one-component-at-a-time without separate time updates so that $\mathbf{K}_i$ and $\mathbf{D}$ in equation (8.1-13) or (8.1-14) are vectors and no matrix inversion is required. This change slightly reduces computations and may improve accuracy. When one-component-at-a-time processing is used, each individual measurement and row of $\mathbf{H}_i$ may be divided by the square root of the corresponding diagonal element of $\mathbf{R}_i$ so that the scaled measurement noise is 1.0; that is, the scaled scalar measurement y'_{i_j} for component $j = 1, 2,\ldots, m$ is formed as

$$\begin{aligned} y'_{i\_j} &= \frac{y_{i\_j}}{\sqrt{R_{i\_jj}}} = \frac{\mathbf{h}_{i\_j}\mathbf{x}_i + r_{i\_j}}{\sqrt{R_{i\_jj}}} \\ &= \mathbf{h}'_{i\_j}\mathbf{x}_i + r'_{i\_j} \end{aligned}$$

where $\mathbf{h}_{i\_j}$ is row j of $\mathbf{H}_i$ and R_{i_jj} is the j-th diagonal of $\mathbf{R}_i$. Notice that $E\left[(r'_{i\_j})^2\right] = 1$ so equation (8.1-10) becomes

$$\mathbf{k}_{i\_j} = \mathbf{P}_{a\_j-1}\mathbf{h}_{i\_j}^T / (\mathbf{h}_{i\_j}\mathbf{P}_{a\_j-1}\mathbf{h}_{i\_j}^T + 1)$$

where $\mathbf{P}_{a\_j-1}$ is the covariance matrix from step $j - 1$ of the processing, and $\mathbf{P}_{a\_0} = \mathbf{P}_{i/i-1}$. This scaling is necessary for some square-root filter algorithms to be described in Chapter 10. When $\mathbf{R}_i$ is not diagonal, one-component-at-a-time processing may still be used if $\mathbf{y}_i$ and $\mathbf{H}_i$ are scaled by the inverse "square root" of $\mathbf{R}_i$, as done in earlier chapters on least-squares processing.

Equation (8.1-12) is sometimes referred to as the *Joseph-stabilized* form of the Kalman filter covariance update (Bucy and Joseph 1968)). Since equation (8.1-12) is the sum of two positive definite symmetric matrices rather than a difference, it is less susceptible to numerical errors in computing $\mathbf{K}_i$ than equation (8.1-13). However, in practice it is subject to some of the same numerical problems as equation (8.1-13), and can sometimes be less accurate than equation (8.1-13) or (8.1-14) when they are implemented so that $\mathbf{P}_{i/i-1}$ and $\mathbf{P}_{i/i}$ are forced to be symmetric—done by averaging upper and lower triangular partitions of $\mathbf{P}_{i/i-1}$ and $\mathbf{P}_{i/i}$, or only computing the upper triangular partition. Furthermore, if equations (8.1-10) and (8.1-12) are implemented as shown, they involve approximately $2mn(n + m) + 0.5m^3 + 0.5n^2(3n + 1) + n$ floating-point multiplications, which can be orders of magnitude greater than that of equation (8.1-14) when $n \gg m$. Bierman (1977b, p. 82) observed that equation (8.1-12) can be implemented much more efficiently using the same approach as equation (8.1-14). That is

$$\begin{aligned}\mathbf{A}_i &= \mathbf{P}_{i/i-1} - \mathbf{K}_i\mathbf{D}^T \\ \mathbf{P}_{i/i} &= \mathbf{A}_i - \mathbf{A}_i\mathbf{H}_i^T\mathbf{K}_i^T + \mathbf{K}_i\mathbf{R}_i\mathbf{K}_i^T\end{aligned}. \tag{8.1-15}$$

Bierman recommended that the simplification $\mathbf{P}_{i/i} = \mathbf{A}_i + (\mathbf{K}_i\mathbf{R}_i - \mathbf{A}_i\mathbf{H}_i^T)\mathbf{K}_i^T$ not be used because numerical experience showed that when $\mathbf{K}_i\mathbf{R}_i \cong \mathbf{A}_i\mathbf{H}_i^T$ (i.e., when $\mathbf{K}_i$ is nearly optimal) "serious errors" can occur. Even though equation (8.1-15) is not expressed as a sum of two positive definite symmetric matrices, it usually has the same numerical properties as equation (8.1-12). More will be said in Chapter 10 about numerical errors and computational efficiency of the Kalman filter.

The filter time update equations (8.1-4) and (8.1-5) may also be implemented more efficiently when the state vector includes bias states, which occurs frequently. That is, when equation (8.1-1) can be written as

$$\begin{bmatrix}\mathbf{x}_d \\ \mathbf{x}_b\end{bmatrix}_i = \begin{bmatrix}\mathbf{\Phi}_{dd} & \mathbf{\Phi}_{db} \\ \mathbf{0} & \mathbf{I}\end{bmatrix}_{i,i-1}\begin{bmatrix}\mathbf{x}_d \\ \mathbf{x}_b\end{bmatrix}_{i-1} + \begin{bmatrix}\mathbf{q}_d \\ \mathbf{0}\end{bmatrix}_{i,i-1} + \begin{bmatrix}\mathbf{u}_d \\ \mathbf{0}\end{bmatrix}_{i,i-1}$$

where $\mathbf{x}_d$ represents n_d dynamic states and $\mathbf{x}_b$ represents n_b biases, then it is not necessary to update $\hat{\mathbf{x}}_b$ in equation (8.1-4). Furthermore equation (8.1-5) may be efficiently implemented by first forming a temporary $n_d \times (n_d + n_b)$ array $\mathbf{D}$ as

$$\boxed{\begin{aligned}[\mathbf{D}_{dd} \quad \mathbf{D}_{db}] &= [\mathbf{\Phi}_{dd} \quad \mathbf{\Phi}_{db}]_{i,i-1}\begin{bmatrix}\mathbf{P}_{dd} & \mathbf{P}_{db} \\ \mathbf{P}_{db}^T & \mathbf{P}_{bb}\end{bmatrix}_{i-1/i-1} \\ &= \left[\mathbf{\Phi}_{dd}\mathbf{P}_{dd} + \mathbf{\Phi}_{db}\mathbf{P}_{db}^T \quad \mathbf{\Phi}_{dd}\mathbf{P}_{db} + \mathbf{\Phi}_{db}\mathbf{P}_{bb}\right]\end{aligned}} \tag{8.1-16}$$

where $\mathbf{P}_{dd}$, $\mathbf{P}_{db}$, and $\mathbf{P}_{bb}$ are the appropriate dynamic and bias partitions of $\mathbf{P}_{i-1/i-1}$. Although $\mathbf{D}$ is shown in partitions, it is formed as a single matrix multiplication of a rectangular matrix (upper rows of $\mathbf{\Phi}$) and square $\mathbf{P}_{i-1/i-1}$. This operation requires $n_d(n_d + n_b)^2$ multiplications. Then $\mathbf{P}_{i/i-1}$ is formed as

$$\boxed{\begin{bmatrix}\mathbf{P}_{dd} & \mathbf{P}_{db} \\ \mathbf{P}_{bd} & \mathbf{P}_{bb}\end{bmatrix}_{i/i-1} = \begin{bmatrix}\mathbf{D}_{dd}\mathbf{\Phi}_{dd}^T + \mathbf{D}_{db}\mathbf{\Phi}_{db}^T + \mathbf{Q} & \mathbf{D}_{db} \\ \mathbf{D}_{db}^T & \mathbf{P}_{bb\_i-1/i-1}\end{bmatrix}.} \tag{8.1-17}$$

Notice that the lower right partition is unchanged from $\mathbf{P}_{i-1/i-1}$ values. Formation of equation (8.1-17) requires a total of $(n_d + n_b)n_d(n_d + 1)/2$ multiplications provided that only the upper partition of $\mathbf{P}_{dd}$ is computed. Thus the total number of multiplications required to implement equations (8.1-16) and (8.1-17) is $(n_d + n_b)n_d(1.5n_d + n_b + 0.5)$. Since formation of equation (8.1-5) as shown requires $(n_d + n_d)^2(1.5(n_d + n_d) + 0.5)$ multiplications, the computational reduction of using equations (8.1-16) and (8.1-17) is significant when many biases are present; for example, approximately 58% reduction when $n_d = n_b$. Further savings are also possible when $\mathbf{\Phi}_{dd}$ is sparse (e.g., core model states are decoupled from colored noise measurement error states).

It should be noted that the term "bias state" usually implies that the state is static with no variation due to process noise. Bias states are commonly used to model parameters that are constant for the time period in which the Kalman filter processes data. This is appropriate when the data span is limited in time. Examples include target tracking (missiles, aircraft, tanks, etc.), most types of orbit determination, or system testing. It is less valid when the Kalman filter is used in an operational system for a long period of time without resetting, such as in process control applications. In these cases, filter bias variances eventually become negligibly small if the biases are observable from the given measurement data. Thus the filter ignores new measurement information on biases. For that reason it is sometimes advisable to model small levels of process noise on "bias" states so that the filter is always somewhat responsive to the measurements.

Example 8.1: Single-State Kalman Filter

Many books on Kalman filtering use one-state examples. While not wanting to appear unimaginative, we also use a one-state example because it easily demonstrates important Kalman filter properties. It must be realized, however, that one-state models are almost never used for real-systems, so applicability of the example is very limited.

Consider the time-invariant single-state dynamic model

$$x_i = \Phi x_{i-1} + q_i$$

where the sampling interval T is fixed, $\Phi = e^{-T/\tau}$ and $E[q_i^2] = Q$. If $0 < \Phi < 1$ the sequence of states $\{x_i, x_{i-1}, \ldots, x_0\}$ is a first-order stationary Markov sequence, where the steady-state variance is $\sigma_{ss}^2 = E[x_i^2] = Q/(1-\Phi^2)$. If $\Phi = 1$, the system is a random walk process, which is nonstationary since the variance increases linearly with time. If $\Phi > 1$ the system is unstable.

The measurement model is $y_i = x_i + r_i$, where we have assumed without loss of generality (since the state x can be scaled) that $H = 1$. Also $E[r_i^2] = R$ is assumed constant. Using these definitions, the state and variance time updates are

$$\hat{x}_{i/i-1} = \Phi \hat{x}_{i-1/i-1}$$
$$P_{i/i-1} = \Phi^2 P_{i-1/i-1} + Q\text{'}$$

the measurement gain matrix is

$$K = P_{i/i-1} / (P_{i/i-1} + R),$$

and the state estimate and variance measurement updates are

$$\begin{aligned}\hat{x}_{i/i} &= \hat{x}_{i/i-1} + K(y_i - \hat{x}_{i/i-1}) \\ P_{i/i} &= P_{i/i-1} - K P_{i/i-1} \quad . \\ &= P_{i/i-1} R / (P_{i/i-1} + R)\end{aligned}$$

Figure 8.2 shows the time profile of the *a posteriori* $\sqrt{P_{i/i}}$ for several values of Q when $R = 100$, $P_{0/0} = 10R = 1000$, $T = 1$, and $\Phi = e^{-T/\tau} = 0.9048$ for $\tau = 10$. The line labels are defined as follows:

"sigx-qp1" uses $Q = 0.1$, or $\sigma_{ss} = \sqrt{Q/(1-\Phi^2)} = 0.743$,

"sigx-q1" uses $Q = 0.1$, or $\sigma_{ss} = 2.349$, and

"sigx-q10" uses $Q = 10$, or $\sigma_{ss} = 7.427$.

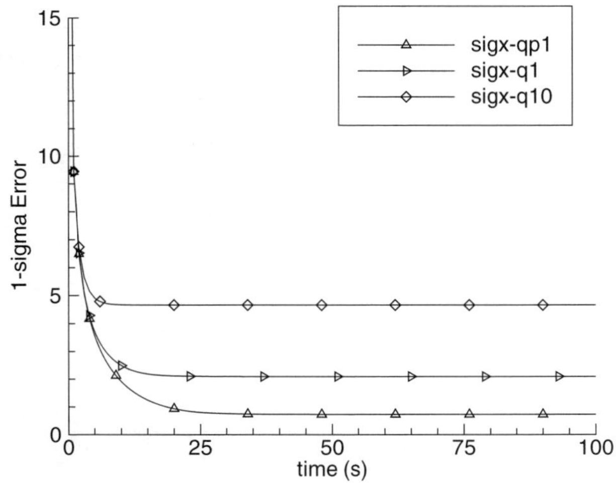

FIGURE 8.2: *A posteriori* $1 - \sigma$ uncertainty for one-state Kalman filter with $\Phi = 0.90$.

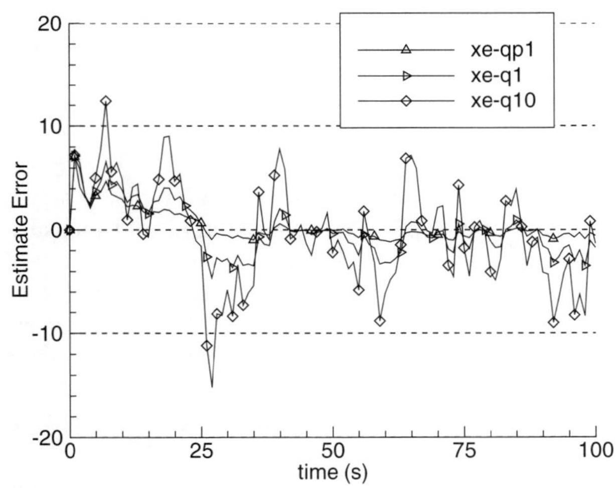

FIGURE 8.3: Sample *a posteriori* estimate errors for one-state Kalman filter with $\Phi = 0.90$.

Notice that $\sqrt{P_{i/i}}$ quickly decays from the initial value of 31.6 to a steady-state value that is a function of Q. The time to reach steady-state is shortest for the largest value of Q. The steady-state $\sqrt{P_{ss}}$ values are 0.732, 2.098, and 4.665 for $Q = 0.1, 1$, and 10, respectively. The corresponding steady-state Kalman gains are 0.00533, 0.0421, and 0.179, respectively. Notice that $\sqrt{P_{ss}} = 0.732$ for $Q = 0.1$ is nearly equal to σ_{ss}, so the measurements are providing little information (as indicated by $K = 0.00533$). However, for $Q = 10$, $\sqrt{P_{ss}} = 4.665$ is 37% smaller than $\sigma_{ss} = 7.427$, so the measurements significantly reduce the *a priori* uncertainty.

Figure 8.3 shows the corresponding *a posteriori* state estimate errors $\hat{x}_{i/i} - x_i$ when using random simulated process and measurement noise with variances defined by Q and R. The three different cases (using different values of Q) all use the same time sequence of random noise values so that the results can be directly compared. Notice that actual errors $\hat{x}_{i/i} - x_i$ occasionally reach $3\sqrt{P_{i/i}}$ at corresponding values of t_i, but they generally follow the pattern shown in Figure 8.2: the model with the largest Q is most influenced by the measurements and thus measurement noise.

The steady-state estimate variance P_{ss} can be computed by solving

$$P_{ss} = \frac{(\Phi^2 P_{ss} + Q)R}{\Phi^2 P_{ss} + Q + R},$$

from (8.1-5) and (8.1-13), which gives

$$\Phi^2 P_{ss}^2 + (Q + R - \Phi^2 R)P_{ss} - QR = 0$$

or

$$\begin{aligned} P_{ss} &= \frac{-(Q + R - \Phi^2 R) + \sqrt{(Q + R - \Phi^2 R)^2 + 4\Phi^2 QR}}{2\Phi^2} \\ &= \frac{\alpha}{\Phi}\left(\sqrt{1 + QR/\alpha^2} - 1\right) \end{aligned}$$

where

$$\alpha = \frac{Q + R(1 - \Phi^2)}{2\Phi}.$$

Notice that P_{ss} is a function of the relative magnitudes Q and R, but the relationship is not simple. When process noise is large and measurement noise is small ($R \ll Q$), the steady-state variance will be approximately equal to the measurement variance: $P_{ss} \cong R$. When $R \approx Q$ and the dynamic model has very short time constants ($\tau \ll T$), $\Phi \cong 0$, and $P_{ss} \cong QR/(Q + R)$. However, this case is unusual because most systems are designed with sampling frequencies well above the Nyquist rate (two samples in the shortest period of the signal), so $0.04 < \Phi \le 1$. In the current example (and many real systems), $R > Q$ and thus P_{ss} is a strong function of Φ.

For the random walk ($\Phi = 1$) case the equation simplifies somewhat:

$$P_{ss} = \frac{-Q + \sqrt{Q^2 + 4QR}}{2} = \frac{Q(\sqrt{1 + 4R/Q} - 1)}{2}.$$

Thus

$$\begin{aligned} Q \gg R &\rightarrow P_{ss} \cong R \\ Q \ll R &\rightarrow P_{ss} \cong \sqrt{QR} \\ Q = R &\rightarrow P_{ss} \cong 0.618R \end{aligned}.$$

Figure 8.4 shows the *a posteriori* $1 - \sigma$ ($\sqrt{P_{ss}}$) values for the random walk case. Even though the process model is nonstationary, measurements keep the estimate error bounded. The steady-state $\sqrt{P_{ss}}$ values are now 1.764, 3.084, and 5.198, respectively for $Q = 0.1$, 1, and 10. Comparing $\sqrt{P_{ss}}$ with the Markov process case of Figure 8.2, $\sqrt{P_{ss}}$ increased by a factor of 142% when using $Q = 0.1$, but it only increased by 11% when using $Q = 10$. Also notice that

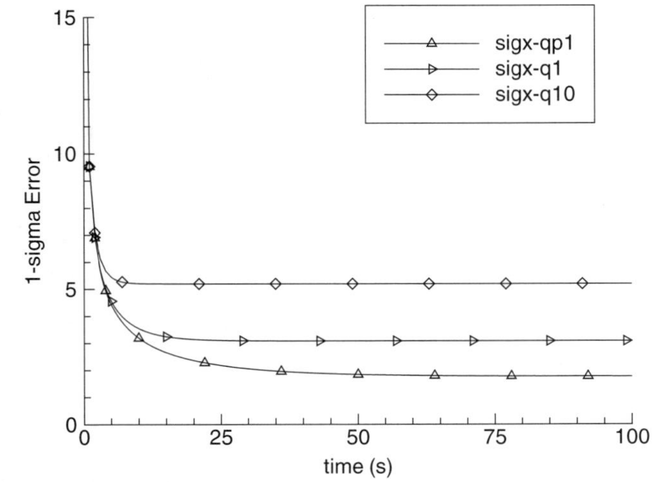

FIGURE 8.4: *A posteriori* $1 - \sigma$ uncertainty for one-state Kalman filter with $\Phi = 1$.

because $Q \ll R$ for these random walk cases, the approximation $P_{ss} \cong \sqrt{QR}$ is fairly accurate.

Even though this Kalman filter example is not representative of real problems, it has characteristics that are applicable to many problems. Specifically:

1. The Kalman gain is initially large—making the estimate sensitive to measurements—but it eventually decays to a steady-state value for most time-invariant systems with fixed sampling intervals.
2. The *a posteriori* variances are also initially large, but stabilize at steady-state values.
3. The time required to reach steady state is a complicated function of Q, R, and Φ, but that time is generally shorter when Q/R and P_0 are large since more weight is given to the measurements.
4. The steady-state variances are also a complicated function of Q, R, and Φ, but the variances generally increase with increasing Q or R.

Example 8.2: Two-State Kalman Filter

This next example is a more realistic problem. We use the same structure but allow the measured "position" state to be driven by another "velocity" state that is modeled as a random walk; that is, $\dot{x}_1 = x_2$, $\dot{x}_2 = q$ where $q(t)$ is white random noise. This is a realistic physical model used for time/frequency errors of rubidium clocks, or one axis of a two-dimensional "constant-velocity" tracking problem. The state transition matrix and state noise matrices are

$$\mathbf{\Phi} = \begin{bmatrix} 1 & T \\ 0 & 1 \end{bmatrix}, \quad \mathbf{Q} = Q_v \begin{bmatrix} T^3/3 & T^2/2 \\ T^2/2 & T \end{bmatrix}$$

where, as before, $Q_v = 0.1$, 1, or 10. This equation for **Q** was obtained using equation (2.2-20), that is,

$$\begin{aligned}\mathbf{Q} &= E\left[\int_0^T \mathbf{\Phi}(\tau)\mathbf{G}\,q(T-\tau)\,dt\int_0^T q(T-\lambda)\mathbf{G}^T\mathbf{\Phi}^T(\lambda)\,d\lambda\right]\\ &= \int_0^T \begin{bmatrix}1 & \tau\\ 0 & 1\end{bmatrix}\begin{bmatrix}0\\1\end{bmatrix} Q_v[0 \quad 1]\begin{bmatrix}1 & 0\\ \tau & 1\end{bmatrix} d\tau\\ &= Q_v\int_0^T \begin{bmatrix}\tau^2 & \tau\\ \tau & 1\end{bmatrix} d\tau\\ &= Q_v\begin{bmatrix}T^3/3 & T^2/2\\ T^2/2 & T\end{bmatrix}\end{aligned}$$

where $E[q(\tau)q(\lambda)] = Q_v\delta(\tau-\lambda)$. In systems with relatively short intervals between measurements, it is not unusual to find the approximation

$$\mathbf{Q} = \begin{bmatrix}0 & 0\\ 0 & Q_vT\end{bmatrix}.$$

Whether or not this is a good approximation depends on the magnitude of T, scaling of the states, and the number of integrators between the process noise and the measurements. We use both expressions in this example, but start with the first. Hence

$$\begin{aligned}\mathbf{P}_{i/i-1} &= \mathbf{\Phi}\mathbf{P}_{i-1/i-1}\mathbf{\Phi}^T + \mathbf{Q}\\ &= \begin{bmatrix}\mathbf{P}_{11} + 2T\mathbf{P}_{11} + T^2\mathbf{P}_{22} + Q_vT^3/3 & \mathbf{P}_{12} + T\mathbf{P}_{22} + Q_vT^2/2\\ \mathbf{P}_{21} + T\mathbf{P}_{22} + Q_vT^2/2 & \mathbf{P}_{22} + Q_vT\end{bmatrix}_{i-1/i-1}\end{aligned}$$

The measured variable is "position":

$$\begin{aligned}y_i &= x_1\\ \mathbf{H} &= [1 \quad 0]\end{aligned}$$

so the Kalman gain matrix is

$$\mathbf{K} = \frac{1}{(P_{11})_{i/i-1} + R}[P_{11} \quad P_{12}]_{i/i-1}.$$

Figure 8.5 shows the *a posteriori* $1-\sigma$ "position" estimate error $\sqrt{(P_{11})_{i/i}}$ when using this model with the same parameter values used for Example 8.2. Notice that after processing the first measurement, $\sqrt{(P_{11})_{i/i}}$ drops to slightly less than the $1-\sigma$ measurement noise of 10. The estimate error quickly reaches a steady-state value, but the time to reach steady state and the steady-state variance are larger than for the comparable one-state problem (Fig. 8.4). Figure 8.6 plots the sample estimate errors on an extended time scale to better show steady-state behavior. Notice that the magnitudes are consistent with Figure 8.5. Figure 8.7 shows the *a posteriori* $1-\sigma$ "velocity" estimate error $\sqrt{(P_{22})_{i/i}}$. Notice that

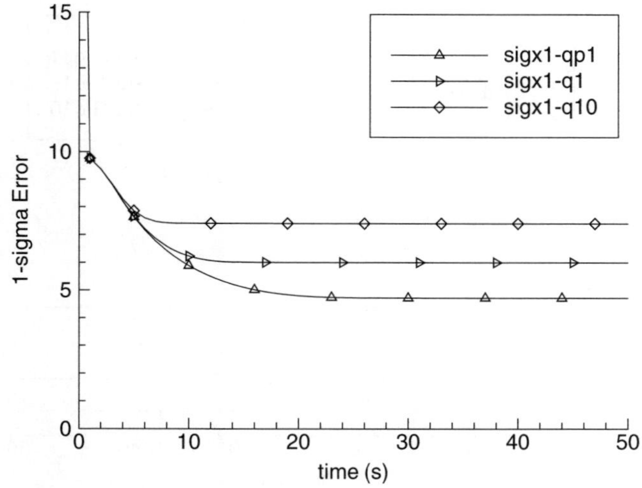

FIGURE 8.5: *A posteriori* $1-\sigma$ position uncertainty of two-state Kalman filter.

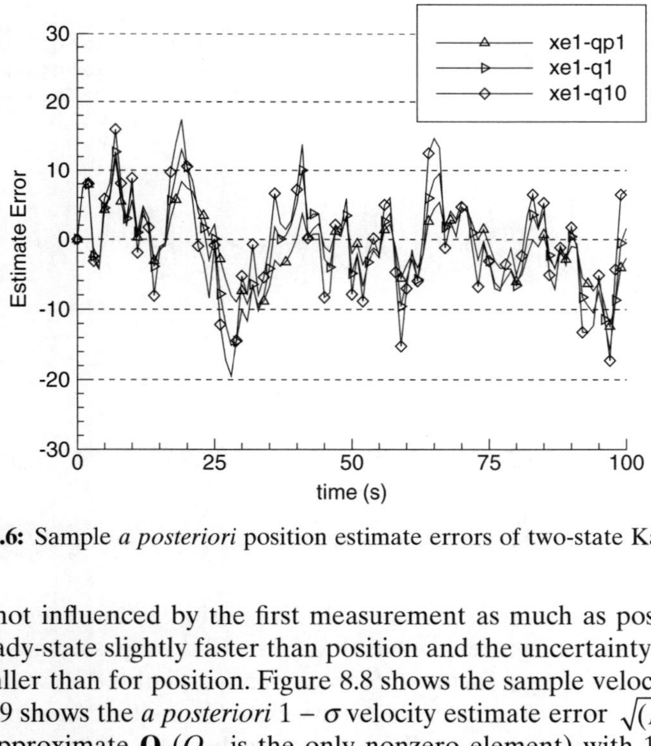

FIGURE 8.6: Sample *a posteriori* position estimate errors of two-state Kalman filter.

velocity is not influenced by the first measurement as much as position, but it reaches steady-state slightly faster than position and the uncertainty for 1 s sampling is smaller than for position. Figure 8.8 shows the sample velocity errors.

Figure 8.9 shows the *a posteriori* $1-\sigma$ velocity estimate error $\sqrt{(P_{22})_{i/i}}$ when using the approximate **Q** (Q_{22} is the only nonzero element) with 1 s sampling. The only significant differences between this plot and Figure 8.7 are the steady-state velocity errors: when using the approximate **Q** the velocity errors are about 12% larger for $Q_v = 10$ and 6% larger for $Q_v = 1$. Differences in the steady-state position errors are less than 1%. Larger velocity errors when **Q** elements are eliminated may seem counterintuitive, but removal of these terms has the effect of decreasing correlation between position and velocity, thus decreasing measurement weighting.

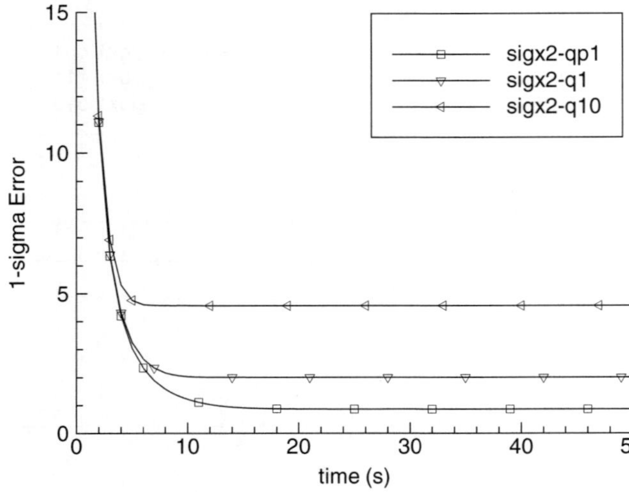

FIGURE 8.7: *A posteriori* $1-\sigma$ velocity uncertainty of two-state Kalman filter.

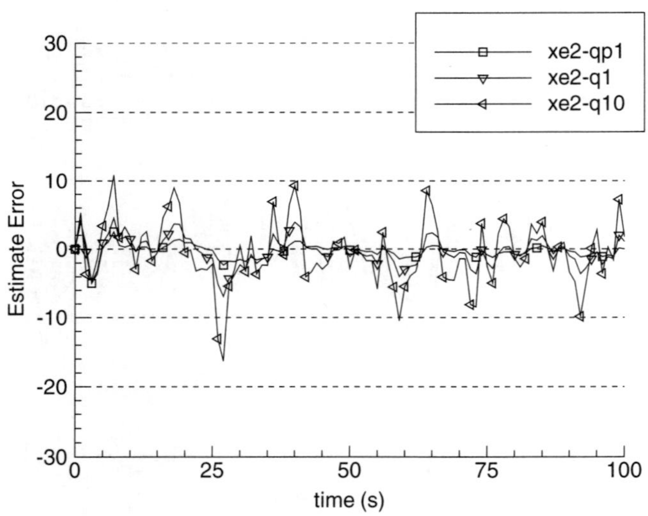

FIGURE 8.8: Sample *a posteriori* velocity estimate errors of two-state Kalman filter.

Figure 8.10 shows the *a posteriori* $1-\sigma$ position estimate error when the measurement sampling rate is reduced to 0.2 s and the measurement noise variance R is increased by a factor of 5. In comparing Figure 8.10 with Figure 8.5, notice that the larger R reduces the filter weight for the first measurements. The time to reach steady state is similar but the steady-state errors are 5%, 9%, and 16% larger, respectively, for $Q_v = 0.1$, 1, and 10. Differences in velocity errors are less than 8%. This result shows that when randomness of the dynamic model is significant compared with the measurement noise, increasing the measurement noise variance by the same ratio as the change in sampling rates is not equivalent. More will be said about this topic when discussing the continuous-time Kalman-Bucy filter.

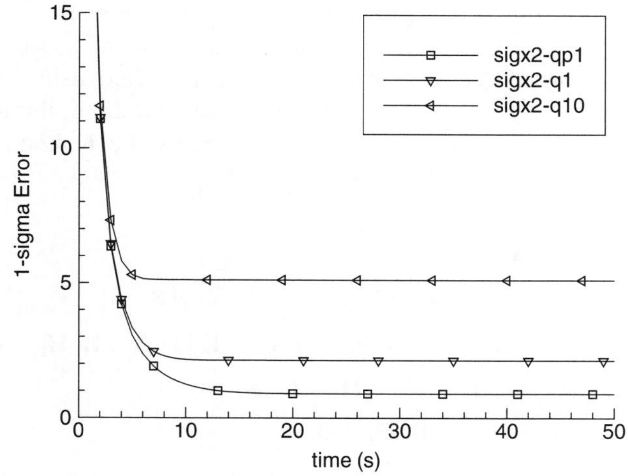

FIGURE 8.9: *A posteriori* $1-\sigma$ velocity uncertainty of two-state Kalman filter: Approximate Q.

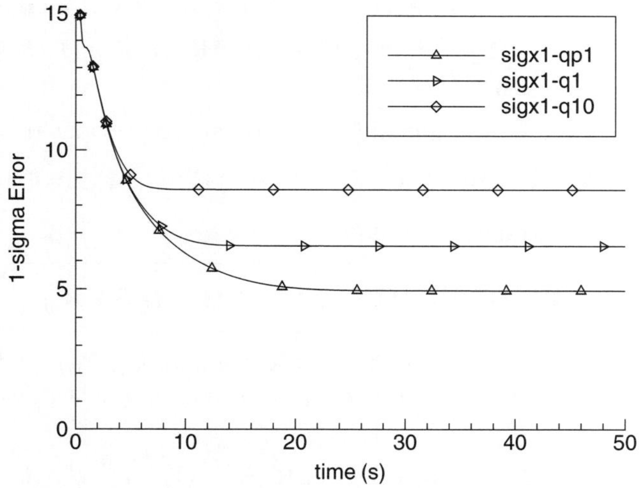

FIGURE 8.10: *A posteriori* $1-\sigma$ position uncertainty of two-state Kalman filter: $T = 0.2$.

8.2 EXTENSIONS OF THE DISCRETE FILTER

We now address two generalizations of the Kalman filter that remove assumptions used in deriving the basic filter. Also discussed are properties of the measurement residuals that allow testing for outlying measurements and computation of the likelihood function.

8.2.1 Correlation between Measurement and Process Noise

Recall that in deriving the Kalman filter it was assumed that measurement and process noise are uncorrelated: $E[\mathbf{q}_{j,j-1}\mathbf{r}_i] = \mathbf{0}$ for all i, j. It is unusual for the correlation to be nonzero when using a physically based model: if correlations are significant they can be handled by recognizing that the measured quantity is a physical phenomenon of the system that can be modeled using state variables. However, for ARMA models with equal numerator and denominator orders (see Section 2.1), the correlation may exist. In these cases the Kalman gain must be modified. We start by computing the *a posteriori* error covariance for this correlated model:

$$\begin{aligned}
\mathbf{P}_{i/i} &= E\left[(\mathbf{x}_i - \hat{\mathbf{x}}_{i/i})(\mathbf{x}_i - \hat{\mathbf{x}}_{i/i})^T\right] \\
&= E\left[((\mathbf{I}-\mathbf{K}_i\mathbf{H}_i)\tilde{\mathbf{x}}_{i/i-1} - \mathbf{K}_i\mathbf{r}_i))((\tilde{\mathbf{x}}_{i/i-1}^T(\mathbf{I}-\mathbf{K}_i\mathbf{H}_i)^T - \mathbf{r}_i^T\mathbf{K}_i^T))\right] \\
&= (\mathbf{I}-\mathbf{K}_i\mathbf{H}_i)\mathbf{P}_{i/i-1}(\mathbf{I}-\mathbf{K}_i\mathbf{H}_i)^T - \mathbf{B}_i\mathbf{K}_i^T - \mathbf{K}_i\mathbf{B}_i^T + \mathbf{K}_i(\mathbf{H}_i\mathbf{B}_i + \mathbf{B}_i^T\mathbf{H}_i^T + \mathbf{R}_i)\mathbf{K}_i^T \\
&= \mathbf{P}_{i/i-1} - \mathbf{K}_i(\mathbf{H}_i\mathbf{P}_{i/i-1} + \mathbf{B}_i^T) - (\mathbf{P}_{i/i-1}\mathbf{H}_i^T + \mathbf{B}_i)\mathbf{K}_i^T \\
&\quad + \mathbf{K}_i(\mathbf{H}_i\mathbf{P}_{i/i-1}\mathbf{H}_i^T + \mathbf{H}_i\mathbf{B}_i + \mathbf{B}_i^T\mathbf{H}_i^T + \mathbf{R}_i)\mathbf{K}_i^T
\end{aligned} \tag{8.2-1}$$

where $\mathbf{B}_i \triangleq E[\mathbf{q}_{i,i-1}\mathbf{r}_i^T]$. The $\mathbf{K}_i$ minimizing tr($\mathbf{P}_{i/i}$) is computed using the orthogonality principle

$$\begin{aligned}
E\left[(\mathbf{x}_i - \hat{\mathbf{x}}_{i/i})\mathbf{y}^T\right] &= E\left[(\tilde{\mathbf{x}}_{i/i-1} - \mathbf{K}_i(\mathbf{H}_i\tilde{\mathbf{x}}_{i/i-1} + \mathbf{r}_i)(\mathbf{H}_i\mathbf{x}_i + \mathbf{r}_i)^T\right] \\
&= E\left[((\mathbf{I}-\mathbf{K}_i\mathbf{H}_i)\tilde{\mathbf{x}}_{i/i-1} - \mathbf{K}_i\mathbf{r}_i)((\hat{\mathbf{x}}_{i/i-1} + \tilde{\mathbf{x}}_{i/i-1})^T\mathbf{H}_i^T + \mathbf{r}_i^T)\right] \\
&= (\mathbf{I}-\mathbf{K}_i\mathbf{H}_i)\mathbf{P}_{i/i-1}\mathbf{H}_i^T - \mathbf{K}_i\mathbf{B}_i^T\mathbf{H}_i^T - \mathbf{K}_i\mathbf{R}_i + (\mathbf{I}-\mathbf{K}_i\mathbf{H}_i)\mathbf{B}_i \\
&= \mathbf{0}
\end{aligned} \tag{8.2-2}$$

where we again used the orthogonality condition at the previous time step:

$$E\left[\tilde{\mathbf{x}}_{i/i-1}\hat{\mathbf{x}}_{i/i-1}^T\right] = E\left[(\mathbf{\Phi}_{i,i-1}\tilde{\mathbf{x}}_{i-1/i-1} + \mathbf{q}_{i,i-1})\hat{\mathbf{x}}_{i-1/i-1}^T\right]\mathbf{\Phi}_{i,i-1}^T = \mathbf{0}.$$

Equation (8.2-2) is rearranged to obtain

$$\boxed{\mathbf{K}_i = \left(\mathbf{P}_{i/i-1}\mathbf{H}_i^T + \mathbf{B}_i\right)\left(\mathbf{H}_i\mathbf{P}_{i/i-1}\mathbf{H}_i^T + \mathbf{B}_i^T\mathbf{H}_i^T + \mathbf{H}_i\mathbf{B}_i + \mathbf{R}_i\right)^{-1}.} \tag{8.2-3}$$

(Again readers may wish to verify that this $\mathbf{K}_i$ minimizes tr($\mathbf{P}_{i/i}$) by differentiation of equation [8.2-1].) Notice that when $\mathbf{K}_i$ is substituted in the last term of equation (8.2-1), that term becomes

$$\mathbf{K}_i(\mathbf{H}_i\mathbf{P}_{i/i-1}\mathbf{H}_i^T + \mathbf{H}_i\mathbf{B}_i + \mathbf{B}_i^T\mathbf{H}_i^T + \mathbf{R}_i)\mathbf{K}_i^T = (\mathbf{P}_{i/i-1}\mathbf{H}_i^T + \mathbf{B}_i)\mathbf{K}_i^T.$$

Thus the *a posteriori* error covariance is

$$\boxed{\mathbf{P}_{i/i} = \mathbf{P}_{i/i-1} - \mathbf{K}_i(\mathbf{H}_i\mathbf{P}_{i/i-1} + \mathbf{B}_i^T).} \tag{8.2-4}$$

Equations (8.2-3) and (8.2-4) are the only modifications of the standard equations required to handle correlated process and measurement noise.

Bryson and Henrikson (1968) present the filter equations resulting when $E[\mathbf{q}_{i,i-1}\mathbf{r}_{i-1}^T] \neq 0$ rather than $E[\mathbf{q}_{i,i-1}\mathbf{r}_i^T] \neq 0$. The modifications involve the filter time update rather than the measurement update. Redefining $\mathbf{B}_{i-1} \triangleq E[\mathbf{q}_{i,i-1}\mathbf{r}_{i-1}^T]$, the time update without control input is

$$\boxed{\begin{aligned}
&\mathbf{D}_i \triangleq E[\mathbf{q}_{i,i-1} \mid \mathbf{r}_{i-1}] = \mathbf{B}_{i-1}\mathbf{R}_{i-1}^{-1} \\
&\hat{\mathbf{x}}_{i/i-1} = \mathbf{\Phi}_{i,i-1}\hat{\mathbf{x}}_{i-1/i-1} + \mathbf{D}_i(\mathbf{y}_{i-1} - \mathbf{H}_{i-1}\hat{\mathbf{x}}_{i-1/i-1}) \\
&\mathbf{P}_{i/i-1} = (\mathbf{\Phi}_{i,i-1} - \mathbf{D}_i\mathbf{H}_{i-1})\mathbf{P}_{i-1/i-1}(\mathbf{\Phi}_{i,i-1} - \mathbf{D}_i\mathbf{H}_{i-1})^T + \mathbf{Q}_{i,i-1} - \mathbf{D}_i\mathbf{R}_{i-1}\mathbf{D}_i^T
\end{aligned}\ .} \tag{8.2-5}$$

The state update is derived by noting that

$$E[\mathbf{r}_{i-1} \mid \mathbf{y}_{i-1}, \hat{\mathbf{x}}_{i-1/i-2}] = \mathbf{y}_{i-1} - \mathbf{H}_{i-1} E[\mathbf{x}_{i-1} \mid \mathbf{y}_{i-1}, \hat{\mathbf{x}}_{i-1/i-2}] = \mathbf{y}_{i-1} - \mathbf{H}_{i-1}\hat{\mathbf{x}}_{i-1/i-1},$$

which gives

$$\begin{aligned}\hat{\mathbf{x}}_{i/i-1} &= E[(\mathbf{\Phi}_{i,i-1}\mathbf{x}_{i-1} + \mathbf{q}_{i,i-1}) \mid \mathbf{y}_{i-1}, \hat{\mathbf{x}}_{i-1/i-2}] \\ &= \mathbf{\Phi}_{i,i-1}\hat{\mathbf{x}}_{i-1/i-1} + \mathbf{D}_i(\mathbf{y}_{i-1} - \mathbf{H}_{i-1}\hat{\mathbf{x}}_{i-1/i-1})\end{aligned}.$$

The covariance update $\mathbf{P}_{i/i-1} \triangleq E\left[(\mathbf{x}_i - \hat{\mathbf{x}}_{i/i-1})(\mathbf{x}_i - \hat{\mathbf{x}}_{i/i-1})^T\right]$ is then obtained using substitution and noting that $E\left[(\mathbf{x}_{i-1} - \hat{\mathbf{x}}_{i-1/i-1})(\mathbf{q}_{i,i-1}^T - \mathbf{r}_{i-1}^T\mathbf{D}_i^T)\right] = \mathbf{0}$ since the measurement update equations are unchanged from equations (8.1-7) to (8.1-13). Equation (8.2-5) is referenced in Chapter 12 when describing implementation of ARMA models in Kalman filters.

8.2.2 Time-Correlated (Colored) Measurement Noise

Another assumption used in the basic filter derivation is $E[\mathbf{r}_j\mathbf{r}_i^T] = \mathbf{0}$ for $j \neq i$. There are several reasons why measurement noise may actually be time-correlated. For example, some systems measure range by tracking the phase of known signals (sinusoids or pseudo-random codes) using phase-locked or delay-locked loops. Range is determined by comparing phases between transmitted and received signals to compute signal delay, which is then divided by the speed of light to compute range. The bandwidth of the tracking loop must be small to suppress measurement noise, but this introduces a loop delay that makes errors in successive range samples correlated when the sampling interval is shorter than loop time constants. Another example occurs in systems that continuously measure delta range by differencing successive phases at adjacent sampling times. Even when sampling intervals are large compared with loop time constants, the correlation of measurement errors for successive phase differences is

$$\frac{E[(r_i - r_{i-1})(r_{i-1} - r_{i-2})]}{E[(r_i - r_{i-1})^2]} = \frac{-\sigma_r^2}{2\sigma_r^2} = -0.5.$$

Bryson and Henrikson (1968) developed two methods for handling time-correlated measurements. Their methods and an alternate are discussed below.

8.2.2.1 *Colored Measurement Noise States* The first approach simply models the correlation as part of the dynamic model by adding a state. We demonstrate this method using a scalar measurement, but the extension to vector measurements is obvious. For a scalar measurement the dynamic model is

$$\begin{bmatrix}\mathbf{x} \\ r\end{bmatrix}_i = \begin{bmatrix}\mathbf{\Phi} & \mathbf{0} \\ \mathbf{0}^T & \rho_r\end{bmatrix}\begin{bmatrix}\mathbf{x} \\ r\end{bmatrix}_{i-1} + \begin{bmatrix}\mathbf{q}_x \\ q_r\end{bmatrix}_i \tag{8.2-6}$$

and the measurement model is

$$y_i = [\mathbf{H} \quad 1]\begin{bmatrix}\mathbf{x}_i \\ r_i\end{bmatrix} \tag{8.2-7}$$

where $\rho_r = E[r_i r_{i-1}]/\sigma_r^2$, $\sigma_r^2 = E[r_i^2]$ and $Q_r = E[q_r^2] = \sigma_r^2(1-\rho_r^2)$. For irregularly sampled continuous systems, use of $\rho_r = e^{-T/\tau}$, where τ is the correlation time constant, will allow the approach to work for variable time steps. Notice that

$$\begin{aligned} E[(y_i - \mathbf{H}\mathbf{x}_i)(y_{i-1} - \mathbf{H}\mathbf{x}_{i-1})] &= E[r_i r_{i-1}] \\ &= \sigma_r^2 \rho_r \end{aligned}$$

which is the desired property.

One potential problem of this augmented state approach is caused by modeling the random measurement noise in equation (8.2-7) as zero. For this model the Kalman gain is

$$\mathbf{K} = \begin{bmatrix} \mathbf{P}_{xx}\mathbf{H}^T + \mathbf{P}_{xr} \\ \mathbf{P}_{rx}\mathbf{H}^T + P_{rr} \end{bmatrix}_{i/i-1} / (\mathbf{H}\mathbf{P}_{xx}\mathbf{H}^T + \mathbf{P}_{rx}\mathbf{H}^T + \mathbf{H}\mathbf{P}_{xr} + P_{rr})_{i/i-1} \tag{8.2-8}$$

where the covariance terms $\mathbf{P}_{xx}$, $\mathbf{P}_{xr}$, P_{rr} are from $\mathbf{P}_{i/i-1}$. Notice that a measurement noise variance term is not included in the denominator. Maybeck (1979, section 5.10) notes that this can potentially cause numerical problems. Random measurement errors are instead accounted for by the Q_r term in the covariance time update

$$\begin{bmatrix} \mathbf{P}_{xx} & \mathbf{P}_{xr} \\ \mathbf{P}_{rx} & P_{rr} \end{bmatrix}_{i/i-1} = \begin{bmatrix} \mathbf{\Phi} & \mathbf{0} \\ \mathbf{0}^T & \rho_r \end{bmatrix} \begin{bmatrix} \mathbf{P}_{xx} & \mathbf{P}_{xr} \\ \mathbf{P}_{rx} & P_{rr} \end{bmatrix}_{i-1/i-1} \begin{bmatrix} \mathbf{\Phi}^T & \mathbf{0} \\ \mathbf{0}^T & \rho_r \end{bmatrix} + \begin{bmatrix} \mathbf{Q}_x & \mathbf{0} \\ \mathbf{0}^T & Q_r \end{bmatrix}_{i,i-1}. \tag{8.2-9}$$

If numerical accuracy is a concern, it may be better to assume that measurement errors are the sum of correlated and random components—often a good assumption. Measurements are thus modeled as $y_i = \mathbf{H}\mathbf{x}_i + r_i + v_i$ where v_i is white noise and $\rho_r\sigma_r^2/(\sigma_r^2 + \sigma_v^2)$ is the desired temporal correlation in measurement noise. With this change, σ_v^2 is added to the denominator of the Kalman gain in equation (8.2-8).

8.2.2.2 Measurement Differencing and Delay State Modeling of Colored Noise The second approach avoids adding colored noise states by differencing the measurements to remove the correlation, and then adding delay states. For time-correlated scalar measurement errors that can be modeled as

$$r_i = \rho_r r_{i-1} + v_i$$

(where v_i is white noise), differencing of successive measurement errors as

$$r_i - \rho_r r_{i-1} = v_i$$

gives an uncorrelated sequence of variables with $\sigma_v^2 = E[v_i^2] = \sigma_r^2(1-\rho_r^2)$. (As before, variable time intervals between measurements can be accommodated if ρ_r is modified appropriately.) This concept is applied to actual vector measurements modeled as

$$\begin{aligned} \mathbf{y}_i &= \mathbf{H}_i\mathbf{x}_i + \mathbf{r}_i \\ &= \mathbf{H}_i\mathbf{x}_i + \mathbf{\Omega}_i\mathbf{r}_{i-1} + \mathbf{v}_i \end{aligned}, \tag{8.2-10}$$

where $\mathbf{\Omega}_i$ is a diagonal matrix that allows for possibly different ρ_r values for different measurements. Measurement differencing produces the modified measurements

$$\boxed{\begin{aligned}\mathbf{z}_i &= \mathbf{y}_i - \mathbf{\Omega}_i\mathbf{y}_{i-1} \\ &= (\mathbf{H}_i\mathbf{x}_i - \mathbf{\Omega}_i\mathbf{H}_{i-1}\mathbf{x}_{i-1}) + \mathbf{v}_i\end{aligned}} \tag{8.2-11}$$

with uncorrelated measurement errors $\mathbf{v}_i$. However, these $\mathbf{z}_i$ variables cannot be directly used as filter measurements because $\mathbf{z}_i$ is a function of the state $\mathbf{x}$ at two different points in time. One approach to this problem uses *delay states* that are a delayed copy of selected filter states at the previous measurement time.

Measurements in real systems are typically a direct function of only a small number of states. For example, position and sometimes velocity measurements are typically used in navigation filters that may have dozens of states. Hence the position and velocity states are the only ones that need be delayed. This state selection is implicitly defined by writing $\mathbf{H}_i = \mathbf{N}_i\mathbf{T}$, where $\mathbf{N}_i$ is created from $\mathbf{H}_i$ by deleting zero columns, and $\mathbf{T}$ is a selection matrix having only one nonzero element (equal to 1) in each row. If there are $k \le n$ nonzero columns in the $m \times n$ matrix $\mathbf{H}_i$, $\mathbf{N}_i$ will be $m \times k$ and $\mathbf{T}$ will be $k \times n$. Hence the measurement equation (8.2-10) may be written as

$$\mathbf{y}_i = \mathbf{N}_i\mathbf{T}\,\mathbf{x}_i + \mathbf{r}_i\,.$$

An augmented state vector is formed as

$$\boxed{\mathbf{x}'_i \triangleq \begin{bmatrix}\mathbf{T}\mathbf{x}_{i-1} \\ \mathbf{x}_i\end{bmatrix} = \begin{bmatrix}\mathbf{d}_i \\ \mathbf{x}_i\end{bmatrix}} \tag{8.2-12}$$

with delay states $\mathbf{d}_i = \mathbf{T}\mathbf{x}_{i-1}$. The time update of this augmented state is

$$\begin{bmatrix}\mathbf{d}_i \\ \mathbf{x}_i\end{bmatrix} = \begin{bmatrix}\mathbf{0} & \mathbf{T} \\ \mathbf{0} & \mathbf{\Phi}_i\end{bmatrix}\begin{bmatrix}\mathbf{d}_{i-1} \\ \mathbf{x}_{i-1}\end{bmatrix} + \begin{bmatrix}\mathbf{0} \\ \mathbf{q}_i\end{bmatrix} \tag{8.2-13}$$

or

$$\mathbf{x}'_i = \mathbf{\Phi}'\mathbf{x}'_{i-1} + \mathbf{q}'_i$$

using the augmented state transition matrix

$$\mathbf{\Phi}' = \begin{bmatrix}\mathbf{0} & \mathbf{T} \\ \mathbf{0} & \mathbf{\Phi}_i\end{bmatrix}.$$

(If the state vector $\mathbf{x}_i$ is partitioned into core dynamic states and biases as described in Section 8.1.2.2, it is usually most efficient to locate the delay states with the dynamic states near the top of the state vector.) Notice that equation (8.2-13) ignores the $\mathbf{d}_{i-1}$ states at each time update. Using this modified version of the state transition matrix, the covariance time update for this augmented system is

$$\boxed{\begin{aligned}\mathbf{P}'_{i/i-1} &\triangleq \begin{bmatrix}\mathbf{P}_{dd} & \mathbf{P}_{dx} \\ \mathbf{P}_{xd} & \mathbf{P}_{xx}\end{bmatrix}_{i/i-1} \\ &= \mathbf{\Phi}'\mathbf{P}'_{i-1/i-1}\mathbf{\Phi}'^T + \mathbf{Q}' \\ &= \begin{bmatrix}\mathbf{T}\mathbf{P}_{xx}\mathbf{T}^T & \mathbf{T}\mathbf{P}_{xx}\mathbf{\Phi}_i^T \\ \mathbf{\Phi}_i\mathbf{P}_{xx}\mathbf{T}^T & \mathbf{\Phi}_i\mathbf{P}_{xx}\mathbf{\Phi}_i^T + \mathbf{Q}_i\end{bmatrix}_{i-1/i-1}\end{aligned}} \tag{8.2-14}$$

If the number of states in $\mathbf{d}_i$ is small, this update can be efficiently implemented without computing separate partitions. Now the differenced measurement is written in terms of the augmented state as

$$\begin{aligned}\mathbf{z}_i &= \mathbf{y}_i - \mathbf{\Omega}_i \mathbf{y}_{i-1} \\ &= \begin{bmatrix} -\mathbf{\Omega}_i \mathbf{N}_{i-1} & \mathbf{N}_i \mathbf{T} \end{bmatrix} \begin{bmatrix} \mathbf{d}_i \\ \mathbf{x}_i \end{bmatrix} + \mathbf{v}_i \end{aligned}. \tag{8.2-15}$$

With this definition, the augmented measurement sensitivity matrix $\mathbf{H}' = [-\Omega_i \mathbf{N}_{i-1} \quad \mathbf{N}_i \mathbf{T}]$ and measurement error covariance

$$\mathbf{R}' = \begin{bmatrix} \sigma_{v1}^2 & \mathbf{0} & \cdots & \mathbf{0} \\ \mathbf{0} & \sigma_{v2}^2 & \cdots & \mathbf{0} \\ \vdots & \vdots & \ddots & \vdots \\ \mathbf{0} & \mathbf{0} & \cdots & \sigma_{vm}^2 \end{bmatrix} = \begin{bmatrix} \sigma_{r1}^2(1-\rho_{r1}^2) & \mathbf{0} & \cdots & \mathbf{0} \\ \mathbf{0} & \sigma_{r2}^2(1-\rho_{r2}^2) & \cdots & \mathbf{0} \\ \vdots & \vdots & \ddots & \vdots \\ \mathbf{0} & \mathbf{0} & \cdots & \sigma_{rm}^2(1-\rho_{rm}^2) \end{bmatrix}$$

are used in the Kalman filter measurement update equations (8.1-10), (8.1-11), and (8.1-13) for the augmented state $\hat{\mathbf{x}}'_{i/i-1}$ and covariance $\mathbf{P}'_{i/i-1}$. These first two methods for handling correlated noise have the advantage that the Kalman filter equations can be used without modification provided that measurements, states, and associated model arrays are defined appropriately.

Implementation of these delay state equations can be made slightly more efficient by carrying the delay states vectors and arrays as separate equations rather than as part of an augmented state. However, this increases the coding effort. The delay state method can be applied when measurements are sampled at varying intervals and measurements are of varying types, but the delay state for the given measurement type must be available at the time of the previous measurement. If the delay state is not available, the measurement must be skipped or processed without differencing.

8.2.2.3 Measurement Differencing and Correlated Measurement-Process Noise

The implementation of Bryson and Henrikson (1968) is equivalent to the above approach, but rather than using delay states, they write the differenced measurements as

$$\begin{aligned}\mathbf{z}_i &= \mathbf{y}_i - \mathbf{\Omega}_i \mathbf{y}_{i-1} \\ &= (\mathbf{H}_i \mathbf{\Phi}_{i,i-1} - \mathbf{\Omega}_i \mathbf{H}_{i-1})\mathbf{x}_{i-1} + (\mathbf{H}_i \mathbf{q}_{i,i-1} + \mathbf{v}_i)\end{aligned}. \tag{8.2-16}$$

Notice that the differenced measurement at time t_i is now dependent on the state vector at time t_{i-1}, so there is a one-step delay in obtaining state estimates. Also notice that the measurement sensitivity matrix in equation (8.2-16) is

$$\mathbf{M}_i = \mathbf{H}_i \mathbf{\Phi}_{i,i-1} - \mathbf{\Omega}_i \mathbf{H}_{i-1} \tag{8.2-17}$$

and the effective measurement noise $\mathbf{H}_i \mathbf{q}_{i,i-1} + \mathbf{v}_i$ involves both random process and measurement noise. Hence the covariance of measurement and process noise is $E[\mathbf{q}_{i,i-1}(\mathbf{H}_i \mathbf{q}_{i,i-1} + \mathbf{v}_i)^T] = \mathbf{Q}_{i,i-1}\mathbf{H}_i^T$. Bryson and Henrikson use $\mathbf{M}_i$, covariance

$\mathbf{S}_i = \mathbf{Q}_{i,i-1}\mathbf{H}_i^T$ and $\mathbf{R}_i = E[\mathbf{v}_i\mathbf{v}_i^T] + \mathbf{H}_i\mathbf{Q}_{i,i-1}\mathbf{H}_i^T$ in a correlated process-measurement noise version of the Kalman filter. The algorithm is somewhat more complicated than equations (8.2-3) to (8.2-4) because backward smoothing is required to handle the one-step delay in time: the equations are summarized below as they would be implemented. To be consistent with our previous usage, the notation below is slightly different from that of Bryson and Henrikson, and we have added dual time subscripts on $\mathbf{\Phi}$ and $\mathbf{Q}$ (i.e., $\mathbf{\Phi}_{i,i-1}, \mathbf{Q}_{i,i-1}$) to avoid confusion caused by use of multiple time indices.

$$
\begin{aligned}
\mathbf{S}_{i-1} &= \mathbf{Q}_{i,i-1}\mathbf{H}_i^T \\
\mathbf{R}_{i-1} &= E[\mathbf{v}_i\mathbf{v}_i^T] + \mathbf{H}_i\mathbf{Q}_{i,i-1}\mathbf{H}_i^T = E[\mathbf{v}_i\mathbf{v}_i^T] + \mathbf{H}_i\mathbf{S}_{i-1} \\
\mathbf{M}_{i-1} &= \mathbf{H}_i\mathbf{\Phi}_{i,i-1} - \mathbf{\Omega}_i\mathbf{H}_{i-1} \\
\mathbf{K}_{i-1} &= \mathbf{P}_{i-1/i-1}\mathbf{M}_{i-1}^T(\mathbf{M}_{i-1}\mathbf{P}_{i-1/i-1}\mathbf{M}_{i-1}^T + \mathbf{R}_{i-1})^{-1} \\
\hat{\mathbf{x}}_{i-1/i} &= \hat{\mathbf{x}}_{i-1/i-1} + \mathbf{K}_{i-1}(\mathbf{z}_i - \mathbf{M}_{i-1}\hat{\mathbf{x}}_{i-1/i-1}) \\
\hat{\mathbf{x}}_{i/i} &= \mathbf{\Phi}_{i,i-1}\hat{\mathbf{x}}_{i-1/i} + \mathbf{S}_{i-1}(\mathbf{M}_{i-1}\mathbf{P}_{i-1/i-1}\mathbf{M}_{i-1}^T + \mathbf{R}_{i-1})^{-1}(\mathbf{z}_i - \mathbf{M}_{i-1}\hat{\mathbf{x}}_{i-1/i-1}) \\
\mathbf{P}_{i-1/i} &= (\mathbf{I} - \mathbf{K}_{i-1}\mathbf{M}_{i-1})\mathbf{P}_{i-1/i-1}(\mathbf{I} - \mathbf{K}_{i-1}\mathbf{M}_{i-1})^T + \mathbf{K}_{i-1}\mathbf{R}_{i-1}\mathbf{K}_{i-1}^T \\
\mathbf{P}_{i/i} &= \mathbf{\Phi}_{i,i-1}\mathbf{P}_{i-1/i-1}\mathbf{\Phi}_{i,i-1}^T + \mathbf{Q}_{i,i-1} - \mathbf{S}_{i-1}(\mathbf{M}_{i-1}\mathbf{P}_{i-1/i-1}\mathbf{M}_{i-1}^T + \mathbf{R}_{i-1})^{-1}\mathbf{S}_{i-1}^T \\
&\quad - \mathbf{\Phi}_{i,i-1}\mathbf{K}_{i-1}\mathbf{S}_{i-1}^T - \mathbf{S}_{i-1}\mathbf{K}_{i-1}^T\mathbf{\Phi}_{i,i-1}^T
\end{aligned}
\tag{8.2-18}
$$

This approach requires more coding than an augmented state vector with delay states, but the computational effort may be comparable. Assumptions of the two methods are similar so flexibility in measurement processing is also comparable. With appropriate compensation, all three methods can handle varying intervals between measurements. However, all are difficult to use when different measurement types appear at different points in time. The following example demonstrates these methods for a simple problem.

Example 8.3: Two-State Model with Correlated Measurement Noise

This example uses the two-state model of Example 8.2, but the measurement errors of adjacent samples are correlated. As before, the state transition and state noise matrices are

$$
\mathbf{\Phi} = \begin{bmatrix} 1 & T \\ 0 & 1 \end{bmatrix}, \quad \mathbf{Q} = Q_v \begin{bmatrix} T^3/3 & T^2/2 \\ T^2/2 & T \end{bmatrix}
$$

and the measurement model is

$$
y_i = [1 \quad 0] \begin{bmatrix} x_1 \\ x_2 \end{bmatrix}_i + r_i
$$

where $E[r_i^2] = 1$. However, in this example $E[r_i r_{i-1}] = 0.5$ rather than zero. Figure 8.11 shows the error in estimated position for a standard Kalman filter and the

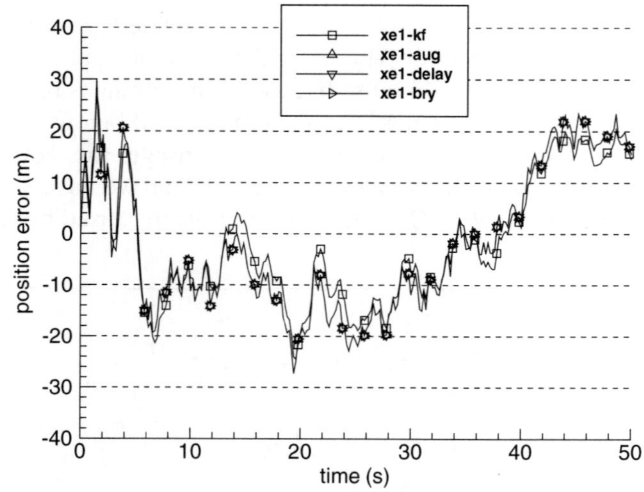

FIGURE 8.11: Error in estimated position for two-state filter with correlated noise.

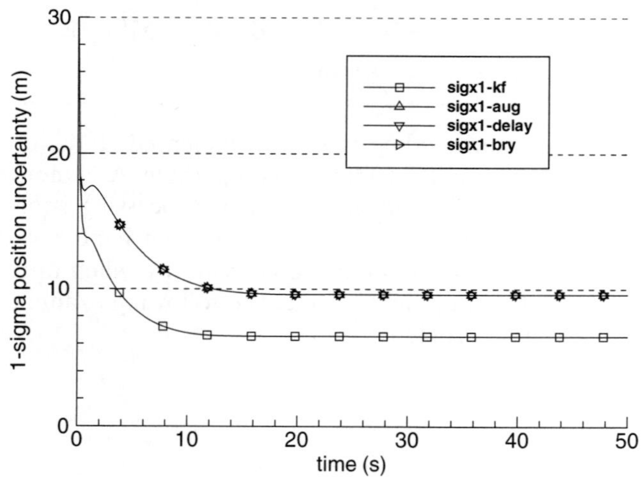

FIGURE 8.12: *A posteriori* $1 - \sigma$ uncertainty in estimated position for two-state filter with correlated noise.

three different methods of handling correlated measurement noise. The labels are defined as:

"xe1-kf" is error in the first state (position) for the standard Kalman filter,

"xe1-aug" uses an augmented state modeling the correlation,

"xe1-delay" uses a delay state, and

"xe1-bry" uses the Bryson-Henrikson method.

Notice that the error in estimated position for the standard Kalman filter is slightly larger than that of the other methods, which are identical. Figure 8.12 shows the $1 - \sigma$ uncertainty computed from the *a posteriori* error covariance matrix for each method. As expected, the standard Kalman filter underestimates actual errors (because it has an incorrect model of measurement noise), while estimates of the other three methods—again identical—are more realistic. Although actual errors of Figure 8.11 are occasionally more than two times the $1-\sigma$ uncertainty of Figure 8.12, statistically this is not significant.

8.2.3 Innovations, Model Validation, and Editing

Recall that the Kalman filter *a priori* m-element measurement residual vector is defined as

$$\begin{aligned}\tilde{\mathbf{y}}_i &\triangleq \mathbf{y}_i - \mathbf{H}_i\hat{\mathbf{x}}_{i/i-1} \\ &= \mathbf{H}_i\mathbf{x}_i + \mathbf{r}_i - \mathbf{H}_i\hat{\mathbf{x}}_{i/i-1}, \\ &= \mathbf{H}_i\tilde{\mathbf{x}}_{i/i-1} + \mathbf{r}_i\end{aligned}$$

with $E[\tilde{\mathbf{y}}_i] = \mathbf{0}$. Because the residuals represent new information not present in the current filter state, they are often referred to as *innovations*. (The term "residuals" is sometimes used to mean *a posteriori* residuals $\mathbf{y}_i - \mathbf{H}_i\tilde{\mathbf{x}}_{i/i}$, but that meaning is not used in this book.) When the Kalman filter accurately models the true system, the measurement residuals will be random variables with covariance

$$\begin{aligned}\mathbf{C}_i &= E[\tilde{\mathbf{y}}_i\tilde{\mathbf{y}}_i^T] \\ &= E[(\mathbf{H}\tilde{\mathbf{x}}_{i/i-1} + \mathbf{r}_i)(\mathbf{H}_i\tilde{\mathbf{x}}_{i/i-1} + \mathbf{r}_i)^T] \\ &= \mathbf{H}\mathbf{P}_{i/i-1}\mathbf{H}^T + \mathbf{R}_i\end{aligned} \tag{8.2-19}$$

for the standard case when $E[\mathbf{q}_{i,i-1}\mathbf{r}_i] = \mathbf{0}$. Furthermore they will be white (uncorrelated in time). To demonstrate this property we first note that

$$\begin{aligned}\tilde{\mathbf{x}}_{i/i-1} &= \mathbf{x}_i - \hat{\mathbf{x}}_{i/i-1} \\ &= \mathbf{\Phi}_i\tilde{\mathbf{x}}_{i-1/i-1} + \mathbf{q}_i\end{aligned}$$

and

$$\begin{aligned}\tilde{\mathbf{x}}_{i/i} &= \mathbf{x}_i - \hat{\mathbf{x}}_{i/i} \\ &= \mathbf{x}_i - \hat{\mathbf{x}}_{i/i-1} - \mathbf{K}_i\tilde{\mathbf{y}}_i \\ &= \tilde{\mathbf{x}}_{i/i-1} - \mathbf{K}_i(\mathbf{H}_i\tilde{\mathbf{x}}_{i/i-1} + \mathbf{r}_i) \\ &= (\mathbf{I} - \mathbf{K}_i\mathbf{H}_i)\tilde{\mathbf{x}}_{i/i-1} - \mathbf{K}_i\mathbf{r}_i\end{aligned}.$$

Then

$$\begin{aligned}E[\tilde{\mathbf{y}}_i\tilde{\mathbf{y}}_{i-1}^T] &= E[(\mathbf{r}_i + \mathbf{H}_i\tilde{\mathbf{x}}_{i/i-1})(\mathbf{r}_{i-1} + \mathbf{H}_{i-1}\tilde{\mathbf{x}}_{i-1/i-2})^T] \\ &= E[(\mathbf{r}_i + \mathbf{H}_i(\mathbf{\Phi}_i\tilde{\mathbf{x}}_{i-1/i-1} + \mathbf{q}_i))(\mathbf{r}_{i-1} + \mathbf{H}_{i-1}\tilde{\mathbf{x}}_{i-1/i-2})^T] \\ &= E[(\mathbf{r}_i + \mathbf{H}_i\mathbf{\Phi}_i[(\mathbf{I} - \mathbf{K}_{i-1}\mathbf{H}_{i-1})\tilde{\mathbf{x}}_{i-1/i-2} - \mathbf{K}_{i-1}\mathbf{r}_{i-1}] + \mathbf{H}_i\mathbf{q}_i)(\mathbf{r}_{i-1} + \mathbf{H}_{i-1}\tilde{\mathbf{x}}_{i-1/i-2})^T] \\ &= -\mathbf{H}_i\mathbf{\Phi}_i\mathbf{K}_{i-1}\mathbf{R}_{i-1} + \mathbf{H}_i\mathbf{\Phi}_i[(\mathbf{I} - \mathbf{K}_{i-1}\mathbf{H}_{i-1})\mathbf{P}_{i-1/i-2}\mathbf{H}_{i-1}^T \\ &= \mathbf{H}_i\mathbf{\Phi}_i[-\mathbf{K}_{i-1}(\mathbf{H}_{i-1}\mathbf{P}_{i-1/i-2}\mathbf{H}_{i-1}^T + \mathbf{R}_{i-1}) + \mathbf{P}_{i-1/i-2}\mathbf{H}_{i-1}^T] \\ &= \mathbf{H}_i\mathbf{\Phi}_i[-\mathbf{P}_{i-1/i-2}\mathbf{H}_{i-1}^T + \mathbf{P}_{i-1/i-2}\mathbf{H}_{i-1}^T] \\ &= \mathbf{0}\end{aligned}$$

The fact that the residuals should be white with known covariance can be used to determine whether the filter is accurately modeling the system. Note that for

Gaussian errors, $E[\tilde{\mathbf{y}}_i^T \mathbf{C}_i^{-1} \tilde{\mathbf{y}}_i]$ should be a χ^2-distributed variables with degrees-of-freedom equal to the number of measurements in $\tilde{\mathbf{y}}_i$. Hence if

$$\boxed{\tilde{\mathbf{y}}_i^T \mathbf{C}_i^{-1} \tilde{\mathbf{y}}_i - m > k\sqrt{2m}} \tag{8.2-20}$$

where k is the editing threshold (typically 4 or 5), then there is reason to believe that either the filter is "diverging" (not following the measurements) or the measurement set is an outlier that should be edited. More will be said about divergence later, but for the moment we assume that when this test is violated the measurement update of state estimate and covariance (eqs. [8.1-11] and [8.1-13]) should be bypassed. This editing algorithm is very effective and should be used whenever a filter processes real data. Multiple levels of editing should be employed in critical operational systems: specifically, a preprocessor should attempt to edit anomalous measurements before they are input to the Kalman filter.

Since the innovations are white when all models are correct, the residuals should be plotted and/or tested for whiteness. When significant patterns appear in the residuals, modeling errors should be suspected. Equations for residual signatures due to bias errors will be derived in a later section, and these curves can be used to help identify error sources.

It is not unusual to encounter problems where the measurement noise (or even process noise) is non-Gaussian, and then questions arise as to whether the Kalman formulation is valid and whether tests such as equation (8.2-20) should be used. The Kalman filter can be derived for linear systems without assuming Gaussian noise, so the derivation is valid provided that the distribution is mean zero, symmetric and the variance is correctly specified. Notice however that the innovation covariance equation (8.2-19) consists of two terms: $\mathbf{HP}_{i/i-1}\mathbf{H}^T$ maps the errors in the state *a priori* estimate into the measurement space, and $\mathbf{R}_i$ accounts for random measurement noise. If process noise is small and the filter has reached steady state, $\mathbf{HP}_{i/i-1}\mathbf{H}^T$ is often much smaller than $\mathbf{R}_i$ so the innovations are primarily measurement noise. In this case non-Gaussian measurement errors will cause the innovations to also be non-Gaussian, and this should be accounted for when using equation (8.2-20). However, if the filter gain is still "open," then the innovations are often close to Gaussian even when $\mathbf{r}_i$ is not. In this case equation (8.2-20) can be used.

Beware however of nonsymmetric noise distributions: they can significantly distort Kalman filter behavior. In these cases a nonlinear transformation on the measurements should be used to make the measurement errors more symmetric. Nonlinear systems will be discussed in the next chapter.

8.2.3.1 The Kalman Filter as a Whitening Transformation The fact that the Kalman filter innovations are uncorrelated with a known covariance is not only useful for detecting outlying measurements, but also allows easy computation of the likelihood function. Recall that in Chapter 4 the measurement vector at time t_i was deterministically modeled for least-square estimation as

$$\mathbf{y}_i = \mathbf{H}_i(\mathbf{\Phi}_i \mathbf{\Phi}_{i-1} \cdots \mathbf{\Phi}_1)\mathbf{x}_{T0} + \mathbf{r}_i$$

where $\mathbf{x}_{T0}$ is the true value of an unknown state vector at an epoch time t_0. These measurements were differenced with predicted measurements $\hat{\mathbf{y}}_i = \mathbf{H}_i(\mathbf{\Phi}_i \mathbf{\Phi}_{i-1} \cdots \mathbf{\Phi}_1)\mathbf{x}_0$ based on a hypothesized epoch state $\mathbf{x}_0$ to compute measurement residuals

$$\begin{aligned}\tilde{\mathbf{y}}_i &= \mathbf{y}_i - \mathbf{H}_i(\boldsymbol{\Phi}_i\boldsymbol{\Phi}_{i-1}\cdots\boldsymbol{\Phi}_1)\mathbf{x}_0\\ &= \mathbf{H}_i(\boldsymbol{\Phi}_i\boldsymbol{\Phi}_{i-1}\cdots\boldsymbol{\Phi}_1)(\mathbf{x}_{T0}-\mathbf{x}_0)+\mathbf{r}_i^{\cdot}\end{aligned}$$

The probability density of the measurement residuals for the Bayesian case was defined using the residual covariance

$$E\left(\begin{bmatrix}\tilde{\mathbf{y}}_1\\ \tilde{\mathbf{y}}_2\\ \vdots\\ \tilde{\mathbf{y}}_m\end{bmatrix}\begin{bmatrix}\tilde{\mathbf{y}}_1^T & \tilde{\mathbf{y}}_2^T & \cdots & \tilde{\mathbf{y}}_m^T\end{bmatrix}\right)=$$

$$\begin{bmatrix}\mathbf{H}_1\\ \mathbf{H}_2\\ \vdots\\ \mathbf{H}_m\end{bmatrix}\mathbf{P}_o\begin{bmatrix}\mathbf{H}_1^T & \mathbf{H}_2^T & \cdots & \mathbf{H}_m^T\end{bmatrix}+\begin{bmatrix}\mathbf{R}_1 & \mathbf{0} & \cdots & \mathbf{0}\\ \mathbf{0} & \mathbf{R}_2 & \cdots & \mathbf{0}\\ \vdots & \vdots & \ddots & \vdots\\ \mathbf{0} & \mathbf{0} & \cdots & \mathbf{R}_m\end{bmatrix}$$

where $\mathbf{P}_o = E[(\mathbf{x}_0 - \mathbf{x}_{T0})(\mathbf{x}_0 - \mathbf{x}_{T0})^T]$. For simplicity we assumed that $E[\mathbf{r}_i\mathbf{r}_j^T]=\mathbf{R}_i\delta_{ij}$, which is the typical case. Use of this expression allowed the minimum variance estimate of $\mathbf{x}_0$ to be computed by maximizing the log of the probability density.

A similar approach based on an epoch state cannot be directly used for the stochastic model of the Kalman filter because process noise $\mathbf{q}_i$ in the time update

$$\begin{aligned}\mathbf{x}_i &= \boldsymbol{\Phi}_i\mathbf{x}_{i-1}+\mathbf{q}_i\\ &= \boldsymbol{\Phi}_i(\boldsymbol{\Phi}_{i-1}\mathbf{x}_{i-2}+\mathbf{q}_{i-1})+\mathbf{q}_i\\ &= \boldsymbol{\Phi}_i(\boldsymbol{\Phi}_{i-1}(\boldsymbol{\Phi}_{i-2}(\cdots\mathbf{x}_0)+\mathbf{q}_{i-2})+\mathbf{q}_{i-1})+\mathbf{q}_i\\ &= (\boldsymbol{\Phi}_i\boldsymbol{\Phi}_{i-1}\boldsymbol{\Phi}_{i-2}\cdots\boldsymbol{\Phi}_1)\mathbf{x}_0+(\boldsymbol{\Phi}_i(\boldsymbol{\Phi}_{i-1}(\boldsymbol{\Phi}_{i-2}(\cdots+\mathbf{q}_1)+\mathbf{q}_{i-2})+\mathbf{q}_{i-1})+\mathbf{q}_i)\end{aligned}$$

greatly complicates the covariance calculation. Notice that the state $\mathbf{x}_i$ is composed of two terms: one propagates the epoch state $\mathbf{x}_0$ through the product of state transition matrices (just as with the least-squares model), and the other, $(\boldsymbol{\Phi}_i(\boldsymbol{\Phi}_{i-1}(\boldsymbol{\Phi}_{i-2}(\cdots + \mathbf{q}_1) + \mathbf{q}_{i-2}) + \mathbf{q}_{i-1}) + \mathbf{q}_i)$, propagates a sequence of discrete process noises. Hence measurement residuals formed from an epoch state will contain an extra term $\tilde{\mathbf{x}}_i^q$ that is the collective sum of all process noise effects:

$$\begin{aligned}\tilde{\mathbf{y}}_i &= \mathbf{y}_i - \mathbf{H}_i\mathbf{x}_0\\ &= \mathbf{H}_i(\boldsymbol{\Phi}_i\boldsymbol{\Phi}_{i-1}\cdots\boldsymbol{\Phi}_1)(\mathbf{x}_0-\mathbf{x}_{T0})+\mathbf{H}_i\tilde{\mathbf{x}}_i^q+\mathbf{r}_i^{\cdot}\end{aligned}$$

The covariance of the stacked $\tilde{\mathbf{y}}_i$ vectors will contain additional highly correlated terms due to $\tilde{\mathbf{x}}_i^q$. Not only is this covariance difficult to compute, but inversion of the large dimension correlated matrix would be computationally expensive and subject to numerical errors.

Fortunately the sequentially computed innovations from the Kalman filter do not have this problem because they are uncorrelated and the covariance is known. The Kalman filter can be viewed as a whitening transformation $\mathbf{T}$ on the correlated measurements $\mathbf{y}_i$; that is,

$$\begin{bmatrix}\tilde{\mathbf{y}}_1\\ \tilde{\mathbf{y}}_2\\ \vdots\\ \tilde{\mathbf{y}}_m\end{bmatrix}=\mathbf{T}\begin{bmatrix}\mathbf{y}_1\\ \mathbf{y}_2\\ \vdots\\ \mathbf{y}_m\end{bmatrix}$$

where

$$E\left(\begin{bmatrix}\tilde{\mathbf{y}}_1\\ \tilde{\mathbf{y}}_2\\ \vdots\\ \tilde{\mathbf{y}}_m\end{bmatrix}\begin{bmatrix}\tilde{\mathbf{y}}_1^T & \tilde{\mathbf{y}}_2^T & \cdots & \tilde{\mathbf{y}}_m^T\end{bmatrix}\right)=\begin{bmatrix}\mathbf{C}_1 & \mathbf{0} & \cdots & \mathbf{0}\\ \mathbf{0} & \mathbf{C}_2 & \cdots & \mathbf{0}\\ \vdots & \vdots & \ddots & \vdots\\ \mathbf{0} & \mathbf{0} & \cdots & \mathbf{C}_m\end{bmatrix}$$

and $\mathbf{C}_i$ is obtained from equation (8.1-8). Thus the log likelihood of the conditional density for k measurement residuals is easily computed as

$$\ln L(\mathbf{y}\,|\,\mathbf{x}_0,\boldsymbol{\theta})=-\frac{1}{2}\left[\sum_{i=1}^{k} m_i\ln(2\pi)+\ln|\mathbf{C}_i|+\tilde{\mathbf{y}}_i^T\mathbf{C}_i^{-1}\tilde{\mathbf{y}}_i\right] \tag{8.2-21}$$

where $\boldsymbol{\theta}$ is a vector representing all the model parameters—such as Markov time constants, process noise power spectral densities (PSDs), and measurement noise variances—used in the Kalman filter. The log likelihood is a very useful metric that can be used to compute optimal values of elements in $\boldsymbol{\theta}$. It also allows models with different structures and numbers of states to be compared, and the optimal model to be selected. Maximum likelihood parameter estimation is discussed in Chapter 11.

8.3 CONTINOUS-TIME KALMAN-BUCY FILTER

The Kalman-Bucy filter (Kalman and Bucy 1961) assumes that measurements are obtained continuously rather than at discrete times, and the dynamic model is defined using stochastic differential equations rather than difference equations. The possibly time-varying system dynamic model is

$$\dot{\mathbf{x}}(t)=\mathbf{F}(t)\mathbf{x}(t)+\mathbf{G}(t)\mathbf{q}_c(t) \tag{8.3-1}$$

for n-vector $\mathbf{x}(t)$ where $\mathbf{q}_c(t)$ is zero-mean white noise. This equation is essentially the same one used in Chapter 2 when discussing discrete sampling of continuous system dynamics. Known control inputs can also be included in the model but are eliminated in this derivation because they do not affect the error covariance or measurement weighting. The measurement m-vector

$$\mathbf{y}(t)=\mathbf{H}(t)\mathbf{x}(t)+\mathbf{r}_c(t) \tag{8.3-2}$$

is continuously available and processed in the filter. In addition to differences in availability, this measurement model also differs from that of the discrete Kalman filter in that the types of measurements included in $\mathbf{y}(t)$ are not normally allowed to change with time. (A change would introduce a step discontinuity in the estimates.) The process and measurement noises are assumed to be independent (white) zero-mean random processes with covariances

$$\left.\begin{aligned}E[\mathbf{q}_c(t)\mathbf{q}_c^T(\tau)]&=\mathbf{Q}_s(t)\delta(t-\tau)\\ E[\mathbf{r}_c(t)\mathbf{r}_c^T(\tau)]&=\mathbf{R}_s(t)\delta(t-\tau)\\ E[\mathbf{q}_c(t)\mathbf{r}_c^T(\tau)]&=\mathbf{0}\end{aligned}\right\}\quad \text{for all } t,\tau$$

where $\delta(t-\tau)$ is the Dirac delta function. Since

$$\int_{-\infty}^{\infty} \delta(t)dt = 1,$$

δ has units of 1/unit-of-time and $\mathbf{Q}_s(t)$ and $\mathbf{R}_s(t)$ are PSD matrices with units equal to units of $\mathbf{q}_c\mathbf{q}_c^T$ or $\mathbf{r}_c\mathbf{r}_c^T$, respectively, multiplied by units of time. Also $\mathbf{R}_s(t)$ is assumed to be positive definite for all time, implying that no component or linear combination of $\mathbf{y}$ can be measured perfectly.

We digress briefly to discuss differences between continuous and discrete measurement noise. If measurements are measured in meters, the discrete measurement covariance $\mathbf{R}_i = E[\mathbf{r}_i\mathbf{r}_i^T]$ at time t_i will have units of m^2. If time is measured in seconds, the continuous-time PSD $\mathbf{R}_s(t)$ will have units of $m^2 \cdot s$. Based on this dimensional analysis, one may suspect that a continuous filter using $\mathbf{R}_s(t_i) = \mathbf{R}_i T$ for "small" $T = t_i - t_{i-1}$ will have approximately the same performance as a discrete filter using measurement noise covariance $\mathbf{R}_i$. This is in fact the case, and it can be justified by considering the averaging effect of multiple measurements. It is easily shown that the variance of the mean of p random independent zero-mean samples is equal to the variance of the individual samples divided by p, that is,

$$\begin{aligned}
E\left(\frac{1}{p}\sum_{i=1}^{p} r_i\right)^2 &= \frac{1}{p^2} E[r_1 + r_2 + \ldots + r_p]^2 \\
&= \frac{1}{p^2} E[r_1^2 + r_2^2 + \ldots + r_p^2] \\
&= \frac{1}{p}\sigma_r^2
\end{aligned}$$

where $E[r_i] = 0$ and $E[r_i r_j] = \sigma_r^2 \delta_{ij}$. Hence if the state $\mathbf{x}$ is nearly constant over the measurement update interval, then a discrete Kalman filter using a measurement update interval of $\Delta t = T/p$ should use a discrete measurement noise covariance $\mathbf{R}_i(\Delta t) = \mathbf{R}_i(T)(T/\Delta t)$ for the measurements to have the equivalent effect as a discrete filter using an update interval of T with measurement noise covariance $\mathbf{R}_i(T)$. In the limit as $\Delta t \to 0$,

$$\mathbf{R}_s(t_i) = \lim_{\Delta t \to 0} (\mathbf{R}_i(\Delta t)\Delta t) \cong \mathbf{R}_i(T)T.$$

Simon (2006, section 8.1.2) provides another argument justifying this relationship.

We now return to the continuous-time filter. The optimal estimate for $\mathbf{x}(t_i)$ is defined as a weighted integral of measurements from t_0 to t:

$$\hat{\mathbf{x}}(t_1 \mid t) \triangleq \int_{t_0}^{t} \mathbf{A}(t_1, \tau)\mathbf{y}(\tau)d\tau,$$

such that

$$E\left[(\mathbf{x}(t_1) - \hat{\mathbf{x}}(t_1))(\mathbf{x}(t_1) - \hat{\mathbf{x}}(t_1))^T\right]$$

is minimum. In this section we are primarily concerned with the filter solution at $t_1 = t$, and the $n \times m$ matrix $\mathbf{A}(t_1,\tau)$ is assumed to be continuously differentiable in both arguments.

Weiner computed a solution for $\mathbf{A}(t_1,\tau)$ when $\mathbf{F}$, $\mathbf{G}$, $\mathbf{H}$, $\mathbf{Q}_s$, and $\mathbf{R}_s$ are fixed (time-invariant system) and the system dynamics are stable (eigenvalues of $\mathbf{F}$ in the left-half complex plane). The solution is obtained using the Weiner-Hopf equation

$$E[\mathbf{x}(t_1)\mathbf{y}^T(\lambda)] = \int_{t_0}^{t} \mathbf{A}(t_1,\tau)E[\mathbf{y}(\tau)\mathbf{y}^T(\lambda)]d\tau.$$

Weiner's solution is described later in Section 8.6. Kalman and Bucy developed a recursive solution to the more general time-varying (with possibly unstable dynamics) filter problem where the optimal estimate is computed as

$$\boxed{\frac{d\hat{\mathbf{x}}(t\,|\,t)}{dt} = \mathbf{F}(t)\hat{\mathbf{x}}(t\,|\,t) + \mathbf{K}_c(t)[\mathbf{y}(t) - \mathbf{H}(t)\hat{\mathbf{x}}(t\,|\,t)].} \tag{8.3-3}$$

The difficulty comes in computing $\mathbf{K}_c(t)$. Kalman and Bucy solved the problem using "advanced and unconventional methods" including calculus of variations, Hamiltonian functions, and co-states (for definitions, see Athans and Falb (1966) or other texts on optimal control). A simplified version of that derivation is also included in Section 8.6. However, in this section we use the more common derivation: the continuous filter is defined as a limiting case of the discrete filter when the interval T between measurements approaches zero.

For small T the state transition matrix may be approximated as $\mathbf{\Phi}(T) \cong \mathbf{I} + \mathbf{F}T$. For simplicity we assume that $\mathbf{F}$, $\mathbf{G}$, and $\mathbf{Q}_s$ are constant over the integration interval, although they may change between intervals. The state at time $t_{i+1} = t_i + T$ is

$$\mathbf{x}(t_{i+1}) = (\mathbf{I} + \mathbf{F}T)\mathbf{x}(t_i) + \mathbf{q}_{i+1,i}$$

where

$$\begin{aligned}\mathbf{q}_{i+1,i} &= \int_{t_i}^{t_{i+1}} \mathbf{\Phi}(t_{i+1},\lambda)\mathbf{G}\,\mathbf{q}_c(\lambda)\,d\lambda \\ &= \int_{t_i}^{t_{i+1}} (\mathbf{I} + \mathbf{F}(t_{i+1} - \lambda))\,\mathbf{G}\,\mathbf{q}_c(\lambda)\,d\lambda\end{aligned}.$$

Using results from Section 2.2.1, the discrete process noise matrix $\mathbf{Q}_{i+1,i} = E[\mathbf{q}_{i+1,i}\mathbf{q}_{i+1,i}^T]$ is computed as

$$\begin{aligned}\mathbf{Q}_{i+1,i} &= \int_{t_i}^{t_{i+1}} [\mathbf{I} + \mathbf{F}(t_{i+1} - \lambda)]\mathbf{G}\,\mathbf{Q}_s\mathbf{G}^T[\mathbf{I} + \mathbf{F}^T(t_{i+1} - \lambda)]\,d\lambda \\ &= \int_0^T [\mathbf{I} + \mathbf{F}\tau]\mathbf{G}\,\mathbf{Q}_s\mathbf{G}^T[\mathbf{I} + \mathbf{F}^T\tau]\,d\tau \\ &= \mathbf{G}\,\mathbf{Q}_s\mathbf{G}^T T + O(T^2)\end{aligned}$$

where $Q(T^2)$ denotes terms of order T^2. This result is then used in equation (8.1-5) to compute the *a priori* covariance matrix

$$\begin{aligned}\mathbf{P}_{i+1/i} &\cong (\mathbf{I} + \mathbf{F}T)\mathbf{P}_{i/i}(\mathbf{I} + \mathbf{F}^T T) + \mathbf{G}\,\mathbf{Q}_s\mathbf{G}^T T + O(T^2) \\ &= \mathbf{P}_{i/i} + (\mathbf{F}\mathbf{P}_{i/i} + \mathbf{P}_{i/i}\mathbf{F}^T)T + \mathbf{G}\,\mathbf{Q}_s\mathbf{G}^T T + O(T^2)\end{aligned}.$$

Hence to first order,

$$\frac{\mathbf{P}_{i+1/i} - \mathbf{P}_{i/i}}{T} \cong \mathbf{F}\,\mathbf{P}_{i/i} + \mathbf{P}_{i/i}\mathbf{F}^T + \mathbf{G}\,\mathbf{Q}_s\mathbf{G}^T.$$

The normalized difference between successive *a priori* covariances is thus

$$\frac{\mathbf{P}_{i+1/i}-\mathbf{P}_{i/i-1}}{T}=\frac{1}{T}((\mathbf{P}_{i+1/i}-\mathbf{P}_{i/i})+(\mathbf{P}_{i/i}-\mathbf{P}_{i/i-1}))$$
$$=\mathbf{F}\mathbf{P}_{i/i}+\mathbf{P}_{i/i}\mathbf{F}^T+\mathbf{G}\mathbf{Q}_s\mathbf{G}^T-\frac{1}{T}\mathbf{K}_i\mathbf{H}\mathbf{P}_{i/i-1} \tag{8.3-4}$$

where we have used $\mathbf{P}_{i/i}-\mathbf{P}_{i/i-1}=-\mathbf{K}_i\mathbf{H}\mathbf{P}_{i/i-1}$ from equation (8.1-13). Notice the term

$$\frac{1}{T}\mathbf{K}_i\mathbf{H}\mathbf{P}_{i/i-1}=\mathbf{P}_{i/i-1}\mathbf{H}^T\left(T\,\mathbf{H}\mathbf{P}_{i/i-1}\mathbf{H}^T+T\,\mathbf{R}\right)^{-1}\mathbf{H}\mathbf{P}_{i/i-1}.$$

In the limit as $T\to 0$, $T\mathbf{H}\mathbf{P}_{i/i-1}\mathbf{H}^T\to 0$, but under our assumption that the discrete $\mathbf{R}$ must be scaled by $1/T$ as T is reduced, $T\mathbf{R}$ is a constant equal to $\mathbf{R}_s$. Hence defining

$$\dot{\mathbf{P}}=\lim_{T\to 0}\left(\frac{\mathbf{P}_{i+1/i}-\mathbf{P}_{i/i-1}}{T}\right)$$

we obtain the differential Riccati equation

$$\boxed{\dot{\mathbf{P}}(t)=\mathbf{F}\mathbf{P}(t)+\mathbf{P}(t)\mathbf{F}^T+\mathbf{G}\mathbf{Q}_s\mathbf{G}^T-\mathbf{K}_c(t)\mathbf{R}_s\mathbf{K}_c^T(t)} \tag{8.3-5}$$

where the continuous-time gain matrix is defined as

$$\boxed{\mathbf{K}_c(t)=\mathbf{P}(t)\mathbf{H}^T\mathbf{R}_s^{-1}.} \tag{8.3-6}$$

Although not explicitly shown, $\mathbf{F}$, $\mathbf{H}$, $\mathbf{Q}_s$, and $\mathbf{R}_s$ may be functions of time. Notice that process noise, represented by $\mathbf{G}\mathbf{Q}_s\mathbf{G}^T$, increases the error covariance while measurements reduce the covariance through the effect of $\mathbf{K}_c(t)\mathbf{R}_s\mathbf{K}_c^T(t)$. Large values of $\mathbf{K}_c(t)$ indicate that the measurements have much information compared with the current $\mathbf{P}$ and thus reduce the covariance significantly.

Equation (8.3-6) for $\mathbf{K}_c(t)$ is substituted in the state differential equation (8.3-3), or using simpler notation

$$\boxed{\dot{\hat{\mathbf{x}}}(t)=\mathbf{F}\,\hat{\mathbf{x}}(t)+\mathbf{K}_c(t)[\mathbf{y}(t)-\mathbf{H}\hat{\mathbf{x}}(t)].} \tag{8.3-7}$$

Figure 8.13 shows the data flow of the continuous-time Kalman filter. As with the discrete filter, the continuous filter must be initialized with an *a priori* estimate $\hat{\mathbf{x}}(0)$ and corresponding error covariance $\mathbf{P}(0)=E\left[(\mathbf{x}(0)-\hat{\mathbf{x}}(0))(\mathbf{x}(0)-\hat{\mathbf{x}}(0))^T\right]$. The continuous filter can also be derived using other approaches (see Johnson 1969).

The covariance differential equations (8.3-5) and (8.3-7) can be numerically integrated using any standard method such as Runge-Kutta, Bulirsch-Stoer, or predictor-corrector. Notice that only $n(n+1)/2$ elements in $\mathbf{P}$ must be integrated since $\mathbf{P}$ is symmetric. However, integration can be computationally intensive when time constants in $\mathbf{F}$ are small. In fact, it may be necessary to integrate equation (8.3-5) using step sizes much smaller than the shortest time constants in $\mathbf{F}$. Two alternative methods attempt to reduce computational effort: the state transition approach and the Chandrasekhar algorithm (see Kailath 2000; Simon 2006). Another concern is numerical round-off since, just like the discrete Kalman filter, the covariance update is obtained as the difference between symmetric positive semi-definite matrices. Furthermore, the condition number of the covariance is double that of the measurement sensitivity matrix, just as with least-squares estimation. Morf

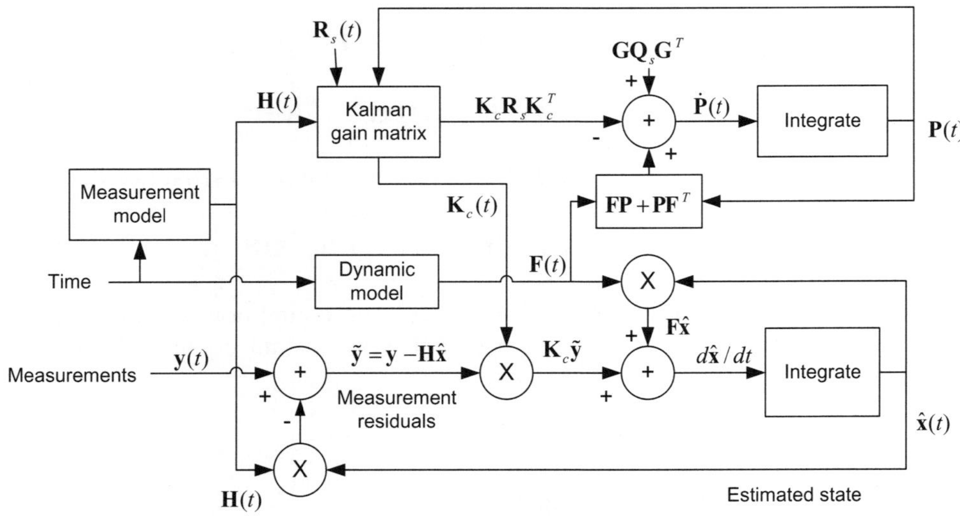

FIGURE 8.13: Continuous-time Kalman-Bucy filter data flow.

et al. (1978) present a square-root version of the continuous filter guaranteeing that $\mathbf{P}$ remains positive semi-definite, and unlike earlier methods, it does not contain any antisymmetric matrix. Simon (2006, section 8.8.3) presents another square-root algorithm, but it requires formation of the covariance derivative equation (8.3-5), so it is subject to some of the same numerical problems as straight integration of equation (8.3-5).

As noted previously, the continuous Kalman-Bucy filter is rarely implemented in practice, so we defer to other sources for more information on practical implementations and generalizations of the filter that relax assumptions. The primary reasons for studying the continuous filter are to understand properties of the discrete filter, to derive extensions of the discrete filter, and to compute the steady-state covariance $\mathbf{P}$, which is usually easier than for the discrete filter. This is demonstrated in the following example.

Example 8.4: Continuous Two-State Filter

This example uses the two-state model of Example 8.2, but measurements are continuous rather than discrete. The continuous dynamic model is defined by the arrays

$$\mathbf{F} = \begin{bmatrix} 0 & 1 \\ 0 & 0 \end{bmatrix}, \quad \mathbf{G}\mathbf{Q}_s\mathbf{G}^T = \begin{bmatrix} 0 & 0 \\ 0 & Q_{sv} \end{bmatrix},$$

where Q_{sv} is equal to either 0.1, 1, or 10 to match the results of Example 8.2 since the measurement update interval T of Example 8.2 was 1 s. The measurement sensitivity matrix and noise PSD are

$$\mathbf{H} = [1 \quad 0], \quad \mathbf{R}_s = 100.$$

Again $\mathbf{R}_s$ is equal to $\mathbf{R}$ of the discrete model because $T = 1$. Hence

$$\mathbf{K}_c(t) = \mathbf{P}(t)\mathbf{H}^T\mathbf{R}_s^{-1} = \begin{bmatrix} P_{11}/R_s \\ P_{21}/R_s \end{bmatrix}$$

and

$$\begin{aligned}\dot{\mathbf{P}}(t) &= \mathbf{F}\mathbf{P}(t) + \mathbf{P}(t)\mathbf{F}^T + \mathbf{G}\mathbf{Q}_s\mathbf{G}^T - \mathbf{K}_c(t)\mathbf{R}_s\mathbf{K}_c^T(t) \\ &= \begin{bmatrix} P_{21} & P_{22} \\ 0 & 0 \end{bmatrix} + \begin{bmatrix} P_{12} & 0 \\ P_{22} & 0 \end{bmatrix} + \begin{bmatrix} 0 & 0 \\ 0 & Q_{sv} \end{bmatrix} - \begin{bmatrix} P_{11}^2/R_s & P_{11}P_{21}/R_s \\ P_{11}P_{21}/R_s & P_{21}^2/R_s \end{bmatrix}. \\ &= \begin{bmatrix} 2P_{12} - P_{11}^2/R_s & P_{22} - P_{11}P_{12}/R_s \\ P_{22} - P_{11}P_{12}/R_s & Q_{sv} - P_{12}^2/R_s \end{bmatrix}\end{aligned}$$

Notice that $P_{21} = P_{12}$ (because of symmetry) so that only the upper triangular portion of $\mathbf{P}$ need be computed. From this equation we immediately see that the steady-state solution $\dot{\mathbf{P}} = \mathbf{0}$ is defined by

$$\begin{aligned} P_{12} &= \sqrt{Q_{sv}R_s} \\ P_{11} &= \sqrt{2P_{12}R_s} \\ P_{22} &= P_{11}P_{12}/R_s \end{aligned}$$

Table 8.1 compares the steady-state standard deviations for the continuous and discrete filters. Notice that the steady-state standard deviations of the continuous filter and discrete filter using $T = 0.2$ s compare closely. For the discrete filter using $T = 1$, the $\sqrt{P_{22}}$ values agree within 9%, but $\sqrt{P_{11}}$ is 6% to 17% smaller in the discrete filter than the continuous filter. This same behavior was noted in Example 8.2 when comparing the $T = 0.2$ and $T = 1$ cases. Even though the dynamic model of this example uses a free integrator with no feedback (and thus the time constant is infinite), the presence of large magnitude process noise invalidates the assumption that the "state $\mathbf{x}$ is nearly constant over the measurement update interval," which allows scaling the discrete $\mathbf{R}$ as $\mathbf{R}_i(\Delta t) = \mathbf{R}_i(\mathrm{T})(T/\Delta t)$. As observed, the problem is greater when $\mathbf{Q}$ is largest and $T = 1$. However, when using $T = 0.2$, the state variation due to $q_c(t)$ at these levels is sufficiently small that there is little difference between the discrete and continuous filters.

Figure 8.14 shows $\sqrt{P_{11}}$ versus time when using adaptive Bulirsch-Stoer integration of the continuous $\dot{\mathbf{P}}$. Notice that Figure 8.14 is nearly identical to Figure 8.10 for the discrete filter with $T = 0.2$. Using an integrator accuracy tolerance of

TABLE 8.1: Steady-State σ of Two-State Model for Continuous and Discrete Filter

	Continuous Filter		Discrete Filter *A Posteriori* ($T = 0.2$ s)		Discrete Filter *A Posteriori* ($T = 1$ s)	
Q_{sv}	$\sqrt{P_{11}}$	$\sqrt{P_{22}}$	$\sqrt{P_{11}}$	$\sqrt{P_{22}}$	$\sqrt{P_{11}}$	$\sqrt{P_{22}}$
0.1	5.01	0.89	4.95	0.89	4.72	0.86
1.0	6.69	2.11	6.54	2.09	6.00	2.00
10	8.92	5.01	8.57	4.92	7.41	4.56

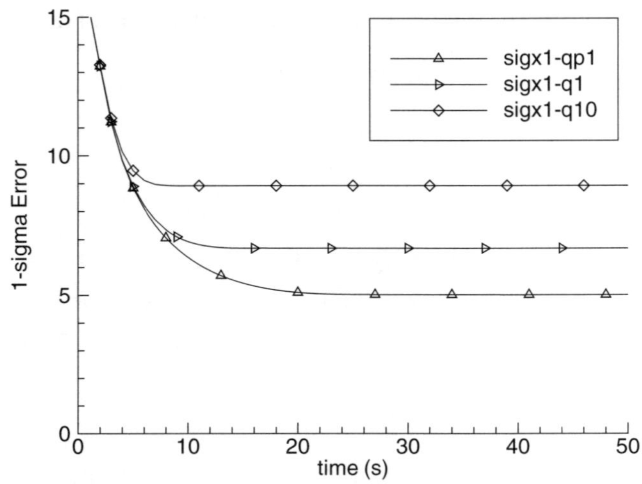

FIGURE 8.14: *A posteriori* $1-\sigma$ uncertainty for position state of two-state continuous Kalman filter.

10^{-8} (a typical "loose" number for orbit determination problems), 308 derivative evaluations were required in the first second of the simulation, and the number dropped to 106, 57, 43, and 43, respectively, for the next 4 s. Only seven evaluations per second were required after 31 s when $Q_{sv} = 0.1$ (17 s when $Q_{sv} = 10$) because of lower weight on the measurements. As noted above, integration of $\dot{\mathbf{P}}$ often requires a step size much smaller than the shortest time constants of the dynamic model. When relaxing the integrator accuracy tolerance to 10^{-6}, the number of derivative evaluations dropped to 199 in the first second. Use of the square-root algorithm listed by Simon (2006, section 8.8.3) decreased the number of function calls in the first 5 s by approximately one-third with essentially the same accuracy.

This result demonstrates why the continuous filter is rarely used: in addition to fact that most measurements are sampled discretely rather than continuously, the computational burden and coding effort of the continuous filter is not competitive with the discrete filter. However, there are two reasons why the continuous Riccati equation is of interest in filtering applications. The first is the relative ease with which the steady-state covariance and gain may be computed. Computation of the discrete steady-state covariance is more difficult than for the continuous system demonstrated above. The second reason is closely related to the first. The Riccati equation allows easy identification of "smoothable" versus "nonsmoothable" states. By smoothable we mean states that are directly or indirectly driven by process noise such that the filter estimate at the final time cannot be simply integrated backward in time to obtain the optimal estimate at previous times. In this example both states are driven by process noise, so both are smoothable. However, if process noise were present only on the first state as

$$\mathbf{G}\mathbf{Q}_s\mathbf{G}^T = \begin{bmatrix} Q_{sp} & 0 \\ 0 & 0 \end{bmatrix},$$

the continuous Riccati equation becomes

$$
\begin{aligned}
\dot{\mathbf{P}}(t) &= \mathbf{FP}(t)+\mathbf{P}(t)\mathbf{F}^T+\mathbf{G}\mathbf{Q}_s\mathbf{G}^T-\mathbf{K}_c(t)\mathbf{R}_s\mathbf{K}_c^T(t) \\
&= \begin{bmatrix} P_{21} & P_{22} \\ 0 & 0 \end{bmatrix} + \begin{bmatrix} P_{12} & 0 \\ P_{22} & 0 \end{bmatrix} + \begin{bmatrix} Q_{sp} & 0 \\ 0 & 0 \end{bmatrix} - \begin{bmatrix} P_{11}^2/R_s & P_{11}P_{21}/R_s \\ P_{11}P_{21}/R_s & P_{21}^2/R_s \end{bmatrix}. \\
&= \begin{bmatrix} 2P_{12}-P_{11}^2/R_s+Q_{sp} & P_{22}-P_{11}P_{12}/R_s \\ P_{22}-P_{11}P_{12}/R_s & -P_{12}^2/R_s \end{bmatrix}
\end{aligned}
$$

It is evident that the steady-state solutions for P_{22} and P_{12} are zero, that is, $\hat{x}_2$ is not smoothable.

8.4 MODIFICATIONS OF THE DISCRETE KALMAN FILTER

The discrete Kalman filter of Section 8.1 or linearizations applicable to nonlinear problems (described in Chapter 9) are by far the most commonly used versions of the Kalman filter. However, there are two modifications that are worthy of note. The first is used when there are a significant number of bias states in the filter. Those bias states can be removed from the full state vector to create a bias-free Kalman filter that also carries equations modeling the sensitivity of the bias-free state estimates to errors in assumed values of bias parameters. Because the bias states are ignored, the bias-free filter is suboptimal, but full optimality is obtained using a bias-restoring filter based on the sensitivity equations. The method was developed primarily to reduce computations, but it is now mostly of interest because the approach allows other extensions to be easily implemented. The second filter modification is a suboptimal version of the Kalman filter used to account for uncertainty in unadjusted states when computing measurement weighting for adjusted states.

8.4.1 Friedland Bias-Free/Bias-Restoring Filter

Friedland (1969) derived a version of the Kalman filter that was intended to reduce computations by separating solutions for nonbias and bias states. He suggested that partitioning of the solution would also make the filter less susceptible to numerical errors, but did not present results demonstrating this. A third advantage of the approach is that it makes the effect of biases on $\hat{\mathbf{x}}$ more readily apparent, and this is a useful starting point for other estimation algorithms. The method is implemented using a bias-free filter that does not include biases in the vector of estimated states, and a bias-restoring filter that only estimates biases. Measurement residuals of the bias-free filter are used as measurements for the bias-restoring filter.

We start by computing errors in estimated bias-free filter states due to errors in unadjusted biases. This development is significantly different from Friedland's, but the result is the same. The total state vector is assumed to consist of n_d time-varying dynamic states $\mathbf{x}_i$ and n_b bias states $\mathbf{b}$. Time propagation of the true state at time t_i is modeled as

$$
\begin{bmatrix} \mathbf{x}_i \\ \mathbf{b} \end{bmatrix} = \begin{bmatrix} \mathbf{\Phi}_{xx} & \mathbf{\Phi}_{xb} \\ \mathbf{0} & \mathbf{I} \end{bmatrix} \begin{bmatrix} \mathbf{x}_{i-1} \\ \mathbf{b} \end{bmatrix} + \begin{bmatrix} \mathbf{q}_i \\ \mathbf{0} \end{bmatrix} \tag{8.4-1}
$$

where as in a normal Kalman filter, $E[\mathbf{q}_i]=\mathbf{0}$ and $E[\mathbf{q}_i\mathbf{q}_j^T]=\mathbf{Q}\delta_{ij}$. To simplify notation we have eliminated time subscripts from $\mathbf{\Phi}_{xx}$, $\mathbf{\Phi}_{xu}$ and $\mathbf{Q}$, but these arrays may vary with time. The bias-free filter models the state estimate time update as

$$\boxed{\hat{\mathbf{x}}_{i/i-1}=\mathbf{\Phi}_{xx}\hat{\mathbf{x}}_{i-1/i-1}+\mathbf{\Phi}_{xb}\mathbf{b}_f} \tag{8.4-2}$$

where $\mathbf{b}_f$ is the fixed value of $\mathbf{b}$ assumed by the bias-free filter ($\mathbf{b}_f$ is often $\mathbf{0}$). Assuming that linearity and superposition apply, errors propagate as

$$\begin{aligned}\tilde{\mathbf{x}}_{i/i-1}&=\mathbf{x}_i-\hat{\mathbf{x}}_{i/i-1}\\&=\mathbf{\Phi}_{xx}(\tilde{\mathbf{x}}^{nb}_{i-1/i-1}+\tilde{\mathbf{x}}^{b}_{i-1/i-1})+\mathbf{\Phi}_{xb}\tilde{\mathbf{b}}+\mathbf{q}_i\end{aligned} \tag{8.4-3}$$

where

$\tilde{\mathbf{b}}=\mathbf{b}-\mathbf{b}_f$ is the error in $\mathbf{b}_f$,

$\tilde{\mathbf{x}}^{b}_{i-1/i-1}$ is the component of $\tilde{\mathbf{x}}_{i-1/i-1}$ due to $\tilde{\mathbf{b}}$, and

$\tilde{\mathbf{x}}^{nb}_{i-1/i-1}$ is the component of $\tilde{\mathbf{x}}_{i-1/i-1}$ due to all errors other than $\tilde{\mathbf{b}}$ (i.e., errors in the prior estimate $\tilde{\mathbf{x}}^{f}_{0/0}$, measurement noise and process noise).

It is further assumed that $\tilde{\mathbf{x}}^{b}_{i-1/i-1}$ can be defined using a $n_d\times n_b$ linear sensitivity matrix $\mathbf{S}^{b}_{i-1/i-1}$,

$$\boxed{\tilde{\mathbf{x}}^{b}_{i-1/i-1}\triangleq\mathbf{S}^{b}_{i-1/i-1}\tilde{\mathbf{b}}} \tag{8.4-4}$$

where $\tilde{\mathbf{b}}$ is assumed to be fixed, zero-mean, independent of other errors, and has covariance $E[\tilde{\mathbf{b}}\tilde{\mathbf{b}}^T]=\mathbf{P}_b$. Provided that $E[\tilde{\mathbf{x}}_{0/0}\tilde{\mathbf{b}}^T]=\mathbf{0}$, $\mathbf{S}^{b}_{0/0}=\mathbf{0}$. Thus

$$\boxed{\begin{aligned}\tilde{\mathbf{x}}_{i/i-1}&=\mathbf{\Phi}_{xx}\tilde{\mathbf{x}}^{nb}_{i-1/i-1}+\mathbf{S}^{b}_{i/i-1}\tilde{\mathbf{b}}+\mathbf{q}_i\\&=\tilde{\mathbf{x}}^{nb}_{i/i-1}+\mathbf{S}^{b}_{i/i-1}\tilde{\mathbf{b}}\end{aligned}} \tag{8.4-5}$$

where

$$\boxed{\mathbf{S}^{b}_{i/i-1}\triangleq\frac{\partial\tilde{\mathbf{x}}_{i/i-1}}{\partial\tilde{\mathbf{b}}}=\mathbf{\Phi}_{xx}\mathbf{S}^{b}_{i-1/i-1}+\mathbf{\Phi}_{xb}} \tag{8.4-6}$$

and

$$\tilde{\mathbf{x}}^{nb}_{i/i-1}=\mathbf{\Phi}_{xx}\tilde{\mathbf{x}}^{nb}_{i-1/i-1}+\mathbf{q}_i\,. \tag{8.4-7}$$

Because the bias-free filter treats $\mathbf{b}_f$ as known, the filter model of the *a priori* covariance is

$$\boxed{\begin{aligned}\mathbf{P}^{xnb}_{i/i-1}&\triangleq E[\tilde{\mathbf{x}}^{nb}_{i/i-1}(\tilde{\mathbf{x}}^{nb}_{i/i-1})^T]\\&=\mathbf{\Phi}_{xx}\mathbf{P}^{xnb}_{i-1/i-1}\mathbf{\Phi}^T_{xx}+\mathbf{Q}\end{aligned}} \tag{8.4-8}$$

which is smaller than the true error covariance. The true and bias-free measurement models are

$$\boxed{\begin{aligned}\mathbf{y}_i&=\mathbf{H}_x\mathbf{x}+\mathbf{H}_b\mathbf{b}+\mathbf{r}_i\\\hat{\mathbf{y}}_i&=\mathbf{H}_x\hat{\mathbf{x}}_{i/i-1}+\mathbf{H}_b\mathbf{b}_f\end{aligned}} \tag{8.4-9}$$

so the measurement residual is

$$\begin{aligned}\tilde{\mathbf{y}}_i &= \mathbf{y}_i - \hat{\mathbf{y}}_i \\ &= \mathbf{H}_x\tilde{\mathbf{x}}_{i/i-1} + \mathbf{H}_b\tilde{\mathbf{b}} + \mathbf{r}_i \\ &= \mathbf{H}_x(\tilde{\mathbf{x}}^{nb}_{i/i-1} + \mathbf{S}^b_{i/i-1}\tilde{\mathbf{b}}) + \mathbf{H}_b\tilde{\mathbf{b}} + \mathbf{r}_i \\ &= \mathbf{H}_x\tilde{\mathbf{x}}^{nb}_{i/i-1} + \mathbf{T}_i\tilde{\mathbf{b}} + \mathbf{r}_i\end{aligned} \tag{8.4-10}$$

where

$$\boxed{\mathbf{T}_i = \mathbf{H}_x\mathbf{S}^b_{i/i-1} + \mathbf{H}_b} \tag{8.4-11}$$

is an $m \times n_b$ matrix. Notice that the $\mathbf{T}_i\tilde{\mathbf{b}}$ term gives the residuals $\tilde{\mathbf{y}}_i$ a systematic (nonwhite) signature. The bias-free filter models the measurement residual covariance as

$$\boxed{\mathbf{C}^{nb}_i \triangleq \mathbf{H}_x\mathbf{P}^{xnb}_{i/i-1}\mathbf{H}^T_x + \mathbf{R}} \tag{8.4-12}$$

where $\mathbf{R} = E[\mathbf{r}_i\mathbf{r}_i^T]$, but the true residual error covariance is

$$\begin{aligned}\mathbf{C}_i &\triangleq E[\tilde{\mathbf{y}}_i\tilde{\mathbf{y}}_i^T] \\ &= \mathbf{H}_x\mathbf{P}^{xnb}_{i/i-1}\mathbf{H}^T_x + \mathbf{R} + \mathbf{T}_i\mathbf{P}_b\mathbf{T}_i^T\end{aligned}.$$

The bias-free gain matrix is computed as

$$\boxed{\mathbf{K}_i = \mathbf{P}^{xnb}_{i/i-1}\mathbf{H}^T_x(\mathbf{C}^{nb}_i)^{-1}} \tag{8.4-13}$$

and the *a posteriori* state estimate and covariance are

$$\boxed{\begin{aligned}\hat{\mathbf{x}}_{i/i} &= \hat{\mathbf{x}}_{i/i-1} + \mathbf{K}_i\tilde{\mathbf{y}}_i \\ \mathbf{P}^{xnb}_{i/i} &= (\mathbf{I} - \mathbf{K}_i\mathbf{H}_x)\mathbf{P}^{xnb}_{i/i-1}\end{aligned}}. \tag{8.4-14}$$

However, the true estimate error is

$$\begin{aligned}\tilde{\mathbf{x}}_{i/i} &\triangleq \mathbf{x}_i - \hat{\mathbf{x}}_{i/i} \\ &= \tilde{\mathbf{x}}^{nb}_{i/i-1} + \tilde{\mathbf{x}}^{b}_{i/i-1} - \mathbf{K}_i\left(\mathbf{H}_x\tilde{\mathbf{x}}^{nb}_{i/i-1} + \mathbf{T}_i\tilde{\mathbf{b}} + \mathbf{r}_i\right) \\ &= (\mathbf{I} - \mathbf{K}_i\mathbf{H}_x)\tilde{\mathbf{x}}^{nb}_{i/i-1} + \mathbf{S}^b_{i/i}\tilde{\mathbf{b}} - \mathbf{K}_i\mathbf{r}_i\end{aligned} \tag{8.4-15}$$

where

$$\boxed{\mathbf{S}^b_{i/i} = \mathbf{S}^b_{i/i-1} - \mathbf{K}_i\mathbf{T}_i.} \tag{8.4-16}$$

Systematic signatures in the bias-free measurement residuals $\tilde{\mathbf{y}}_i$, associated covariance $\mathbf{C}_i$ and sensitivity arrays $\mathbf{T}_i$ provide the information necessary to estimate the bias parameters. Since the bias-free filter is only suboptimal because the error in $\mathbf{b}_f$ is neglected, a properly designed bias-restoring filter can compute optimal estimates $\hat{\mathbf{b}}_{i/i}$ that are then used to correct $\hat{\mathbf{x}}_{i/i}$ to produce optimal estimates $\hat{\mathbf{x}}^o_{i/i}$; that is, $\hat{\mathbf{b}}_{i/i}$ and $\hat{\mathbf{x}}^o_{i/i}$ are the same as those computed from a single optimal filter.

The bias-restoring filter processes each bias-free measurement residual $\tilde{\mathbf{y}}_i$ as it becomes available to compute an updated estimate of the biases $\hat{\mathbf{b}}_{i/i}$ and associated error covariance $\mathbf{P}^b_{i/i}$. The bias-restoring filter is initialized as $\hat{\mathbf{b}}_{0/0} = \mathbf{b}_f$ and $\mathbf{P}^b_{0/0} = \mathbf{P}_b$, and the bias-restoring gain matrix is computed as

$$\boxed{\mathbf{K}^b_i = \left(\mathbf{P}^b_{i-1/i-1}\mathbf{T}^T_i\right)\left(\mathbf{T}_i\mathbf{P}^b_{i-1/i-1}\mathbf{T}^T_i + \mathbf{C}_i\right)^{-1}.} \tag{8.4-17}$$

TABLE 8.2: Multiplication Count for Partitioned Kalman versus Bias-Free/Bias-Restoring Filter (Covariance Equations Only)

Filter Step	Kalman Filter Using Equations (8.1-14), (8.1-16), and (8.1-17)	Bias-Free/Bias-Restoring Filter
Time update	$1.5n_d^3 + 2.5n_d^2 n_b + 0.5n_x^2$	$1.5n_d^3 + n_d^2 n_b + 0.5n_d^2$
Measurement update	$m[1.5n_d^2 + 3n_d n_b + 1.5n_b^2 + 0.5m(n_d + n_b) + 0.5m^2]$	$m[2.5n_d^2 + 2n_d n_b + 1.5n_b^2 + m(n_d + n_b) + m^2]$
Equation (8.4-19)	—	$1.5n_d^2 n_b + n_d n_b^2 + 0.5n_d n_b$

The above equation is somewhat simpler than the equivalent equations given by Friedland. Notice that there is no time update in the bias-restoring filter. The bias-restoring *a posteriori* state estimate and covariance are

$$\begin{aligned} \hat{\mathbf{b}}_{i/i} &= \hat{\mathbf{b}}_{i-1/i-1} + \mathbf{K}_i^b[\tilde{\mathbf{y}}_i - \mathbf{T}_i(\hat{\mathbf{b}}_{i-1/i-1} - \mathbf{b}_f)] \\ \mathbf{P}_{i/i}^b &= \mathbf{P}_{i-1/i-1}^b - \mathbf{K}_i^b(\mathbf{T}_i\, \mathbf{P}_{i-1/i-1}^b) \end{aligned} . \tag{8.4-18}$$

After computing $\hat{\mathbf{b}}_{i/i}$ and $\mathbf{P}_{i/i}^b$, an optimal estimate of the bias-free state and error covariance can then be computed as

$$\begin{aligned} \hat{\mathbf{x}}_{i/i}^o &= \hat{\mathbf{x}}_{i/i} + \mathbf{S}_{i/i}^b(\hat{\mathbf{b}}_{i/i} - \mathbf{b}_f) \\ \mathbf{P}_{i/i}^{xo} &= \mathbf{P}_{i/i}^{xnb} + \mathbf{S}_{i/i}^b \mathbf{P}_{i/i}^b (\mathbf{S}_{i/i}^b)^T \end{aligned} . \tag{8.4-19}$$

The equation for $\mathbf{P}_{i/i}^{xo}$ is obtained by noting that

$$\begin{aligned} \mathbf{x}_i - \hat{\mathbf{x}}_{i/i}^o &= \tilde{\mathbf{x}}_{i/i}^{nb} + \mathbf{S}_{i/i}^b(\mathbf{b} - \mathbf{b}_f) + \mathbf{S}_{i/i}^b(\mathbf{b}_f - \hat{\mathbf{b}}_{i/i}) \\ &= \tilde{\mathbf{x}}_{i/i}^{nb} + \mathbf{S}_{i/i}^b(\mathbf{b} - \hat{\mathbf{b}}_{i/i}) \end{aligned}$$

and computing $\mathbf{P}_{i/i}^{xo} = E\left[(\mathbf{x}_i - \hat{\mathbf{x}}_{i/i}^o)(\mathbf{x}_i - \hat{\mathbf{x}}_{i/i}^o)^T\right]$.

The difference in total floating point operations between Friedland's method and a carefully implemented full-state Kalman filter is usually small. The total number of multiplications required to implement the covariance updates of the Kalman filter in equations (8.1-14), (8.1-16), and (8.1-17) versus Friedland's method (as described above) are shown in Table 8.2. Although computational savings does not justify use of Friedland's method, the insight into effects of biases is used for other purposes in Chapters 9 and 11.

In some applications it is not necessary to compute $\hat{\mathbf{b}}_{i/i}$ and $\hat{\mathbf{x}}_{i/i}$ after each measurement is processed. When it is only necessary to compute these values at the final measurement time t_m, $\hat{\mathbf{b}}_{m/m}$ may be computed as a least-squares solution using all the concatenated measurements; that is,

$$\begin{bmatrix} \tilde{\mathbf{y}}_1 \\ \tilde{\mathbf{y}}_2 \\ \vdots \\ \tilde{\mathbf{y}}_m \end{bmatrix} = \begin{bmatrix} \mathbf{T}_1 \\ \mathbf{T}_2 \\ \vdots \\ \mathbf{T}_m \end{bmatrix} \mathbf{b}_{m/m} + \begin{bmatrix} \mathbf{v}_1 \\ \mathbf{v}_2 \\ \vdots \\ \mathbf{v}_m \end{bmatrix} \tag{8.4-20}$$

or

$$\tilde{\mathbf{y}}_T = \mathbf{T}_T \mathbf{b}_{m/m} + \mathbf{v}_T$$

where $E(\mathbf{v}_i \mathbf{v}_j^T) = \mathbf{C}_i \delta_{ij}$ (the previously computed residual covariance). Using the normal equations,

$$\hat{\mathbf{b}}_{m/m} = \left(\sum_{i=1}^{m} \mathbf{T}_i^T \mathbf{C}_i^{-1} \mathbf{T}_i \right)^{-1} \left(\sum_{i=1}^{m} \mathbf{T}_i^T \mathbf{C}_i^{-1} \tilde{\mathbf{y}}_i \right), \tag{8.4-21}$$

although QR solutions are generally preferred. Then state estimates at all previous times can be computed as

$$\hat{\mathbf{x}}_{i/i}^{o} = \hat{\mathbf{x}}_{i/i}^{f} + \mathbf{S}_{i/i}^{b} (\hat{\mathbf{b}}_{m/m} - \mathbf{b}_f) \tag{8.4-22}$$

for $i = 1, 2, \ldots, m$ provided that $\hat{\mathbf{x}}_{i/i}$ and $\mathbf{S}_{i/i}^{b}$ were saved. This approach can reduce computations in some cases, but that is not the primary reason for interest in the approach. The fact that serial correlation in the bias-free $\tilde{\mathbf{y}}_i$ is only due to errors in $\mathbf{b}_f$ allows easy hypothesis testing for different model error possibilities. This will be discussed again in Chapter 11.

8.4.2 Kalman-Schmidt Consider Filter

In the early 1960s Stanley Schmidt developed a general-purpose filter error analysis program for the National Aeronautics and Space Administration (NASA) Goddard Space Flight Center. As described in McGee and Schmidt (1985), one option in that software provided a "means of optimally compensating for modeling errors" where the algorithm "includes the effects of (but does not solve for) selected error states." Schmidt (1966, p. 333) described it as a method that improved convergence for poorly observable problems with a large number of unknowns, and also reduced computations for these problems. The approach includes the effect of unknown parameters on the state estimates without estimating the parameters. Schmidt's modification became known as the Kalman-Schmidt consider filter.

Somewhat oddly, later descriptions of the algorithm differ both in the implementation and the motivation. Some references describe it as a method designed to reduce computations by not estimating "nuisance" colored (Markov process) measurement noise states that do not affect the more important core dynamic states. These descriptions also emphasize the computational reduction resulting from the lack-of-coupling between core and nuisance states in the state transition matrix. Other references describe Schmidt's approach as a way to compensate for bias parameters in either the dynamic or measurement models, where models using those parameters are poor and estimation of the parameters could causes erroneous estimates of other states. Divergence control is another possible reason. For Kalman filters that operate over long periods of time without resetting, the variance of non-smoothable states eventually approaches zero, thus making the filter unresponsive to new measurements. If filter models are imperfect, the estimated states may then diverge from the true states.

Schmidt's approach accounts for uncertainty of the "nuisance" or "consider" parameters in weighting of the measurements when updating state estimates. (Notice that this use of the term "consider" is different from that used in Chapter 6 on least squares.) This procedure is equivalent to setting rows of the Kalman gain matrix for specified consider states to zero, although Schmidt did not directly use that approach. The value and covariance of the consider states either remain at the initial values (for biases) or steady-state values (for Markov processes), and the parameter uncertainty is accounted for when computing measurement weights for the remaining states. Whether or not this is the best solution to the problem is sometimes debated. In cases where "bias" states are actually somewhat time-varying, it may be better to model process noise on the state, but that approach is not always applicable.

The presentation of this section is general and allows the consider parameters to be dynamic model biases, measurement model biases, or colored noise measurement errors. It has been this author's experience that the consider filter is primarily used to handle poorly modeled bias parameters, but this derivation makes no such assumption. To explain the Kalman-Schmidt filter, let the state vector be partitioned into adjustable states and consider or "nuisance" parameters as

$$\begin{bmatrix} \mathbf{x} \\ \mathbf{c} \end{bmatrix},$$

where $\mathbf{c}$ represents the true values of the consider parameters and $\hat{\mathbf{c}}$ is the estimate used in the filter. The dynamic model for this augmented state vector is

$$\begin{bmatrix} \mathbf{x} \\ \mathbf{c} \end{bmatrix}_i = \begin{bmatrix} \mathbf{\Phi}_{xx} & \mathbf{\Phi}_{xc} \\ \mathbf{0} & \mathbf{\Phi}_{cc} \end{bmatrix} \begin{bmatrix} \mathbf{x} \\ \mathbf{c} \end{bmatrix}_{i-1} + \begin{bmatrix} \mathbf{q}_x \\ \mathbf{q}_c \end{bmatrix}.$$

When process noise is not included on consider parameters, time evolution of consider parameters can be modeled in the $\mathbf{\Phi}_{xc}$ term (if it exists) and it can be assumed that $\mathbf{\Phi}_{cc} = \mathbf{I}$. Time updates of the state and covariance are the same as for a normal Kalman filter. If written in partitions the filter state and covariance time updates are

$$\begin{bmatrix} \hat{\mathbf{x}} \\ \hat{\mathbf{c}} \end{bmatrix}_{i/i-1} = \begin{bmatrix} \mathbf{\Phi}_{xx} & \mathbf{\Phi}_{xc} \\ \mathbf{0} & \mathbf{\Phi}_{cc} \end{bmatrix} \begin{bmatrix} \hat{\mathbf{x}} \\ \hat{\mathbf{c}} \end{bmatrix}_{i-1/i-1} = \begin{bmatrix} \mathbf{\Phi}_{xx}\hat{\mathbf{x}} + \mathbf{\Phi}_{xc}\hat{\mathbf{c}} \\ \mathbf{\Phi}_{cc}\hat{\mathbf{c}} \end{bmatrix}_{i-1/i-1} \tag{8.4-23}$$

and

$$\begin{aligned}
\begin{bmatrix} \mathbf{P}_{xx} & \mathbf{P}_{xc} \\ \mathbf{P}_{xc}^T & \mathbf{P}_{cc} \end{bmatrix}_{i/i-1}
&= \begin{bmatrix} \mathbf{\Phi}_{xx} & \mathbf{\Phi}_{xc} \\ \mathbf{0} & \mathbf{\Phi}_{cc} \end{bmatrix} \begin{bmatrix} \mathbf{P}_{xx} & \mathbf{P}_{xc} \\ \mathbf{P}_{xc}^T & \mathbf{P}_{cc} \end{bmatrix}_{i-1/i-1} \begin{bmatrix} \mathbf{\Phi}_{xx}^T & \mathbf{0} \\ \mathbf{\Phi}_{xc}^T & \mathbf{\Phi}_{cc}^T \end{bmatrix} + \begin{bmatrix} \mathbf{Q}_{xx} & \mathbf{0} \\ \mathbf{0} & \mathbf{Q}_{cc} \end{bmatrix} \\
&= \begin{bmatrix} \mathbf{\Phi}_{xx}\mathbf{P}_{xx} + \mathbf{\Phi}_{xc}\mathbf{P}_{xc}^T & \mathbf{\Phi}_{xx}\mathbf{P}_{xc} + \mathbf{\Phi}_{xc}\mathbf{P}_{cc} \\ \mathbf{\Phi}_{cc}\mathbf{P}_{xc}^T & \mathbf{\Phi}_{cc}\mathbf{P}_{cc} \end{bmatrix}_{i-1/i-1} \begin{bmatrix} \mathbf{\Phi}_{xx}^T & \mathbf{0} \\ \mathbf{\Phi}_{xc}^T & \mathbf{\Phi}_{cc}^T \end{bmatrix} + \begin{bmatrix} \mathbf{Q}_{xx} & \mathbf{0} \\ \mathbf{0} & \mathbf{Q}_{cc} \end{bmatrix} \\
&= \begin{bmatrix} (\mathbf{\Phi}_{xx}\mathbf{P}_{xx} + \mathbf{\Phi}_{xc}\mathbf{P}_{xc}^T)\mathbf{\Phi}_{xx}^T + (\mathbf{\Phi}_{xx}\mathbf{P}_{xc} + \mathbf{\Phi}_{xc}\mathbf{P}_{cc})\mathbf{\Phi}_{xc}^T + \mathbf{Q}_{xx} & (\mathbf{\Phi}_{xx}\mathbf{P}_{xc} + \mathbf{\Phi}_{xc}\mathbf{P}_{cc})\mathbf{\Phi}_{cc}^T \\ \mathbf{\Phi}_{cc}(\mathbf{P}_{xc}^T\mathbf{\Phi}_{xx}^T + \mathbf{P}_{cc}\mathbf{\Phi}_{xc}^T) & \mathbf{\Phi}_{cc}\mathbf{P}_{cc}\mathbf{\Phi}_{cc}^T + \mathbf{Q}_{cc} \end{bmatrix}
\end{aligned} \tag{8.4-24}$$

where $\mathbf{Q}_{xx} = E[\mathbf{q}_x\mathbf{q}_x^T]$ and $\mathbf{Q}_{cc} = E[\mathbf{q}_c\mathbf{q}_c^T]$. Notice that it is not necessary to compute the updates in partitions if a temporary array

$$\begin{aligned}
[\mathbf{D}_{xx} \quad \mathbf{D}_{xc}] &= [\mathbf{\Phi}_{xx} \quad \mathbf{\Phi}_{xc}] \begin{bmatrix} \mathbf{P}_{xx} & \mathbf{P}_{xc} \\ \mathbf{P}_{xc}^T & \mathbf{P}_{cc} \end{bmatrix}_{i-1/i-1} \\
&= \left[\mathbf{\Phi}_{xx}\mathbf{P}_{xx} + \mathbf{\Phi}_{xc}\mathbf{P}_{xc}^T \quad \mathbf{\Phi}_{xx}\mathbf{P}_{xc} + \mathbf{\Phi}_{xc}\mathbf{P}_{cc}\right]_{i-1/i-1}
\end{aligned} \tag{8.4-25}$$

is computed as a single matrix multiplication of rectangular and square arrays. Then

$$\begin{bmatrix} \mathbf{P}_{xx} & \mathbf{P}_{xc} \\ \mathbf{P}_{xc}^T & \mathbf{P}_{cc} \end{bmatrix}_{i/i-1} = \begin{bmatrix} \mathbf{D}_{xx}\boldsymbol{\Phi}_{xx}^T + \mathbf{D}_{xc}\boldsymbol{\Phi}_{xc}^T + \mathbf{Q}_{xx} & \mathbf{D}_{xc}\boldsymbol{\Phi}_{cc}^T \\ \boldsymbol{\Phi}_{cc}\mathbf{D}_{xc}^T & \boldsymbol{\Phi}_{cc}\mathbf{P}_{cc}\boldsymbol{\Phi}_{cc}^T + \mathbf{Q}_{cc} \end{bmatrix} \tag{8.4-26}$$

where $\mathbf{D}_{xx}\boldsymbol{\Phi}_{xx}^T + \mathbf{D}_{xc}\boldsymbol{\Phi}_{xc}^T$ is also computed as a single matrix multiplication. If $\mathbf{c}$ is composed of only biases, $\hat{\mathbf{c}}$ and $\mathbf{P}_{cc}$ are fixed and not updated. If $\mathbf{c}$ represents stable low-order Markov process models, the initial value of $\mathbf{P}_{cc}$ is usually set to the steady-state value and thus $\boldsymbol{\Phi}_{cc}\mathbf{P}_{cc}\boldsymbol{\Phi}_{cc}^T + \mathbf{Q}_{cc} = \mathbf{P}_{cc}$ so that it is not necessary to update this partition. Further computations can be eliminated if $\boldsymbol{\Phi}_{xc} = \mathbf{0}$.

The measurement model for this system is $\mathbf{y}_i = \mathbf{H}_x\mathbf{x}_i + \mathbf{H}_c\mathbf{c}_i + \mathbf{r}_i$ and the measurement residual is

$$\begin{aligned} \tilde{\mathbf{y}}_i &= \mathbf{y}_i - (\mathbf{H}_x\hat{\mathbf{x}}_{i/i-1} + \mathbf{H}_c\hat{\mathbf{c}}_{i/i-1}) \\ &= \mathbf{H}_x\tilde{\mathbf{x}}_i + \mathbf{H}_c\tilde{\mathbf{c}}_i + \mathbf{r}_i \end{aligned}. \tag{8.4-27}$$

The residual covariance is

$$\mathbf{C}_i \triangleq E[\tilde{\mathbf{y}}_i\tilde{\mathbf{y}}_i^T] = \mathbf{H}_x\mathbf{P}_{xx}\mathbf{H}_x^T + \mathbf{H}_c\mathbf{P}_{xc}^T\mathbf{H}_x^T + \mathbf{H}_x\mathbf{P}_{xc}\mathbf{H}_c^T + \mathbf{H}_c\mathbf{P}_{cc}\mathbf{H}_c^T + \mathbf{R} \tag{8.4-28}$$

where all covariance partitions are from $\mathbf{P}_{i/i-1}$. The modified gain matrix at time t_i is the normal Kalman filter gain with rows corresponding to the consider states set to zero. The measurement covariance update is simpler when the consider states are located at the bottom, which implies that $\mathbf{K}$ is

$$\begin{bmatrix} \mathbf{K}_x \\ \mathbf{0} \end{bmatrix} = \begin{bmatrix} (\mathbf{P}_{xx}\mathbf{H}_x^T + \mathbf{P}_{xc}\mathbf{H}_c^T)\mathbf{C}_i^{-1} \\ \mathbf{0} \end{bmatrix}_{i/i-1}. \tag{8.4-29}$$

Again the upper equation can be implemented without separate matrix multiplications. Because this gain matrix is suboptimal, one might believe it necessary to use the long form of the covariance measurement update, equation (8.1-12), rather than the short form, equation (8.1-13). However, a modified version of the short form can be used if the consider states are at the bottom. This greatly reduces computations as can be seen from the modified state and covariance measurement update

$$\begin{bmatrix} \hat{\mathbf{x}} \\ \hat{\mathbf{c}} \end{bmatrix}_{i/i} = \begin{bmatrix} \hat{\mathbf{x}} \\ \hat{\mathbf{c}} \end{bmatrix}_{i/i-1} + \begin{bmatrix} \mathbf{K}_x[\mathbf{y}_i - (\mathbf{H}_x\hat{\mathbf{x}} + \mathbf{H}_c\hat{\mathbf{c}})] \\ \mathbf{0} \end{bmatrix}_{i/i-1} \tag{8.4-30}$$

$$\begin{aligned} \begin{bmatrix} \mathbf{P}_{xx} & \mathbf{P}_{xc} \\ \sim & \mathbf{P}_{cc} \end{bmatrix}_{i/i} &= \begin{bmatrix} \mathbf{P}_{xx} & \mathbf{P}_{xc} \\ \sim & \mathbf{P}_{cc} \end{bmatrix}_{i/i-1} - \begin{bmatrix} \mathbf{K}_x \\ \mathbf{0} \end{bmatrix} [\mathbf{H}_x\mathbf{P}_{xx} + \mathbf{H}_c\mathbf{P}_{xc}^T \quad \mathbf{H}_x\mathbf{P}_{xc} + \mathbf{H}_c\mathbf{P}_{cc}]_{i/i-1} \\ &= \begin{bmatrix} \mathbf{P}_{xx} - \mathbf{K}_x(\mathbf{H}_x\mathbf{P}_{xx} + \mathbf{H}_c\mathbf{P}_{xc}^T) & \mathbf{P}_{xc} - \mathbf{K}_x(\mathbf{H}_x\mathbf{P}_{xc} + \mathbf{H}_c\mathbf{P}_{cc}) \\ \sim & \mathbf{P}_{cc} \end{bmatrix}_{i/i-1} \end{aligned}. \tag{8.4-31}$$

Notice that the $[\mathbf{P}_{cx}]_{i/i}$ partition is not computed but is set (if needed) as $[\mathbf{P}_{cx}]_{i/i} = [\mathbf{P}_{xc}^T]_{i/i}$. When consider parameters are either biases or Markov process that do not drive the core dynamic states (i.e., $\boldsymbol{\Phi}_{xc} = \mathbf{0}$), the computational savings may be significant when the number of consider states is large. Also notice that the measurement update of partitions $\mathbf{P}_{xx}$ and $\mathbf{P}_{xc}$ is optimal for that one step. This explains why the long form of the measurement update is not required. However, because $\mathbf{P}_{cc}$ is not updated, filter processing of all measurements is suboptimal. Use of consider states is demonstrated in Example 9.7.

8.5 STEADY-STATE SOLUTION

As shown in several time-invariant, fixed sampling interval examples, the *a posteriori* covariance of the Kalman filter will often reach a steady-state value after a sufficient number of measurements have been processed. This implies that the gain matrix has a fixed value. The steady-state covariance is of interest because performance requirements are sometimes specified for steady-state behavior. Also, when filters are implemented on embedded microprocessors with limited computational capabilities, it is often desirable to precompute the steady-state gain so that the expensive covariance updates need not be calculated. One drawback of this approach is the difficulty in handling filter initialization transients in cases where the initial state estimate is poorly known. This problem can be addressed by computing several different gain values that are used sequentially during filter start-up.

It must be mentioned that not all problems have steady-state solutions. For example, unobservable random walk states have a variance that grows linearly with time. Time-varying systems or systems with varying measurement sampling also do not have steady-state gains. Furthermore multiple steady-state solutions are possible when results depend on the initial conditions and solutions do not result in a stable filter. While conditions for a steady-state solution can be expressed in terms of stabilizability and detectability conditions on filter arrays, the behavior of the Kalman filter covariance over an extended period of time demonstrates whether or not a particular problem has a steady-state solution. Note that bias states not driven directly or indirectly by process noise will have a steady-state variance of zero, but that is a steady-state solution.

While the steady-state covariance can be determined by processing a sufficient number of real or simulated measurements, a more analytic approach is sometimes needed for greater accuracy. Provided that a steady-state solution does exist, the Riccati equation for the continuous system can be used to determine that solution. That is, from equations (8.3-5) and (8.3-6), the $n \times n$ equation

$$\mathbf{FP}(\infty)+\mathbf{P}(\infty)\mathbf{F}^T+\mathbf{G}\mathbf{Q}_s\mathbf{G}^T-\mathbf{P}(\infty)\mathbf{H}^T\mathbf{R}_s^{-1}\mathbf{HP}(\infty)=\mathbf{0} \tag{8.5-1}$$

defines the steady-state solution $\mathbf{P}(\infty)$, and the steady-state gain is computed as equation (8.3-6) using $\mathbf{P}(\infty)$. Equation (8.5-1) is called the *Continuous Algebraic Riccati Equation* (CARE). A discrete form (DARE) can also be developed directly from the discrete Kalman filter equations by equating the *a priori* covariances for two time steps. When measurement and process noise are uncorrelated, the result is

$$\begin{aligned}\lim_{i\to\infty}(\mathbf{P}_{i/i-1}) &= \mathbf{P}(\infty)\\ &= \mathbf{\Phi}\left(\mathbf{P}(\infty)-\mathbf{P}(\infty)\mathbf{H}^T[\mathbf{HP}(\infty)\mathbf{H}^T+\mathbf{R}]^{-1}\mathbf{HP}(\infty)\right)\mathbf{\Phi}^T+\mathbf{Q}_d\end{aligned}. \tag{8.5-2}$$

Simon (2006, section 7.3) lists the result for correlated measurement and process noise. Solution of the DARE is similar to that of the CARE, but numerical issues are slightly different.

There are a surprising number of methods that have been developed to obtain symmetric positive semi-definite solutions to algebraic Riccati problems. One of the earliest methods—described by Potter (1966) and others—was based on eigen decomposition of a $2n \times 2n$ matrix equivalent to equation (8.5-1). That approach has several numerical problems, including difficulty in calculating eigenvectors for repeated eigenvalues. Alternately Newton iterations can be used to solve equation (8.5-1) if a suitably accurate initial guess is available. Newton's method is sometimes used to improve a solution obtained by another method. An iterative method based on the matrix sign function computes eigenvectors without computing eigenvalues. The basic method due to Denman and Beavers (1976) is easy to implement and works well for many problems, but more robust implementations (e.g., Byers 1987) are somewhat complex. Sima (2005) compares newer methods for solving algebraic Riccati equations. Arnold and Laub (1984) describe the algorithms implemented in MATLAB.

Provided that access to a good linear algebra package (such as LAPACK) is available, one of the easiest to implement CARE solution methods is based on Schur decomposition of the $2n \times 2n$ Hamiltonian matrix

$$\mathbf{Z} = \begin{bmatrix} \mathbf{F}^T & -\mathbf{H}^T\mathbf{R}_s^{-1}\mathbf{H} \\ -\mathbf{G}\mathbf{Q}_s\mathbf{G}^T & -\mathbf{F} \end{bmatrix} \tag{8.5-3}$$

The algorithm is due to Laub (1979). He explains that for a unique positive semi-definite solution $\mathbf{P}(\infty)$ to exist, $(\mathbf{F}^T, \mathbf{H}^T\mathbf{R}_s^{-1/2})$ must be a "stabilizable pair" and $(\mathbf{G}\mathbf{Q}^{1/2}, \mathbf{F}^T)$ must be a "detectable pair." These terms are defined (see Simon 2006, section 1.7.3; Wonham 1968; or Kalman 1960) as follows:

1. A system is stabilizable if it is either controllable or stable, or if the uncontrollable modes are stable.
2. A system is detectable if it is either observable or stable, or the unobservable modes are stable.
3. A system is observable if its initial state can be uniquely determined given a time history of input and output over a period of time. For a continuous system it must apply for any final time.
4. A system is controllable if for any initial condition there exists a control that will drive the state to any specified condition at a final time. For a continuous system it must apply for any final time.
5. A continuous system is stable if all eigenvalues lie within the left half complex plane, and a discrete system is stable if all eigenvalues lie within the unit circle.

Also, $\mathbf{H}^T\mathbf{R}_s^{-1}\mathbf{H}$ and $\mathbf{G}\mathbf{Q}_s\mathbf{G}^T$ must be positive semi-definite for a solution to exist. With these conditions no eigenvalues of $\mathbf{Z}$ will be purely imaginary. An orthogonal similarity transformation is applied to $\mathbf{Z}$ to produce a right Schur form $\mathbf{S}$, or equivalently

$$\mathbf{Z} = \mathbf{U}\begin{bmatrix} \mathbf{S}_{11} & \mathbf{S}_{12} \\ \mathbf{0} & \mathbf{S}_{22} \end{bmatrix}\mathbf{U}^T \tag{8.5-4}$$

where real $n \times n$ arrays $\mathbf{S}_{11}$ and $\mathbf{S}_{22}$ are nearly upper triangular (have at most one elements below the diagonal). The diagonal 1×1 or 2×2 blocks of $\mathbf{S}$ are the eigenvalues of $\mathbf{Z}$, and when 2×2 blocks occur, the eigenvalues are complex conjugate. The Schur decomposition can be arranged so that the real parts of the diagonals of $\mathbf{S}_{11}$ are negative and the diagonals of $\mathbf{S}_{22}$ are positive. With the $n \times n$ partitions of

$$\mathbf{U} = \begin{bmatrix} \mathbf{U}_{11} & \mathbf{U}_{12} \\ \mathbf{U}_{21} & \mathbf{U}_{22} \end{bmatrix} \tag{8.5-5}$$

defined to match $\mathbf{S}_{11}$ and $\mathbf{S}_{22}$, $\mathbf{U}_{11}$ is invertible and the steady-state covariance is computed as

$$\boxed{\mathbf{P}(\infty) = \mathbf{U}_{21}\mathbf{U}_{11}^{-1}} \tag{8.5-6}$$

where $\mathbf{P}(\infty) = \mathbf{P}^T(\infty) \geq 0$.

If solving the DARE problem defined by equation (8.5-2),

$$\mathbf{Z} = \begin{bmatrix} \mathbf{\Phi}^T + (\mathbf{H}^T\mathbf{R}^{-1}\mathbf{H})\mathbf{\Phi}^{-1}\mathbf{Q} & -(\mathbf{H}^T\mathbf{R}^{-1}\mathbf{H})\mathbf{\Phi}^{-1} \\ -\mathbf{\Phi}^{-1}\mathbf{Q} & \mathbf{\Phi}^{-1} \end{bmatrix} \tag{8.5-7}$$

and eigenvalues $a + jb$ in $\mathbf{S}_{11}$ must be inside the unit circle (i.e., $|a + jb| = \sqrt{a^2 + b^2} < 1$). When comparing values from different sources, notice that equations (8.5-2) and (8.5-7) are defined for the *a priori* covariance.

LAPACK routine DGEESX can be used to perform the Schur decomposition. Set DGEESX input flags JOBVS = 'V' to compute the Schur vectors, and SORT = 'S' to sort eigenvalues. It is also necessary to provide a separate logical function SENSE (WR,WJ) that returns TRUE only if WR < 0 (real part of eigenvalue is negative) for the CARE, or $\mathrm{WR}^2 + \mathrm{WI}^2 < 1$ for the DARE. Then define input array $\mathbf{A}$ as equation (8.5-2) or (8.5-7). Finally compute equation (8.5-6) using the partitions of the output **VS** array.

If using MATLAB, the CARE function solves the continuous Riccati problem and DARE solves the discrete problem.

Example 8.5: Steady-State Position-Velocity-Acceleration Problem

Since nonlinear problems do not usually have steady-state solutions, we demonstrate computation of the steady-state solution using a linear six-state position-velocity-acceleration problem. With the filter state vector $\mathbf{x}^T = [x \;\; y \;\; \dot{x} \;\; \dot{y} \;\; \ddot{x} \;\; \ddot{y}]$, the continuous dynamic model $\dot{\mathbf{x}} = \mathbf{F}\mathbf{x} + \mathbf{G}\mathbf{q}_s$ is defined by

$$\mathbf{F} = \begin{bmatrix} 0 & 0 & 1 & 0 & 0 & 0 \\ 0 & 0 & 0 & 1 & 0 & 0 \\ 0 & 0 & 0 & 0 & 1 & 0 \\ 0 & 0 & 0 & 0 & 0 & 1 \\ 0 & 0 & 0 & 0 & 0 & 0 \\ 0 & 0 & 0 & 0 & 0 & 0 \end{bmatrix}, \quad \mathbf{G}\mathbf{Q}_s\mathbf{G}^T = \begin{bmatrix} 0 & 0 & 0 & 0 & 0 & 0 \\ 0 & 0 & 0 & 0 & 0 & 0 \\ 0 & 0 & 0 & 0 & 0 & 0 \\ 0 & 0 & 0 & 0 & 0 & 0 \\ 0 & 0 & 0 & 0 & 10^{-6} & 0 \\ 0 & 0 & 0 & 0 & 0 & 10^{-6} \end{bmatrix}$$

TABLE 8.3: Steady-State Filter Solutions

Source	Position $1-\sigma$	Velocity $1-\sigma$	Acceleration $1-\sigma$
Discrete Kalman filter *a priori*: $\sqrt{\mathbf{P}_{1000/999}}$	0.47054	0.057033	0.0045516
Discrete Kalman filter *a posteriori*: $\sqrt{\mathbf{P}_{1000/1000}}$	0.42576	0.053367	0.0044404
Geometric mean of *a priori* and *a posteriori*: $\sqrt{\sqrt{\mathbf{P}_{1000/999}\mathbf{P}_{1000/1000}}}$	0.44759	0.055170	0.0044957
CARE solution: $\sqrt{\mathbf{P}(\infty)}$	0.44721	0.054772	0.0044721
DARE solution for $\lim_{i\to\infty}\sqrt{\mathbf{P}_{i/i-1}}$	0.47054	0.057033	0.0045516

and the continuous measurement model $\mathbf{y} = \mathbf{Hx} + \mathbf{r}_s$ is defined by

$$\mathbf{H} = \begin{bmatrix} 1 & 0 & 0 & 0 & 0 & 0 \\ 0 & 1 & 0 & 0 & 0 & 0 \end{bmatrix}, \quad \mathbf{R}_S = E[\mathbf{r}_s\mathbf{r}_s^T] = \begin{bmatrix} 1 & 0 \\ 0 & 1 \end{bmatrix}$$

or

$$\mathbf{H}^T\mathbf{R}_s^{-1}\mathbf{H} = \begin{bmatrix} 1 & 0 & 0 & 0 & 0 & 0 \\ 0 & 1 & 0 & 0 & 0 & 0 \\ 0 & 0 & 0 & 0 & 0 & 0 \\ 0 & 0 & 0 & 0 & 0 & 0 \\ 0 & 0 & 0 & 0 & 0 & 0 \\ 0 & 0 & 0 & 0 & 0 & 0 \end{bmatrix}.$$

Notice that measurements only observe the first two states, process noise only affects the last two states, and the dynamics matrix $\mathbf{F}$ couples measurements with the states driven by process noise. The free integrators make the dynamics unstable, but measurements make the filter stable. Also notice that the model does not couple motion in the x and y axes, so the problem could be handled as two independent filters. Use of two axes verifies that repeated eigenvalues to not cause numerical problems in the Schur method.

The discrete Kalman filter is nearly in steady state after processing 60 measurements at 1 s intervals, and appears to be fully in steady state at 1000 s. The *a priori* standard deviations for each axis of position, velocity, and acceleration, computed as the square roots of the diagonals of $\mathbf{P}_{1000/999}$, are listed in the first line of Table 8.3. The next line lists *a posteriori* values from $\mathbf{P}_{1000/1000}$. To five digits the values do not change at 10,000 s. The corresponding Schur solution values of $\mathbf{P}(\infty)$ computed for the CARE and $\mathbf{P}_{\infty/\infty-1}$ for the DARE are also listed. Notice that the DARE values agree with $\mathbf{P}_{1000/999}$ from the Kalman filter and the CARE values are close to the geometric mean of $\mathbf{P}_{1000/999}$ and $\mathbf{P}_{1000/1000}$. This verifies that the CARE and DARE solutions computed using the Schur method match actual discrete filter covariances.

8.6 WIENER FILTER

Recall from Chapter 1 that the Wiener filter is the minimum variance or minimum mean-squared error (MMSE) solution to the time-invariant, statistically stationary linear filtering problem. Since the steady-state Kalman filter solution presented in the previous section also solves the MMSE time-invariant, statistically stationary linear filtering problem, the solutions should be identical even if the methods are different. That is in fact the case. To better understand the connection, we expand on the development of the Wiener filter.

As described in Chapter 1, Wiener was interested in estimating a stochastic time-varying signal where measurements were obtained as linear operations on that signal. Figure 8.15 shows the model of Wiener's vector estimation problem, where we have designated the variable to be estimated as the system state n-vector $\mathbf{x}(t)$ rather than a signal $\mathbf{s}(t)$. A linear time-invariant $m \times n$ matrix operation $\mathbf{D}_f(u - t)$ is performed on $\mathbf{x}(t)$ to obtain m-vector $\mathbf{z}(u)$ at time u, where that operation only depends on the time difference $u - t$. A modulation function $\mathbf{c}(u)$ multiplies $\mathbf{z}(u)$. In continuous systems $c(u) = 1$ is typical, but in fixed-interval discretely sampled systems $c(u) = \delta(\text{mod}(t,T))$ where $\delta(t)$ is the Dirac delta function, mod is the modulus function, and T is the sampling interval. Random measurement noise $\mathbf{r}(u)$ is then added to obtain the measurements $\mathbf{y}(u)$ over the interval $t_i \le u \le t_f$. It is assumed that $\mathbf{x}(t)$ and $\mathbf{y}(u)$ are jointly stationary. The desired output of the estimator is designated as $\mathbf{d}(v)$ and it is assumed that this can be obtained as another linear operation $\mathbf{D}_d(v - t)$ on $\mathbf{x}(t)$. For the filtering problem $\mathbf{d}(t) = \mathbf{x}(t)$, while $\mathbf{d}(t) = \mathbf{x}(t + \tau)$ is a prediction if $\tau > 0$ and a smoothed estimate if $\tau < 0$. Other linear operations—such as differentiation—could be specified for $\mathbf{d}(v)$. It is also assumed that $\mathbf{x}(t)$ and $\mathbf{d}(v)$ are jointly stationary.

For the purposes of this section, we assume that a filter solution is desired ($\mathbf{d}(t) = \mathbf{x}(t)$), the system is continuous with no modulating signal ($c(u) = 1$), and $\mathbf{x}(t)$ completely defines the system state so that $\mathbf{D}_f(u - t)$ need not have any memory. In fact, we assume that $\mathbf{D}_f$ is a constant.

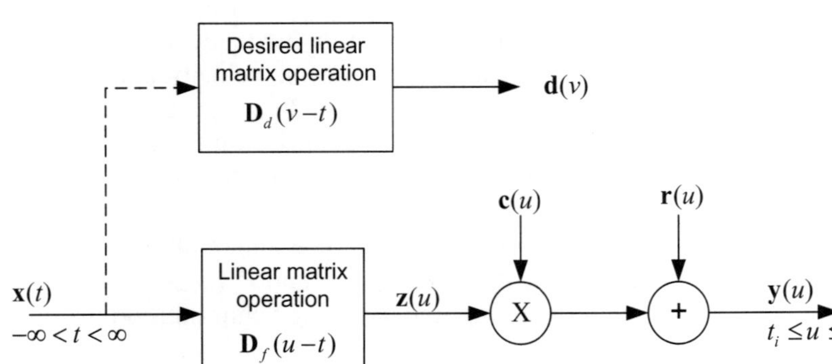

FIGURE 8.15: Vector time-invariant linear estimation problem.

8.6.1 Wiener-Hopf Equation

The following development is based on the derivation by Van Trees (1968, sections 3.4.5, 6.1, and 6.2), but with modifications to make it more compatible with the Kalman filter problem. It should be noted however that Wiener's (and Kalman-Bucy's) approach is more general than summarized here.

The goal is to define a linear weighting function of past measurements that produces the MMSE estimate of $\mathbf{x}(t)$ as shown in Figure 8.16.

Van Trees shows that a MAP estimate of $\mathbf{x}(t)$ over $t_i \le t \le t_f$ for the filtering-smoothing problem can be computed as

$$\hat{\mathbf{x}}(t) = \int_{t_i}^{t_f} \mathbf{A}_o(t-u)\mathbf{y}(u)\,du, \qquad t_i \le t \le t_f \tag{8.6-1}$$

where the $n \times m$ matrix $\mathbf{A}_o(t-u)$ is the "optimum" impulse response of the processor. For linear modulation $c(u)$, $\hat{\mathbf{x}}(t)$ is also an MMSE estimate. To find the optimal impulse response we first define the estimate error for any arbitrary impulse response $\mathbf{A}(t-u)$ as

$$\begin{aligned}\tilde{\mathbf{x}}(t) &= \mathbf{x}(t) - \hat{\mathbf{x}}(t) \\ &= \mathbf{x}(t) - \int_{t_i}^{t_f} \mathbf{A}(t-u)\mathbf{y}(u)\,du\end{aligned} \tag{8.6-2}$$

which has the mean-squared value

$$J = E[\tilde{\mathbf{x}}(t)^T \tilde{\mathbf{x}}(t)] = \mathrm{tr}\, E[\tilde{\mathbf{x}}(t)\tilde{\mathbf{x}}^T(t)]. \tag{8.6-3}$$

After substituting equation (8.6-2) and bringing the expectation inside the integral, the estimate error covariance is

$$\begin{aligned}E[\tilde{\mathbf{x}}(t)\tilde{\mathbf{x}}^T(t)] = &\int_{t_i}^{t_f}\int_{t_i}^{t_f} \mathbf{A}(t-u)E[\mathbf{y}(u)\mathbf{y}^T(v)]\mathbf{A}^T(t-v)\,du\,dv + E[\mathbf{x}(t)\mathbf{x}^T(t)] \\ &- \int_{t_i}^{t_f} \mathbf{A}(t-u)E[\mathbf{y}(u)\mathbf{x}^T(t)]\,du - \int_{t_i}^{t_f} E[\mathbf{x}(t)\mathbf{y}^T(u)]\mathbf{A}^T(t-u)\,du\end{aligned}. \tag{8.6-4}$$

Now suppose that the given impulse response is the sum of an optimum impulse response and a perturbation:

$$\mathbf{A}(t-u) = \mathbf{A}_o(t-u) + \varepsilon \mathbf{A}_\varepsilon(t-u), \tag{8.6-5}$$

so that $J = J_o + \Delta J(\varepsilon)$ where J_o is defined using $\hat{\mathbf{x}}(t)$ from equation (8.6-1). Substituting equaton (8.6-5) in equation (8.6-4) yields

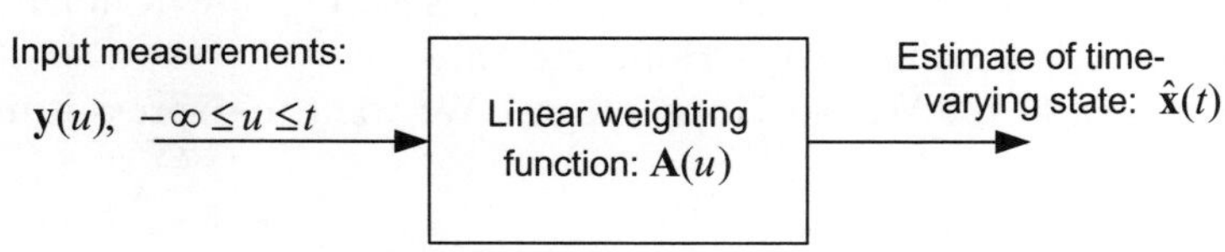

FIGURE 8.16: Wiener filter.

$$
\begin{aligned}
E[\tilde{\mathbf{x}}(t)\tilde{\mathbf{x}}^T(t)] &= \int_{t_i}^{t_f}\int_{t_i}^{t_f} \mathbf{A}_o(t-u)E[\mathbf{y}(u)\mathbf{y}^T(v)]\mathbf{A}_o^T(t-v)\,du\,dv + E[\mathbf{x}(t)\mathbf{x}^T(t)] \\
&\quad - \int_{t_i}^{t_f} \mathbf{A}_o(t-u)E[\mathbf{y}(u)\mathbf{x}^T(t)]\,du - \int_{t_i}^{t_f} E[\mathbf{x}(t)\mathbf{y}^T(u)]\mathbf{A}_o^T(t-u)\,du \\
&\quad + \varepsilon \begin{pmatrix} -\int_{t_i}^{t_f} \mathbf{A}_\varepsilon(t-u)E[\mathbf{y}(u)\mathbf{x}^T(t)]\,du - \int_{t_i}^{t_f} E[\mathbf{x}(t)\mathbf{y}^T(u)]\mathbf{A}_\varepsilon^T(t-u)\,du \\ +\int_{t_i}^{t_f}\int_{t_i}^{t_f} \mathbf{A}_o(t-u)E[\mathbf{y}(u)\mathbf{y}^T(v)]\mathbf{A}_\varepsilon^T(t-v)\,du\,dv \\ +\int_{t_i}^{t_f}\int_{t_i}^{t_f} \mathbf{A}_\varepsilon(t-u)E[\mathbf{y}(u)\mathbf{y}^T(v)]\mathbf{A}_o^T(t-v)\,du\,dv \end{pmatrix}. \\
&\quad + \varepsilon^2 \int_{t_i}^{t_f}\int_{t_i}^{t_f} \mathbf{A}_\varepsilon(t-u)E[\mathbf{y}(u)\mathbf{y}^T(v)]\mathbf{A}_\varepsilon^T(t-v)\,du\,dv
\end{aligned}
\tag{8.6-6}
$$

If $\mathbf{A}_o(t-u)$ is the optimum filter impulse response, $\Delta J(\varepsilon)$ must be positive for all values of ε. A necessary and sufficient condition that $\Delta J(\varepsilon) > 0$ for $\varepsilon \neq 0$ is that the terms multiplying ε be zero, or

$$
\int_{t_i}^{t_f} E[\mathbf{x}(t)\mathbf{y}^T(u)]\mathbf{A}_\varepsilon^T(t-u)\,du = \int_{t_i}^{t_f}\int_{t_i}^{t_f} \mathbf{A}_o(t-v)E[\mathbf{y}(v)\mathbf{y}^T(u)]\mathbf{A}_\varepsilon^T(t-u)\,dv\,du
$$

where we have interchanged the definitions of u and v in the right-hand-side. Since $\mathbf{A}_\varepsilon(t-u)$ is arbitrary and not a function of v, the condition becomes

$$
\boxed{E[\mathbf{x}(t)\mathbf{y}^T(u)] = \int_{t_i}^{t_f} \mathbf{A}_o(t-v)E[\mathbf{y}(v)\mathbf{y}^T(u)]\,dv, \qquad t_i \le u \le t_f}
\tag{8.6-7}
$$

This is the Wiener-Hopf integral. Kalman and Bucy (1961) noted Pugachev's observation that this is a special case of the orthogonal projection principle. This can be understood by using equation (8.6-1) to compute

$$
\boxed{E\left[\hat{\mathbf{x}}(t)\mathbf{y}^T(u)\right] = \int_{t_i}^{t_f} \mathbf{A}_o(t-v)E\left[\mathbf{y}(v)\mathbf{y}^T(u)\right]dv, \qquad t_i \le u \le t_f.}
\tag{8.6-8}
$$

Since this is the same equation as equation (8.6-7), we have

$$
\boxed{E\left[(\mathbf{x}(t)-\hat{\mathbf{x}}(t))\mathbf{y}^T(u)\right] = \mathbf{0}}
\tag{8.6-9}
$$

which is the orthogonal projection principle. Since $\hat{\mathbf{x}}(t)$ is obtained as a linear operation on $\mathbf{y}(u)$, equation (8.6-9) can also be expressed as

$$
\boxed{E\left[(\mathbf{x}(t)-\hat{\mathbf{x}}(t))\hat{\mathbf{x}}^T(t)\right] = \mathbf{0}.}
\tag{8.6-10}
$$

These last two equations will be used later when showing how Kalman and Bucy derived their filter from the Wiener-Hopf integral.

We now return to the Wiener-Hopf integral. We want to express it in terms of the covariances

$$
\begin{aligned}
\mathbf{P}_{xy}(t-u) &\triangleq E[(\mathbf{x}(t)-\bar{\mathbf{x}})(\mathbf{y}^T(u)-\bar{\mathbf{y}}^T)] = E[\mathbf{x}(t)\mathbf{y}^T(u)] - \overline{\mathbf{x}\mathbf{y}}^T \\
\mathbf{P}_{yy}(t-u) &\triangleq E[(\mathbf{y}(t)-\bar{\mathbf{y}})(\mathbf{y}^T(u)-\bar{\mathbf{y}}^T)] = E[\mathbf{y}(t)\mathbf{y}^T(u)] - \overline{\mathbf{y}\mathbf{y}}^T
\end{aligned}
$$

where $\bar{\mathbf{x}} \triangleq E[\mathbf{x}(t)]$ and $\bar{\mathbf{y}} \triangleq E[\mathbf{y}(t)]$ for the stationary processes. Assuming that $\mathbf{x}(t)$ and the measurement noise $\mathbf{r}(t)$ are zero-mean processes, the Wiener-Hopf integral becomes

$$\mathbf{P}_{xy}(t-u) = \int_{t_i}^{t_f} \mathbf{A}_o(t-v)\mathbf{P}_{yy}(v-u)\,dv, \qquad t_i \le u \le t \tag{8.6-11}$$

Since we are interested in the steady-state filter solution, we set $t_f = t$ and allow $t_i = -\infty$ (or at least set t_i much farther in the past than the "memory" of the system.) We also change variables, $\tau = t - u$ and $\lambda = t - v$, to obtain

$$\boxed{\mathbf{P}_{xy}(\tau) = \int_0^{\infty} \mathbf{A}_o(\lambda)\mathbf{P}_{yy}(\tau-\lambda)\,d\lambda, \qquad 0 \le \tau \le t - t_i.} \tag{8.6-12}$$

This expression defines the optimum linear filter in terms of the impulse response $\mathbf{A}_o(\lambda)$ given covariances $\mathbf{P}_{xy}(\tau)$ and $\mathbf{P}_{yy}(\tau - \lambda)$. Wiener obtained a solution for $\mathbf{A}_o(\lambda)$ in the time domain, but we use the frequency domain solution due to Bode and Shannon (1950) because it is more easily understood.

8.6.2 Solution for the Optimal Weighting Function

Since the power spectral density is the Fourier transform of the autocorrelation function, we compute the Fourier transform of the convolution integral equation (8.6-12)

$$\boxed{\begin{aligned}
\mathbf{S}_{xy}(j\omega) &= \int_{-\infty}^{\infty} \mathbf{P}_{xy}(\tau)e^{-j\omega\tau}d\tau \\
&= \int_{-\infty}^{\infty}\left(\int_0^{\infty} \mathbf{A}_o(\lambda)\mathbf{P}_{yy}(\tau-\lambda)\,d\lambda\right)e^{-j\omega\tau}d\tau \\
&= \int_0^{\infty} \mathbf{A}_o(\lambda)\left(\int_{-\infty}^{\infty} \mathbf{P}_{yy}(\tau-\lambda)e^{-j\omega\tau}d\tau\right)d\lambda \\
&= \int_0^{\infty} \mathbf{A}_o(\lambda)e^{-j\omega\lambda}\left(\int_{-\infty}^{\infty} \mathbf{P}_{yy}(\gamma)e^{-j\omega\gamma}d\gamma\right)d\lambda \\
&= \left(\int_0^{\infty} \mathbf{A}_o(\lambda)e^{-j\omega\lambda}\,d\lambda\right)\mathbf{S}_{yy}(\omega)
\end{aligned}} \tag{8.6-13}$$

where $j = \sqrt{-1}$, $\omega = 2\pi f$, and f is frequency in Hz. Notice that $\mathbf{S}_{yy}(\omega)$ is real-valued because the covariance $\mathbf{P}_{yy}(\tau)$ is symmetric about $\tau = 0$. $\mathbf{P}_{xy}(\tau)$ is normally time-symmetric so $\mathbf{S}_{xy}(\omega)$ should be real, but this cannot be guaranteed. To be realizable (i.e., causal), $\mathbf{A}_o(\lambda)$ must be zero for negative time, so the lower limit of the above integration is zero rather than $-\infty$.

Equation (8.6-13) can be rearranged to obtain

$$\mathbf{A}_o^F(j\omega) \triangleq \mathbf{S}_{xy}(j\omega)[\mathbf{S}_{yy}(\omega)]^{-1} \tag{8.6-14}$$

provided that $\mathbf{S}_{yy}(\omega)$ is nonsingular. However, the inverse Fourier transform

$$\mathbf{A}_o(t) = \mathscr{F}^{-1}[\mathbf{A}_o^F(j\omega)] \tag{8.6-15}$$

will not result in a function $\mathbf{A}_o(t)$ that is zero for negative time. Provided that $\mathbf{S}_{xy}(\omega)$ is a real-valued function, the computed $\mathbf{A}_o(t)$ will be symmetric about $t = 0$ and thus is the solution for a nonrealizable (noncausal) filter. Equation (8.6-15) can be used for post-processing applications or to compute an approximate solution (using a truncated $\mathbf{A}_o(t)$) when a long lag between filter input and output is allowed.

To obtain the transform of a realizable filter, the spectral density $\mathbf{S}_{yy}(\omega)$ is factored as

$$\begin{aligned}\mathbf{S}_{yy}(\omega) &= \mathbf{S}_{yy}^{+}(j\omega)\mathbf{S}_{yy}^{-}(j\omega) \\ &= \mathbf{S}_{yy}^{+}(j\omega)[\mathbf{S}_{yy}^{+}(j\omega)]^{*}\end{aligned} \tag{8.6-16}$$

where $\mathbf{S}_{yy}^{+}(j\omega)$ is zero for negative time (poles in left-half complex plane), $\mathbf{S}_{yy}^{-}(j\omega)$ is zero for positive time (poles in right-half complex plane), and $[\mathbf{S}(j\omega)]^*$ denotes the complex conjugate of $\mathbf{S}(j\omega)$. Then the transform of the realizable filter response is computed as

$$\boxed{\mathbf{A}_o^F(j\omega) = \left[\mathbf{S}_{xy}(j\omega)\left([\mathbf{S}_{yy}^{+}(j\omega)]^{*}\right)^{-1}\right]^{+}\left[\mathbf{S}_{yy}^{+}(j\omega)\right]^{-1}} \tag{8.6-17}$$

where $[\cdot]^+$ denotes that only the realizable portion (corresponding to $t \ge 0$) of the term in brackets is used. This is obtained from the inverse Fourier transform of the term in brackets:

$$\boxed{\mathbf{A}'(t) = \mathscr{F}^{-1}\left[\mathbf{S}_{xy}(j\omega)\left([\mathbf{S}_{yy}^{+}(j\omega)]^{*}\right)^{-1}\right]} \tag{8.6-18}$$

and setting $\mathbf{A}'(t) = \mathbf{0}$ for $t < 0$. The inverse transform

$$\mathbf{W}(j\omega) = \left[\mathbf{S}_{yy}^{+}(j\omega)\right]^{-1} \tag{8.6-19}$$

is sometimes referred to as a whitening transformation because time-correlated measurements processed by it are uncorrelated. To obtain the filter impulse response $\mathbf{A}_o(t)$, the inverse transform $\mathscr{F}^{-1}[\mathbf{W}(j\omega)]$ can be convolved with $[\mathbf{A}'(t)]^+$. Alternately $[\mathbf{A}'(t)]^+$ can be re-transformed to the frequency domain, multiplied by $\mathbf{W}(j\omega)$, and the inverse transform of the result computed. Unless the system is very low order and the covariance functions are simple, this is not a particularly easy method for analytic computation. A scalar example at the end of this section demonstrates the procedure more clearly.

8.6.3 Filter Input Covariances

The previous sections described computation of the Wiener filter weighting function, but did not define how the input covariances $\mathbf{P}_{xy}(\tau)$ and $\mathbf{P}_{yy}(\tau)$ are computed from the system model equations. The model equations are

$$\mathbf{x}(t+\tau) = \mathbf{\Phi}(\tau)\mathbf{x}(t) + \mathbf{q}(t+\tau,\tau), \quad E[\mathbf{q}(t+\tau,t)\mathbf{q}^T(t+\tau,t)] = \mathbf{Q}(\tau) \tag{8.6-20}$$

$$\mathbf{y}(t) = \mathbf{H}\,\mathbf{x}(t) + \mathbf{r}(t), \qquad E[\mathbf{r}(t)\mathbf{r}(\tau)] = \mathbf{R}\delta(t-\tau) \tag{8.6-21}$$

where all variables are zero mean, $E[\mathbf{q}(u+\tau,u)\mathbf{r}^T(t)] = \mathbf{0}$ for all t and u, and $E[\mathbf{q}(u+\tau,u)\mathbf{q}^T(t+\tau,t)] = \mathbf{0}$ for non-overlapping time intervals. Hence the covariances are

$$\boxed{\begin{aligned}\mathbf{P}_{xy}(\tau) &= E\left\{(\mathbf{x}(t+\tau)-\bar{\mathbf{x}})\left[(\mathbf{x}(t)-\bar{\mathbf{x}})^T\mathbf{H}^T + \mathbf{r}^T(t)\right]\right\} \\ &= \mathbf{\Phi}(\tau)\mathbf{P}_{xx}\mathbf{H}^T\end{aligned}} \tag{8.6-22}$$

$$\mathbf{P}_{yy}(\tau-\lambda) = E\{[\mathbf{H}(\mathbf{x}(t+\tau-\lambda)-\bar{\mathbf{x}})+\mathbf{r}^T(t+\tau-\lambda)][(\mathbf{x}(t)-\bar{\mathbf{x}})^T\mathbf{H}^T+\mathbf{r}^T(t)]\}$$
$$= \mathbf{H}\mathbf{\Phi}(\tau-\lambda)\mathbf{P}_{xx}\mathbf{H}^T+\mathbf{R}_s\delta(\tau-\lambda) \tag{8.6-23}$$

where

$$\mathbf{P}_{xx} = E\left[(\mathbf{x}(t)-\bar{\mathbf{x}})(\mathbf{x}(t)-\bar{\mathbf{x}})^T\right] = E[\mathbf{x}(t)\mathbf{x}^T(t)] \tag{8.6-24}$$

since $\bar{\mathbf{x}} = \mathbf{0}$ is assumed. The Wiener-Hopf equation for this model is

$$\mathbf{\Phi}(\tau)\mathbf{P}_{xx}\mathbf{H}^T = \int_0^\infty \mathbf{A}_o(\lambda)[\mathbf{H}\mathbf{\Phi}(\tau-\lambda)\mathbf{P}_{xx}\mathbf{H}^T+\mathbf{R}_s\delta(\tau-\lambda)]\,d\lambda. \tag{8.6-25}$$

This equation does not appear similar to the Kalman filter equations. This is partly due to computation of $\hat{\mathbf{x}}$ in equation (8.6-1) as the time convolution of a weighting function and measurements, versus the recursive processing of individual measurements in the Kalman filter. Also the covariance is defined as $\mathbf{P}_{xx} \triangleq E\left[(\bar{\mathbf{x}}-\hat{\mathbf{x}})(\bar{\mathbf{x}}-\hat{\mathbf{x}})^T\right]$ rather than $\mathbf{P} \triangleq E\left[(\mathbf{x}-\hat{\mathbf{x}})(\mathbf{x}-\hat{\mathbf{x}})^T\right]$ as done in the Kalman filter. The Wiener filter error covariance from equation (8.6-4) for the optimal $\mathbf{A}_o$ can be written as

$$\begin{aligned}
E[\tilde{\mathbf{x}}(t)\tilde{\mathbf{x}}^T(t)] &= \int_{ti}^{tf}\left[\int_{ti}^{tf}\mathbf{A}_o(t-u)\mathbf{P}_{yy}(u-v)\,du\right]\mathbf{A}_o^T(t-v)\,dv+\mathbf{P}_{xx} \\
&\quad -\int_{ti}^{tf}\mathbf{A}_o(t-u)\mathbf{P}_{yx}(u-t)]\,du-\int_{ti}^{tf}\mathbf{P}_{xy}(t-u)\mathbf{A}_o^T(t-u)\,du \\
&= \int_0^\infty\left[\int_0^\infty\mathbf{A}_o(\tau)\mathbf{P}_{yy}(\lambda-\tau)\,d\tau\right]\mathbf{A}_o^T(\lambda)\,d\lambda+\mathbf{P}_{xx} \\
&\quad -\int_0^\infty\mathbf{A}_o(\tau)\mathbf{P}_{yx}(-\tau)]\,d\tau-\int_0^\infty\mathbf{P}_{xy}(\tau)\mathbf{A}_o^T(\tau)\,d\tau \\
&= \mathbf{P}_{xx}+\int_0^\infty\mathbf{P}_{xy}(\lambda)\mathbf{A}_o^T(\lambda)\,d\lambda-\int_0^\infty\mathbf{A}_o(\tau)\mathbf{P}_{yx}(\tau)]\,d\tau-\int_0^\infty\mathbf{P}_{xy}(\tau)\mathbf{A}_o^T(\tau)\,d\tau \\
&= \mathbf{P}_{xx}-\int_0^\infty\mathbf{A}_o(\tau)\mathbf{P}_{yx}(\tau)]\,d\tau
\end{aligned} \tag{8.6-26}$$

but this does not provide any additional insight in comparing the Wiener and Kalman filters.

8.6.4 Equivalence of Weiner and Steady-State Kalman-Bucy Filters

To show how the Kalman-Bucy filter can be derived from the Wiener-Hopf integral, we follow the derivation of Kalman and Bucy (1961), except in the last steps. The derivation is based on the model

$$\dot{\mathbf{x}}(t) = \mathbf{F}\mathbf{x}(t)+\mathbf{G}\mathbf{q}_c(t), \quad E[\mathbf{x}(t)\mathbf{q}_c^T(t)] = \mathbf{0}, \quad E[\mathbf{q}_c(t)\mathbf{q}_c^T(\tau)] = \mathbf{Q}_s\delta(t-\tau) \tag{8.6-27}$$

$$\mathbf{y}(t) = \mathbf{H}\mathbf{x}(t)+\mathbf{r}_c(t), \quad E[\mathbf{x}(t)\mathbf{r}_c^T(t)] = \mathbf{0}, \quad E[\mathbf{r}_c(t)\mathbf{r}_c^T(\tau)] = \mathbf{R}_s\delta(t-\tau) \tag{8.6-28}$$

and $E[\mathbf{r}_c(t)\mathbf{q}_c^T(\tau)] = \mathbf{0}$. We first differentiate the Wiener-Hopf equation (8.6-11) with respect to time. Using equation (8.6-27), the left side of the derivative is

$$\frac{\partial}{\partial t}[\mathbf{P}_{xy}(t-u)] = \mathbf{F}\mathbf{P}_{xy}(t-u) = \int_{ti}^{t}\mathbf{F}\mathbf{A}_o(t-v)\mathbf{P}_{yy}(v-u)\,dv$$

and the right side is

$$\begin{aligned}\frac{\partial}{\partial t}\int_{t_i}^{t}\mathbf{A}_o(t-v)\mathbf{P}_{yy}(v-u)\,dv &= \int_{t_i}^{t}\frac{\partial \mathbf{A}_o(t-v)}{\partial t}\mathbf{P}_{yy}(v-u)dv + \mathbf{A}_o(0)\mathbf{H}\,\mathbf{P}_{xy}(t-u)\\ &= \int_{t_i}^{t}\left[\frac{\partial \mathbf{A}_o(t-v)}{\partial t}\mathbf{P}_{yy}(v-u) + \mathbf{A}_o(0)\mathbf{H}\,\mathbf{A}_o(t-v)\mathbf{P}_{yy}(v-u)\right]dv\end{aligned},$$

where we have used equation (8.6-28) and obtained $\mathbf{P}_{xy}(t-u)$ from equation (8.6-11). Combining the two terms yields

$$\int_{t_i}^{t}\left[\mathbf{F}\mathbf{A}_o(t-v) - \frac{\partial \mathbf{A}_o(t-v)}{\partial t} - \mathbf{A}_o(0)\mathbf{H}\,\mathbf{A}_o(t-v)\right]\mathbf{P}_{yy}(v-u)dv = \mathbf{0}. \quad (8.6\text{-}29)$$

This is satisfied if the term in brackets is zero, so any $\mathbf{A}_o(t-v)$ satisfying

$$\frac{\partial \mathbf{A}_o(t-v)}{\partial t} = [\mathbf{F} - \mathbf{A}_o(0)\mathbf{H}]\mathbf{A}_o(t-v)$$

for all $t_i \le v \le t$ must be an optimal solution. Now using equaton (8.6-1)

$$\boxed{\begin{aligned}\frac{d}{dt}\hat{\mathbf{x}}(t) &= \int_{t_i}^{t_f}\frac{\partial}{\partial t}\mathbf{A}_o(t-u)\mathbf{y}(u)\,du + \mathbf{A}_o(0)\mathbf{y}(t)\\ &= \int_{t_i}^{t_f}[\mathbf{F} - \mathbf{A}_o(0)\mathbf{H}]\mathbf{A}_o(t-u)\mathbf{y}(u)\,du + \mathbf{A}_o(0)\mathbf{y}(t)\\ &= \mathbf{F}\hat{\mathbf{x}}(t) + \mathbf{K}[\mathbf{y}(t) - \mathbf{H}\hat{\mathbf{x}}(t)]\end{aligned}} \quad (8.6\text{-}30)$$

where the gain is defined as $\mathbf{K} \triangleq \mathbf{A}_o(0)$. Kalman and Bucy allowed this to be time-varying, but we now assume that steady-state conditions have been reached so that $\mathbf{K}$ is constant. This equation shows that the Weiner filter can be implemented in the recursive form of the Kalman-Bucy filter. To compute the optimal gain, Kalman and Bucy use the orthogonal projection theorem, but the derivation is not intuitive. We deviate slightly from their derivation. First use equation (8.6-28) and the orthogonality conditions equations (8.6-9) and (8.6-10) to compute

$$\begin{aligned}E\left[\tilde{\mathbf{x}}(t)\mathbf{y}^T(t)\right] &= E\left[\tilde{\mathbf{x}}(t)(\mathbf{x}^T(t)\mathbf{H}^T + \mathbf{r}_c^T(t))\right]\\ &= E\left[\tilde{\mathbf{x}}(t)(\hat{\mathbf{x}}^T(t)\mathbf{H}^T + \tilde{\mathbf{x}}^T(t)\mathbf{H}^T + \mathbf{r}_c^T(t))\right]\\ &= \mathbf{P}(\infty)\mathbf{H}^T + E\left[\tilde{\mathbf{x}}(t)\mathbf{r}_c^T(t)\right]\\ &= \mathbf{0}\end{aligned} \quad (8.6\text{-}31)$$

where as before $\tilde{\mathbf{x}} = \mathbf{x} - \hat{\mathbf{x}}$ and $\mathbf{P}(\infty) = E[\tilde{\mathbf{x}}(t)\tilde{\mathbf{x}}^T(t)]$ is the steady-state error covariance. The last term in the above sum is computed using equation (8.6-1):

$$\begin{aligned}E\left[\tilde{\mathbf{x}}(t)\mathbf{r}^T(t)\right] &= -E\left[\hat{\mathbf{x}}(t)\mathbf{r}^T(t)\right]\\ &= -\int_{t_i}^{t}\mathbf{A}_o(t-u)E\left[\mathbf{y}(u)\mathbf{r}^T(t)\right]du\\ &= -\mathbf{A}_o(0)\mathbf{R}_s\\ &= -\mathbf{K}\mathbf{R}_s\end{aligned} \quad (8.6\text{-}32)$$

where

$$\begin{aligned}E[\mathbf{y}(u)\mathbf{r}_c^T(t)] &= E[\mathbf{r}_c(u)\mathbf{r}_c^T(t)]\\ &= \mathbf{R}_s\delta(u-t)\end{aligned}.$$

Combining equations (8.6-31) and (8.6-32), assuming that $\mathbf{R}_s$ is nonsingular,

$$\boxed{\mathbf{K} = \mathbf{P}(\infty)\mathbf{H}^T\mathbf{R}_s^{-1}.} \tag{8.6-33}$$

Equations (8.6-30) and (8.6-33) are the same equations presented in Section 8.3, except that a constant $\mathbf{P}(\infty)$ is used rather than the time-varying $\mathbf{P}(t)$. Hence the state error covariance is also the same. Kalman and Bucy demonstrated that the continuous filter can be derived from the Wiener-Hopf integral. Since Wiener's solution to the Wiener-Hopf equation is for a time-invariant system, the steady-state Kalman-Bucy filter is equivalent to the Wiener filter.

Example 8.6: Wiener Filter for a First-Order Markov Process

Consider the first-order Markov process and scalar measurement

$$\dot{x}(t) = -a\,x(t) + q_c(t), \quad E[q_c(t)q_c(u)] = Q_s\delta(t-u)$$
$$y(t) = x(t) + r_c(t), \qquad E[r_c(t)r_c(u)] = R_s\delta(t-u)$$

with $a > 0$. Using the above equations and methods of Chapter 2, the covariances are

$$P_{xx} = \frac{Q_s}{2a}, \qquad P_{xy}(\lambda) = \frac{Q_s}{2a}e^{-a|\lambda|}, \qquad P_{yy}(\lambda) = \frac{Q_s}{2a}e^{-a|\lambda|} + R_s\delta(\lambda)$$

and the power spectral densities are

$$S_{xy}(\omega) = \frac{Q_s}{\omega^2 + a^2}, \qquad S_{yy}(\omega) = \frac{Q_s}{\omega^2 + a^2} + R_s = R_s\left(\frac{\omega^2 + b^2}{\omega^2 + a^2}\right)$$

where $b = \sqrt{a^2 + Q_s / R_s}$. The factor of $S_{yy}(\omega)$ having an inverse transform that is zero for $t < 0$ is

$$S_{yy}^{+}(j\omega) = \sqrt{R_s}\left(\frac{j\omega + b}{j\omega + a}\right).$$

Then

$$\begin{aligned} S_{xy}(j\omega)\left([S_{yy}^{+}(j\omega)]^*\right)^{-1} &= \frac{1}{\sqrt{R_s}}\left(\frac{Q_s}{\omega^2 + a^2}\right)\left(\frac{-j\omega + a}{-j\omega + b}\right) \\ &= \frac{Q_s / \sqrt{R_s}}{(j\omega + a)(-j\omega + b)} \end{aligned}$$

and

$$\begin{aligned} A'(t) &= \mathscr{F}^{-1}\left[S_{xy}(j\omega)\left([S_{yy}^{+}(j\omega)]^*\right)^{-1}\right] \\ &= \frac{Q_s}{2\pi\sqrt{R_s}}\int_{-\infty}^{\infty}\left(\frac{e^{j\omega t}}{(j\omega + a)(-j\omega + b)}\right)d\omega \\ &= \frac{Q_s}{2\pi(a+b)\sqrt{R_s}}\int_{-\infty}^{\infty}\left(\frac{1}{j\omega + a} + \frac{1}{-j\omega + b}\right)e^{j\omega t}d\omega \\ &= \frac{Q_s}{(a+b)\sqrt{R_s}}\begin{cases} e^{-at} & t \ge 0 \\ e^{bt} & t < 0 \end{cases} \end{aligned}$$

$A'(t)$ is the sum of two terms: one is zero for $t < 0$ and the other is zero for $t > 0$. We now use the realizable portion that is zero for $t < 0$:

$$\mathscr{F}\left([A'(t)]^{+}\right)=\frac{Q_s}{(a+b)\sqrt{R_s}}\left(\frac{1}{j\omega+a}\right).$$

Hence

$$\begin{aligned}A_o^F(j\omega)&=\left[S_{xy}(j\omega)\left([S_{yy}^{+}(j\omega)]^{*}\right)^{-1}\right]^{+}\left[S_{yy}^{+}(j\omega)\right]^{-1}\\&=\frac{Q_s}{(a+b)\sqrt{R_s}}\left(\frac{1}{j\omega+a}\right)\frac{1}{\sqrt{R_s}}\left(\frac{j\omega+a}{j\omega+b}\right)\\&=\frac{Q_s}{(a+b)R_s}\left(\frac{1}{j\omega+b}\right)\end{aligned}$$

and

$$\boxed{A_o(t)=\frac{Q_s}{(a+b)R_s}\begin{cases}e^{-bt} & t\geq 0\\ 0 & t<0\end{cases}}$$

This solution has been verified using Monte Carlo simulation. When integration step sizes are less than $0.05/b$, the Wiener filter using the above weighting function in equation (8.6-1) generates the same state estimates as a steady-state Kalman-Bucy filter within 3%. The steady-state covariance of the Kalman-Bucy filter is obtained by solving the Riccati equation

$$-2a\,P(\infty)+Q_s-P^2(\infty)/R_s=0$$

for $P(\infty)$. This yields the steady-state gain

$$\begin{aligned}K&=P(\infty)R_s^{-1}=-a+\sqrt{a^2+Q_s/R_s}\\&=\frac{(-a+b)(a+b)}{(a+b)}\\&=\frac{-a^2+a^2+Q_s/R_s}{(a+b)}\\&=\frac{Q_s}{(a+b)R_s}\\&=A_o(0)\end{aligned}.$$

As expected, the weighting given to the current measurement is the same for both the steady-state Kalman-Bucy and Wiener filters. The difficulty in computing the Wiener-Hopf weighting function and the ease of implementing the steady-state Kalman-Bucy filter for this simple example demonstrates why the Wiener filter is rarely used.

8.7 SUMMARY

This chapter has presented the basics of Kalman filtering. For the discrete-time case, the time propagation and measurement models are

$$\mathbf{x}_i = \mathbf{\Phi}_{i,i-1}\mathbf{x}_{i-1} + \mathbf{q}_{i,i-1} + \mathbf{u}_{i,i-1}, \qquad E[\mathbf{q}_{i,i-1}] = \mathbf{0}, \quad E[\mathbf{q}_{i,i-1}\mathbf{q}_{j,j-1}^T] = \mathbf{Q}_{i,i-1}\delta_{ij}$$

$$\mathbf{y}_i = \mathbf{H}_i\mathbf{x}_i + \mathbf{r}_i, \qquad E[\mathbf{r}_i] = \mathbf{0}, \qquad E[\mathbf{r}_i\mathbf{r}_j^T] = \mathbf{R}_i\delta_{ij}.$$

It is usually assumed that $E[\mathbf{r}_i\mathbf{q}_{j,j-1}^T] = \mathbf{0}$ for all i, j, but the basic filter equations can be modified to handle this correlation. The filter *a posteriori* filter state estimate and covariance at time t_{i-1}, based on measurements up to and including time t_{i-1}, are designated as $\hat{\mathbf{x}}_{i-1/i-1}$ and $\mathbf{P}_{i-1/i-1}$, respectively. The *a priori* state and covariance at time t_i are computed as

$$\hat{\mathbf{x}}_{i/i-1} = \mathbf{\Phi}_{i,i-1}\hat{\mathbf{x}}_{i-1/i-1} + \mathbf{u}_{i,i-1}, \qquad \mathbf{P}_{i/i-1} = \mathbf{\Phi}_{i,i-1}\mathbf{P}_{i-1/i-1}\mathbf{\Phi}_{i,i-1}^T + \mathbf{Q}_{i,i-1}.$$

The state estimate and covariance are updated to include the measurements at time t_i as

$$\begin{aligned}\hat{\mathbf{x}}_{i/i} &= \hat{\mathbf{x}}_{i/i-1} + \mathbf{K}_i(\mathbf{y}_i - \mathbf{H}_i\hat{\mathbf{x}}_{i/i-1})\\ \mathbf{P}_{i/i} &= (\mathbf{I} - \mathbf{K}_i\mathbf{H}_i)\mathbf{P}_{i/i-1}\end{aligned}$$

where the $n \times m$ gain matrix is

$$\mathbf{K}_i = \mathbf{P}_{i/i-1}\mathbf{H}_i^T(\mathbf{H}_i\mathbf{P}_{i/i-1}\mathbf{H}_i^T + \mathbf{R}_i)^{-1}.$$

A longer form of the covariance update—called the Joseph form—can also be used.

Section 8.2 showed how the discrete filter equations are modified when the measurement noise and process noise are correlated or the when the measurement noise is time-correlated. The white characteristic of the measurement residuals (innovations) and the relevance for data editing and model validation were discussed.

The continuous-time Kalman-Bucy filter uses the dynamic and measurement models

$$\dot{\mathbf{x}}(t) = \mathbf{F}(t)\mathbf{x}(t) + \mathbf{G}(t)\mathbf{q}_c(t), \qquad E[\mathbf{q}_c(t)] = \mathbf{0}, \quad E[\mathbf{q}_c(t)\mathbf{q}_c^T(\tau)] = \mathbf{Q}_s(t)\delta(t-\tau),$$

$$\mathbf{y}(t) = \mathbf{H}(t)\mathbf{x}(t) + \mathbf{r}_c(t), \qquad E[\mathbf{r}_c(t)] = \mathbf{0}, \quad E[\mathbf{r}_c(t)\mathbf{r}_c^T(\tau)] = \mathbf{R}_s(t)\delta(t-\tau)$$

with $E[\mathbf{q}_c(t)\mathbf{r}_c^T(\tau)] = \mathbf{0}$ usually assumed. The filter state estimate and covariance obey the differential equations

$$\begin{aligned}\dot{\hat{\mathbf{x}}}(t) &= \mathbf{F}\hat{\mathbf{x}}(t) + \mathbf{K}_c(t)[\mathbf{y}(t) - \mathbf{H}\hat{\mathbf{x}}(t)]\\ \dot{\mathbf{P}}(t) &= \mathbf{F}\mathbf{P}(t) + \mathbf{P}(t)\mathbf{F}^T + \mathbf{G}\mathbf{Q}_s\mathbf{G}^T - \mathbf{K}_c(t)\mathbf{R}_s\mathbf{K}_c^T(t)\end{aligned}$$

where the continuous-time gain matrix is

$$\mathbf{K}_c(t) = \mathbf{P}(t)\mathbf{H}^T\mathbf{R}_s^{-1}.$$

The equation for $\dot{\mathbf{P}}(t)$ is a Riccati equation.

Section 8.4 discussed two extensions of the discrete Kalman filter. The first, due to Friedland, is equivalent to the standard Kalman filter but it is implemented as a bias-free filter followed by a bias-restoring filter. It was originally developed to reduce computations and improve filter accuracy, but these benefits are negligible compared with good implementations of the Kalman filter. Friedland's filter is primarily important as a basis for understanding the effects of biases and for development of other algorithms. The second extension of the discrete Kalman filter is due to Schmidt, and is called the Kalman-Schmidt consider filter. It has the advantage that the effects of poorly known "nuisance parameters" are accounted for in computing the gain matrix for the adjusted states. The consider filter is suboptimal but it can sometimes be more robust than the Kalman filter (with all parameters estimated) when models are uncertain.

Section 8.5 discussed computation of the steady-state gain matrix for the continuous and discrete Kalman filters. Most solution methods compute the steady-state covariance using a $2n \times 2n$ Hamiltonian matrix, where n is the number if states. Standard solution methods compute eigenvalues of the Hamiltonian matrix. Methods based on Schur decomposition or the matrix sign function are considered the most reliable.

Finally, Section 8.6 described implementation of the Wiener filter and showed that it is equivalent to the steady-state solution of the continuous Kalman-Bucy filter.

CHAPTER 9

FILTERING FOR NONLINEAR SYSTEMS, SMOOTHING, ERROR ANALYSIS/MODEL DESIGN, AND MEASUREMENT PREPROCESSING

This chapter covers extensions of Kalman filtering that are routinely used. Specific topics considered in this chapter are:

1. *Kalman filtering for nonlinear systems*: The standard techniques include linearization about a reference trajectory (linearized filtering), linearization about the current estimate (extended Kalman filtering), and iterated linearized filtering (iterated extended Kalman filter or iterated linear filter-smoother). The extended Kalman filter is probably the most frequently used Kalman filter implementation.
2. *Smoothing*: Smoothers compute the minimum mean-squared error (MMSE) estimate of a state in past time based on measurements up to a later time. Smoothing options include fixed point, fixed lag, and fixed interval.
3. *Error analysis and model state selection*: Error analysis should be a necessary step in most Kalman filter implementations. As with least-squares estimation, Monte Carlo and covariance error analysis can be used. Options for linear covariance error analysis include perturbation analysis of independent error sources (measurement noise, process noise, and errors in unadjusted or "considered" model parameters), error analysis for *reduced-order models* (ROMs) defined as transformations on detailed models, and error analysis for truth and filter models that are structurally different but a subset of important states are common to both. Analysis of different model structures is important when attempting to develop a high-accuracy filter that is insensitive to model structure and parameter assumptions.
4. *Measurement preprocessing*: A Kalman filter that has reached steady-state conditions can be very effective in identifying anomalous measurements. During the initialization period, however, it is much less effective. Because the filter has an *infinite impulse response* (IIR) characteristic, it is important

Advanced Kalman Filtering, Least-Squares and Modeling: A Practical Handbook, by Bruce P. Gibbs

that anomalous measurements be edited during preprocessing. The available preprocessing methods are essentially the same as described for least-squares estimation.

9.1 NONLINEAR FILTERING

Although Kalman's original (1960) paper used linear models, the first application was for a nonlinear system: navigation of the Apollo spacecraft. As described by McGee and Schmidt (1985), Stanley Schmidt worked in the Dynamics Analysis Branch at Ames Research Center, and in 1960 that group had the task of designing the midcourse navigation and guidance for the Apollo circumlunar mission. It was assumed that the spacecraft crew would use an inertially referenced optical sensor to measure angles to the earth and moon. However, the method of computing the navigation solution was unclear. Iterative weighted least squares was too complex for state-of-the-art onboard computers. Wiener filtering could not be applied without making approximations that would restrict the observation system or degrade accuracy. Other alternatives could not meet accuracy requirements.

Fortunately Kalman visited Ames in the fall of 1960, and the staff immediately recognized the usefulness of Kalman's sequential solution. However, it was not at first clear how the linear filter would be used for the nonlinear system. Partly because the Ames staff had already been using linear perturbation concepts in guidance studies, Schmidt realized that the nonlinear trajectory equations could be linearized about a reference trajectory if the filter equations were broken into separate discrete time and measurement updates. It was also suspected that better results would be obtained if the linearization was recomputed using the current state solution at each step. Although many details and practical problems remained, the approach allowed successful navigation to the moon. Because of this experience and others, Schmidt enthusiastically promoted use of Kalman's filter.

Schmidt's modification was at first called the Kalman-Schmidt filter by some authors, but it came to be accepted as the *extended Kalman filter* (EKF). Because the vast majority of applications are for nonlinear systems, it is often just called the Kalman filter.

9.1.1 Linearized and Extended Kalman Filters

The *extended Kalman filter* (EKF) assumes that the nonlinear system dynamic and measurement models can be expanded in first-order Taylor series about the current estimate. That is, the filter time update becomes

$$\boxed{\begin{aligned} \hat{\mathbf{x}}_{i/i-1} &= \boldsymbol{\varphi}(\hat{\mathbf{x}}_{i-1/i-1}, \mathbf{u}_i, t_i, t_{i-1}) \\ \boldsymbol{\Phi}_{i,i-1} &= \frac{\partial \boldsymbol{\varphi}(\hat{\mathbf{x}}_{i-1/i-1}, \mathbf{u}_{i-1}, t_i, t_{i-1})}{\partial \mathbf{x}} \\ \mathbf{P}_{i/i-1} &= \boldsymbol{\Phi}_{i,i-1}\mathbf{P}_{i-1/i-1}\boldsymbol{\Phi}_{i,i-1}^T + \mathbf{Q}_{i,i-1} \end{aligned}} \tag{9.1-1}$$

where $\boldsymbol{\varphi}(\hat{\mathbf{x}}_{i-1/i-1}, \mathbf{u}_i, t_i, t_{i-1})$ is usually obtained by numerically integrating the state differential equation $\dot{\mathbf{x}}(t) = \mathbf{f}[\mathbf{x}(t), \mathbf{u}(t), t]$ as described in Section 2.2, and we again use the simplified notation

$$\frac{\partial \boldsymbol{\varphi}(\boldsymbol{\xi},\mathbf{u},t_i,t_{i-1})}{\partial \mathbf{x}} = \left.\frac{\partial \boldsymbol{\varphi}(\mathbf{x},\mathbf{u},t_i,t_{i-1})}{\partial \mathbf{x}}\right|_{\mathbf{x}=\boldsymbol{\xi}}.$$

The state noise covariance $\mathbf{Q}_{i,j-1}$ is also obtained using the methods of Section 2.2.1.

The EKF measurement update is then

$$\boxed{\begin{aligned}\hat{\mathbf{y}}_i &= \mathbf{h}(\hat{\mathbf{x}}_{i/i-1})\\ \mathbf{H}_i &= \frac{\partial \mathbf{h}(\hat{\mathbf{x}}_{i/i-1})}{\partial \mathbf{x}}\\ \mathbf{K}_i &= \mathbf{P}_{i/i-1}\mathbf{H}_i^T(\mathbf{H}_i\mathbf{P}_{i/i-1}\mathbf{H}_i^T+\mathbf{R})^{-1}.\\ \hat{\mathbf{x}}_{i/i} &= \hat{\mathbf{x}}_{i/i-1}-\mathbf{K}_i(\mathbf{y}_i-\hat{\mathbf{y}}_i)\\ \mathbf{P}_{i/i} &= \mathbf{P}_{i/i-1}-\mathbf{K}_i\mathbf{H}_i\mathbf{P}_{i/i-1}\end{aligned}} \tag{9.1-2}$$

The Joseph form of the $\mathbf{P}_{i/i}$ update can also be used.

Early users of the EKF reported that filters sometimes became unresponsive to measurements and/or the state estimate diverged from true values much more than predicted by the filter covariance matrix. These problems were often caused by modeling errors, outlying measurements, or numerical instability resulting from single precision implementations and/or failure to keep $\mathbf{P}_{i/i}$ symmetric. The divergence problem is more serious in an extended filter than a linear filter because partial derivatives are evaluated at the current state estimate, and if that deviates from truth, partials will be incorrect causing further divergence.

For some applications in which the process and measurement noise is small and models are good, an accurate reference trajectory can be computed prior to filtering. Then the filter partial derivatives $\mathbf{\Phi}_{i,i-1}$ and $\mathbf{H}_i$ are evaluated along the reference trajectory. Satellite orbit determination for mid and high earth orbit spacecraft is one example where a limited tracking span allows fairly accurate ephemeris predictions for periods of days. For low altitude satellites, batch least-squares solutions can be used as a reference trajectory for a Kalman filter/smoother that improves the accuracy of the orbit (Gibbs 1978, 1979). Use of an accurate reference trajectory may help avoid accelerated divergence due to incorrect partials, although it does not alleviate effects of measurement outlier, modeling, or numerical errors. The linearized Kalman filter is nearly identical to an extended filter but $\mathbf{\Phi}_{i,i-1}$ and $\mathbf{H}_i$ are computed as

$$\boxed{\mathbf{\Phi}_{i,i-1} = \frac{\partial \boldsymbol{\varphi}(\mathbf{x}_{i-1}^{ref},\mathbf{u}_{i-1},t_i,t_{i-1})}{\partial \mathbf{x}}} \tag{9.1-3}$$

and

$$\boxed{\mathbf{H}_i = \left.\frac{\partial \mathbf{h}(\mathbf{x}_i^{ref})}{\partial \mathbf{x}}\right|} \tag{9.1-4}$$

where $\mathbf{x}_i^{ref}$ is the reference trajectory.

Most Kalman filter applications involve nonlinear systems and use the EKF described above. The EKF would not be so widely accepted if the approach was not successful. However, our first example is intended to make you cautious when working with nonlinear systems. Later examples show more positive results.

Example 9.1: Nonlinear Measurement Model for Bias State

A trivial one-state, one-measurement model demonstrates some of the problems encountered when working with nonlinear models and the EKF. This example uses a bias state x with a scalar measurement that is a nonlinear function of the state:

$$y = x + \alpha x^2 + r$$

where $E[r] = 0$, $R = E[r^2] = 1$ and $x = 0$. To make the effect of the nonlinearity significant, we set $\alpha = 0.5$. The *a priori* filter estimate is $\hat{x}_{0/0} = 0$ (optimistically equal to the true value) and $P_{0/0} = E[(x - \hat{x}_{0/0})^2] = 1$. The EKF model of the measurement and partial derivative is

$$\hat{y} = \hat{x}_{0/0} + \alpha \hat{x}_{0/0}^2 = 0$$
$$H = \frac{\delta y}{\delta x} = 1 + 2\alpha \hat{x}_{0/0} = 1 .$$

Since x is a bias, the time update is simply $\hat{x}_{1/0} = \hat{x}_{0/0}$ and $P_{1/0} = P_{0/0}$. Using the above values, the Kalman gain, *a posteriori* state estimate and covariance are computed as

$$K = P_{0/0} H / (P_{0/0} H^2 + R) = 0.5$$
$$\hat{x}_{1/1} = \hat{x}_{0/0} + K(y - \hat{y}) = 0.5y \quad .$$
$$P_{1/1} = P_{0/0} - K\,H P_{0/0} = 0.5$$

Now we consider the problem using probability density functions. Recall from Chapter 4 that the minimum variance/MMSE estimate for linear or nonlinear models is the conditional mean; that is, the mean x of the density function

$$p_{X|Y}(x \mid y) = p_{Y|X}(y \mid x)\, p_X(x) / p_Y(y)$$

where $p_{Y|X}(y|x)$ is the conditional probability density of receiving the measurement given the state x, $p_X(x)$ is the prior density function of x, and

$$p_Y(y) = \int_{-\infty}^{\infty} p_{Y|X}(y \mid x)\, p_X(x)\, dx.$$

Assuming that the one-dimensional distributions are Gaussian,

$$p_{Y|X}(y \mid x) = \frac{1}{(2\pi)^{1/2} R^{1/2}} \exp(-[y - h(x)]^2/(2R))$$

and

$$p_X(x) = \frac{1}{(2\pi)^{1/2} P_{0/0}^{1/2}} \exp(-(x - \hat{x}_{0/0})^2/(2P_{0/0}))$$

for the measurement model $y = h(x) + r$. The conditional mean is computed as

$$\hat{x} = \left(\int_{-\infty}^{\infty} x\, p_{X|Y}(x \mid y) dx\right) \Big/ \left(\int_{-\infty}^{\infty} p_{X|Y}(x \mid y) dx\right)$$

Figure 9.1 shows $p_{X|Y}(x|y)$ for the given example model with y computed using different values of the measurement noise r. The lines are labeled as

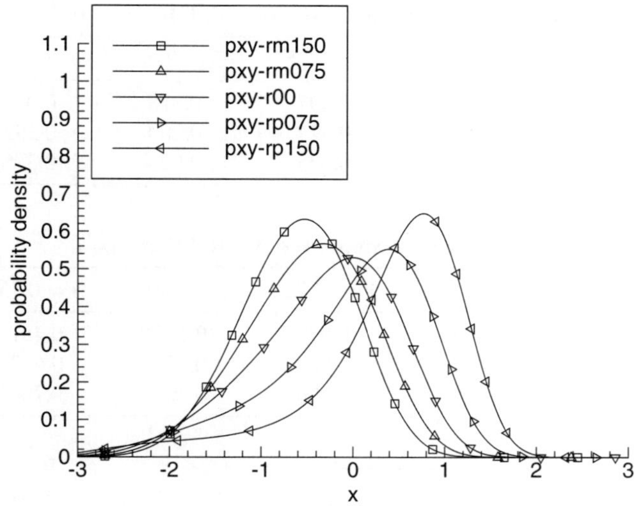

FIGURE 9.1: Conditional probability density for one-state model with filter $x_0 = 0$.

pxy_rm150	uses $r = -1.50$
pxy_rm075	uses $r = -0.75$
pxy_r00	uses $r = 0.0$
pxy_rp075	uses $r = 0.75$
pxy_rp150	uses $r = 1.5$.

The largest r value is only 1.5 times the specified noise standard deviation, so these are typical noise magnitudes. Notice that the conditional distributions are highly skewed for all cases (even $r = 0$). Table 9.1 lists the mean and mode of $p_{X|Y}(x|y)$ and the EKF estimate. Notice that the EKF estimate does not match either the mean or the mode, but it is closer to the mode for all cases except $r = -1.5$. Neither the conditional mean nor the mode are symmetric about $r = 0$, but the mode is closer to symmetric. The EKF estimate is symmetric about $r = 0$ because $\hat{x}_{1/1} = 0.5y$ when $\hat{x}_{0/0} = 0$. As expected, the root-mean-squared (RMS) error from the true $x = 0$ is slightly smaller for the conditional mean estimate (0.412) than for the conditional mode (0.475) or the EKF (0.530). (Note: The nonlinear equations for $\hat{y}$ and H can be substituted in equation (4.4-21) to compute Maximum *A Posteriori* (MAP) solutions for x. Those agree with the mode $p_{X|Y}(x|y)$ values listed in Table 9.1.)

Figure 9.2 and Table 9.2 show the results when $\hat{x}_{0/0} = 1$ rather than the true value ($x = 0$). Again the conditional distributions are highly skewed and the EKF estimate is closer to the mode than the mean. In this case the RMS error of the conditional mean is larger than either the conditional mode or the EKF estimate. However, Monte Carlo simulation using random samples of the true x distributed as $N(\hat{x}_{0/0}, P_{0/0})$ shows that the RMS error of the conditional mean is about 4–7% smaller than either the mode or the EKF estimates, which supports the assertion that the conditional mean is the minimum variance estimate. In fact, the conditional mean is essentially unbiased, while the mode and EKF estimates have a significant bias error (~0.23).

TABLE 9.1: State Estimates for One-State Model with Filter $x_0 = 0$

Estimate Type	$r = -1.5$	$r = -0.75$	$r = 0.0$	$r = 0.75$	$r = 1.5$
Conditional mean	−0.613	−0.497	−0.318	−0.040	0.353
Conditional mode	−0.527	−0.324	0.000	0.392	0.769
EKF estimate	−0.750	−0.375	0.000	0.375	0.750

TABLE 9.2: State Estimates for One-State Model with Filter $x_0 = 1$

Estimate Type	$r = -1.5$	$r = -0.75$	$r = 0.0$	$r = 0.75$	$r = 1.5$
Conditional mean	−0.264	−0.078	0.173	0.487	0.836
Conditional mode	−0.152	0.087	0.379	0.692	1.000
EKF estimate	−0.200	0.100	0.400	0.700	1.000

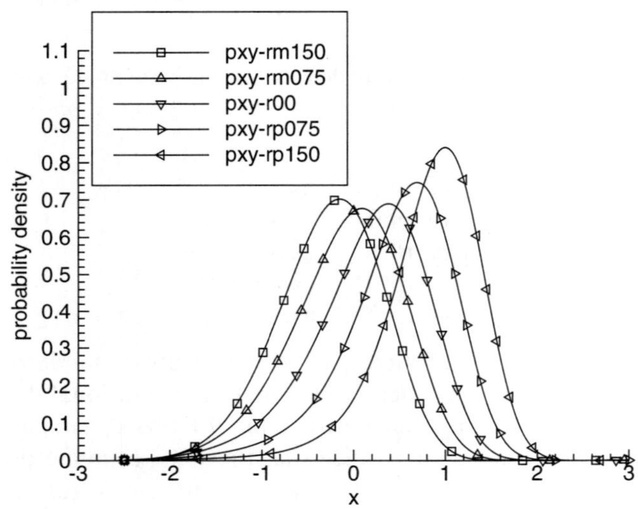

FIGURE 9.2: Conditional probability density for one-state model with filter $x_0 = 1$.

The bias in EKF or conditional mode estimates is a function of the magnitude of the nonlinearity compared with the measurement noise. When the effect of the nonlinearity for expected variations in state estimates is small compared with the measurement noise standard deviation, the EKF estimate should be reasonably accurate and unbiased. More precisely, for a Taylor series expansion of the measurement model about the current state estimate,

$$y(x) = y(\hat{x}) + \frac{dy}{dx}(x-\hat{x}) + 0.5\frac{d^2y}{dx^2}(x-\hat{x})^2 + \ldots,$$

the effect of the quadratic and higher order terms for the given state uncertainty specified by P_x should be small compared with the specified measurement noise; that is,

$$\frac{d^2y}{dx^2}P_x \ll 2\sqrt{R}$$

where $P_x = E[(x-\hat{x})^2]$. This relationship should apply for every processed measurement. In the current example, when $\sqrt{R}=5$ rather than 1, the conditional mean, mode, and EKF estimates are all nearly unbiased and match within a few percent.

To summarize, when nonlinearities for the expected error in state estimates are significant relative to measurement noise, the EKF estimate will be biased and the computed error covariance will underestimate actual errors. In some cases the bias can cause the filter to diverge from the true state. The EKF estimate for each measurement is not minimum variance. However, because the EKF is a feedback loop, it will usually (but not always) converge to the correct answer after processing many measurements, even if convergence takes longer than it should. In this particular example, the bias in the EKF estimate decayed at about the same rate as the uncertainty computed from the *a posteriori* covariance, and the behavior was not significantly different from that of the linear case with $\alpha = 0$. Figure 9.3 shows the state estimate ("xf") and $1-\sigma$ uncertainty ("sig_xf") for one Monte Carlo sample using 10,000 measurements.

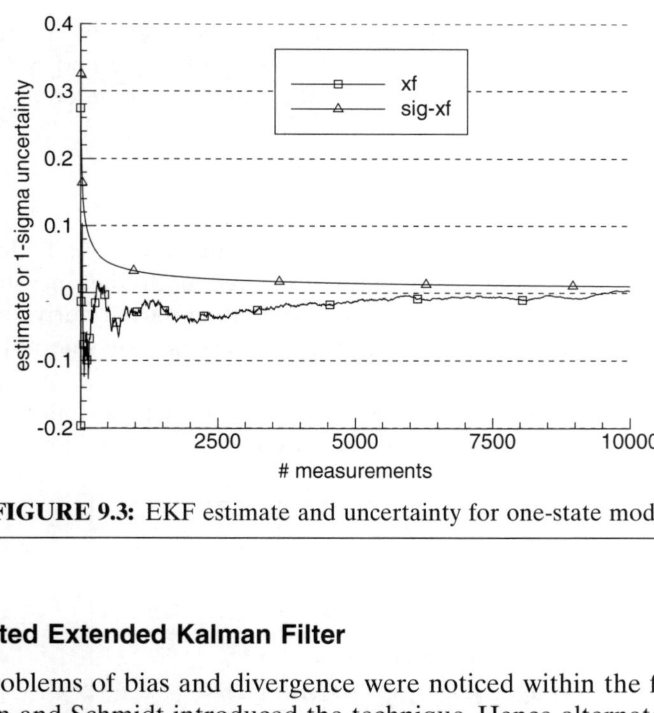

FIGURE 9.3: EKF estimate and uncertainty for one-state model.

9.1.2 Iterated Extended Kalman Filter

The EKF problems of bias and divergence were noticed within the first few years after Kalman and Schmidt introduced the technique. Hence alternate methods for handling nonlinearities were developed. Some approaches include higher order terms in the filter design, but these require higher order partial derivatives and they are generally difficult to compute and implement. Jazwinski (1970) summarizes early research in these areas, and compares second-order filters with two approaches that are fairly easy to implement: the *iterated extended Kalman filter* (IEKF) and the *iterated linear filter-smoother*. Gelb (1974, sections 6.1 and 6.2) compares the iterated filter with higher order filters and a statistically linearized filter that characterizes the nonlinearity using describing functions.

The IEKF was apparently developed by J. Breakwell, and then analyzed by Denham and Pines (1966). The *a posteriori* state estimate obtained after processing a measurement is used in the IEKF as the reference state for recomputing $\mathbf{H}_i$. In the iterated linear filter-smoother due to Wishner et al. (1968), the *a posteriori* state estimate is used to recompute both $\mathbf{\Phi}_{i,i-1}$ and $\mathbf{H}_i$. That is, the time and measurement updates are recomputed using the new $\mathbf{\Phi}_{i,i-1}$ and $\mathbf{H}_i$, and this is repeated until the change in state estimates is small. Both algorithms are implemented to avoid weighting the measurements in the solution more than once.

For the IEKF, the following measurement update equations are repeated:

$$\boxed{\begin{aligned}
\hat{\mathbf{y}}_i^j &= \mathbf{h}(\boldsymbol{\eta}^j) \\
\mathbf{H}_i^j &= \frac{\partial \mathbf{h}(\boldsymbol{\eta}^j)}{\partial \mathbf{x}} \\
\mathbf{K}_i^j &= \mathbf{P}_{i/i-1}(\mathbf{H}_i^j)^T[\mathbf{H}_i^j\mathbf{P}_{i/i-1}(\mathbf{H}_i^j)^T + \mathbf{R}_i]^{-1} \\
\boldsymbol{\eta}^{j+1} &= \hat{\mathbf{x}}_{i/i-1} + \mathbf{K}_i^j[\mathbf{y}_i - \hat{\mathbf{y}}_i^j - \mathbf{H}_i^j(\hat{\mathbf{x}}_{i/i-1} - \boldsymbol{\eta}^j)]
\end{aligned}} \quad . \qquad (9.1\text{-}5)$$

(In these equations the superscript j is the iteration counter, not an exponent.) The iterations are initialized with $\boldsymbol{\eta}^1 = \hat{\mathbf{x}}_{i/i-1}$. When converged, the updated state for the final iteration j is set as $\hat{\mathbf{x}}_{i/i} = \boldsymbol{\eta}^{j+1}$, and the covariance is computed as

$$\mathbf{P}_{i/i} = (\mathbf{I} - \mathbf{K}_i^j\mathbf{H}_i^j)\mathbf{P}_{i/i-1}$$

or using the Joseph form.

The IEKF only attempts to remove the effects of measurement model nonlinearities. The iterated linear filter-smoother improves both the measurement and dynamic model reference trajectory when computing partial derivatives. Because the method requires modifying $\hat{\mathbf{x}}_{i-1/i-1}$ after updating $\hat{\mathbf{x}}_{i/i}$, the smoothing iterations are more complicated. The iterations are initialized with $\xi^1 = \hat{\mathbf{x}}_{i-1/i-1}$ and $\eta^1 = \hat{\mathbf{x}}_{i/i-1}$ for the state dynamic model $\hat{\mathbf{x}}_{i/i-1} = \boldsymbol{\varphi}(\hat{\mathbf{x}}_{i-1/i-1}, t_i, t_{i-1})$. Then the following equations are repeated in the order shown until changes in η^j are negligible:

$$\boxed{\begin{aligned}
\mathbf{\Phi}_{i,i-1}^j &= \frac{\partial \boldsymbol{\varphi}(\boldsymbol{\xi}^j, t_i, t_{i-1})}{\partial \mathbf{x}} \\
\mathbf{x}_{i/i-1} &= \boldsymbol{\varphi}(\boldsymbol{\xi}^j, t_i, t_{i-1}) + \mathbf{\Phi}_{i,i-1}^j(\hat{\mathbf{x}}_{i-1/i-1} - \boldsymbol{\xi}^j) \\
\mathbf{P}_{i/i-1}^j &= \mathbf{\Phi}_{i,i-1}^j\mathbf{P}_{i-1/i-1}(\mathbf{\Phi}_{i,i-1}^j)^T + \mathbf{Q}_{i,i-1} \\
\mathbf{S}^j &= \mathbf{P}_{i-1/i-1}(\mathbf{\Phi}_{i,i-1}^j)^T(\mathbf{P}_{i/i-1}^j)^{-1} \\
\hat{\mathbf{y}}_i^j &= \mathbf{h}(\boldsymbol{\eta}^j) \\
\mathbf{H}_i^j &= \frac{\partial \mathbf{h}(\boldsymbol{\eta}^j)}{\partial \mathbf{x}} \\
\mathbf{K}_i^j &= \mathbf{P}_{i/i-1}^j(\mathbf{H}_i^j)^T[\mathbf{H}_i^j\mathbf{P}_{i/i-1}^j(\mathbf{H}_i^j)^T + \mathbf{R}_i]^{-1} \\
\boldsymbol{\eta}^{j+1} &= \hat{\mathbf{x}}_{i/i-1} + \mathbf{K}_i^j[\mathbf{y}_i - \hat{\mathbf{y}}_i^j - \mathbf{H}_i^j(\hat{\mathbf{x}}_{i/i-1} - \boldsymbol{\eta}^j)] \\
\boldsymbol{\xi}^{j+1} &= \hat{\mathbf{x}}_{i-1/i-1} + \mathbf{S}^j(\boldsymbol{\eta}^{j+1} - \hat{\mathbf{x}}_{i/i-1})
\end{aligned}} \quad . \qquad (9.1\text{-}6)$$

After convergence compute

$$\begin{aligned}\hat{\mathbf{x}}_{i/i} &= \boldsymbol{\eta}^{j+1} \\ \mathbf{P}_{i/i} &= (\mathbf{I}-\mathbf{K}_i^j\mathbf{H}_i^j)\mathbf{P}_{i/i-1}^j\end{aligned} \tag{9.1-7}$$

The Joseph form of the covariance update can also be used. Jazwinski (1970, chapter 9) shows that the IEKF solution is always equal to the mode of the conditional density. Hence while an IEKF solution may be more accurate than an EKF solution, it is not minimum variance and does not eliminate the estimate bias. Depending upon the type of nonlinearity, the iterations are not guaranteed to converge. However, the methods are easy to implement and are computationally efficient. The statistically linearized filter (Gelb 1974, section 6.2) is slightly more difficult to implement and possibly slower than the iterated filters, but may be useful for some problems. Newer techniques that evaluate the nonlinear equations at selected points (allowing computation of first- and second-order moments) are discussed in Chapter 11.

Example 9.2: Continuation of Example 9.1 Using the IEKF

Example 9.1 did not include a time update model because the state is a bias. Hence we use the IEKF rather than the iterated filter-smoother for this example. In each case the IEKF solutions are identical to the conditional mode values of Tables 9.1 and 9.2, which are slightly different from the EKF solutions. However, the number of iterations required for state convergence within 0.001 varies greatly. With $r = -1.5$ and $\hat{x}_{0/0} = 0$, the iterations are barely stable (62 iterations are required), which is a reason for caution when using this method. For $r = 0$ to 1.5, only two to three iterations are required.

Using a scalar example with quadratic dynamic and cubic measurement model nonlinearities, Jazwinski shows that IEKF solutions closely match both the mean of the conditional probability density and estimates obtained using a modified Gaussian second-order filter. The EKF estimates are biased, although they eventually converge to the true state. Jazwinski's result may or may not be more representative of typical real applications than our bias example. Gelb (1974, example 6.1-3) shows that the IEKF is more accurate than either the EKF or a second-order filter for range tracking of a falling ball—similar to our Example 7.3 but with an unknown atmospheric drag coefficient. Gelb also shows that the statistically linearized filter is superior to the EKF or higher order filters for a scalar nonlinear measurement problem, but the iterated filters were not tested for this example. Wishner et al. (1969) show that iterated filters can reduce estimation errors compared with the EKF.

It is surprisingly difficult to find real-world problems for which the performance of the iterated filters is superior to that of the EKF. This statement is generally true for most approaches to nonlinear filtering. For mildly nonlinear problems the EKF works well and iterated (or other nonlinear filters) do not improve the solution significantly. For more significant nonlinearities, the iterated filters sometimes fail to converge in cases when the EKF provides a satisfactory (if not optimal) solution. There are some problems for which the iterated filters work better than the EKF, but general characteristics of these problems are not clear.

We compared the EKF with iterated filters for the passive ship tracking problem of Section 3.1 (also used as Example 7.4), and the maneuvering tank and aircraft problems of Section 3.2. Unfortunately the poor observability and nonlinearity of the ship passive tracking problem does not allow any recursive filter method making only one pass through the measurements to be competitive with Gauss-Newton iterations that re-linearize about the current trajectory at each iteration. To make the problem more suitable for the EKF, the measurement sampling rate was increased, measurement accuracy was improved, and the initial state covariance was set to a smaller value than that of Example 7.4. Even so, the IEKF often failed to converge (usually with very erratic steps) in cases where the EKF worked satisfactorily. For the tank tracking problem with measurement sampling intervals of 0.1, and the aircraft tracking problem with 1 s measurement sampling, the performance of the EKF and iterated filter-smoother was essentially the same. Results for the tank and aircraft tracking problems are shown in Section 9.2 when discussing smoothing.

As stated at the beginning of this book, you should evaluate different algorithms for your particular application because results obtained for simple examples do not necessarily apply to other problems.

9.2 SMOOTHING

As explained in Chapter 1, Kalman and Wiener filters can be used to not only provide the optimal estimate of the system state at the time of the last measurement, but also to compute the optimal estimate at future or past times. When modeling errors do not exist, the optimum linear predictor is obtained by simply the integrating the filter state to the time of interest. However, when the system under consideration is stochastic (subject to random perturbations), the effect of process noise must be accounted for when computing optimal *smoothed* state estimates for times in the past. The smoothing operation effectively gives more weight to measurements near the time of the smoothed estimates than those further away. Smoothed estimates are computed using a time history of filter estimates, or information that is equivalent.

Since smoothing requires significant computation and coding effort, you may wonder if it necessary to compute a smoothed estimate rather than simply using the filtered state at the desired time. Smoothed estimates are usually much more accurate than filtered estimates because both past and future measurements are used to compute the smoothed estimate at each time point. In some cases smoothing will also average the effects of modeling errors, thus further improving the estimate.

There are three general categories of smoothing:

1. *Fixed point*: the smoothed state is only computed at a fixed point in time, or possibly several fixed points. This can be implemented as a real-time processor.
2. *Fixed lag*: the smoothed state is computed at a fixed lag in time from the current measurement; that is, the smoothed epoch moves as new measurements become available. The smoothed epoch must match a measurement time. This can also be implemented for real-time processing.

3. *Fixed interval*: smoothed state estimates are computed at every measurement time from the start to the end. Fixed-interval smoothing is generally only used for post-processing since all the data must be available.

We now consider each of these cases. Although both continuous-time and discrete-time smoothing algorithms exist, we only describe the discrete-time versions since these are the primary methods implemented. The continuous-time versions are more useful when analyzing smoother behavior, and when deriving other algorithms.

9.2.1 Fixed-Point Smoother

Equations defining the fixed-point smoother may be found in many references (e.g., Gelb 1974; Maybeck 1982a; Simon 2006), but it is often easier to augment the filter state with a copy of the current state at the desired smoothing time. This can be repeated multiple times to obtain multiple fixed-point smoothed estimates.

Mathematically this is implemented as follows. Let $\hat{\mathbf{x}}_{i/i}$ be the *a posteriori* filter state n-vector at time t_i, and $\mathbf{P}_{xx\_i/i}$ be the corresponding $n \times n$ error covariance matrix. Time t_i is a desired point at which a smoothed estimate is required. We assume that the state is composed of n_d dynamic and n_b bias states:

$$\hat{\mathbf{x}}_{i/i} = \begin{bmatrix} \hat{\mathbf{x}}_d \\ \hat{\mathbf{x}}_b \end{bmatrix}_{i/i} \tag{9.2-1}$$

with associated covariance

$$\mathbf{P}_{xx\_i/i} = \begin{bmatrix} \mathbf{P}_{dd} & \mathbf{P}_{db} \\ \mathbf{P}_{db}^T & \mathbf{P}_{bb} \end{bmatrix}_{i/i} \tag{9.2-2}$$

expressed in partitions of dynamic and bias states. Since bias states are not driven by process noise and are not smoothable, it is not necessary to include them as smoothed states. We define the smoothed estimate of the dynamic states at time t_i based on measurements up to and including t_j (where $j \geq i$) as

$$\hat{\mathbf{s}}_{i/j} \triangleq [\hat{\mathbf{x}}_d]_{i/j} = \hat{\mathbf{x}}_{d\_i/j}\,. \tag{9.2-3}$$

It is not necessary for $\hat{\mathbf{s}}$ to include all n_d states, but for this discussion we assume that it does. After processing measurements at the desired smoothing time t_i, an augmented state vector of dimension $2n_d + n_b$ is created by appending $\hat{\mathbf{s}}_{i/i} = \hat{\mathbf{x}}_{d\_i/i}$ to the bottom of the filter state vector:

$$\hat{\mathbf{x}}_{i/i}^a = \begin{bmatrix} \hat{\mathbf{x}} \\ \hat{\mathbf{s}} \end{bmatrix}_{i/i} = \begin{bmatrix} \hat{\mathbf{x}}_d \\ \hat{\mathbf{x}}_b \\ \hat{\mathbf{x}}_d \end{bmatrix}_{i/i}. \tag{9.2-4}$$

Notice that the smoothed state $\hat{\mathbf{s}}_{i/i}$ is initialized with a copy of $\hat{\mathbf{x}}_{d\_i/i}$. The error covariance is likewise augmented and initialized as

$$\mathbf{P}_{i/i}^a = \begin{bmatrix} \mathbf{P}_{xx} & \mathbf{P}_{xs} \\ \mathbf{P}_{xs}^T & \mathbf{P}_{ss} \end{bmatrix}_{i/i} = \begin{bmatrix} \mathbf{P}_{dd} & \mathbf{P}_{db} & \mathbf{P}_{dd} \\ \mathbf{P}_{db}^T & \mathbf{P}_{bb} & \mathbf{P}_{bd} \\ \mathbf{P}_{dd} & \mathbf{P}_{db}^T & \mathbf{P}_{dd} \end{bmatrix}_{i/i} \tag{9.2-5}$$

where $\mathbf{P}_{xs} = E[(\mathbf{x}-\hat{\mathbf{x}})(\mathbf{s}-\hat{\mathbf{s}})^T]$ and $\mathbf{P}_{ss} = E[(\mathbf{s}-\hat{\mathbf{s}})(\mathbf{s}-\hat{\mathbf{s}})^T]$ are initialized as

$$\mathbf{P}_{xs\_i/i} = [\mathbf{P}_{xs}]_{i/i} = \begin{bmatrix} \mathbf{P}_{dd} \\ \mathbf{P}_{bd} \end{bmatrix}_{i/i}. \qquad (9.2\text{-}6)$$
$$\mathbf{P}_{ss\_i/i} = [\mathbf{P}_{dd}]_{i/i}$$

Notice that the covariance $\mathbf{P}^a_{i/i}$ is singular at this point, but that has no impact on subsequent processing. Since the added "smoothed" states are effectively biases, they can be handled as other biases, but the smoothed states do not appear in either the dynamic or measurement models. As future measurements are processed, the smoothed states $\hat{\mathbf{s}}_{i/j}$ for $j > i$ are automatically updated because the covariance term $\mathbf{P}_{xs\_i/j}$ couples changes in $\hat{\mathbf{x}}_{j/j}$ to $\hat{\mathbf{s}}_{i/j}$.

The state transition matrix for the next measurement time becomes

$$\mathbf{\Phi}^a_{i+1,i} = \begin{bmatrix} \mathbf{\Phi} & \mathbf{0} \\ \mathbf{0} & \mathbf{I} \end{bmatrix}_{i+1,i} \qquad (9.2\text{-}7)$$

and the state process noise matrix is

$$\mathbf{Q}^a_{i+1,i} = \begin{bmatrix} \mathbf{Q} & \mathbf{0} \\ \mathbf{0} & \mathbf{0} \end{bmatrix}_{i+1,i} \qquad (9.2\text{-}8)$$

where

$$\mathbf{\Phi} = \begin{bmatrix} \mathbf{\Phi}_{dd} & \mathbf{\Phi}_{db} \\ \mathbf{0} & \mathbf{I} \end{bmatrix} \quad \text{and} \quad \mathbf{Q} = \begin{bmatrix} \mathbf{Q}_{dd} & \mathbf{0} \\ \mathbf{0} & \mathbf{0} \end{bmatrix}$$

are the state transition and process noise matrices for the original n-state system $\hat{\mathbf{x}}$. The augmented measurement sensitivity matrix is equal to the original $\mathbf{H}$ matrix with n_d added zero columns:

$$\mathbf{H}^a_{i+1} = [\mathbf{H} \quad \mathbf{0}]_{i+1}. \qquad (9.2\text{-}9)$$

Provided that the filter is implemented to handle bias states efficiently (avoiding multiplications by 0 or 1 and taking advantage of symmetry), this implementation of the fixed-point smoother requires approximately the same number of floating point operations as explicit implementation of the fixed-point smoothing equations, but it is easier to implement. It also has the advantage that general-purpose Kalman filter libraries can be used.

It is not difficult, however, to express the time and measurement update for the separate partitions using the above matrix equations. The time update is

$$\boxed{\begin{aligned} \hat{\mathbf{x}}_{i+1/i} &= \mathbf{\Phi}\hat{\mathbf{x}}_{i/i} \\ \mathbf{P}_{xx\_i+1/i} &= \mathbf{\Phi}\mathbf{P}_{xx\_i/i}\mathbf{\Phi}^T + \mathbf{Q}. \\ \mathbf{P}_{xs\_i+1/i} &= \mathbf{\Phi}\mathbf{P}_{xs\_i/i} \end{aligned}} \qquad (9.2\text{-}10)$$

Notice that time updates are not required for $\hat{\mathbf{s}}$ and $\mathbf{P}_{ss}$, and the multiplication $\mathbf{\Phi}\mathbf{P}_{xs\_i/i}$ is the only extra calculation required beyond the normal filter time update. The measurement update is

$$\begin{aligned}
\hat{\mathbf{x}}_{i+1/i+1} &= \hat{\mathbf{x}}_{i+1/i} + \mathbf{K}_x(\mathbf{y}_{i+1} - \mathbf{H}\hat{\mathbf{x}}_{i+1/i}) \\
\hat{\mathbf{s}}_{i/i+1} &= \hat{\mathbf{s}}_{i/i} + \mathbf{K}_s(\mathbf{y}_{i+1} - \mathbf{H}\hat{\mathbf{x}}_{i+1/i}) \\
\mathbf{P}_{xx_i+1/i+1} &= \mathbf{P}_{xx_i+1/i} - \mathbf{K}_x\mathbf{H}\mathbf{P}_{xx_i+1/i} \\
\mathbf{P}_{xs_i+1/i+1} &= \mathbf{P}_{xs_i+1/i} - \mathbf{K}_x\mathbf{H}\mathbf{P}_{xs_i+1/i} \\
\mathbf{P}_{ss_i/i+1} &= \mathbf{P}_{ss_i/i} - \mathbf{K}_s\mathbf{H}\mathbf{P}_{ss_i/i}
\end{aligned} \tag{9.2-11}$$

for gain matrices

$$\begin{aligned}
\mathbf{K}_x &= \mathbf{P}_{xx_i+1/i}\mathbf{H}^T\left(\mathbf{H}\mathbf{P}_{xx_i+1/i}\mathbf{H}^T + \mathbf{R}\right)^{-1} \\
\mathbf{K}_s &= \mathbf{P}_{xs_i+1/i}^T\mathbf{H}^T\left(\mathbf{H}\mathbf{P}_{xx_i+1/i}\mathbf{H}^T + \mathbf{R}\right)^{-1}
\end{aligned}. \tag{9.2-12}$$

The equations for $\mathbf{K}_x$, $\hat{\mathbf{x}}$, and $\mathbf{P}_{xx}$ are the normal filter measurement update. As with all Kalman filter implementations, the covariances $\mathbf{P}_{xx}$ and $\mathbf{P}_{ss}$ should be forced to be symmetric. We have expressed the state update using linear operations, but the nonlinear form of the EKF may also be used. As expected, neither the smoothed state nor the smoothed covariance affect the original filter state $\hat{\mathbf{x}}$. Calculation of smoothed states requires additional computations but does not alter the original filter equations for $\hat{\mathbf{x}}$ and $\mathbf{P}_{xx}$. As before, significant computational savings are possible by storing intermediate products, although that may not be important unless the state size is large.

Notice that gain equation for the smoothed state can also be written as

$$\mathbf{K}_s = \mathbf{P}_{xs\_i/i}^T\mathbf{\Phi}^T\mathbf{P}_{xx\_i+1/i}^{-1}\mathbf{K}_x. \tag{9.2-13}$$

This suggests the fixed-point smoother equations can be written in other forms that do not require separate calculation of $\mathbf{P}_{xs}$ and $\mathbf{P}_{ss}$. The equations for the full state vector (including nonsmoothable biases) are

$$\begin{aligned}
\hat{\mathbf{x}}_{i/j} &= \hat{\mathbf{x}}_{i/j-1} + \mathbf{B}_j(\hat{\mathbf{x}}_{j/j} - \hat{\mathbf{x}}_{j/j-1}) \\
&= \hat{\mathbf{x}}_{i/j-1} + \mathbf{B}_j\mathbf{K}_j(\mathbf{y}_j - \mathbf{H}_j\hat{\mathbf{x}}_{j/j-1}) \\
\mathbf{P}_{i/j} &= \mathbf{P}_{i/j-1} + \mathbf{B}_j(\mathbf{P}_{j/j} - \mathbf{P}_{j/j-1})\mathbf{B}_j^T \\
&= \mathbf{P}_{i/j-1} - \mathbf{B}_j\mathbf{K}_j\mathbf{H}_j\mathbf{P}_{j/j-1}\mathbf{B}_j^T
\end{aligned} \tag{9.2-14}$$

for $j > i$ where

$$\mathbf{B}_j = \mathbf{B}_{j-1}\left(\mathbf{P}_{j-1/j-1}\mathbf{\Phi}_{j,j-1}^T\mathbf{P}_{j/j-1}^{-1}\right) \tag{9.2-15}$$

is initialized with $\mathbf{B}_i = \mathbf{I}$, and recursively updated. This form of the fixed-point smoother requires the same storage as the previous equations—the upper n_d rows of $\mathbf{B}_j$ are the same size as $\mathbf{P}_{xs}$ plus $\mathbf{P}_{ss}$. However, equation (9.2-15) requires inversion of the $(n_d + n_b) \times (n_d + n_b)$ symmetric matrix $\mathbf{P}_{xx\_j/j-1}$ and several other matrix products that involve at least $n_d(n_d + n_b)^2$ multiplications. Since equations (9.2-14) and (9.2-15) do not reduce storage or computations and it is usually easier to implement the smoother by augmenting the state at the desired smoothing point, this form is infrequently used.

9.2.2 Fixed-Lag Smoother

The augmented state fixed-point smoother method above can be used to implement a fixed-lag smoother when the number of lags is small. For example, if the smoothing lag corresponds to three measurement time intervals, the augmented *a posteriori* state vector at t_i is defined as

$$\hat{\mathbf{x}}^a_{i/i} = \begin{bmatrix} \hat{\mathbf{x}}_{i/i} \\ \hat{\mathbf{s}}_{i-1/i} \\ \hat{\mathbf{s}}_{i-2/i} \\ \hat{\mathbf{s}}_{i-3/i} \end{bmatrix}.$$

Before processing the time update of the next measurement, the elements of $\hat{\mathbf{x}}^a_{i/i}$ corresponding to $\hat{\mathbf{s}}_{i-3/i}$ are replaced with $\hat{\mathbf{s}}_{i-2/i}$ and the original $\hat{\mathbf{s}}_{i-2/i}$ elements are replaced with $\hat{\mathbf{s}}_{i-1/i}$. Finally the $\hat{\mathbf{s}}_{i-1/i}$ elements are replaced with the dynamic states $\hat{\mathbf{x}}_{d\_i/i}$. Hence the *a priori* state vector used to process the next measurement becomes

$$\hat{\mathbf{x}}^a_{i+1/i} = \begin{bmatrix} \hat{\mathbf{x}}_{i+1/i} \\ \hat{\mathbf{s}}_{i/i} \\ \hat{\mathbf{s}}_{i-1/i} \\ \hat{\mathbf{s}}_{i-2/i} \end{bmatrix}$$

where $\hat{\mathbf{s}}_{i/i} = \hat{\mathbf{x}}_{d\_i/i}$. The corresponding elements of the covariance must also be shifted and the new covariance terms set as $\mathbf{P}_{ss\_i/i} = \mathbf{P}_{dd\_i/i}$ and $\mathbf{P}_{ds\_i/i} = \mathbf{P}_{dd\_i/i}$. This process of shifting and replacing is repeated at every measurement time. When the desired smoothing lag is significant, this results in a large state vector and covariance with consequent impact on the computational burden. However, it is relatively easy to implement using standard filter software.

One form of the discrete-time fixed-lag smoother appears in Gelb (1974, section 5.4) and Maybeck (1982a, section 8.6). It uses the full state vector rather than just dynamic states. Denoting N as the smoothing lag, the equations are:

$$\boxed{\begin{aligned} \hat{\mathbf{x}}_{i+1/i+1+N} &= \mathbf{\Phi}_{i+1,i}\,\hat{\mathbf{x}}_{i/i+N} + \mathbf{Q}_{i+1,i}\mathbf{\Phi}^{-T}_{i+1,i}\mathbf{P}^{-1}_{i/i}(\hat{\mathbf{x}}_{i/i+N} - \hat{\mathbf{x}}_{i/i}) \\ &\quad + \mathbf{B}_{i+1+N}\mathbf{K}_{i+1+N}(\mathbf{y}_{i+1+N} - \mathbf{H}_{i+1+N}\hat{\mathbf{x}}_{i+1+N/i+N}) \\ \mathbf{P}_{i+1/i+1+N} &= \mathbf{P}_{i+1/i} - \mathbf{B}_{i+1+N}\mathbf{K}_{i+1+N}\mathbf{H}_{i+1+N}\mathbf{P}_{i+1+N/i+N}\mathbf{B}^T_{i+1+N} \\ &\quad - \mathbf{A}_i^{-1}(\mathbf{P}_{i/i} - \mathbf{P}_{i/i+N})\mathbf{A}_i^{-T} \end{aligned}} \tag{9.2-16}$$

where

$$\boxed{\begin{aligned} \mathbf{A}_i &= \mathbf{P}_{i/i}\mathbf{\Phi}^T_{i+1,i}\mathbf{P}^{-1}_{i+1/i} \\ \mathbf{B}_{i+1+N} &= \mathbf{A}_{i+1}\mathbf{A}_{i+2}\cdots\mathbf{A}_{i+N}\,. \\ &= \mathbf{A}_i^{-1}\mathbf{B}_{i+N}\mathbf{A}_{i+N} \end{aligned}} \tag{9.2-17}$$

All other variables are the output of the normal Kalman filter. The smoothed estimate and covariance are initialized with $\hat{\mathbf{x}}_{0/N}$ and $\mathbf{P}_{0/N}$ obtained from observation time N of a fixed-point smoother. Notice that this version also requires that arrays ($\mathbf{A}_i$) be stored over the smoothing lag interval, with a new array added and an old one dropped from storage at each measurement step. A matrix inversion is also required. Simon (2006, section 9.3) presents a different form of the smoother that does not require matrix inversion.

9.2.3 Fixed-Interval Smoother

While the fixed-point and fixed-lag smoothers can be used for real-time processing, the fixed interval smoother is used after all measurements have been processed by the filter. Fraser and Potter (1969) describe the optimal linear smoother as a combination of an optimal filter operating forward in time and another filter operating backward in time, as shown in Figure 9.4.

Since the measurement and process noises used in the two filters are independent if the current point is only included in one filter, the filter estimate errors are also independent. Hence the optimum smoothed estimate at a given point is obtained by appropriate weighting of estimates with independent errors; that is,

$$\boxed{\hat{\mathbf{x}}_{i/N} = \mathbf{A}_i \hat{\mathbf{x}}_{i/i} + (\mathbf{I} - \mathbf{A}_i)\hat{\mathbf{x}}^b_{i/i+1:N}} \tag{9.2-18}$$

where $\mathbf{A}_i$ is the weighting matrix to be determined and $\hat{\mathbf{x}}^b_{i/i+1:N}$ is the state estimate from the backward filter at time t_i using measurements from t_{i+1} to t_N. Notice that weighting of $\hat{\mathbf{x}}^b_{i/i+1:N}$ is $(\mathbf{I} - \mathbf{A}_i)$ so that the smoothed estimate is unbiased if the individual filter estimates are unbiased:

$$\begin{aligned} E[\hat{\mathbf{x}}_{i/N}] &= \mathbf{A}_i E[\hat{\mathbf{x}}_{i/i}] + (\mathbf{I} - \mathbf{A}_i) E[\hat{\mathbf{x}}^b_{i/i+1:N}] \\ &= \mathbf{A}_i \mathbf{x}_i + (\mathbf{I} - \mathbf{A}_i)\mathbf{x}_i \\ &= \mathbf{x}_i \end{aligned}.$$

The minimum variance smoother solution is obtained by minimizing the trace of the error covariance for $\hat{\mathbf{x}}_{i/N}$; that is,

$$\begin{aligned} J &= \mathrm{tr}(E[(\mathbf{x}_i - \hat{\mathbf{x}}_{i/N})(\mathbf{x}_i - \hat{\mathbf{x}}_{i/N})^T]) \\ &= \mathrm{tr}(E[(\mathbf{A}_i \tilde{\mathbf{x}}_{i/i} + (\mathbf{I} - \mathbf{A}_i)\tilde{\mathbf{x}}^b_{i/i+1:N})(\tilde{\mathbf{x}}^T_{i/i}\mathbf{A}^T_i + (\tilde{\mathbf{x}}^b_{i/i+1:N})^T(\mathbf{I} - \mathbf{A}^T_i))]) \\ &= \mathrm{tr}(\mathbf{A}_i \mathbf{P}_{i/i}\mathbf{A}^T_i + (\mathbf{I} - \mathbf{A}_i)\mathbf{P}^b_{i/i+1:N}(\mathbf{I} - \mathbf{A}^T_i)) \end{aligned}$$

where $\tilde{\mathbf{x}}_{i/i} = \mathbf{x}_i - \hat{\mathbf{x}}_{i/i}$ and $\tilde{\mathbf{x}}^b_{i/i+1:N} = \mathbf{x}_i - \hat{\mathbf{x}}^b_{i/i+1:N}$. The minimum is obtained by setting

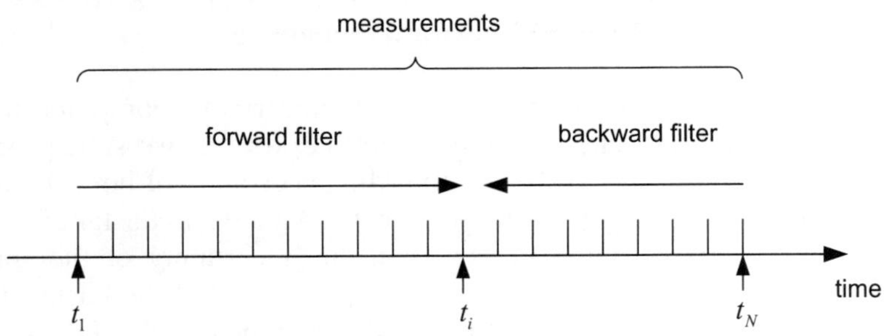

FIGURE 9.4: Fixed-interval smoother as combination of forward and backward filters.

$$\frac{\partial J}{\partial \mathbf{A}_i} = 2\left(\mathbf{A}_i \mathbf{P}_{i/i} - (\mathbf{I} - \mathbf{A}_i)\mathbf{P}^b_{i/i+1:N}\right) = \mathbf{0}$$

or

$$\boxed{\mathbf{A}_i = \mathbf{P}^b_{i/i+1:N}(\mathbf{P}_{i/i} + \mathbf{P}^b_{i/i+1:N})^{-1}.} \tag{9.2-19}$$

Thus

$$\boxed{\begin{aligned}\hat{\mathbf{x}}_{i/N} &= \left(\mathbf{P}^b_{i/i+1:N}\hat{\mathbf{x}}_{i/i} + \mathbf{P}_{i/i}\hat{\mathbf{x}}^b_{i/i+1:N}\right)\left(\mathbf{P}_{i/i} + \mathbf{P}^b_{i/i+1:N}\right)^{-1} \\ \mathbf{P}_{i/N} &= \mathbf{A}_i\mathbf{P}_{i/i}\mathbf{A}_i^T + (\mathbf{I} - \mathbf{A}_i)\mathbf{P}^b_{i/i+1:N}(\mathbf{I} - \mathbf{A}_i^T) \\ &= \mathbf{P}_{i/i} - \mathbf{P}_{i/i}\left(\mathbf{P}_{i/i} + \mathbf{P}^b_{i/i+1:N}\right)^{-1}\mathbf{P}_{i/i} \\ &= \left(\mathbf{P}^{-1}_{i/i} + (\mathbf{P}^b_{i/i+1:N})^{-1}\right)^{-1}\end{aligned}} \quad . \tag{9.2-20}$$

are the smoothed estimate and covariance at time t_i given the estimates from the forward and backward filters. Notice that covariance $\mathbf{P}_{i/N}$ will be smaller than either $\mathbf{P}_{i/i}$ or $\mathbf{P}^b_{i/i+1:N}$, which shows the benefit of smoothing. Also notice that it is necessary to store all state estimates and covariance matrices from the forward filter (or alternatively from the backward filter) in order to compute the smoothed estimates.

The forward-backward filter solution to the smoothing problem is often used to derive other smoothing properties, but it is seldom implemented. Notice that the backward filter cannot use prior information because that would introduce correlation between forward and backward filter errors. The requirement for a prior-free filter operating backward in time implies that it should be implemented as an information filter using $(\mathbf{P}^b_{i/i+1:N})^{-1}$, rather than the covariance $\mathbf{P}^b_{i/i+1:N}$. Then equation (9.2-20) is modified to avoid inversion of $\mathbf{P}^b_{i/i+1:N}$. See Fraser and Potter (1969) for the details, but note that the listed equation for the backward filter time update of the information matrix should be

$$\begin{aligned}\mathbf{U}'_{k-1} &= \mathbf{\Phi}^T_{k-1}[(\mathbf{I} - \mathbf{J}_k)\mathbf{U}_k(\mathbf{I} - \mathbf{J}_k)^T + \mathbf{J}_k\mathbf{Q}^{-1}_{k-1}\mathbf{J}_k]\,\mathbf{\Phi}_{k-1} \\ &= \mathbf{\Phi}^T_{k-1}(\mathbf{I} - \mathbf{J}_k)\mathbf{U}_k\,\mathbf{\Phi}_{k-1}\end{aligned} .$$

Other sources for information on the forward-backward smoother derivation include Maybeck (1982a, section 8.4) and Simon (2006, section 9.4.1).

Example 9.3: Two-State Forward-Backward Smoothing

Figure 9.5 shows the $1-\sigma$ "position" (state #1) uncertainty computed from the *a posteriori* filter and smoother covariance matrices for the two-state problem of Example 8.2 when using $Q = 1$. Notice that the uncertainty of both the forward and backward filters is initially large, but decays to a nearly constant value after processing about 10 s of data. As expected, the uncertainty of the smoother matches the forward filter uncertainty at the end time, and the backward filter at the initial time. The uncertainty of the smoother in the middle of the data span is much less than that of either the forward or backward filters.

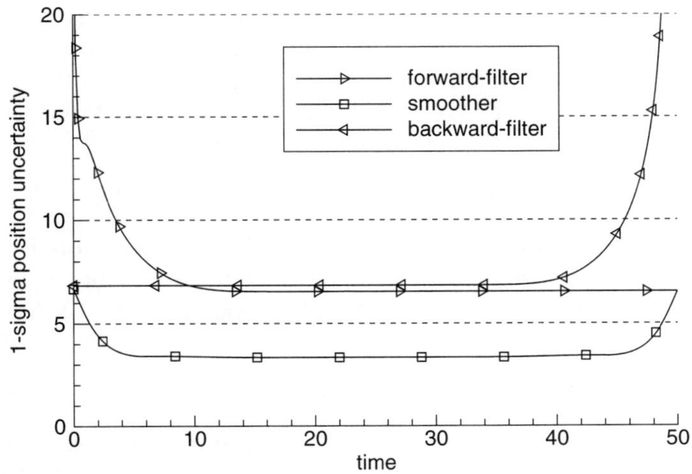

FIGURE 9.5: Two-state forward filter, backward filter, and smoother 1-sigma position uncertainty.

Creation of a separate backward information filter with all the associated model structure makes the forward-backward smoother approach somewhat unattractive. Two more popular fixed-interval covariance-based smoothers are:

1. Rauch-Tung-Striebel (RTS) smoother: The state estimates, *a priori* covariance and *a posteriori* covariances (or equivalent information) from the forward filter are saved and used in a backward state/covariance recursion that is initialized with the final filter solution.
2. Modified Bryson-Frazier (mBF) smoother: The algorithm derived from the work of Rauch et al. is a more efficient form of a result reported by Bryson and Frazier (1963). Bierman (1977b) refers to it as mBF. Rather than using filter covariances to compute smoother gains, this approach uses a backward recursion of adjoint variables that avoids matrix inversions. Although the mBF smoother can be derived from the RTS algorithm, the information saved from the forward filter involves measurements rather than process noise.

These two smoothers are described in following sections. Chapter 10 discusses three "square-root" smoothers: the square-root information smoother (SRIS), the Dyer and McReynolds (1969) covariance smoother, and a square-root version of the RTS. A number of other smoothing algorithms are available but infrequently used. Surveys of smoothing algorithms are found in Meditch (1969, 1973), Kailath and Frost (1968), Kailath (1975), and Bierman (1974, 1977b).

In should be noted that the Kalman filter used to generate inputs for fixed-interval smoothers must be optimal. That is, a Kalman-Schmidt consider filter cannot be used because the state estimates are suboptimal, thus introducing correlations between forward and backward solutions that are not accounted for. There is a critical loss of information in using a consider filter that cannot be reconstructed in a backward sweep smoother.

9.2.3.1 RTS Smoother Rauch et al. (1965) elegantly derived a backwardly recursive smoother using conditional probabilities and maximization of the likelihood function for assumed Gaussian densities. They showed that the joint probability density for the states at t_i and t_{i+1} and measurements from t_0 to the final time t_N can be expressed using conditional probability densities as

$$p(\mathbf{x}_i,\mathbf{x}_{i+1},\mathbf{y}_0,\mathbf{y}_1,\ldots,\mathbf{y}_N)=p(\mathbf{x}_{i+1}|\mathbf{x}_i)\,p(\mathbf{x}_i,\mathbf{y}_0,\mathbf{y}_1,\ldots,\mathbf{y}_i)\,p(\mathbf{y}_{i+1},\ldots,\mathbf{y}_N|\mathbf{x}_{i+1}),$$

where subscripts on p have been omitted to simplify the notation. By maximizing the logarithm of the above density using the time update model equations (8.1-4) and (8.1-5) to compute conditional densities, they found that $\hat{\mathbf{x}}_{i/N}$ is the solution that minimizes

$$J=(\hat{\mathbf{x}}_{i+1/N}-\mathbf{\Phi}_{i+1,i}\mathbf{x}_i)^T\mathbf{Q}_{i+1,i}^{-1}(\hat{\mathbf{x}}_{i+1/N}-\mathbf{\Phi}_{i+1,i}\mathbf{x}_i)+(\mathbf{x}_i-\hat{\mathbf{x}}_{i/i})^T\mathbf{P}_{i/i}^{-1}(\mathbf{x}_i-\hat{\mathbf{x}}_{i/i})$$

That solution is obtained by setting the gradient to zero and using the matrix inversion identity of Appendix A. The RTS algorithm can also be derived from the forward-backward smoothing equations, as demonstrated by Simon (2006, section 9.4.2), or from the SRIS of Chapter 10.

The RTS backward recursion is

$$\boxed{\begin{aligned}\hat{\mathbf{x}}_{i/N}&=\hat{\mathbf{x}}_{i/i}+\mathbf{G}_i(\hat{\mathbf{x}}_{i+1/N}-\hat{\mathbf{x}}_{i+1/i})\\ \mathbf{P}_{i/N}&=\mathbf{P}_{i/N}+\mathbf{G}_i(\mathbf{P}_{i+1/N}-\mathbf{P}_{i+1/i})\mathbf{G}_i^T\end{aligned}} \tag{9.2-21}$$

where the gain matrix is

$$\boxed{\mathbf{G}_i=\mathbf{P}_{i/i}\mathbf{\Phi}_{i+1,i}^T\mathbf{P}_{i+1/i}^{-1}} \tag{9.2-22}$$

and $\hat{\mathbf{x}}_{i/i}$, $\hat{\mathbf{x}}_{i+1/i}$, $\mathbf{P}_{i/i}$, $\mathbf{P}_{i+1/i}$, $\mathbf{\Phi}_{i+1,i}$, $\mathbf{Q}_{i+1,i}$ are generated from the Kalman filter and associated computations. The backward recursions are initialized with $\hat{\mathbf{x}}_{N/N}$ and $\mathbf{P}_{N/N}$ from the forward filter and repeated for $i = N-1, N-2,\ldots, 1$. Although the gain matrix is usually written as equation (9.2-22), by rearranging the equation for the filter covariance time update equation (8.1-5) as

$$\mathbf{P}_{i/i}=\mathbf{\Phi}_{i+1,i}^{-1}(\mathbf{P}_{i+1/i}-\mathbf{Q}_{i+1,i})\mathbf{\Phi}_{i+1,i}^{-T},$$

the gain can be written as

$$\boxed{\mathbf{G}_i=\mathbf{\Phi}_{i+1,i}^{-1}(\mathbf{I}-\mathbf{Q}_{i+1,i}\mathbf{P}_{i+1/i}^{-1}).} \tag{9.2-23}$$

This equation directly shows that in the absence of process noise $\mathbf{Q}$, smoothing is just backward integration. Furthermore, bias states that are not driven by process noise are not changed from the final filter estimate; that is, they are not smoothable. If the bias states are located at the bottom of the state vector, the gain matrix is

$$\boxed{\begin{aligned}\begin{bmatrix}\mathbf{G}_d\\ \mathbf{I}\end{bmatrix}_i&=\begin{bmatrix}\mathbf{\Phi}_{dd}^{-1} & -\mathbf{\Phi}_{dd}^{-1}\mathbf{\Phi}_{db}\\ \mathbf{0} & \mathbf{I}\end{bmatrix}_{i+1,i}\begin{bmatrix}\mathbf{I}-\mathbf{Q}_{dd}(\mathbf{P}_{i+1/i}^{-1})_{dd} & -\mathbf{Q}_{dd}(\mathbf{P}_{i+1/i}^{-1})_{db}\\ \mathbf{0} & \mathbf{I}\end{bmatrix}\\ &=\begin{bmatrix}\mathbf{\Phi}_{dd}^{-1}[\mathbf{I}-\mathbf{Q}_{dd}(\mathbf{P}_{i+1/i}^{-1})_{dd}] & -\mathbf{\Phi}_{dd}^{-1}[\mathbf{\Phi}_{db}+\mathbf{Q}_{dd}(\mathbf{P}_{i+1/i}^{-1})_{db}]\\ \mathbf{0} & \mathbf{I}\end{bmatrix}\end{aligned}} \tag{9.2-24}$$

where we have used the same partitioning of dynamic and bias states shown in Section 9.2.1. The notation $(\mathbf{P}_{i+1/i}^{-1})_{dd}$ and $(\mathbf{P}_{i+1/i}^{-1})_{db}$ refers to the dynamic-dynamic and dynamic-bias partitions of $\mathbf{P}_{i+1/i}^{-1}$. When the fraction of bias states is significant and $\mathbf{Q}_{dd}$ is sparse, this form of the gain matrix can greatly reduce computations, even though it does require inversion of $\mathbf{\Phi}_{dd}$.

Since the required arrays can be computed in different forms, there are several options for saving output from the forward Kalman filter: the best choice for a particular problem depends on the number of dynamic and bias states and the sparseness of $\mathbf{\Phi}_{dd}$, $\mathbf{\Phi}_{db}$ and $\mathbf{Q}_{dd}$. It is generally efficient to write

$$\hat{\mathbf{x}}_{i/i},\ \hat{\mathbf{x}}_{i+1/i},\ \mathbf{\Phi}_{i+1,i},\ \mathbf{Q}_{i+1,i},\ \mathbf{P}_{i/i}$$

from the forward filter to direct access disk after performing the filter time update at t_{i+1}, but before performing the measurement update. When arrays are small and disk storage size is unimportant, $\mathbf{P}_{i+1/i}$ can also be saved. Otherwise it can be computed as $\mathbf{\Phi}_{i+1,i}\mathbf{P}_{i/i}\mathbf{\Phi}_{i+1,i}^T + \mathbf{Q}_{i+1,i}$. If equation (9.2-23) is used to compute the gain, $\mathbf{P}_{i+1/i}$ rather than $\mathbf{P}_{i/i}$ can be saved. Notice that $\mathbf{\Phi}_{i+1,i}^{-1} = \mathbf{\Phi}_{i,i+1}$, so it may be more efficient when models are analytic to compute $\mathbf{\Phi}_{i,i+1}$ for a negative time step rather than to invert a matrix. To conserve storage, only the upper triangular portion of $\mathbf{P}_{i/i}$ or $\mathbf{P}_{i+1/i}$ should be written to the file. If the number of bias states is significant, only the nonzero partitions of $\mathbf{\Phi}_{i+1,i}$ and $\mathbf{Q}_{i+1,i}$ should be saved.

The RTS smoother is probably the most popular type of smoother. It is relatively easy to understand and implement, and the algorithm can be used to derive other types of smoothers. It does not require storage of measurement information, which makes it flexible: measurements may be processed in the filter one-component-at-a-time, or measurements types and sizes may be variable without any impact to the smoother. It can also be very efficient when the number of dynamic states is comparable to or smaller than the number of bias states.

One possible disadvantage of the RTS is numerical accuracy. The state and covariance differencing of equation (9.2-21) can result in a loss of precision when the quantities differenced are comparable in magnitude. However, there are two reasons why loss of numerical accuracy is rarely a problem. First, since fixed-interval smoothing is a post-processing activity, it is usually executed on computers having double precision floating point registers—such as the PC. Second, errors in computing the smoothed covariance $\mathbf{P}_{i/N}$ have no impact on the accuracy of the smoothed state $\hat{\mathbf{x}}_{i/N}$ since the gain matrix only uses quantities from the Kalman filter. In fact, $\mathbf{P}_{i/N}$ is sometimes not computed to reduce computational run time. Because of covariance differencing, there is the potential for the computed $\mathbf{P}_{i/N}$ to have negative variances, but this is rarely a problem when using double precision.

Another problem is caused by the need to invert $\mathbf{P}_{i+1/i}$ or $\mathbf{\Phi}$ or both. The iterations can lose accuracy if $\mathbf{\Phi}$ has a negligibly small value on the diagonal—possibly because of data gaps causing propagation over a long time interval. To prevent the problem for long time integrations, either limit the maximum time step or set a lower limit on diagonal elements of $\mathbf{\Phi}$ for Markov process states (when $e^{-FT/\tau}$ is negligibly small).

9.2.3.2 mBF Smoother

This algorithm uses the adjoint variable approach from optimal control theory. The adjoint n-vector λ used in the mBF is equivalent to

$$\boldsymbol{\lambda}_{i/i} = \mathbf{\Phi}_{i+1,i}^T \mathbf{P}_{i+1/i}^{-1}(\hat{\mathbf{x}}_{i+1/N} - \hat{\mathbf{x}}_{i+1/i}). \qquad (9.2\text{-}25)$$

(Rauch et al. showed that the mBF smoother can be derived from the RTS smoother using the above substitution: notice that the first line of equation (9.2-21) can be written as $\hat{\mathbf{x}}_{i/N} = \hat{\mathbf{x}}_{i/i} + \mathbf{P}_{i/i}\boldsymbol{\lambda}_{i/i}$) The mBF also computes an $n \times n$ adjoint matrix $\boldsymbol{\Lambda}$ that is used to calculate the smoothed covariance.

Starting at the final time t_N with $\boldsymbol{\lambda}_{N/N} = \mathbf{0}$ and $\boldsymbol{\Lambda}_{N/N} = \mathbf{0}$, a measurement update is performed to remove the effect of the last measurement from the adjoint variables for $i = N$:

$$\begin{aligned} \boldsymbol{\lambda}_{i/i-1} &= \boldsymbol{\lambda}_{i/i} + \mathbf{H}_i^T[\mathbf{C}_i^{-1}\tilde{\mathbf{y}}_i - \mathbf{K}_i^T\boldsymbol{\lambda}_{i/i}] \\ &= (\mathbf{I} - \mathbf{K}_i\mathbf{H}_i)^T\boldsymbol{\lambda}_{i/i} + \mathbf{H}_i^T\mathbf{C}_i^{-1}\tilde{\mathbf{y}}_i \\ \boldsymbol{\Lambda}_{i/i-1} &= (\mathbf{I} - \mathbf{K}_i\mathbf{H}_i)^T\boldsymbol{\Lambda}_{i/i}(\mathbf{I} - \mathbf{K}_i\mathbf{H}_i) + \mathbf{H}_i^T\mathbf{C}_i^{-1}\mathbf{H}_i \end{aligned} \quad . \tag{9.2-26}$$

where $\mathbf{H}_i$, $\mathbf{C}_i = \mathbf{H}_i\mathbf{P}_{i/i-1}\mathbf{H}_i^T + \mathbf{R}_i$, $\tilde{\mathbf{y}}_i = \mathbf{y}_i - \mathbf{H}_i\hat{\mathbf{x}}_{i/i-1}$, $\mathbf{K}_i$, and $\boldsymbol{\Phi}_{i,i-1}$ are generated from the Kalman filter. Then the backward time update

$$\begin{aligned} \boldsymbol{\lambda}_{i-1/i-1} &= \boldsymbol{\Phi}_{i,i-1}^T\boldsymbol{\lambda}_{i/i-1} \\ \boldsymbol{\Lambda}_{i-1/i-1} &= \boldsymbol{\Phi}_{i,i-1}^T\boldsymbol{\Lambda}_{i/i-1}\boldsymbol{\Phi}_{i,i-1} \end{aligned} \tag{9.2-27}$$

is evaluated, and smoothed estimates and covariances at t_{i-1} are computed as

$$\begin{aligned} \hat{\mathbf{x}}_{i-1/N} &= \hat{\mathbf{x}}_{i-1/i-1} + \mathbf{P}_{i-1/i-1}\boldsymbol{\lambda}_{i-1/i-1} \\ \mathbf{P}_{i-1/N} &= \mathbf{P}_{i-1/i-1} - \mathbf{P}_{i-1/i-1}\boldsymbol{\Lambda}_{i-1/i-1}\mathbf{P}_{i-1/i-1} \end{aligned} \quad . \tag{9.2-28}$$

Note: The $\bar{\boldsymbol{\lambda}}$ variable used by Bierman is equal to the negative of our $\boldsymbol{\lambda}_{i/i}$, and the $\mathbf{w}_i$ variable of Rauch et al. is equal to our $\boldsymbol{\lambda}_{i+1/i}$. The formulations presented by Bierman (1977b) and Rauch et al. (1965) both have minor errors, such as missing transpose operators or incorrect symbols. These or similar errors also appear in other papers. Equations (9.2-26) to (9.2-28) above have been numerically verified to match the RTS and forward-backward smoothers.

As with the RTS smoother, these equations are repeated for $i = N, N - 1, \ldots, 2$. Because measurements are explicitly processed in equation (9.2-26), this time index is shifted by +1 compared with the index used in the previous section, but the number of steps is the same. Notice that the mBF formulation uses the filter variables

$$\hat{\mathbf{x}}_{i-1/i-1},\ \mathbf{P}_{i-1/i-1}, \boldsymbol{\Phi}_{i,i-1}, \mathbf{H}_i, \mathbf{C}_i, \mathbf{K}_i, \tilde{\mathbf{y}}_i$$

but does not use $\mathbf{Q}_{i+1,i}$ (as required in the RTS algorithm). The filter implementation should save these variables to direct access disk after computing $\mathbf{H}_i$, $\mathbf{C}_i$, $\mathbf{K}_i$, and $\tilde{\mathbf{y}}_i$, but before measurement updating to compute $\hat{\mathbf{x}}_{i/i}$ and $\mathbf{P}_{i/i}$. Notice that the mBF smoother does not use matrix inversions, but does require more $n \times n$ matrix multiplications than the RTS algorithm. As with the RTS algorithm, numerical errors in calculating the covariance $\mathbf{P}_{i/N}$ do not have any impact on the smoothed state estimate $\hat{\mathbf{x}}_{i/N}$. Numerical accuracy may or may not be better than for the RTS algorithm. It is somewhat more difficult to take advantage of dynamic model structure in the mBF than in the RTS. Handling of measurements processed one-component-at-at-time in the Kalman filter is also more difficult with the mBF. This may explain why the RTS is more frequently used.

Example 9.4: Maneuvering Tank Tracking

This example uses the maneuvering tank data and six-state model of Section 3.2.1. Since both the measurement and dynamic models are nonlinear, the example demonstrates use of both the EKF and RTS smoothing. It was also used to test the iterated-filter-smoother, but results were essentially the same as those of the EKF.

The reconstructed position data for the Scout vehicle is shown in Figure 9.6. A sensor measuring azimuth and range to the target tank was located at $x = 600$ m, $y = -600$ m. Also shown in the figure are lines from the sensor to the target at 190 and 300 s of tracking. (Although the filter and smoother processed the entire tracking span, performance is only shown for this shorter span so that plots are less cluttered.) The simulated $1-\sigma$ noise was 0.3 milliradians for azimuth and 2 m for range. Process noise was only modeled on the alongtrack and crosstrack acceleration states, where the assumed power spectral density (PSD) was 1.0 and 0.5 m^2/s^5, respectively.

Figure 9.7 shows the error in EKF-estimated velocity states for the time span 190 to 300 s. Figure 9.8 is the comparable plot for the smoother velocity error. Notice that the smoother error is indeed "smoother" and also much smaller. The improvement of the smoother is much greater than would be expected from the approximate doubling of the number of measurements used for each state estimate. The relative filter/smoother reduction in peak position errors is approximately the same (4:1) as for velocity. Smoother alongtrack acceleration errors are about 40% smaller than those of the filter, but crosstrack acceleration error magnitudes are similar to those of the filter.

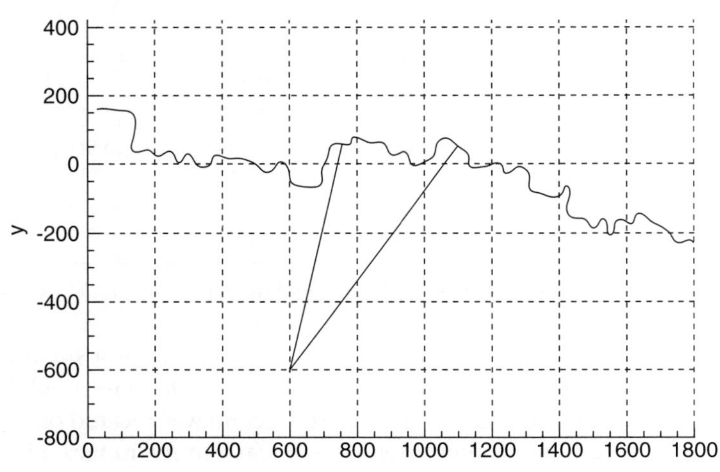

FIGURE 9.6: Tank ground track and sensor location.

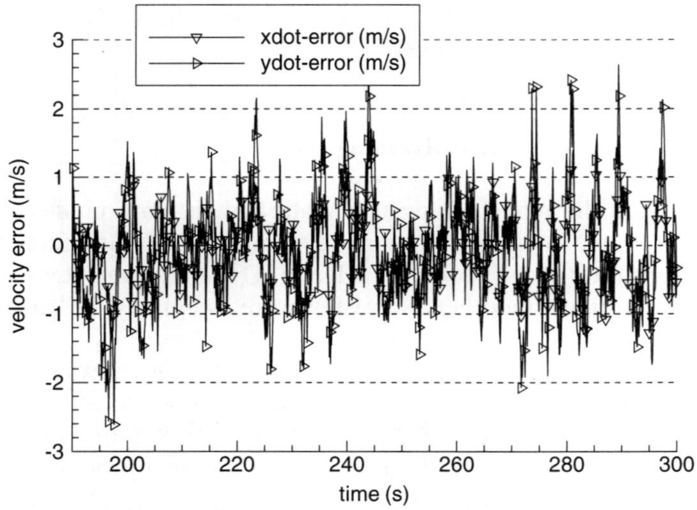

FIGURE 9.7: Error in filter-estimated tank velocity.

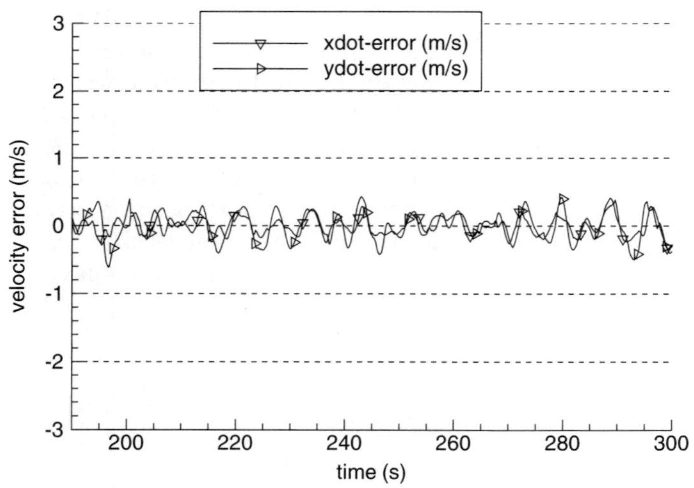

FIGURE 9.8: Error in smoother-estimated tank velocity.

Example 9.5: Tracking of Aircraft in Landing Pattern

A second example was described in Section 3.2.2. Motion of a small aircraft in the airport landing pattern was simulated using an eight-state model. A radar measuring range, azimuth, and elevation every second was located on the airport, where the simulated measurement noise $1-\sigma$ was 1 m for ranges and 0.2 milliradian for angles. Vertical velocity, crosstrack acceleration, and alongtrack acceleration were modeled in the filter as random walk processes with input white process noise PSDs of $0.08\,m^2/s^3$, $1.0\,m^2/s^5$, and $0.4\,m^2/s^5$, respectively. This model for random perturbations is time-invariant, but actual perturbations on these parameters only occur for short intervals. These PSD values were selected to give "good" average performance.

Figures 9.9 and 9.10 show the error in EKF-estimated position and velocity versus time. Notice that distinct error patterns occur during the turns (at 45–55 and 76–87 s), changes in vertical velocity (at 76 and 90 s) and during deceleration (90–95 s). Figures 9.11 and 9.12 are the comparable plots for the smoother. The errors are much smaller, but significant error patterns remain during maneuvers. This demonstrates the difficulty in using a time-invariant process noise model to characterize state perturbations that are really step jumps. Use of a first-order

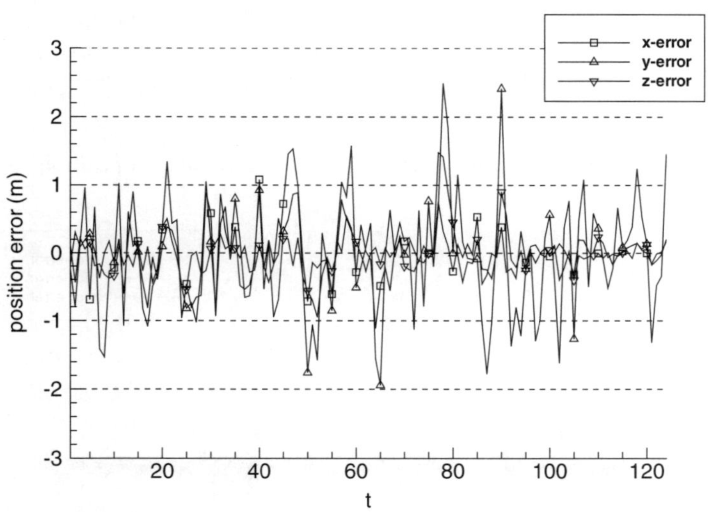

FIGURE 9.9: Error in filter-estimated aircraft position.

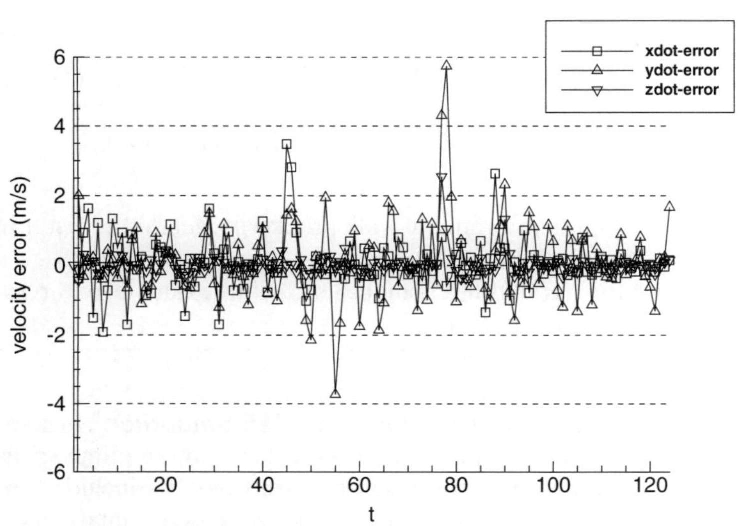

FIGURE 9.10: Error in filter-estimated aircraft velocity.

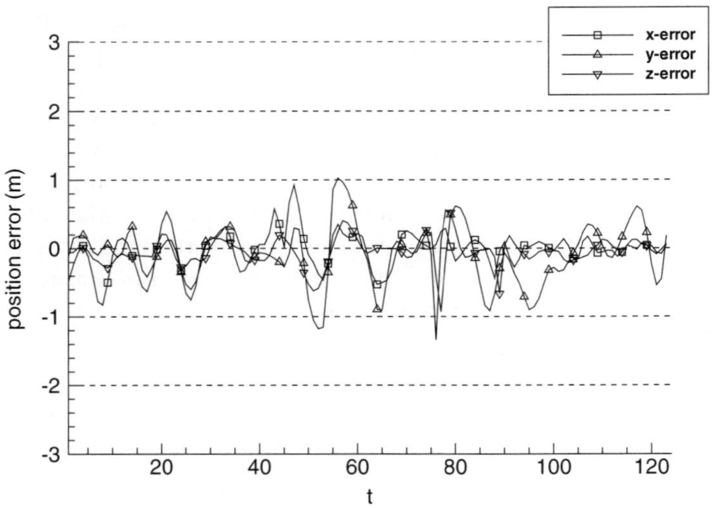

FIGURE 9.11: Error in smoother-estimated aircraft position.

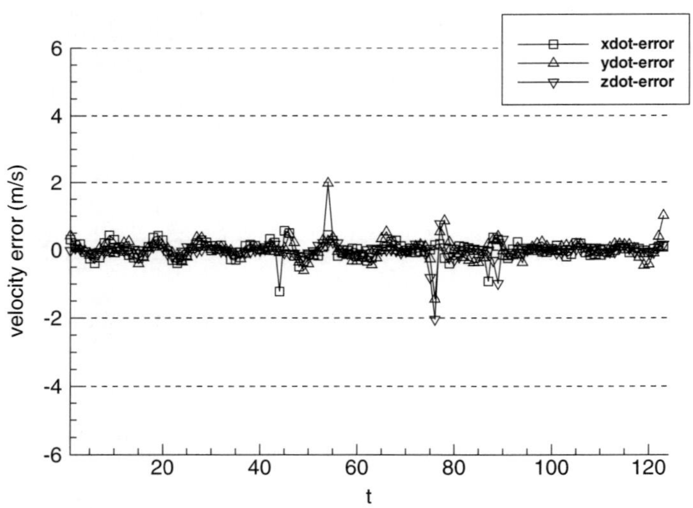

FIGURE 9.12: Error in smoother-estimated aircraft velocity.

Markov process rather than a random walk process model for the last three states did not improve performance significantly.

We also tested the iterated filter-smoother for this example, and results were essentially the same as for the EKF.

9.2.3.3 Pass Disposable Bias States in the RTS Smoother In some applications individual sensors only acquire measurements for limited time spans, so measurement biases associated with these sensors need only be included in the filter state vector when the sensor is receiving data. Orbit determination is one such example because ground-based sensors only view low-altitude satellites for short

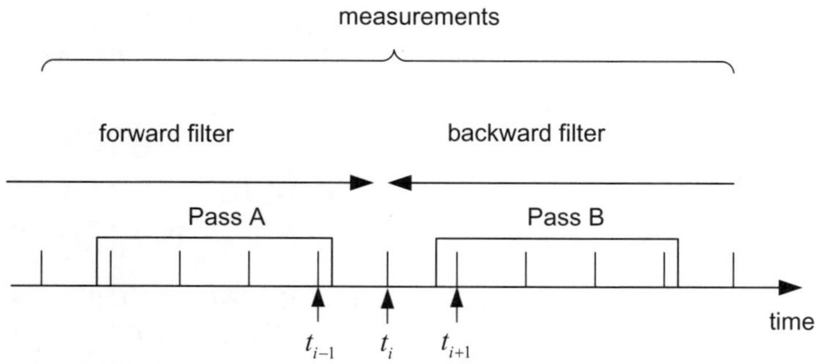

FIGURE 9.13: Estimation of smoothed pass-parameters.

periods of time, and may not see a satellite again for a day or more. Since a large component of the measurement bias is due to local tropospheric and ionospheric atmospheric effects or other time-varying errors, the biases can be treated as independent for different passes. The RTS algorithm has the ability to reconstruct pass-disposable bias states without the need to carry them continuously in the Kalman filter. Other smoothers—particularly information-based smoothers—probably have a comparable ability, but we have not seen algorithms for reconstructing the pass-parameters. Implementation of pass-disposable biases greatly reduces the filter/smoother computational and storage burdens since a few filter bias states can optimally handle hundreds of pass-disposable biases.

Tanenbaum (1977) discovered that information for bias states dropped by the filter can be reconstructed in the backward recursion of the RTS algorithm because the bias states are not smoothable. Tanenbaum derived the algorithm using the forward-backward smoother algorithm of Fraser and Potter, and then applied it to the RTS. The derivation has not been published, but because of space limitations we only summarize the results.

We first consider the state estimates obtained from forward and backward filters at a time point located between two station passes, as shown in Figure 9.13. The *a priori* information for the pass *B* biases is treated as a measurement that does not enter the forward filter until the pass begins. Hence the forward filter *a priori* estimate and covariance at t_i have the form

$$\hat{\mathbf{x}}_{i/i-1} = \begin{bmatrix} \hat{\mathbf{x}}_c \\ \hat{\mathbf{x}}_a \\ \mathbf{0} \end{bmatrix}_{i/i-1}, \quad \mathbf{P}_{i/i-1} = \begin{bmatrix} \mathbf{P}_{cc} & \mathbf{P}_{ca} & \mathbf{0} \\ \mathbf{P}_{ac} & \mathbf{P}_{aa} & \mathbf{0} \\ \mathbf{0} & \mathbf{0} & \infty \end{bmatrix}_{i/i-1},$$

where subscript *c* denotes states that are common to both the forward and backward filters and subscripts *a* and *b* refer to pass biases for the two passes. Since information on the pass *B* states is not available, the covariance is represented as infinity. The backward filter *a posteriori* estimate and covariance at t_i are of the form

$$\hat{\mathbf{x}}^b_{i/i:N} = \begin{bmatrix} \hat{\mathbf{x}}_c \\ \mathbf{0} \\ \hat{\mathbf{x}}_b \end{bmatrix}_{i/i:N}, \quad \mathbf{P}^b_{i/i:N} = \begin{bmatrix} \mathbf{P}^b_{cc} & \mathbf{0} & \mathbf{P}^b_{cb} \\ \mathbf{0} & \infty & \mathbf{0} \\ \mathbf{P}^b_{bc} & \mathbf{0} & \mathbf{P}^b_{cb} \end{bmatrix}_{i/i:N},$$

where *a prior* information is never used in the backward filter. Since the smoothed covariance is the inverse of the sum of information matrices for the two independent estimates, it has the form

$$\mathbf{P}_{i/N} = \left(\mathbf{P}_{i/i-1}^{-1} + (\mathbf{P}_{i/i:N}^{b})^{-1}\right)^{-1} = \begin{bmatrix} \mathbf{D}^{-1} & \mathbf{P}_{cc}^{b}\mathbf{D}^{-1}\mathbf{P}_{ca} & \mathbf{P}_{cc}\mathbf{D}^{-1}\mathbf{P}_{cb}^{b} \\ \sim & \mathbf{P}_{aa} - \mathbf{P}_{ac}\mathbf{D}^{-1}\mathbf{P}_{ca} & \mathbf{P}_{ac}\mathbf{D}^{-1}\mathbf{P}_{cb}^{b} \\ \sim & \sim & \mathbf{P}_{bb}^{b} - \mathbf{P}_{bc}^{b}\mathbf{D}^{-1}\mathbf{P}_{cb}^{b} \end{bmatrix}$$

where $\mathbf{D} = \left(\mathbf{P}_{cc\_i/i-1}^{-1} + (\mathbf{P}_{cc\_i/i:N}^{b})^{-1}\right)$, and the symmetric lower triangular partitions are not shown. (Inversion of matrices containing infinity must be done with great care, but the result can be derived more rigorously starting with information arrays.) Notice that the solution for common parameters does not depend on pass-parameters, and the solution for pass A parameters does not depend on pass B (and vice versa). This verifies that it is not necessary to carry pass-parameters outside the pass, but does not yet show that they can be reconstructed in the backward sweep of the RTS smoother.

We now consider what happens at point t_{i-1} which we assume is the time of the last measurement in pass A. The RTS algorithm uses the covariance difference at t_i, which after some manipulation can be shown to have the form

$$\mathbf{P}_{i/N} - \mathbf{P}_{i/i-1} = \begin{bmatrix} \Delta\mathbf{P}_{cc} & \Delta\mathbf{P}_{cc}(\mathbf{P}_{cc}^{-1}\mathbf{P}_{ca}) & \mathbf{P}_{cc}(\mathbf{P}_{cc} + \mathbf{P}_{cc}^{b})^{-1}\mathbf{P}_{cb}^{b} \\ \sim & (\mathbf{P}_{ac}\mathbf{P}_{cc}^{-1})\Delta\mathbf{P}_{cc}(\mathbf{P}_{cc}^{-1}\mathbf{P}_{ca}) & \mathbf{P}_{ac}(\mathbf{P}_{cc} + \mathbf{P}_{cc}^{b})^{-1}\mathbf{P}_{cb}^{b} \\ \sim & \sim & -\infty \end{bmatrix}_i \qquad (9.2\text{-}29)$$

where $\Delta\mathbf{P}_{cc} = -\mathbf{P}_{cc}(\mathbf{P}_{cc} + \mathbf{P}_{cc}^{b})^{-1}\mathbf{P}_{cc}$, but the actual common state covariance difference $\Delta\mathbf{P}_{cc} = (\mathbf{P}_{cc})_{i/N} - (\mathbf{P}_{cc})_{i/i-1}$ is used in implementing the algorithm. (Covariances $\mathbf{P}_{ac}$ and $\mathbf{P}_{cc}$ in equation [9.2-29] are obtained from the filter $\mathbf{P}_{i/i-1}$, and $\mathbf{P}_{cb}^{b}, \mathbf{P}_{cc}^{b}$ are obtained from $\mathbf{P}_{i/i:N}^{b}$ from the backward filter.) Values in the last row and column of equation (9.2-29) are irrelevant because they will not be used again in the backward sweep of the RTS. To avoid a growing smoother state size, these states should be eliminated from the smoother when they no longer appear in the filter state vector.

The above equation verifies that "missing" terms of $\mathbf{P}_{i/N} - \mathbf{P}_{i/i-1}$ can be reconstructed by pre- or post-multiplying $\Delta\mathbf{P}_{cc}$ by the factor $\mathbf{P}_{cc}^{-1}\mathbf{P}_{ca}$ obtained from the forward filter $\mathbf{P}_{i/i-1}$. Hence pass-parameters should be discarded in the filter after writing $\hat{\mathbf{x}}_{i/i-1}, \mathbf{P}_{i/i-1}, \ldots$ at t_i to the disk. By a similar procedure it is shown that the difference in state estimates can be reconstructed as

$$\hat{\mathbf{x}}_{i/N} - \hat{\mathbf{x}}_{i/i-1} = \begin{bmatrix} \Delta\hat{\mathbf{x}}_{cc} \\ (\mathbf{P}_{ac}\mathbf{P}_{cc}^{-1})\Delta\hat{\mathbf{x}}_{cc} \end{bmatrix}_i, \qquad (9.2\text{-}30)$$

where the pass B states have been eliminated.

The last step in the reconstruction fills in the pass A terms of the RTS gain matrix. Again after some manipulation using equation (9.2-23), it can be shown that

$$\boxed{\mathbf{G}' = \begin{bmatrix} \mathbf{\Phi}_{cc}^{-1}(\mathbf{I} - \mathbf{Q}_{cc}\mathbf{P}_{cc}^{-1}) \\ \mathbf{P}_{ac}\mathbf{P}_{cc}^{-1} \end{bmatrix}_{i/i-1}} \qquad (9.2\text{-}31)$$

is the gain matrix to be used in

$$\boxed{\begin{aligned}\hat{\mathbf{x}}_{i-1/N} &= \hat{\mathbf{x}}_{i-1/i-1} + \mathbf{G}'\Delta\hat{\mathbf{x}}_{cc} \\ \mathbf{P}_{i-1/N} &= \mathbf{P}_{i-1/i-1} + \mathbf{G}'\Delta\mathbf{P}_{cc}(\mathbf{G}')^T\end{aligned}} \tag{9.2-32}$$

to reconstruct the missing pass A bias states from the *a priori* forward filter common states and covariance at time t_i. The pass-disposable bias states to be added in the backward smoother sweep will not in general be located at the bottom of the filter state vector, and the smoother state vector will be smaller than the filter state when pass-parameters are added. Hence it is necessary to shift state and covariance ordering when implementing the reconstruction. This is greatly simplified if state labels are saved with the filter variables written to disk so that the location of common filter-smoother states can be identified. Software that implements this method is described in Chapter 13.

This algorithm was used operationally in orbit determination filter-smoothers at the U.S. Naval Surface Weapons Center and at the National Aeronautics and Space Administration (Gibbs 1979, 1982) to estimate pass-disposable bias states for hundreds of tracking passes using storage and computations for only a few biases.

Example 9.6: Global Positioning System (GPS) Satellite Navigation

The previous examples are of relatively low order, but it is also practical to use smoothers for large complex problems. Navigation for the GPS constellation of satellites is one such example. GPS and the navigation filter were described in Section 3.4.10.

A Kalman filter similar to the Master Control Station (MCS) navigation filter was used to process pseudo-range data from 19 ground receivers obtained during days 238 to 243 of 2005. An RTS smoother was also implemented and used for a 5 day span. To minimize disk storage and computations, it was desirable to explicitly take advantage of the sparseness in the state transition and process noise matrices, and to use pass-disposable monitor station biases, as described above. For the newest space vehicles in service at that time—designated as Block IIR—the smoothed satellite ephemeris typically agreed within 5–14 cm of the ephemeris computed by the International GNSS Service (IGS) using a much larger set of ground stations. We used the smoothed ephemeris as a reference to calculate the user range error (URE) of the filter-calculated ephemeris. The URE is defined for GPS as

$$URE = \sqrt{(\Delta r - \Delta t)^2 + (\Delta i^2 + \Delta c^2)/50}$$

where $\Delta r, \Delta i, \Delta c$ are the radial, intrack, and crosstrack components of the ephemeris error, and Δt is the error in the calculated clock time multiplied by the speed of light. Figure 9.14 shows the zero-age-of-data (no prediction) URE calculated using the difference between filter and smoother ephemeris/clock states for Block IIR space vehicles. Notice that peak URE values are 23 cm, but average UREs are usually less than 10 cm. Certain space vehicles tend to have slightly greater UREs than others because of differences in clock stability.

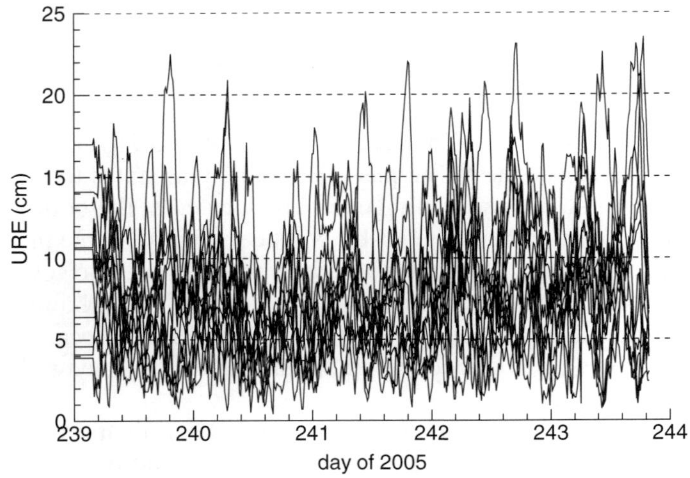

FIGURE 9.14: Computed GPS user range error for 18 block IIR space vehicles.

Since this example uses actual rather than simulated GPS measurements, the true errors in filter and smoother state estimates are unknown. Based on computed filter and smoother covariances, and on the smoother comparison with IGS results listed above, it is believed that most of the URE shown in Figure 9.14 is filter error.

9.3 FILTER ERROR ANALYSIS AND REDUCED-ORDER MODELING

Just as with least-squares processing, the formal error covariance $\mathbf{P}_{i/i}$ of the Kalman filter is a lower bound on estimate errors $\tilde{\mathbf{x}}_{i/i}$ when process noise, measurement noise, and initial condition errors are correctly modeled in the filter. Actual errors may be larger when the filter model does not match reality, some states are deliberately not estimated, the gain matrix is suboptimal, or the effects of nonlinearities are not included. A more realistic estimate of $E[\tilde{\mathbf{x}}_{i/i}\tilde{\mathbf{x}}_{i/i}^T]$—or even a distribution function—is important for several reasons. First, it is very useful in the design stage when attempting to develop a filter that meets accuracy requirements under realistic operating conditions. Availability of a realistic error covariance—particularly when contributions of different error sources are separately available—allows comparison of different designs and selection of the "best". Second, knowledge of the true error covariance is of interest for its own sake. For example, it is often necessary to list expected accuracy when providing specifications for customers or when "proving" that the system meets specifications. Example accuracy specifications include $3-\sigma$ error limits, confidence bounds, or probability of "impact." Finally in systems where the filter output is used to make decisions on possible control actions, the error covariance or distribution can be a required input. For example, the estimate error covariance is sometimes required for risk analysis associated with potential response actions.

As with least squares, there are two basic methods by which the statistical characteristics of the estimation errors may be determined: linear covariance analysis and Monte Carlo simulation. In covariance analysis, the effects of various error sources are propagated through the filter equations assuming that the error effects are additive and linear. The approach can also be applied to nonlinear systems, but error models are linearizations about the state trajectory. Covariance error analysis has the advantages that the effects of error sources may be separately computed, and it is computationally fast. However, coding effort may be significant. Monte Carlo error analysis is more rigorous because (1) it includes the effects of nonlinearities; (2) it can handle more complicated cases where there are significant structural differences between the filter model and a detailed "truth" model; and (3) it is easily coded. However, it is computationally slow since many random samples are necessary to properly characterize the error distribution. It is even slower when it is necessary to run separate simulations for each error source to separately determine the effects. The remainder of this section discusses options for linear covariance analysis.

Covariance error analysis is used when filter models of system dynamics and measurements are different from those of detailed truth models. In most cases the differences arise because the filter uses a ROM. There are a variety of reasons why filter models are of lower order than detailed simulation models. Since computations increase as the square or cube of filter order for most implementations, computational reduction is one of the primary reasons for using a ROM. Other reasons include improved computational accuracy (due to improved observability), imprecise knowledge of parameters used in detailed models, and improved robustness (in some cases).

Methods used to develop a ROM from a detailed model include the following:

1. Do not estimate (i.e., do not include in the filter state vector) various fixed parameters that are reasonably well known.
2. Eliminate detailed models of low-amplitude effects and instead increase $\mathbf{Q}$ or $\mathbf{R}$ if the errors are nearly random.
3. Same as above, but also modify coefficients in $\boldsymbol{\Phi}$ to allow the filter to have a broader frequency response than the detailed model.
4. Average or subsample measurements so that the filter model need not include states that model high-frequency effects.
5. Replace complex random force models with colored noise (low-order Markov process) models.
6. Combine states having similar measurement signatures since observability of the separate parameters will be poor.
7. Replace distributed parameter models with lumped parameter models.

When a ROM simply deletes states of a detailed model, linear perturbation methods may be used to either analyze the effects of errors in unadjusted parameters on filter accuracy, or to "consider" the uncertainty of the parameter values when weighting the measurements. However, when a filter model uses states and parameters different from those of a detailed model, the above approach is not directly applicable. Alternate forms of covariance error analysis can be used when the filter model states are a linear transformation of detailed model states, or when the filter model and detailed model have a subset of states in common. When the filter model has a different structure than a detailed model, Monte Carlo simulation is often the only option for error analysis.

We present three approaches for linear covariance error analysis of the Kalman filter. The first is used when the filter model structure closely matches reality, but some parameters of the model may be unadjusted or treated as consider parameters. Linearized sensitivity and covariance matrices for the effects of five possible error sources are summed to compute the total error covariance. The approach also computes derivatives of the measurement residuals with respect to unadjusted parameters. The second approach is used when the filter ROM is a linear transformation on a more accurate representation of the system. It is assumed that the transformation is used to generate reduced-order filter versions of the $\mathbf{\Phi}$, $\mathbf{Q}$, and $\mathbf{H}$ matrices. Then the adjusted states and unadjusted parameters of the transformed model are analyzed using the first approach. The third approach is used when the filter values and possibly structure of the $\mathbf{\Phi}$, $\mathbf{Q}$, $\mathbf{H}$, and $\mathbf{R}$ arrays are different from those of the detailed model, but a subset of important states are common to both models. In this case error propagation can be analyzed using an augmented state vector that includes both the detailed and reduced-order state vectors.

Extended discussions of ROM philosophy appear in Maybeck (1979, chapter 6), Grewal and Andrews (2001, section 7.5), Gelb (1974, chapter 7), and to a lesser extent in Anderson and Moore (1979, chapter 11) and Simon (2006, section 10.3). We encourage you to read one or more of these books. To summarize, ROM is a trade-off between developing a model that

1. accurately characterizes the system and provides excellent estimation performance when all parameters of the model are correct, but may deteriorate rapidly when model parameters are incorrect, versus
2. a model that provides somewhat lower accuracy under the best conditions but deteriorates less rapidly when model parameters are wrong.

The trade-off between computations and performance may also be a factor. A ROM may be less sensitive to parameter errors than a higher order model, but that is not necessarily true. Covariance error analysis can be invaluable in evaluating the trade-offs, but ultimately Monte Carlo analysis should be used to verify the performance.

9.3.1 Linear Analysis of Independent Error Sources

This section derives linear error effects in the discrete-time Kalman filter due to un-estimated bias parameters, Schmidt-type consider bias parameters, measurement noise, process noise, and the initial estimate. The following derivation is mostly due to Aldrich (1971) and Hatch (Gibbs and Hatch 1973), partly based on work by Jazwinski (1970, p. 246) and Schmidt (1966). This presentation assumes that the unadjusted and consider parameters are biases, so errors can be modeled using linear sensitivity arrays. The same approach can also handle unadjusted and consider states that are colored noise (rather than biases), but it is necessary to instead carry error covariance components rather than sensitivity arrays.

The error in the *a posteriori* state estimate at time t_{i-1} is assumed to be a linear function of five statistically independent error sources:

$$\begin{aligned}\tilde{\mathbf{x}}_{i-1/i-1} &= \mathbf{x}_{i-1} - \hat{\mathbf{x}}_{i-1/i-1} \\ &= \tilde{\mathbf{x}}^u_{i-1/i-1} + \tilde{\mathbf{x}}^c_{i-1/i-1} + \tilde{\mathbf{x}}^0_{i-1/i-1} + \tilde{\mathbf{x}}^r_{i-1/i-1} + \tilde{\mathbf{x}}^q_{i-1/i-1}\end{aligned} \tag{9.3-1}$$

where

$\tilde{\mathbf{x}}^u_{i-1/i-1}$ is the state error due to errors in filter values ($\hat{\mathbf{u}}$) of static unadjusted (but analyzed) parameters **u**,

$\tilde{\mathbf{x}}^c_{i-1/i-1}$ is the state error due to errors in filter values ($\hat{\mathbf{c}}$) of static consider parameters **c**,

$\tilde{\mathbf{x}}^0_{i-1/i-1}$ is the state error due to errors in the initial filter estimate $\hat{\mathbf{x}}_{0/0}$,

$\tilde{\mathbf{x}}^r_{i-1/i-1}$ is the state error due to cumulative errors in measurement noise, and

$\tilde{\mathbf{x}}^q_{i-1/i-1}$ is the state error due to cumulative errors in process noise.

(Superscripts are used in this section to denote components of a total vector or array.) There are no restrictions on the effects of unadjusted-analyze and consider parameters: they may appear in the force model, the measurement model, or both.

It is further assumed that the error in $\hat{\mathbf{x}}_{i-1/i-1}$ due to static parameters can be represented using linear sensitivity arrays:

$$\begin{aligned}\tilde{\mathbf{x}}^u_{i-1/i-1} &= \mathbf{S}^u_{i-1/i-1}\tilde{\mathbf{u}} \\ \tilde{\mathbf{x}}^c_{i-1/i-1} &= \mathbf{S}^c_{i-1/i-1}\tilde{\mathbf{c}} \\ \tilde{\mathbf{x}}^0_{i-1/i-1} &= \mathbf{S}^0_{i-1/i-1}\tilde{\mathbf{x}}_{0/0}\end{aligned} \tag{9.3-2}$$

where $\tilde{\mathbf{u}} = \mathbf{u} - \hat{\mathbf{u}}$, $\tilde{\mathbf{c}} = \mathbf{c} - \hat{\mathbf{c}}$, and $\tilde{\mathbf{x}}_{0/0} = \mathbf{x}_0 - \hat{\mathbf{x}}_{0/0}$. All error sources are assumed to be zero mean and independent.

Time propagation of the true state at time t_i is modeled as

$$\mathbf{x}_i = \mathbf{\Phi}_{xx}\,\mathbf{x}_{i-1} + \mathbf{\Phi}_{xu}\,\mathbf{u} + \mathbf{\Phi}_{xc}\,\mathbf{c} + \mathbf{q}_i\,. \tag{9.3-3}$$

The time subscripts have been dropped from partitions of $\mathbf{\Phi}$ to simplify notation. The corresponding filter time update is

$$\hat{\mathbf{x}}_{i/i-1} = \mathbf{\Phi}_{xx}\hat{\mathbf{x}}_{i-1/i-1} + \mathbf{\Phi}_{xu}\hat{\mathbf{u}} + \mathbf{\Phi}_{xc}\hat{\mathbf{c}} \tag{9.3-4}$$

so errors propagate as

$$\begin{aligned}\tilde{\mathbf{x}}_{i/i-1} &= \mathbf{x}_i - \hat{\mathbf{x}}_{i/i-1} \\ &= \mathbf{\Phi}_{xx}(\tilde{\mathbf{x}}^u_{i-1/i-1} + \tilde{\mathbf{x}}^c_{i-1/i-1} + \tilde{\mathbf{x}}^0_{i-1/i-1} + \tilde{\mathbf{x}}^r_{i-1/i-1} + \tilde{\mathbf{x}}^q_{i-1/i-1}) + \mathbf{\Phi}_{xu}\tilde{\mathbf{u}} + \mathbf{\Phi}_{xc}\tilde{\mathbf{c}} + \mathbf{q}_i\,. \\ &= \mathbf{S}^u_{i/i-1}\tilde{\mathbf{u}} + \mathbf{S}^c_{i/i-1}\tilde{\mathbf{c}} + \mathbf{S}^0_{i/i-1}\tilde{\mathbf{x}}_{0/0} + \mathbf{\Phi}_{xx}\tilde{\mathbf{x}}^r_{i-1/i-1} + \mathbf{\Phi}_{xx}\tilde{\mathbf{x}}^q_{i-1/i-1} + \mathbf{q}_i\end{aligned} \tag{9.3-5}$$

where

$$\begin{aligned}\mathbf{S}^u_{i/i-1} &\triangleq \frac{\partial \tilde{\mathbf{x}}_{i/i-1}}{\partial \mathbf{u}} = \mathbf{\Phi}_{xx}\mathbf{S}^u_{i-1/i-1} + \mathbf{\Phi}_{xu} \\ \mathbf{S}^c_{i/i-1} &\triangleq \frac{\partial \tilde{\mathbf{x}}_{i/i-1}}{\partial \mathbf{c}} = \mathbf{\Phi}_{xx}\mathbf{S}^c_{i-1/i-1} + \mathbf{\Phi}_{xc} \\ \mathbf{S}^0_{i/i-1} &\triangleq \frac{\partial \tilde{\mathbf{x}}_{i/i-1}}{\partial \tilde{\mathbf{x}}_{0/0}} = \mathbf{\Phi}_{xx}\mathbf{S}^0_{i-1/i-1}\end{aligned} \tag{9.3-6}$$

Defining the portions of the total error covariance due to measurement and process noises as

$$\begin{aligned}\mathbf{P}^r_{i-1/i-1} &\triangleq E[\tilde{\mathbf{x}}^r_{i-1/i-1}(\tilde{\mathbf{x}}^r_{i-1/i-1})^T]\\ \mathbf{P}^q_{i-1/i-1} &\triangleq E[\tilde{\mathbf{x}}^q_{i-1/i-1}(\tilde{\mathbf{x}}^q_{i-1/i-1})^T]\end{aligned},$$

and noting that the error sources are assumed to be independent, the true error covariance is

$$\boxed{\begin{aligned}\mathbf{P}_{i/i-1} = \mathbf{S}^u_{i/i-1}\mathbf{P}_u(\mathbf{S}^u_{i/i-1})^T + \mathbf{S}^c_{i/i-1}\mathbf{P}_c(\mathbf{S}^c_{i/i-1})^T + \mathbf{S}^0_{i/i-1}\mathbf{P}_{0/0}(\mathbf{S}^0_{i/i-1})^T\\ + \mathbf{\Phi}_{xx}\mathbf{P}^r_{i-1/i-1}\mathbf{\Phi}^T_{xx} + (\mathbf{\Phi}_{xx}\mathbf{P}^q_{i-1/i-1}\mathbf{\Phi}^T_{xx} + \mathbf{Q})\end{aligned}} \tag{9.3-7}$$

where $\mathbf{P}_u = E[\tilde{\mathbf{u}}\tilde{\mathbf{u}}^T]$, $\mathbf{P}_c = E[\tilde{\mathbf{c}}\tilde{\mathbf{c}}^T]$, $\mathbf{P}_{0/0} = E[\tilde{\mathbf{x}}_{0/0}\tilde{\mathbf{x}}^T_{0/0}]$, and $\mathbf{Q} = E[\mathbf{q}_i\mathbf{q}_i^T]$. Again we drop the time subscript from $\mathbf{Q}$ and subsequent $\mathbf{H}$ and $\mathbf{R}$ arrays for simplicity in notation, not because they are assumed constant. From equation (9.3-7) the portions of $\mathbf{P}_{i/i-1}$ due to measurement and process noise are, respectively,

$$\boxed{\begin{aligned}\mathbf{P}^r_{i/i-1} &= \mathbf{\Phi}_{xx}\mathbf{P}^r_{i-1/i-1}\mathbf{\Phi}^T_{xx}\\ \mathbf{P}^q_{i/i-1} &= \mathbf{\Phi}_{xx}\mathbf{P}^q_{i-1/i-1}\mathbf{\Phi}^T_{xx} + \mathbf{Q}\end{aligned}}. \tag{9.3-8}$$

The sensitivity and covariance arrays are initialized as follows:

$$\boxed{\mathbf{S}^u_{0/0} = \mathbf{0},\quad \mathbf{S}^c_{0/0} = \mathbf{0},\quad \mathbf{S}^0_{0/0} = \mathbf{I},\quad \mathbf{P}^r_{0/0} = \mathbf{0},\quad \mathbf{P}^q_{0/0} = \mathbf{0}.} \tag{9.3-9}$$

The filter model of the state error covariance $\mathbf{P}^f_{i/i-1}$ is the same as equation (9.3-7) without the $\mathbf{S}^u_{i/i-1}\mathbf{P}_u(\mathbf{S}^u_{i/i-1})^T$ term. (Superscript "f" denotes filter model.)

The true measurement model is

$$\mathbf{y}_i = \mathbf{H}_x\mathbf{x} + \mathbf{H}_u\mathbf{u} + \mathbf{H}_c\mathbf{c} + \mathbf{r}_i \tag{9.3-10}$$

and the filter prediction is

$$\hat{\mathbf{y}}_i = \mathbf{H}_x\hat{\mathbf{x}}_{i/i-1} + \mathbf{H}_u\hat{\mathbf{u}} + \mathbf{H}_c\hat{\mathbf{c}} \tag{9.3-11}$$

so the measurement residual is

$$\begin{aligned}\tilde{\mathbf{y}}_i &= \mathbf{y}_i - \hat{\mathbf{y}}_i\\ &= \mathbf{H}_x\tilde{\mathbf{x}}_{i/i-1} + \mathbf{H}_u\tilde{\mathbf{u}} + \mathbf{H}_c\tilde{\mathbf{c}} + \mathbf{r}_i\\ &= \mathbf{H}_x(\mathbf{S}^0_{i/i-1}\tilde{\mathbf{x}}_{0/0} + \mathbf{S}^c_{i/i-1}\tilde{\mathbf{c}} + \mathbf{S}^u_{i/i-1}\tilde{\mathbf{u}} + \tilde{\mathbf{x}}^r_{i/i-1} + \tilde{\mathbf{x}}^q_{i/i-1}) + \mathbf{H}_u\tilde{\mathbf{u}} + \mathbf{H}_c\tilde{\mathbf{c}} + \mathbf{r}_i\\ &= \mathbf{T}^0_i\tilde{\mathbf{x}}^0_{i/i-1} + \mathbf{T}^u_i\tilde{\mathbf{u}} + \mathbf{T}^c_i\tilde{\mathbf{c}} + \mathbf{H}_x(\tilde{\mathbf{x}}^r_{i/i-1} + \tilde{\mathbf{x}}^q_{i/i-1}) + \mathbf{r}_i\end{aligned} \tag{9.3-12}$$

where the measurement residual derivatives for the static errors are

$$\boxed{\begin{aligned}\mathbf{T}^0_i &\triangleq \frac{\partial\tilde{\mathbf{y}}_i}{\partial\tilde{\mathbf{x}}_{0/0}} = \mathbf{H}_x\mathbf{S}^0_{i/i-1}\\ \mathbf{T}^u_i &\triangleq \frac{\partial\tilde{\mathbf{y}}_i}{\partial\tilde{\mathbf{u}}} = \mathbf{H}_x\mathbf{S}^u_{i/i-1} + \mathbf{H}_u\\ \mathbf{T}^c_i &\triangleq \frac{\partial\tilde{\mathbf{y}}_i}{\partial\tilde{\mathbf{c}}} = \mathbf{H}_x\mathbf{S}^c_{i/i-1} + \mathbf{H}_c\end{aligned}}. \tag{9.3-13}$$

These residual derivatives can be very helpful when trying to determine types of model errors that can cause particular signatures in measurement residuals. However, the **T** arrays should be multiplied by the expected standard deviation for each parameter (e.g., $\mathbf{T}_i^u \mathbf{P}_u^{1/2}$ for diagonal $\mathbf{P}_u$) before plotting as a function of time. In other words, compare the observed residual signatures with expected perturbations (not derivatives) due to each unadjusted parameter.

The true residual error covariance is

$$\begin{aligned} \mathbf{C}_i &\triangleq E[\tilde{\mathbf{y}}_i \tilde{\mathbf{y}}_i^T] \\ &= \mathbf{T}_i^0 \mathbf{P}_{0/0} (\mathbf{T}_i^0)^T + \mathbf{T}_i^u \mathbf{P}_u (\mathbf{T}_i^u)^T + \mathbf{T}_i^c \mathbf{P}_c (\mathbf{T}_i^c)^T + \mathbf{H}_x (\mathbf{P}_{i/i-1}^r + \mathbf{P}_{i/i-1}^q) \mathbf{H}_x^T + \mathbf{R} \end{aligned} \quad (9.3\text{-}14)$$

However, the filter does not know about unadjusted error parameters, and thus models the residual covariance as

$$\boxed{\mathbf{C}_i^f = \mathbf{T}_i^0 \mathbf{P}_{0/0} (\mathbf{T}_i^0)^T + \mathbf{T}_i^c \mathbf{P}_c (\mathbf{T}_i^c)^T + \mathbf{H}_x (\mathbf{P}_{i/i-1}^r + \mathbf{P}_{i/i-1}^q) \mathbf{H}_x^T + \mathbf{R}.} \quad (9.3\text{-}15)$$

The Kalman gain matrix for adjusted states (with consider parameter) is computed as

$$\boxed{\mathbf{K}_i = \left(\mathbf{P}_{i/i-1}^f \mathbf{H}_x^T + \mathbf{S}_{i/i-1}^c \mathbf{P}_c \mathbf{H}_c^T\right) (\mathbf{C}_i^f)^{-1}} \quad (9.3\text{-}16)$$

where

$$\boxed{\mathbf{P}_{i/i-1}^f = \mathbf{S}_{i/i-1}^0 \mathbf{P}_{0/0} (\mathbf{S}_{i/i-1}^0)^T + \mathbf{S}_{i/i-1}^c \mathbf{P}_c (\mathbf{S}_{i/i-1}^c)^T + \mathbf{P}_{i/i-1}^r + \mathbf{P}_{i/i-1}^q} \quad (9.3\text{-}17)$$

is the filter-modeled covariance of $\tilde{\mathbf{x}}_{i/i-1}$. (Note: significant computational saving may be obtained by grouping and saving matrix products for use in multiple equations.) The *a posteriori* state estimate is

$$\hat{\mathbf{x}}_{i/i} = \hat{\mathbf{x}}_{i/i-1} + \mathbf{K}_i \tilde{\mathbf{y}}_i \quad (9.3\text{-}18)$$

and the error is

$$\boxed{\begin{aligned} \tilde{\mathbf{x}}_{i/i} &= \mathbf{x}_i - \hat{\mathbf{x}}_{i/i} \\ &= \left(\tilde{\mathbf{x}}_{i/i-1}^u + \tilde{\mathbf{x}}_{i/i-1}^c + \tilde{\mathbf{x}}_{i/i-1}^r + \tilde{\mathbf{x}}_{i/i-1}^q + \tilde{\mathbf{x}}_{i/i-1}^0\right) - \mathbf{K}_i \left(\mathbf{T}_i^0 \tilde{\mathbf{x}}_{i/i-1}^0 + \mathbf{T}_i^u \tilde{\mathbf{u}} + \mathbf{T}_i^c \tilde{\mathbf{c}} + \mathbf{H}_x (\tilde{\mathbf{x}}_{i/i-1}^r + \tilde{\mathbf{x}}_{i/i-1}^q) + \mathbf{r}_i\right) \\ &= \mathbf{S}_{i/i}^0 \tilde{\mathbf{x}}_{0/0} + \mathbf{S}_{i/i}^u \tilde{\mathbf{u}} + \mathbf{S}_{i/i}^c \tilde{\mathbf{c}} + (\mathbf{I} - \mathbf{K}_i \mathbf{H}_x) \tilde{\mathbf{x}}_{i/i-1}^q + (\mathbf{I} - \mathbf{K}_i \mathbf{H}_x) \tilde{\mathbf{x}}_{i/i-1}^r + \mathbf{K}_i \mathbf{r}_i \end{aligned}}$$

(9.3-19)

where the *a posteriori* sensitivity arrays are

$$\boxed{\begin{aligned} \mathbf{S}_{i/i}^0 &= \mathbf{S}_{i/i-1}^0 - \mathbf{K}_i \mathbf{T}_i^0 \\ \mathbf{S}_{i/i}^u &= \mathbf{S}_{i/i-1}^u - \mathbf{K}_i \mathbf{T}_i^u . \\ \mathbf{S}_{i/i}^c &= \mathbf{S}_{i/i-1}^c - \mathbf{K}_i \mathbf{T}_i^c \end{aligned}} \quad (9.3\text{-}20)$$

The recursions for those portions of the covariance matrix due to measurement and process noise are

$$\boxed{\begin{aligned}\mathbf{P}^r_{i/i} &= (\mathbf{I}-\mathbf{K}_i\mathbf{H}_i)\mathbf{P}^r_{i/i-1}(\mathbf{I}-\mathbf{K}_i\mathbf{H}_i)^T + \mathbf{K}_i\mathbf{R}_i\mathbf{K}_i^T \\ \mathbf{P}^q_{i/i} &= (\mathbf{I}-\mathbf{K}_i\mathbf{H}_i)\mathbf{P}^q_{i/i-1}(\mathbf{I}-\mathbf{K}_i\mathbf{H}_i)^T\end{aligned}} \tag{9.3-21}$$

and the total error covariance $E[\tilde{\mathbf{x}}_{i/i}\tilde{\mathbf{x}}^T_{i/i}]$ (including effects of unadjusted-analyze parameter errors) is

$$\boxed{\mathbf{P}_{i/i} = \mathbf{S}^u_{i/i}\mathbf{P}_u(\mathbf{S}^u_{i/i})^T + \mathbf{S}^c_{i/i}\mathbf{P}_c(\mathbf{S}^c_{i/i})^T + \mathbf{S}^0_{i/i}\mathbf{P}_{0/0}(\mathbf{S}^0_{i/i})^T + \mathbf{P}^r_{i/i} + \mathbf{P}^q_{i/i}\,.} \tag{9.3-22}$$

The filter model of the *a posteriori* error covariance is the same without the $\mathbf{S}^u_{i/i}\mathbf{P}_u(\mathbf{S}^u_{i/i})^T$ term.

Notice that when the number of measurements is smaller than the number of states, significant computational reduction is obtained by implementing equation (9.3-21) as

$$\begin{aligned}\mathbf{A} &= (\mathbf{I}-\mathbf{K}_i\mathbf{H}_i)\mathbf{P}^r_{i/i-1} \\ \mathbf{P}^r_{i/i} &= \mathbf{A} + (\mathbf{K}_i\mathbf{R}_i - \mathbf{A}\mathbf{H}_i^T)\mathbf{K}_i^T\end{aligned}$$

and

$$\begin{aligned}\mathbf{A} &= (\mathbf{I}-\mathbf{K}_i\mathbf{H}_i)\mathbf{P}^q_{i/i-1} \\ \mathbf{P}^q_{i/i} &= \mathbf{A} - (\mathbf{A}\mathbf{H}_i^T)\mathbf{K}_i^T\end{aligned}.$$

The effort required to code the above equations expressed in matrix partitions is considerable. (Chapter 13 describes flexible code developed for this purpose—module COVAR_ANAL—where all static parameters may be selected by input as either adjusted, consider, or unadjusted-analyze.) If it is not necessary to know the separate contributions of measurement noise, process noise, and initial estimate errors, these terms may be combined into a single covariance, as normally done in a Kalman filter. That is, delete the $\mathbf{S}^0_{i/i}$ and $\mathbf{P}^q_{i/i}$ terms, rename the $\mathbf{P}^r_{i/i}$ term as $\mathbf{P}^{0qr}_{i/i}$ (initialized with $\mathbf{P}^{0qr}_{0/0} = \mathbf{I}$), replace equation (9.3-8) with

$$\mathbf{P}^{0qr}_{i/i-1} = \mathbf{\Phi}_{xx}\mathbf{P}^{0qr}_{i-1/i-1}\mathbf{\Phi}^T_{xx} + \mathbf{Q}\,,$$

and replace equation (9.3-21) with

$$\mathbf{P}^{0qr}_{i/i} = (\mathbf{I}-\mathbf{K}_i\mathbf{H}_i)\mathbf{P}^{0qr}_{i/i-1}(\mathbf{I}-\mathbf{K}_i\mathbf{H}_i)^T + \mathbf{K}_i\mathbf{R}_i\mathbf{K}_i^T\,.$$

Also, coding is simpler and potentially more efficient if states in the global model are ordered as dynamic-core, dynamic-bias, and measurement-bias. With this ordering, the state transition, process noise, and measurement model matrices have the forms

$$\begin{bmatrix}\mathbf{\Phi}_{core} & \mathbf{\Phi}_{core-bias} & \mathbf{0} \\ \mathbf{0} & \mathbf{I} & \mathbf{0} \\ \mathbf{0} & \mathbf{0} & \mathbf{I}\end{bmatrix}, \quad \begin{bmatrix}\mathbf{Q}_{core} & \mathbf{0} & \mathbf{0} \\ \mathbf{0} & \mathbf{0} & \mathbf{0} \\ \mathbf{0} & \mathbf{0} & \mathbf{0}\end{bmatrix}, \quad [\mathbf{H}_{core} \quad \mathbf{0} \quad \mathbf{H}_{meas}].$$

Then

$\mathbf{\Phi}_{xx}$ is formed as $\mathbf{\Phi}_{core}$ plus selected columns of $\mathbf{\Phi}_{core-bias}$, with added zero columns and 1's placed on the diagonal,

$\mathbf{Q}_{xx}$ is formed as $\mathbf{Q}_{core}$ plus zero row/columns,

$\mathbf{\Phi}_{xu}$ and $\mathbf{\Phi}_{xc}$ are formed by mapping columns of $\mathbf{\Phi}_{core-bias}$ and adding zero columns,

$\mathbf{H}_{xx}$ is formed as $\mathbf{H}_{core}$ plus selected columns of $\mathbf{H}_{meas}$ and zero columns,

$\mathbf{H}_u$ and $\mathbf{H}_c$ are formed by mapping columns of $\mathbf{H}_{meas}$ and adding zero columns.

Example 9.7: Strapdown Inertial Navigation System

This example uses the 27-state model of a strapdown inertial navigation system discussed in Section 3.3. Recall that this model includes three states each for nine groups: errors in computed position, velocity, gyro tilt, accelerometer bias, accelerometer scale factor error, gyro bias, gyro scale factor error, gyro G-sensitive errors, and G-squared-sensitive errors. Simulated error magnitudes were listed in Section 3.3. The acceleration profile shown in Figure 3.3 was obtained for a simulated 790 s trajectory of a ballistic missile. There are multiple periods in which the acceleration changes because of booster staging or intentional bus maneuvers.

Position measurements generated by the simulation were processed by a Kalman filter that implemented the error equations of this section. The software allowed each of the 27 states to be individually selected as ignore, unadjusted-analyze, consider, or estimated. Initially only the first 15 states (position, velocity, tilt, accelerometer bias, and accelerometer scale factor) were estimated, with all remaining states set as unadjusted-analyze. Figure 9.15 shows the covariance-computed uncertainty in the x-component of position error due to each error source. Uncertainty signatures for the y and z components are similar. The lines are labeled as follows:

"sig0-delPx" is the $1-\sigma$ uncertainty due to errors in the initial state estimate

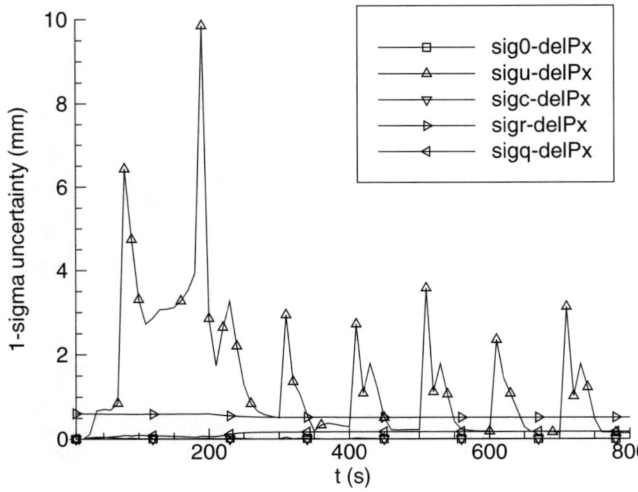

FIGURE 9.15: x-Position uncertainty versus time with 15 estimated states (12 unadjusted).

"sigu-delPx" is the $1 - \sigma$ uncertainty due to errors in unadjusted-analyze parameters

"sigc-delPx" is the $1 - \sigma$ uncertainty due to errors in consider parameters

"sigr-delPx" is the $1 - \sigma$ uncertainty due to measurement noise

"sigq-delPx" is the $1 - \sigma$ uncertainty due to process noise

Since consider parameters were not included in this filter scenario, consider error effects are zero. Although difficult to see in the plot, the contribution of initial condition errors is negligible because the accuracy of the first measurement (0.61 mm $1 - \sigma$) is much greater than the initial position error (4000 mm). The effects of process noise are also very small for the assumed process noise magnitudes. Uncertainty due to measurement noise (about 0.5 mm) is only slightly smaller than the measurement noise for a single measurement. However, the effect of errors in unadjusted-analyze parameters is quite large. Figure 9.16 is a similar plot for the x-component of velocity. The error behavior for velocity is similar to that of position, although the effects of process noise (0.06 mm/s after 200 s) are slightly larger than those of measurement noise (0.04 mm/s).

Filter errors were next analyzed with various sets of unadjusted parameters changed to consider. The effects of errors in individual unadjusted parameters can be determined by examining columns of the $\mathbf{S}^{u}_{i/i}$ array multiplied by the corresponding uncertainty of $\sqrt{\mathbf{P}_u}$ for each parameter. Gyro biases had the largest effect, so they were made consider parameters. Figures 9.17 and 9.18 show the results: the error magnitudes are greatly reduced because inclusion of gyro bias consider parameters reduced the error sensitivity. In Figures 9.19 and 9.20 the gyro biases were estimated rather than considered. Notice that the errors *increased* greatly compared with Figures 9.17 and 9.18, and this result was confirmed by actual estimation errors of limited Monte Carlo trials. In other words, the position and velocity error magnitudes are actually larger when the gyro biases are estimated rather than considered. This counterintuitive result only

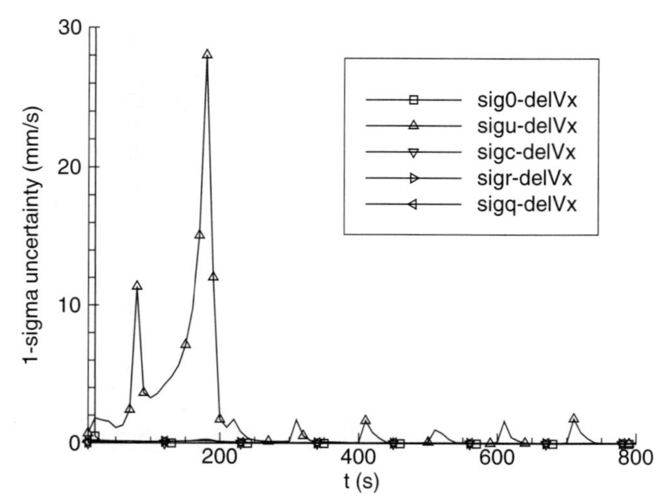

FIGURE 9.16: x-Velocity uncertainty versus time with 15 estimated states (12 unadjusted).

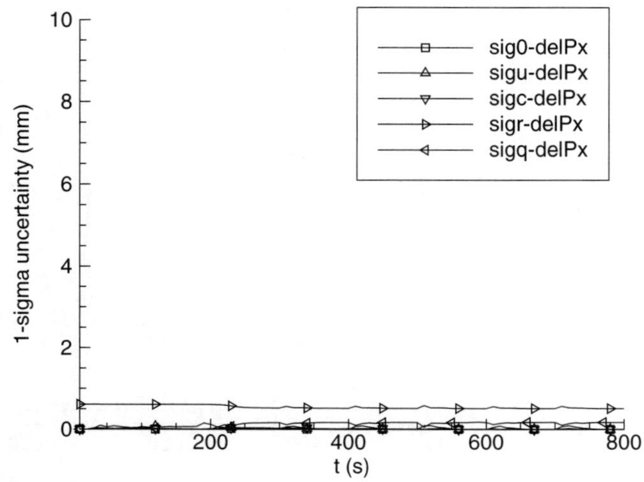

FIGURE 9.17: x-Position uncertainty versus time with 15 estimated states and 3 consider parameters (9 unadjusted).

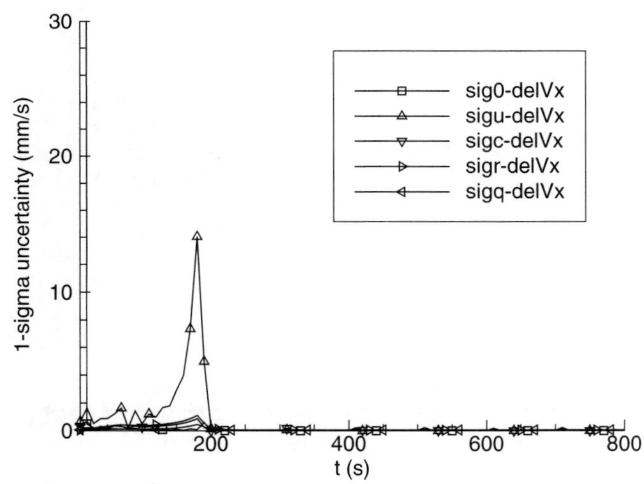

FIGURE 9.18: x-Velocity uncertainty versus time with 15 estimated states and 3 consider parameters (9 unadjusted).

occurs because other gyro errors (scale factor and acceleration-sensitive parameters) are unadjusted. When all 27 variables are adjusted (not plotted), the maximum errors after the boost phase are reduced to about 0.5 mm for position and 0.07 m/s for velocity. This demonstrates that the Kalman filter is "optimal" (meaning that errors are smallest) when all variables are estimated, *provided that no other modeling errors exist.* When modeling errors are present, errors in estimated states may not necessarily be smallest when all variables are estimated. In this example, treating the gyro bias parameters as consider parameters—rather than estimated—reduced the sensitivity of other estimated states to modeling errors caused by ignoring unadjusted parameters.

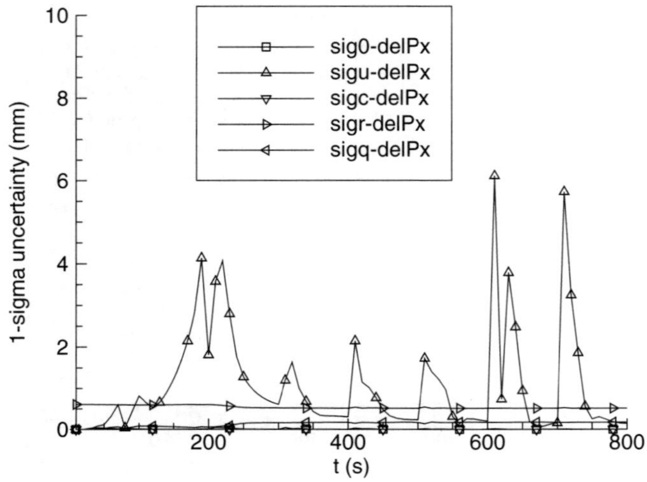

FIGURE 9.19: x-Position uncertainty versus time with 18 estimated states (9 unadjusted).

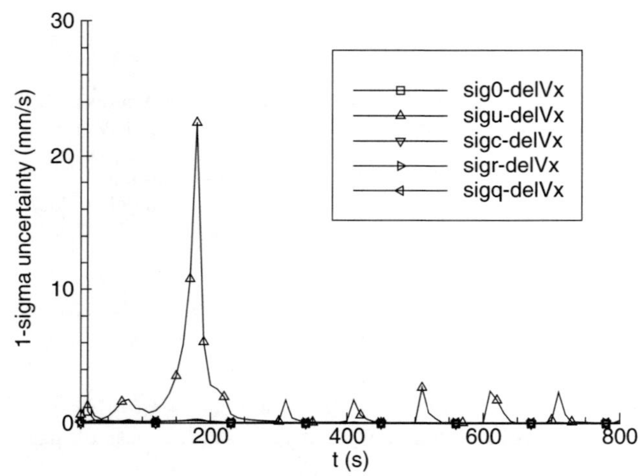

FIGURE 9.20: x-Velocity uncertainty versus time with 18 estimated states (9 unadjusted).

This example shows how the error analysis technique of this section can be applied to a realistic problem, and used to determine the best choice of estimated states and consider parameters for a ROM.

9.3.2 Error Analysis for ROM Defined as a Transformed Detailed Model

When the ROM simply deletes states of the detailed model, methods of the previous section may be used to either analyze the effects of errors in unadjusted parameters on filter accuracy, or to "consider" the uncertainty of the parameter values when weighting the measurements. However, when the filter model uses states and parameters different from those of the detailed model, the above approach is not directly applicable. The more general case will be discussed in the next section, or when that does not apply, Monte Carlo analysis can be used. When filter model states are a linear transformation of detailed model states, the covariance error analysis of the previous section can still be used provided that a preprocessing step is first applied, as will now be shown.

Let the detailed model states be denoted by $\mathbf{x}_d$ of dimension n_d and the filter states be denoted by $\mathbf{x}_f$ of dimension $n_f < n_d$, where $\mathbf{x}_f = \mathbf{T}_f\mathbf{x}_d$ and $\mathbf{T}_f$ is $n_f \times n_d$. An invertible transformation between the two state vectors must exist to analyze error propagation, so we instead assume that an augmented state vector $\mathbf{x}_t$ of dimension n_d is defined as

$$\mathbf{x}_t = \begin{bmatrix} \mathbf{x}_f \\ \mathbf{a} \end{bmatrix} = \begin{bmatrix} \mathbf{T}_f \\ \mathbf{T}_a \end{bmatrix} \mathbf{x}_d \tag{9.3-23}$$

where $(n_d - n_f) \times n_d$ matrix $\mathbf{T}_a$ is chosen to make

$$\mathbf{T} = \begin{bmatrix} \mathbf{T}_f \\ \mathbf{T}_a \end{bmatrix} \tag{9.3-24}$$

full rank. Also $\mathbf{T}_a$ must *not* transform states driven by process noise if it is used in the error analysis of the previous section. Otherwise the definitions of $\mathbf{a}$ and $\mathbf{T}_a$ are nearly arbitrary. There are a variety of approaches for chosing $\mathbf{T}_f$. For example, it may sum $\mathbf{x}_d$ states that have similar effects, or combine states to compute variables of interest (perhaps for control purposes), or may be composed of eigenvectors of the steady-state filter covariance for $\mathbf{x}_d$ corresponding to the largest (poorly observable) eigenvalues.

Using the state transformation of equation (9.3-24), the state transition, state noise, measurement sensitivity matrices, and initial covariance for a model based on $\mathbf{x}_t$ become

$$\begin{aligned} \mathbf{\Phi}_t &= \mathbf{T}\mathbf{\Phi}_d\mathbf{T}^{-1} \\ \mathbf{Q}_t &= \mathbf{T}\mathbf{Q}_d\mathbf{T}^T \\ \mathbf{H}_t &= \mathbf{H}_d\mathbf{T}^{-1} \\ (\mathbf{P}_t)_{0/0} &= \mathbf{T}(\mathbf{P}_d)_{0/0}\mathbf{T}^T \end{aligned} \tag{9.3-25}$$

where $\mathbf{\Phi}_d$, $\mathbf{Q}_d$, and $\mathbf{H}_d$ are the model matrices for the detailed model. It is now conceptually simple to use $\mathbf{x}_t$ as the total parameter set from which adjusted states ($\mathbf{x}_f$), unadjusted-analyze, and consider states are selected. These selections are then used in the error analysis equations of the previous section; that is, the appropriate partitions of the $\mathbf{\Phi}_t$, $\mathbf{Q}_t$, $\mathbf{H}_t$, and $(\mathbf{P}_t)_{0/0}$ arrays for the corresponding states or parameters are substituted in the error equations. Since those error equations assume that unadjusted-analyze and consider parameters are biases (rather than dynamic states driven by process noise), $\mathbf{a}$ must include only biases, which explains why $\mathbf{T}_a$ must only transform biases. If the equations of the previous section are modified to include colored noise unadjusted-analyze and consider states, the restrictions may be relaxed.

9.3.3 Error Analysis for Different Truth and Filter Models

When the filter model has states and structure different from those of a detailed model, Monte Carlo simulation is often the only option for error analysis. However, if a subset of states are common to both filter and detailed models—even if the dynamic or measurement models have different structures and coefficients—it is possible to implement covariance error analysis using a dual-state model approach. (See Gelb 1974, or Grewal and Andrews 2001, for other examples.)

It is assumed that the detailed model is of the form

$$\begin{bmatrix} \mathbf{x}_c \\ \mathbf{x}_e \end{bmatrix}_i = \begin{bmatrix} \mathbf{\Phi}_{cc} & \mathbf{\Phi}_{ce} \\ \mathbf{0} & \mathbf{\Phi}_{ee} \end{bmatrix} \begin{bmatrix} \mathbf{x}_c \\ \mathbf{x}_e \end{bmatrix}_{i-1} + \begin{bmatrix} \mathbf{q}_c \\ \mathbf{q}_e \end{bmatrix}_i$$
$$\mathbf{y}_i = [\mathbf{H}_c \quad \mathbf{H}_e] \begin{bmatrix} \mathbf{x}_c \\ \mathbf{x}_e \end{bmatrix}_i + \mathbf{r}_i \tag{9.3-26}$$

where $\mathbf{x}_c$ represents a combination of dynamic and static states that are also used in the filter, and $\mathbf{x}_e$ represents dynamic and static states that do not appear in the Kalman filter. For simplicity we assume that $\mathbf{\Phi}_{ec} = \mathbf{0}$ ($\mathbf{x}_c$ states do not drive the dynamics of $\mathbf{x}_e$), but that is not an inherent limitation of the method. It is also assumed that

$$E\left(\begin{bmatrix} \mathbf{q}_c \\ \mathbf{q}_e \end{bmatrix} [\mathbf{q}_c^T \quad \mathbf{q}_e^T] \right) = \begin{bmatrix} \mathbf{Q}_{cc} & \mathbf{Q}_{ce} \\ \mathbf{Q}_{ce}^T & \mathbf{Q}_{ee} \end{bmatrix}, \tag{9.3-27}$$

$E[\mathbf{q}_c] = \mathbf{0}$, $E[\mathbf{q}_e] = \mathbf{0}$, $E[\mathbf{r}_i] = \mathbf{0}$, $E[\mathbf{r}_i \mathbf{r}_i^T] = \mathbf{R}$, and $E[\mathbf{q}_{ci} \mathbf{r}_i^T] = \mathbf{0}$. The $\mathbf{Q}_{ce}$ term may be zero in some problems.

The corresponding filter models are

$$\begin{bmatrix} \hat{\mathbf{x}}_c \\ \hat{\mathbf{x}}_d \end{bmatrix}_{i/i-1} = \begin{bmatrix} \mathbf{\Phi}_{cc}^f & \mathbf{\Phi}_{cd}^f \\ \mathbf{0} & \mathbf{\Phi}_{dd}^f \end{bmatrix} \begin{bmatrix} \hat{\mathbf{x}}_c \\ \hat{\mathbf{x}}_d \end{bmatrix}_{i-1/i-1}$$
$$\hat{\mathbf{y}}_i = \begin{bmatrix} \mathbf{H}_c^f & \mathbf{H}_d^f \end{bmatrix} \begin{bmatrix} \hat{\mathbf{x}}_c \\ \hat{\mathbf{x}}_d \end{bmatrix}_i \tag{9.3-28}$$
$$\begin{bmatrix} \hat{\mathbf{x}}_c \\ \hat{\mathbf{x}}_d \end{bmatrix}_{i/i} = \begin{bmatrix} \hat{\mathbf{x}}_c \\ \hat{\mathbf{x}}_d \end{bmatrix}_{i/i-1} + \begin{bmatrix} \mathbf{K}_c \\ \mathbf{K}_d \end{bmatrix}_i (\mathbf{y}_i - \hat{\mathbf{y}}_i)$$

where $\mathbf{K}_c$ and $\mathbf{K}_d$ are the Kalman gain for the given system model defined by $\mathbf{\Phi}^f$, $\mathbf{Q}^f$, $\mathbf{H}^f$, $\mathbf{R}^f$, and $\mathbf{P}_{0/0}$. It is further assumed that $\hat{\mathbf{x}}_c$ has the same dimension and state definitions as $\mathbf{x}_c$, but no relationships between $\mathbf{\Phi}_{cc}^f$ and $\mathbf{\Phi}_{cc}$, or $\mathbf{H}_{cc}^f$ and $\mathbf{H}_{cc}$, are assumed. There are also no assumptions about relationships between $\hat{\mathbf{x}}_d$ and $\mathbf{x}_e$: they may have different dimensions and state definitions. This allows for the possibility that $\hat{\mathbf{x}}_d$ is a colored noise and bias model of a larger model based on states in $\mathbf{x}_e$.

The two separate models are concatenated and "common" states are differenced to give an augmented state vector represented as $\mathbf{z}$. The dynamic model for $\mathbf{z}$ is

$$\mathbf{z}_{i/i-1} = \begin{bmatrix} \mathbf{x}_c - \hat{\mathbf{x}}_c \\ \hat{\mathbf{x}}_d \\ \mathbf{x}_c \\ \mathbf{x}_e \end{bmatrix}_{i/i-1} = \begin{bmatrix} \mathbf{\Phi}_{cc}^f & -\mathbf{\Phi}_{cd}^f & \mathbf{\Phi}_{cc} - \mathbf{\Phi}_{cc}^f & \mathbf{\Phi}_{ce} \\ \mathbf{0} & \mathbf{\Phi}_{dd}^f & \mathbf{0} & \mathbf{0} \\ \mathbf{0} & \mathbf{0} & \mathbf{\Phi}_{cc} & \mathbf{\Phi}_{ce} \\ \mathbf{0} & \mathbf{0} & \mathbf{0} & \mathbf{\Phi}_{ee} \end{bmatrix} \begin{bmatrix} \mathbf{x}_c - \hat{\mathbf{x}}_c \\ \hat{\mathbf{x}}_d \\ \mathbf{x}_c \\ \mathbf{x}_e \end{bmatrix}_{i-1/i-1} + \begin{bmatrix} \mathbf{q}_c \\ \mathbf{0} \\ \mathbf{q}_c \\ \mathbf{q}_e \end{bmatrix}_i \tag{9.3-29}$$
$$= \mathbf{A}_i \mathbf{z}_{i-1/i-1} + \mathbf{w}_i$$

where

$$
\mathbf{A}_i = \begin{bmatrix} \boldsymbol{\Phi}_{cc}^f & -\boldsymbol{\Phi}_{cd}^f & \boldsymbol{\Phi}_{cc} - \boldsymbol{\Phi}_{cc}^f & \boldsymbol{\Phi}_{ce} \\ \mathbf{0} & \boldsymbol{\Phi}_{dd}^f & \mathbf{0} & \mathbf{0} \\ \mathbf{0} & \mathbf{0} & \boldsymbol{\Phi}_{cc} & \boldsymbol{\Phi}_{ce} \\ \mathbf{0} & \mathbf{0} & \mathbf{0} & \boldsymbol{\Phi}_{ee} \end{bmatrix}.
$$
$$
\mathbf{w}_i = \begin{bmatrix} \mathbf{q}_c \\ \mathbf{0} \\ \mathbf{q}_c \\ \mathbf{q}_e \end{bmatrix}_i, \quad \mathbf{Q}_w = E(\mathbf{w}_i \mathbf{w}_i^T) = \begin{bmatrix} \mathbf{Q}_{cc} & \mathbf{0} & \mathbf{Q}_{cc} & \mathbf{Q}_{ce} \\ \mathbf{0} & \mathbf{0} & \mathbf{0} & \mathbf{0} \\ \mathbf{Q}_{cc} & \mathbf{0} & \mathbf{Q}_{cc} & \mathbf{Q}_{ce} \\ \mathbf{Q}_{ce}^T & \mathbf{0} & \mathbf{Q}_{ce}^T & \mathbf{Q}_{ee} \end{bmatrix} \tag{9.3-30}
$$

Notice that process noise does not drive the $\hat{\mathbf{x}}_c$ or $\hat{\mathbf{x}}_d$ states since the filter state time update does not include a random noise term. (It is of course modeled in the covariance time update used in computing the Kalman gain.) If we treat the initial states of the true model ($\mathbf{x}_{c\_0}$ and $\mathbf{x}_{e\_0}$) as fixed-but-unknown (nonrandom) parameters, the expected value of $\mathbf{z}_{0/0}$ is

$$
E(\mathbf{z}_{0/0}) = \bar{\mathbf{z}}_{0/0} = \begin{bmatrix} \mathbf{x}_{c\_0} - \mathbf{x}_{c\_0} \\ \bar{\mathbf{x}}_{d\_0} \\ \mathbf{x}_{c\_0} \\ \mathbf{x}_{e\_0} \end{bmatrix} = \begin{bmatrix} \mathbf{0} \\ \bar{\mathbf{x}}_{d\_0} \\ \mathbf{x}_{c\_0} \\ \mathbf{x}_{e\_0} \end{bmatrix}.
$$

Since $\mathbf{x}_{d\_0}$ is not required to represent physical variables, a "true" value of $\mathbf{x}_{d\_0}$ does not really exist. However, we can assume that $\bar{\mathbf{x}}_{d\_0}$ represents a reasonable guess for the mean value of those states. Then $E(\mathbf{z}_{i/i-1}) = \bar{\mathbf{z}}_{i/i-1}$ can be computed by propagating $\bar{\mathbf{z}}_{0/0}$ without the effects of process noise or measurement noise. Thus the covariance of $\mathbf{z}_{i/i-1}$ is

$$
\begin{aligned} \mathbf{P}_{i/i-1}^z &= E[(\mathbf{z}_{i/i-1} - \bar{\mathbf{z}}_{i/i-1})(\mathbf{z}_{i/i-1} - \bar{\mathbf{z}}_{i/i-1})^T] \\ &= \mathbf{A}\mathbf{P}_{i-1/i-1}^z \mathbf{A}^T + \mathbf{Q}_i^w \end{aligned} \tag{9.3-31}
$$

initialized with

$$
\mathbf{P}_{0/0}^z = \begin{bmatrix} \mathbf{P}_{cc\_0/0} & \mathbf{0} & \mathbf{0} & \mathbf{0} \\ \mathbf{0} & \mathbf{P}_{dd\_0/0} & \mathbf{0} & \mathbf{0} \\ \mathbf{0} & \mathbf{0} & \mathbf{0} & \mathbf{0} \\ \mathbf{0} & \mathbf{0} & \mathbf{0} & \mathbf{0} \end{bmatrix}. \tag{9.3-32}
$$

The arrays $\mathbf{P}_{cc\_0/0}$ and $\mathbf{P}_{dd\_0/0}$ should normally be equal to the state covariances used to initialize the filter. Again notice that $\mathbf{P}_{dd\_0/0}$ does not really exist since $\hat{\mathbf{x}}_d$ is not a physical state, but $\mathbf{P}_{dd\_0/0}$ must be specified to initialize the Kalman filter, so one hopes that it reasonably characterizes the effective errors in the filter model.

The detailed measurement model is represented using state vector $\mathbf{z}_{i/i-1}$ as

$$\mathbf{y}_i = [\mathbf{0} \quad \mathbf{0} \quad \mathbf{H}_c \quad \mathbf{H}_e] \begin{bmatrix} \mathbf{x}_c - \hat{\mathbf{x}}_c \\ \hat{\mathbf{x}}_d \\ \mathbf{x}_c \\ \mathbf{x}_e \end{bmatrix}_{i/i-1} + \mathbf{r}_i$$

and the filter measurement model is

$$\hat{\mathbf{y}}_i = \begin{bmatrix} -\mathbf{H}_c^f & \mathbf{H}_d^f & \mathbf{H}_c^f & \mathbf{0} \end{bmatrix} \begin{bmatrix} \mathbf{x}_c - \hat{\mathbf{x}}_c \\ \hat{\mathbf{x}}_d \\ \mathbf{x}_c \\ \mathbf{x}_e \end{bmatrix}_{i/i-1},$$

so the measurement residual used in the Kalman filter can be written as

$$\tilde{\mathbf{y}}_i = \mathbf{y}_i - \hat{\mathbf{y}}_i = \begin{bmatrix} \mathbf{H}_c^f & -\mathbf{H}_d^f & \mathbf{H}_c - \mathbf{H}_c^f & \mathbf{H}_e \end{bmatrix} \begin{bmatrix} \mathbf{x}_c - \hat{\mathbf{x}}_c \\ \hat{\mathbf{x}}_d \\ \mathbf{x}_c \\ \mathbf{x}_e \end{bmatrix}_{i/i-1} + \mathbf{r}_i. \tag{9.3-33}$$

Now using $\tilde{\mathbf{y}}_i$ in the Kalman filter update equation (8.1-11) yields the *a posteriori* state $\mathbf{z}_{i/i}$:

$$\begin{bmatrix} \mathbf{x}_c - \hat{\mathbf{x}}_c \\ \hat{\mathbf{x}}_d \\ \mathbf{x}_c \\ \mathbf{x}_e \end{bmatrix}_{i/i} = \begin{bmatrix} \mathbf{x}_c - \hat{\mathbf{x}}_c \\ \hat{\mathbf{x}}_d \\ \mathbf{x}_c \\ \mathbf{x}_e \end{bmatrix}_{i/i-1} + \begin{bmatrix} -\mathbf{K}_c \\ \mathbf{K}_d \\ \mathbf{0} \\ \mathbf{0} \end{bmatrix}_i \begin{bmatrix} \mathbf{H}_c^f & -\mathbf{H}_d^f & \mathbf{H}_c - \mathbf{H}_c^f & \mathbf{H}_e \end{bmatrix} \begin{bmatrix} \mathbf{x}_c - \hat{\mathbf{x}}_c \\ \hat{\mathbf{x}}_d \\ \mathbf{x}_c \\ \mathbf{x}_e \end{bmatrix}_{i/i-1} + \begin{bmatrix} -\mathbf{K}_c \\ \mathbf{K}_d \\ \mathbf{0} \\ \mathbf{0} \end{bmatrix}_i \mathbf{r}_i$$

or

$$\mathbf{z}_{i/i} = \mathbf{D}_i \mathbf{z}_{i/i-1} + \mathbf{E}_i \mathbf{r}_i \tag{9.3-34}$$

where

$$\boxed{\mathbf{D}_i = \begin{bmatrix} \mathbf{I} - \mathbf{K}_c \mathbf{H}_c^f & \mathbf{K}_c \mathbf{H}_d^f & -\mathbf{K}_c(\mathbf{H}_c - \mathbf{H}_c^f) & -\mathbf{K}_c \mathbf{H}_e \\ \mathbf{K}_d \mathbf{H}_c^f & \mathbf{I} - \mathbf{K}_d \mathbf{H}_d^f & \mathbf{K}_d(\mathbf{H}_c - \mathbf{H}_c^f) & \mathbf{K}_d \mathbf{H}_e \\ \mathbf{0} & \mathbf{0} & \mathbf{I} & \mathbf{0} \\ \mathbf{0} & \mathbf{0} & \mathbf{0} & \mathbf{I} \end{bmatrix}, \quad \mathbf{E}_i = \begin{bmatrix} -\mathbf{K}_c \\ \mathbf{K}_d \\ \mathbf{0} \\ \mathbf{0} \end{bmatrix}_i.} \tag{9.3-35}$$

Thus the error covariance for $\mathbf{z}_{i/i}$ is

$$\boxed{\begin{aligned} \mathbf{P}_{i/i}^z &\triangleq E[(\mathbf{z}_{i/i} - \bar{\mathbf{z}}_{i/i})(\mathbf{z}_{i/i} - \bar{\mathbf{z}}_{i/i})^T]. \\ &= \mathbf{D}_i \mathbf{P}_{i/i-1}^z \mathbf{D}_i^T + \mathbf{E}_i \mathbf{R}_i \mathbf{E}_i^T \end{aligned}} \tag{9.3-36}$$

Equations (9.3-30) and (9.3-36) give the time and measurement updates of the $\mathbf{z}$ covariance. The upper left $n_f \times n_f$ partition gives the error covariance $E[(\mathbf{x}_c - \hat{\mathbf{x}}_c)(\mathbf{x}_c - \hat{\mathbf{x}}_c)^T]$ for the common states, which is the desired output. Since "true" values of $\hat{\mathbf{x}}_d$ do not exist, it not possible to compute an error covariance for these filter states.

9.4 MEASUREMENT PREPROCESSING

As noted in Chapter 8, hypothesis tests on filter innovations can be quite effective in identifying and removing outlying measurements. However, these residual tests are relatively insensitive during the initialization period when the filter gain is high. Another reason for concern is the IIR characteristic of the Kalman filter. Unedited anomalous measurements can cause significant errors in the filter estimates for an extended period of time. For these reasons it is important that an operational system edit measurements before processing with a Kalman filter.

The methods available for measurement preprocessing and editing are mostly the same as described in Section 7.5 for least squares. High-rate data can be averaged or smoothed, and measurement editing logic can be included as part of the averaging/smoothing operation. Alternately, batch least squares can be used to compute parameter estimates (e.g., polynomial coefficients) and covariances that model an assumed deterministic evolution of the measurements over a short time span. The parameter estimates and covariance are used as measurements in the Kalman filter. This is called a *mini-batch* approach.

A low-order filter operating on raw measurement data can detect jumps and either edit individual measurements or alert the Kalman filter that parameter values may have jumped (thus allowing the filter to switch to a high-Q model or otherwise compensate for the jump.) Known measurements errors can be removed or known transformations applied in the preprocessing. When multiple measurements provide some redundancy, measurements may be compared to determine if all are consistent. This method is used in GPS ground system preprocessing to detect unexpected satellite and ground clock timing errors.

9.5 SUMMARY

Most Kalman filtering applications are for nonlinear systems. The EKF handles the nonlinearities by linearizing about the current estimate at each measurement update. The EKF is by far the most commonly used nonlinear method, and it works well for most applications. However, the estimate may be biased because the nonlinearities can cause the conditional probability density to become nonsymmetric. The linearized Kalman filter linearizes the model about a reference trajectory that is not updated after each measurement. It is used in cases (such as orbit determination) when it is possible to compute an accurate reference trajectory over an extended period of time. The IEKF is an extension of the EKF that relinearizes and iterates the measurement update step when the measurement model is nonlinear. The iterated linear filter-smoother works similarly, but iterates on both the time update and measurement update steps when both models are nonlinear. These iterated filters are designed to compute the conditional mode rather than the conditional mean. They have the potential to work better than the EKF but there are few examples demonstrating improved performance. Also the iterations may not converge or may converge slowly. Other nonlinear techniques include higher order derivatives of the models when attempting to compute the mean of the conditional probability. Unfortunately these methods are only feasible for simple models and they seldom work well. Statistically linearized methods use describing functions to model the effective linear behavior of the model. Two other nonlinear filtering methods (described in Chapter 11) evaluate the model states at multiple points to better characterize the conditional probability density function.

Smoothing solutions compute the MMSE state at specific times using measurements both before and after the time of the smoothed state. Fixed-point smoothers only compute the state estimate at a fixed time. Several implementations are possible: the simplest augments the smoothed state (or a subset of states) with a copy of the current state at the desired smoothing time. Fixed-lag smoothers compute the smoothed state at a fixed time interval behind the measurements, and can operate as real-time processors. They are implemented either by augmenting and shifting the state vector, or using a chain of matrix multiplications over the lag time. These fixed-lag smoothers can be computationally expensive for real-time applications. Fixed-interval smoothers operate as post-processors and are implemented using a backward recursion to process outputs generated by the forward filter. The forward-backward filter due to Fraser and Potter is useful for theoretical purposes but is rarely implemented. The RTS and mBF methods are the most popular covariance-based fixed-interval smoothers. The state estimates and covariances (or equivalent information) from the forward filter are saved and used in a backward state/covariance recursion that is initialized with the final filter solution. The mBF uses adjoint variables in the backward recursion to avoid matrix inversions, but it is somewhat less flexible than the RTS. Both methods usually work well with high (but not excessive) computational burdens. An implementation of the RTS method was shown in Example 9.6 to work with disposable pass-dependent biases. Other types of smoothers will be discussed in the next chapter.

There are several reasons for analyzing the effects of various error sources on filter accuracy. Error analysis is very useful when designing filters to meet accuracy requirements under realistic operating conditions. A breakdown of the effects of the different error sources can help identify parameters to be adjusted in a ROM. This analysis is also helpful when attempting to design filters that are relatively insensitive to assumptions about model and filter parameter values. In some cases realistic error analysis is a required output when "proving" to a customer that the system meets accuracy specifications, or when providing input that is used for subsequent actions, such as control.

As with least-squares estimation, Monte Carlo and linear covariance error analysis can be used. Covariance error analysis can separately compute the effects of error sources and is computationally fast. Monte Carlo error analysis is more rigorous because it includes the effects of nonlinearities, can handle more complicated cases where there are significant structural differences between the filter model and a detailed "truth" model, and it is easily coded. Options for linear covariance error analysis include perturbation analysis of independent error sources (measurement noise, process noise, and errors in unadjusted or "considered" model parameters), analysis of ROM models defined as transformations on more accurate models, and analysis of truth and filter models that are structurally different but have a common subset of important states. In the latter case error propagation can be analyzed using an augmented state vector that includes state vectors for both models.

Section 9.3 also discussed methods for developing a ROM, and design considerations that impact selection of states. The goal of ROM development is usually to develop a high-accuracy filter that is insensitive to model parameter assumptions, or has greatly reduced computational requirements with minimal accuracy loss. Covariance error analysis can be invaluable when analyzing the trade-offs, but ultimately Monte Carlo analysis should be used to verify the performance for nonlinear systems.

CHAPTER 10

FACTORED (SQUARE-ROOT) FILTERING

From the first application of Kalman filtering—midcourse navigation and guidance for the Apollo moon mission (see McGee and Schmidt 1985)—engineers were concerned about the numerical accuracy of the Kalman filter covariance measurement update

$$\mathbf{P}_{i/i} = \mathbf{P}_{i/i-1} - \mathbf{P}_{i/i-1}\mathbf{H}_i^T(\mathbf{H}_i\mathbf{P}_{i/i-1}\mathbf{H}_i^T + \mathbf{R}_i)^{-1}\mathbf{H}_i\mathbf{P}_{i/i-1}$$

since it differences two positive definite arrays. This was a potential problem for the Apollo mission because the filter was implemented on a 16-bit fixed-point arithmetic flight computer. This project provided the incentive for Potter to develop the first square-root implementation of the Kalman filter (see Battin 1964), although this version could not handle process noise. Within the next few years other square-root variants or extensions were developed by Bellantoni and Dodge (1967), Andrews (1968), Dyer and McReynolds (1969), Schmidt (1970), Carlson (1973), Bierman (1974, 1977b), and others. In the same time period investigators compared the numerical accuracy and execution speed of the various algorithms with the standard and Joseph forms of the Kalman equations. Results of some studies were published (Gura and Bierman 1971; Kaminski et al. 1971; Bierman and Thornton 1977; Verhaegen and Van Dooren 1986) while other studies remained as internal white papers. To this author's knowledge the results were fairly consistent: the covariance forms of the Kalman filter measurement update—equations (8.1-12) and (8.1-13)—are not reliable when implemented in single precision, and any of the square-root algorithms implemented in single precision provide nearly the same accuracy as the covariance forms implemented in double precision. Bierman and Thornton's U-D algorithm was found to be as fast or faster than other square-root algorithms, and relatively easy to implement. In fact, the measurement update portion of the algorithm is not only comparable to the short form of the Kalman filter equation (8.1-13) in speed, but in some cases it is faster. Another square-root algorithm, the *square-root information filter* (SRIF), has advantages for certain types of problems, but is often slower than the U-D filter and somewhat more difficult to use.

Advanced Kalman Filtering, Least-Squares and Modeling: A Practical Handbook, by Bruce P. Gibbs

It should be noted that square-root filtering algorithms are generally defined by the method used to implement the measurement update, with several options available for the time update of each algorithm.

Even though improvements have been made in other square-root algorithms, the U-D filter is still the dominant type of factored filter algorithm, with the SRIF a good alternate for some applications. Three factors were responsible for the popularity of the U-D filter: results of study comparisons, the book and papers written by Bierman, and the Fortran Estimation Subroutine Package (ESP) created by Bierman and Nead (1978) that the NASA/Jet Propulsion Laboratory provided for free to interested parties. Later Bierman formed his own company (Factorized Estimation Applications Inc.) and created a commercial version of the estimation subroutines—called the Estimation Subroutine Library (ESL; Bierman and Bierman 1984)—that is currently available at http://netlib.org/a/esl.tgz. The ESP/ESL subroutine that implements the Householder transformations (THHC) was used in the least-squares examples of Chapters 5 and 7. This particular implementation is equivalent to the Modified Gram-Schmidt (MGS) algorithm.

This chapter begins with a summary of Kalman filter numerical issues, and then introduces the U-D filter. The measurement update is defined, and several options for the time update are then described. Bierman's U-D version of the Rauch-Tung-Striebel (RTS) smoother is also presented. The SRIF, *square-root information smoother* (SRIS), and *Dyer-McReynolds covariance smoother* (DMCS) are next discussed. Since the SRIF operates in the information domain rather than the covariance domain, the properties are very different from those of the U-D filter. The SRIF is derived using the "data equation," which is a useful concept for some problems and is also a better starting point for developing other algorithms. The SRIF preserves much of the sparseness present in measurement partial derivative matrices, and this is a great benefit for some problems. An example shows how the SRIF/SRIS can be used to apply statistical continuity constraints to a static three-dimensional (3-D) geophysical/hydrogeological modeling problem defined as finite-difference flow equations on a grid of many thousand nodes. Methods for factorized error analysis and SRIF estimation for large sparse systems are also discussed in this chapter.

10.1 FILTER NUMERICAL ACCURACY

A simple example demonstrates the loss of numerical accuracy in the measurement covariance update of the Kalman filter.

Example 10.1: Numerical Accuracy of One-State Kalman Filter

Consider the filter problem with a scalar state x_i and measurement

$$y_i = x_i + r_i, \quad P_{i/i-1} = 1, \quad R = \sigma_r^2 \triangleq E[r_i^2].$$

The Kalman gain for this model is $K_i = P_{i/i-1}/(P_{i/i-1} + R) = 1/(1 + R)$, and the covariance measurement update is

$$\begin{aligned} P_{i/i} &= P_{i/i-1} - K_i P_{i/i-1} \\ &= P_{i/i-1} - P_{i/i-1}^2 / (P_{i/i-1} + R). \\ &= 1 - 1/(1+R) \end{aligned}$$

If this is implemented in single precision on a PC with optimization disabled (to prevent execution entirely in internal 80-bit floating-point stack registers), any value of $\sigma_r < 2.44 \times 10^{-4}$ will result in $P_{i/i} = 0$. This implies that all precision is lost when the measurement noise standard deviation is less than approximately 2.44×10^{-4} times the *a priori* state standard deviation. Even values of σ_r near $0.01\sqrt{P_{i/i-1}}$ can result in a relative loss of accuracy in the fourth digit.

Notice that the Joseph form of the covariance update only partially solves the numerical problem because it still forms $1 - KH$:

$$\begin{aligned} P_{i/i} &= (1 - K_i)^2 P_{i/i-1} + K_i^2 R \\ &= (1 - 1/(1+R))^2 P_{i/i-1} + R/(1+R)^2 \end{aligned}.$$

For this particular scalar model the Joseph form is fairly accurate because the $R/(1 + R)^2$ term restores most of the error in the $(1 - 1/(1 + R))^2 P_{i/i-1}$ term. However, for larger state sizes the extra computations of the Joseph form result in an accumulation of numerical errors. For some problems the Joseph measurement update can actually have larger errors than the "short" Kalman form.

When using multi-state filters where correlations between states are significant, even larger values of σ_r can cause numerical precision loss. The problem becomes worse when additional measurements are processed because numerical errors accumulate. One example that has appeared in the literature is a two-state filter with a scalar measurement. The prior covariance and measurement equation are

$$\mathbf{P}_{i/i-1} = \begin{bmatrix} 1 & 1-\varepsilon \\ 1-\varepsilon & 1 \end{bmatrix}, \quad y_i = [1 \quad 1] \begin{bmatrix} x_1 \\ x_2 \end{bmatrix}_i + r_i, \quad R = \sigma_r^2$$

where ε may vary from a "small" positive number (e.g., 0.0001) to 1.0. When ε is small the two states are highly correlated—a common situation. However, the numerical deterioration of this example is essentially similar to that of our one-state example.

The goal of the "square-root" filter algorithms is to either avoid the subtraction in the above covariance update—at least for the diagonal terms—or to only difference square roots of variables so that the precision loss is much smaller. In the U-D algorithm the D term is equal to the variance P of our scalar example, and the U-D update directly computes

$$P_{i/i} = P_{i/i-1}R/(P_{i/i-1} + R).$$

There is still a small loss of accuracy due to the fact that $P_{i/i-1} + R$ is numerically equal to $P_{i/i-1}$ for "small" R, but the relative error of this effect is small. True "square-root" algorithms compute the square root of $P_{i/i}$, while the U-D filter avoids the square roots by saving squared elements separately from the off-diagonal factors of $P_{i/i}$.

Numerical problems with the conventional Kalman measurement update may occur when the covariance matrix is rapidly reduced by processing very accurate measurements, or when there is a great variation in observability of different linear combinations of variables (i.e., when the range of covariance eigenvalues is great). However, Bierman and Thornton (1977) show that numerical errors can also occur for benign problems when working in single precision. It is somewhat unusual to have numerical problems when implementing "covariance" Kalman filters in double precision. This author routinely implements filters in both the standard covariance form (because it is often easier to validate these implementations) and as U-D or SRIF filters. Differences in results have only been observed for two applications, and these cases were unusually stressful in that measurements were much more accurate than the prior estimate. As Bierman and Thornton note, one does not know in advance when numerical problems will occur, so it is always best to use a factored filter.

Finally we note that it is also possible to have numerical problems with factorized filters. As shown in Example 7.1, attempts to enforce constraints in least-squares estimators by using very high measurement weighting can result in numerical problems—even when using the Householder or MGS algorithms. Similar problems can occur in factorized filters because they also use Householder or MGS transformations, although the conditions under which these problems occur are unusual. This is another reason for implementing filters using more than one algorithm and cross-checking the results.

10.2 U-D FILTER

Bierman and Thornton's U-D filter is called a factored filter rather than a "square-root" filter because square-root operations are not required. It factors the covariance as

$$\mathbf{P} = \mathbf{U}\mathbf{D}\mathbf{U}^T \qquad (10.2\text{-}1)$$

where $\mathbf{U}$ is a unit upper triangular matrix with 1's along the diagonal and $\mathbf{D}$ is a diagonal matrix. To conserve storage the $\mathbf{D}$ elements are stored on the diagonals of $\mathbf{U}$ with all elements stored in a vector, that is,

$$\begin{bmatrix} D_1 & U_{12} & U_{13} & \cdots & U_{1n} \\ 0 & D_2 & U_{23} & \cdots & U_{2n} \\ 0 & 0 & D_3 & \cdots & U_{3n} \\ \vdots & \vdots & \vdots & \ddots & \vdots \\ 0 & 0 & 0 & \cdots & D_n \end{bmatrix}$$

is stored as vector

$$[D_1 \quad U_{12} \quad D_2 \quad U_{13} \quad U_{23} \quad D_3 \quad U_{14} \quad \cdots \quad D_n]^T.$$

A 3 × 3 positive semi-definite symmetric covariance matrix is thus factored as

$$\begin{bmatrix} P_{11} & P_{12} & P_{13} \\ P_{12} & P_{22} & P_{23} \\ P_{13} & P_{23} & P_{33} \end{bmatrix} = \begin{bmatrix} 1 & U_{12} & U_{13} \\ 0 & 1 & U_{23} \\ 0 & 0 & 1 \end{bmatrix} \begin{bmatrix} D_1 & 0 & 0 \\ 0 & D_2 & 0 \\ 0 & 0 & D_3 \end{bmatrix} \begin{bmatrix} 1 & 0 & 0 \\ U_{12} & 1 & 0 \\ U_{13} & U_{23} & 1 \end{bmatrix}$$
$$= \begin{bmatrix} D_1 & U_{12}D_2 & U_{13}D_3 \\ 0 & D_2 & U_{23}D_3 \\ 0 & 0 & D_3 \end{bmatrix} \begin{bmatrix} 1 & 0 & 0 \\ U_{12} & 1 & 0 \\ U_{13} & U_{23} & 1 \end{bmatrix}$$
$$= \begin{bmatrix} D_1 + U_{12}^2 D_2 + U_{13}^2 D_3 & U_{12}D_2 + U_{13}U_{23}D_3 & U_{13}D_3 \\ \sim & D_2 + U_{23}^2 D_3 & U_{23}D_3 \\ \sim & \sim & D_3 \end{bmatrix}$$

where the symmetric lower elements are not shown. Hence

$$\begin{aligned} &D_3 = P_{33}, \quad U_{23} = P_{23} / D_3, \quad U_{13} = P_{13} / D_3 \\ &D_2 = P_{22} - U_{23}^2 D_3, \quad U_{12} = (P_{12} - U_{13}U_{23}D_3) / D_2. \\ &D_1 = P_{11} - U_{12}^2 D_2 - U_{13}^2 D_3 \end{aligned}$$

The COV2UD subroutine in Bierman's ESP/ESL computes the U-D factors for a positive definite symmetric matrix. The code computes the U-D factors "in place" where it is assumed that upper triangular **P** matrix has been loaded into the **U** array before calling COV2UD. Then COV2UD overwrites the computed U-D factors into the **U** array. The algorithm is

$$\boxed{\begin{array}{ll} U_{11} = P_{11} & \text{(implicitly set)} \\ \text{for } j = n \text{ to 2 by } -1 & \text{(column loop)} \\ \quad \text{if } \mathrm{U}_{jj} > 0,\ \{\alpha = 1/\mathrm{U}_{jj}\}, \text{else } \{\alpha = 0\} & \\ \quad \text{for } k = 1 \text{ to } j-1 & \text{(row loop)} \\ \quad\quad \beta = \mathrm{U}_{kj} & \\ \quad\quad \mathrm{U}_{kj} \Leftarrow \alpha \mathrm{U}_{kj} & \text{(scale elements in } j \text{ column)} \\ \quad\quad \mathrm{U}_{1:k,k} \Leftarrow \mathrm{U}_{1:k,k} - \beta \mathrm{U}_{1:k,j} & \text{(modify column } k) \\ \quad \text{end } k \text{ loop} & \\ \text{end } j \text{ loop} & \end{array}} \quad . \qquad (10.2\text{-}2)$$

(The symbol ⇐ means replace.) Other useful ESP/ESL conversion routines are UD2COV and U2SIG, which convert a U-D matrix to a covariance or just the standard deviations.

As with the Kalman filter, the U-D filter consists of separate time and measurement updates. We first discuss the measurement update because there are several options for the time update.

10.2.1 U-D Filter Measurement Update

The U-D filter measurement update is equivalent to the measurement update of the Kalman filter listed in Chapter 9:

$$
\begin{aligned}
\mathbf{K}_i &= \mathbf{P}_{i/i-1}\mathbf{H}_i^T\left(\mathbf{H}_i\mathbf{P}_{i/i-1}\mathbf{H}_i^T+\mathbf{R}_i\right)^{-1}\\
\hat{\mathbf{x}}_{i/i} &= \hat{\mathbf{x}}_{i/i-1}+\mathbf{K}_i(\mathbf{y}_i-\mathbf{H}_i\hat{\mathbf{x}}_{i/i-1})\\
\mathbf{P}_{i/i} &= \mathbf{P}_{i/i-1}-\mathbf{K}_i\mathbf{H}_i\mathbf{P}_{i/i-1}\\
&= \mathbf{P}_{i/i-1}-\mathbf{P}_{i/i-1}\mathbf{H}_i^T(\mathbf{H}_i\mathbf{P}_{i/i-1}\mathbf{H}_i^T+\mathbf{R}_i)^{-1}\mathbf{H}_i\mathbf{P}_{i/i-1}
\end{aligned} \tag{10.2-3}
$$

The U-D filter combines measurement information with the *a priori* **U** matrix using rank-1 updates, so the measurements must be processed as scalars. If multiple measurements are available at a given time, the vector measurement $\mathbf{y}_i = \mathbf{H}_i\mathbf{x}_i + \mathbf{r}_i$, or

$$
\begin{bmatrix} y_1\\ y_2\\ \vdots\\ y_m \end{bmatrix}_i = \begin{bmatrix} \mathbf{h}_1\\ \mathbf{h}_2\\ \vdots\\ \mathbf{h}_m \end{bmatrix}_i \mathbf{x}_i + \begin{bmatrix} r_1\\ r_2\\ \vdots\\ r_m \end{bmatrix}_i
$$

where each $\mathbf{h}_j$ is a row vector, must be processed separately and sequentially as $y_j = \mathbf{h}_j\mathbf{x}_i + r_j$ for $j = 1, 2,\ldots, m$. For this to work, the individual measurements must be uncorrelated, so if $E[\mathbf{r}_i\mathbf{r}_i^T]=\mathbf{R}_i$ is not diagonal, it should be de-correlated using the method of the previous chapters; that is, compute $\mathbf{R}_i=\mathbf{R}_i^{1/2}\mathbf{R}_i^{T/2}$ (for any factorization) and then transform the measurement as $\mathbf{z}_i=\mathbf{R}_i^{-1/2}\mathbf{y}_i=(\mathbf{R}_i^{-1/2}\mathbf{H}_i)\mathbf{x}_i+(\mathbf{R}_i^{-1/2}\mathbf{r}_i)$ such that $E[\mathbf{R}_i^{-1/2}\mathbf{r}_i\mathbf{r}_i^T\mathbf{R}_i^{-T/2}]=\mathbf{I}$. Alternately the factorization $\mathbf{R}_i=\mathbf{U}_i\mathbf{D}_i\mathbf{U}_i^T$ can be used to yield $E[\mathbf{U}_i^{-1}\mathbf{r}_i\mathbf{r}_i^T\mathbf{U}_i^{-T}]=\mathbf{D}_i$. If the measurement model is nonlinear, the transformation is applied to measurement residuals, $\mathbf{z}_i=\mathbf{R}_i^{-1/2}(\mathbf{y}_i-\mathbf{H}_i\mathbf{x}_i)$, rather than directly to the measurements.

The remainder of this section assumes that only a single scalar measurement is processed. When vector measurements are available, the processing is repeated for each element. The Kalman measurement update for the covariance can be implemented in U-D form as

$$
\begin{aligned}
\mathbf{U}_{i/i}\mathbf{D}_{i/i}\mathbf{U}_{i/i}^T &= \mathbf{U}_{i/i-1}\mathbf{D}_{i/i-1}\mathbf{U}_{i/i-1}^T\\
&\quad -\mathbf{U}_{i/i-1}\mathbf{D}_{i/i-1}\mathbf{U}_{i/i-1}^T\mathbf{h}_i^T(\mathbf{h}_i\mathbf{U}_{i/i-1}\mathbf{D}_{i/i-1}\mathbf{U}_{i/i-1}^T\mathbf{h}_i^T+R_i)^{-1}\mathbf{h}_i\mathbf{U}_{i/i-1}\mathbf{D}_{i/i-1}\mathbf{U}_{i/i-1}^T.
\end{aligned} \tag{10.2-4}
$$

Defining

$$
\boxed{\begin{aligned}
\mathbf{f} &\triangleq \mathbf{U}_{i/i-1}^T\mathbf{h}_i^T\\
\mathbf{g} &\triangleq \mathbf{D}_{i/i-1}\mathbf{f}\\
\alpha &\triangleq \mathbf{h}_i\mathbf{P}_{i/i-1}\mathbf{h}_i^T+R_i\\
&= \mathbf{f}^T\mathbf{g}+R_i
\end{aligned}}, \tag{10.2-5}
$$

the update can be written as

$$
\mathbf{U}_{i/i}\mathbf{D}_{i/i}\mathbf{U}_{i/i}^T=\mathbf{U}_{i/i-1}\left(\mathbf{D}_{i/i-1}-\mathbf{g}\mathbf{g}^T/\alpha\right)\mathbf{U}_{i/i-1}^T. \tag{10.2-6}
$$

The U-D measurement update is implemented by factoring the middle term as

$$\boxed{\bar{\mathbf{U}}\bar{\mathbf{D}}\bar{\mathbf{U}}^T \triangleq \mathbf{D}_{i/i-1} - \mathbf{g}\mathbf{g}^T/\alpha.} \tag{10.2-7}$$

(The over bar denotes an intermediate result, not a mean value.) Then the result is computed as

$$\boxed{\mathbf{U}_{i/i}\mathbf{D}_{i/i}\mathbf{U}_{i/i}^T = (\mathbf{U}_{i/i-1}\bar{\mathbf{U}})\bar{\mathbf{D}}(\bar{\mathbf{U}}^T\mathbf{U}_{i/i-1}^T)} \tag{10.2-8}$$

or

$$\boxed{\begin{aligned}\mathbf{D}_{i/i} &= \bar{\mathbf{D}}\\ \mathbf{U}_{i/i} &= \mathbf{U}_{i/i-1}\bar{\mathbf{U}}\end{aligned}\;.} \tag{10.2-9}$$

Because of the special structure of the $\mathbf{U}$ arrays, $\mathbf{U}_{i/i-1}\bar{\mathbf{U}}$ is also unit upper triangular and can be computed very efficiently. The vector Kalman gain needed for the state update in equation (10.2-3) is computed as

$$\boxed{\mathbf{k} = \mathbf{U}_{i/i-1}\mathbf{g}/\alpha.} \tag{10.2-10}$$

The primary "trick" of the update involves the method used to perform the rank-1 update equation (10.2-7). The method is a numerically more stable modification of an algorithm due to Agee and Turner (1972). To understand the problem, consider the 3 × 3 case

$$\begin{aligned}&\begin{bmatrix} D_1 - g_1^2/\alpha & -g_1g_2/\alpha & -g_1g_3/\alpha \\ -g_1g_2/\alpha & D_2 - g_2^2/\alpha & -g_2g_3/\alpha \\ -g_1g_3/\alpha & -g_2g_3/\alpha & D_3 - g_3^2/\alpha \end{bmatrix}\\
&\quad = \begin{bmatrix} 1 & \bar{U}_{12} & \bar{U}_{13} \\ 0 & 1 & \bar{U}_{23} \\ 0 & 0 & 1 \end{bmatrix}\begin{bmatrix} \bar{D}_1 & 0 & 0 \\ 0 & \bar{D}_2 & 0 \\ 0 & 0 & \bar{D}_3 \end{bmatrix}\begin{bmatrix} 1 & 0 & 0 \\ \bar{U}_{12} & 1 & 0 \\ \bar{U}_{13} & \bar{U}_{23} & 1 \end{bmatrix}\\
&\quad = \begin{bmatrix} \bar{D}_1 + \bar{U}_{12}^2\bar{D}_2 + \bar{U}_{13}^2\bar{D}_3 & \bar{U}_{12}\bar{D}_2 + \bar{U}_{13}\bar{U}_{23}\bar{D}_3 & \bar{U}_{13}\bar{D}_3 \\ \sim & \bar{D}_2 + \bar{U}_{23}^2\bar{D}_3 & \bar{U}_{23}\bar{D}_3 \\ \sim & \sim & \bar{D}_3 \end{bmatrix}\end{aligned}$$

Hence the factors are

$$\begin{aligned}&\bar{D}_3 = D_3 - g_3^2/\alpha, \quad \bar{U}_{23} = -g_2g_3/(\alpha\bar{D}_3), \quad \bar{U}_{13} = -g_1g_3/(\alpha\bar{D}_3)\\ &\bar{D}_2 = D_2 - g_2^2/\alpha - \bar{U}_{23}^2\bar{D}_3, \quad \bar{U}_{12} = -(g_1g_2/\alpha + \bar{U}_{13}\bar{U}_{23}\bar{D}_3)/\bar{D}_2\;.\\ &\bar{D}_1 = D_1 - g_1^2/\alpha - (\bar{U}_{12}^2\bar{D}_2 + \bar{U}_{13}^2\bar{D}_3)\end{aligned}$$

Notice that the diagonal $\bar{D}$ elements are computed as the difference of two or more positive numbers. Because this differencing can lead to a loss of precision, the algorithm is modified. Recalling that $g_j = D_jf_j$, it is observed that

$$\left.\begin{aligned}\bar{D}_j &= D_j(1 + c_jg_jf_j)\\ 1/c_{j-1} &= 1/c_j + g_jf_j\end{aligned}\right\} \quad j = n, n-1, \ldots, 1 \tag{10.2-11}$$

where $1/c_n = -\alpha$. Although this recursion does not directly eliminate the subtraction (notice that $c_n < 0$), it can be rewritten as a forward recursion that avoids differencing for diagonals (see Bierman 1977b, section V.3 for the details.) The final steps of the measurement update—including the multiplication $\mathbf{U}_{i/i} = \mathbf{U}_{i/i-1}\bar{\mathbf{U}}$—are computed as

$$
\boxed{
\begin{aligned}
&\gamma_1 = R + g_1 f_1 \\
&\bar{D}_1 = D_1 R / \gamma_1 \\
&\mathbf{k} = [g_1 \quad 0 \quad \ldots \quad 0]^T \\
&\text{for } j = 2 \text{ to } n \qquad\qquad \text{(column loop)} \\
&\quad \gamma_j = \gamma_{j-1} + g_j f_j \\
&\quad \bar{D}_j = D_j \gamma_{j-1} / \gamma_j \\
&\quad \left.\begin{aligned} &\bar{\mathbf{u}}_j = \mathbf{u}_j - (f_j / \gamma_{j-1})\mathbf{k} \\ &\mathbf{k} \Leftarrow \mathbf{k} + g_j \mathbf{u}_j \end{aligned}\right\} \text{(only compute vector elements } 1{:}j-1) \\
&\text{end loop}
\end{aligned}
}
\tag{10.2-12}
$$

where

$$\mathbf{U} = [\mathbf{u}_1 \quad \mathbf{u}_2 \quad \ldots \quad \mathbf{u}_n], \quad \bar{\mathbf{U}} = [\bar{\mathbf{u}}_1 \quad \bar{\mathbf{u}}_2 \quad \ldots \quad \bar{\mathbf{u}}_n]$$

and the columns $\mathbf{u}_j$ and $\bar{\mathbf{u}}_j$ only have $j - 1$ nonzero elements above the unit diagonal. The loops can be arranged so that $\bar{\mathbf{U}}$ and $\mathbf{U}$ occupy the same storage, and $\mathbf{g}$ and $\mathbf{k}$ occupy the same storage. At loop completion, $\mathbf{k} = \mathbf{U}_{i/i-1}\mathbf{g}$, so the Kalman gain can be computed as

$$\boxed{\mathbf{k} \Leftarrow \mathbf{k} / \alpha.} \tag{10.2-13}$$

Because of the manner in which $\mathbf{k}$ is updated, the final $\bar{\mathbf{U}}$ is equal to $\mathbf{U}_{i/i} = \mathbf{U}_{i/i-1}\bar{\mathbf{U}}$, so no further processing is needed. The simplicity and efficiency of this algorithm is impressive.

These equations are implemented in the ESP/ESL subroutine UDMES. The ESP UDMES does not directly allow for data outlier hypothesis testing, but the ESL version does. In the ESL UDMEAS, equations (10.2-5) are computed first and then

$$\eta = \tilde{y} / \sqrt{\alpha}$$

is formed. This should be $N(0,1)$ if all models are correct and the measurement noise is Gaussian. When η^2 is greater than a specified input editing threshold (e.g., 5^2), equation (10.2-12) is bypassed to edit the measurement.

10.2.2 U-D Filter Time Update

There are at least three methods for computing the U-D time update equivalent to the covariance update

$$\mathbf{P}_{i/i-1} = \mathbf{\Phi} \mathbf{P}_{i-1/i-1} \mathbf{\Phi}^T + \mathbf{Q}. \tag{10.2-14}$$

(We have eliminated subscripts from $\mathbf{\Phi}$ and $\mathbf{Q}$ to simplify notation, but it is recognized that they are often time-dependent.) The options are:

1. Form an augmented rectangular array using $\mathbf{\Phi U}_{i-1/i-1}$ and $\mathbf{Q}^{1/2}$, and transform it to upper triangular using *modified weighted Gram-Schmidt* (MWGS) orthogonalization.
2. Compute $\mathbf{\Phi U}_{i-1/i-1}$, use Householder transformations to re-triangularize, and incorporate of $\mathbf{Q}$ as a sequence of rank-1 updates.
3. Compute $\mathbf{\Phi U}_{i-1/i-1}$ for the dynamic states only, form $(\mathbf{\Phi U}_{i-1/i-1})\mathbf{D}_{i-1/i-1}(\mathbf{U}_{i-1/i-1}^T\mathbf{\Phi}^T)+\mathbf{Q}$ for the dynamic states, and re-factor as $\mathbf{U}_{i/i-1}\mathbf{D}_{i/i-1}\mathbf{U}_{i/i-1}^T$.

10.2.2.1 MWGS The most general method for computing the U-D filter covariance time update is due to Thornton (Thornton and Bierman 1975; Thornton 1976). It is related to a method originally developed by Schmidt to work with the Potter square-root covariance filter. Schmidt discovered that equation (10.2-14) can be written as

$$\begin{aligned}\mathbf{P}_{i/i-1} &= [\mathbf{\Phi L}_{i-1/i-1} \quad \mathbf{G}]\begin{bmatrix}\mathbf{L}_{i-1/i-1}^T\mathbf{\Phi}^T\\ \mathbf{G}^T\end{bmatrix}\begin{matrix}\}n\times n\\ \}p\times n\end{matrix}\\ &= [\mathbf{\Phi L}_{i-1/i-1} \quad \mathbf{G}]\mathbf{T}^T\mathbf{T}\begin{bmatrix}\mathbf{L}_{i-1/i-1}^T\mathbf{\Phi}^T\\ \mathbf{G}^T\end{bmatrix}\\ &= [\mathbf{L}_{i/i-1} \quad \mathbf{0}]\begin{bmatrix}\mathbf{L}_{i/i-1}^T\\ \mathbf{0}\end{bmatrix}\end{aligned} \tag{10.2-15}$$

where

$\mathbf{P}_{i-1/i-1} = \mathbf{L}_{i-1/i-1}\mathbf{L}_{i-1/i-1}^T$ is $n \times n$,
$\mathbf{L}$ is the $n \times n$ lower triangular Cholesky factor of $\mathbf{P}$,
$\mathbf{Q} = \mathbf{G}\mathbf{G}^T$ is the $n \times n$ state noise covariance,
$\mathbf{G}$ is $n \times p$ where $p \le n$, and
$\mathbf{T}$ is an $(n + p) \times (n + p)$ orthogonal matrix ($\mathbf{T}^T\mathbf{T} = \mathbf{I}$).

Matrix $\mathbf{G}$ can either be obtained directly from the model, or by Cholesky (or U-D) factorization of a $p \times p$ copy of $\mathbf{Q}$ obtained by deleting zero rows and columns. Matrix $\mathbf{T}$ is computed as a sequence of orthogonal transformations (usually MGS) to zero-out the lower left block of

$$\begin{bmatrix}\mathbf{L}_{i-1/i-1}^T\mathbf{\Phi}^T\\ \mathbf{G}^T\end{bmatrix}.$$

The method used to compute the individual transformations is the same as described in Chapter 5 when discussing the QR algorithm. The covariance factors are never squared, so numerical precision of the factors is retained.

The presence of a diagonal matrix in the U-D factorization of

$$\boxed{\mathbf{P}_{i/i-1} = [\mathbf{\Phi U}_{i-1/i-1} \quad \mathbf{G}]\begin{bmatrix}\mathbf{D}_{i-1/i-1} & \mathbf{0}\\ \mathbf{0} & \mathbf{D}_q\end{bmatrix}\begin{bmatrix}\mathbf{U}_{i/i-1}^T\mathbf{\Phi}^T\\ \mathbf{G}^T\end{bmatrix}\begin{matrix}\}n\times n\\ \}p\times n\end{matrix}} \tag{10.2-16}$$

(where $\mathbf{Q} = \mathbf{G}\mathbf{D}_q\mathbf{G}^T$) complicates the triangularization process. One approach forms

$$\begin{bmatrix}\mathbf{\Phi U}_{i-1/i-1}\mathbf{D}_{i-1/i-1}^{1/2} & \mathbf{G}\mathbf{D}_q^{1/2}\end{bmatrix},$$

uses post-multiplying orthogonal transformations to make the above array upper triangular, and then scales the columns by their root sum-of-squares to make the array unit upper triangular. The column sums-of-squares form the output diagonal matrix. However, Thornton discovered a better method. The algorithm uses MWGS orthogonalization to compute a $n \times (n+p)$ matrix $\mathbf{V}$ such that

$$[\mathbf{\Phi U}_{i-1/i-1} \quad \mathbf{G}] = \mathbf{U}_{i/i-1}\mathbf{V} \tag{10.2-17}$$

and

$$\mathbf{V}\begin{bmatrix}\mathbf{D}_{i-1/i-1} & \mathbf{0} \\ \mathbf{0} & \mathbf{D}_q\end{bmatrix}\mathbf{V}^T = \mathbf{D}_{i/i-1} \tag{10.2-18}$$

where $\mathbf{U}_{i/i-1}$ is $n \times n$ unit upper triangular and $\mathbf{D}_{i/i-1}$ is $n \times n$ diagonal. Equation (10.2-18) explains why matrix $\mathbf{V}$ is considered D-orthogonal. Since the Gram-Schmidt algorithm sequentially operates on previously formed orthogonal row vectors to form each new orthogonal row vector in $\mathbf{V}$, the vector weighting can be expressed as multiplication by a triangular matrix, which is the desired result ($\mathbf{U}_{i/i-1}$).

The MWGS algorithm is implemented in ESP subroutine WGS, or ESL routine WGSG. The inputs to the subroutine are

$$\boxed{\begin{aligned}\mathbf{W} &= [\mathbf{\Phi U}_{i-1/i-1} \quad \mathbf{G}]\ \}n\times(n+p)\\ \mathbf{D}_w &= \begin{bmatrix}\mathbf{D}_{i-1/i-1} & \mathbf{0} \\ \mathbf{0} & \mathbf{D}_q\end{bmatrix}\Big\}(n+p)\times(n+p)\end{aligned}}, \tag{10.2-19}$$

where the diagonals of $\mathbf{D}_w$ are stored as a vector. The output $\mathbf{U}_{i/i-1}$ and $\mathbf{D}_{i/i-1}$ satisfy

$$\boxed{\begin{aligned}\mathbf{W}\mathbf{D}_w\mathbf{W}^T &= \mathbf{U}_{i/i-1}\mathbf{V}\mathbf{D}_w\mathbf{V}^T\mathbf{U}_{i/i-1}^T\\ &= \mathbf{U}_{i/i-1}\mathbf{D}_{i/i-1}\mathbf{U}_{i/i-1}^T\end{aligned}}. \tag{10.2-20}$$

It should be noted that the execution speed of these routines can be improved significantly when $\mathbf{W}$ is sparse (as it usually is) by saving the indices of the nonzero elements for each row as it is processed. This modification is included in the WGS_M subroutine associated with this book.

Since the number of flops required by the MWGS method is proportional to $(n+p)n(n+1)$, it is very inefficient to compute $\mathbf{T}$ for the full $n \times (n+p)$ $\mathbf{W}$ matrix. If the state vector includes many biases such that

$$\boxed{\mathbf{P}_{i/i-1} = \begin{bmatrix}\mathbf{\Phi}_{dd} & \mathbf{\Phi}_{db} \\ \mathbf{0} & \mathbf{I}\end{bmatrix}\begin{bmatrix}\mathbf{P}_{dd} & \mathbf{P}_{db} \\ \mathbf{P}_{db}^T & \mathbf{P}_{bb}\end{bmatrix}_{i-1/i-1}\begin{bmatrix}\mathbf{\Phi}_{dd}^T & \mathbf{0} \\ \mathbf{\Phi}_{db}^T & \mathbf{I}\end{bmatrix} + \begin{bmatrix}\mathbf{Q}_{dd} & \mathbf{0} \\ \mathbf{0} & \mathbf{0}\end{bmatrix}}, \tag{10.2-21}$$

it is easily shown that

$$\boxed{\begin{aligned}\mathbf{U}_{bb\_i/i-1} &= \mathbf{U}_{bb\_i-1/i-1}\\ \mathbf{D}_{b\_i/i-1} &= \mathbf{D}_{b\_i-1/i-1}\\ \mathbf{U}_{db\_i/i-1} &= \mathbf{\Phi}_{dd}\mathbf{U}_{db\_i-1/i-1} + \mathbf{\Phi}_{db}\mathbf{U}_{bb\_i-1/i-1}\end{aligned}}. \tag{10.2-22}$$

Hence the MWGS method should only be applied to the dynamic portion

$$\mathbf{U}_{dd\_i/i-1}\mathbf{D}_{dd\_i/i-1}\mathbf{U}^T_{dd\_i/i-1} = [\mathbf{\Phi}_{dd}\mathbf{U}_{dd} \quad \mathbf{G}]_{i-1/i-1}\begin{bmatrix}\mathbf{D}_{dd} & \mathbf{0}\\ \mathbf{0} & \mathbf{D}_q\end{bmatrix}_{i-1/i-1}\begin{bmatrix}\mathbf{U}^T_{dd}\mathbf{\Phi}^T_{dd}\\ \mathbf{G}^T\end{bmatrix}_{i-1/i-1} \tag{10.2-23}$$

where $\mathbf{\Phi}_{dd}$, $\mathbf{U}_{dd}$, $\mathbf{D}_{dd}$ are $n_d \times n_d$ and $\mathbf{G}$ is $n_d \times p$ where $p \le n_d$.

10.2.2.2 Rank-1 Updates In cases where the number of bias and colored noise (Markov) states are significant, Bierman (1977b, section VII.4) suggests that the time update take advantage of the assumed structure

$$\begin{bmatrix}\mathbf{x}_d\\ \mathbf{x}_m\\ \mathbf{x}_b\end{bmatrix}_i = \begin{bmatrix}\mathbf{\Phi}_{dd} & \mathbf{\Phi}_{dm} & \mathbf{\Phi}_{db}\\ \mathbf{0} & \mathbf{\Phi}_{mm} & \mathbf{0}\\ \mathbf{0} & \mathbf{0} & \mathbf{I}\end{bmatrix}\begin{bmatrix}\mathbf{x}_d\\ \mathbf{x}_m\\ \mathbf{x}_b\end{bmatrix}_{i-1} + \begin{bmatrix}\mathbf{q}_d\\ \mathbf{q}_m\\ \mathbf{0}\end{bmatrix}_{i,i-1} \tag{10.2-24}$$

where $\mathbf{\Phi}_{mm}$ is the diagonal block for first-order Markov process (colored noise) states. In Bierman's book $\mathbf{q}_d = \mathbf{0}$, which only allows process noise to impact the dynamic states $\mathbf{x}_d$ through $\mathbf{\Phi}_{dm}$ for the next time step. This gives the impression that the algorithm does not work when process noise directly affects $\mathbf{x}_d$. Later writings by Bierman include the $\mathbf{q}_d$ term, but examples often ignore it because presence of $\mathbf{q}_d$ can significantly increase computations.

Bierman suggests that the deterministic time update (forming $\mathbf{\Phi}\mathbf{U}_{i-1/i-1}$ and refactoring the "*dd*" partition) be performed first without using the $\mathbf{\Phi}_{mm}$ term. Then the stochastic update (inclusion of process noise) is implemented either using one-component-at-a-time processing with the Agee-Turner algorithm, or using the MWGS method on the full matrix as in the previous section. The Agee-Turner method is sometimes more efficient, but the difference is usually not significant unless the number of Markov process states is large.

There are several objections to this method when implemented assuming that $\mathbf{q}_d = \mathbf{0}$. First, the model is overly restrictive because it cannot handle systems where process noise directly drives the dynamic states or where $\mathbf{x}_m$ includes second-order Markov process states. The lack of a $\mathbf{q}_d$ term driving $\mathbf{x}_m$ implies that the measurement update interval must be much smaller than the time constants in $\mathbf{\Phi}_{dd}$, which implies that this method cannot handle data gaps or moderately long measurement intervals without artificially forcing multiple time updates at short time intervals. Another disadvantage is the coding effort required to implement processing in partitions. Finally, it has been this author's experience that when a filter implementation takes advantage of model structure, someone (often you) invariably discovers that it is necessary to change the model. The lack of flexibility in this approach is a serious objection. However, the general approach does work when $\mathbf{q}_d \ne \mathbf{0}$. For simplicity in equations to follow, it is assumed that Markov process states are included in $\mathbf{x}_d$, so the state vector just consists of $\mathbf{x}_d$ and $\mathbf{x}_b$.

The system is partitioned into dynamic and biases states as shown in equation (10.2-21), and the stochastic update is implemented using one-component-at-a-time processing. The deterministic update is first performed by computing

$$\mathbf{A} = \mathbf{\Phi}\mathbf{U}_{i-1/i-1} \tag{10.2-25}$$

and then using MWGS to compute $\mathbf{U}^0, \mathbf{D}^0$ such that

$$\boxed{\mathbf{U}^0\mathbf{D}^0(\mathbf{U}^0)^T = \mathbf{P}^0 = \mathbf{A}\mathbf{D}_{i-1/i-1}\mathbf{A}^T.} \tag{10.2-26}$$

The stochastic update is implemented for each of the p nonzero diagonal elements in $\mathbf{D}_q$ of the model

$$\mathbf{Q}_{dd} = \mathbf{G}\mathbf{D}_q\mathbf{G}^T = [\mathbf{g}_1 \quad \mathbf{g}_2 \quad \cdots \quad \mathbf{g}_p]\begin{bmatrix} D_{q1} & 0 & \cdots & 0 \\ 0 & D_{q2} & \cdots & 0 \\ \vdots & \vdots & \ddots & \vdots \\ 0 & 0 & \cdots & D_{qp} \end{bmatrix}\begin{bmatrix} \mathbf{g}_1^T \\ \mathbf{g}_2^T \\ \vdots \\ \mathbf{g}_p^T \end{bmatrix} \tag{10.2-27}$$

where the $\mathbf{g}_i$ are column vectors. Since typically $p < n$, $\mathbf{G}$ is a rectangular matrix. The $\mathbf{G}$ and $\mathbf{D}_q$ arrays can be obtained by U-D factoring $\mathbf{Q}_{dd}$ and then eliminating zero columns.

The stochastic update is implemented by executing

$$\mathbf{P}^j = \mathbf{P}^{j-1} + D_{qj}\mathbf{g}_j\mathbf{g}_j^T \tag{10.2-28}$$

or

$$\boxed{\mathbf{U}^j\mathbf{D}^j(\mathbf{U}^j)^T = \mathbf{U}^{j-1}\mathbf{D}^{j-1}(\mathbf{U}^{j-1})^T + D_{qj}\mathbf{g}_j\mathbf{g}_j^T} \tag{10.2-29}$$

for $j = 1, 2, \ldots, p$. The $\mathbf{U}^j$, $\mathbf{D}^j$ arrays are computed using the Agee-Turner rank-1 update algorithm for the general equation

$$\bar{\mathbf{U}}\bar{\mathbf{D}}\bar{\mathbf{U}}^T = \mathbf{U}\mathbf{D}\mathbf{U}^T + c\mathbf{v}\mathbf{v}^T$$

where $\bar{\mathbf{U}} = \mathbf{U}^j, \bar{\mathbf{D}} = \mathbf{D}^j, \mathbf{U} = \mathbf{U}^{j-1}, \mathbf{D} = \mathbf{D}^{j-1}, c = D_{qj}$, and $\mathbf{v} = \mathbf{g}_j$. The rank-1 update is implemented in the ESP/ESL RANK1 subroutine as:

$$\boxed{\begin{array}{ll} b_n = c & \\ \text{for } j = n \text{ to } 2 \text{ by } -1 & \text{(column loop)} \\ \quad \bar{D}_j = D_j + b_j v_j^2 & \\ \left.\begin{array}{l} \quad \mathbf{v} \Leftarrow \mathbf{v} - v_j \mathbf{u}_j \\ \quad \bar{\mathbf{u}}_j = \mathbf{u}_j + (b_j v_j / \bar{D}_j)\mathbf{v} \end{array}\right\} & \text{(only compute elements } 1 : j-1) \\ \quad b_{j-1} = (D_j / \bar{D}_j) b_j & \\ \text{end loop} & \\ \bar{D}_1 = D_1 + b_1 v_1^2 & \end{array}} \tag{10.2-30}$$

where v_j is the j-th element of $\mathbf{v}$. As with the U-D measurement update, the vectors are defined as

$$\mathbf{U} = [\mathbf{u}_1 \quad \mathbf{u}_2 \quad \ldots \quad \mathbf{u}_n], \quad \bar{\mathbf{U}} = [\bar{\mathbf{u}}_1 \quad \bar{\mathbf{u}}_2 \quad \ldots \quad \bar{\mathbf{u}}_n]$$

and the columns $\mathbf{u}_j$ and $\bar{\mathbf{u}}_j$ only have $j - 1$ nonzero elements above the unit diagonal. The loops are arranged so that $\bar{\mathbf{U}}$ and $\mathbf{U}$ can occupy the same storage. To minimize numerical problems the logic is slightly different when $\bar{D}_j > 16D_j$.

For mostly full $\mathbf{G}$ arrays this sequence of rank-1 updates is comparable in speed to the MWGS method, particularly if MWGS is implemented as a sparse matrix algorithm. However, when **G** is sparse—as when explicitly using model equation (10.2-24) with $\mathbf{q}_d = \mathbf{0}$ and diagonal elements of $\mathbf{\Phi}_{mm}$ scale the **U** rows—the rank-1 approach can be faster than MWGS.

10.2.2.3 Square and Re-Factor Another alternative for the U-D time update explicitly forms the covariance matrix for the dynamic states as

$$\mathbf{P}_{dd} = (\mathbf{\Phi}_{dd}\mathbf{U}_{dd})\mathbf{D}_{dd}(\mathbf{\Phi}_{dd}\mathbf{U}_{dd})^T + \mathbf{Q}_{dd} \tag{10.2-31}$$

and then factors $\mathbf{P}_{dd}$ as $\mathbf{U}_{dd\_i/i-1}\mathbf{D}_{dd\_i/i-1}\mathbf{U}^T_{dd\_i/i-1}$ The dynamic-bias terms are obtained from equation (10.2-22). Since most of the precision loss in the Kalman equations occurs in the measurement update, this squaring and factoring approach is usually accurate when the number of dynamic states is smaller than the number of biases. (The addition of $\mathbf{Q}_{dd}$ tends to compensate for numerical errors.) This method is generally faster than the MWGS method because factoring requires less computation than MWGS. Bierman did not approve of this approach, but it is often used successfully.

10.2.3 RTS Smoother for U-D Filter

In response to complaints that the SRIF/SRIS was too slow for some filter/smoothing problems (e.g., Gibbs 1979), Bierman developed two different versions of RTS smoothers that work with output from the U-D filter (Bierman 1983, 1988). We summarize the first method because the second is more complicated and requires extended discussion. The advantages of the second method are described at the end of this section, but conditions under which the first method has problems are rarely encountered. Bierman intended that the smoother would be used with the colored noise model of equation (10.2-24) where the stochastic update was implemented using one-component-at-a-time processing. However, the approach also works (although less efficiently) for a more general model with one-component-at-a-time stochastic update processing.

The assumed filter time update model is

$$\mathbf{P}_{i/i-1} = (\mathbf{\Phi}\mathbf{U}_{i-1/i-1})\mathbf{D}_{i-1/i-1}(\mathbf{\Phi}\mathbf{U}_{i-1/i-1})^T + \sum_{j=1}^{p} D_{qj}\mathbf{g}_j\mathbf{g}_j^T \tag{10.2-32}$$

where

$$\mathbf{Q}_{dd} = \sum_{j=1}^{p} D_{qj}\mathbf{g}_j\mathbf{g}_j^T$$

is defined by equation (10.2-27). Recall from Chapter 9 that the RTS backward recursion is

$$\begin{aligned}\hat{\mathbf{x}}_{i/N} &= \hat{\mathbf{x}}_{i/i} + \mathbf{G}_i(\hat{\mathbf{x}}_{i+1/N} - \hat{\mathbf{x}}_{i+1/i})\\ \mathbf{P}_{i/N} &= \mathbf{P}_{i/N} + \mathbf{G}_i(\mathbf{P}_{i+1/N} - \mathbf{P}_{i+1/i})\mathbf{G}_i^T\end{aligned} \tag{10.2-33}$$

where the gain matrix can be expressed as

$$\begin{aligned}\mathbf{G}_i &= \mathbf{P}_{i/i}\mathbf{\Phi}^T\mathbf{P}_{i+1/i}^{-1} \\ &= \mathbf{\Phi}^{-1}(\mathbf{I}-\mathbf{Q}\mathbf{P}_{i+1/i}^{-1}) \\ &= \mathbf{\Phi}^{-1}\left(\mathbf{I}-\sum_{j=1}^{p} D_{qj}\mathbf{g}_j(\mathbf{g}_j^T\mathbf{P}_{i+1/i}^{-1})\right)\end{aligned} \quad . \tag{10.2-34}$$

Rather than process the backward update in a single step, the U-D smoother assumes that the filter performs the time update using one deterministic update followed by a sequence of rank-1 stochastic updates. Then the smoother executes p stochastic updates (executed in reverse order from the filter) followed by one deterministic update. To avoid confusion in notation, we drop time indices in the following steps and just use the stochastic update index j as a superscript. We retain j as a subscript on D_{qj} and $\mathbf{g}_j$ since they are an element/vector of an array. Since each of the individual filter stochastic updates is modeled as equation (10.2-29), the smoother gain corresponding to that update is

$$\boxed{\begin{aligned}\mathbf{G}^j &= \mathbf{P}^{j-1}\left(\mathbf{P}^{j-1}+D_{qj}\mathbf{g}_j\mathbf{g}_j^T\right)^{-1} \\ &= \mathbf{P}^{j-1}\left[(\mathbf{P}^{j-1})^{-1}-(\mathbf{P}^{j-1})^{-1}\mathbf{g}_j\left(\mathbf{g}_j^T(\mathbf{P}^{j-1})^{-1}\mathbf{g}_j+D_{qj}\right)^{-1}\mathbf{g}_j^T(\mathbf{P}^{j-1})^{-1}\right] \\ &= \mathbf{I}-\lambda^j\mathbf{g}_j(\mathbf{v}^j)^T\end{aligned}} \tag{10.2-35}$$

where

$$\boxed{\begin{aligned}\mathbf{v}^j &= (\mathbf{P}^{j-1})^{-1}\mathbf{g}_j \\ \lambda^j &= 1/\left(\mathbf{g}_j^T\mathbf{v}^j+D_{qj}\right)\end{aligned}} . \tag{10.2-36}$$

(The second step in equation (10.2-35) uses the matrix inversion identity from Appendix B.) Vector $\mathbf{v}^j$ can be computed efficiently when using the U-D filter by back-solving

$$\mathbf{U}^{j-1}\mathbf{a}=\mathbf{g}_j, \qquad (\mathbf{U}^{j-1})^T\mathbf{v}^j=(\mathbf{D}^{j-1})^{-1}\mathbf{a} \tag{10.2-37}$$

in two steps taking advantage of the triangular structure of $\mathbf{U}^{j-1}$. This can be implemented in either the filter or the smoother depending upon the arrays that are saved to disk by the filter. The solution for $\mathbf{v}^j$ is one of the more computationally expensive steps of the smoother for the general problem. Bierman notes that the computations are significantly reduced by using the colored noise model equation (10.2-24) when $\mathbf{\Phi}_{mm}$ is diagonal.

The stochastic updates of smoother state and covariance at time t_{i+1} are initialized with

$$\boxed{\begin{aligned}\bar{\mathbf{x}}^p &= \hat{\mathbf{x}}_{i+1/N} \\ \bar{\mathbf{P}}^p &= \mathbf{P}_{i+1/N}\end{aligned}} ,$$

and for $j = p, p-1, \ldots, 1$ compute

$$\boxed{\begin{aligned}\bar{\mathbf{x}}^{j-1} &= \hat{\mathbf{x}}^{j-1}+\mathbf{G}^j(\bar{\mathbf{x}}^j-\hat{\mathbf{x}}^j) \\ &= \hat{\mathbf{x}}^{j-1}+\left(\mathbf{I}-\lambda^j\mathbf{g}_j(\mathbf{v}^j)^T\right)(\bar{\mathbf{x}}^j-\hat{\mathbf{x}}^j) \\ &= \hat{\mathbf{x}}^j-\lambda^j\mathbf{g}_j(\mathbf{v}^j)^T(\bar{\mathbf{x}}^j-\hat{\mathbf{x}}^j) \\ &= \hat{\mathbf{x}}^j-\mathbf{g}_j\left(\lambda^j(\mathbf{v}^j)^T(\bar{\mathbf{x}}^j-\hat{\mathbf{x}}^j)\right)\end{aligned}} \tag{10.2-38}$$

$$\begin{aligned}\bar{\mathbf{P}}^{j-1} &= \mathbf{P}^{j-1} + \mathbf{G}^j(\bar{\mathbf{P}}^j - \mathbf{P}^j)(\mathbf{G}^j)^T \\ &= \mathbf{G}^j\bar{\mathbf{P}}^j(\mathbf{G}^j)^T + \left(\mathbf{P}^{j-1} - \mathbf{P}^{j-1}(\mathbf{P}^j)^{-1}\mathbf{P}^j(\mathbf{G}^j)^T\right) \\ &= \mathbf{G}^j\bar{\mathbf{P}}^j(\mathbf{G}^j)^T + \mathbf{P}^{j-1}\lambda^j\mathbf{v}^j\mathbf{g}_j^T \\ &= \mathbf{G}^j\bar{\mathbf{P}}^j(\mathbf{G}^j)^T + \lambda^j\mathbf{g}_j\mathbf{g}_j^T\end{aligned} \tag{10.2-39}$$

where $\hat{\mathbf{x}}^j$ and $\mathbf{P}^j$ are outputs of the filter stochastic update j, and we have used $\mathbf{G}^j = \mathbf{P}^{j-1}(\mathbf{P}^j)^{-1}$ since $\mathbf{\Phi} = \mathbf{I}$ for the stochastic updates. Finally the deterministic update is computed using $\mathbf{G}^0 = \mathbf{\Phi}^{-1}$, which gives

$$\begin{aligned}\hat{\mathbf{x}}_{k/N} &= \hat{\mathbf{x}}_{k/k} + \mathbf{G}^0(\bar{\mathbf{x}}^0 - \hat{\mathbf{x}}^0) \\ &= \hat{\mathbf{x}}_{k/k} + \mathbf{\Phi}^{-1}(\bar{\mathbf{x}}^0 - \hat{\mathbf{x}}_{k+1/k}). \\ &= \mathbf{\Phi}^{-1}\bar{\mathbf{x}}^0\end{aligned} \tag{10.2-40}$$

In other words, $\hat{\mathbf{x}}_{k/N}$ just backward integrates $\bar{\mathbf{x}}^0$ which was obtained from $\hat{\mathbf{x}}_{k+/N}$ with the effects of process noise removed. The smoothed deterministic covariance update is

$$\begin{aligned}\mathbf{P}_{k/N} &= \mathbf{P}_{k/k} + \mathbf{G}^0(\bar{\mathbf{P}}^0 - \mathbf{P}^0)(\mathbf{G}^0)^T \\ &= \mathbf{\Phi}^{-1}\bar{\mathbf{P}}^0\mathbf{\Phi}^{-T} + \left(\mathbf{P}_{k/k} - \mathbf{\Phi}^{-1}\mathbf{P}^0\mathbf{\Phi}^{-T}\right). \\ &= \mathbf{\Phi}^{-1}\bar{\mathbf{P}}^0\mathbf{\Phi}^{-T}\end{aligned} \tag{10.2-41}$$

Notice that computation of $\mathbf{P}_{k/N}$ does not difference covariances, so numerical accuracy is improved compared with the normal RTS smoother. As with the RTS smoother, these iterations can lose accuracy if $\mathbf{\Phi}$ has a negligibly small value on the diagonal (e.g., because of data gaps causing propagation over a long time interval.) Bierman's use of the colored noise model equation (10.2-24) and associated changes to the smoothing algorithm solve this problem for the first-order Markov process states, but impose restrictions on the system structure. His second version of the RTS smoother (published in 1988 after his death) is better able to handle correlated process noise terms and ill-conditioned filter covariances, and does not difference state estimates. However, the implementation is more complex. Prvan and Osborne (1988) developed another variant of the RTS smoother that claims to better handle correlated process noise.

Finally we note that for the general case where states are only partitioned into dynamic and bias subsets—without using the colored noise model of equation (10.2-24)—the computational burden of this smoothing algorithm is comparable to that of the partitioned RTS solution using equation (9.2-20). This comparison assumes that both algorithms are implemented using a U-D filter and similar coding structures. The main advantage of Bierman's 1983 RTS smoother is the improved accuracy of the covariance calculation due to elimination of covariance differences.

10.2.4 U-D Error Analysis

The covariance error analysis for unadjusted bias parameters presented in Chapter 9 can also be used with the U-D filter. Notice that propagation of the bias sensitivity and covariance arrays ($\mathbf{S}^u$, $\mathbf{S}^0$, $\mathbf{P}^q$, $\mathbf{P}^r$) in Section 9.3.1 only depend on inputs to the filter ($\mathbf{\Phi}_{xx}$, $\mathbf{\Phi}_{xu}$, $\mathbf{Q}$, $\mathbf{H}_x$, $\mathbf{H}_u$, $\mathbf{R}$) and on the Kalman gain matrix $\mathbf{K}$. Since the U-D filter also computes **K** (for one-component-at-a-time processing), the equations listed in Section 9.3.1 can be directly applied using U-D outputs. Alternately, error propagation for arbitrary Kalman gains can be implemented using the U-D filter, as shown by Thornton (1976) and Thornton and Bierman (1978). A U-D implementation is generally not required for numerical accuracy because error analysis is rarely implemented in single precision.

Error analysis for Kalman-Schmidt "consider states" is a problem because the U-D measurement update assumes that the filter is optimal. Bierman and Vandergraft (1986, unpublished) developed an approach for making the U-D filter operate as a consider filter: the optimal measurement update of the U-D factors is first computed, and then the addition to reset the covariance for the consider parameters is incorporated as a series of rank-1 updates (similar to the rank-1 approach for process noise). The consider parameter "states" are also reset to the original values. Because there is limited need for a U-D consider filter, we omit the derivation and caution the reader that the consider parameter analysis of Section 9.3.1 cannot be used with the U-D filter.

Bierman (1977b, chapter 8) presents an extended discussion of error analysis, and points out that there are at least four categories of errors that can affect filter performance:

1. incorrect *a priori* covariance on the estimated states,
2. incorrect measurement noise covariance,
3. incorrect process noise covariance, and
4. unestimated bias or colored noise (Markov) states,

Bierman shows how the effects of these error sources can be analyzed: the discussion is general and does not explicitly depend on the U-D filter. The approach of our Section 9.3.1 primarily addresses the last category, but gives insight on the effects of errors in the *a priori* estimate and of measurement and process noise. As noted in our Chapter 9, incorrect model structure and incorrect model parameters (in $\mathbf{\Phi}$ and $\mathbf{H}$) are also potential sources of error that must be analyzed using other approaches.

10.3 SQUARE ROOT INFORMATION FILTER (SRIF)

While the U-D filter and previously mentioned square-root filters factor the state error covariance, the SRIF operates in the information domain. We have already used the measurement update step of the SRIF in Chapters 5, 6, and 7: it is the MGS version of the QR algorithm. Bierman recognized that this measurement update could be generalized in a *data equation* representation that would also apply to the time update of the Kalman filter. A data equation has the form

$$\mathbf{A}\mathbf{x} = \mathbf{y} + \mathbf{v} \tag{10.3-1}$$

where $\mathbf{y}$ is a vector of data (Bierman uses $\mathbf{z}$), $\mathbf{A}$ is a general matrix, $\mathbf{x}$ is a vector to be estimated and $\mathbf{v}$ is zero mean, unit covariance "noise." To use this concept in a filter, the variables to be estimated and the measurements are partitioned as

$$\begin{bmatrix} \mathbf{B} & \mathbf{C} \\ \mathbf{D} & \mathbf{E} \end{bmatrix} \begin{bmatrix} \mathbf{w} \\ \mathbf{x} \end{bmatrix} = \begin{bmatrix} \mathbf{z} \\ \mathbf{y} \end{bmatrix} + \begin{bmatrix} \mathbf{v}_z \\ \mathbf{v}_y \end{bmatrix}. \tag{10.3-2}$$

We will shortly define **w** as process noise, but for the moment that is irrelevant. As with the QR algorithm, we multiply both sides of the data equation by an orthogonal transformation

$$\mathbf{T}\begin{bmatrix} \mathbf{B} & \mathbf{C} \\ \mathbf{D} & \mathbf{E} \end{bmatrix} \begin{bmatrix} \mathbf{w} \\ \mathbf{x} \end{bmatrix} = \mathbf{T}\begin{bmatrix} \mathbf{z} \\ \mathbf{y} \end{bmatrix} + \mathbf{T}\begin{bmatrix} \mathbf{v}_z \\ \mathbf{v}_y \end{bmatrix} \tag{10.3-3}$$

where $\mathbf{T}\mathbf{T}^T = \mathbf{I}$ and

$$\mathbf{T}E\left(\begin{bmatrix} \mathbf{v}_z \\ \mathbf{v}_y \end{bmatrix} \begin{bmatrix} \mathbf{v}_z^T & \mathbf{v}_y^T \end{bmatrix}\right)\mathbf{T}^T = \mathbf{T}\mathbf{I}\mathbf{T}^T = \mathbf{I}.$$

The properties $E[\mathbf{v}\mathbf{v}^T] = \mathbf{I}$ and $\mathbf{T}E[\mathbf{v}\mathbf{v}^T]\mathbf{T}^T = \mathbf{I}$ are critical to the SRIF. We select **T** to make the transformed left-side matrix upper triangular; that is, equation (10.3-3) becomes

$$\begin{bmatrix} \mathbf{W} & \mathbf{V} \\ \mathbf{0} & \mathbf{R} \end{bmatrix} \begin{bmatrix} \mathbf{w} \\ \mathbf{x} \end{bmatrix} = \begin{bmatrix} \mathbf{z}_w \\ \mathbf{z}_x \end{bmatrix} + \begin{bmatrix} \mathbf{v}'_z \\ \mathbf{v}'_y \end{bmatrix} \tag{10.3-4}$$

where **W** and **R** are upper triangular. As with the QR method, matrix **T** is obtained using elementary Householder or Givens transformations, or MGS orthogonalization, but **T** is never explicitly computed. Then the minimum variance/minimum mean-squared error (MMSE) solution for **x** is obtained by solving

$$\boxed{\mathbf{R}\hat{\mathbf{x}} = \mathbf{z}_x} \tag{10.3-5}$$

and it is not necessary to explicitly solve for **w** since it is usually of little interest. **R** will be singular if insufficient measurement information has been obtained to make **x** observable. When **R** is nonsingular, $\hat{\mathbf{x}}$ is computed by back-solving equation (10.3-5), and the error covariance is

$$\boxed{\mathbf{P} = \mathbf{R}^{-1}\mathbf{R}^{-T}.} \tag{10.3-6}$$

The following descriptions of the SRIF and smoothers are based on the work of Bierman and others, but generally follow a particularly clear summary of Vandergraft (1991).

10.3.1 SRIF Time Update

As with the Kalman filter, the SRIF time update starts with an *a posteriori* state estimate $\hat{\mathbf{x}}_{i-1/i-1}$ at time t_{i-1}. For the SRIF that estimate is implicitly defined by

$$\mathbf{R}_{i-1/i-1}\hat{\mathbf{x}}_{i-1/i-1} = \mathbf{z}_{i-1/i-1} \tag{10.3-7}$$

where $\mathbf{R}_{i-1/i-1}$ is upper triangular such that the error in $\mathbf{z}_{i-1/i-1}$,

$$\begin{aligned} \mathbf{v}_{i-1}^x &\triangleq \mathbf{R}_{i-1/i-1}\mathbf{x}_{i-1} - \mathbf{z}_{i-1/i-1} \\ &= \mathbf{R}_{i-1/i-1}(\mathbf{x}_{i-1} - \hat{\mathbf{x}}_{i-1/i-1}) \end{aligned}, \tag{10.3-8}$$

is zero mean with unit covariance. Hence $\mathbf{R}_{i-1/i-1}^T\mathbf{R}_{i-1/i-1}$ is the information matrix for $\hat{\mathbf{x}}_{i-1/i-1}$ Provided that $\mathbf{R}_{i-1/i-1}$ is nonsingular, the estimate error covariance is

$$\mathbf{P}_{i-1/i-1} \triangleq E[(\mathbf{x}_{i-1} - \hat{\mathbf{x}}_{i-1/i-1})(\mathbf{x}_{i-1} - \hat{\mathbf{x}}_{i-1/i-1})^T] = \mathbf{R}_{i-1/i-1}^{-1}\mathbf{R}_{i-1/i-1}^{-T}, \tag{10.3-9}$$

but inversion of $\mathbf{R}_{i-1/i-1}$ is not required for the derivation. A data equation for $\mathbf{x}_{i-1}$ is obtained as

$$\begin{aligned}\mathbf{R}_{i-1/i-1}\mathbf{x}_{i-1} &= \mathbf{R}_{i-1/i-1}\hat{\mathbf{x}}_{i-1/i-1} + \mathbf{R}_{i-1/i-1}(\mathbf{x}_{i-1} - \hat{\mathbf{x}}_{i-1/i-1}) \\ &= \mathbf{z}_{i-1/i-1} + \mathbf{v}_{i-1}^x\end{aligned}. \tag{10.3-10}$$

The state time update model from equation (8.1-1) is

$$\mathbf{x}_i = \mathbf{\Phi}\mathbf{x}_{i-1} + \mathbf{G}\mathbf{q}_{i,i-1}, \quad E[\mathbf{q}_{i,i-1}\mathbf{q}_{i,i-1}^T] = \mathbf{I} \tag{10.3-11}$$

where we have inserted the $n \times p$ $\mathbf{G}$ matrix so that $\mathbf{q}_{i,i-1}$ has unit covariance and does not include zero elements. $\mathbf{G}$ may be obtained as $\mathbf{G} = \mathbf{U}\mathbf{D}^{1/2}$ for the U-D factorization $\mathbf{Q} = \mathbf{U}\mathbf{D}\mathbf{U}^T$, where zero columns of $\mathbf{G}$ are eliminated. Also the $\mathbf{u}_{i,i-1}$ "control" term has been dropped from equation (8.1-1) because it does not impact the information or covariance equations. Thus equation (10.3-11) may be written as

$$\mathbf{x}_{i-1} = \mathbf{\Phi}^{-1}\mathbf{x}_i - \mathbf{\Phi}^{-1}\mathbf{G}\mathbf{q}_{i,i-1}. \tag{10.3-12}$$

Multiplying by $\mathbf{R}_{i-1/i-1}$ and using equation (10.3-10) gives

$$\mathbf{R}_{i-1/i-1}\mathbf{\Phi}^{-1}\mathbf{x}_i - \mathbf{R}_{i-1/i-1}\mathbf{\Phi}^{-1}\mathbf{G}\mathbf{q}_{i,i-1} = \mathbf{z}_{i-1/i-1} + \mathbf{v}_{i-1}^x. \tag{10.3-13}$$

Since the mean value of $\mathbf{q}_{i,i-1}$ is zero, we can write another data equation

$$\mathbf{q}_{i,i-1} = \mathbf{0} + \mathbf{v}_{i-1}^q \tag{10.3-14}$$

where $E[\mathbf{v}_{i-1}^q(\mathbf{v}_{i-1}^q)^T] = \mathbf{I}$ Combining these two equations gives

$$\begin{bmatrix} \mathbf{I} & \mathbf{0} \\ -\mathbf{R}_{i-1/i-1}\mathbf{\Phi}^{-1}\mathbf{G} & \mathbf{R}_{i-1/i-1}\mathbf{\Phi}^{-1} \end{bmatrix}\begin{bmatrix} \mathbf{q}_{i,i-1} \\ \mathbf{x}_i \end{bmatrix} = \begin{bmatrix} \mathbf{0} \\ \mathbf{z}_{i-1/i-1} \end{bmatrix} + \begin{bmatrix} \mathbf{v}_{i-1}^q \\ \mathbf{v}_{i-1}^x \end{bmatrix} \tag{10.3-15}$$

which can be multiplied by an orthogonal transformation as

$$\boxed{\mathbf{T}\begin{bmatrix} \mathbf{I} & \mathbf{0} \\ -\mathbf{R}_{i-1/i-1}\mathbf{\Phi}^{-1}\mathbf{G} & \mathbf{R}_{i-1/i-1}\mathbf{\Phi}^{-1} \end{bmatrix}\begin{bmatrix} \mathbf{q}_{i,i-1} \\ \mathbf{x}_i \end{bmatrix} = \mathbf{T}\begin{bmatrix} \mathbf{0} \\ \mathbf{z}_{i-1/i-1} \end{bmatrix} + \mathbf{T}\begin{bmatrix} \mathbf{v}_{i-1}^q \\ \mathbf{v}_{i-1}^x \end{bmatrix}} \tag{10.3-16}$$

to obtain

$$\boxed{\begin{bmatrix} \mathbf{W}_{i/i-1} & \mathbf{V}_{i/i-1} \\ \mathbf{0} & \mathbf{R}_{i/i-1} \end{bmatrix}\begin{bmatrix} \mathbf{q}_{i,i-1} \\ \mathbf{x}_i \end{bmatrix} = \begin{bmatrix} \mathbf{z}_{i/i-1}^q \\ \mathbf{z}_{i/i-1}^x \end{bmatrix} + \begin{bmatrix} \mathbf{v}_{i-1}'^q \\ \mathbf{v}_{i-1}'^x \end{bmatrix}} \tag{10.3-17}$$

where $\mathbf{W}_{i/i-1}$ and $\mathbf{R}_{i/i-1}$ are upper triangular and

$$E\begin{bmatrix} \mathbf{v}_{i-1}'^q \\ \mathbf{v}_{i-1}'^x \end{bmatrix} = \mathbf{0}, \quad E\left(\begin{bmatrix} \mathbf{v}_{i-1}'^q \\ \mathbf{v}_{i-1}'^x \end{bmatrix}\begin{bmatrix} \mathbf{v}_{i-1}'^q \\ \mathbf{v}_{i-1}'^x \end{bmatrix}^T\right) = \mathbf{I}.$$

Provided that $\hat{\mathbf{x}}_{i-1/i-1}$ is MMSE and the system is linear, $\hat{\mathbf{x}}_{i/i-1}$ defined by

$$\mathbf{R}_{i/i-1}\hat{\mathbf{x}}_{i/i-1} = \mathbf{z}_{i/i-1}^x \tag{10.3-18}$$

is also MMSE. However, equation (10.3-18) is not used to compute $\hat{\mathbf{x}}_{i/i-1}$ since it is either defined by $\hat{\mathbf{x}}_{i/i-1} = \mathbf{\Phi}\hat{\mathbf{x}}_{i-1/i-1}$ for linear systems, or by integration of $\hat{\mathbf{x}}_{i-1/i-1}$ for nonlinear systems. Rather, equation (10.3-16) is used to define the time update to the square-root information matrix and vector. This is easily implemented by appending the right-side **z**-vector to the information array of the left side to create an augmented $(n + p) \times (n + p + 1)$ array that is multiplied by **T**; that is,

$$\mathbf{T}\begin{bmatrix} \mathbf{I} & \mathbf{0} & \mathbf{0} \\ -\mathbf{R}_{i-1/i-1}\mathbf{\Phi}^{-1}\mathbf{G} & \mathbf{R}_{i-1/i-1}\mathbf{\Phi}^{-1} & \mathbf{z}_{i-1/i-1} \end{bmatrix} = \begin{bmatrix} \mathbf{W}_{i/i-1} & \mathbf{V}_{i/i-1} & \mathbf{z}^{q}_{i/i-1} \\ \mathbf{0} & \mathbf{R}_{i/i-1} & \mathbf{z}^{x}_{i/i-1} \end{bmatrix}. \tag{10.3-19}$$

Triangularization of the rectangular matrix in equation (10.3-19) can be performed by ESL subroutine TDHHT. Bierman suggests that the SRIF time update be performed one-component-at-a-time (using the columns of **G**) because this is somewhat more efficient than a vector update, but vector processing can also be used.

The upper rows of equation (10.3-17) provide information on $\mathbf{q}_{i,i-1}$. Notice that an estimate of $\mathbf{q}_{i,i-1}$ can be obtained as

$$\hat{\mathbf{q}}_{(i,i-1)/i-1} = \mathbf{W}^{-1}_{i/i-1}(\mathbf{z}^{q}_{i/i-1} - \mathbf{V}_{i/i-1}\hat{\mathbf{x}}_{i/i-1}). \tag{10.3-20}$$

This relationship is used for computing smoothed estimates, so $\mathbf{W}_{i/i-1}$, $\mathbf{V}_{i/i-1}$, and $\mathbf{z}^{q}_{i/i-1}$ should be saved when smoothing is to be performed. Otherwise the upper rows of equation (10.3-17) can be discarded after computation.

10.3.2 SRIF Measurement Update

The SRIF measurement update combines *a priori* information represented by the data equation

$$\mathbf{R}_{i/i-1}\mathbf{x}_i = \mathbf{z}^{x}_{i/i-1} + \mathbf{v}'^{x}_{i-1} \tag{10.3-21}$$

with scaled measurement information represented by

$$\mathbf{H}_i\mathbf{x}_i = \mathbf{y}_i - \mathbf{r}_i, \tag{10.3-22}$$

where the measurement error $\mathbf{r}_i$ is assumed to be zero mean and unit covariance. If $E[\mathbf{r}_i\mathbf{r}_i^T] \neq \mathbf{I}$ for the unscaled measurement equation $\mathbf{y}_i = \mathbf{H}_i\mathbf{x}_i + \mathbf{r}_i$, then $\mathbf{y}_i$ and $\mathbf{H}_i$ should be multiplied by the inverse square root of $E[\mathbf{r}_i\mathbf{r}_i^T]$ before use in equation (10.3-22).

The prior information and measurement equations are combined in the single data equation

$$\begin{bmatrix} \mathbf{R}_{i/i-1} \\ \mathbf{H}_i \end{bmatrix}\mathbf{x}_i = \begin{bmatrix} \mathbf{z}^{x}_{i/i-1} \\ \mathbf{y}_i \end{bmatrix} + \begin{bmatrix} \mathbf{v}'^{x}_{i-1} \\ -\mathbf{r}_i \end{bmatrix}. \tag{10.3-23}$$

Multiplying by an orthogonal transformation,

$$\mathbf{T}\begin{bmatrix} \mathbf{R}_{i/i-1} \\ \mathbf{H}_i \end{bmatrix}\mathbf{x}_i = \mathbf{T}\begin{bmatrix} \mathbf{z}^{x}_{i/i-1} \\ \mathbf{y}_i \end{bmatrix} + \mathbf{T}\begin{bmatrix} \mathbf{v}'^{x}_{i-1} \\ -\mathbf{r}_i \end{bmatrix}, \tag{10.3-24}$$

the transformed data equation becomes

$$\begin{bmatrix} \mathbf{R}_{i/i} \\ \mathbf{0} \end{bmatrix}\mathbf{x}_i = \begin{bmatrix} \mathbf{z}^{x}_{i/i} \\ \mathbf{e}_i \end{bmatrix} + \begin{bmatrix} \mathbf{v}^{x}_{i} \\ \mathbf{v}^{y}_{i} \end{bmatrix} \tag{10.3-25}$$

where $\mathbf{R}_{i/i}$ is upper triangular. As with the QR algorithm, $\mathbf{T}$ is computed using elementary Householder transformations of the form $\mathbf{T} = \mathbf{I} - (2/\mathbf{u}^T\mathbf{u})\mathbf{u}\mathbf{u}^T$. Triangularization of the combined $\mathbf{R}_{i/i-1}$, $\mathbf{z}^x_{i/i-1}$ and $\mathbf{H}_i$, $\mathbf{y}_i$ to compute $\mathbf{R}_{i/i}$, $\mathbf{z}^x_{i/i}$ can be performed by the ESL subroutine THHC.

The *a posteriori* MMSE is computed as

$$\mathbf{R}_{i/i}\hat{\mathbf{x}}_{i/i} = \mathbf{z}^x_{i/i}. \tag{10.3-26}$$

Notice that the Bayesian cost function for this problem is

$$\begin{aligned} J &= \frac{1}{2}[(\mathbf{z}^x_{i/i-1} - \mathbf{R}_{i/i-1}\mathbf{x}_i)^T(\mathbf{z}^x_{i/i-1} - \mathbf{R}_{i/i-1}\mathbf{x}_i) + (\mathbf{y}_i - \mathbf{H}\mathbf{x}_i)^T(\mathbf{y}_i - \mathbf{H}\mathbf{x}_i)] \\ &= \frac{1}{2}[(\mathbf{z}^x_{i/i-1} - \mathbf{R}_{i/i-1}\mathbf{x}_i)^T \quad (\mathbf{y}_i - \mathbf{H}\mathbf{x}_i)^T]\mathbf{T}^T\mathbf{T}\begin{bmatrix} \mathbf{z}^x_{i/i-1} - \mathbf{R}_{i/i-1}\mathbf{x}_i \\ \mathbf{y}_i - \mathbf{H}\mathbf{x}_i \end{bmatrix} \\ &= \frac{1}{2}[(\mathbf{z}^x_{i/i} - \mathbf{R}_{i/i}\mathbf{x}_i)^T \quad \mathbf{e}_i^T]\begin{bmatrix} \mathbf{z}^x_{i/i} - \mathbf{R}_{i/i}\mathbf{x}_i \\ \mathbf{e}_i \end{bmatrix} \end{aligned} \tag{10.3-27}$$

Since $\mathbf{z}^x_{i/i-1} - \mathbf{R}_{i/i-1}\mathbf{x}_i$ is the only term that is a function of $\mathbf{x}_i$, the minimum J is obtained with $\hat{\mathbf{x}}_{i/i}$ defined by equation (10.3-26), which yields the *a posteriori* cost function

$$J = \frac{\mathbf{e}_i^T\mathbf{e}_i}{2}. \tag{10.3-28}$$

In other words, the *a posteriori* weighted residual sum-of-squares is $\mathbf{e}_i^T\mathbf{e}_i$

10.3.3 Square Root Information Smoother (SRIS)

The SRIS uses the discarded process noise information from the SRIF time update to compute the smoothed estimates. Bierman (1976) credits Kaminski (Kaminski and Bryson 1972) with the algorithm. Christensen (Christensen and Reinbold 1974) provided early insight on smoothing, and used the SRIS for processing Mariner 10 data. As with the RTS and mBF smoothers, the final filter estimate is the starting value for the backward recursions of the SRIS. The SRIS combines the upper data equation for process noise of the final time step, equation (10.3-17), with the data equation for the final measurement update, equation (10.3-25), as

$$\begin{bmatrix} \mathbf{W}_{N/N-1} & \mathbf{V}_{N/N-1} \\ \mathbf{0} & \mathbf{R}_{N/N} \end{bmatrix}\begin{bmatrix} \mathbf{q}_{N,N-1} \\ \mathbf{x}_N \end{bmatrix} = \begin{bmatrix} \mathbf{z}^q_{N/N-1} \\ \mathbf{z}^x_{N/N} \end{bmatrix} + \begin{bmatrix} \mathbf{v}'^q_{N-1} \\ \mathbf{v}^x_N \end{bmatrix}. \tag{10.3-29}$$

We substitute the time update model equation (10.3-11) for the final time,

$$\mathbf{x}_N = \mathbf{\Phi}_{N,N-1}\mathbf{x}_{N-1} + \mathbf{G}_{N,N-1}\mathbf{q}_{N,N-1},$$

and rearrange to obtain

$$\begin{bmatrix} \mathbf{W}_{N/N-1} + \mathbf{V}_{N/N-1}\mathbf{G}_{N,N-1} & \mathbf{V}_{N/N-1}\mathbf{\Phi}_{N,N-1} \\ \mathbf{R}_{N/N}\mathbf{G}_{N,N-1} & \mathbf{R}_{N/N}\mathbf{\Phi}_{N,N-1} \end{bmatrix}\begin{bmatrix} \mathbf{q}_{N,N-1} \\ \mathbf{x}_{N-1} \end{bmatrix} = \begin{bmatrix} \mathbf{z}^q_{N/N-1} \\ \mathbf{z}^x_{N/N} \end{bmatrix} + \begin{bmatrix} \mathbf{v}'^q_{N-1} \\ \mathbf{v}^x_N \end{bmatrix}. \tag{10.3-30}$$

(We have again added time subscripts on $\mathbf{G}$ and $\mathbf{\Phi}$ to avoid confusion in subsequent equations.) An orthogonal transformation is applied to the above equation to triangularize the left array as

$$\begin{bmatrix} \mathbf{W}_{N-1/N} & \mathbf{V}_{N-1/N} \\ \mathbf{0} & \mathbf{R}_{N-1/N} \end{bmatrix} \begin{bmatrix} \mathbf{q}_{N,N-1} \\ \mathbf{x}_{N-1} \end{bmatrix} = \begin{bmatrix} \mathbf{z}^q_{N-1/N} \\ \mathbf{z}^x_{N-1/N} \end{bmatrix} + \begin{bmatrix} \mathbf{v}''^q_{N-1} \\ \mathbf{v}''^x_{N-1} \end{bmatrix}, \tag{10.3-31}$$

and the smoother estimate is computed as

$$\hat{\mathbf{x}}_{N-1/N} = \mathbf{R}^{-1}_{N-1/N} \mathbf{z}^x_{N-1/N}. \tag{10.3-32}$$

Since $E[\mathbf{v}''^x_{N-1}(\mathbf{v}''^x_{N-1})^T] = \mathbf{I}$, the covariance of $\hat{\mathbf{x}}_{N-1/N}$ is

$$\mathbf{P}_{N-1/N} = \mathbf{R}^{-1}_{N-1/N} \mathbf{R}^{-T}_{N-1/N}. \tag{10.3-33}$$

This procedure can then be repeated for earlier time points. The general expression for the SRIS recursion is represented as

$$\mathbf{T}_i \begin{bmatrix} \mathbf{W}_{i/i-1} + \mathbf{V}_{i/i-1}\mathbf{G}_{i,i-1} & \mathbf{V}_{i/i-1}\mathbf{\Phi}_{i,i-1} & \mathbf{z}^q_{i/i-1} \\ \mathbf{R}_{i/N}\mathbf{G}_{i,i-1} & \mathbf{R}_{i/N}\mathbf{\Phi}_{i,i-1} & \mathbf{z}^x_{i/N} \end{bmatrix} = \begin{bmatrix} \mathbf{W}_{i-1/N} & \mathbf{V}_{i-1/N} & \mathbf{z}^q_{i-1/N} \\ \mathbf{0} & \mathbf{R}_{i-1/N} & \mathbf{z}^x_{i-1/N} \end{bmatrix} \tag{10.3-34}$$

for $i = N, N-1, \ldots, 1$. Notice that at each time update the SRIF must save $\mathbf{W}_{i/i-1}$, $\mathbf{V}_{i/i-1}$, $\mathbf{z}^q_{i/i-1}$. Also $\mathbf{G}_{i,i-1}$ and $\mathbf{\Phi}_{i/i-1}$ must either be saved or recomputed from the model. The quantities in the left side of equation (10.3-34) are formed and saved in a $(n+p) \times (n+p+1)$ array, where $\mathbf{R}_{i/N}$, $\mathbf{z}^x_{i/N}$ are obtained from the previous (later in time) SRIF step. Triangularizing Householder transformations are applied to produce the right-hand side of equation (10.3-34). Alternately it may be more efficient to implement the stochastic update of equation (10.3-34) one-component-at-a-time using columns of $\mathbf{G}_{i,i-1}$. The estimate and covariance are computed when needed as

$$\begin{aligned} \hat{\mathbf{x}}_{i-1/N} &= \mathbf{R}^{-1}_{i-1/N} \mathbf{z}^x_{i-1/N} \\ \mathbf{P}_{i-1/N} &= \mathbf{R}^{-1}_{i-1/N} \mathbf{R}^{-T}_{i-1/N} \end{aligned}. \tag{10.3-35}$$

While the storage requirements are slightly less than for the RTS smoother, approximately $pn(2p + 1.5n) + (n+p)^2(n+p-1)/3$ flops (multiply-add) operations are required to form and triangularize equation (10.3-34) and another $n^2(4n+9)/6$ flops are required to compute $\hat{\mathbf{x}}_{i-1/N}$ and $\mathbf{P}_{i-1/N}$. Hence the total is about $n^3 + 3p^2n + 2.5pn^2 + p^3/3$. This compares with about $0.5n^3 + 3.5p^2n + pn^2 + p^3$ for the RTS smoother if implemented to avoid computations for biases. Because of the high SRIS computational burden, an alternate form of the smoother (Dyer-McReynolds) is often used with the SRIF.

It should be noted that SRIS computations are greatly reduced if the smoother covariance is not needed. The process noise can be computed from equation (10.3-17) using the filter outputs $\mathbf{W}^{-1}_{i/i-1}$, $\mathbf{V}_{i/i-1}$, $\mathbf{z}^q_{i/i-1}$ and the state $\mathbf{x}_i$ as

$$\mathbf{q}_{i,i-1} = \mathbf{W}^{-1}_{i/i-1}\left(\mathbf{z}^q_{i/i-1} - \mathbf{V}_{i/i-1}\mathbf{x}_i + \mathbf{v}'^q_{i-1}\right). \tag{10.3-36}$$

Since $E[\mathbf{v}'^q_{i-1}] = \mathbf{0}$ and $E[\hat{\mathbf{x}}_{i/N}] = \mathbf{x}_i$, the (conditional) expected value of the process noise is

$$\hat{\mathbf{q}}_{(i,i-1)/N} = \mathbf{W}^{-1}_{i/i-1}(\mathbf{z}^q_{i/i-1} - \mathbf{V}_{i/i-1}\hat{\mathbf{x}}_{i/N}). \tag{10.3-37}$$

The smoothed state estimate at t_{i-1} is thus computed using equation (10.3-12) as

$$\begin{aligned}\hat{\mathbf{x}}_{i-1/N} &= \mathbf{\Phi}_{i,i-1}^{-1}(\hat{\mathbf{x}}_{i/N} - \mathbf{G}_{i,i-1}\hat{\mathbf{q}}_{(i,i-1)/N}) \\ &= \mathbf{\Phi}_{i,i-1}^{-1}\left[(\mathbf{I}+\mathbf{G}_{i,i-1}\mathbf{W}_{i/i-1}^{-1}\mathbf{V}_{i/i-1})\hat{\mathbf{x}}_{i/N} - \mathbf{G}_{i,i-1}\mathbf{W}_{i/i-1}^{-1}\mathbf{z}_{i/i-1}^{q}\right].\end{aligned} \tag{10.3-38}$$

Since this is a backward recursion, rounding errors in $\hat{\mathbf{x}}_{i/N}$ are passed on to $\hat{\mathbf{x}}_{i-1/N}$, but it is not clear if this recursion causes a growth in the rounding error. Although the SRIS estimate from equation (10.3-35) is also computed by a backward recursion, variables are computed using orthogonal transformations, so error growth may be smaller.

10.3.4 Dyer-McReynolds Covariance Smoother (DMCS)

The DMCS also computes the smoother covariance using outputs from the SRIF, but it does this without triangularizing a rectangular array. To derive the algorithm we substitute equation (10.3-36) in equation (10.3-12) to obtain a backward recursion for the state vector,

$$\mathbf{x}_{i-1} = \mathbf{\Phi}_{i,i-1}^{-1}\left[(\mathbf{I}+\mathbf{G}_{i,i-1}\mathbf{W}_{i/i-1}^{-1}\mathbf{V}_{i/i-1})\mathbf{x}_i - \mathbf{G}_{i,i-1}\mathbf{W}_{i/i-1}^{-1}(\mathbf{z}_{i/i-1}^{q} + \mathbf{v}_{i-1}'^{q})\right], \tag{10.3-39}$$

based on measurement and dynamic model information. Since $E[\mathbf{v}_{i-1}'^{q}] = \mathbf{0}$, the backward Dyer-McReynolds recursion for the smoother estimate is

$$\hat{\mathbf{x}}_{i-1/N} = \mathbf{\Phi}_{i,i-1}^{-1}\left[(\mathbf{I}+\mathbf{G}_{i,i-1}\mathbf{W}_{i/i-1}^{-1}\mathbf{V}_{i/i-1})\hat{\mathbf{x}}_{i/N} - \mathbf{G}_{i,i-1}\mathbf{W}_{i/i-1}^{-1}\mathbf{z}_{i/i}^{q}\right], \tag{10.3-40}$$

which is equation (10.3-38). The recursion is initialized with $\hat{\mathbf{x}}_{N/N}$ from the filter. The smoother estimate error at t_i is the difference between equations (10.3-39) and (10.3-40):

$$\mathbf{x}_{i-1} - \hat{\mathbf{x}}_{i-1/N} = \mathbf{\Phi}_{i,i-1}^{-1}\left[(\mathbf{I}+\mathbf{G}_{i,i-1}\mathbf{W}_{i/i-1}^{-1}\mathbf{V}_{i/i-1})(\mathbf{x}_i - \hat{\mathbf{x}}_{i/N}) - \mathbf{G}_{i,i-1}\mathbf{W}_{i/i-1}^{-1}\mathbf{v}_{i-1}'^{q}\right]. \tag{10.3-41}$$

It is easily shown from equation (10.3-17) that $E[(\mathbf{x}_i - \hat{\mathbf{x}}_{i/i})(\mathbf{v}_{i-1}'^{q})^T] = \mathbf{0}$, but it is less obvious that $E[(\mathbf{x}_i - \hat{\mathbf{x}}_{i/N})(\mathbf{v}_{i-1}'^{q})^T] = \mathbf{0}$. Since $E[\mathbf{v}_{i-1}'^{q}(\mathbf{v}_{i-1}'^{q})^T] = \mathbf{I}$, the smoother error covariance recursion is

$$\begin{aligned}\mathbf{P}_{i-1/N} &\triangleq E[(\mathbf{x}_{i-1} - \hat{\mathbf{x}}_{i-1/N})(\mathbf{x}_{i-1} - \hat{\mathbf{x}}_{i-1/N})^T] \\ &= \mathbf{\Phi}_{i,i-1}^{-1}\left((\mathbf{I}+\mathbf{A}_i\mathbf{V}_{i/i-1})\mathbf{P}_{i/N}(\mathbf{I}+\mathbf{A}_i\mathbf{V}_{i/i-1})^T + \mathbf{A}_i\mathbf{A}_i^T\right)\mathbf{\Phi}_{i,i-1}^{-T}\end{aligned} \tag{10.3-42}$$

where $\mathbf{A}_i = \mathbf{G}_{i,i-1}\mathbf{W}_{i/i-1}^{-1}$. If implemented to minimize computations for bias states, with stochastic updates applied one-component-at-a-time in the SRIF, the total flops required to compute the smoother covariance are approximately $1.5p^2n + 2pn^2 + 2p^3$. This is closer to that of the RTS smoother implemented with similar considerations. DCMS computations are less than those of the SRIS because it is not necessary to re-triangularize a rectangular array. However, actual execution time on modern computers may not be proportional to flop counts because of pipelining and parallel operations.

10.3.5 SRIF Error Analysis

Because the SRIF works in the information domain, it has a different method for computing sensitivity to errors in unadjusted parameters (Bierman 1977b, chapter

9). Bierman's chapter 9 discussion is extensive and covers both incorrect *a priori* information and errors in unadjusted biases. Bierman shows that selection of unadjusted biases in a linear system can be decided *a posteriori* after processing with all parameters adjusted. This is similar to the error analysis method previously used for least-squares estimation. In this section we only consider the effects of unadjusted biases.

The state vector is partitioned into n_x adjusted ($\mathbf{x}$) and n_u unadjusted ($\mathbf{u}$) bias parameters where the unadjusted parameters are located at the bottom of the state vector. With this partitioning the normalized measurement equation is

$$\begin{bmatrix}\mathbf{H}_i^x & \mathbf{H}_i^u\end{bmatrix}\begin{bmatrix}\mathbf{x}\\ \mathbf{u}\end{bmatrix}_i = \mathbf{y}_i - \mathbf{r}_i \tag{10.3-43}$$

where $\mathbf{r}_i$ is zero mean and unit covariance. Hence the SRIF measurement data equation has the form

$$\begin{bmatrix}\mathbf{R}_{i/i-1}^x & \mathbf{S}_{i/i-1}^{xu}\\ \mathbf{0} & \mathbf{R}^u\\ \mathbf{H}_i^x & \mathbf{H}_i^u\end{bmatrix}\begin{bmatrix}\mathbf{x}_i\\ \mathbf{u}\end{bmatrix} = \begin{bmatrix}\mathbf{z}_{i/i-1}^x\\ \mathbf{0}\\ \mathbf{y}_i\end{bmatrix} + \begin{bmatrix}\mathbf{v}_{i-1}'^x\\ \mathbf{v}^u\\ -\mathbf{r}_i\end{bmatrix} \tag{10.3-44}$$

where $\mathbf{S}_{i/i-1}^{xu}$ is the cross-information between adjusted and unadjusted parameters and $\mathbf{R}^u$ defines the prior uncertainty on $\mathbf{u}$. (For a nonlinear systems $\mathbf{u}$ represents the perturbation from the nominal values $\hat{\mathbf{u}}$ used in the filter to compute the measurement residual $\mathbf{y}_i - \mathbf{h}(\hat{\mathbf{x}},\hat{\mathbf{u}})$). After multiplying by triangularizing orthogonal transformations $\mathbf{T}$, the data equation becomes

$$\begin{bmatrix}\mathbf{R}_{i/i}^x & \mathbf{S}_{i/i}^{xu}\\ \mathbf{0} & \mathbf{R}^u\\ \mathbf{0} & \mathbf{H}_i'^u\end{bmatrix}\begin{bmatrix}\mathbf{x}_i\\ \mathbf{u}\end{bmatrix} = \begin{bmatrix}\mathbf{z}_{i/i}^x\\ \mathbf{0}\\ \mathbf{e}_i\end{bmatrix} + \begin{bmatrix}\mathbf{v}_i^x\\ \mathbf{v}^u\\ \mathbf{v}_i^y\end{bmatrix}. \tag{10.3-45}$$

Since it is not necessary (nor desirable) for $\mathbf{T}$ to triangularize the $\mathbf{H}_i'^u$ array, we have assumed that the orthogonal transformations stop after n_x columns so that $\mathbf{R}^u$ is unchanged. Hence the middle row of the above equation could be eliminated. The estimate of adjusted states is computed as

$$\hat{\mathbf{x}}_{i/i} = (\mathbf{R}_{i/i}^x)^{-1}\mathbf{z}_{i/i}^x \tag{10.3-46}$$

and the estimate error is

$$\boxed{\mathbf{x}_i - \hat{\mathbf{x}}_{i/i} = (\mathbf{R}_{i/i}^x)^{-1}(\mathbf{v}_i^x - \mathbf{S}_{i/i}^{xu}\mathbf{u})} \tag{10.3-47}$$

so the sensitivity of the estimate error to $\mathbf{u}$ is $-(\mathbf{R}_{i/i}^x)^{-1}\mathbf{S}_{i/i}^{xu}$. The total error covariance is

$$\boxed{\begin{aligned}\mathbf{P}_{i/i}^a &= E[(\mathbf{x}_i - \hat{\mathbf{x}}_{i/i})(\mathbf{x}_i - \hat{\mathbf{x}}_{i/i})^T]\\ &= (\mathbf{R}_{i/i}^x)^{-1}(\mathbf{R}_{i/i}^x)^{-T} + \mathbf{A}_{i/i}\mathbf{A}_{i/i}^T\end{aligned}} \tag{10.3-48}$$

where

$$\boxed{\mathbf{A}_{i/i} = (\mathbf{R}_{i/i}^x)^{-1}\mathbf{S}_{i/i}^{xu}(\mathbf{R}^u)^{-1}.} \tag{10.3-49}$$

The SRIF time update is also altered when unadjusted biases are included. Equation (10.3-11) becomes

$$\begin{bmatrix} \mathbf{x}_i \\ \mathbf{u} \end{bmatrix} = \begin{bmatrix} \mathbf{\Phi}_{xx} & \mathbf{\Phi}_{xu} \\ \mathbf{0} & \mathbf{I} \end{bmatrix} \begin{bmatrix} \mathbf{x}_{i-1} \\ \mathbf{u} \end{bmatrix} + \begin{bmatrix} \mathbf{G}_x \\ \mathbf{0} \end{bmatrix} \mathbf{q}_{i,i-1} \tag{10.3-50}$$

and equation (10.3-15) becomes

$$\begin{bmatrix} \mathbf{I} & \mathbf{0} & \mathbf{0} \\ -\mathbf{R}_{i-1/i-1}\mathbf{\Phi}_{xx}^{-1}\mathbf{G} & \mathbf{R}_{i-1/i-1}\mathbf{\Phi}_{xx}^{-1} & \mathbf{S}_{i-i/i-i}^{xu} - \mathbf{R}_{i-1/i-1}\mathbf{\Phi}_{xx}^{-1}\mathbf{\Phi}_{xu} \end{bmatrix} \begin{bmatrix} \mathbf{q}_{i,i-1} \\ \mathbf{x}_i \\ \mathbf{u} \end{bmatrix} = \begin{bmatrix} \mathbf{0} \\ \mathbf{z}_{i-1/i-1} \end{bmatrix} + \begin{bmatrix} \mathbf{v}_{i-1}^{q} \\ \mathbf{v}_{i-1}^{x} \end{bmatrix}. \tag{10.3-51}$$

After applying triangularizing orthogonal transformations up to but not including the last n_u columns, the result is

$$\begin{bmatrix} \mathbf{W}_{i/i-1} & \mathbf{V}_{i/i-1}^{x} & \mathbf{V}_{i/i-1}^{u} \\ \mathbf{0} & \mathbf{R}_{i/i-1} & \mathbf{S}_{i/i-1}^{xu} \end{bmatrix} \begin{bmatrix} \mathbf{q}_{i,i-1} \\ \mathbf{x}_i \\ \mathbf{u} \end{bmatrix} = \begin{bmatrix} \mathbf{z}_{i/i-1}^{q} \\ \mathbf{z}_{i/i-1}^{x} \end{bmatrix} + \begin{bmatrix} \mathbf{v}_{i-1}^{\prime q} \\ \mathbf{v}_{i-1}^{\prime x} \end{bmatrix}. \tag{10.3-52}$$

The total error covariance for $\hat{\mathbf{x}}_{i/i-1}$ is

$$\begin{aligned} \mathbf{P}_{i/i-1}^{a} &= E[(\mathbf{x}_i - \hat{\mathbf{x}}_{i/i-1})(\mathbf{x}_i - \hat{\mathbf{x}}_{i/i-1})^T] \\ &= \mathbf{R}_{i/i-1}^{-1}\mathbf{R}_{i/i-1}^{-T} + \mathbf{A}_{i/i-1}\mathbf{A}_{i/i-1}^{T} \end{aligned} \tag{10.3-53}$$

where

$$\mathbf{A}_{i/i-1} = \mathbf{R}_{i/i-1}^{-1}\mathbf{S}_{i/i-1}^{xu}(\mathbf{R}^{u})^{-1}. \tag{10.3-54}$$

10.4 INERTIAL NAVIGATION SYSTEM (INS) EXAMPLE USING FACTORED FILTERS

The INS error model of Section 3.3 is used to compare the factored filter algorithms. Recall that the state vector includes 27 states of which 18 are biases. The modeled process noise for the nine dynamic states (position, velocity, and gyro tilt) is relatively small and the position measurements are very accurate (0.61 mm $1-\sigma$). This example was selected for algorithm comparisons because numerical errors are a potential problem.

Three filters were compared:

1. The standard Kalman covariance filter: implemented using either Kalman's measurement covariance update (with averaging of off-diagonal covariance elements), Bierman's version of the Joseph form, or the full Joseph form.
2. The U-D filter implemented using either Thornton's MWGS algorithm, or the MWGS method for the deterministic update and a series of rank-1 updates for the stochastic (process noise) update.
3. The SRIF implemented as described previously.

The U-D and SRIF were implemented using modified versions of the ESP or ESL utility routines that are associated with this book. The modifications included upgrading to Fortran 90 syntax, implementing many vector operations using implicit MATLAB/Fortran 90 vector operators (thus facilitating porting to MATLAB), specifying input/output INTENT on all calling arguments, rearranging calling arguments as "output, input/output, and input" to make the subroutines appear more like functions, using a separate module to specify whether arithmetic operations are single or double precision, and modifying several routines so that they optionally test for matrix sparseness and only operate on nonzero elements. All modified routines retain comments indicating the ESP/ESL origins of the code, and the subroutine names all end in "_M" to indicate that the routine has been modified and the calling argument order is different than that of the original. Chapter 13 provides more details about the supplied routines.

In each case the filters were implemented to take advantage of the problem structure: 9 dynamic states and 18 biases. The U-D filter implementation used slightly less matrix partitioning than the other two filters because the WGS_M subroutine has the option to test for sparseness of the input $\mathbf{W}$ matrix and bypass multiplications by zero. Likewise the PHIU_M subroutine also tests for nonzero blocks of $\mathbf{\Phi}$ and avoids operations on zeros. Hence the U-D filter code is simpler than code for the other algorithms. In fact, approximately twice as many lines-of-code are required for the SRIF than for the U-D filter. Further computational reductions in both the U-D and SRIF code could be obtained by additional partitioning.

Table 10.1 lists the root-sum-squared (RSS) differences between the filter state estimates and a reference solution summed over the 800 s (80 time steps) of the flight. All filters are implemented in double precision on a 3.0-GHz Pentium 4 PC. The reference estimates are the U-D filter estimates computed using Thornton's MWGS time update. This may or may not be the fastest algorithm, but it is certainly one of the most accurate. The results of Table 10.1 are listed as RSS values for each block of states summed over the 80 time steps. As expected, the U-D filter using rank-1 stochastic updates agrees most closely with the reference U-D MWGS

TABLE 10.1: RSS Difference in Double Precision Filter State Estimates from U-D MWGS

State Group	Short Kalman	Bierman's Joseph Update	Joseph Update	U-D (Rank-1 Q Update)	SRIF
Position error (mm)	8.2e-8	9.2e-7	9.0e-7	6.5e-10	4.0e-9
Velocity error(mm/s)	1.6e-7	2.0e-7	1.9e-7	8.5e-11	7.4e-10
Gyro tilt (arc sec)	1.3e-6	4.0e-5	3.8e-5	6.5e-9	2.8e-8
Accelerometer bias (micro-G)	6.6e-6	1.9e-6	1.8e-6	2.6e-10	6.8e-10
Accelerometer SFE (PPM)	4.3e-5	5.2e-5	4.6e-5	1.6e-8	6.5e-8
Gyro bias drift (arc-sec/s)	4.0e-8	1.0e-7	9.6e-8	1.6e-11	6.7e-11
Gyro SFE(PPM)	5.7e-5	1.1e-5	1.0e-5	3.7e-9	1.3e-8
G-sensitive error (arc-sec/s/G)	3.0e-8	2.2e-8	2.0e-8	4.4e-12	1.7e-11
G^2-sensitive error (arc-sec/s/G^2)	3.3e-9	3.1e-8	2.6e-9	3.4e-13	1.2e-12

solution, and the difference between the SRIF and reference solutions is about five times larger. Considering the large magnitude of many states, the relative differences are less than 10^{-15}, which is approximately the double precision accuracy of the PC. The differences for all covariance-based Kalman filters, however, are factors of 20 to 9700 times larger than those of the SRIF. Even so, the accuracy is still quite good when operating in double precision. These differences are all much smaller than the computed *a posteriori* $1 - \sigma$ uncertainty. None of the three covariance forms is significantly more accurate than the others.

The results are quite different when operating in single precision. All covariance-based Kalman filters failed to complete measurement processing because they eventually computed negative measurement residual variances. This problem is avoided in the two Joseph forms of the measurement update when position and velocity variances in the state covariance matrix are constrained to be greater than $10\,mm^2$ and $10\,mm^2/s^2$, respectively. This primarily affects the variance of position and velocity states, but also impacts other states indirectly. This artificial constraint is a kludge that changes the behavior of the filter. Furthermore it only works for the Joseph forms, not for the standard Kalman filter.

Table 10.2 lists the differences between the single precision filter state estimates and the reference U-D (double precision) solution. These differences are approximately 10^7 times larger than the values in Table 10.1. The estimate errors for the covariance-based filters are much larger than those of the factored filters, but the factored filters do not—at first glance—appear to offer accuracy comparable to the double precision covariance filters. This is misleading because most differences in Table 10.2 are caused by the single precision accuracy of the measurements, measurement partials, and the filter state vector. The position error measurements are about 2×10^7 mm at the end of the data span, so the position errors in the factored filter columns of Table 10.2 are approximately equal in magnitude to the single precision round-off of the measurements. In fact, the round-off errors are

TABLE 10.2: RSS Difference in Single Precision Filter State Estimates from U-D MWGS

State Group	Bierman's Joseph Update	Joseph Update	U-D (MWGS)	U-D (Rank1 Q Update)	SRIF
Position error (mm)	0.48	0.48	0.37	0.37	1.2
Velocity error(mm/s)	0.52	0.53	0.05	0.04	0.12
Gyro tilt (arc sec)	87.1	86.9	2.3	4.3	9.5
Accelerometer bias (micro-G)	17.6	17.5	0.08	0.10	0.14
Accelerometer SFE (PPM)	60.4	60.5	16.7	16.0	6.9
Gyro bias drift (arc-sec/s)	0.24	0.24	0.006	0.010	0.020
Gyro SFE (PPM)	19.9	19.9	0.65	1.1	2.3
G-sensitive error (arc-sec/s/G)	0.089	0.089	0.002	0.002	0.006
G^2-sensitive error (arc-sec/s/G^2)	0.0063	0.0063	0.0001	0.0002	0.0004

almost as large as the position measurement noise (0.61 mm $1-\sigma$). When the measurement magnitudes are reduced by differencing them with a reference state, differences between the state estimates of the three factored filters and a double precision U-D reference are less than a factor of 10 larger than the numbers in Table 10.1 for the covariance filters. This demonstrates that single precision factored filters are nearly capable of accuracy comparable to double precision covariance filters when numerical round-off of measurements and states is small compared with the state corrections computed by the filter measurement update. Unfortunately an accurate reference state for measurement differencing would not be available in a real system, so the better approach (when "single precision" coding is necessary) is to implement processing of the measurements and state vector in double precision, and implement covariance and gain processing in single precision.

Another indication of the accuracy of the factored filters is given in Table 10.3 where the errors in computing the $1-\sigma$ uncertainty (square roots of the covariance diagonals) are about 10^7 times smaller for the factored filters than for the covariance filters. Of course a large part of the error in the covariance filters is caused by the need to constrain the position and velocity variances greater than 10.

Execution time of the various filter implementations is a concern in many applications, particularly when implemented on flight computers or imbedded microprocessors, or for systems with very large state orders. In this particular application the execution time is quite short, but the relative timing of the different algorithms provides general guidance on expected performance for more time-critical applications.

Table 10.4 lists the execution time for a single pass through all the measurements for various implementations of the filters. As indicated previously, all time updates were implemented to take advantage—at least partially—of the differences in behavior of dynamic versus bias states. As expected, the standard Kalman filter is the fastest, with the Joseph update forms much slower. The U-D filter using either

TABLE 10.3: RSS Difference in Single Precision Filter State $1-\sigma$ from U-D MWGS

State Group	Bierman's Joseph Update	Joseph Update	U-D (MWGS)	U-D (Rank-1 Q Update)	SRIF
Position error (mm)	4.3	4.3	5.0e-8	5.6e-8	5.0e-8
Velocity error(mm/s)	5.0	5.0	1.1e-5	3.6e-6	3.6e-4
Gyro tilt (arc sec)	79.6	79.4	1.5e-5	1.3e-5	6.2e-3
Accelerometer bias (micro-G)	22.7	22.6	8.6e-6	8.4e-6	9.5e-6
Accelerometer SFE (PPM)	39.5	39.5	6.3e-6	7.3e-6	4.7e-6
Gyro bias drift (arc-sec/s)	0.18	0.18	3.0e-8	3.3e-8	9.4e-8
Gyro SFE (PPM)	40.6	40.9	1.2e-5	1.3e-5	3.0e-5
G-sensitive error (arc-sec/s/G)	0.060	0.060	3.5e-8	3.4e-8	4.1e-8
G^2-sensitive error (arc-sec/s/G^2)	0.011	0.011	7.3e-10	1.0e-9	2.8e-9

TABLE 10.4: Double Precision Filter Algorithm Timing

Filter	Time (Millisecond)
Short Kalman update	5.6
Bierman's Joseph update	9.1
Joseph update	16.7
U-D: sparse MWGS (σ only)	6.6
U-D: sparse MWGS (full covariance)	8.3
U-D: sparse rank-1 Q update (σ only)	6.9
U-D: normal MWGS (σ only)	10.1
U-D: normal rank-1 Q update (σ only)	9.4
SRIF (σ only)	18.9
SRIF (full *a posteriori* covariance)	20.9
SRIF (without computing σ or covariance)	16.7

the MWGS or rank-1 time updates is almost as fast as the standard Kalman filter provided that it uses the sparse matrix versions of the WGS_M and PHIU_M subroutines and only computes *a posteriori* $1 - \sigma$ values rather than the full covariance. When using the normal (nonsparse) utility routines the execution time increases 35% to 50%, which indicates that the sparse algorithm is effective. Timing of the U-D MWGS and rank-1 time updates is comparable, with the MWGS method better able to benefit from use of the sparse matrix utilities. Computation of the full *a posteriori* covariance at every step increases the execution time by about 26% for the sparse MWGS method. The U-D time updates can be made somewhat faster if further partitioning is used.

The SRIF is two to nearly three times slower than the U-D filters, and use of a sparse matrix THHC_M subroutine for the SRIF measurement update did not improve times significantly. The TDHHT_M subroutine (used for triangularizing a rectangular matrix in the time update) was partially modified to test for nonzero elements, but this did not reduce execution times. Execution time may also be reduced by implementing the stochastic time update one-component-at-a-time, but this is unlikely to be significant. As with the U-D filter, computation of the full covariance significantly increases SRIF execution time.

Although these results were obtained for one specific problem that contained a large number of biases, the relative timing behavior of the algorithms has been observed by this author for other applications.

To summarize, the U-D filter is nearly as fast as the standard Kalman filter for this problem and provides nearly the same accuracy as a standard double precision Kalman filter when implemented in single precision. The single precision covariance-based Kalman filters failed completely for this problem, and the Joseph forms could only be made to work by artificially constraining state variances. The SRIF has approximately the same accuracy as the U-D filter, but is much slower. When implemented in double precision, the factored filters provide about three digits more precision than the covariance-based filters, but even the covariance-based filters provide acceptable accuracy for this problem.

10.5 LARGE SPARSE SYSTEMS AND THE SRIF

We previously mentioned that there are computational benefits when working in the information domain for certain types of problems. One such problem occurs when subsets of measurements only observe subsets of states plus common states. An example is the multi-satellite-arc orbit determination problem where the goal is to compute coefficients of the geopotential field. Ground tracking allows estimation of each satellite arc's orbit elements and possibly drag or solar radiation pressure coefficients. Since the geopotential field affects all satellite orbits, coefficients of a spherical harmonic expansion can also be estimated using multiple satellite arcs. For a system using three satellite arcs (most geopotential solutions use thousands) the filter state vector will be $\begin{bmatrix}\mathbf{x}_1^T & \mathbf{x}_2^T & \mathbf{x}_3^T & \mathbf{x}_c^T\end{bmatrix}^T$, where $\mathbf{x}_1, \mathbf{x}_2, \mathbf{x}_3$ are the satellite arc-specific states (orbit elements and optional force model parameters) and $\mathbf{x}_c$ represents the "common" spherical harmonic states. The measurement partial derivative matrix for this problem will have the form

$$\mathbf{H} = \begin{bmatrix} \mathbf{H}_{11} & \mathbf{0} & \mathbf{0} & \mathbf{H}_{1c} \\ \mathbf{0} & \mathbf{H}_{22} & \mathbf{0} & \mathbf{H}_{2c} \\ \mathbf{0} & \mathbf{0} & \mathbf{H}_{22} & \mathbf{H}_{3c} \end{bmatrix} \tag{10.5-1}$$

where $\mathbf{H}_{11}$, $\mathbf{H}_{22}$, and $\mathbf{H}_{33}$ contain measurement information on the arc-specific states for ground tracking of each of the three satellite arcs, and $\mathbf{H}_{1c}$, $\mathbf{H}_{2c}$, $\mathbf{H}_{3c}$ contain information on the common parameters. We assume that ground tracking does not include relay measurements involving multiple satellites, so there is no cross-information on arc-specific states for different arcs.

When the above measurements are processed in the SRIF, it is found that the resulting solution has the form

$$\begin{bmatrix} \mathbf{R}_{11} & \mathbf{0} & \mathbf{0} & \mathbf{R}_{1c} \\ \mathbf{0} & \mathbf{R}_{22} & \mathbf{0} & \mathbf{R}_{2c} \\ \mathbf{0} & \mathbf{0} & \mathbf{R}_{33} & \mathbf{R}_{3c} \\ \mathbf{0} & \mathbf{0} & \mathbf{0} & \mathbf{R}_{cc} \end{bmatrix} \begin{bmatrix} \hat{\mathbf{x}}_1 \\ \hat{\mathbf{x}}_2 \\ \hat{\mathbf{x}}_3 \\ \hat{\mathbf{x}}_c \end{bmatrix} = \begin{bmatrix} \mathbf{z}_1 \\ \mathbf{z}_2 \\ \mathbf{z}_3 \\ \mathbf{z}_c \end{bmatrix} \tag{10.5-1}$$

provided that process noise is not modeled on the common parameters. Notice that the sparseness in $\mathbf{H}$ between states for different arcs is retained in the $\mathbf{R}$ matrix. (This can be verified by forming the normal equation information matrix $\mathbf{H}^T\mathbf{H}$ and then factoring it as $\mathbf{R}^T\mathbf{R}$.) Hence the solution is

$$\begin{aligned} \hat{\mathbf{x}}_c &= \mathbf{R}_{cc}^{-1}\mathbf{z}_c \\ \hat{\mathbf{x}}_3 &= \mathbf{R}_{33}^{-1}(\mathbf{z}_3 - \mathbf{R}_{3c}\hat{\mathbf{x}}_c) \\ \hat{\mathbf{x}}_2 &= \mathbf{R}_{22}^{-1}(\mathbf{z}_2 - \mathbf{R}_{2c}\hat{\mathbf{x}}_c)\cdot \\ \hat{\mathbf{x}}_1 &= \mathbf{R}_{11}^{-1}(\mathbf{z}_1 - \mathbf{R}_{1c}\hat{\mathbf{x}}_c) \end{aligned} \tag{10.5-3}$$

where all diagonal $\mathbf{R}$ partitions are upper triangular and easily inverted. Notice that a solution for the common parameters can be obtained without solving for the arc-dependent states. Furthermore, transformation of measurement information $\mathbf{H}$ to form $\mathbf{R}$ can be done in arc-dependent partitions without ever forming the full $\mathbf{R}$ matrix. That is, when processing arc #1, the required array partitions are

$$\mathbf{T}\begin{bmatrix}\mathbf{R}_{11} & \mathbf{R}_{1c} & \mathbf{z}_1 \\ \mathbf{0} & \mathbf{R}_{cc} & \mathbf{z}_c \\ \mathbf{H}_{11} & \mathbf{H}_{1c} & \mathbf{y}_1\end{bmatrix}_{1/0} = \begin{bmatrix}\mathbf{R}_{11} & \mathbf{R}_{1c} & \mathbf{z}_1 \\ \mathbf{0} & \mathbf{R}_{cc} & \mathbf{z}_c \\ \mathbf{0} & \mathbf{0} & \mathbf{e}_1\end{bmatrix}_{1/1}. \tag{10.5-4}$$

After all arc #1 measurements are processed, the *a posteriori* $\mathbf{R}_{11}$, $\mathbf{H}_{1c}$, $\mathbf{z}_1$ are saved to disk and replaced with the prior information for arc #2. The arc #2 measurement information is loaded and processed, and this is repeated until all arcs have been processed. The $\hat{\mathbf{x}}_c$ solution is computed and arc-dependent solutions are computed by reloading data from disk. Very large interconnected systems can be handled using this approach. More details may be found in Bierman (1977a).

The general approach can be extended to systems where information on individual "subsystem" states is obtained from multiple sets of measurements. In the above example this could happen if satellite-to-satellite relay measurements were included. It can also happen in static spatial processing problems where measurements involve multiple subsets of states (e.g., spatial interpolation). A spatial example is presented in the next section. The measurement partial matrix for a system of five subsystems where each measurement is a function of two or three state subsystems is

$$\mathbf{H} = \begin{bmatrix}\mathbf{H}_{11} & \mathbf{H}_{12} & \mathbf{0} & \mathbf{0} & \mathbf{0} & \mathbf{H}_{1c} \\ \mathbf{H}_{21} & \mathbf{H}_{22} & \mathbf{H}_{23} & \mathbf{0} & \mathbf{0} & \mathbf{H}_{2c} \\ \mathbf{0} & \mathbf{H}_{32} & \mathbf{H}_{33} & \mathbf{H}_{34} & \mathbf{0} & \mathbf{H}_{3c} \\ \mathbf{0} & \mathbf{0} & \mathbf{H}_{43} & \mathbf{H}_{44} & \mathbf{H}_{45} & \mathbf{H}_{4c} \\ \mathbf{0} & \mathbf{0} & \mathbf{0} & \mathbf{H}_{54} & \mathbf{H}_{55} & \mathbf{H}_{5c}\end{bmatrix}. \tag{10.5-5}$$

After SRIF processing the state equations are

$$\begin{bmatrix}\mathbf{R}_{11} & \mathbf{R}_{12} & \mathbf{R}_{13} & \mathbf{0} & \mathbf{0} & \mathbf{R}_{1c} \\ \mathbf{0} & \mathbf{R}_{22} & \mathbf{R}_{23} & \mathbf{R}_{24} & \mathbf{0} & \mathbf{R}_{2c} \\ \mathbf{0} & \mathbf{0} & \mathbf{R}_{33} & \mathbf{R}_{34} & \mathbf{R}_{35} & \mathbf{R}_{3c} \\ \mathbf{0} & \mathbf{0} & \mathbf{0} & \mathbf{R}_{44} & \mathbf{R}_{45} & \mathbf{R}_{4c} \\ \mathbf{0} & \mathbf{0} & \mathbf{0} & \mathbf{0} & \mathbf{R}_{55} & \mathbf{R}_{5c} \\ \mathbf{0} & \mathbf{0} & \mathbf{0} & \mathbf{0} & \mathbf{0} & \mathbf{R}_{cc}\end{bmatrix}\begin{bmatrix}\hat{\mathbf{x}}_1 \\ \hat{\mathbf{x}}_2 \\ \hat{\mathbf{x}}_3 \\ \hat{\mathbf{x}}_4 \\ \hat{\mathbf{x}}_5 \\ \hat{\mathbf{x}}_c\end{bmatrix} = \begin{bmatrix}\mathbf{z}_1 \\ \mathbf{z}_2 \\ \mathbf{z}_3 \\ \mathbf{z}_4 \\ \mathbf{z}_5 \\ \mathbf{z}_c\end{bmatrix}. \tag{10.5-6}$$

Notice that with (at most) three nonzero partitions in each row block of the **H** matrix, there are also three nonzero partitions in each row block of the **R** matrix. Again this can be implemented by using only nonzero matrix partitions at each stage, and swapping partitions to disk between stages. The state estimate for the above problem is

$$\begin{aligned}\hat{\mathbf{x}}_c &= \mathbf{R}_{cc}^{-1}\mathbf{z}_c \\ \hat{\mathbf{x}}_5 &= \mathbf{R}_{55}^{-1}(\mathbf{z}_5 - \mathbf{R}_{5c}\hat{\mathbf{x}}_c) \\ \hat{\mathbf{x}}_4 &= \mathbf{R}_{44}^{-1}(\mathbf{z}_4 - \mathbf{R}_{45}\hat{\mathbf{x}}_5 - \mathbf{R}_{4c}\hat{\mathbf{x}}_c) \\ \hat{\mathbf{x}}_3 &= \mathbf{R}_{33}^{-1}(\mathbf{z}_3 - \mathbf{R}_{34}\hat{\mathbf{x}}_4 - \mathbf{R}_{35}\hat{\mathbf{x}}_5 - \mathbf{R}_{3c}\hat{\mathbf{x}}_c), \\ \hat{\mathbf{x}}_2 &= \mathbf{R}_{22}^{-1}(\mathbf{z}_2 - \mathbf{R}_{23}\hat{\mathbf{x}}_3 - \mathbf{R}_{24}\hat{\mathbf{x}}_4 - \mathbf{R}_{2c}\hat{\mathbf{x}}_c) \\ \hat{\mathbf{x}}_1 &= \mathbf{R}_{11}^{-1}(\mathbf{z}_1 - \mathbf{R}_{12}\hat{\mathbf{x}}_2 - \mathbf{R}_{13}\hat{\mathbf{x}}_3 - \mathbf{R}_{1c}\hat{\mathbf{x}}_c)\end{aligned} \tag{10.5-7}$$

which can be computed using a backward recursion by reading array partitions from disk as needed. The state error covariance $\mathbf{P} = \mathbf{R}^{-1}\mathbf{R}^{-T}$ for each subset partition can also be computed using this approach, but computations and storage rapidly become quite large as the number of subset states grows.

10.6 SPATIAL CONTINUITY CONSTRAINTS AND THE SRIF DATA EQUATION

The flexibility and usefulness of the SRIF data equation approach is demonstrated by applying it to a static (time-invariant) spatial problem where measurements introduce correlations between spatial elements. More specifically, the data equation can be applied to 2-D or 3-D geophysical or hydrological "data fusion" modeling problems where spatial continuity of geophysical parameters is a soft constraint. We demonstrate the approach for a steady-state subsurface groundwater hydrological flow calibration problem. This time-invariant spatial model is very different than typical time-dependent models used for filtering, but the data equation approach is an ideal application of the concept. The basic idea for this approach was due to Porter (Porter et al. 1993, 1997), but key elements were also developed by Gibbs, Yancey, and Vandergraft (Porter et al. 1994; Gibbs et al. 1998).

The hydrological flow model is defined by finite difference equations on a 3-D grid as shown in Figure 10.1. (The approach can also be applied to finite-element models.) The "MODFLOW row" refers to a popular hydrological flow modeling software package (see Harbaugh and McDonald 1996).

Hydrological data fusion (Porter et al. 1997; Gibbs 1998) uses a Bayesian estimation method that combines information from

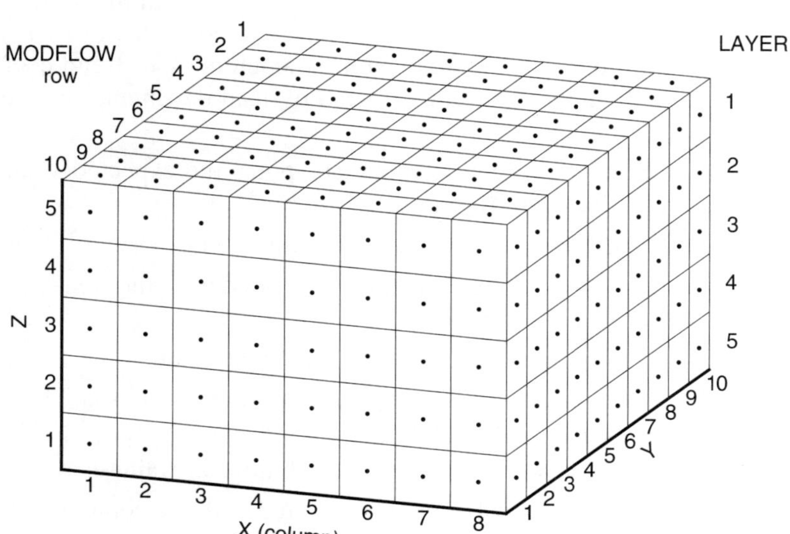

FIGURE 10.1: Flow model grid column/row/layer numbering.

1. Hydraulic head and log conductivity measurements
2. Steady-state groundwater flow constraints
3. Constraints on the spatial variability (heterogeneity) of the hydraulic conductivity field
4. Prior information on model boundary condition parameters

to compute a flow model that best fits all information. Each source of information is weighted in the solution according to the expected accuracy of the data model and constraints. The spatial variability constraint model is incorporated in the solution by treating the constraint within separate hydrological layers as a pseudo-measurement. That is, the output of the spatial continuity model at each node is treated somewhat like an ordinary measurement in that it has an expected value and random error for which the variance is known. The groundwater flow constraint is either treated as a pseudo-measurement or as a hard constraint. Using this approach, all measurements and pseudo-measurements are processed as ordinary measurements in a data equation.

The hydrological data fusion attempts to minimize the residuals (difference between the data and the model output) for all four sources of information described above by minimizing a nonlinear Bayesian penalty function representing the weighted sum-of-squared residuals; that is,

$$J=\frac{1}{2}\sum_i\left[\frac{y_{mi}-\hat{y}_{mi}(\mathbf{x})}{w_{mi}}\right]^2+\frac{1}{2}\sum_i\left[\frac{0-f_i(\mathbf{x})}{w_{fi}}\right]^2+\frac{1}{2}\sum_i\left[\frac{0-g_i(\mathbf{x})}{w_{gi}}\right]^2+\frac{1}{2}\sum_i\left[\frac{\theta_{pi}-\theta_i}{w_{\theta i}}\right]^2 \tag{10.6-1}$$

where:

y_{mi} is measurement i,

$\mathbf{x}$ is the state vector containing all model parameters to be estimated; that is, $\mathbf{x}^T=[\mathbf{h}^T \quad \mathbf{c}^T \quad \boldsymbol{\theta}^T]$ where $\mathbf{h}$ represents hydraulic head at the nodes, $\mathbf{c}$ represents the natural log of y-direction hydraulic conductivity ($\ln K_y$) at the nodes, and $\boldsymbol{\theta}$ represents the common parameters ($\ln K_y$ trend coefficients, anisotropy, recharge, source/drain head of head-dependent flux boundary conditions, etc.).

$\hat{y}_{mi}(\mathbf{x})$ is the nonlinear function which computes the expected value of the measurement based on the state,

w_{mi} is the standard deviation of the measurement noise,

$f_i(\mathbf{x})$ is the nonlinear function that computes, based on $\mathbf{x}$, the steady-state flow constraint error at each node,

w_{fi} is the standard deviation of the flow constraint error,

$g_i(\mathbf{x})$ is the linear function that computes, based on $\mathbf{x}$, the spatial variability interpolation error of $\ln K_y$ at each node,

w_{gi} is the standard deviation of spatial variability constraint error,

θ_{pi} is the prior expected value of component i of $\boldsymbol{\theta}$, the common parameter states,

$w_{\theta i}$ is the standard deviation of the prior estimate error.

Because the flow and measurement models are nonlinear, J is minimized using an iterative Gauss-Newton method that linearizes about the estimated $\mathbf{x}$ at each iteration.

10.6.1 Flow Model

The differential equation describing flow in a variably saturated porous medium model (see Huyakorn and Pinder 1983, section 4.5; Anderson and Woessner 1992) is

$$\frac{\partial}{\partial x_i}\left(K_{ij}k_{rw}\frac{\partial h}{\partial x_j}\right)=\eta\frac{\partial h}{\partial t}-q \qquad (10.6\text{-}2)$$

where

x_i is the i-th component of an orthogonal coordinate system,

K_{ij} is the saturated hydraulic conductivity tensor for the porous medium,

k_{rw} is the relative permeability of water, which is a function of water saturation,

h is hydraulic head,

q is volumetric source/sink flow rate into the system per unit volume of the medium,

η is a composite nonlinear function of h that defines storage of the system.

For this problem we assume steady-state conditions ($\partial h/\partial t = 0$). Hence in a finite difference grid, the flow (L^3/T) to cell i, j, k in the x-direction from cells $i-1,j,k$, and $i+1,j,k$ (i,j,k are, respectively, the grid x,y,z directions) is

$$f_{i\_x}(\mathbf{x})=\begin{pmatrix}\bar{K}_{x:i-1/2,j,k}\bar{S}_{w:i-1/2,j,k}\dfrac{h_{i-1,j,k}-h_{i,j,k}}{(\Delta x_{i-1}+\Delta x_{i+1})/2}\Delta z_{i-1/2,j,k}\\ +\bar{K}_{x:i+1/2,j,k}\bar{S}_{w:i+1/2,j,k}\dfrac{h_{i+1,j,k}-h_{i,j,k}}{(\Delta x_{i+1}+\Delta x_{i+1})/2}\Delta z_{i+1/2,j,k}\end{pmatrix}\Delta y_j \qquad (10.6\text{-}3)$$

where

$\bar{K}_{x:i-1/2,j,k}$ is the harmonic average of saturated conductivity in the x-direction between cells $i-1, j, k$ and i, j, k, and $\bar{K}_{x:i+1/2,j,k}$ is defined accordingly,

$\bar{S}_{w:i-1/2,j,k}$ is the average water saturation between cells $i-1, j, k$ and i, j, k, and $\bar{S}_{w:i+1/2,j,k}$ is defined accordingly ($\bar{S}_w$ is a linear function of h and flow direction),

Δx_{i-1} is the cell x-spacing (same for all j, k indices) between cells $i-1, j, k$ and i, j, k, and Δx_{i+1} is defined accordingly,

Δy_j is the cell y-spacing (same for all i, k indices),

$\Delta z_{i-1/2,j,k}$ is the cell z-spacing between cells $i-1, j, k$ and i, j, k, and $\Delta z_{i-1/2,j,k}$ is defined accordingly.

Flow in the y-direction is similar while flow in the z-direction depends on whether the model is quasi-2-D or fully 3-D. Also the cell z-spacing depends on the i, j, k indices because cell dimensions must follow the hydrological layers. The flow model

$f_i(\mathbf{x})$ of equation (10.6-1) is the sum of flow in all three directions with source/sink flow added. That source/sink flow is zero for all cells except those defined as boundary conditions. Those boundary conditions include fixed head, fixed flow, and head-dependent flow (river/stream, drain, general head-boundary, and evapotranspiration). Head-dependent flow is modeled as

$$q_i = R_i A_i (h_r - h_{\max}) \tag{10.6-4}$$

where R_i is leakance, A_i is cross-sectional area, h_r is the reference head and $h_{\max} = \max(h_i, z_r)$ with z_r as a reference height. Parameters such R_i, h_r and fixed flow at a cell may be included in the parameter vector $\boldsymbol{\theta}$.

To summarize, the nonlinear flow model at each cell depends on hydraulic head and hydraulic conductivity at adjacent cells, and on source/sink model parameters.

10.6.2 Log Conductivity Spatial Continuity Model

Log hydraulic conductivity in the y-direction ($\beta = \ln K_y$) within a hydrological unit (layer) is modeled as a trend in the x-y directions with a "soft" spatial continuity constraint on spatial variability. Conductivity in the x-direction (K_x) is modeled as an anisotropy factor times K_y, and vertical conductivity is modeled as constant within a hydrological unit. The x-y trend coefficients and anisotropy factor are included in the common parameters $\boldsymbol{\theta}$. Since horizontal hydraulic conductivity within a hydrological unit tends to be smooth, deviation of $\ln K_y$ from the trend is treated as a *Markov random field* (MRF) and handled by spatial continuity constraint $g_i(\mathbf{x})$. The MRF is modeled as a 2-D or 3-D autoregressive model (Whittle 1954) of the form

$$u_{i,j,k} = a_x(u_{i-1,j,k} + u_{i+1,j,k}) + a_y(u_{i,j-1,k} + u_{i,j+1,k}) + a_z(u_{i,j,k-1} + u_{i,j,k+1}) + \zeta_{i,j,k} \tag{10.6-5}$$

where

$u_{i,j,k}$ is deviation of $\beta = \ln K_y$ from the trend model at grid location i, j, k,

a_x, a_y, a_z are the autoregressive coefficients ($a_x = a_y = a_z$ is isotropic),

$\zeta_{i,j,k}$ is zero-mean white noise input with variance computed from the specified variance on u and a_x, a_y, a_z.

Notice that the spatial variability model is defined by the constants a_x, a_y, a_z and the variance of $\zeta_{i,j,k}$. The "a" variables are defined as a function of the average grid spacing at the given location divided by specified correlation distances τ_x, τ_y, τ_z; that is,

$$\begin{aligned} d_x &= (\Delta x_{i-1,j,k} + \Delta x_{i+1,j,k})/(2\tau_x) \\ d_y &= (\Delta y_{i,j-1,k} + \Delta y_{i,j+1,k})/(2\tau_y)\,. \\ d_z &= (\Delta z_{i,j,k-1} + \Delta z_{i,j,k+1})/(2\tau_z) \end{aligned} \tag{10.6-6}$$

Unfortunately the relationship between these constants and the behavior of an anisotropic 3-D autoregressive model is nontrivial: the derivation is quite complex (Yancey 1994). It turns out that the continuous version of equation (10.6-5) has an

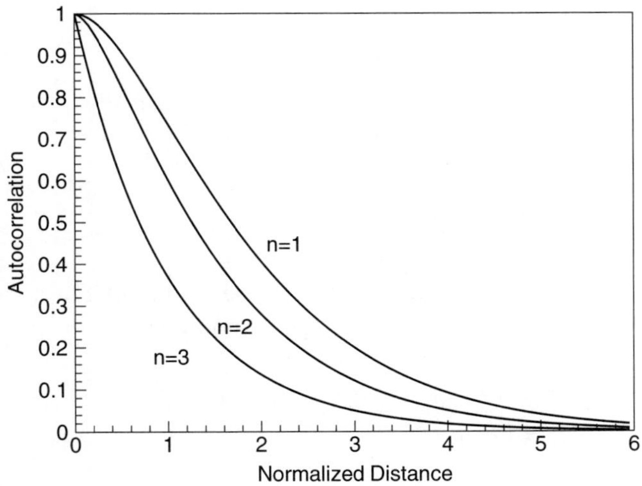

FIGURE 10.2: Continuous autocorrelation functions in one, two, or three dimensions (isotropic).

autocorrelation function that is exponential in three dimensions but deviates from exponential in one or two dimensions, as shown in Figure 10.2.

For a 3-D isotropic model ($\tau = \tau_x = \tau_y = \tau_z$)

$$E[u_{i,j,k}u_{l,m,n}] = \sigma_u^2 \exp\left(-\sqrt{(x_i - x_l)^2 + (y_j - y_m)^2 + (z_k - z_n)^2} \Big/ \tau\right). \tag{10.6-7}$$

The correlation distances τ_x, τ_y, τ_z are specified as the distances at which the correlation function drops to $1/e = 0.37$. The user also specifies the desired standard deviation of u (σ_u). For a 3-D model, the variance of $\zeta_{i,j,k}$ required to obtain the specified σ_u is then computed as

$$\begin{aligned}
\phi &= (d_x d_y d_z)^2 + 2[(d_x d_y)^2 + (d_y d_z)^2 + (d_x d_z)^2] \\
a_x &= (d_y d_z)^2 / \phi \\
a_y &= (d_x d_z)^2 / \phi \\
a_z &= (d_x d_y)^2 / \phi \\
\alpha &= \sqrt{\frac{16 a_y a_z}{(1+2a_x)^2 - 4(a_y - a_z)^2}} \\
\beta &= \sqrt{\frac{16 a_y a_z}{(1-2a_x)^2 - 4(a_y - a_z)^2}} \\
\sigma_\xi^2 &= \frac{2\pi^2 \sqrt{a_y a_z (1-4a_x^2)}}{\int_\alpha^\beta g(\lambda) d\lambda} \sigma_u^2
\end{aligned} \tag{10.6-8}$$

where $g(\lambda)$ is defined as

$$
\begin{aligned}
f &= \frac{1+(a_y-a_z)^2\lambda^2}{4a_y a_z} \\
u &= \left(\frac{4\sqrt{a_y a_z f}}{1-2a_x}-\lambda\right)\left(\lambda-\frac{4\sqrt{a_y a_z f}}{1+2a_x}\right) \\
g(\lambda) &= \frac{\lambda E(\lambda)}{(1-\lambda^2)\sqrt{u}}
\end{aligned}
\tag{10.6-9}
$$

and $E(\lambda)$ is the complete elliptic integral of the second kind. For a 2-D autoregressive model the equivalent relationship is approximately

$$
\begin{aligned}
\phi &= (1.6^2 d_x d_y)^2 + 2(1.6^2)(d_x^2 + d_y^2) \\
a_x &= 1.6^2 d_y^2 / \phi \\
a_y &= 1.6^2 d_x^2 / \phi \\
\lambda &= 4\sqrt{\frac{a_x a_y}{1-4(a_x-a_y)^2}} \\
\sigma_\xi^2 &= \sigma_u^2\left(\frac{\pi\sqrt{(1-4(a_x-a_y)^2)}(1-4(a_x+a_y)^2)}{2E(\lambda)}\right)
\end{aligned}
\tag{10.6-10}
$$

The factor of 1.6 compensates for the fact that the $1/e$ point of the 2-D case occurs at about 1.6 normalized distance.

To summarize, equation (10.6-5) defines the spatial autoregressive model at each grid cell that represents $g_i(\mathbf{x})$ in equation (10.6-1). The "a" coefficients and the variance σ_ξ^2 required to use the model are defined by equations (10.6-8) to (10.6-10).

10.6.3 Measurement Models

Two types of measurement are used for the hydrological data fusion: hydraulic head obtained as well water levels, and log hydraulic conductivity computed from well pumping or drawdown tests. Flow velocity measurements can also be used, but tend to be inaccurate. Since the head and log conductivity measurements are point measurements, the "computed" $\hat{y}_{mi}(\mathbf{x})$ values required for the residuals in equation (10.6-1) are obtained by 2-D or 3-D linear interpolation of the values at the cell centers. Thus all computed measurements are a function of the values at the four or eight nearest cells.

10.6.4 SRIF Processing

The above descriptions show how the four elements of equation (10.6-1) can be rearranged as data equations. Since the flow and measurement equations are nonlinear, the equations must be linearized about the estimate at each Gauss-Newton iteration. Hence the data equations for this system can be written as

$$\begin{bmatrix} \delta\mathbf{y}_m \\ \delta\mathbf{y}_g \\ \delta\mathbf{y}_\theta \\ \delta\mathbf{y}_f \end{bmatrix} = \begin{bmatrix} \mathbf{M}_h & \mathbf{M}_\beta & \mathbf{M}_\theta \\ \mathbf{0} & \mathbf{G}_\beta & \mathbf{0} \\ \mathbf{0} & \mathbf{0} & \mathbf{W}_\theta^{-1} \\ \mathbf{B}_h & \mathbf{B}_\beta & \mathbf{B}_\theta \end{bmatrix} \begin{bmatrix} \delta\mathbf{h} \\ \delta\boldsymbol{\beta} \\ \delta\boldsymbol{\theta} \end{bmatrix} + \begin{bmatrix} \mathbf{v}_m \\ \mathbf{v}_g \\ \mathbf{v}_\theta \\ \mathbf{v}_f \end{bmatrix} \tag{10.6-11}$$

where $\delta\mathbf{y}_m$, $\delta\mathbf{y}_g$, $\delta\mathbf{y}_\theta$, $\delta\mathbf{y}_f$ are the measurement/pseudo-measurement residuals computed as

$$\delta\mathbf{y} = \begin{bmatrix} \delta\mathbf{y}_m \\ \delta\mathbf{y}_g \\ \delta\mathbf{y}_\theta \\ \delta\mathbf{y}_f \end{bmatrix} = \begin{bmatrix} \mathbf{W}_m^{-1}[\mathbf{y}_m - \hat{\mathbf{y}}_m(\mathbf{h},\boldsymbol{\beta},\boldsymbol{\theta})] \\ \mathbf{W}_g^{-1}[\mathbf{0} - \mathbf{g}(\boldsymbol{\beta})] \\ \mathbf{W}_\theta^{-1}(\boldsymbol{\theta}_p - \boldsymbol{\theta}) \\ \mathbf{W}_f^{-1}[\mathbf{0} - \mathbf{f}(\mathbf{h},\boldsymbol{\beta},\boldsymbol{\theta})] \end{bmatrix}, \tag{10.6-12}$$

$\delta\mathbf{x} = [\delta\mathbf{h}^T \;\; \delta\boldsymbol{\beta}^T \;\; \delta\boldsymbol{\theta}^T]^T$ is the correction state vector, and the $\mathbf{v}$ terms are measurement/pseudo-measurement error. Also

$$\mathbf{M}_h = \mathbf{W}_m^{-1}\frac{\partial\hat{\mathbf{y}}_m}{\partial\mathbf{h}}, \quad \mathbf{M}_\beta = \mathbf{W}_m^{-1}\frac{\partial\hat{\mathbf{y}}_m}{\partial\boldsymbol{\beta}}, \quad \mathbf{M}_\theta = \mathbf{W}_m^{-1}\frac{\partial\hat{\mathbf{y}}_m}{\partial\boldsymbol{\theta}}, \tag{10.6-13}$$

$$\mathbf{G}_\beta = \mathbf{W}_g^{-1}\frac{\partial\mathbf{g}}{\partial\boldsymbol{\beta}}, \tag{10.6-14}$$

$$\mathbf{B}_h = \mathbf{W}_f^{-1}\frac{\partial\mathbf{f}}{\partial\mathbf{h}}, \quad \mathbf{B}_\beta = \mathbf{W}_f^{-1}\frac{\partial\mathbf{f}}{\partial\boldsymbol{\beta}}, \quad \mathbf{B}_\theta = \mathbf{W}_f^{-1}\frac{\partial\mathbf{f}}{\partial\boldsymbol{\theta}}. \tag{10.6-15}$$

The **W** matrices are diagonal and contain the squares of the weights in equation (10.6-1). To solve this Bayesian least-squares problem using the SRIF, the **h** and **β** states are stored alternating for each node, rather than as separate blocks of states. Because the measurement, flow, and spatial continuity equations depend only on states at adjacent cells, the "bandwidth" of individual data equations is limited to the number of states in three grid column (i.e., three times the product of number of states in each layer, number of layers, and number of grid rows per column). If all the measurements/pseudo-measurements between two grid columns are processed in one batch, it is found that the matrix triangularization step of the SRIF will nearly fill the **R** matrix for the states in three grid columns plus the common parameters. Thus it is only necessary to store a matrix of this size while processing the measurements. After all measurements/pseudo-measurements for a given grid column are processed, the partitions of the SRIF **R** matrix are swapped to disk and then the partitions are initialized for the next grid column. This is the partitioning approach described in Section 10.5. After processing all grid columns, a back-solving (i.e., SRIS) step reads the stored partitions from disk in reverse order to compute "smoothed" state estimates and covariance partitions. This approach is practical for moderately sized grids (e.g., <100,000 cells) but computations and storage increase rapidly with grid size.

10.6.5 Steady-State Flow Constrained Iterative Solution

Unfortunately the approach outlined above often works poorly in that the Gauss-Newton iterations converge very slowly (if at all), and the "converged" solution often has large mass balance errors at individual boundary condition cells. This is a particular problem for head-dependent boundary conditions. The difficulty is due to the numerical "stiffness" caused by mixing a heavily weighted flow model error (necessary to meet flow mass continuity requirements for subsequent flow transport analysis) with "soft" measurements and spatial continuity pseudo-measurements.

This problem is solved by removing the steady-state flow constraint as a pseudo-measurement, and instead treating it as a hard constraint. That is, the $f_i(\mathbf{x})$ term in equation (10.6-1) is eliminated and replaced with $f_i(\mathbf{x}) = 0$ so that the modified least-squares cost function is

$$J = \frac{1}{2}\sum_i \left[\frac{y_{mi} - \hat{y}_{mi}(\mathbf{x})}{w_{mi}}\right]^2 + \frac{1}{2}\sum_i \left[\frac{0 - g_i(\mathbf{x})}{w_{gi}}\right]^2 + \frac{1}{2}\sum_i \left[\frac{\theta_{pi} - \theta_i}{w_{\theta i}}\right]^2 \qquad (10.6\text{-}16)$$

with constraint $\mathbf{f}(\mathbf{x}) = \mathbf{0}$. Hence the relation $\mathbf{B}_h\delta\mathbf{h} + \mathbf{B}_\beta\delta\boldsymbol{\beta} + \mathbf{B}_\theta\delta\boldsymbol{\theta} = \mathbf{0}$ is inverted to obtain

$$\delta\mathbf{h} = -\mathbf{B}_h^{-1}(\mathbf{B}_\beta\delta\boldsymbol{\beta} + \mathbf{B}_\theta\delta\boldsymbol{\theta}) \qquad (10.6\text{-}17)$$

which is then substituted in the remaining equations to obtain

$$\begin{bmatrix} \delta\mathbf{y}_m \\ \delta\mathbf{y}_g \\ \delta\mathbf{y}_\theta \end{bmatrix} = \begin{bmatrix} \mathbf{M}'_\beta & \mathbf{M}'_\theta \\ \mathbf{G}_\beta & \mathbf{0} \\ \mathbf{0} & \mathbf{W}_\theta^{-1} \end{bmatrix} \begin{bmatrix} \delta\boldsymbol{\beta} \\ \delta\boldsymbol{\theta} \end{bmatrix} + \begin{bmatrix} \mathbf{v}_m \\ \mathbf{v}_g \\ \mathbf{v}_\theta \end{bmatrix}. \qquad (10.6\text{-}18)$$

where

$$\begin{aligned} \mathbf{M}'_\beta &= \mathbf{M}_\beta - \mathbf{M}_h\mathbf{B}_h^{-1}\mathbf{B}_\beta \\ \mathbf{M}'_\theta &= \mathbf{M}_\theta - \mathbf{M}_h\mathbf{B}_h^{-1}\mathbf{B}_\theta \end{aligned}.$$

The steady-state flow Jacobian $\mathbf{B}_h$ is square and full rank, thus allowing the inversion of equation (10.6-17) to be computed. Equation (10.6-18) is in the form of a data equation that potentially could be used in the SRIF to compute the $\delta\boldsymbol{\beta}$ and $\delta\boldsymbol{\theta}$ states minimizing penalty function equation (10.6-16). Unfortunately the structure of $\mathbf{M}'_\beta$ and $\mathbf{M}'_\theta$ is no longer limited to blocks of three grid columns, so the storage and computational requirements of the SRIF are prohibitive. Instead the solution is obtained using the iterative conjugate gradient on the normal equations to minimize the residual (CGNR) method that works with vectors rather than matrices. This approach has been used with grid sizes of 500,000 to 1,000,000 cells. Explicit inversion of $\mathbf{B}_h$ is not required because it is only necessary to solve for $\boldsymbol{\lambda}$ in equations of the form $\mathbf{B}_h\boldsymbol{\lambda} = \mathbf{z}$ or $\mathbf{B}_h^T\boldsymbol{\lambda} = \mathbf{z}$ These equations are solved using the bi-conjugate gradient stabilized linear solver described in Chapter 5. It is also found that convergence of the CGNR iterations is greatly improved by estimating $\mathbf{G}_\beta\delta\boldsymbol{\beta}$ rather than $\delta\boldsymbol{\beta}$. This is effectively a form of preconditioning.

Figure 10.3 shows contours of the estimated log conductivity field in the fifth hydrological unit for a site in Cape Cod, MA. Well locations are shown as shaded squares.

While this example appears to have little connection with the Kalman filter that is the subject of these chapters, it does demonstrate how a concept developed for the SRIF—the data equation—can be used to solve spatial problems for MRF

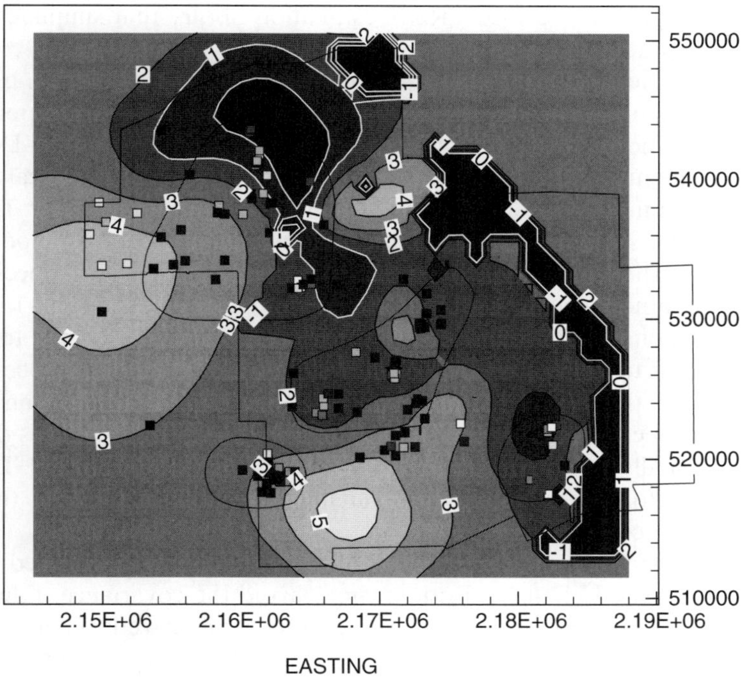

FIGURE 10.3: Data fusion estimated ln ky in layer 18 (Cape Cod, MA).

systems. The data equation and autoregressive model can be applied to many other types of MRF spatial problems.

10.7 SUMMARY

The potential numerical problems of the standard Kalman filter were known from the earliest applications. The differencing of positive semi-definite symmetric matrices in the measurement update of the covariance matrix can cause loss of accuracy. The Joseph form of the covariance update attempts to minimize sensitivity to numerical errors, but it is subject to the some of the same numerical problems as the standard form. More accurate Kalman filter algorithms work with a factored form of the covariance because the condition number of the covariance square root is one-half that of the covariance. Numerous numerical studies show that a factored filter implemented in single precision can achieve nearly the same accuracy as a covariance form implemented in double precision. This was very important for early filter applications on limited precision flight computers.

Studies also showed that the U-D filter is as computationally fast as the covariance Kalman filter, and faster than most other square-root filters. While the U-D filter is similar in concept to the covariance Kalman filter, the SRIF operates in the information domain, and has advantages for certain types of problems. The U-D filter and SRIF are the most widely used factorized filters. This is partly due to

Bierman's development of a FORTRAN estimation library that simplifies implementation of U-D and SRIF filters.

The U-D filter factors the symmetric positive semi-definite covariance matrix as $\mathbf{UDU}^T$ where $\mathbf{U}$ is unit upper triangular and $\mathbf{D}$ is diagonal with positive elements. The U-D measurement update is implemented as rank-1 updates to the U-D factors, where measurements must be processed one-component-at-a-time scalars. The rank-1 update is a modified version of an algorithm due to Agee-Turner. The multiplication $\mathbf{\Phi U}$ and subsequent re-triangularization is the deterministic portion of the time update. There are several options for the stochastic step (process noise inclusion) of the time update. It may be implemented as a series of rank-1 updates, or MWGS orthogonalization may be used to triangularize $[\mathbf{\Phi U} \quad \mathbf{G}]$ and to update $\mathbf{D}$. A third option is to form $\mathbf{\Phi UDU}^T\mathbf{\Phi}^T$ for only those states driven by process noise, and then re-factor the result. This is less desirable than the other options as the squaring operation can introduce numerical errors. The MWGS method is the most flexible and is usually comparable in speed to the rank-1 update method. The U-D filter can also perform error analysis for unadjusted parameters using the same methods as for the covariance filter.

Bierman developed two versions of an RTS smoother that are designed to work with outputs from the U-D filter. Both versions avoid the covariance differencing present in the standard RTS, so numerical accuracy is improved. When used in a general form for systems composed of dynamic states and biases, the computational burden of the U-D RTS smoother is similar to that of the standard RTS implemented with similar considerations. However, Bierman shows that for systems with first-order Markov process states, further partitioning can make the new algorithm faster than the standard RTS.

The SRIF is introduced by showing that the Kalman filter measurement and time updates can both be generalized as data equations. The data equation approach is used to process measurement and time update information by triangularizing the information matrix using orthogonal transforms. This is similar to the method previously presented for least-squares problems solved using the QR method. An example demonstrates how the SRIF can be efficiently used for large interconnected systems. SRIF error analysis for effects of unadjusted biases is also described.

It is shown that outputs of the SRIF time update can be used in a SRIS to compute smoothed estimates and covariances. However, the computational burden of the SRIS covariance step is high because triangulaization of a rectangular array is required. The DMCS algorithm uses the same SRIF outputs to compute the smoothed covariance with less computation than the SRIS.

An INS example is used to compare the numerical accuracy and execution speed of the standard and modified Kalman, U-D, and SRIF filters using enhanced versions of Bierman's ESL routines. The U-D and SRIF filters implemented in single precision are shown to have nearly the accuracy of double precision covariance-based Kalman filters. The covariance-based filters do not execute in single precision for this problem unless the position and velocity variances are arbitrarily constrained to be greater than 10. Execution time for the sparse matrix U-D filter is comparable to that of the standard Kalman filter, while other filter versions are much slower. The SRIF has approximately the same accuracy as the U-D filter but is much slower. When implemented in double precision, differences in accuracy are much less significant, and even the covariance-based filters provide acceptable accuracy for this problem.

The general utility of the data equation approach for solving other types of problems is discussed. A hydrological modeling example shows how the data equation is applied to spatial MRF problems: soft spatial continuity constraints are imposed on hydraulic conductivity within hydrological units using 2-D or 3-D autoregressive models.

CHAPTER 11

ADVANCED FILTERING TOPICS

This chapter covers a variety of topics that are extensions of the basic Kalman filtering theory. Since model parameters are often poorly known, we first discuss maximum likelihood estimation (MLE) of parameters such as initial conditions, process noise variances, measurement noise variances, and dynamic model constants. System characteristics often change with time, so we next discuss methods that allow filters to adapt to model changes. The simplest adaptive filters use statistics on filter innovations to adjust the process noise covariance, and this approach can work well when changes in system "noise" levels are slow. When system states change value abruptly, a jump detection/estimation method based on hypothesis testing is usually a better approach. A multiple model filter may be appropriate when systems transition between a finite number of models.

Methods for enforcing equality or inequality constraints on filter estimates are also of interest. The methods employed are similar to those used for least-squares estimation, but the problem is somewhat more difficult. Robust estimation is another least-squares topic applicable to filtering. H-infinity filters are designed to minimize estimation errors when input errors are larger than expected. Thus they can track more accurately than Kalman filters when anomalous conditions occur.

The final two topics address alternate methods for nonlinear filtering: unscented Kalman filters and particle filters. Both methods approximate the state conditional mean and covariance by directly evaluating the nonlinear equations at multiple points, rather than by linearization at the current estimate (as done in the extended Kalman filter (EKF)). Unscented filters use a limited number of carefully selected evaluation points to accurately compute the state and covariance up to third-order terms of the Taylor series. Particle filters use a cloud of randomly selected evaluation points to approximate the probability density function. Particle filters are potentially more accurate, but require much more computation than unscented filters.

Unlike most of the other techniques discussed in this book, this author has no application experience with H-infinity, unscented or particle filters. This is mostly due to being unaware of the methods at the time that relevant filter work was

Advanced Kalman Filtering, Least-Squares and Modeling: A Practical Handbook, by Bruce P. Gibbs

completed. Even in hindsight, however, it is not clear that these techniques would have been a better choice than the method that was implemented at the time. Of the three new filtering approaches, it is this author's opinion that unscented filters are the most generally useful. The number of published papers on these topics indicates high interest and continuing development. Thus you should be aware of the technology and consider all options.

11.1 MAXIMUM LIKELIHOOD PARAMETER ESTIMATION

Several texts on Kalman filtering include extended discussions of error effects due to incorrect *a priori* estimates and variances, dynamic model parameters, process noise variances, measurement model parameters, and measurement noise variances. We have tended to ignore these topics, and now explain why: these modeling errors can be eliminated with only slightly more effort than required to analyze the effects. The errors are minimized by selecting model parameters that maximize the likelihood function. As shown previously in Section 8.2.3.1, the nominally white characteristic of the Kalman filter measurement innovations allows simple calculation of the log likelihood function for assumed Gaussian distributions. Hence it is only necessary to compute derivatives of the log likelihood function with respect to the unknown model parameters in order to solve for the parameters using an iterative method called *Fisher scoring*, or simply *scoring*. The method, suggested by Rao (1968), is a generalization of Gauss-Newton iterations used for nonlinear least-squares problems. Both methods approximate the Hessian matrix of the log likelihood by its expected value. However, the measurement covariance matrix used in *maximum likelihood* (ML) parameter identification is a function of the estimated parameters, so terms that do not appear in the nonlinear least squares Hessian matrix appear in the ML Hessian. Details of the algorithm are presented below.

Use of scoring with Kalman filter generation of the required partial derivatives is conceptually shown in Figure 11.1. For purposes of this section it is assumed that parameter estimation is performed off-line using a "training" measurement set. Then the optimized parameters are used in a filter that processes operational data. One method for online adaptive estimation of parameters in a single filter is mentioned in the next section on adaptive filtering.

Inclusion of the parameters as states in an EKF is another alternative for estimation of model parameters. That approach sometimes works for bias parameters when nonlinearities are mild, but for some problems the EKF may not converge to the correct solution. It also does not work when process noise and measurement noise variances are unknown.

Although conceptually simple, ML parameter identification is rarely used when developing Kalman filters. That is unfortunate because parameter optimization can greatly improve filter performance and can also make it less sensitive to other types of modeling errors. There are two major reasons why ML parameter identification is not used more frequently. First, the coding effort is significant because computation of partial derivatives is nontrivial and construction of flexible software allowing analysis of different modeling hypotheses is time-consuming. However, once

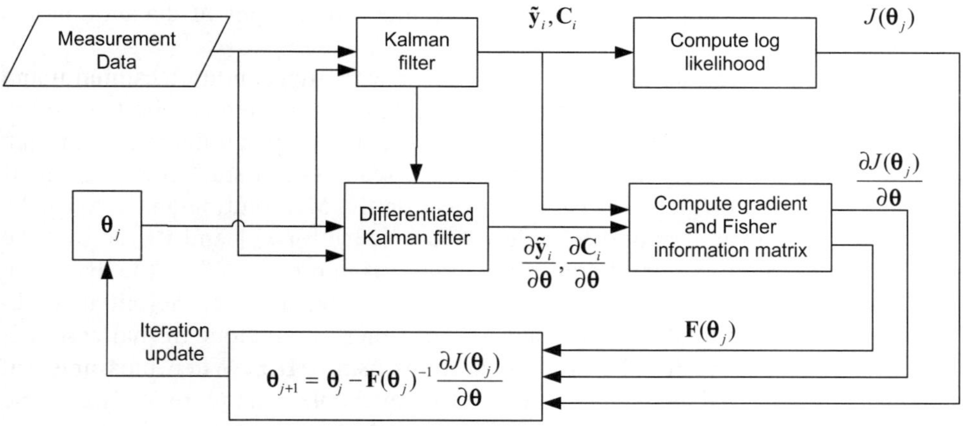

FIGURE 11.1: MLE data flow.

general-purpose software for this purpose has been created (see Chapter 13 for an example), it can be used for different projects with relatively minor modifications. The second problem is the computational burden. The computations required for partial derivatives of the filter innovations with respect to the unknown model parameters are slightly greater than those of one Kalman filter for every unknown parameter. That is, solution for 10 parameters requires slightly more computation than 11 Kalman filters. That problem greatly limited use of MLE several decades ago, but the issue is much less important with today's computers. An example later in this section discusses a problem with 371 states and 103,505 measurements where an ML solution for 17 unknown model parameters was obtained in several hours on a PC.

We start by discussing the types of parameters that contribute to model error and may be estimated via MLE. The possible error sources include incorrect

1. *A priori* estimate $\hat{\mathbf{x}}_{0/0}$
2. *A priori* covariance $\mathbf{P}_{0/0}$
3. Dynamic model parameters $\boldsymbol{\alpha}_d$ in $\dot{\mathbf{x}}(t) = \mathbf{f}(\mathbf{x}, \boldsymbol{\alpha}_d)$
4. Measurement model parameters $\boldsymbol{\alpha}_m$ in $\mathbf{y}(t) = \mathbf{h}(\mathbf{x}, \boldsymbol{\alpha}_m)$
5. Process noise covariances $\mathbf{Q}$, or power spectral densities (PSDs) ($\mathbf{Q}_s$) when using a continuous model
6. Measurement noise variances $\mathbf{R}$

It is unusual to have errors in measurement model parameter that are not states in the model (e.g., biases, colored noise), so $\boldsymbol{\alpha}_m$ parameters are generally not present. Dynamic model parameters $\boldsymbol{\alpha}_d$ are usually the constants of Markov process models (e.g., a in $\dot{x} = -ax + q$) since these colored noise models are mostly used to account for random effects that are difficult to characterize. Core state dynamic models are frequently based on physical variables and less likely to have unknown parameters, but the following development does not preclude other types of dynamic model parameters in $\boldsymbol{\alpha}_d$.

The first two categories of model error—incorrect initial condition estimates and covariances—are inherently different from the other types and can be handled in MLE using a different approach. If the measurement data span is much greater than the time required for the Kalman filter to nearly reach steady state, then the effects of initial condition errors will have little effect on the ML solution and can usually be ignored. However, it may still be desirable to solve for $\bar{\mathbf{x}}_{0/0}$ and $\mathbf{P}_{0/0}$ so that the start-up response of an operational Kalman filter is improved. When the data span is not much greater than the time to reach steady state, errors in the initial conditions may significantly affect ML solutions for other parameters. In that case it is also desirable to solve for the initial conditions using MLE. When measurement data from multiple independent tests (trials) are available, it is theoretically possible to solve for both the mean ($\bar{\mathbf{x}}_{0/0}$) and covariance ($\mathbf{P}_{0/0}$) of the initial states, but there is often insufficient data to compute $\mathbf{P}_{0/0}$. Hence for single trials MLE can only estimate $\mathbf{x}_{0/0}$. Although the filter response may be affected by incorrect $\mathbf{P}_{0/0}$, the impact on errors in ML estimates of other parameters should be minimal if $\mathbf{x}_{0/0}$ is estimated prior-free in the ML solution.

To summarize, the parameters estimated in single trials via MLE may be generally defined as two groups,

$$\boldsymbol{\theta} = \begin{bmatrix} \boldsymbol{\beta} \\ \boldsymbol{\alpha} \end{bmatrix}, \tag{11.1-1}$$

where $\boldsymbol{\beta}$ defines the filter state initial conditions $\hat{\mathbf{x}}_{0/0}$ for a single trial, and $\boldsymbol{\alpha}$ includes other types of model parameters. Often $\boldsymbol{\beta} = \hat{\mathbf{x}}_{0/0}$ but it may desirable to initialize Markov process model states to zero if the initial condition is not readily observable from the available measurement data. Four types of model parameters are included in $\boldsymbol{\alpha}$:

$$\boldsymbol{\alpha} = \begin{bmatrix} \boldsymbol{\alpha}_d^T & \boldsymbol{\alpha}_m^T & \boldsymbol{\alpha}_q^T & \boldsymbol{\alpha}_r^T \end{bmatrix}^T \tag{11.1-2}$$

where

- $\boldsymbol{\alpha}_d$ are parameters that define the dynamic model $\dot{\mathbf{x}}(t) = \mathbf{f}(\mathbf{x}, \boldsymbol{\alpha}_d)$
- $\boldsymbol{\alpha}_m$ are parameters that define the measurement model $\mathbf{y}(t) = \mathbf{h}(\mathbf{x}, \boldsymbol{\alpha}_m)$,
- $\boldsymbol{\alpha}_q$ are parameters that define the continuous process noise PSD $\mathbf{Q}_s$, which is usually diagonal,
- $\boldsymbol{\alpha}_r$ are parameters that define the measurement noise variance $\mathbf{R}$.

In many cases the number of elements in $\boldsymbol{\alpha}_q$ will be less than the number of diagonal entries in $\mathbf{Q}_s$ because the same value can be used for different channels or coordinates. That may also be true for $\boldsymbol{\alpha}_d$.

11.1.1 Calculation of the State Transition Partial Derivatives

To compute the gradient of the log likelihood required for the ML scoring iterations, the partial derivatives of the measurement innovations and innovation covariance with respect to $\boldsymbol{\theta}$ must be computed. Those partials are computed from partial derivatives of Kalman filter quantities. We start with the dynamic model and the filter time update equations. As in Chapter 2, the true stochastic differential equation for the state is assumed to be

$$\dot{\mathbf{x}}(t) = \mathbf{f}(\mathbf{x}(t), \boldsymbol{\alpha}) + \mathbf{q}(t) \tag{11.1-3}$$

where $E[\mathbf{q}(t)] = \mathbf{0}$ and $E[\mathbf{q}(t)\mathbf{q}(\tau)^T] = \mathbf{Q}_s\,\delta(t-\tau)$. This leads to the filter time update

$$\hat{\mathbf{x}}_{i/i-1} = \hat{\mathbf{x}}_{i-1/i-1} + \int_{t_{i-1}}^{t_i} \mathbf{f}(\hat{\mathbf{x}}(\lambda), \boldsymbol{\alpha})\,d\lambda. \tag{11.1-4}$$

We again ignore the effects of known control inputs, but those terms can be added if needed. Partial derivatives of the state time update with respect to $\boldsymbol{\alpha}$ are a required input when calculating derivatives of the Kalman filter. Derivatives of $\hat{\mathbf{x}}$ with respect to $\boldsymbol{\beta}$ will be discussed later. The partial derivatives $\partial\hat{\mathbf{x}}_{i/i-1}/\partial\boldsymbol{\alpha}$, $\partial\boldsymbol{\Phi}_{i,i-1}/\partial\boldsymbol{\alpha}$, $\partial\mathbf{Q}_{i,i-1}/\partial\boldsymbol{\alpha}$ can be computed using the same methods described in Chapter 2 for computing $\hat{\mathbf{x}}_{i/i-1}$, $\boldsymbol{\Phi}_{i,i-1}$ and $\mathbf{Q}_{i,i-1}$. We now summarize several of these methods. Notice that some approaches work adequately for short time intervals, but it is often necessary to use scaling and squaring to obtain the desired transition arrays when the interval between measurements is large.

11.1.1.1 Taylor Series The homogenous solution to equation (11.1-3) for a general nonlinear system is

$$\begin{aligned}
\mathbf{x}_i &= \mathbf{x}_{i-1} + \int_{t_{i-1}}^{t_i} \mathbf{f}(\mathbf{x}(\lambda), \boldsymbol{\alpha})\,d\lambda \\
&= \mathbf{x}_{i-1} + \mathbf{f}(\mathbf{x}_{i-1}, \boldsymbol{\alpha})T + \mathbf{F}_x(\mathbf{x}_{i-1}, \boldsymbol{\alpha})\mathbf{f}(\mathbf{x}_{i-1}, \boldsymbol{\alpha})\frac{T^2}{2} \\
&\quad + \left(\frac{\partial \mathbf{F}_x(\mathbf{x}_{i-1}, \boldsymbol{\alpha})}{\partial \mathbf{x}} + \mathbf{F}_x(\mathbf{x}_{i-1}, \boldsymbol{\alpha})\mathbf{F}_x(\mathbf{x}_{i-1}, \boldsymbol{\alpha})\right)\mathbf{f}(\mathbf{x}_{i-1}, \boldsymbol{\alpha})\frac{T^3}{6} + \ldots
\end{aligned} \tag{11.1-5}$$

where $\mathbf{x}_i = \mathbf{x}(t_i)$, $T = t_i - t_{i-1}$ and

$$\boxed{\mathbf{F}_x(\mathbf{x}, \boldsymbol{\alpha}) \triangleq \frac{\partial \mathbf{f}(\mathbf{x}, \boldsymbol{\alpha})}{\partial \mathbf{x}}.} \tag{11.1-6}$$

The notation $\partial\mathbf{F}_x/\partial\mathbf{x}$ means $\partial\mathbf{F}_x/\partial x_j$ for $j = 1$ to n; that is, it is a set of matrices. Hence

$$\boxed{\begin{aligned}
\frac{\partial \hat{\mathbf{x}}_{i/i-1}}{\partial \boldsymbol{\alpha}} &= \frac{\partial \hat{\mathbf{x}}_{i-1/i-1}}{\partial \boldsymbol{\alpha}} + \left(\mathbf{F}_\alpha(\hat{\mathbf{x}}_{i-1/i-1}, \boldsymbol{\alpha}) + \mathbf{F}_x(\hat{\mathbf{x}}_{i-1/i-1}, \boldsymbol{\alpha})\frac{\partial \hat{\mathbf{x}}_{i-1/i-1}}{\partial \boldsymbol{\alpha}}\right)T \\
&\quad + \left(\begin{aligned}&\frac{\partial \mathbf{F}_x(\hat{\mathbf{x}}_{i-1/i-1}, \boldsymbol{\alpha})}{\partial \pm}\mathbf{f}(\hat{\mathbf{x}}_{i-1/i-1}, \boldsymbol{\alpha}) + \mathbf{F}_x(\hat{\mathbf{x}}_{i-1/i-1}, \boldsymbol{\alpha})\mathbf{F}_\alpha(\hat{\mathbf{x}}_{i-1/i-1}, \boldsymbol{\alpha}) \\ &+ \left(\frac{\partial \mathbf{F}_x(\hat{\mathbf{x}}_{i-1/i-1}, \boldsymbol{\alpha})}{\partial \hat{\mathbf{x}}}\mathbf{f}(\hat{\mathbf{x}}_{i-1/i-1}, \boldsymbol{\alpha}) + \mathbf{F}_x(\hat{\mathbf{x}}_{i-1/i-1}, \boldsymbol{\alpha})\mathbf{F}_x(\hat{\mathbf{x}}_{i-1/i-1}, \boldsymbol{\alpha})\right)\frac{\partial \hat{\mathbf{x}}_{i-1/i-1}}{\partial \boldsymbol{\alpha}}\end{aligned}\right)\frac{T^2}{2} \\
&\quad + \ldots
\end{aligned}} \tag{11.1-7}$$

where

$$\boxed{\mathbf{F}_\alpha(\mathbf{x}, \boldsymbol{\alpha}) \triangleq \frac{\partial \mathbf{f}(\mathbf{x}, \boldsymbol{\alpha})}{\partial \boldsymbol{\alpha}},} \tag{11.1-8}$$

$\partial\hat{\mathbf{x}}_{i/i-1}/\partial\boldsymbol{\alpha}$ means $\partial\hat{\mathbf{x}}_{i/i-1}/\partial\alpha_j$, and $\partial\mathbf{F}_x/\partial\boldsymbol{\alpha}$ means $\partial\mathbf{F}_x/\partial\alpha_j$ for $j = 1, 2,\ldots, p$ if there are p parameters in $\boldsymbol{\alpha}$. Now from equation (11.1-5)

$$\begin{aligned}\boldsymbol{\Phi}_{i,i-1} &\triangleq \frac{\partial\mathbf{x}_i}{\partial\mathbf{x}_{i-1}}\\ &= \mathbf{I} + \mathbf{F}_x(\mathbf{x}_{i-1},\boldsymbol{\alpha})T + \left(\frac{\partial\mathbf{F}_x(\mathbf{x}_{i-1},\boldsymbol{\alpha})}{\partial\mathbf{x}}\mathbf{f}(\mathbf{x}_{i-1},\boldsymbol{\alpha}) + \mathbf{F}_x(\mathbf{x}_{i-1},\boldsymbol{\alpha})\mathbf{F}_x(\mathbf{x}_{i-1},\boldsymbol{\alpha})\right)\frac{T^2}{2} + \ldots\end{aligned} \tag{11.1-9}$$

and

$$\boxed{\begin{aligned}\frac{\partial\boldsymbol{\Phi}_{i,i-1}}{\partial\boldsymbol{\alpha}} &\cong \frac{\partial\mathbf{F}_x(\mathbf{x}_{i-1},\boldsymbol{\alpha})}{\partial\boldsymbol{\alpha}}T\\ &+ \begin{pmatrix}\dfrac{\partial^2\mathbf{F}_x(\mathbf{x}_{i-1},\boldsymbol{\alpha})}{\partial\boldsymbol{\alpha}\partial\mathbf{x}^T}\mathbf{f}(\mathbf{x}_{i-1},\boldsymbol{\alpha}) + \dfrac{\partial\mathbf{F}_x(\mathbf{x}_{i-1},\boldsymbol{\alpha})}{\partial\mathbf{x}}\mathbf{F}_\alpha(\mathbf{x}_{i-1},\boldsymbol{\alpha})\\ + \dfrac{\partial\mathbf{F}_x(\mathbf{x}_{i-1},\boldsymbol{\alpha})}{\partial\boldsymbol{\alpha}}\mathbf{F}_x(\mathbf{x}_{i-1},\boldsymbol{\alpha}) + \mathbf{F}_x(\mathbf{x}_{i-1},\boldsymbol{\alpha})\dfrac{\partial\mathbf{F}_x(\mathbf{x}_{i-1},\boldsymbol{\alpha})}{\partial\boldsymbol{\alpha}}\end{pmatrix}\frac{T^2}{2} + \ldots\end{aligned}} \tag{11.1-10}$$

This expression ignores terms of the form $(\partial\mathbf{f}/\partial\mathbf{x})(\partial\mathbf{x}/\partial\boldsymbol{\alpha})$ and $(\partial\mathbf{F}_x/\partial\mathbf{x})(\partial\mathbf{x}/\partial\boldsymbol{\alpha})$, but they may be significant in some cases. From equation (2.3-17), the discrete process noise covariance can be expressed as

$$\begin{aligned}\mathbf{Q}_{i,i-1} = \mathbf{Q}_s T &+ [\mathbf{F}_x(\mathbf{x}_{i-1},\boldsymbol{\alpha})\mathbf{Q}_s + \mathbf{Q}_s\mathbf{F}_x^T(\mathbf{x}_{i-1},\boldsymbol{\alpha})]T^2/2\\ &+ \mathbf{F}_x(\mathbf{x}_{i-1},\boldsymbol{\alpha})\mathbf{Q}_s\mathbf{F}_x^T(\mathbf{x}_{i-1},\boldsymbol{\alpha})T^3/3 + \ldots\end{aligned} \tag{11.1-11}$$

so

$$\boxed{\begin{aligned}\frac{\partial\mathbf{Q}_{i,i-1}}{\partial\boldsymbol{\alpha}} = \frac{\partial\mathbf{Q}_s}{\partial\boldsymbol{\alpha}}T &+ \left(\frac{\partial\mathbf{F}_x(\mathbf{x}_{i-1},\boldsymbol{\alpha})}{\partial\boldsymbol{\alpha}}\mathbf{Q}_s + \mathbf{F}_x(\mathbf{x}_{i-1},\boldsymbol{\alpha})\frac{\partial\mathbf{Q}_s}{\partial\boldsymbol{\alpha}}\right)\frac{T^2}{2}\\ &+ \left(\frac{\partial\mathbf{F}_x(\mathbf{x}_{i-1},\boldsymbol{\alpha})}{\partial\boldsymbol{\alpha}}\mathbf{Q}_s + \mathbf{F}_x(\mathbf{x}_{i-1},\boldsymbol{\alpha})\frac{\partial\mathbf{Q}_s}{\partial\boldsymbol{\alpha}}\right)^T\frac{T^2}{2} + \ldots\end{aligned}}. \tag{11.1-12}$$

Notice that $\mathbf{Q}_s$ is a function of only $\boldsymbol{\alpha}_q$, while $\mathbf{F}_x$ is a function of only $\boldsymbol{\alpha}_d$, so the different dependencies should be utilized when computing the derivatives. Also some terms disappear when the system is linear in $\mathbf{x}$.

Provided that the integration step T is much smaller than the shortest time constants in $\mathbf{F}_x$, it is often adequate to truncate the partial derivatives $\partial\boldsymbol{\Phi}_{i,i-1}/\partial\boldsymbol{\alpha}$, $\partial\mathbf{Q}_{i,i-1}/\partial\boldsymbol{\alpha}$ after the first or second terms. When actual time steps between measurements are greater than the shortest time constants in $\mathbf{F}_x$, the above partial derivatives are evaluated for a short time interval, and scaling and squaring is used to reach the desired time step.

11.1.1.2 Integration of $\partial\dot{\mathbf{x}}/\partial\boldsymbol{\alpha}$ and $\partial\dot{\boldsymbol{\Phi}}/\partial\boldsymbol{\alpha}$ The state partial derivative $\partial\hat{\mathbf{x}}_{i/i-1}/\partial\boldsymbol{\alpha}$ can be computed by numerically integrating $\partial\dot{\mathbf{x}}(t)/\partial\boldsymbol{\alpha}$ initialized with $\partial\hat{\mathbf{x}}_{i-1/i-1}/\partial\boldsymbol{\alpha}$.

Since the time derivative of the state transition matrix obeys

$$\frac{d\boldsymbol{\Phi}(\mathbf{x}_{i-1},\boldsymbol{\alpha},t-t_{i-1})}{dt}=\mathbf{F}_x(\mathbf{x}(t),\boldsymbol{\alpha})\boldsymbol{\Phi}(\mathbf{x}_{i-1},\boldsymbol{\alpha},t-t_{i-1}), \tag{11.1-13}$$

the partial derivatives of $\dot{\boldsymbol{\Phi}}$ can be obtained by integration of

$$\boxed{\frac{\partial\dot{\boldsymbol{\Phi}}(\mathbf{x}_{i-1},\boldsymbol{\alpha},t-t_{i-1})}{\partial\boldsymbol{\alpha}}\cong\frac{\partial\mathbf{F}_x(\mathbf{x}(t),\boldsymbol{\alpha})}{\partial\boldsymbol{\alpha}}\boldsymbol{\Phi}(\mathbf{x}_{i-1},\boldsymbol{\alpha},t-t_{i-1})+\mathbf{F}_x(\mathbf{x}(t),\boldsymbol{\alpha})\frac{\partial\boldsymbol{\Phi}(\mathbf{x}_{i-1},\boldsymbol{\alpha},t-t_{i-1})}{\partial\boldsymbol{\alpha}}} \tag{11.1-14}$$

initialized with

$$\frac{\partial\dot{\boldsymbol{\Phi}}(\hat{\mathbf{x}}_{i-1/i-1},\boldsymbol{\alpha},0)}{\partial\boldsymbol{\alpha}}=\frac{\partial\mathbf{F}_x(\hat{\mathbf{x}}_{i-1/i-1},\boldsymbol{\alpha}))}{\partial\boldsymbol{\alpha}}$$

and using the same numerical integrator used for computing $\hat{\mathbf{x}}_{i/i-1}$ and $\boldsymbol{\Phi}_{i,i-1}$. We have again ignored terms of the form $(\partial\mathbf{F}_x/\partial\mathbf{x})(\partial\mathbf{x}/\partial\boldsymbol{\alpha})$.

Unfortunately the discrete state noise matrix does not obey a similarly simple differential equation, so it is necessary to use Taylor series integration or another method to compute $\partial\mathbf{Q}_{i,i-1}/\partial\boldsymbol{\alpha}$.

11.1.1.3 Numeric Partial Derivatives In cases where complex dynamics make analytic computation of partial derivatives difficult, numeric partial derivatives using central differences may be used to compute

$$\frac{\partial\hat{\mathbf{x}}_{i/i-1}}{\partial\boldsymbol{\alpha}}\quad\text{and}\quad\frac{\partial\boldsymbol{\Phi}_{i,i-1}}{\partial\boldsymbol{\alpha}}$$

where $\hat{\mathbf{x}}_{i/i-1}$ is computed by numerical integration. Sometimes numeric partial derivatives are only used for the portion of the system with complex dynamics, and analytic or Taylor series integration is used for simpler dynamics in the same system. Numeric partials cannot be used to compute $\partial\mathbf{Q}_{i,i-1}/\partial\boldsymbol{\alpha}$, so truncated Taylor series is often used for short time intervals because $\partial\mathbf{Q}_{i,i-1}/\partial\boldsymbol{\alpha}$ need not be as accurate as $\partial\boldsymbol{\Phi}_{i,i-1}/\partial\boldsymbol{\alpha}$.

11.1.1.4 Scaling and Squaring (Interval Doubling) Since the above methods may not be accurate when evaluated for the full interval between measurements, it may be necessary to use scaling and squaring to reach the desired measurement interval. Recall that for a time-invariant process, the discrete process noise matrix for an interval of $2T$ is obtained from equation (2.3-19) as

$$\mathbf{Q}(2T)=\boldsymbol{\Phi}(T)\mathbf{Q}(T)\boldsymbol{\Phi}^T(T)+\mathbf{Q}(T)$$

and the state transition matrix is

$$\boldsymbol{\Phi}(2T)=\boldsymbol{\Phi}(T)\boldsymbol{\Phi}(T).$$

Hence

$$\boxed{\begin{aligned}\frac{\partial\mathbf{Q}(2T)}{\partial\boldsymbol{\alpha}}&=\frac{\partial\boldsymbol{\Phi}(T)}{\partial\boldsymbol{\alpha}}\mathbf{Q}(T)\boldsymbol{\Phi}^T(T)+\boldsymbol{\Phi}(T)\frac{\partial\mathbf{Q}(T)}{\partial\boldsymbol{\alpha}}\boldsymbol{\Phi}^T\\&\quad+\boldsymbol{\Phi}(T)\mathbf{Q}(T)\frac{\partial\boldsymbol{\Phi}^T(T)}{\partial\boldsymbol{\alpha}}+\frac{\partial\mathbf{Q}(T)}{\partial\boldsymbol{\alpha}}\end{aligned}} \tag{11.1-15}$$

and

$$\boxed{\frac{\partial \boldsymbol{\Phi}(2T)}{\partial \boldsymbol{\alpha}} = \frac{\partial \boldsymbol{\Phi}(T)}{\partial \boldsymbol{\alpha}} \boldsymbol{\Phi}(T) + \boldsymbol{\Phi}(T) \frac{\partial \boldsymbol{\Phi}(T)}{\partial \boldsymbol{\alpha}}.} \tag{11.1-16}$$

11.1.2 Derivatives of the Filter Time Update

The filter state time update is listed above as equation (11.1-4), and the corresponding covariance update is

$$\mathbf{P}_{i/i-1} = \boldsymbol{\Phi}_{i,i-1} \mathbf{P}_{i-1/i-1} \boldsymbol{\Phi}_{i,i-1}^T + \mathbf{Q}_{i,i-1}. \tag{11.1-17}$$

For the moment we only consider partial derivatives of these equations with respect to $\boldsymbol{\alpha}$: partials with respect to $\mathbf{x}_{0/0}$ will be computed later using the methods of Chapter 9. Partial derivatives of the *a priori* state estimate can be computed using one of the methods described in the last section. For linear systems

$$\boxed{\frac{\partial \hat{\mathbf{x}}_{i/i-1}}{\partial \hat{\boldsymbol{\alpha}}} = \boldsymbol{\Phi}_{i,i-1} \frac{\partial \hat{\mathbf{x}}_{i-1/i-1}}{\partial \hat{\boldsymbol{\alpha}}} + \frac{\partial \boldsymbol{\Phi}_{i,i-1}}{\partial \hat{\boldsymbol{\alpha}}} \hat{\mathbf{x}}_{i-1/i-1}} \tag{11.1-18}$$

where $\partial \hat{\mathbf{x}}_{i/i-1} / \partial \hat{\boldsymbol{\alpha}}$ means $\partial \hat{\mathbf{x}}_{i/i-1} / \partial \hat{\alpha}_j$ for $j = 1$ to p. For linear or nonlinear systems the partial derivatives of the covariance are

$$\boxed{\frac{\partial \mathbf{P}_{i/i-1}}{\partial \hat{\boldsymbol{\alpha}}} = \left(\frac{\partial \boldsymbol{\Phi}_{i,i-1}}{\partial \hat{\boldsymbol{\alpha}}} \mathbf{P}_{i-1/i-1} \boldsymbol{\Phi}_{i,i-1}^T \right) + \left(\frac{\partial \boldsymbol{\Phi}_{i,i-1}}{\partial \hat{\boldsymbol{\alpha}}} \mathbf{P}_{i-1/i-1} \boldsymbol{\Phi}_{i,i-1}^T \right)^T + \boldsymbol{\Phi}_{i,i-1} \frac{\partial \mathbf{P}_{i-1/i-1}}{\partial \hat{\boldsymbol{\alpha}}} \boldsymbol{\Phi}_{i,i-1}^T + \frac{\partial \mathbf{Q}_{i,i-1}}{\partial \hat{\boldsymbol{\alpha}}}.} \tag{11.1-19}$$

Notice that the derivatives are with respect to $\hat{\boldsymbol{\alpha}}$ rather than $\boldsymbol{\alpha}$. We use $\hat{\boldsymbol{\alpha}}$ to denote the value of $\boldsymbol{\alpha}$ used in the Kalman filter model for the current ML scoring iteration.

Since computation of the above equations plus $\partial \boldsymbol{\Phi}_{i,i-1} / \partial \hat{\boldsymbol{\alpha}}$ is usually the most numerically intensive step of the entire process, it is important that those equations be implemented efficiently. Terms should only be calculated for partials that are nonzero, for example, $\boldsymbol{\Phi}_{i,i-1}$ is only a function of $\boldsymbol{\alpha}_d$ and $\mathbf{Q}_{i,i-1}$ is only a function of $\boldsymbol{\alpha}_q$. Further, if there are a significant number of biases in the state vector such that

$$\boldsymbol{\Phi} = \begin{bmatrix} \boldsymbol{\Phi}_{dd} & \boldsymbol{\Phi}_{db} \\ \mathbf{0} & \mathbf{I} \end{bmatrix},$$

it is often true that

$$\frac{\partial \boldsymbol{\Phi}}{\partial \boldsymbol{\alpha}} = \begin{bmatrix} \dfrac{\partial \boldsymbol{\Phi}_{dd}}{\partial \boldsymbol{\alpha}} & \mathbf{0} \\ \mathbf{0} & \mathbf{0} \end{bmatrix}$$

so the zero partitions should not be computed and matrix multiplications should ignore these partitions. One approach that minimizes unnecessary calculations implements the derivatives using Friedland's bias-free/bias-restoring filters, where the two filters are separately differentiated (see Shuster et al. 1983 or Shuster and Porter 1984 for an example based on multi-test accuracy analysis). However, a single filter implemented in dynamic and bias partitions can be just as efficient, as demonstrated in Chapter 8 for the filter. Also, sparse matrix multiplication utilities (see the included SMULT and SMULT_BLOCK routines) should be used since the above partial arrays tend to be very sparse.

11.1.3 Derivatives of the Filter Measurement Update

Given the nonlinear measurement model $\hat{\mathbf{y}}_i = \mathbf{h}_i(\hat{\mathbf{x}}_{i/i-1}, \hat{\boldsymbol{\alpha}})$, the Kalman filter measurement covariance update is

$$\begin{aligned}
\mathbf{H}_i &= \frac{\partial \mathbf{h}_i(\hat{\mathbf{x}}_{i/i-1}, \hat{\boldsymbol{\alpha}})}{\partial \hat{\mathbf{x}}_{i/i-1}} \\
\mathbf{C}_i &= \mathbf{H}_i \mathbf{P}_{i/i-1} \mathbf{H}_i^T + \mathbf{R}_i \\
\mathbf{K}_i &= \mathbf{P}_{i/i-1} \mathbf{H}_i^T \mathbf{C}_i^{-1} \\
\mathbf{P}_{i/i} &= \mathbf{P}_{i/i-1} - \mathbf{K}_i \mathbf{H}_i \mathbf{P}_{i/i-1}
\end{aligned} \tag{11.1-20}$$

and the state update is

$$\begin{aligned}
\tilde{\mathbf{y}}_i &= \mathbf{y}_i - \mathbf{h}_i(\hat{\mathbf{x}}_{i/i-1}, \hat{\boldsymbol{\alpha}}) \\
\hat{\mathbf{x}}_{i/i} &= \hat{\mathbf{x}}_{i/i-1} + \mathbf{K}_i \tilde{\mathbf{y}}_i
\end{aligned}. \tag{11.1-21}$$

The partial derivatives of these equations are thus

$$\boxed{\begin{aligned}
\frac{\partial \mathbf{C}_i}{\partial \hat{\boldsymbol{\alpha}}} &= \left(\mathbf{H}_i \mathbf{P}_{i/i-1} \frac{\partial \mathbf{H}_i^T}{\partial \hat{\boldsymbol{\alpha}}} \right) + \left(\mathbf{H}_i \mathbf{P}_{i/i-1} \frac{\partial \mathbf{H}_i^T}{\partial \hat{\boldsymbol{\alpha}}} \right)^T + \mathbf{H}_i \frac{\partial \mathbf{P}_{i/i-1}}{\partial \hat{\boldsymbol{\alpha}}} \mathbf{H}_i^T + \frac{\partial \mathbf{R}_i}{\partial \hat{\boldsymbol{\alpha}}} \\
\frac{\partial \mathbf{K}_i}{\partial \hat{\boldsymbol{\alpha}}} &= \left(\frac{\partial \mathbf{P}_{i/i-1}}{\partial \hat{\boldsymbol{\alpha}}} \mathbf{H}_i^T + \mathbf{P}_{i/i-1} \frac{\partial \mathbf{H}_i^T}{\partial \hat{\boldsymbol{\alpha}}} - \mathbf{K}_i \frac{\partial \mathbf{C}_i}{\partial \hat{\boldsymbol{\alpha}}} \right) \mathbf{C}_i^{-1} \\
\frac{\partial \mathbf{P}_{i/i}}{\partial \hat{\boldsymbol{\alpha}}} &= \frac{\partial \mathbf{P}_{i/i-1}}{\partial \hat{\boldsymbol{\alpha}}} - \mathbf{K}_i \mathbf{H}_i \frac{\partial \mathbf{P}_{i/i-1}}{\partial \hat{\boldsymbol{\alpha}}} - \left(\frac{\partial \mathbf{K}_i}{\partial \hat{\boldsymbol{\alpha}}} \mathbf{H}_i + \mathbf{K}_i \frac{\partial \mathbf{H}_i}{\partial \hat{\boldsymbol{\alpha}}} \right) \mathbf{P}_{i/i-1}
\end{aligned}} \tag{11.1-22}$$

and

$$\boxed{\begin{aligned}
\frac{\partial \tilde{\mathbf{y}}_i}{\partial \hat{\boldsymbol{\alpha}}} &= -\left(\frac{\partial \mathbf{h}_i(\hat{\mathbf{x}}_{i/i-1}, \hat{\boldsymbol{\alpha}})}{\partial \hat{\boldsymbol{\alpha}}} + \mathbf{H}_i \frac{\partial \hat{\mathbf{x}}_{i/i-1}}{\partial \hat{\boldsymbol{\alpha}}} \right) \\
\frac{\partial \hat{\mathbf{x}}_{i/i}}{\partial \hat{\boldsymbol{\alpha}}} &= \frac{\partial \hat{\mathbf{x}}_{i/i-1}}{\partial \hat{\boldsymbol{\alpha}}} + \frac{\partial \mathbf{K}_i}{\partial \hat{\boldsymbol{\alpha}}} \tilde{\mathbf{y}}_i + \mathbf{K}_i \frac{\partial \tilde{\mathbf{y}}_i}{\partial \hat{\boldsymbol{\alpha}}}
\end{aligned}}, \tag{11.1-23}$$

where we have used the equation for the derivative of the matrix inverse:

$$\frac{\partial \mathbf{C}^{-1}}{\partial \hat{\alpha}} = -\mathbf{C}^{-1} \frac{\partial \mathbf{C}}{\partial \hat{\alpha}} \mathbf{C}^{-1}.$$

The equations are actually simpler than indicated because $\partial \mathbf{H}_i / \partial \hat{\boldsymbol{\alpha}}$ is normally zero and several intermediate matrix products are repeated. Implementation is almost trivial when measurements are processed one-component-at-a-time so that $\mathbf{C}_i$ and $\tilde{\mathbf{y}}_i$ are scalars and $\mathbf{H}_i$ is a row vector. Since $\mathbf{H}_i$ is usually very sparse, it may be beneficial to use sparse matrix multiplication utilities when the state dimension is large.

Although the above partial derivatives for the time and measurement updates were computed using the filter covariance equations, it is also possible to compute partials when using the U-D filter. Bierman et al. (1990) obtained partial derivatives of the U-D filter using line-by-line differentiation of the individual U-D filter algorithms. A similar approach was also used by Kulikova (2009). However, the method is more difficult to implement efficiently than the above equations and may not benefit as much from matrix sparseness. Note that the cost gradient used in the scoring iterations need not be as accurate as the cost function since small gradient errors primarily affect the rate of convergence rather than preventing convergence. Large gradient errors may cause convergence problems, but significant precision loss is rarely encountered when working in double precision.

11.1.4 Partial Derivatives for Initial Condition Errors

As previously indicated, it is very inefficient to compute filter partial derivatives with respect to initial condition parameters using the differentiated filter equations above. Instead we use sensitivity equations from Section 9.3.1, but modify the definition to be consistent with usage for $\boldsymbol{\alpha}$. Provided that any dependence of $\boldsymbol{\Phi}$ on $\hat{\boldsymbol{\beta}}$ can be ignored, the *a priori* sensitivity is

$$\boxed{\mathbf{S}^0_{i/i-1} \triangleq \frac{\partial \tilde{\mathbf{x}}_{i/i-1}}{\partial \hat{\boldsymbol{\beta}}} = \boldsymbol{\Phi}_{i,i-1}\mathbf{S}^0_{i-1/i-1}} \tag{11.1-24}$$

where $\tilde{\mathbf{x}}_{i/i-1} = \mathbf{x}_i - \hat{\mathbf{x}}_{i/i-1}$ and $\mathbf{S}^0_{0/0}$ is initialized as the $n \times n$ matrix

$$\boxed{\mathbf{S}^0_{0/0} = -\frac{\partial \hat{\mathbf{x}}_{0/0}}{\partial \hat{\boldsymbol{\beta}}}} \tag{11.1-25}$$

if there n elements in $\boldsymbol{\beta}$. Although it appears redundant, we retain the superscript 0 on $\mathbf{S}$ (indicating sensitivity to initial conditions) from Chapter 9. The sensitivity of the filter innovations is

$$\boxed{\mathbf{T}^0_i \triangleq \frac{\partial \tilde{\mathbf{y}}_i}{\partial \hat{\boldsymbol{\beta}}} = \mathbf{H}_i\mathbf{S}^0_{i/i-1}} \tag{11.1-26}$$

and the *a posteriori* state sensitivity is

$$\boxed{\mathbf{S}^0_{i/i} = \mathbf{S}^0_{i/i-1} - \mathbf{K}_i\mathbf{T}^0_i.} \tag{11.1-27}$$

The quantity of interest for subsequent MLE calculations is $\mathbf{T}^0_i$. Notice that for linear systems $\mathbf{K}_i$ is not a function of $\hat{\mathbf{x}}_{0/0}$, and for nonlinear systems the functional dependence is ignored because the sensitivity of $\mathbf{K}_i$ to $\hat{\mathbf{x}}_{0/0}$ is often small. If this not the case, the $\hat{\boldsymbol{\beta}}$ elements can be included as part of $\hat{\boldsymbol{\alpha}}$ and the equations of the previous section can be used.

11.1.5 Computation of the Log Likelihood and Scoring Step

Recall from Chapter 8 that the Kalman filter innovations $\tilde{\mathbf{y}}_i$ are white when all models are correct, and this allows the log likelihood for k measurement residuals to be computed as

$$\ln L(\mathbf{y}\,|\,\mathbf{x}_0,\boldsymbol{\theta}) = -\frac{1}{2}\left[\sum_{i=1}^{k} m_i \ln(2\pi) + \ln|\mathbf{C}_i| + \tilde{\mathbf{y}}_i^T \mathbf{C}_i^{-1}\tilde{\mathbf{y}}_i\right]. \qquad (11.1\text{-}28)$$

We define the cost function for the MLE as the negative of the log likelihood so that algorithms for cost function minimization can be used:

$$\begin{aligned} J &= -\ln L(\mathbf{y}\,|\,\mathbf{x}_0,\boldsymbol{\theta}) \\ &= \frac{1}{2}\left[\sum_{i=1}^{k} m_i \ln(2\pi) + \ln|\mathbf{C}_i| + \tilde{\mathbf{y}}_i^T \mathbf{C}_i^{-1}\tilde{\mathbf{y}}_i\right], \end{aligned} \qquad (11.1\text{-}29)$$

Using the matrix derivative

$$\frac{\partial \ln|\mathbf{C}|}{\partial \hat{\boldsymbol{\alpha}}} = \frac{\partial \ln|\mathbf{C}|}{\partial |\mathbf{C}|}\frac{\partial |\mathbf{C}|}{\partial \hat{\boldsymbol{\alpha}}} = \frac{1}{|\mathbf{C}|}\frac{\partial |\mathbf{C}|}{\partial \hat{\boldsymbol{\alpha}}} = tr\left[\mathbf{C}^{-1}\frac{\partial \mathbf{C}}{\partial \hat{\boldsymbol{\alpha}}}\right]$$

the gradients of J with respect to the unknown parameters are

$$\boxed{\frac{\partial J}{\partial \hat{\boldsymbol{\alpha}}} = \sum_{i=1}^{k} \frac{1}{2} tr\left(\mathbf{C}_i^{-1}\frac{\partial \mathbf{C}_i}{\partial \hat{\boldsymbol{\alpha}}}\right) - \frac{1}{2}\tilde{\mathbf{y}}_i^T \mathbf{C}_i^{-1}\frac{\partial \mathbf{C}_i}{\partial \hat{\boldsymbol{\alpha}}}\mathbf{C}_i^{-1}\tilde{\mathbf{y}}_i + \tilde{\mathbf{y}}_i^T \mathbf{C}_i^{-1}\frac{\partial \tilde{\mathbf{y}}_i}{\partial \hat{\boldsymbol{\alpha}}}} \qquad (11.1\text{-}30)$$

$$\boxed{\begin{aligned} \frac{\partial J}{\partial \hat{\boldsymbol{\beta}}} &= \sum_{i=1}^{k} \tilde{\mathbf{y}}_i^T \mathbf{C}_i^{-1}\frac{\partial \tilde{\mathbf{y}}_i}{\partial \hat{\boldsymbol{\beta}}} \\ &= \sum_{i=1}^{k} \tilde{\mathbf{y}}_i^T \mathbf{C}_i^{-1}\mathbf{T}_i^0 \end{aligned}}\,. \qquad (11.1\text{-}31)$$

The second equation assumes that $\mathbf{C}_i$ and $\mathbf{T}_i^0$ are not a function of $\hat{\boldsymbol{\beta}}$, which implies that the dependence of $\mathbf{H}_i(\hat{\mathbf{x}}_{i/i-1},\hat{\boldsymbol{\alpha}})$ and $\boldsymbol{\Phi}_{i,i-1}(\hat{\mathbf{x}}_{i-1/i-1},\hat{\boldsymbol{\alpha}})$ on $\hat{\boldsymbol{\beta}}$ is either zero or negligible. If this is not true, additional terms must be included so that equation is effectively the same as equation (11.1-30).

The above gradients compose the *score vector* $\partial J/\partial \boldsymbol{\theta}$. Using our previous convention for derivatives with respect to vectors, the partial derivatives are treated as row vectors.

The corresponding *Fisher information matrix* **F** (not to be confused with the partial derivative matrices $\mathbf{F}_x$ and $\mathbf{F}_\alpha$ for the dynamic model equation [11.1-7]) is

$$\mathbf{F} = E\left[\frac{\partial^2 J}{\partial \boldsymbol{\theta}\partial \boldsymbol{\theta}^T}\right] = E\left[\left(\frac{\partial J}{\partial \boldsymbol{\theta}}\right)^T\left(\frac{\partial J}{\partial \boldsymbol{\theta}}\right)\right]. \qquad (11.1\text{-}32)$$

For $\mathbf{F}^{-1}$ to be used as a lower bound on the error covariance for $\hat{\boldsymbol{\theta}}_{ML}$, the gradients in the above expression should be computed at the true $\boldsymbol{\theta}$, but since this is unknown, the model must be evaluated at $\hat{\boldsymbol{\theta}}_{ML}$. (During the scoring iterations the model must be evaluated at the current estimate $\hat{\boldsymbol{\theta}}$ for the iteration.)

The partial derivative of equation (11.1-30) is

$$E\left[\frac{\partial^2 J}{\partial\hat{\alpha}_l\partial\hat{\alpha}_m}\right]=E\left[\sum_{i=1}^{k}\left(\begin{array}{l}\frac{1}{2}tr\left(-\mathbf{C}_i^{-1}\frac{\partial\mathbf{C}_i}{\partial\hat{\alpha}_l}\mathbf{C}_i^{-1}\frac{\partial\mathbf{C}_i}{\partial\hat{\alpha}_m}+\mathbf{C}_i^{-1}\frac{\partial^2\mathbf{C}_i}{\partial\hat{\alpha}_l\partial\hat{\alpha}_m}\right)-\frac{\partial\tilde{\mathbf{y}}_i^T}{\partial\hat{\alpha}_l}\mathbf{C}_i^{-1}\frac{\partial\mathbf{C}_i}{\partial\hat{\alpha}_m}\mathbf{C}_i^{-1}\tilde{\mathbf{y}}_i\\+\tilde{\mathbf{y}}_i^T\mathbf{C}_i^{-1}\frac{\partial\mathbf{C}_i}{\partial\hat{\alpha}_l}\mathbf{C}_i^{-1}\frac{\partial\mathbf{C}_i}{\partial\hat{\alpha}_m}\mathbf{C}_i^{-1}\tilde{\mathbf{y}}_i-\frac{1}{2}\tilde{\mathbf{y}}_i^T\mathbf{C}_i^{-1}\frac{\partial^2\mathbf{C}_i}{\partial\hat{\alpha}_l\partial\hat{\alpha}_m}\mathbf{C}_i^{-1}\tilde{\mathbf{y}}_i\\+\frac{\partial\tilde{\mathbf{y}}_i^T}{\partial\hat{\alpha}_l}\mathbf{C}_i^{-1}\frac{\partial\tilde{\mathbf{y}}_i}{\partial\hat{\alpha}_m}-\tilde{\mathbf{y}}_i^T\mathbf{C}_i^{-1}\frac{\partial\mathbf{C}_i}{\partial\hat{\alpha}_l}\mathbf{C}_i^{-1}\frac{\partial\tilde{\mathbf{y}}_i}{\partial\hat{\alpha}_m}+\tilde{\mathbf{y}}_i^T\mathbf{C}_i^{-1}\frac{\partial^2\tilde{\mathbf{y}}_i}{\partial\hat{\alpha}_l\partial\hat{\alpha}_m}\end{array}\right)\right].\tag{11.1-33}$$

This expression includes two types of components that are analyzed differently, so we write $\mathbf{F}=\mathbf{F}^{(1)}+\mathbf{F}^{(2)}$ where individual elements are

$$F_{l,m}^{(1)}=E\left[\sum_{i=1}^{k}\left(\begin{array}{l}\frac{1}{2}tr\left(-\mathbf{C}_i^{-1}\frac{\partial\mathbf{C}_i}{\partial\hat{\alpha}_l}\mathbf{C}_i^{-1}\frac{\partial\mathbf{C}_i}{\partial\hat{\alpha}_m}+\mathbf{C}_i^{-1}\frac{\partial^2\mathbf{C}_i}{\partial\hat{\alpha}_l\partial\hat{\alpha}_m}\right)\\+\tilde{\mathbf{y}}_i^T\mathbf{C}_i^{-1}\frac{\partial\mathbf{C}_i}{\partial\hat{\alpha}_l}\mathbf{C}_i^{-1}\frac{\partial\mathbf{C}_i}{\partial\hat{\alpha}_m}\mathbf{C}_i^{-1}\tilde{\mathbf{y}}_i-\frac{1}{2}\tilde{\mathbf{y}}_i^T\mathbf{C}_i^{-1}\frac{\partial^2\mathbf{C}_i}{\partial\hat{\alpha}_l\partial\hat{\alpha}_m}\mathbf{C}_i^{-1}\tilde{\mathbf{y}}_i+\frac{\partial\tilde{\mathbf{y}}_i^T}{\partial\hat{\alpha}_l}\mathbf{C}_i^{-1}\frac{\partial\tilde{\mathbf{y}}_i}{\partial\hat{\alpha}_m}\end{array}\right)\right]\tag{11.1-34}$$

$$F_{l,m}^{(2)}=E\left[\sum_{i=1}^{k}\left(-\frac{\partial\tilde{\mathbf{y}}_i^T}{\partial\hat{\alpha}_l}\mathbf{C}_i^{-1}\frac{\partial\mathbf{C}_i}{\partial\hat{\alpha}_m}\mathbf{C}_i^{-1}\tilde{\mathbf{y}}_i-\tilde{\mathbf{y}}_i^T\mathbf{C}_i^{-1}\frac{\partial\mathbf{C}_i}{\partial\hat{\alpha}_l}\mathbf{C}_i^{-1}\frac{\partial\tilde{\mathbf{y}}_i}{\partial\hat{\alpha}_m}+\tilde{\mathbf{y}}_i^T\mathbf{C}_i^{-1}\frac{\partial^2\tilde{\mathbf{y}}_i}{\partial\hat{\alpha}_l\partial\hat{\alpha}_m}\right)\right].\tag{11.1-35}$$

We first address $\mathbf{F}^{(1)}$. Noting that $\mathbf{a}^T\mathbf{b}=tr[\mathbf{a}\mathbf{b}^T]=tr[\mathbf{b}\mathbf{a}^T]$ for general vectors **a**, **b**, we compute

$$\begin{aligned}F_{l,m}^{(1)}&=\sum_{i=1}^{k}\left(\begin{array}{l}\frac{1}{2}tr\left(-\mathbf{C}_i^{-1}\frac{\partial\mathbf{C}_i}{\partial\hat{\alpha}_l}\mathbf{C}_i^{-1}\frac{\partial\mathbf{C}_i}{\partial\hat{\alpha}_m}\right)+\frac{1}{2}tr\left(\mathbf{C}_i^{-1}\frac{\partial^2\mathbf{C}_i}{\partial\hat{\alpha}_l\partial\hat{\alpha}_m}\right)\\+E\left[\begin{array}{l}tr\left(\mathbf{C}_i^{-1}\frac{\partial\mathbf{C}_i}{\partial\hat{\alpha}_m}\mathbf{C}_i^{-1}\tilde{\mathbf{y}}_i\tilde{\mathbf{y}}_i^T\mathbf{C}_i^{-1}\frac{\partial\mathbf{C}_i}{\partial\hat{\alpha}_l}\right)-\frac{1}{2}tr\left(\mathbf{C}_i^{-1}\tilde{\mathbf{y}}_i\tilde{\mathbf{y}}_i^T\mathbf{C}_i^{-1}\frac{\partial^2\mathbf{C}_i}{\partial\hat{\alpha}_l\partial\hat{\alpha}_m}\right)\\+tr\left(\mathbf{C}_i^{-1}\frac{\partial\tilde{\mathbf{y}}_i}{\partial\hat{\alpha}_m}\frac{\partial\tilde{\mathbf{y}}_i^T}{\partial\hat{\alpha}_l}\right)\end{array}\right]\end{array}\right)\\&=\sum_{i=1}^{k}\frac{1}{2}tr\left(\mathbf{C}_i^{-1}\frac{\partial\mathbf{C}_i}{\partial\hat{\alpha}_l}\mathbf{C}_i^{-1}\frac{\partial\mathbf{C}_i}{\partial\hat{\alpha}_m}\right)+E\left[tr\left(\mathbf{C}_i^{-1}\frac{\partial\tilde{\mathbf{y}}_i}{\partial\hat{\alpha}_l}\frac{\partial\tilde{\mathbf{y}}_i^T}{\partial\hat{\alpha}_m}\right)\right]\end{aligned}.\tag{11.1-36}$$

Unfortunately there is no simple method for calculating the expected value of the second term, so in practice the sample value is used because this converges asymptotically to the expected value.

We now examine $\mathbf{F}^{(2)}$, which can be written as

$$F_{l,m}^{(2)}=E\left[\sum_{i=1}^{k}tr\left(-\left(\frac{\partial\mathbf{C}_i}{\partial\hat{\alpha}_m}\mathbf{C}_i^{-1}\tilde{\mathbf{y}}_i\frac{\partial\tilde{\mathbf{y}}_i^T}{\partial\hat{\alpha}_l}\mathbf{C}_i^{-1}\right)-\left(\mathbf{C}_i^{-1}\frac{\partial\tilde{\mathbf{y}}_i}{\partial\hat{\alpha}_m}\tilde{\mathbf{y}}_i^T\mathbf{C}_i^{-1}\frac{\partial\mathbf{C}_i}{\partial\hat{\alpha}_l}\right)+\left(\mathbf{C}_i^{-1}\frac{\partial^2\tilde{\mathbf{y}}_i}{\partial\hat{\alpha}_l\partial\hat{\alpha}_m}\tilde{\mathbf{y}}_i^T\right)\right)\right].\tag{11.1-37}$$

Notice from equation (11.1-23) that $\partial\tilde{\mathbf{y}}_i^T/\partial\hat{\boldsymbol{\alpha}}$ is a function of $\hat{\mathbf{x}}_{i/i-1}$ and $\partial\hat{\mathbf{x}}_{i/i-1}/\partial\hat{\boldsymbol{\alpha}}$, and both of these terms are a function of $\tilde{\mathbf{y}}_{i-1}, \tilde{\mathbf{y}}_{i-2},\ldots, \tilde{\mathbf{y}}_1$ and $\hat{\mathbf{x}}_{0/0}$, but not $\tilde{\mathbf{y}}_i$. Since $E[\tilde{\mathbf{y}}_i\tilde{\mathbf{y}}_{i-j}^T]=\mathbf{0}$ for all $j>0$, $E[(\partial\tilde{\mathbf{y}}_i/\partial\hat{\alpha}_l)\tilde{\mathbf{y}}_i^T]$ must be zero. This same approach can be applied to higher derivatives such as $E\left[(\partial^2\tilde{\mathbf{y}}_i/\partial\hat{\alpha}_l\partial\hat{\alpha}_m)\tilde{\mathbf{y}}_i^T\right]$, so all three terms of $\mathbf{F}^{(2)}$ are zero. (Numerical experience on Example 11.1 below demonstrates that the first two terms of $\mathbf{F}^{(2)}$ are less than 2% of the values in $\mathbf{F}^{(1)}$.) Thus we obtain

$$\boxed{\mathrm{F}_{l,m}=\sum_{i=1}^{k}\frac{1}{2}tr\left(\mathbf{C}_i^{-1}\frac{\partial\mathbf{C}_i}{\partial\hat{\theta}_l}\mathbf{C}_i^{-1}\frac{\partial\mathbf{C}_i}{\partial\hat{\theta}_m}\right)+tr\left(\mathbf{C}_i^{-1}\frac{\partial\tilde{\mathbf{y}}_i}{\partial\hat{\theta}_l}\frac{\partial\tilde{\mathbf{y}}_i^T}{\partial\hat{\theta}_m}\right).} \tag{11.1-38}$$

The same result is obtained when $\mathbf{F}_{l,m}$ is computed as $E[(\partial J/\partial\theta_l)(\partial J/\partial\theta_m)]$. This equation is written using variable $\hat{\boldsymbol{\theta}}$ because it applies to both $\hat{\boldsymbol{\alpha}}$ and $\hat{\boldsymbol{\beta}}$, but we have assumed that terms involving $\partial\mathbf{C}_i/\partial\hat{\boldsymbol{\beta}}$ are zero. Since the second term uses sample values, this expression for $\mathbf{F}$ is a mixture of the true and expected Hessians. Asymptotically this sample value converges to the true Fisher information matrix. Use of the sample values in equation (11.1-38) to compute the scoring iterations should not cause problems, but one should be somewhat cautious when using it to compute the lower bound on the error covariance matrix:

$$E\left[(\boldsymbol{\theta}-\hat{\boldsymbol{\theta}}_{ML})(\boldsymbol{\theta}-\hat{\boldsymbol{\theta}}_{ML})^T\right]\geq\mathbf{F}^{-1}.$$

The scoring iteration is

$$\boxed{\hat{\boldsymbol{\theta}}_j=\hat{\boldsymbol{\theta}}_{j-1}-\left(\mathbf{F}(\hat{\boldsymbol{\theta}}_{j-1})\right)^{-1}\left(\frac{\partial J(\hat{\boldsymbol{\theta}}_{j-1})}{\partial\hat{\boldsymbol{\theta}}}\right)^T} \tag{11.1-39}$$

where j is the iteration number. Due to the nonlinear nature of the problem, it is often better to use Levenburg-Marquardt or back-solving optimization algorithms, rather than just simple scoring steps.

The portion of $\mathbf{F}$ for initial condition parameters $\boldsymbol{\beta}$ is more easily computed as

$$\mathbf{F}_{\beta\beta}=E\left[\frac{\partial^2 J}{\partial\hat{\boldsymbol{\beta}}\partial\hat{\boldsymbol{\beta}}^T}\right]=\sum_{i=1}^{k}(\mathbf{T}_i^0)^T\mathbf{C}_i^{-1}\mathbf{T}_i^0. \tag{11.1-40}$$

The cross-term $\mathbf{F}_{\beta\alpha}$ can be obtained from equation (11.1-38) or by differentiating equation (11.1-31) with respect to $\boldsymbol{\alpha}$:

$$\begin{aligned}\mathbf{F}_{\beta\alpha}&=E\left[\frac{\partial^2 J}{\partial\hat{\boldsymbol{\alpha}}\partial\hat{\boldsymbol{\beta}}^T}\right]\\&=E\left[\sum_{i=1}^{k}\frac{\partial(\mathbf{T}_i^0)^T}{\partial\hat{\boldsymbol{\alpha}}}\mathbf{C}_i^{-1}\tilde{\mathbf{y}}_i-(\mathbf{T}_i^0)^T\mathbf{C}_i^{-1}\frac{\partial\mathbf{C}_i^{-1}}{\partial\hat{\boldsymbol{\alpha}}}\mathbf{C}_i^{-1}\tilde{\mathbf{y}}_i+(\mathbf{T}_i^0)^T\mathbf{C}_i^{-1}\frac{\partial\tilde{\mathbf{y}}_i}{\partial\hat{\boldsymbol{\alpha}}}\right]\\&=E\left[\sum_{i=1}^{k}(\mathbf{T}_i^0)^T\mathbf{C}_i^{-1}\frac{\partial\tilde{\mathbf{y}}_i}{\partial\hat{\boldsymbol{\alpha}}}\right]\end{aligned} \tag{11.1-41}$$

provided that $\partial\mathbf{C}_i/\partial\hat{\boldsymbol{\beta}}=\mathbf{0}$ and $\partial\mathbf{T}_i^0/\partial\hat{\boldsymbol{\beta}}=\mathbf{0}$. Since $E[\partial\tilde{\mathbf{y}}_i/\partial\hat{\boldsymbol{\alpha}}]=\mathbf{0}$ and $\mathbf{T}_i^0$ is deterministic, $\mathbf{F}_{\beta\alpha}$ is zero.

It can also be shown that $\hat{\boldsymbol{\beta}}_j$ for multiple tests j are sufficient statistics for estimating $\bar{\boldsymbol{\beta}}$ and $\mathbf{P}_{\beta\beta}$ across tests. The method is described in Shuster et al. (1983) and Shuster and Porter (1984), and it can be very useful for test and evaluation purposes. Other general references on ML parameter estimation are Gupta and Mehra (1974), Maybeck (1982a), Schweppe (1973), Rao (1968), Ljung (1999), and Levine (1996, section 58).

Finally we note that MLE is not the only method that can be used to determine model parameters. When models are nonlinear and the purpose of a Kalman filter is to predict future values of states, it is sometimes possible to obtain more accurate predictions when directly using the root-mean-squared (RMS) prediction error as the metric for determination of model parameters. This was the method used by Cruz et al. (2005) to compute measurement and process noise variances of Global Positioning System (GPS) system models.

For more information on the MLE implementation, refer to the associated software (TST_FIS2.F) used for the next example below. A more general modular version that uses sparse matrix multiplication routines and is designed to flexibly handle a variety of system models is also provided (MLE_PARAMETERID.F)

Example 11.1: Estimation of Markov Process Constants, Process Noise, and Measurement Noise

A four-state model demonstrates the application of ML parameter identification. Although the model order is low, this problem has attributes of many realistic filtering problems because colored noise (first-order Markov) processes are driving forces for other states. The continuous dynamic model is

$$\begin{bmatrix} \dot{x}_1 \\ \dot{x}_2 \\ \dot{x}_3 \\ \dot{x}_4 \end{bmatrix} = \begin{bmatrix} 0 & 1 & 0 & 0 \\ 0 & -a_1 & 0 & 0 \\ 0 & 0 & 0 & 1 \\ 0 & 0 & 0 & -a_2 \end{bmatrix} \begin{bmatrix} x_1 \\ x_2 \\ x_3 \\ x_4 \end{bmatrix} + \begin{bmatrix} 0 \\ w_2 \\ 0 \\ w_4 \end{bmatrix}$$

with true $a_1 = 0.1$ and $a_2 = 0.2$. The process noises w_2 and w_2 are zero mean and white. The discrete model $\mathbf{x}_i = \mathbf{\Phi}\mathbf{x}_{i-1} + \mathbf{q}_i$ for this system uses

$$\mathbf{\Phi} = \begin{bmatrix} 1 & (1-e^{-a_1 T})/a_1 & 0 & 0 \\ 0 & e^{-a_1 T} & 0 & 0 \\ 0 & 0 & 1 & (1-e^{-a_2 T})/a_2 \\ 0 & 0 & 0 & e^{-a_2 T} \end{bmatrix}, \quad \mathbf{Q} = E\left[\mathbf{q}_i \mathbf{q}_i^T\right] = \begin{bmatrix} 0 & 0 & 0 & 0 \\ 0 & Q_2 & 0 & 0 \\ 0 & 0 & 1 & 0 \\ 0 & 0 & 0 & Q_4 \end{bmatrix},$$

initialized with $\mathbf{x}_0^T = [10 \quad 0 \quad 0 \quad 0]$. Notice that only two diagonal elements of $\mathbf{Q}$ are nonzero even though other elements should be nonzero if the discrete model is the integral of the given continuous model. This approximation is used because the measurement sampling interval of $T = 0.25$ s is much smaller than the time constants of the two Markov processes (10 and 5 s). The true values of $\mathbf{Q}$ are set so that $E\left[x_2^2\right] = E\left[x_4^2\right] \cong 4^2$; that is, $Q_2 = (1-e^{-2a_1 T})4.05^2 = 0.80$ and $Q_4 = (1-e^{-2a_2 T})4.10^2 = 1.60$.

Two different measurement model were used for this example: the linear model

$$\begin{bmatrix} y_1 \\ y_1 \end{bmatrix}_i = \begin{bmatrix} 1 & 0 & 0 & 0 \\ 0 & 0 & 1 & 0 \end{bmatrix} \mathbf{x}_i + \begin{bmatrix} r_1 \\ r_2 \end{bmatrix}_i$$

where r_1 and r_2 are $N(0,0.5^2)$, and the nonlinear model

$$\begin{bmatrix} y_1 \\ y_1 \end{bmatrix}_i = \begin{bmatrix} \sqrt{x_1^2 + x_3^2} \\ \tan^{-1}(x_1 / x_3) \end{bmatrix}_i + \begin{bmatrix} r_1 \\ r_2 \end{bmatrix}_i$$

where r_1 is $N(0,0.5^2)$ and r_2 is $N(0,0.032^2)$. The six parameters to be estimated via MLE are

$$\boldsymbol{\theta}^T = [a_1 \quad a_2 \quad Q_2 \quad Q_4 \quad R_1 \quad R_2]$$

where $R_1 = E[r_1^2]$ and $R_2 = E[r_2^2]$. The partial derivatives $\partial\boldsymbol{\Phi}/\partial\boldsymbol{\theta}$ are

$$\frac{\partial \Phi_{12}}{\partial a_1} = \left(Te^{-a_1T} - (1-e^{-a_1T})/a_1\right)/a_1, \quad \frac{\partial \Phi_{22}}{\partial a_1} = -Te^{-a_1T}$$

$$\frac{\partial \Phi_{34}}{\partial a_2} = \left(Te^{-a_2T} - (1-e^{-a_2T})/a_2\right)/a_2, \quad \frac{\partial \Phi_{44}}{\partial a_2} = -Te^{-a_2T}$$

and partial derivatives of $\mathbf{Q}$ and $\mathbf{R}$ with respect to $\boldsymbol{\theta}$ elements are either zero or one. Simulated measurements over the time interval 0 to 100 s (400 time samples) were generated using the above models and random samples for $\mathbf{q}$ and $\mathbf{r}$. A Kalman filter initialized at the true $\mathbf{x}_0$ (to avoid estimation of initial conditions) processed the measurements. J, $\partial J/\partial\boldsymbol{\theta}$, and the Fisher information matrix $\mathbf{F}$ were evaluated at the true $\boldsymbol{\theta}$. These values were compared with the numerically computed central-difference values of the gradient and $\mathbf{F}$ to verify coding and the accuracy of the approximation equation (11.1-38). As expected, the single-sample, numerically computed gradient for the linear measurement model matches the analytic gradient

$$\partial J / \partial\boldsymbol{\theta} = [2.5439 \quad 13.658 \quad -5.3302 \quad -2.8885 \quad -21.765 \quad 7.4462]$$

to five or more digits. The single-sample numerically computed Hessian is

$$\left[\frac{\partial^2 J}{\partial\boldsymbol{\theta}\partial\boldsymbol{\theta}^T}\right]_{1-sample} = \begin{bmatrix} 514.06 & 0 & -62.043 & 0 & -1.2085 & 0 \\ & 317.35 & 5.7\times10^{-4} & -36.885 & -5.7\times10^{-4} & -1.6485 \\ & & 45.362 & -5.7\times10^{-4} & 19.292 & 0 \\ & & & 13.893 & -5.7\times10^{-4} & 9.4377 \\ & & & & 740.16 & -5.7\times10^{-4} \\ & & & & & 601.51 \end{bmatrix}.$$

Values less than 1×10^{-4} are shown as zero, and only the upper triangular partition is listed. Notice that because the measurement and dynamic models for the two channels are uncoupled, the cross elements of the Hessian are nearly zero. This Hessian is compared with the Fisher information from equation (11.1-38) for the same set of random numbers:

$$\mathbf{F}(\boldsymbol{\theta}_{true}) = \begin{bmatrix} 514.84 & 0 & -62.212 & 0 & 4.4226 & 0 \\ & 313.94 & 0 & -31.946 & 0 & 9.9175 \\ & & 35.411 & 0 & 13.744 & 0 \\ & & & 10.215 & 0 & 9.5640 \\ & & & & 660.69 & 0 \\ & & & & & 629.66 \end{bmatrix}.$$

Since the Fisher information matrix is partially an expected value, the numbers should not be identical to the single-sample Hessian, but most elements are close. The 1000-sample average of the numerically computed Hessian for the linear measurement model is

$$\left[\frac{\partial^2 J}{\partial \boldsymbol{\theta} \partial \boldsymbol{\theta}^T}\right]_{1000-sample} = \begin{bmatrix} 468.01 & 0 & -57.411 & 0 & -1.3863 & 0 \\ & 244.57 & 0 & -28.238 & 0 & -2.0452 \\ & & 32.954 & 0 & 18.895 & 0 \\ & & & 9.7011 & 0 & 11.147 \\ & & & & 652.53 & 0 \\ & & & & & 626.82 \end{bmatrix}.$$

Notice that the (2,2) diagonal element is significantly different from the corresponding element of **F**, which shows the effects of computing **F** using sample values in equation (11.1-38) rather than expected values. Ignoring terms in **F** (because they are zero at the ML estimate) is another source of error in equation (11.1-38). The above **F** was evaluated at the true value of $\boldsymbol{\theta}$ because this is necessary for $\mathbf{F}^{-1}(\hat{\boldsymbol{\theta}})$ to be interpreted as a lower bound on the error covariance $E\left[(\boldsymbol{\theta}-\hat{\boldsymbol{\theta}})(\boldsymbol{\theta}-\hat{\boldsymbol{\theta}})^T\right]$. Since the true $\boldsymbol{\theta}$ is generally unknown, it is necessary to evaluate equation (11.1-4) at $\hat{\boldsymbol{\theta}}_{ML}$. For the first sample of the Monte Carlo simulation,

$$\hat{\boldsymbol{\theta}}_{ML} = [0.1145 \quad 0.1758 \quad 0.9702 \quad 1.8077 \quad 0.5290 \quad 0.4852]^T$$

and

$$\mathbf{F}(\hat{\boldsymbol{\theta}}_{ML}) = \begin{bmatrix} 427.40 & 0 & -47.462 & 0 & -2.3958 & 0 \\ & 283.75 & 0 & -25.148 & 0 & -2.8499 \\ & & 22.590 & 0 & 15.416 & 0 \\ & & & 7.8435 & 0 & 11.088 \\ & & & & 578.05 & 0 \\ & & & & & 653.43 \end{bmatrix}.$$

Notice that the (2,2), (3,3), (4,4), and (5,5) elements differ by more than 10% from those of the average 1000-sample Hessian. This variability demonstrates the limitations of using $\mathbf{F}^{-1}(\hat{\boldsymbol{\theta}}_{ML})$ to characterize the error covariance for problems where dynamic model parameters are estimated.

We next check whether $\mathbf{F}^{-1}(\hat{\boldsymbol{\theta}}_{ML})$ accurately characterizes the estimate errors. The standard deviations computed from $\mathbf{F}^{-1}(\hat{\boldsymbol{\theta}}_{ML})$ for this problem are

$$\sigma_{\boldsymbol{\theta}} = [0.0554 \quad 0.0705 \quad 0.2431 \quad 0.4289 \quad 0.0421 \quad 0.0398]^T,$$

where

$$\boldsymbol{\theta}_{true} = [0.10 \quad 0.20 \quad 0.80 \quad 1.60 \quad 0.50 \quad 0.50]^T,$$

so the $1-\sigma$ uncertainties of a_1 and a_2 are, respectively, 35% and 55% of the actual values. (These uncertainties would be smaller if the measurements were more accurate.) Observability of these parameters is good because 800 scalar measurements were processed, the sampling interval is 20 times smaller than the shortest time constant, and the total data span is 10 times longer than the longest time constant. In comparing $\hat{\boldsymbol{\theta}}_{ML}$ with $\boldsymbol{\theta}_{true}$, it is found that the maximum $|\boldsymbol{\theta}_{true} - \hat{\boldsymbol{\theta}}_{ML}|$ is only 70% of σ_θ, so $\mathbf{F}^{-1}(\hat{\boldsymbol{\theta}}_{ML})$ does seem to be a reasonably accurate bound on the estimate error for this case.

When using the nonlinear range and azimuth measurement model, the two channels are coupled and elements of the above **F** that were previously zero become nonzero. Otherwise the characteristics of the problem are not significantly changed by use of coupled nonlinear measurements. Experiments were also conducted using alternate parameterizations for $\boldsymbol{\theta}$: $\log(a_1)$, $\log(a_2)$ or $1/a_1$, $1/a_2$ were tested as replacements for a_1, a_2, and $\log(Q_2)$, $\log(Q_4)$ or $Q_2/(2a_1)$, $Q_4/(2a_2)$ were tested as replacements for Q_2, Q_4. In all cases unconstrained scoring converged in three iterations when initialized at truth, so for this particular problem no parameterization was superior to the others when using 100 s of data and 0.25 s sampling.

For good parameter observability it is generally desirable to use data spans 10 times (or more) longer than the longest time constants of the system, and to have at least five measurement samples in the shortest time constants. To test behavior with marginal data, the total data span was shortened to 50 s, which caused degraded observability for the 10 s Markov process parameters. It was found that convergence of the iterations was fastest when using a_1, a_2 in $\boldsymbol{\theta}$, and much slower when using $1/a_1$, $1/a_2$. Convergence when using $Q_2/(2a_1)$, $Q_4/(2a_2)$ was slower than when using Q_2, Q_4 or $\log(Q_2)$, $\log(Q_4)$. Use of $\log(Q_2)$, $\log(Q_4)$ has the added benefit of automatically constraining $Q > 0$. Although not used in this example, it is necessary on some problems to use backtracking or Levenberg-Marquardt optimization when observability of dynamic model constants is poor.

In summary, ML parameter identification worked well for this problem, and $\mathbf{F}^{-1}(\hat{\boldsymbol{\theta}}_{ML})$ did characterize the actual estimation error.

Example 11.2: Estimation of Process Noise PSD for the GPS Constellation

ML parameter identification was used for the GPS constellation described in Section 3.4.10. This is the same filter problem used in Example 9.6 to demonstrate fixed-interval smoothers. The navigation filter processed pseudo-range (PR) and PR minus accumulated delta range (ADR) measurements between 29 GPS satellites and 19 ground receivers. The ADR measurements are less noisy than PR, while PR-ADR is not influenced by tropospheric delay modeling errors but does contain a pass-dependent bias. The Kalman filter estimated either 10 or 13 orbital, solar radiation and clock states for each satellite, and a total of three clock and tropospheric error states for each ground station. Six earth orientation states were also estimated. The satellite velocity, solar radiation and clock states, and ground clock and tropospheric error states were the only states driven by process noise. All were modeled as either random walk or integrated random walk—none were colored noise states. The smoothed PR and PR-ADR measurements available for testing were obtained during 2005 days 238 to 243. A total of 371 permanent states were used in the filter, with pass-disposable station-satellite PR-ADR bias states added and removed as satellites were

TABLE 11.1: Process Noise Parameters Estimated via MLE

State Type	Process Noise State Components		
Satellite velocity	Alongtrack velocity	Crosstrack velocity	Radial velocity
Satellite solar radiation force	Scale factor	*y*-body component	—
Satellite cesium clocks	Time	Frequency	—
"Stable" satellite rubidium clocks	Time	Frequency	Frequency drift
"Less stable" satellite rubidium clocks	Time	Frequency	Sin and cos for frequency drift
Monitor station cesium clocks	Time	Frequency	—
Tropospheric refraction	Zenith path delay	—	—

observed from the monitor stations. For efficiency, the state and covariance time updates of each satellite were computed separately, and sparse matrix multiplication routines were extensively used in filter and filter derivative computations. $\hat{\mathbf{x}}$ and $\mathbf{\Phi}$ were numerically integrated for the orbital states, and a Taylor series was used for the clock and other random walk states. $\partial\mathbf{Q}/\partial\boldsymbol{\alpha}$ was computed as $\mathbf{Q} \approx \mathbf{Q}_s T$ for intervals of 5 min or less, and then interval doubling was used to obtain $\hat{\mathbf{x}}, \mathbf{\Phi}, \mathbf{Q}$ and $\partial\mathbf{Q}/\partial\boldsymbol{\alpha}$ for the 15 min intervals of smoothed measurements. $\partial\hat{\mathbf{x}}/\partial\boldsymbol{\alpha}$ and $\partial\mathbf{\Phi}/\partial\boldsymbol{\alpha}$ were not required for this problem.

ML parameter identification was used to determine 17 process noise PSD values as listed in Table 11.1. Because raw measurements are available at a relatively high sampling rate, measurement noise variances can be accurately determined simply by plotting measurement deviations from a smooth curve. Several references also provided information on the noise characteristics. No dynamic model parameters ($\boldsymbol{\alpha}_d$) were estimated because colored noise states are not normally used in the model. To avoid ML estimation of initial condition states, a Kalman filter first processed 2 days of data using nominal initial condition states and parameter values. The filter state estimates at the final time were then used to initialize the MLE filter, which processed 103,505 scalar PR and PR-ADR measurements from days 240.0 to 244.0. Initial values of the process noise PSD values were obtained from published values for clock states, or as educated guesses for the other states.

The MLE was executed in several individual runs of three to five scoring iterations that typically executed in about 3 to 4 h per run on a 3.8-GHz PC. After completion of each scoring solution, the filter and convergence behavior was analyzed and adjustments were made to the filter setup. In one case an alternate clock model was tested. The final solution did not reduce the convergence metric of equation (7.3-20) below 0.11, which indicates the presence of small modeling errors—this was expected because the ADR measurements are very accurate (1.7 cm noise), and the solar radiation, clock, and tropospheric delay models are imperfect. The success of the ML approach was demonstrated when a separate Kalman filter using the ML-estimated process noise PSD then predicted user range error (URE) for zero to 24-h intervals: accuracy improved significantly compared with results using the initial parameter estimates. The zero age-of-data (no prediction) URE statistics were presented in Figure 9.14 of Example 9.6.

This exercise demonstrated the practicality of implementing MLE for large-scale problems. This case included 371 states, a variable number of pass-disposable station-satellite biases, 103,505 scalar measurements, and 17 MLE parameters.

11.2 ADAPTIVE FILTERING

Example 9.5 demonstrated that a Kalman filter using fixed **Q** has difficulty in tracking an aircraft that only occasionally makes turns or otherwise maneuvers. A better model would recognize that either **Q** should be increased or the dynamic model should be modified when the aircraft maneuvers. Tracking of maneuvering vehicles is one type of problem in which filters that adjust model characteristics in response to observed performance can track more accurately than filters using a fixed model. Many other examples exist because it is not unusual for either measurement noise or process noise to change magnitude as external conditions change. Both dynamic and measurement models may also vary with time, and these changes may be either abrupt or slow.

These problems led to the development of various adaptive Kalman filters. Finite memory and exponentially weighted filters are examples of one approach to the problem. Other filter types compute statistics on the squared measurement residuals over a sliding window, and modify the process noise covariance **Q** as the residual variance changes. This change in **Q** may be done in steps or as a linear function of the residual variance. One simple adaptive filter of this type uses the algorithm

$$\begin{aligned}\eta_i &= \sum_{j=0}^{k} \tilde{\mathbf{y}}_{i-j}^T \mathbf{C}_{i-j}^{-1} \tilde{\mathbf{y}}_{i-j} \\ \mathbf{Q}_{i+1,i} &= \eta_i \mathbf{Q}_0\end{aligned} \tag{11.2-1}$$

where k is the length of the sliding window and $\mathbf{Q}_0$ is a "base" value. Jazwinski (1970, section 8.11) developed a method for adaptive noise estimation that is more sophisticated. It assumes that any excess of the actual residual sum-of-squares minus the predicted residual variance without the effects of $\mathbf{Q}_{i,i-1}$, that is,

$$\tilde{\mathbf{y}}_i^T \tilde{\mathbf{y}}_i - \left(\mathbf{H}_i^T \mathbf{\Phi}_{i,i-1} \mathbf{P}_{i-1/i-1} \mathbf{\Phi}_{i,i-1}^T \mathbf{H}_i + \mathbf{R}_i\right),$$

is due to the effects of process noise and can be used to estimate the diagonal elements of **Q**. To minimize the sensitivity of the adaptive filter to measurement noise, the estimate of **Q** diagonals is computed using all residuals over a sliding time window in a Bayesian least-squares estimator. Jazwinski presents results showing that the approach improves the prediction accuracy for a difficult satellite orbit determination problem. Bar-Shalom et al. (2001) describes techniques of this type as *noise level adjustment* for continuous systems, or *noise level switching* for discrete noise changes.

While these adaptive approaches work well for some problems with "clean" measurements, they also increase **Q** in response to measurement outliers, and residual smoothing causes a slow response when the system dynamics suddenly change. To overcome the problem of erroneously responding to measurement outliers, some adaptive filters add logic to edit several (e.g., three) successive measurements before using residuals in the **Q** estimation. While this makes the filter more robust, it also increases the filter lag when responding to actual changes in dynamics.

When system inputs abruptly change or additional states (e.g., acceleration) suddenly become active, it is better to explicitly model the step change in input and estimate that step via least squares. Bar-Shalom et al. refer to these methods as *input estimation*. The basic approach only models a single jump at a fixed time located at the beginning of a sliding window. The method is not designed to handle multiple jumps at unknown times. Bar-Shalom et al. note that multiple hypotheses can be used, but this becomes "quite expensive."

The next sections present two other adaptive filtering approaches that the author has used successfully. The first method is an extension of input estimation: it assumes that the system may suddenly change dynamic or measurement characteristics at unknown times. This is sometimes called the *jump or fault detection* problem, depending upon the application. Hence the algorithm continuously performs hypothesis testing to determine whether jumps have occurred, and then the filter may either modify the model for future measurements, or the filter may use recent measurements to instantly update the filter state estimate. While the latter approach is more complicated, it has a much shorter response lag.

The other approach assumes that the system at any point in time may be characterized by one of a finite number of models. This adaptive filtering approach continuously operates multiple Kalman filters in parallel, and the optimal estimate is determined using tracking metrics computed from measurement innovations. Many other types of adaptive filters have been developed over the years. For example, Maybeck (1982, section 10.4) uses an MLE approach in an online mode: innovations are accumulated over a sliding window and parameter estimates are updated periodically.

11.3 JUMP DETECTION AND ESTIMATION

Many systems exhibit sudden changes in behavior. Vehicles such as ships, aircraft, and missiles often operate in constant velocity modes for extended periods of time, but then suddenly maneuver. Some devices such as *inertial measuring units* (IMUs) suddenly change internal characteristics, possibly in response to external forces. Individual components of complex systems (e.g., process control) may suddenly exhibit a partial or complete failure and thus affect behavior of the system. In most cases neither the time nor the mode change is known. Thus the jump detection algorithm must detect the change, and then determine both the jump time and magnitude.

Willsky (1976) describes most jump detection techniques as being either multiple model approaches, innovations-based techniques, or jump process formulations. In addition there are also off-line local likelihood (scoring) techniques based on the gradient of the likelihood function (Morgan et al. 1985; Basseville and Benveniste 1986). Blackman (1986) describes three generic types of maneuver detection and adaptive filtering algorithms used for online tracking applications:

1. Modify filter parameters such as the process noise covariance after the maneuver is detected, but retain the same filter form.
2. Augment the filter state model to include additional parameters upon maneuver detection.
3. Use multiple filter models, each with different maneuver models. The filter with the "best model" can be used, or the filter outputs can be linearly combined using *a posteriori* probabilities.

To that list we add another option: use recent measurements to both detect the maneuver and estimate the maneuver parameters, and then update the filter state to include the maneuver.

One widely referenced method for jump detection and estimation is based on the *generalized likelihood ratio* (GLR) test, as documented in Willsky and Jones (1974, 1976), Willsky (1976), Basseville (1988), and Basseville and Benveniste (1986). Van Trees (1968) provides more background on the GLR test. GLR-based jump detection is an innovations approach where the probability of the no-jump hypothesis and various candidate jump hypotheses are continuously evaluated to determine if a jump has occurred. In the online GLR implementation by Willsky and Jones, an unknown jump vector is hypothesized to exist at each measurement time within a finite window of past time. The innovations from a "jump-free" filter are used for hypothesis testing to determine whether a jump occurred within the window. If a jump is detected, the jump-free filter state and, optionally, covariance are updated and the filter proceeds. As proposed, the method is directed toward the detection of only one jump within a narrow window of time. If multiple jumps occur within the window, the correlation between the jump estimates is not accounted for. This is a significant limitation of many jump detection approaches because multiple jumps are not uncommon in post-processing applications, and the best results are obtained when using the entire data span. Another problem with the GLR method is the selection of hypotheses based upon past data only—if, at one stage, it makes an incorrect decision about the presence of jumps, it may not recover properly from the error.

We now discuss an innovations-based approach related to GLR that does not suffer when multiple jumps occur within the data window. It uses stepwise regression to systematically test all of the potential jump candidates. This method of Gibbs (1992) was motivated by postflight processing of missile test data in which the IMU may exhibit as many as 40 to 50 jumps in a single missile flight. Various forms of the techniques described above had been tested on simulated data, and the results were not satisfactory. The stepwise regression method was found to work well on this problem for single or multiple jumps. In contrast to off-line scoring techniques, convergence of the regression technique is guaranteed, and it automatically computes the number of jump parameters. Also, the method has been extended to real-time applications in a manner much like GLR (i.e., using a sliding data window), but the correlation between jumps within the window is correctly included in the calculations.

The stepwise regression approach assumes a model and filter structure similar to that commonly used for GLR, and the implementation of the hypothesis test for a single jump is similar. The measurement data are processed using a "jump-free filter" that contains all the dynamic states but does not model the jumps in those states. Sensitivity of the innovations with respect to all possible previous bias jumps is computed using an approach similar to Friedland's bias-free filter. The jump-free filter innovations and innovation sensitivities are processed using forward stepwise regression to compute the minimum set of required jump states. These jump state estimates can then be used to recompute the total state estimate at each time point, if desired.

For simplicity in the discussion to follow, we assume that all of the potential jump states are hypothesized to jump at each measurement time. However, this is not a restriction. Implementation can assume that jumps occur at discrete intervals not synchronized with the measurement times, or it can be assumed that a restricted set of states jump within a limited time span.

Since the number of jump parameters may grow linearly with time, and the sensitivity of the measurement innovations with respect to all previous jump parameters must be computed and stored, the computational and storage burdens increase quadratically with the number of measurement times included in the window of measurements. This is the primary disadvantage of the method. Thus it may be necessary to make approximations so that the number of jump parameters is limited.

11.3.1 Jump-Free Filter Equations

Development of the jump detection/estimation algorithm begins with the standard Kalman filter equations (11.1-4), (11.1-17), (11.1-20), and (11.1-21). Now define $\mathbf{s}_i$ of dimension $l \leq n$ as the subset of the state vector that is assumed to jump. In other words, the portion of $\mathbf{x}$ due to the jumps has been separated from $\mathbf{x}$ and included as an additive linear term in the time update equation as

$$\mathbf{x}_i = \mathbf{x}_{i-1} + \int_{t_{i-1}}^{t_i} (\mathbf{f}(\mathbf{x}(\lambda), \boldsymbol{\alpha}) + \mathbf{q}(\lambda))\, d\lambda + \mathbf{D}_i \mathbf{s}_i \tag{11.3-1}$$

where $\mathbf{D}_i$ is a mapping matrix (composed of only 1's and 0's) of dimension $n \times l$. Since jumps are assumed to be rare events, the $\mathbf{s}_i$ vector will usually be zero, and will only be nonzero during time intervals in which jumps occur. By a suitable choice of states, this model can be used to handle dynamic model jumps or steps, and sensor jumps or steps. Hence the linearized error in the filter state estimate is the difference between equations (11.3-1) and (11.1-4):

$$\boxed{\mathbf{x}_i - \hat{\mathbf{x}}_{i/i-1} = \boldsymbol{\Phi}_{i,i-1}(\mathbf{x}_{i-1} - \hat{\mathbf{x}}_{i-1/i-1}) + \mathbf{q}_{i,i-1} + \mathbf{D}_i \mathbf{s}_i .} \tag{11.3-2}$$

We now define a jump parameter vector, $\boldsymbol{\theta}_i$, at time t_i as the concatenation of all $\mathbf{s}_j$ for $j = 1$ to i; that is,

$$\boxed{\boldsymbol{\theta}_i \triangleq [\mathbf{s}_1^T \quad \mathbf{s}_2^T \quad \cdots \quad \mathbf{s}_i^T]^T .} \tag{11.3-3}$$

The dimension of $\boldsymbol{\theta}_i$ is il, which increases with the time index i. Also define $\mathbf{T}_{i/i}$ as the sensitivity of the error in $\hat{\mathbf{x}}_{i/i}$ with respect to $\boldsymbol{\theta}_i$:

$$\boxed{\mathbf{T}_{i/k}(j) \triangleq \frac{\partial(\mathbf{x}_i - \hat{\mathbf{x}}_{i/k})}{\partial \boldsymbol{\theta}_i(j)}} \tag{11.3-4}$$

where $\boldsymbol{\theta}_i(j)$ denotes the j-th element of $\boldsymbol{\theta}_i$, and $\mathbf{T}_{i/k}(j)$ denotes the j-th column of the matrix at time t_i based upon measurements up to and including time t_k. Note that the column dimension of $\mathbf{T}_{i/k}$ increases as the size of $\boldsymbol{\theta}_i$ increases. Also note that these jump times need not be limited to measurement times if filter time updates are inserted between measurements.

From equation (11.3-2), the time update of $\mathbf{T}$ is

$$\boxed{\mathbf{T}_{i/i-1} = \boldsymbol{\Phi}_{i,i-1}\mathbf{T}_{i-1/i-1}} \tag{11.3-5}$$

for all $\mathbf{s}_j$, $1 \le j \le i$. At measurement time t_i, $\mathbf{s}_i$ must be appended to $\boldsymbol{\theta}_i$, so the last l columns (corresponding to $\mathbf{s}_i$) are appended to $\mathbf{T}_{i/i-1}$ and those columns are initialized as $\mathbf{D}_i$.

From equation (11.1-21) the sensitivity of the innovations $\tilde{\mathbf{y}}_i$ (of dimension m) to jump states j is

$$\boxed{\frac{\partial \tilde{\mathbf{y}}_i}{\partial \boldsymbol{\theta}_i(j)} = \mathbf{H}_i\mathbf{T}_{i/i-1}(j)} \tag{11.3-6}$$

and the measurement update for $\mathbf{T}$ is

$$\boxed{\mathbf{T}_{i/i}(j) = \mathbf{T}_{i/i-1}(j) - \mathbf{K}_i \frac{\partial \tilde{\mathbf{y}}_i}{\partial \boldsymbol{\theta}_i(j)}.} \tag{11.3-7}$$

Since the jump-free filter is optimal except for the effect of the jumps, the innovations consist of two components: a white Gaussian component with covariance $\mathbf{C}_i$, and a component due to the jumps ($\mathbf{H}_i\mathbf{T}_{i/i-1}\boldsymbol{\theta}_i$). Thus these innovations, innovation covariance, and innovation sensitivity equation (11.3-7) are all that are required to compute the jump states using weighted least squares. Equations (11.3-5) to (11.3-7) are simpler than the equivalent GLR equations in Willsky and Jones (1976), and only require about one-half as many flops.

In implementing this approach, it is desirable that all measurements be processed one-component-at-a-time within the Kalman filter. If the measurement covariance $\mathbf{R}_i$ is not diagonal, it is necessary to multiply $\mathbf{y}_i$ and $\mathbf{H}_i$ by $\mathbf{R}_i^{-1/2}$. Then each scalar measurement innovation and measurement sensitivity equation (11.3-6) can be scaled by $C_i^{-1/2}$ so that the resulting scaled innovation has unity variance. This simplifies the regression calculations to follow. The result is

$$\boxed{\mathbf{z} = \begin{bmatrix} \tilde{y}_1/\sqrt{C_1} \\ \tilde{y}_2/\sqrt{C_2} \\ \vdots \\ \tilde{y}_i/\sqrt{C_i} \end{bmatrix}, \quad \mathbf{M} = \begin{bmatrix} (\partial\tilde{y}_1/\partial\theta_1)/\sqrt{C_1} \\ (\partial\tilde{y}_2/\partial\theta_2)/\sqrt{C_2} \\ \vdots \\ (\partial\tilde{y}_i/\partial\theta_i)/\sqrt{C_i} \end{bmatrix}} \tag{11.3-8}$$

that form the linearized measurement equation

$$\boxed{\mathbf{z} = \mathbf{M}\boldsymbol{\theta} + \mathbf{v}} \tag{11.3-9}$$

where $\mathbf{v}$ is $N(\mathbf{0},\mathbf{I})$ measurement noise. Because this method calculates and stores the sensitivity of each measurement with respect to all jump states preceding the measurement, the computational burden and storage requirements can be high. However, with modern computers this is less of an issue than in the past.

11.3.2 Stepwise Regression

The normalized innovations **z** and innovation sensitivities **M** contain all the information needed to compute the least-squares estimate of the jump parameters. If it is hypothesized that all possible jumps occur at every measurement time, the number of jump states is large. For example, if the number of measurement times $k = 200$ and $l = 20$ states are allowed to jump, there are $kl = 4000$ possible jump candidates. Depending upon the relative values of l and m, the number of scalar measurements may be more or less than the number of jump states. In general, all these jump states will not be observable from the available measurement data. Even if the number of measurements exceeds the number of jump states, the signatures of the jump states will be so similar that observability problems are nearly guaranteed.

Fortunately, by the nature of the jump problem it is assumed that jumps are rare events—otherwise the jumps could be treated as continuous process noise. Thus we wish to find the minimal set of jump states that reduce the measurement innovations sum-of-squares by a statistically significant amount; that is, find the subset $\hat{\boldsymbol{\theta}}_x$ of dimension p that minimizes the sum-of-squares (S_s) of residuals $\mathbf{z} - \mathbf{M}\hat{\boldsymbol{\theta}}_x$. Forward stepwise regression described in Section 6.3.2 provides a systematic means to search through all the jump possibilities. In Section 6.3.2 the state vector of estimated variables at regression step j is denoted as $\hat{\mathbf{x}}_{x\_j}$, but this is replaced with $\hat{\boldsymbol{\theta}}_{x\_j}$ when used for jump detection. Hence the F-statistic of equation (6.3-11) becomes

$$\frac{\left[S_s(\hat{\boldsymbol{\theta}}_{x\_j-1}) - S_s(\hat{\boldsymbol{\theta}}_{x\_j})\right] / \sigma_r^2}{\left[S_s(\hat{\boldsymbol{\theta}}_{x\_j-1})\right] / \sigma_r^2}, \tag{11.3-10}$$

which is used when the measurement noise variance is unknown. However, in our jump application the measurements **z** were normalized to have unity variance for the assumed measurement and dynamic models. Thus the test can be based on just the numerator of equation (11.3-10), which is χ^2 distributed with one degree-of-freedom when only one parameter is added at each regression stage. In comparing stepwise regression with GLR, we note that stepwise regression recursively tests many hypotheses that a scalar parameter has jumped, while GLR tests for a vector jump. Also, stepwise regression includes a test on "tolerance," and can delete parameters that have already been included in the regression. Gibbs (1992) shows that under the following conditions the regression test for a single hypothesis is equivalent to the log likelihood ratio statistic used in GLR:

1. The measurement noise is assumed to be known, which is usually true.
2. Only one jump is allowed in the data window.
3. The GLR implementation only allows a scalar parameter to jump.

Gibbs shows that differences in assumptions of the two algorithms can have a significant affect on performance. Other important points of the paper are:

1. There is no inherent reason to believe that the GLR test is in some sense better than the regression test—neither is a *universally most powerful* (UMP) test.
2. Implicit estimation of the measurement noise variance in stepwise regression gives it a variable threshold feature that may, or may not, be desirable. The residual RMS of the first few iterations is usually large and thus the residual test effectively uses a large threshold. As the dominant jumps are removed,

the residual RMS decreases closer to the actual noise level and thus the residual test approximates the likelihood test. This characteristic is also important when the true noise variance is unknown.

3. The best results are obtained for the test example when prior information on the jump magnitudes is included as a pseudo-measurement in the data equation (11.3-9). Because of the large number of hypothesized jump possibilities, prior-free estimation of jump parameters sometimes produced over-fitting of data in response to measurement noise and possibly modeling errors; that is, an excessive number of jump parameters were estimated where one jump estimate was cancelling another. Inclusion of a reasonable level of prior on the jump magnitudes eliminated this problem. However, use of prior information on the jump magnitudes requires that the d_f calculation be modified for the additional measurements; that is, $d_f = (km) - p + p = km$.
4. Prior information about the times of the jumps can be included by restricting the hypothesized times of the jumps; that is, by limiting the augmentation of columns to $\mathbf{T}_{i/k}$.
5. Calculations for a two-step staircase signature in the innovations show that stepwise regression will always select one of the actual jumps as the first jump, rather than using an average of the two jumps. The generality of this result is unknown, but experience has tended to confirm this behavior.
6. For problems where the filter state includes many bias states, jump detection improves significantly when the bias states are removed from the jump-free filter and instead included as permanently estimated states in the stepwise regression. Depending on the implementation, algorithm execution time can also be greatly reduced because of the reduction in the number of partial derivatives. However, this approach is only of use in post-processing (not real-time) applications. The improvement in detection performance is due to the nonlinear behavior of the hypothesis test rather than a difference in state estimates. For linear systems the state estimates must be the same whether bias states are included in the jump-free or jump-restoring filters.

11.3.3 Correction of Jump-Free Filter State

Given a set of jump states estimated by stepwise regression, it is desirable to correct the jump-free filter state taking into account the estimated jumps. This is similar to the correction in Friedland's bias-restoring filter. In some applications it is sufficient to compute the total state vector at the final time as

$$\boxed{\hat{\mathbf{x}}_{k/k}(corrected) = \hat{\mathbf{x}}_{k/k} + \mathbf{T}_{x\_k/k}\hat{\boldsymbol{\theta}}_x} \tag{11.3-11}$$

where t_k is the final time and $\mathbf{T}_{x\_k/k}$ includes all jump states estimated by the regression. Likewise, the corrected error covariance is:

$$\boxed{\mathbf{P}_{k/k}(corrected) = \mathbf{P}_{k/k} + \mathbf{T}_{x\_k/k}(\mathbf{M}_x^T\mathbf{M}_x + \mathbf{P}_{ax}^{-1})^{-1}\mathbf{T}_{x\_k/k}^T} \tag{11.3-12}$$

where $\mathbf{P}_{ax}$ is the prior covariance for the jump parameters. Since the prior estimate is zero, an extra term for the prior does not appear in equation (11.3-11). Note that the computed state covariance does not include any uncertainty due to errors in computing the time of the jumps. If this is needed, the jump parameters estimated by the regression can be used as initial conditions in a scoring method: scoring will automatically include the uncertainty on the jump times in the Fisher information matrix.

If the entire filtered state time history is required, it is necessary to save the entire filtered state estimates and $\mathbf{T}_{i/i}$ matrix at each measurement time, and correct the state using the equivalent of equation (11.3-11). However, the covariance is more difficult to compute since the filter state is conditioned on past data while the jump parameter estimate is conditioned on the entire data span. If the smoothed state time history is desired, it is possible to derive an equivalent form of equations (11.3-11) and (11.3-12) for the smoothed jump-free states, but it is simpler to rerun the filter and add an increment to the covariance matrix for the specific jump states at the computed jump times. That covariance increment is set equal to the prior uncertainty on the jump state. Then standard smoothing methods can be used.

11.3.4 Real-Time Jump Detection Using Stepwise Regression

Although the jump-detection algorithm was originally developed as an off-line post-processing method, it can also be used for real-time operation in much the same manner as GLR. However, we implement the estimate updating procedure quite differently from that described by Willsky and Jones (1974). These modifications appear to greatly improve the jump estimation.

The success of the approach depends upon whether the signatures of the jumps are well defined over a finite period of time. As with GLR, the algorithm is implemented to estimate jumps within a finite window of past time. As each new jump state is appended to the $\mathbf{T}_{i/i}$ matrix, the oldest jump state is dropped from $\mathbf{T}_{i/i}$. Thus the number of hypothesized jump states remains fixed, rather than growing without limit. Stepwise regression is performed at each new measurement time. However, the detected jumps are not immediately used to update the jump-free state vector, covariance, and the past jump-free innovations. Rather, the algorithm keeps track of the estimated jumps within the window. For the purposes of real-time estimation, the effect of jumps is added to the filter estimate using equation (11.3-11), but the jump-free filter state is not modified. The filter state is only updated when the full signature of the jump has been observed (i.e., when the trailing edge of the sliding data window coincides with the jump time). Because of this delay in implementing the filter update, the jump detection algorithm must be capable of handling multiple jumps, which basic GLR cannot. When the filter state is updated for an estimated jump, the past filter residuals and derivatives within the window are also modified using equations (11.3-6) and (11.3-7).

This delay in implementing the filter update insures that the entire data window is used to detect and estimate each jump, thus greatly improving jump estimation. To satisfy the need for minimal detection lag in critical real-time applications, a temporary filter state representing the best current estimate of the state (including the effect of jumps) is created at each measurement time. However, it is recognized that the estimate of the jump parameter set may change from one time point to the next. The temporary state is used for real-time prediction purposes (e.g., fire control in a tracking application). This hybrid scheme satisfies both the requirement for minimal detection lag and good jump estimation. Provided that the correct jump is

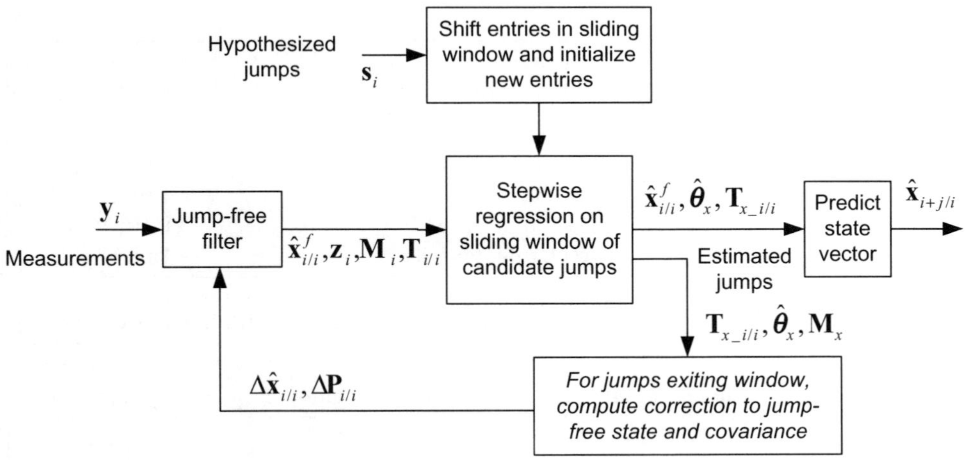

FIGURE 11.2: Processing flow for real-time jump detection/estimation.

detected, the state estimate produced by the prediction is optimal at each measurement time.

To reduce sensitivity of the jump detection to measurement outliers and noise, the real-time algorithm is implemented with a lag in hypothesizing jumps. That is, stepwise regression will not consider a candidate jump until several measurements are available after the candidate jump time. The lag must take into account the time required to observe a jump signature larger than the measurement noise. Thus the desired lag is problem-dependent. Since the signature of measurement outliers will generally not match the signature of candidate jumps, stepwise regression should not detect a jump when outliers occur. This may or may not actually happen. Use of the lag does, however, provide some capability to separate measurement outliers from jumps.

Figure 11.2 shows the processing flow of the real-time jump detection/estimation algorithm. Flow for the off-line post-processing algorithm is similar, but the window of candidate jumps does not slide (since it includes the entire data span) and stepwise regression is only performed once rather than as each new measurement is received.

The following example demonstrates that the performance of this online method is almost as good as that of the post-processing off-line technique. Refer to the associated software (ACTRKFIL_JUMP.f, and JUMPCOM.f) for the aircraft tracking example described below for more detailed information on the implementation.

Example 11.3: Multiple Jumps in IMU

This example uses the 27-state inertial navigation system (INS) model described in Section 3.3 and also used in Example 9.7. More details of this example and results are provided in Gibbs (1992). Table 11.2 is a copy of Table 3.3 that lists the simulated jumps for the missile flight. Table 11.3 lists the jumps estimated using stepwise regression on the entire data span when the 18 biases are moved

TABLE 11.2: Simulated INS Jumps

Simulation Jump Time (s)	Jump Parameter	Jump Magnitude
25	*y*-gyro SFE	–60 PPM
50	*y*-gyro g-sensitive error	+0.10 arc-sec/g
100	z-gyro bias drift	–0.10 arc-sec/s
120	*x*-accelerometer SFE	+1.0 PPM
400	*y*-accelerometer bias	–4 micro-g

TABLE 11.3: Estimated INS Jumps Using Off-Line Regression with Biases Removed

Jump Time (s)	Jump Parameter	Jump Magnitude	$1-\sigma$ Jump Uncertainty
30	*y*-gyro SFE	–56.5 PPM	17.6 PPM
50	*y*-gyro g-sensitive error	+0.088 arc-sec/g	0.010 arc-sec/g
100	z-gyro bias drift	–0.080 arc-sec/s	0.016 arc-sec/s
120	*x*-accelerometer SFE	+1.29 PPM	0.15 PPM
400	*y*-accelerometer bias	–4.09 micro-g	0.20 micro-g

from the jump-free filter to the jump-restoring filter. The regression F-to-add was 9.0, F-to-remove was 4.0, and tolerance threshold was 0.01. Notice in Table 11.3 that all jump parameters were correctly determined and all errors in jump estimates were within 1.9 times the computed $1-\sigma$ uncertainty. In fact, the *x*-accelerometer scale factor error (SFE) was the only parameter having an error greater than 1.3 times the $1-\sigma$ uncertainty. The *y*-gyro SFE jump at 25 s was computed to occur at 30 s because measurements were only sampled every 10 s. The results were nearly identical when the 18 biases were included in the jump-free filter, except that the jump in *y*-gyro SFE at 25 s was instead estimated as a –0.044 arc-sec/g jump in *y*-gyro g-sensitive error. The mis-modeling caused by the 5 s error in allowable jump times was mostly responsible for the incorrect jump parameter selection.

The GLR method was also tested on this data using a variety of algorithm options. A window width of 250 s, lag in hypothesizing jumps of 20 s, and detection threshold of 0.9 produced results as good as other GLR options. GLR was implemented so that the filter state and covariance were reset after a jump was detected, which was thought to be the best approach. Unfortunately GLR results were poor. Only the y-accelerometer bias at 400 s was estimated correctly (–4.07 micro-g): since the jump appears as a step function in the residuals, it would be surprising if this was incorrectly estimated. The x-accelerometer SFE jump at 120 s was calculated to occur at 130 s, and the other three jumps were estimated as two jumps at the wrong times, parameters, and channels. Another problem with GLR is a lack of robustness: the estimated parameters varied greatly from one Monte Carlo sample to the next. This was not a problem with stepwise regression.

Table 11.4 lists the estimated jump parameters when the stepwise regression method was implemented as a real-time sliding window algorithm as described previously. The window width, lag in hypothesizing jumps, and detection threshold were the same as used for GLR. Notice that all five estimated jumps were at the correct times, but the jumps at 30 and 100 s were determined for the wrong

TABLE 11.4: Estimated INS Jumps Using "Real-Time" Regression

Jump Time (s)	Jump Parameter	Jump Magnitude	$1-\sigma$ Jump Uncertainty
30	*y*-gyro g-sensitive error	−0.041 arc-sec/g	0.013 arc-sec/g
50	*y*-gyro g-sensitive error	+0.073 arc-sec/g	0.014 arc-sec/g
100	*z*-gyro g-sensitive error	−0.022 arc-sec/g	0.005 arc-sec/g
120	*x*-accelerometer SFE	+1.25 PPM	0.16 PPM
400	*y*-accelerometer bias	−4.07 micro-g	0.22 micro-g

parameter. Although incorrect, the estimate at 30 s closely matched the estimate based on the full data span when the 18 biases were included in the jump-free filter. Only the z-gyro bias drift jump at 100 s was estimated differently when using the full data span. The error of 10 s predictions was also computed: real-time stepwise regression produced average RSS velocity errors 13% smaller than those of GLR.

The stepwise regression jump detection/estimation method was also used operationally to process missile test flight data. Approximately 40 to 50 jumps were estimated in each flight. Estimation and removal of these jumps reduced the filter residual sum-of-squares by factors of 6 to 15. Independent tracking confirmed that the jump-corrected state was more accurate than the uncorrected state.

Example 11.4: Real-Time Aircraft Tracking

The jump detection/estimation algorithm was also tested on the aircraft tracking problem described in Section 3.2.2 and used in Example 9.5. Recall that the simulated aircraft turned left 90 degrees at 45–55 s and at 76–87 s, changed vertical velocity at 76 and 90 s, and decelerated from 90–95 s. Range, azimuth, and elevation were measured every second with simulated noise standard deviations of 1 m for ranges and 0.2 milliradians for angles.

The real-time jump detection/estimation algorithm was implemented as described above, where the window width was 12 s (12 measurement samples) and the lag in hypothesizing jumps was 1 s. The filter state and covariance were not updated until the jump exited the 12 s sliding window (as previously described), but intermediate jump estimates were used in predicting the aircraft state. The stepwise regression F-to-add was set to 9.0, the F-to-remove was 4.0, and the tolerance threshold was 0.01. Other regression parameters were also tested, but these values worked as well as others. The prior uncertainty on the jump magnitudes used in stepwise regression was set equal to the filter state uncertainty: 6 m/s, 0.5, and 1 m/s^2, respectively, for vertical velocity, crosstrack, and alongtrack acceleration.

In Example 9.5 the filter process noise PSDs for vertical velocity, crosstrack acceleration, and alongtrack acceleration were set to values appropriate for handling the maneuvers. Since the jump detection/estimation algorithm is designed to model maneuvers, the process noise for those parameters should be nearly zero. We used small nonzero values (0.0003 m^2/s^3, 0.0003, and 0.0003 m^2/s^5) for all three components so that remaining modeling errors would not cause problems.

TABLE 11.5: Simulated Aircraft Maneuvers

Maneuver Start Time (s)	Maneuver End Time (s)	Vertical Velocity (m/s)	Tangent (bank angle)	Alongtrack Acceleration (m/s^2)
44.0	54.0	0	–0.53	0
76.0	87.0	–4.0	–0.47	0
89.0	94.0	–6.0	0	–1.0

TABLE 11.6: Estimated Aircraft Maneuvers

Window Exit Time (s)	Estimated Maneuver Time (s)	Vertical Velocity (m/s)	Tangent (bank angle)	Alongtrack Acceleration (m/s^2)
51	44	—	–0.481	—
61	54	—	+0.452	—
83	76	–4.08	–0.420	–0.604
88	81	—	—	+0.750
92	85	—	—	–0.569
94	87	—	+0.475	—
96	89	–2.06	—	—
102	95	—	—	+0.821

Table 11.5 lists the simulated maneuvers. A "perfect" jump detection algorithm should estimate both entry and exit for each of the five maneuvers listed in Table 11.5 (10 jumps total). Notice that this is a particularly difficult test of the jump algorithm because there are only 11 measurements during the longest maneuver, multiple states jump during the maneuvers, and the gap between the last two maneuvers is only 9 s.

Table 11.6 lists the maneuvers estimated by the jump detection algorithm. A total of 8 jumps were actually estimated and most (but not all) of these jumps approximated simulated jumps. The jump algorithm tended to believe that alongtrack acceleration changed slightly more often than actual. Even so, predicted states that used the intermediate jump estimates are fairly accurate. Figures 11.3 and 11.4 show the error in the estimated position and velocity when using the intermediate estimates. When comparing these plots with Figures 9.9 and 9.10 for an EKF using the larger process noise values, it is noted that the RMS position error during nonjump periods is 30% to 50% smaller when using the jump algorithm, although the position errors are briefly larger during the later jumps. However, the velocity errors are much smaller during nonjump periods when using the jump algorithm, and errors during jump periods are only slightly larger. The jump algorithm seems to have the greatest difficulty during the period from 86 to 94 s when multiple parameters are changing. The difficulty may be partly due to the fact that the model is nonlinear and the estimated jump parameters are not updated in the filter until 12 s has elapsed. Shorter window lengths helped in some respects, but caused other problems.

To summarize, use of the real-time jump algorithm greatly improved aircraft tracking accuracy compared with a Kalman filter using fixed process noise.

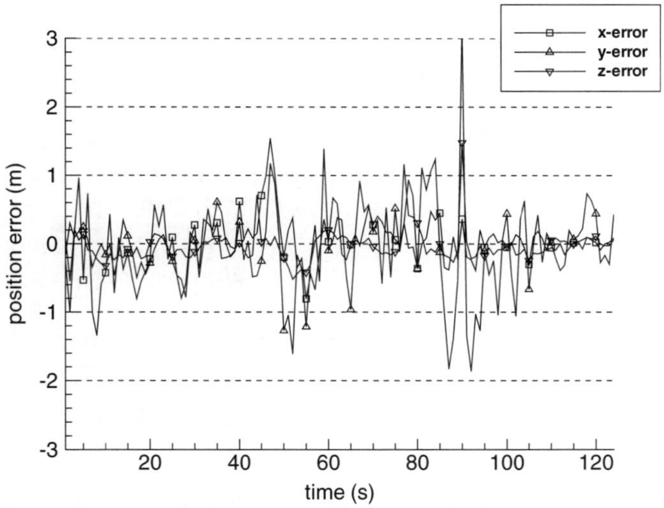

FIGURE 11.3: Error in estimated position using real-time jump algorithm.

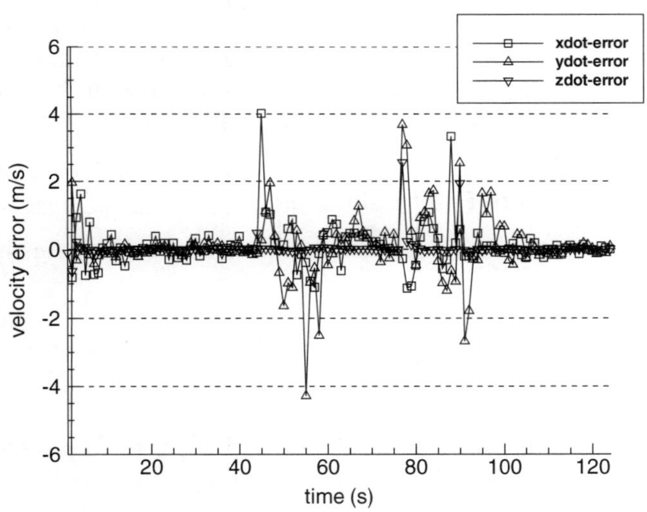

FIGURE 11.4: Error in estimated velocity using real-time jump algorithm.

11.4 ADAPTIVE TARGET TRACKING USING MULTIPLE MODEL HYPOTHESES

While the jump detection/estimation algorithm is a very good solution for problems in which the system model randomly changes in steps, many systems transition abruptly or smoothly between a finite number of behavior models. The types of behavior may include different levels of process noise, different levels of measurement noise, or different dynamic models—perhaps different frequency bands of

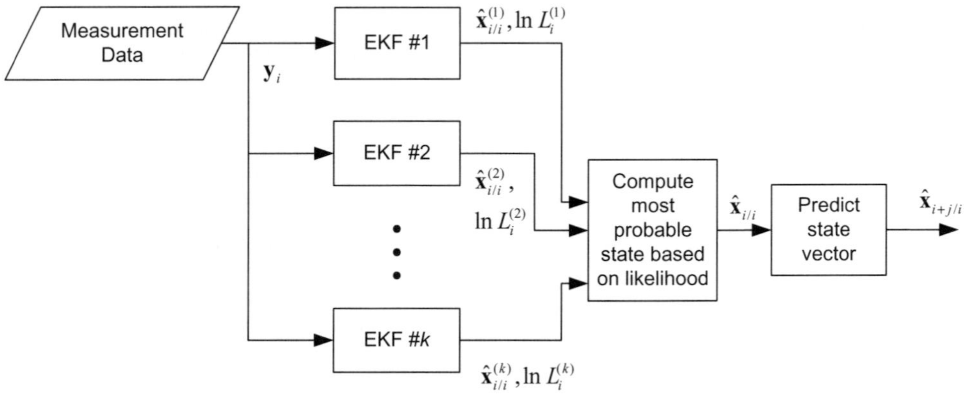

FIGURE 11.5: Multiple model adaptive filter.

colored noise disturbances. This suggests that a bank of Kalman filters using different models should continuously operate in parallel, and the "optimal" state estimate should somehow be determined from the different state estimates. The optimal state estimate may either be computed as a weighted sum of state estimates from all filters, or the state estimate from the filter with the best "tracking metric" can be used. In either case the weighting or tracking metric should be based on the probability that each filter model is correct. This suggests that the likelihood function should be used for this purpose, as shown in Figure 11.5.

11.4.1 Weighted Sum of Filter Estimates

The state estimate can be computed as a weighted sum of state estimates from k filters as

$$\hat{\mathbf{x}}_{i/i}^{optimal} = \sum_{j=1}^{k} w_i^j \hat{\mathbf{x}}_{i/i}^j \tag{11.4-1}$$

where w_i^j is the weighting of filter j at time t_i, and $\hat{\mathbf{x}}_{i/i}^j$ is the *a posteriori* state estimate of filter j at time t_i. The weighting function is computed as

$$w_i^j = \frac{p_{\mathbf{Y|X}}^j(\mathbf{y}_1, \mathbf{y}_2, \ldots, \mathbf{y}_i \mid \mathbf{x}_{0/0}^j)}{\sum_{l=1}^{k} p_{\mathbf{Y|X}}^l(\mathbf{y}_1, \mathbf{y}_2, \ldots, \mathbf{y}_i \mid \mathbf{x}_{0/0}^l)} \tag{11.4-2}$$

where $p_{\mathbf{Y|X}}^j(\mathbf{y}_1, \mathbf{y}_2, \ldots, \mathbf{y}_i \mid \mathbf{x}_{0/0}^j)$ is the joint probability density of receiving measurements $\mathbf{y}_1, \mathbf{y}_2, \ldots, \mathbf{y}_i$ given the prior state estimate $\hat{\mathbf{x}}_{0/0}^j$ and the models of filter j. The probability densities are computed from the calculated log likelihood of each filter,

$$\ln L^j(\mathbf{y}_1, \mathbf{y}_2, \ldots, \mathbf{y}_i \mid \mathbf{x}_{0/0}^j) = -\frac{1}{2}\left[\sum_{l=1}^{i} m_l \ln(2\pi) + \ln\left|\mathbf{C}_l^j\right| + (\tilde{\mathbf{y}}_l^j)^T (\mathbf{C}_l^j)^{-1} \tilde{\mathbf{y}}_l^j\right], \tag{11.4-3}$$

where $\tilde{\mathbf{y}}_l^j$ and $\mathbf{C}_l^j$ are the filter j innovation and innovation covariance at time t_l. Thus the weighting is

$$w_i^j = \frac{\exp\left(\ln L^j(\mathbf{y}_1, \mathbf{y}_2, \ldots, \mathbf{y}_i \mid \mathbf{x}_{0/0}^j)\right)}{\sum_{l=1}^{k} \exp\left(\ln L^l(\mathbf{y}_1, \mathbf{y}_2, \ldots, \mathbf{y}_i \mid \mathbf{x}_{0/0}^l)\right)}. \tag{11.4-4}$$

Since the true system model may change with time, it is usually best to sum the terms in equation (11.4-3) over a sliding window of time, particularly when operating the filters for lengthy time periods.

There is a practical problem in using equation (11.4-4) for weighting. The log likelihood is not limited in magnitude because it is the log of a probability density. Values of equation (11.4-3) may be hundreds or even thousands. This will cause overflows or underflows when computing the exponential, so it is necessary to scale the numerator and denominator by the inverse exponential of the largest $\ln L^j(\mathbf{y}_1, \mathbf{y}_2, \ldots, \mathbf{y}_i \mid \mathbf{x}_{0/0}^j)$. This is done by adding the negative of

$$\max_j \left(\ln L^j(\mathbf{y}_1, \mathbf{y}_2, \ldots, \mathbf{y}_i \mid \mathbf{x}_{0/0}^j)\right)$$

to all values before evaluating the exponentials. All differences less than −40 can be ignored (i.e., set $w_i^j = 0$) because $e^{-40} = 4.2 \times 10^{-18}$ is beyond double precision accuracy on most computers.

Calculation of a weighted sum of filter states is only meaningful when state definitions for all filters are identical. This may happen, for example, when the filter models simply use different levels of process or measurement noise, or different parameters in the state transition matrix.

11.4.2 Maximum Likelihood Filter Selection

The above approach cannot be used when the state definitions of the parallel filters are different. It is also not unusual to find that the likelihood functions computed for different filters differ by orders of magnitude. Hence the filter with the largest likelihood dominates the weighted sum of equation (11.4-1). In these cases it is better to use the state estimate from the filter with the largest likelihood function. If the true system only uses one model at a given time, selection of a single filter may work better than when using the weighted sum, even when the likelihood of one filter is not dominant. An example later in this section demonstrates the effectiveness of this approach for tracking maneuvering tanks.

Notice that the filter state dimension affects the degrees-of-freedom for the filter and this should be taken into account when computing the likelihood function. However, there are two reasons why this is usually not done in practice. First, the effective degrees-of-freedom for different filter state sizes are initially equal because of differences in the number and values of terms included in the filter $\hat{\mathbf{x}}_{0/0}^j$ and $\mathbf{P}_{0/0}^j$ for different j. Also, other differences in filter models are more important than the state size. For example, the process noise magnitude is often increased to account for the larger model error when creating reduced-order models.

11.4.3 Dynamic and Interactive Multiple Models

Selection of the optimal filter based solely on the current likelihood function often works well when the tracking data span is short. When longer data spans are available, it is possible to obtain better results using prior information on the probabilities of mode transitions, and by dynamically selecting a larger set of filters that retain history of the mode transitions. Bar-Shalom et al. (2001, section 11.6) describes several methods using this approach, and the preferred method is called the *interacting multiple model* (IMM) estimator.

As with other multiple model approaches, it is assumed that the system behavior at any given time can be adequately described by one of r models: M^j where $1 \le j \le r$. The model (or mode) at time t_i is defined as M_i^j, and the system model is

$$\boxed{\begin{aligned} \mathbf{x}_i &= \mathbf{\Phi}_{i,i-1}^j \mathbf{x}_{i-1} + \mathbf{q}_{i,i-1} \\ \mathbf{y}_i &= \mathbf{H}_i^j \mathbf{x}_i + \mathbf{r}_i \end{aligned}} \tag{11.4-5}$$

where

$$\begin{aligned} \mathbf{q}_{i,i-1} &\sim N\left(\overline{\mathbf{q}}_{i,i-1}^j, \mathbf{Q}_{i,i-1}^j\right) \\ \mathbf{r}_i &\sim N\left(\overline{\mathbf{r}}_i^j, \mathbf{R}_i^j\right) \end{aligned} \tag{11.4-6}$$

and we use the abbreviated notation

$$\begin{aligned} \mathbf{\Phi}_{i,i-1}^j &= \mathbf{\Phi}_{i,i-1}(\mathbf{M}_i^j), \quad & \mathbf{Q}_{i,i-1}^j &= \mathbf{Q}_{i,i-1}(\mathbf{M}_i^j) \\ \mathbf{H}_i^j &= \mathbf{H}_i(\mathbf{M}_i^j), & \mathbf{R}_i^j &= \mathbf{R}_i(\mathbf{M}_i^j) \end{aligned}.$$

Notice that use of a mode-dependent nonzero mean ($\overline{\mathbf{q}}_{i,i-1} \neq \mathbf{0}$) allows modeling of jumps in input.

At the next measurement time t_{i+1}, the system may change modes to M_{i+1}^k, so in the absence of any prior information on the probability of mode transitions, there are r^2 possible transitions from M_i^j to M_{i+1}^k. Hence if r parallel filters were operating at t_i, it would be necessary to use r^2 parallel filters at t_{i+1} in order to optimally track the system state. This exponential increase in the number of filters makes implementation of an optimal approach prohibitive, so suboptimal methods are used to "prune" the number of filters carried at each time step.

It is first assumed that the mode transition probabilities are time-invariant and independent of the state, which makes it a *homogenous Markov chain*. We now make the notation even more confusing by defining the mode sequence

$$\mathbf{M}_i^{\mathbf{u}} = \left\{\mathbf{M}_{i-1}^{\mathbf{s}}, M_i^j\right\},$$

where superscript **s** denotes the *parent sequence* through t_{i-1} and M_i^j is the mode at t_i; that is, j is the last mode of mode sequence **u**. Then using the Markov property that the system state at one time is sufficient to describe the probability of future states, we have

$$\Pr\left\{M_i^j \mid \mathbf{M}_{i-1}^{\mathbf{s}}\right\} = \Pr\left\{M_i^j \mid M_{i-1}^k\right\} \tag{11.4-7}$$

where k is the last mode of mode sequence **s** and $\Pr\{\cdot\}$ denotes the probability of the given discrete event. The conditional probability of the mode change k to j is denoted as

$$\boxed{p_{kj} \triangleq \Pr\{M_i^j \mid M_{i-1}^k\},} \tag{11.4-8}$$

and values for all possible mode changes are a required input for the IMM method. We now define the conditional probability of a given mode history at t_i given the history of measurements up to t_i as

$$\boxed{\mu_i^{\mathbf{u}} \triangleq \Pr\{\mathbf{M}_i^{\mathbf{u}} \mid \mathbf{y}_i, \mathbf{y}_{i-1}, \ldots, \mathbf{y}_1\}.} \tag{11.4-9}$$

Bar-Shalom et al. show that for a Markov chain the conditional probability can be written as

$$\mu_i^{\mathbf{u}} = \frac{1}{c} p_{\mathbf{Y}i|\mathbf{M},\mathbf{Y}i-}(\mathbf{y}_i \mid \mathbf{M}_i^{\mathbf{u}}, \mathbf{y}_{i-1}, \ldots, \mathbf{y}_1)\, p_{kj} \mu_{i-1}^{\mathbf{s}} \tag{11.4-10}$$

where k is the last mode of sequence $\mathbf{s}$, j is the last mode of sequence $\mathbf{u}$, c is a normalizing constant to make the sum of probabilities for different sequences equal to 1.0, and $p_{\mathbf{Y}i|\mathbf{M},\mathbf{Y}i-}(\mathbf{y}_i \mid \mathbf{M}_i^{\mathbf{u}}, \mathbf{y}_{i-1}, \ldots, \mathbf{y}_1)$ is the conditional probability density of receiving measurements $\mathbf{y}_i$ given the current mode sequence and previous measurement sequence. Equivalently

$$p_{\mathbf{Y}i|\mathbf{M},\mathbf{Y}i-}(\mathbf{y}_i \mid \mathbf{M}_i^{\mathbf{u}}, \mathbf{y}_{i-1}, \ldots, \mathbf{y}_1) = L(\mathbf{y}_i \mid \hat{\mathbf{x}}_{i/i-1}^{(\mathbf{u})}) \tag{11.4-11}$$

where $L(\mathbf{y}_i \mid \hat{\mathbf{x}}_{i/i-1}^{(\mathbf{u})})$ is the likelihood function for measurements at t_i, and $\hat{\mathbf{x}}_{i/i-1}^{(\mathbf{u})}$ denotes the *a priori* filter estimate that was obtaining using mode sequence $\mathbf{u}$ and measurements $\mathbf{y}_{i-1}, \ldots, \mathbf{y}_1$.

To implement the dynamic multiple model algorithms it is necessary to define two conditional probabilities for modes, rather than mode sequences:

$$\boxed{\begin{aligned} \lambda_i^j &\triangleq \Pr\{M_i^j \mid \mathbf{y}_i, \ldots, \mathbf{y}_1\} \\ &= \frac{1}{c} p_{\mathbf{Y}i|M,\mathbf{Y}i-}(\mathbf{y}_i \mid M_i^j, \mathbf{y}_{i-1}, \ldots, \mathbf{y}_1)\bar{c}_j \end{aligned}} \tag{11.4-12}$$

where

$$\boxed{\begin{aligned} \bar{c}_j &= \sum_{k=1}^{r} p_{kj} \lambda_{i-1}^k \\ c &= \sum_{j=1}^{r} p_{\mathbf{Y}i|M,\mathbf{Y}i-}(\mathbf{y}_i \mid M_i^j, \mathbf{y}_{i-1}, \ldots, \mathbf{y}_1)\bar{c}_j \end{aligned}} \;. \tag{11.4-13}$$

and

$$\boxed{\begin{aligned} \lambda_{i-1/i-1}^{k|j} &\triangleq \Pr\{M_{i-1}^k \mid M_i^j, \mathbf{y}_{i-1}, \ldots, \mathbf{y}_1\} \\ &= \frac{1}{\bar{c}_j} p_{kj} \lambda_{i-1}^k \end{aligned}} \;. \tag{11.4-14}$$

There are several options for pruning the mode options at each measurement step. One approach, called *generalized pseudo-Bayesian estimator of first order* (GPB1), uses the λ_i^j probabilities as weighting factors to sum the *a posteriori* state estimates of r filters into a single state estimate $\hat{\mathbf{x}}_{i/i}$, which is approximate. The corresponding covariance is created by similar weighting. Then the summed state is propagated forward in time to the next measurement using the r different mode possibilities, so r filters are always used. The *GPB2* algorithm is similar but all mode possibilities at the current and last time are retained so that the number of filters is r^2.

The IMM algorithm also retains r filters, but the state estimate for each filter uses a different combination of mode-conditioned estimates. This mixing of the estimates, or hypothesis merging, is performed after the *a posteriori* filter estimates are computed. The steps are:

1. *Compute the mixing probabilities* as

$$\begin{aligned}\bar{c}^j &= \sum_{k=1}^{r} p_{kj}\lambda_{i-1}^k \\ \lambda_{i-1/i-1}^{k|j} &= \frac{1}{\bar{c}^j} p_{kj}\lambda_{i-1}^k\end{aligned} \tag{11.4-15}$$

for $j, k = 1,\ldots, r$ where λ_{i-1}^k is the mode probability for mode k at t_{i-1} from the previous time step (see step 4 below). The initial probability is set as $\lambda_0^k = 1/r$ for all k.

2. *Compute the mixed state estimate and covariance initial conditions* for all modes M_i^j as

$$\hat{\mathbf{x}}_{i-1/i-1}^{0j} = \sum_{k=1}^{r} \lambda_{i-1/i-1}^{k|j}\hat{\mathbf{x}}_{i-1/i-1}^{k} \tag{11.4-16}$$

$$\mathbf{P}_{i-1/i-1}^{0j} = \sum_{k=1}^{r} \lambda_{i-1/i-1}^{k|j}\left[\mathbf{P}_{i-1/i-1}^{k} + (\hat{\mathbf{x}}_{i-1/i-1}^{k} - \hat{\mathbf{x}}_{i-1/i-1}^{0j})(\hat{\mathbf{x}}_{i-1/i-1}^{k} - \hat{\mathbf{x}}_{i-1/i-1}^{0j})^T\right] \tag{11.4-17}$$

for $j = 1,\ldots, r$ where $\hat{\mathbf{x}}_{i-1/i-1}^{k}$ and $\mathbf{P}_{i-1/i-1}^{k}$ are the output of the r filters from the previous time step.

3. *Predict the mixed state estimate and covariance to the next measurement time, process measurement, and compute the likelihood for each filter.*

Time update:

$$\begin{aligned}\hat{\mathbf{x}}_{i/i-1}^{0j} &= \mathbf{\Phi}_{i,i-1}^{j}\hat{\mathbf{x}}_{i-1/i-1}^{0j} \\ \tilde{\mathbf{y}}_i^j &= \mathbf{y}_i - \mathbf{H}_i^j\hat{\mathbf{x}}_{i/i-1}^{0j} \\ \mathbf{P}_{i/i-1}^{0j} &= (\mathbf{\Phi}_{i,i-1}^{j}\mathbf{P}_{i-1/i-1}^{0j})(\mathbf{\Phi}_{i,i-1}^{j})^T + \mathbf{Q}_{i,i-1}^{j} \\ \mathbf{C}_i^j &= (\mathbf{H}_i^j\mathbf{P}_{i/i-1}^{0j})(\mathbf{H}_i^j)^T + \mathbf{R}_i^j\end{aligned}. \tag{11.4-18}$$

Measurement update:

$$\boxed{\begin{aligned}\mathbf{K}_i^j &= (\mathbf{P}_{i/i-1}^{0j})(\mathbf{H}_i^j)^T(\mathbf{C}_i^j)^{-1}\\ \hat{\mathbf{x}}_{i/i}^j &= \hat{\mathbf{x}}_{i/i-1}^{0j} + \mathbf{K}_i^j\tilde{\mathbf{y}}_i^j\\ \mathbf{P}_{i/i}^j &= \mathbf{P}_{i/i-1}^{0j} - \mathbf{K}_i^j\mathbf{H}_i^j\mathbf{P}_{i/i-1}^{0j}\end{aligned}} \tag{11.4-19}$$

for $j = 1,\ldots, r$. The Joseph, U-D, or square-root information filter (SRIF) updates can also be used. Then compute the likelihood for each filter as

$$\boxed{L^j(\mathbf{y}_i \mid M_i^j, \hat{\mathbf{x}}_{i-1/i-1}^{0j}) = \frac{1}{(2\pi)^{m/2}\left|\mathbf{C}_i^j\right|^{1/2}}\exp\left(-(\tilde{\mathbf{y}}_i^j)^T(\mathbf{C}_i^j)^{-1}\tilde{\mathbf{y}}_i^j/2\right).} \tag{11.4-20}$$

4. *Update the mode probabilities as*

$$\boxed{\begin{aligned}c &= \sum_{k=1}^{r}\bar{c}^j L^j(\mathbf{y}_i \mid M_i^j, \hat{\mathbf{x}}_{i-1/i-1}^{0j})\\ \lambda_i^j &= \frac{\bar{c}^j}{c}L^j(\mathbf{y}_i \mid M_i^j, \hat{\mathbf{x}}_{i-1/i-1}^{0j})\end{aligned}} \tag{11.4-21}$$

with $\bar{c}^j$ from (11.4-15).

5. *Combine state estimates and covariances* (optional for output only)

$$\boxed{\hat{\mathbf{x}}_{i/i} = \sum_{j=1}^{r}\lambda_i^j\hat{\mathbf{x}}_{i/i}^j} \tag{11.4-22}$$

$$\boxed{\mathbf{P}_{i/i} = \sum_{j=1}^{r}\lambda_i^j\left[\mathbf{P}_{i/i}^j + (\mathbf{x}_{i/i}^j - \hat{\mathbf{x}}_{i/i})(\mathbf{x}_{i/i}^j - \hat{\mathbf{x}}_{i/i})^T\right].} \tag{11.4-23}$$

This completes the IMM summary. See Bar-Shalom et al. (2001, section 11.6) for additional details of the algorithm.

It is interesting to compare the IMM approach with the jump algorithm of Section 11.3.4 because both are dynamic multiple model approaches. IMM is somewhat more general because the jump algorithm only handles one type of model change (step jumps in input), while the jump algorithm considers all jump hypotheses within the sliding window without requiring prior knowledge of mode transition probabilities. To summarize, differences between the IMM and jump detection/estimation algorithm are:

1. Jump detection considers all mode sequences within the sliding window. It does not prune the hypothesis within the window. As estimated jumps leave the window they are assumed to be known and thus used to reset the jump-free filter state. This is the only pruning performed.
2. The jump mode changes are limited to steps in input states. This assumption allows the algorithm to operate without running multiple Kalman filters in parallel because the multiple jump hypotheses are handled as a Bayesian least-squares regression problem. The IMM method does not explicitly model

step changes in input, and some information on jump signatures is lost when state estimates are combined in the mixing step.

3. The jump algorithm uses prior information on the magnitude of the jumps, but does not use information on mode transition probabilities. Compared with IMM this is a weakness when those prior probabilities are known. For example, in the aircraft tracking problem the probability of having a jump to the "no-turn" mode after a turn has been detected is higher when a short period of time has elapsed. This information is not normally used in the jump algorithm, but it could be if the prior estimate for future jumps was set to the negative of the detected turn bank angle. However, it is also difficult to use the turn information in IMM because the transition probabilities are a function of the time elapsed since the last jump, and this violates the assumption that transition probabilities are time-invariant. Even when transition probabilities are fixed, the values may not be known.
4. When F-tests are used in stepwise regression for the jump detection, the algorithm automatically adapts to changes in measurement noise. If stepwise regression uses chi-squared tests, the measurement noise variance must be known. The IMM method has the capability to explicitly estimate changes in measurement noise variances, but this requires that the additional modes be carried in the bank of filters.
5. Provided that both algorithms detect a given jump, both should predict the future behavior with little lag, although the methods used are very different. The jump algorithm may be more sensitive to the jump, but predictions may also be more erratic than IMM because IMM averages the filter response in the mixing step.

Example 11.5: Tank Tracking with Multiple Models

Section 3.2.1 described development of five tank maneuver models, and Example 9.4 demonstrated smoothing for the "Scout" vehicle model. Recall that one motion model is "constant velocity," one uses two first-order Markov process acceleration models, and three motion models use two second-order Markov process models. Since "operational" tank motion tends to be serpentine, use of jump models is not appropriate.

These tank maneuver models are used here to demonstrate fire control tracking and prediction based on the multiple model approach. Of the three multiple model approaches discussed above, selection of the filter with the maximum likelihood is the only method tested. The IMM approach is not ideal because the allowable tracking time is limited to 10 s in combat situations, and there is little basis for defining transition mode probabilities—normally only one mode would apply in a 10-s segment. Also, averaging of state estimates from different models is meaningless when the states represent different physical variables.

The Scout track of Figure 3.3 is used as the test data. This plot is repeated in Figure 11.6 where an "own tank" observer is positioned 2 km from the Scout.

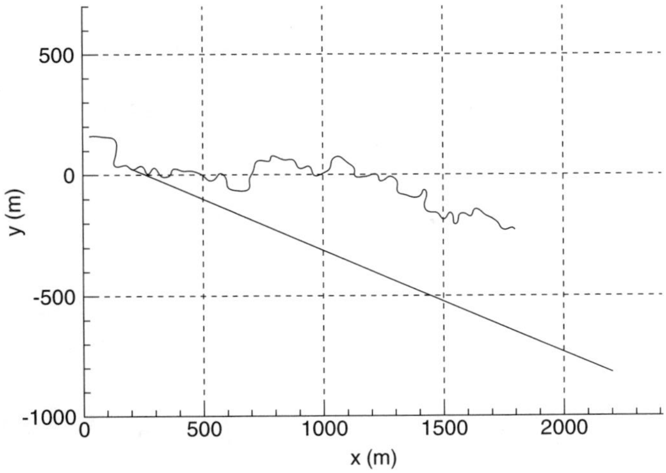

FIGURE 11.6: Scout track and line-of-sight to typical observer location.

For purposes of simulating range and azimuth tracking data, observers were randomly located about 2 km ahead of the Scout for each tracking test. The mean observer location was 25° below the x-axis with a Gaussian-distributed $1 - \sigma$ of 30°; that is, the $1 - \sigma$ limits are +5° and −55°. Ranges of 1 and 3 km were also tested, but results are only listed for 2 km. Simulated observer-to-target ranges and azimuth (bearing) angles were generated at a 10 Hz sampling rate, and Gaussian noise of 2 m and 0.3 milliradians ($1 - \sigma$) was added.

Since 10 s is considered the maximum tracking duration for each encounter, the 568 s track was broken into 109 segments of 10 s that overlap by 5 s. Each segment was treated as a separate encounter and the observer was randomly positioned for each encounter as described above. It is assumed that observers have at least 4 s in which to evaluate tracking performance of the filters. The "optimal" filter state is determined once per second from 4 to 10 s of tracking, and the optimal state vector is used to predict the Scout position at the time of expected projectile impact assuming a projectile speed of 1500 m/s. The error in the predicted azimuth—called the lead prediction error—is computed for each sample, and statistics on whether the error is within the NATO box (±1.15 m) are tabulated. A mean hit probability (p_h) is computed for the seven samples of each encounter, and histograms of hit probability are generated. Additional gun-ammo dispersion errors should also be included for a more realistic assessment of hit probability, but that dispersion is not included in this example. See Gibbs and Porter (1980) for more information about this problem and example.

Figure 11.7 is a histogram of hit probability when using a single constant-velocity filter, which is the type of lead prediction usually employed. Figure 11.8 is the comparable plot when using the five-model adaptive filter where the optimal state is obtained from the filter with the greatest likelihood function. Typically zero or one model change occurred in each encounter, with the "low maneuver" and "M60" models the most commonly selected. This was

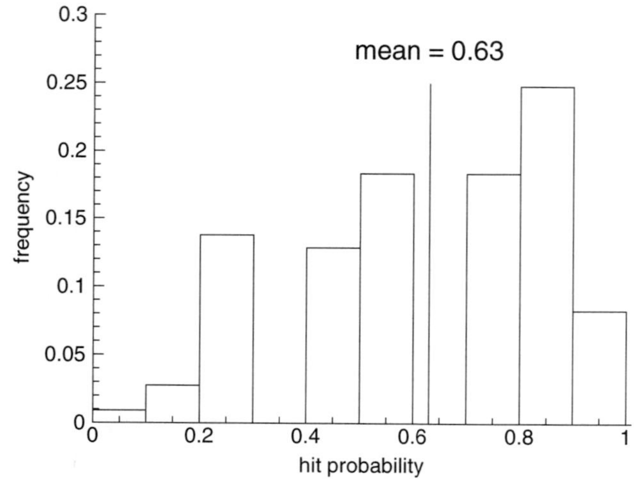

FIGURE 11.7: Hit probability histogram for constant velocity filter.

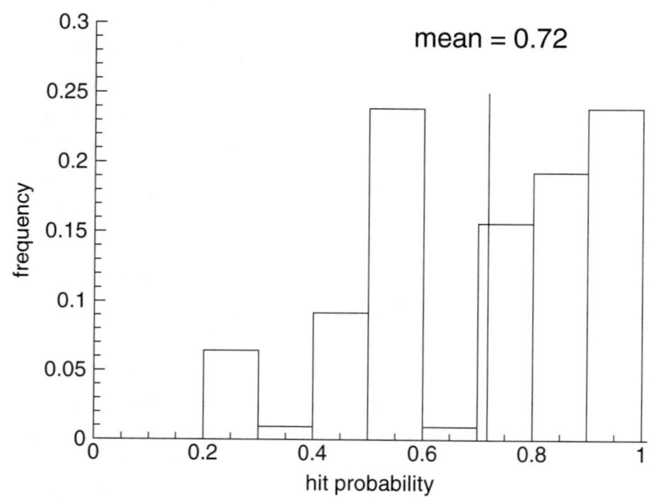

FIGURE 11.8: Hit probability histogram for adaptive filter.

particularly true in the latter half of the scenario when Scout speeds were lower. Notice in the figures that the number of samples with low hit probability is much smaller for the adaptive filter, and the mean p_h is 0.72 compared with 0.63 for the constant velocity filter. This is a significant improvement, but the difference is smaller when gun-ammo dispersion is included. Since some systems only generate range measurements once per second, the difference in performance with reduced range tracking was also tested. This had no effect on p_h for the constant velocity filter, but reduced the average p_h to 0.70 for the adaptive filter. Not surprisingly, higher order filters need more range measurements to accurately estimate the higher derivatives.

11.5 CONSTRAINED ESTIMATION

It is sometimes necessary to enforce equality or inequality constrains on Kalman filter state estimates. The methods employed for this purpose are similar to those used for least-squares estimation as described in Section 7.1. For equality constraints it is generally best to use the constraint equation to change state definitions so that the number of state variables is reduced. However, this is not always possible. The constraint can be enforced by treating the constraint as a heavily weighted pseudo-measurement, but this must be done cautiously because it can cause the numerical problems described in Section 7.1.1. Other methods, such as Lagrange multipliers, can also be used, but they are somewhat more difficult to implement for filters than for least-squares estimation. Simon (2006, section 7.5) provides additional information on how the various constraint methods are used with Kalman filters.

Inequality constraints may be imposed using the same approaches used for least squares (e.g., active set methods), but there is a significant difference. Use of an inequality constraint in batch least-squares estimation can be mathematically defined as the solution to a specific problem, but this is less clear for recursive filters. Since imposition of the constraint may occur when processing some measurements but not others, the filter response is nonlinear and the nonlinear behavior is a function of the sequence of measurement noise values. This may work for some applications, but the state estimates are not optimal in a well-defined sense. Rather than using inequality constraints, it may be better to use a nonlinear change of state variables—such as log function—to ensure that values are greater than zero. This is only acceptable when the added nonlinearity does not cause linearization problems in an EKF.

11.6 ROBUST ESTIMATION: *H*-INFINITY FILTERS

The Kalman filter is a minimum variance estimator. That is, it minimizes the l_2 norm of expected errors in the state estimate. In some cases it is more important to limit the maximum errors of some linear combination of states for worst case assumptions about the process noise, measurement noise, and initial condition errors. Filters designed to minimize a weighted norm of state errors for worst case conditions are called H_∞ or *minimax* filters.

The discrete time H_∞ filter uses the same state model as the Kalman filter, which we repeat here from Chapter 8:

$$\mathbf{x}_i = \mathbf{\Phi}_{i,i-1}\mathbf{x}_{i-1} + \mathbf{q}_{i,i-1} \tag{11.6-1}$$

$$\mathbf{y}_i = \mathbf{H}_i\mathbf{x}_i + \mathbf{r}_i \tag{11.6-2}$$

where $E[\mathbf{q}_{i,i-1}] = \mathbf{0}$, $E[\mathbf{q}_{i,i-1}\mathbf{q}_{j,j-1}^T] = \mathbf{Q}_{i,i-1}\delta_{ij}$, $E[\mathbf{r}_i] = \mathbf{0}$, $E[\mathbf{r}_i\mathbf{r}_j^T] = \mathbf{R}_i\delta_{ij}$, and $E(\mathbf{r}_i\mathbf{q}_{j,j-1}^T) = \mathbf{0}$ for all i, j. In deriving the optimal measurement update for the Kalman filter, we computed the gain matrix $\mathbf{K}_i$ in the filter update

$$\hat{\mathbf{x}}_{i/i} = \hat{\mathbf{x}}_{i/i-1} + \mathbf{K}_i(\mathbf{y}_i - \mathbf{H}_i\hat{\mathbf{x}}_{i/i-1})$$

that minimized

$$J = \frac{1}{2} E\left[\tilde{\mathbf{x}}_{i/i}^T \tilde{\mathbf{x}}_{i/i}\right] = \frac{1}{2} tr(\mathbf{P}_{i/i}),$$

where $\tilde{\mathbf{x}}_{i/i} = \mathbf{x}_i - \hat{\mathbf{x}}_{i/i}$ and $\mathbf{P}_{i/i} = E\left[\tilde{\mathbf{x}}_{i/i} \tilde{\mathbf{x}}_{i/i}^T\right]$. It was noted that the $\mathbf{K}_i$ minimizing J for arbitrary weighting $\mathbf{W}$ in $E\left[\tilde{\mathbf{x}}_{i/i}^T \mathbf{W} \tilde{\mathbf{x}}_{i/i}\right]$ also minimizes J for any weighting, so we conveniently assumed $\mathbf{W} = \mathbf{I}$.

Unlike the Kalman filter, the cost function to be minimized in H_∞ filters is not uniquely defined. Thus there are multiple H_∞ filter approaches. The presentation of this section is based on the approach of Banavar and Speyer (1991) for continuous systems, but adapted for discrete systems as developed by Shen et al. (1996), Shen and Deng (1995, 1997, 1999), and Simon (2006, Chapter 11),

The H_∞ approach first defines a linear transformation of states

$$\boxed{\mathbf{z}_i \triangleq \mathbf{L}_i \mathbf{x}_i} \tag{11.6-3}$$

where $\mathbf{L}_i$ is a full-rank, user-defined matrix. The corresponding transformation on the state estimate is $\hat{\mathbf{z}}_{i/j} = \mathbf{L}_i \hat{\mathbf{x}}_{i/j}$. The residual $\tilde{\mathbf{z}}_{i/j} = \mathbf{z}_i - \hat{\mathbf{z}}_{i/j}$ is used in the cost function of the H_∞ filter, so inclusion of the transformation $\mathbf{L}_i$ allows certain combinations of states to be given more weight than others. For example, it may be desirable in a tracking problem that the error component of predicted position normal to the line-of-sight be smaller than the error component along the line-of-sight.

The cost function of the H_∞ filter is defined to be

$$\boxed{J'_k = \frac{\sum_{i=0}^{k-1} \tilde{\mathbf{z}}_{i+1/i}^T \mathbf{S}_i \tilde{\mathbf{z}}_{i+1/i}}{\tilde{\mathbf{x}}_{0/0}^T \mathbf{P}_{0/0}^{-1} \tilde{\mathbf{x}}_{0/0} + \sum_{i=0}^{k-1} \mathbf{q}_{i+1,i}^T \mathbf{Q}_{i+1,i}^{-1} \mathbf{q}_{i+1,i} + \sum_{i=0}^{k-1} \mathbf{r}_i^T \mathbf{R}_i^{-1} \mathbf{r}_i}} \tag{11.6-4}$$

where $\mathbf{S}_i$ is a user-defined positive definite symmetric weighting matrix. Notice that this cost function differs from that of the Kalman filter in three respects. First, J'_k is computed as a sum of terms over all measurements up to but not including k, rather than as an expected value for each measurement. Second, the denominator of J'_k includes the three random inputs to the system, so if one term is larger than expected, the residual $\tilde{\mathbf{z}}_{i/j}$ is allowed to be larger to obtain the same J'_k. Finally, the weighting matrix $\mathbf{S}_i$ for this J'_k does influence the measurement gain matrix, so both $\mathbf{L}_i$ and $\mathbf{S}_i$ affect the solution $\hat{\mathbf{x}}_{i+1/i}$.

This cost function is not sufficiently defined to allow computation of a unique minimum. The goal is find estimates $\hat{\mathbf{x}}_{i+1/i}$ that keep the cost below a given value $(1/\theta)$ for worst case conditions; that is,

$$J'_k < \frac{1}{\theta}. \tag{11.6-5}$$

Thus by rearranging equation (11.6-4) and using equation (11.6-2) with equation (11.6-3), a modified cost function is defined:

$$\begin{aligned} J''_k &= \frac{-1}{\theta} \tilde{\mathbf{x}}_{0/0}^T \mathbf{P}_{0/0}^{-1} \tilde{\mathbf{x}}_{0/0} + \sum_{i=0}^{k-1} \Gamma_i \\ &< 0 \end{aligned} \tag{11.6-6}$$

where

$$\Gamma_i = (\mathbf{x}_{i+1} - \hat{\mathbf{x}}_{i+1/i})^T \mathbf{W}_i (\mathbf{x}_{i+1} - \hat{\mathbf{x}}_{i+1/i}) - \frac{1}{\theta}\left(\mathbf{q}_{i+1,i}^T \mathbf{Q}_{i+1,i}^{-1} \mathbf{q}_{i+1,i} + (\mathbf{y}_i - \mathbf{H}_i \mathbf{x}_i)^T \mathbf{R}_i^{-1} (\mathbf{y}_i - \mathbf{H}_i \mathbf{x}_i)\right) \tag{11.6-7}$$

and

$$\mathbf{W}_i = \mathbf{L}_i^T \mathbf{S}_i \mathbf{L}_i . \tag{11.6-8}$$

This cost function does not include the dynamic model constraint equation (11.6-1), so this is added using a vector of Lagrange multipliers $\boldsymbol{\lambda}_{i+1}$:

$$J_k = \frac{-1}{\theta} \tilde{\mathbf{x}}_{0/0}^T \mathbf{P}_{0/0}^{-1} \tilde{\mathbf{x}}_{i/i} + \sum_{i=0}^{k-1} \left(\Gamma_i + \frac{2\boldsymbol{\lambda}_{i+1}^T}{\theta} (\boldsymbol{\Phi}_{i+1,i} \mathbf{x}_i + \mathbf{q}_{i+1,i} - \mathbf{x}_{i+1}) \right) + \frac{2\boldsymbol{\lambda}_0^T}{\theta} \mathbf{x}_0 - \frac{2\boldsymbol{\lambda}_0^T}{\theta} \mathbf{x}_0 . \tag{11.6-9}$$

The scaling $(2/\theta)\boldsymbol{\lambda}_{i+1}^T$ is unusual, but it does not affect the solution. Addition and subtraction of $(2/\theta)\boldsymbol{\lambda}_0^T \mathbf{x}_0$ may appear redundant, but it is helpful in following steps to group $+(2/\theta)\boldsymbol{\lambda}_0^T \mathbf{x}_0$ with the sum of terms from $i = 0$ to $k - 1$, and to keep $-(2/\theta)\boldsymbol{\lambda}_0^T \mathbf{x}_0$ separate.

Cost function equation (11.6-9) can now be used as the basis for the solution. The goal is to find equations defining $\hat{\mathbf{x}}_{i+1/i}$—or equivalently a measurement weighting matrix (similar to the Kalman gain matrix)—that minimize the cost for worst case assumptions about $\mathbf{x}_0$, $\mathbf{q}_{i+1,i}$ and $\mathbf{y}_i$. That is, find

$$J_k^* = \min_{\mathbf{x}_i} \max_{\mathbf{x}_0, \mathbf{q}_{i+1,i}, \mathbf{y}_i} J_k .$$

This is a game theoretic problem that is solved in two steps. Notice that the estimator has complete knowledge of $\mathbf{y}_i$ but no knowledge $\mathbf{x}_0$ and $\mathbf{q}_{i+1,i}$. Therefore it should be prepared for worst case values of the unknown variables. In the first step partial derivatives of J_k with respect to $\mathbf{x}_0$, $\mathbf{q}_{i+1,i}$, and $\boldsymbol{\lambda}_i$ are set to zero to obtain relationships for the maximum J_k as a function of $\hat{\mathbf{x}}_{i+1/i}$ and $\mathbf{y}_i$. In the second step, the partial derivatives of J_k with respect to $\hat{\mathbf{x}}_{i+1/i}$ and $\mathbf{y}_i$ are set to zero to obtain an expression for the measurement gain matrix minimizing J_k from the first step.

By setting $\partial J / \partial \mathbf{x}_0 = \mathbf{0}^T$ with J defined by equation (11.6-9), it is found that

$$\mathbf{x}_0 = \hat{\mathbf{x}}_{0/0} + \mathbf{P}_{0/0} \boldsymbol{\lambda}_0 \tag{11.6-10}$$

and

$$(\mathbf{x}_0 - \hat{\mathbf{x}}_{1/0})^T \mathbf{W}_0 = -\frac{1}{\theta}\left(\boldsymbol{\lambda}_0^T + \boldsymbol{\lambda}_1^T \boldsymbol{\Phi}_{1,0} + (\mathbf{y}_0 - \mathbf{H}_0 \mathbf{x}_0)^T \mathbf{R}_0^{-1} \mathbf{H}_0\right). \tag{11.6-11}$$

(Presence of measurement $\mathbf{y}_0$ at the initial time is unusual, but for generality it is allowed.) Based on equation (11.6-10), it is assumed that

$$\mathbf{x}_i = \boldsymbol{\mu}_i + \mathbf{P}_{i/i} \boldsymbol{\lambda}_i \tag{11.6-12}$$

where $\boldsymbol{\mu}_i$ and $\mathbf{P}_{i/i}$ are to be determined. (It is later found that an optimal solution is obtained using this assumption.) Then setting $\partial J / \partial \mathbf{q}_{i+1,i} = \mathbf{0}^T$ and $\partial J / \partial \boldsymbol{\lambda}_i = \mathbf{0}^T$ leads to expressions for J as a function of $\hat{\mathbf{x}}_{i+1/i}$ and $\mathbf{y}_i$. The solution for $\hat{\mathbf{x}}_{i+1/i}$ is obtained (after much manipulation) by setting $\partial J / \partial \hat{\mathbf{x}}_{i+1/i} = \mathbf{0}^T$ and $\partial J / \partial \mathbf{y}_i = \mathbf{0}^T$. See Shen and Deng (1995, 1997) or Simon (2006, section 11.3) for the details. The result is

$$\boxed{\begin{aligned} \mathbf{D}_i &= \left[\mathbf{I} - \theta \mathbf{W}_i \mathbf{P}_{i/i-1} + \mathbf{H}_i^T \mathbf{R}_i^{-1} \mathbf{H}_i \mathbf{P}_{i/i-1}\right]^{-1} \\ \mathbf{K}_i &= \mathbf{P}_{i/i-1} \mathbf{D}_i \mathbf{H}_i^T \mathbf{R}_i^{-1} \\ \hat{\mathbf{x}}_{i+1/i} &= \mathbf{\Phi}_{i+1,i}\left[\hat{\mathbf{x}}_{i/i-1} + \mathbf{K}_i(\mathbf{y}_i - \mathbf{H}_i \hat{\mathbf{x}}_{i/i-1})\right] \\ \mathbf{P}_{i+1/i} &= \mathbf{\Phi}_{i+1,i} \mathbf{P}_{i/i-1} \mathbf{D}_i \mathbf{\Phi}_{i+1,i}^T + \mathbf{Q}_{i+1,i} \end{aligned}} \tag{11.6-13}$$

provided that θ is sufficiently small to maintain

$$\mathbf{P}_{i/i-1}^{-1} - \theta \mathbf{W}_i + \mathbf{H}_i^T \mathbf{R}_i^{-1} \mathbf{H}_i \tag{11.6-14}$$

positive definite. With this restriction, the solution $\hat{\mathbf{x}}_{i+1/i}$ minimizes the cost function of equation (11.6-9) with dynamic model constraint equation (11.6-1). Notice that when $\theta = 0$, equation (11.6-13) is equivalent to the Kalman filter: this is verified using the matrix inversion identity of Appendix A. Also notice that equation (11.6-13) requires inversion of an $n \times n$ matrix when computing $\mathbf{D}_i$. This can be computationally expensive (compared with the standard Kalman filter) when the state size n is large. For some applications the computational burden can be a problem, and it may be necessary to replace the time-varying gain with a precomputed steady-state gain for time-invariant systems.

Notice that in addition to the normal Kalman filter covariance matrices $\mathbf{Q}_{i+1,i}$, $\mathbf{R}_i$, and $\mathbf{P}_{0/0}$, the H_∞ filter also requires input of $\mathbf{L}_i$, $\mathbf{S}_i$, and θ. Even for this one approach to H_∞ filtering, there are an infinite number of possible solutions that depend on the selection of these three parameters. θ must be small so that equation (11.6-14) is positive definite, but it not clear how matrices $\mathbf{L}_i$ and $\mathbf{S}_i$ should be chosen for a given problem. The performance can be highly sensitive to the weights. Testing using both simulated and real data for hypothesized worst case conditions is recommended to verify that the selected $\mathbf{L}_i$, $\mathbf{S}_i$, and θ provide performance meeting requirements. H_∞ filters have the potential to greatly reduce maximum estimation errors compared with Kalman filters, but actual behavior depends on the choice of input parameters and the specific "non-ideal" conditions encountered.

Since performance of H_∞ filters will generally be worse than that of Kalman filters under "normal" (non-worst case) conditions, some minimax filters attempt to merge Kalman and H_∞ filters. These and other implementations are discussed in Simon (2006, chapters 11 and 12).

11.7 UNSCENTED KALMAN FILTER (UKF)

Use of the EKF and iterated EKF for nonlinear systems was discussed in Chapter 10. The limitations of these approaches were also discussed, and it was noted that these linearized methods may not converge to the minimum variance solution when nonlinearities are large and observabiliy is poor. In this section we examine another approach to the problem. Rather than linearize the model about the current estimate, the UKF computes carefully selected perturbations about the current state estimate. Then the perturbed states are propagated through the nonlinear time update and measurement model equations to compute samples of the state estimate and predicted measurement. The distribution of these states and measurements is used to infer the mean and covariance, and these are substituted in the Kalman update equations. The concept is based on the observation that it is easier to approximate a probability distribution than it is to approximate an arbitrary nonlinear function. The UKF requires more computation than a simple EKF implemented using analytic partial derivatives, but UKF computations are generally comparable to those of an EKF that uses central difference numeric partial derivatives for state transition and measurement sensitivity arrays.

The UKF approach was first developed by Julier et al. (1995). Later UKF publications include Julier and Uhlmann (1997, 2002, 2004a,b), Julier et al. (2000), and Julier (2002). Simon (2006, chapter 14) also provides a good discussion of the algorithms. Julier and Uhlmann state that UKF applications have included navigation systems for high-speed road vehicles, public transportation systems, data assimilation systems, and underwater vehicles. Simon notes that the UKF has been used for aircraft engine health estimation, aircraft model estimation, neural network training, financial forecasting, and motor state estimation.

Different papers by Julier and Uhlman present slightly different algorithms for the UKF. Since their papers are often referenced in this section, we use the abbreviation J&U0x to indicate their paper published in 200x. The following description of the UKF covers the scaled version of the unscented transform using an extended symmetric distribution of sigma points. Most of the material is covered in J&U04a, but examination of earlier papers is helpful in understanding details. We start by describing how the unscented transform can be used to compute the mean and covariance of a nonlinear function of a random variable.

11.7.1 Unscented Transform

Suppose that we have a nonlinear vector function of a vector random variable $\mathbf{x}$,

$$\mathbf{z} = \mathbf{f}(\mathbf{x}), \tag{11.7-1}$$

where we want to determine the mean and covariance of $\mathbf{z}$ given the mean $\bar{\mathbf{x}}$ and covariance $\mathbf{P}_x$ of $\mathbf{x}$. This type of nonlinear equation can appear in the EKF time update of the state vector ($\hat{\mathbf{x}}_{i/i-1}$), and/or when computing predicted measurements $\hat{\mathbf{y}}_i = \mathbf{h}_i(\hat{\mathbf{x}}_{i/i-1})$. In the EKF, the equation is expanded in a Taylor series,

$$\mathbf{z}(\mathbf{x}) = \mathbf{z}(\mathbf{x}_0) + \frac{\partial \mathbf{f}}{\partial \mathbf{x}}(\mathbf{x}-\mathbf{x}_0) + \frac{1}{2}(\mathbf{x}-\mathbf{x}_0)^T \frac{\partial^2 \mathbf{f}}{\partial \mathbf{x} \partial \mathbf{x}^T}(\mathbf{x}-\mathbf{x}_0) + \ldots,$$

and truncated after the linear term. The linearization is in error because all second-order and higher terms are ignored. Example 7.3 showed for a nonlinear falling ball problem that contours of constant least-squares cost computed from the linearized covariance may be significantly different from contours of the true cost. These contours reflect the difference between the true and linearized probability density of the state $\mathbf{x}$ for Gaussian errors. J&U04a presents another example for range and bearing target tracking. It shows that the actual distribution is very different from that of the elliptical model computed from the covariance matrix.

Monte Carlo simulation is another method for handling the nonlinearity. Here random samples of $\mathbf{x}^{(i)}$ are generated to have mean $\bar{\mathbf{x}}$ and covariance $\mathbf{P}_x$. Then the sample mean and covariance of $\mathbf{z}^{(i)} = \mathbf{f}(\mathbf{x}^{(i)})$ are computed. The approach is accurate provided that a sufficiently large number of samples are selected.

The UKF uses a different approach. By carefully selecting values of $\mathbf{x}^{(i)}$, called *sigma points*, the mean and covariance of $\mathbf{z}^{(i)}$ can be computed using only $2n + 1$ samples, where n is the dimension of $\mathbf{x}$. The $2n + 1$ samples and corresponding weights are selected as:

$$\begin{aligned}
&\mathbf{x}^{(0)} = \bar{\mathbf{x}} \\
&\left.\begin{aligned}
\mathbf{x}^{(i)} &= \bar{\mathbf{x}} + \sqrt{\frac{n}{1-W^{(0)}}}\left(\mathbf{P}_x^{1/2}\right)_{col_i} \\
\mathbf{x}^{(i+n)} &= \bar{\mathbf{x}} - \sqrt{\frac{n}{1-W^{(0)}}}\left(\mathbf{P}_x^{1/2}\right)_{col_i}
\end{aligned}\right\} \quad i = 1,2,\ldots,n \\
&W^{(i)} = (1-W^{(0)})/(2n), \quad i = 1,2,\ldots,n
\end{aligned} \tag{11.7-2}$$

where $\mathbf{P}_x = \mathbf{P}_x^{1/2}\mathbf{P}_x^{T/2}$ and $\left(\mathbf{P}_x^{1/2}\right)_{col\_i}$ is the i-th column of $\mathbf{P}_x^{1/2}$. Cholesky factorization is typically used, where $\mathbf{P}_x = \mathbf{L}\mathbf{L}^T$ and $\mathbf{P}_x^{1/2} = \mathbf{L}$ is lower triangular. The $W^{(i)}$ are used to weight the samples $\mathbf{z}^{(i)} = \mathbf{f}(\mathbf{x}^{(i)})$ when computing the sample mean and covariance. They must satisfy the constraint

$$\sum_{i=0}^{2n} W^{(i)} = 1. \tag{11.7-3}$$

$W^{(0)}$ is the weight given to $\mathbf{z}^{(0)}$, and it is normally set as $W^{(0)} = 0$. In cases when information about the distribution of $\mathbf{x}$ is known, it may be appropriate to use a nonzero $W^{(0)}$ so that fourth-order moments of the sample distribution for $\mathbf{z}^{(i)}$ match the true distribution. However this requires analysis of the nonlinearity, so $W^{(0)}$ is typically set to zero.

The sample mean and covariance of $\mathbf{z}$ are computed as

$$\begin{aligned}
\bar{\mathbf{z}}_s &= \sum_{i=0}^{2n} W^{(i)}\mathbf{z}^{(i)} \\
\boldsymbol{\Sigma}_z &= \sum_{i=0}^{2n} W^{(i)}(\mathbf{z}^{(i)} - \bar{\mathbf{z}}_s)(\mathbf{z}^{(i)} - \bar{\mathbf{z}}_s)^T
\end{aligned} \; . \tag{11.7-4}$$

These equations are derived using a Taylor series expansion of the nonlinearity,

$$\mathbf{z}^{(i)} = \mathbf{z}(\mathbf{x}^{(i)}) = \mathbf{z}(\bar{\mathbf{x}}) + \frac{\partial \mathbf{f}}{\partial \mathbf{x}}\tilde{\mathbf{x}}^{(i)} + \frac{1}{2}(\tilde{\mathbf{x}}^{(i)})^T\left(\frac{\partial^2 \mathbf{f}}{\partial \mathbf{x}\partial \mathbf{x}^T}\right)\tilde{\mathbf{x}}^{(i)} + \ldots$$

where $\tilde{\mathbf{x}}^{(i)} = \mathbf{x}^{(i)} - \bar{\mathbf{x}}$. Then the Taylor series is substituted in equation (11.7-4). Since the sigma points were selected so that

$$\tilde{\mathbf{x}}^{(i)} = -\tilde{\mathbf{x}}^{(i+n)}, \quad \sum_{i=0}^{2n} \tilde{\mathbf{x}}^{(i)} = \mathbf{0} \quad \text{and} \quad \sum_{i=0}^{2n} \tilde{\mathbf{x}}^{(i)}(\tilde{\mathbf{x}}^{(i)})^T = \frac{2n}{1-W^{(0)}}\mathbf{P}_x,$$

odd powers of the series for $\bar{\mathbf{z}}$ and $\boldsymbol{\Sigma}_z$ vanish and the remaining terms are found to match $E[\mathbf{z}^{(i)}]$ and $E[(\mathbf{z}^{(i)} - \bar{\mathbf{z}})(\mathbf{z}^{(i)} - \bar{\mathbf{z}})^T]$ computed by taking the expected values of the Taylor series through third-order terms. See Julier et al. (2000), J&U02, J&U04a, or Simon (2006, Section 14.2) for the details.

Notice that the perturbation of the *i*-th element of the state vector $\bar{\mathbf{x}}$ in equation (11.7-2) is approximately equal to $\sqrt{nP_{x\_ii}}$ when $\mathbf{P}_x$ is a diagonally dominant matrix. This can be a problem when the state vector is large, and/or when a filter is initialized with large variances. For example, when $n = 400$ (a not unusual state size), the perturbation will be about 20 times larger than the standard deviations of the states. This is may greatly exceed the acceptable range of the model and cause model failure (e.g., argument out-of-range for square root, arc sine, or log functions.) Julier (2002) noted that the range of the distribution can be restricted using a scaling variable α. This scaling was introduced to minimize the errors due to high-order terms when sigma points are not symmetric, but it can also be used to limit the excursions with symmetric sampling. The modified transform is

$$\boxed{\begin{aligned} \mathbf{x}^{(0)} &= \bar{\mathbf{x}} \\ \left.\begin{aligned} \mathbf{x}^{(i)} &= \bar{\mathbf{x}} + \alpha\sqrt{\frac{n}{1-W^{(0)}}}\left(\mathbf{P}_x^{1/2}\right)_{col\_i} \\ \mathbf{x}^{(i+n)} &= \bar{\mathbf{x}} - \alpha\sqrt{\frac{n}{1-W^{(0)}}}\left(\mathbf{P}_x^{1/2}\right)_{col\_i} \end{aligned}\right\} & \quad i = 1, 2, \ldots, n \\ W_s^{(0)} &= (W^{(0)} + \alpha^2 - 1)/\alpha^2 \\ W_s^{(i)} &= (1 - W^{(0)})/(2n\alpha^2), \quad i = 1, 2, \ldots, n \end{aligned}} \tag{11.7-5}$$

and the sample mean and covariance are

$$\boxed{\begin{aligned} \bar{\mathbf{z}}_s &= \sum_{i=0}^{2n} W_s^{(i)}\mathbf{z}^{(i)} \\ \mathbf{\Sigma}_z &= \sum_{i=0}^{2n} W_s^{(i)}(\mathbf{z}^{(i)} - \bar{\mathbf{z}}_s)(\mathbf{z}^{(i)} - \bar{\mathbf{z}}_s)^T + (1+\beta-\alpha^2)(\mathbf{z}^{(0)} - \bar{\mathbf{z}}_s)(\mathbf{z}^{(0)} - \bar{\mathbf{z}}_s)^T \end{aligned}} . \tag{11.7-6}$$

Julier (2002) used $0.01 < \alpha \le 1$ when working with nonsymmetric sigma points, but perhaps setting $\alpha\sqrt{n/(1-W^{(0)})}$ equal to two or three is a better approach for symmetric sigma points. This gives "2-sigma" or "3-sigma" $\mathbf{x}$ perturbations in the directions of $\mathbf{P}_x^{1/2}$. The value of β is chosen to minimize the fourth-order error terms for known distributions, and it is a function of the kurtosis. When $\beta = 0$ and sigma point sampling is symmetric, equations for the mean and covariance are accurate to third-order. If it is known that the distribution of $\mathbf{x}$ is Gaussian, setting $\beta = 2$ can reduce errors in fourth-order terms. However, this can also lead to nonpositive definite $\mathbf{\Sigma}_z$ unless equation (11.7-6) is modified to compute the covariance about $\mathbf{z}^{(0)}$ rather than $\bar{\mathbf{z}}_s$.

Another version of the unscented transform is based on a set of $n + 1$ *simplex sigma points*, rather than the $2n + 1$ symmetric sigma points. The covariance computed from simplex sampling can sometimes have smaller errors when distributions are skewed in an optimal direction, but simplex sampling can also have much larger errors when the skew direction is non-optimal. For that reason simplex sigma sampling should be used with great caution, if at all.

11.7.2 UKF Algorithm

The UKF uses the unscented transform to compute conditional means and covariances for nonlinear models. As with the Kalman filter, the UKF is initialized with an estimate of the state $\hat{\mathbf{x}}_{0/0}$ and covariance $\mathbf{P}_{0/0}$ at time t_0. Then for each new measurement at time t_i, the following steps are executed.

1. Select a set of sigma points $\mathbf{x}^{(j)}_{i-1/i-1}$ and weights $W^{(j)}$ for $0 \le j \le 2n+1$ using algorithm equation (11.7-2) or equation (11.7-5) starting with $\hat{\mathbf{x}}_{i-1/i-1}$ and $\mathbf{P}_{i-1/i-1}$. Normally set $W^{(0)} = 0$.
2. Integrate the $2n+1$ points to obtain $\hat{\mathbf{x}}^{(j)}_{i/i-1} = \mathbf{f}(\mathbf{x}^{(j)}_{i-1/i-1})$.
3. Compute the *a priori* state and covariance at time t_i:

$$\boxed{\hat{\mathbf{x}}_{i/i-1} = \sum_{j=0}^{2n} W^{(j)} \hat{\mathbf{x}}^{(j)}_{i/i-1}} \tag{11.7-7}$$

$$\boxed{\begin{aligned}\boldsymbol{\Sigma}_{i/i-1} = &\sum_{j=0}^{2n} \left(W^{(j)} (\mathbf{x}^{(j)}_{i/i-1} - \hat{\mathbf{x}}_{i/i-1})(\mathbf{x}^{(j)}_{i/i-1} - \hat{\mathbf{x}}_{i/i-1})^T \right) \\ &+ (1-\alpha^2)(\mathbf{x}^{(0)}_{i/i-1} - \hat{\mathbf{x}}_{i/i-1})(\mathbf{x}^{(0)}_{i/i-1} - \hat{\mathbf{x}}_{i/i-1})^T + \mathbf{Q}_{i,i-1}\end{aligned}} . \tag{11.7-8}$$

Equation (11.7-8) assumes that the process noise is sufficiently small so that the contribution can be modeled as an additive linear term $\mathbf{Q}_{i,i-1}$. When $\mathbf{Q}_{i,i-1}$ is a significant fraction of $\boldsymbol{\Sigma}_{i/i-1}$, the process noise $\mathbf{q}_{i,i-1}$ should be handled differently. The method is described at the end of this section.

4. Select a new set of $2n+1$ sigma points $\mathbf{x}^{(j)}_{i/i-1}$ and weights using algorithm equation (11.7-2) or (11.7-5) with $\hat{\mathbf{x}}_{i/i-1}$ and $\boldsymbol{\Sigma}_{i/i-1}$. Normally set $W^{(0)} = 0$. If the measurements are linear in $\mathbf{x}$, the sigma point calculations can be bypassed and the linear model used directly.
5. Compute $2n+1$ predicted measurements $\hat{\mathbf{y}}^{(j)}_i = \mathbf{h}_i(\mathbf{x}^{(j)}_{i/i-1})$.
6. Compute the predicted measurement and covariance at time t_i:

$$\boxed{\hat{\mathbf{y}}_i = \sum_{j=0}^{2n} W^{(j)} \hat{\mathbf{y}}^{(j)}_i} \tag{11.7-9}$$

$$\boxed{\begin{aligned}\boldsymbol{\Sigma}_{yy\_i} = &\sum_{j=0}^{2n} \left(W^{(j)} (\hat{\mathbf{y}}^{(j)}_i - \hat{\mathbf{y}}_i)(\hat{\mathbf{y}}^{(j)}_i - \hat{\mathbf{y}}_i)^T \right) \\ &+ (1-\alpha^2)(\hat{\mathbf{y}}^{(0)}_i - \hat{\mathbf{y}}_i)(\hat{\mathbf{y}}^{(0)}_i - \hat{\mathbf{y}}_i)^T + \mathbf{R}_i \\ \boldsymbol{\Sigma}_{xy\_i} = &\sum_{j=0}^{2n} W^{(j)} (\mathbf{x}^{(j)}_{i/i-1} - \hat{\mathbf{x}}_{i/i-1})(\hat{\mathbf{y}}^{(j)}_i - \hat{\mathbf{y}}_i)^T \\ &+ (1-\alpha^2)(\mathbf{x}^{(0)}_{i/i-1} - \hat{\mathbf{x}}_{i/i-1})(\hat{\mathbf{y}}^{(0)}_i - \hat{\mathbf{y}}_i)^T\end{aligned}} . \tag{11.7-10}$$

Notice that equation (11.7-10) assumes that measurement noise is small so that the contribution can be modeled as a linear addition $\mathbf{R}_i$. The method used when $\mathbf{R}_i$ is a significant fraction of $\boldsymbol{\Sigma}_{yy\_i}$ is described below.

7. Compute the Kalman gain and *a posteriori* state and covariance as

$$\boxed{\begin{aligned}\mathbf{K}_i &= \boldsymbol{\Sigma}_{xy\_i}\boldsymbol{\Sigma}_{yy\_i}^{-1}\\ \hat{\mathbf{x}}_{i/i} &= \hat{\mathbf{x}}_{i/i-1} + \mathbf{K}_i(\mathbf{y}_i - \hat{\mathbf{y}}_i).\\ \mathbf{P}_{i/i} &= \boldsymbol{\Sigma}_{i/i-1} - \mathbf{K}_i\boldsymbol{\Sigma}_{xy\_i}^T\end{aligned}} \tag{11.7-11}$$

The UKF can also be implemented in a factored covariance form for greater numerical accuracy (Van der Merwe and Wan 2001).

When $\mathbf{Q}_{i,i-1}$ is a significant fraction of $\boldsymbol{\Sigma}_{i/i-1}$, the assumption that process noise can be treated as a linear perturbation may not be valid. In that case the nonzero process noise $\mathbf{q}_{i,i-1}$ components should be added to the bottom of the state vector to create an augmented state. Likewise when $\mathbf{R}_i$ is a significant fraction of $\boldsymbol{\Sigma}_{yy\_i}$, the measurement noise $\mathbf{r}_i$ components should be added to the state vector. Thus the augmented *a priori* state vector and covariance at t_i are

$$\begin{aligned}\hat{\mathbf{x}}_{i/i-1}^a &= \begin{bmatrix}\hat{\mathbf{x}}_{i/i-1}^T & \mathbf{q}_{i,i-1}^T & \mathbf{r}_i^T\end{bmatrix}^T\\ \mathbf{P}_{i/i-1}^a &= \begin{bmatrix}\mathbf{P}_{i/i-1} & \mathbf{0} & \mathbf{0}\\ \mathbf{0} & \mathbf{Q}_{i,i-1} & \mathbf{0}\\ \mathbf{0} & \mathbf{0} & \mathbf{R}_i\end{bmatrix}.\end{aligned} \tag{11.7-12}$$

The augmented state should be used in all references to sigma points in the UKF steps above, and the state size should be changed to $n + n_q + m$ where n_q is the number of nonzero elements in $\mathbf{q}_{i,i-1}$, and m is the number of measurements in $\mathbf{y}_i$. When this approach is used, the $\mathbf{Q}_{i,i-1}$ term should be dropped from equation (11.7-8) and $\mathbf{R}_i$ should be dropped from equation (11.7-10).

J&U04a presents an example of a simulated ballistic reentry tracking problem (repeated below) where UKF estimation errors are significantly smaller than those of an EKF. The corrected results (J&U04b) are less dramatic but still favor the UKF. Simon (2006, chapter 14) uses a similar ballistic problem (but with range measurements only) in a limited Monte Carlo simulation to compare different implementations of the UKF and EKF. He also shows that UKF estimates of states can be more accurate than those of the EKF.

J&U00 states that "the new filter should be preferred over the EKF in virtually all nonlinear estimation applications." This statement seems to be partially based on their assertion (in J&U00, but repeated in slightly different form in J&U04a) that "over 30 years of experience with (the EKF) has led to a general consensus that it is difficult to implement, difficult to tune, and only reliable for systems that are almost linear on the time scale of the update intervals." If this statement were true, it is a wonder that the EKF has been used successfully and reliably for probably thousands of very diverse applications over a period approaching five decades. (Recall that the EKF was first implemented in 1961 for Apollo midcourse lunar navigation.) While it is probably true that the UKF is not used as often as it should be, much more experience on a wider range of problems is needed before it can be declared generally preferable to the EKF. The UKF can be a better option than the EKF for some range-only and angle-only tracking problems. The differences are most likely to be important when nonlinearities are large (particularly over the measurement sampling interval) and measurements only sense one direction of a multidimensional problem. Cases that include hard constraints on the states, or discontinuities, are also better handled with the UKF.

Some of the issues to be considered in deciding between use of the EKF and UKF include the following:

1. Some dynamic and measurement models are fairly linear in a limited region and then suddenly break down. While an EKF implemented with numerical partial derivatives can (and should) use very small state perturbations, the UKF uses perturbations comparable to or larger than the $1-\sigma$ estimation uncertainty. Even when using α scaling described previously, there is the potential for model functions to fail catastrophically, particularly at filter initialization when variances are often large.
2. While the UKF only requires one matrix multiplication and inversion to perform the measurement update, it requires integrating the equations of motion $2n+1$ times when performing the time update using symmetric sigma points. In many applications time integration is by far the most computationally intensive step of the filter. Thus for systems with many states, the EKF state vector is often partitioned into dynamic states and bias states, where only dynamic states are integrated. Partial derivatives with respect to bias states are often trivial and computed analytically, while complicated partials with respect to dynamic states are computed numerically. Hence computations required to implement an EKF may be acceptable, while the $2n+1$ function evaluations required for the UKF may be prohibitive.
3. Julier and Uhlmann criticize the EKF because successful implementation requires "tuning" of the process noise matrix $\mathbf{Q}$. The UKF also requires $\mathbf{Q}$ as an input. It can often be smaller than for an EKF because it need not compensate for model nonlinearities. Even so, the true value of $\mathbf{Q}$ is rarely known with certainty and it still must be "tuned." The most general forms of the UKF also require input of α, β, and $W^{(0)}$, and these parameters must be determined for each problem. Hence tuning of a UKF can be more complicated than tuning an EKF.

Example 11.6: Ballistic Reentry

An example often used in comparing nonlinear filtering methods involves range tracking of a ballistic object subject to earth gravitation and aerodynamic drag forces. Sometimes noisy angle measurements are also used. Variants on this problem appear in Simon (2006, chapter 14) and Gelb (1974, example 6.1-2). It is also similar to our Example 7.1 for a falling ball tracked by range measurements.

The following case of two-dimensional ballistic reentry attempts to match the example in J&U04a/b, which is based on a similar problem of Chang et al. (1977). Ground radar is located at earth-centered coordinates $(x_r, y_r) = (6374, 0)$ km, where $r_e = 6374$ km is the assumed radius of the earth. The ballistic body is modeled by the state equations

$$\dot{x}_1(t) = x_3(t)$$
$$\dot{x}_2(t) = x_4(t)$$
$$\dot{x}_3(t) = d(t)x_3(t) + g(t)x_1(t) + q_3(t)$$
$$\dot{x}_4(t) = d(t)x_4(t) + g(t)x_2(t) + q_4(t)$$
$$\dot{x}_5(t) = q_5(t)$$

where

- x_1, x_2 are the body earth-centered (x, y) positions (km),
- x_3, x_4 are the body earth-centered (x, y) velocities (km/s),
- x_5 is the logarithm of a scale variable to account for variations in atmospheric density and drag force,
- $r(t) = \sqrt{x_1^2(t) + x_2^2(t)}$ is body distance from the earth center (km),
- $v(t) = \sqrt{x_3^2(t) + x_4^2(t)}$ is body velocity (km/s),
- $d(t) = -\beta_0 e^{(r_e - r(t))/h_0} e^{x_5(t)} v(t)$ is the drag function (1/s),
- $g(t) = -G_m/r^3(t)$ is the gravitational function (1/s^2),
- $\rho_0 = 1.225$ is atmospheric density (kg/m^3) at the earth's surface,
- $h_0 = 9.3$ is the atmospheric density scale height (km) for the exponential model,
- $\beta_0 = 0.5\rho_0 C_d A/M = 0.59783$/km is the drag proportionality coefficient at zero altitude ($C_d A/M$ is the drag coefficient times area divided by mass),
- $G_m = 3.9860044 \times 10^5$ is the earth's gravitational constant (km^3/s^2),
- $q_3(t)$, $q_4(t)$, $q_5(t)$ are zero-mean process noise with respective simulated variances of 2.4065×10^{-5}, 2.4065×10^{-5}, and 0 for 0.1 s sampling.

The density scale height is listed as $h_0 = 13.406$ km in J&U04a, but that value is not realistic for the "standard atmosphere" and it does not produce a simulated trajectory matching that of J&U04a. Use of the more accurate $h_0 = 9.3$ km provides a reasonable match to the referenced trajectory.

The state vector for each trial of the 100 sample Monte Carlo simulation is initialized with mean

$$\bar{\mathbf{x}}(0) = [6500.4 \quad 349.14 \quad -1.8093 \quad -6.7967 \quad 0.6932]^T$$

and random perturbations having covariance

$$\mathbf{P}(0) = diag[10^{-6} \quad 10^{-6} \quad 10^{-6} \quad 10^{-6} \quad 0];$$

that is, the initial standard deviations of the simulation are 1 m for position and 1 m/s for velocity. Setting $x_5(0) = 0.6932$ makes the true value of drag two times larger than nominal. The body is initially located 371 km from the radar at an altitude of 349 km, with velocity directed toward an impact point past the radar. After 30 s of flight, the body reaches an altitude of about 70 km and is strongly affected by atmospheric drag. The motion rapidly becomes vertical, with the result that the body is only about 2 km horizontally from the radar when it impacts the earth.

The EKF and UKF are initialized with

$$\hat{\mathbf{x}}_{0/0} = [6500.4 \quad 349.14 \quad -1.8093 \quad -6.7967 \quad 0]^T$$
$$\mathbf{P}_{0/0} = diag[10^{-6} \quad 10^{-6} \quad 10^{-6} \quad 10^{-6} \quad 1].$$

Notice that the variance of x_5 is set to one so that it can accommodate the non-nominal simulation initial condition. The discrete filter process noise matrix is set to

$$\mathbf{Q} = diag[0 \quad 0 \quad 2.4065\times10^{-5} \quad 2.4065\times10^{-5} \quad 1\times10^{-6}].$$

The range and depression angle measurements are modeled (in both the simulation and filters) as

$$R(t) = \sqrt{(x_1(t)-x_r)^2 + (x_2(t)-y_r)^2} + w_1$$
$$\theta(t) = \tan^{-1}\left(\frac{x_2(t)-y_r}{x_1(t)-x_r}\right) + w_2$$

where w_1 and w_2 are zero-mean uncorrelated noise with $E[w_1^2] = 1\text{ m}^2$ and $E[w_2^2] = (0.130\text{ radians})^2$. The measurement noises are listed in J&U04a as having "variances of 1 m and 17 mrd, respectively." Chang et al. (1977) list the assumed angle measurement standard deviation as "0.17 mrad" (0.17×10^{-3} radian), which is a reasonable noise error for tracking radars. It was initially assumed that J&U04a used $E[w_2^2] = (17\text{ mrad})^2$, which is 10,000 times larger than the variance used by Chang et al. However, there is little difference in the state estimates of the EKF and UKF when using this noise variance, and the estimate errors of both filters are much smaller than listed in J&U04a or J&U04b. It was found that the $1-\sigma$ w_2 errors had to be increased to about 130 mrad (note that $0.13^2 = 0.017$) in order to approximately match results of J&U04b. At this noise level the effective measurement error is equivalent to a position error of 13 km at 100 km range. This gives the angle measurements little weight in the solution.

The Monte Carlo simulation included Gaussian random errors in state initial conditions, process noise, and measurement noise. The equations of motion were integrated using a fourth-order Runge-Kutta method with a step size of 0.050 s. The EKF partial derivatives were computed analytically for the measurements, and using numeric central differences for the state dynamics.

Figures 11.9–11.11 show the 100-sample RMS estimate errors and *a posteriori* $1-\sigma$ uncertainty (square roots of covariance diagonals) for the EKF position, velocity, and drag correction states. Notice that the actual position errors are much larger than the modeled $1-\sigma$ error during the first 100 s. The errors peak at about 50 s, which occurs at an altitude of 34 km. The velocity errors are also larger than modeled during the first 100 s, but are smaller than modeled in the second half. The drag correction errors after 40 s of simulation are consistently much larger than the computed $1-\sigma$. In fact, the filter is unable to remove bias errors in the drag correction.

Figures 11.12–11.14 show the comparable results for the UKF. Notice that the estimate errors are consistently smaller than those of the EKF, and actual errors better match (or are smaller) than the computed $1-\sigma$. Also note that the error in the estimated drag correction is approximately equal to the computed

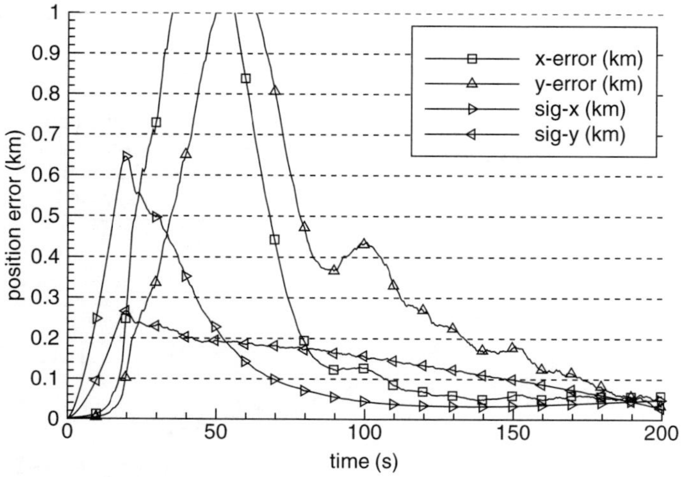

FIGURE 11.9: EKF position errors for ballistic reentry.

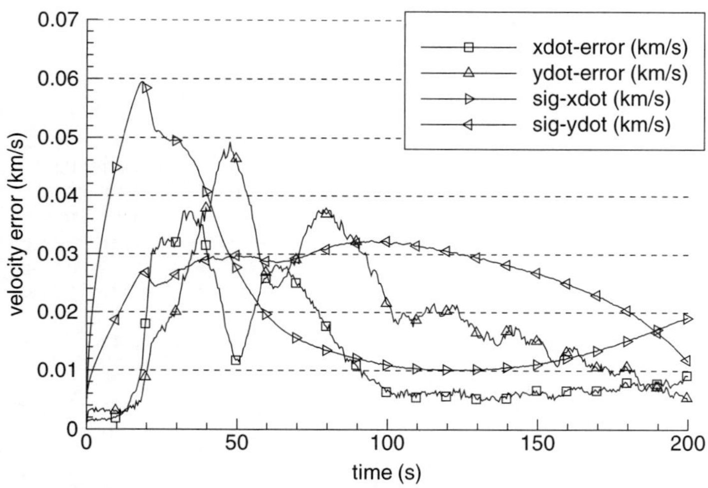

FIGURE 11.10: EKF velocity errors for ballistic reentry.

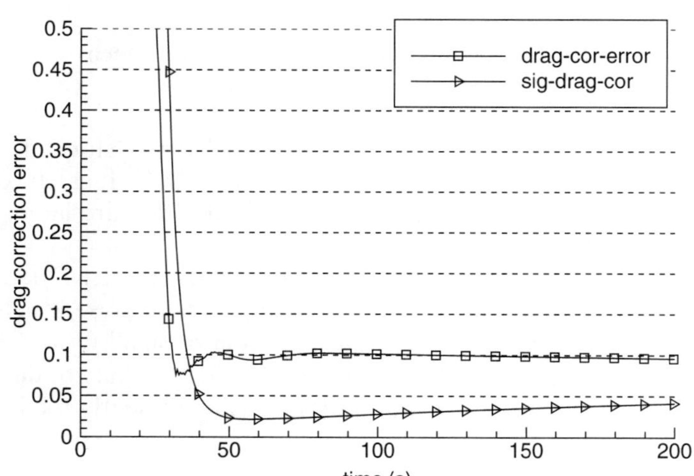

FIGURE 11.11: EKF drag correction errors for ballistic reentry.

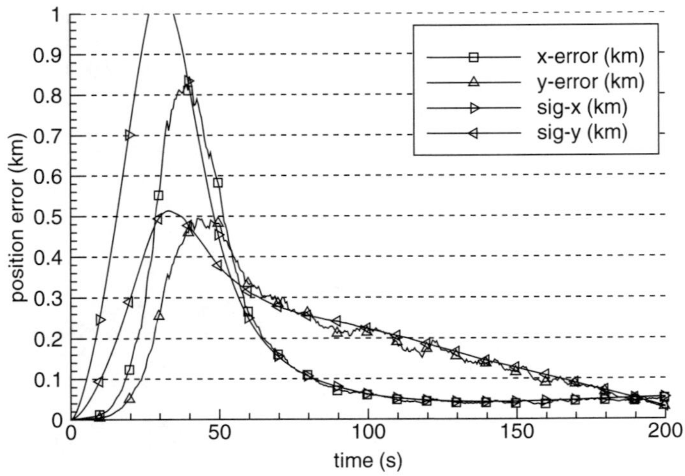

FIGURE 11.12: UKF position errors for ballistic reentry.

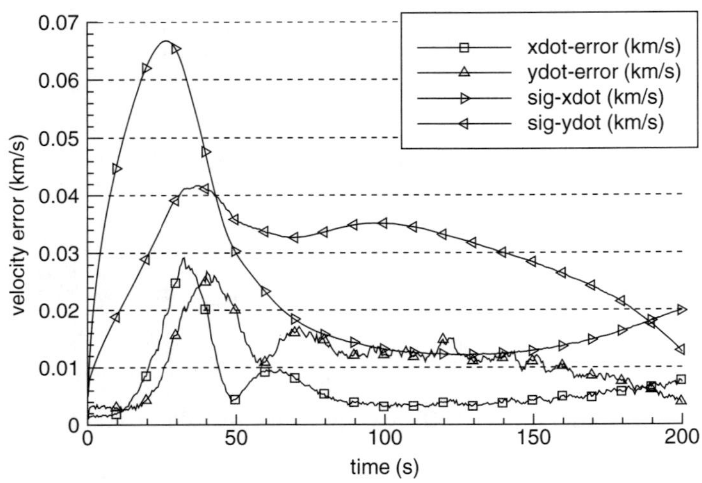

FIGURE 11.13: UKF velocity errors for ballistic reentry.

$1-\sigma$ uncertainty. The results are essentially unchanged (for this five-state problem) when using different values of α and β in the UKF. It can be concluded from these results that the linearization errors of the EKF are large, while the UKF accurately models the system.

There is, however, a relatively simple fix for the EKF. The EKF does not give sufficient weight to the range and angle measurements because it believes that the linearized dynamic model is accurate. Addition of small Q to the position states keeps the filter gain "open," thus allowing better tracking of the measurements. Estimate errors of the EKF drop dramatically to nearly the level of the UKF when using

$$Q_{11} = Q_{22} = 10^{-4}.$$

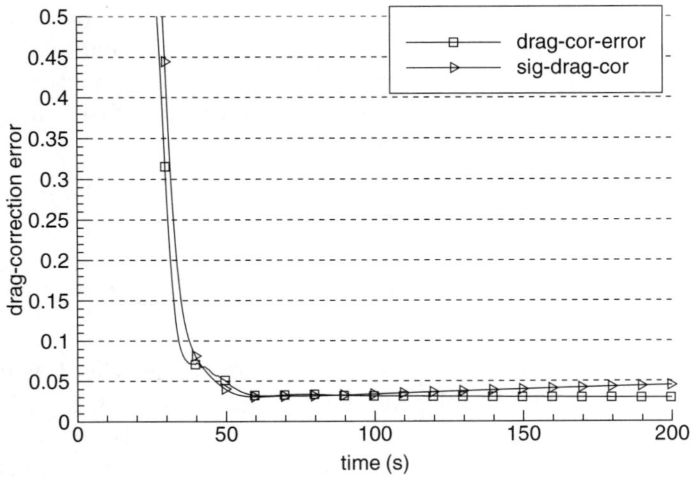

FIGURE 11.14: UKF drag correction errors for ballistic reentry.

To summarize, the UKF is much more accurate than the unmodified EKF for this somewhat unusual tracking problem (where angle measurement errors are more than 700 times larger than typical for tracking radars.) When angle errors are reduced by a factor of 7.6, differences between the EKF and UKF are small. EKF performance can be greatly improved by modeling small process noise on the position states.

11.8 PARTICLE FILTERS

While the UKF carefully selects state evaluation points to compute the sample mean and covariance, particle filters use a Monte Carlo approach. That is, a large number of points are randomly chosen with the goal of approximating the conditional probability density function. This requires much more computation than the UKF, but it has the potential to be more accurate in highly nonlinear problems. Unlike most other filter algorithms, there are many different methods for implementing particle filters. This section does not attempt to describe the algorithms in detail. Rather, the goal is to present an overview of the approach so that readers can decide whether they want to pursue the subject.

It is probably easiest to understand particle filters by examining steps of the filter algorithm. Before describing the operations, however, it is first necessary to explain "particles." The filter uses a large number of individual particles to allow approximate evaluation of the state conditional probability density. Each particle $j = 1, 2,\ldots, L$ at time t_i has two properties: a state vector $\mathbf{x}^j_{i/k}$ and an *importance weight* $w^j_{i/k}$. As before, the subscript i/k denotes that the variable is evaluated at time t_i conditioned on measurements up through time t_k. The weights $w^j_{i/k}$ can be interpreted as conditional probabilities for the associated particle, so the set $\{\mathbf{x}^j_{i/i}, w^j_{i/i} : j = 1,2,\ldots,L\}$ allows approximate computation of the conditional probability density function $p_{X|Y}(\mathbf{x}_i|\mathbf{y}_i, \mathbf{y}_{i-1},\ldots, \mathbf{y}_0)$. (To simplify notation, we allow $\mathbf{y}_0$ to represent prior information on $\mathbf{x}_0$.) The set of posterior particles should accurately characterize all significant moments of the probability density function, particularly when multiple maxima exist. However, at a minimum, the first and second moments of the particle set should be approximately correct:

$$E[\mathbf{x}_i \mid \mathbf{y}_i, \mathbf{y}_{i-1}, \ldots, \mathbf{y}_0] \cong \hat{\mathbf{x}}_{i/i} = \sum_{j=1}^{L} w_{i/i}^j \mathbf{x}_{i/i}^j \tag{11.8-1}$$

$$E[(\mathbf{x}_i - \hat{\mathbf{x}}_{i/i})(\mathbf{x}_i - \hat{\mathbf{x}}_{i/i})^T] \cong \mathbf{P}_{i/i} = \sum_{j=1}^{L} w_{i/i}^j (\mathbf{x}_{i/i}^j - \hat{\mathbf{x}}_{i/i})(\mathbf{x}_{i/i}^j - \hat{\mathbf{x}}_{i/i})^T. \tag{11.8-2}$$

These equations are not exact because the number of particles L is finite. Notice that there are $L(n + 1)$ variables in the set $\{\mathbf{x}_{i/i}^j, w_{i/i}^j : j = 1,2,\ldots,L\}$ and only $n(n + 3)/2$ constraints in the two above equations (where n is the dimension of the state). Thus $\mathbf{x}_{i/k}^j$ and $w_{i/k}^j$ are not uniquely defined by these equations when $L(n + 1) > n(n + 3)/2$, which is the usual case. In addition to these equations, there are also requirements that $0 \le w_{i/k}^j \le 1$ (since it is treated as a probability) and

$$\sum_{j=1}^{L} w_{i/k}^j = 1. \tag{11.8-3}$$

One approach for defining the particles sets $w_{i/k}^j = 1/L$ for $j = 1, 2,\ldots, L$, and uses a random number generator with the desired mean and probability density to set $\mathbf{x}_{i/k}^j$. The result will satisfy equations (11.8-1) and (11.8-2), and this is the method used to initialize particles in the filter. Alternately one could make the weights proportional to the desired probability density function, and then set the points $\mathbf{x}_{i/i}^j$ so that the full set of $\{\mathbf{x}_{i/i}^j, w_{i/i}^j\}$ approximates the conditional probability density. This demonstrates that there are multiple combinations of weights and particle states that can characterize a given probability density function.

Measurements are used in particle filters to modify the weights $w_{i/k}^j$, so measurement processing causes the weights to be unequal, even if initially equal. After processing several measurements, some of the particle weights will be very small while weights for other particles will be large. Particles with small weights become irrelevant to the solution, and resolution suffers because equation (11.8-1) is determined by only a few particles. This problem is corrected by *resampling*, which deletes low-weighted particles and creates additional new particles from the dominant particles. That is, the $\mathbf{x}_{i/i}^j$ values of the dominant particles are duplicated and weights are reduced; for example, $w_{i/k}^j = 1/L$. Other weights can also be used.

With this introduction, we now explain particle filter processing. As with Kalman filters, particle filters have time and measurement updates, but they also have other steps that manipulate particles. The steps are as follows:

1. *Initialization*: Using the initial state estimate $\hat{\mathbf{x}}_{0/0}$ and covariance $\mathbf{P}_{0/0}$, randomly generate L "particles" having the desired distribution (usually Gaussian), for example, $N(\hat{\mathbf{x}}_{0/0}, \mathbf{P}_{0/0})$. The number of particles is generally much larger than the state size, for example, $L = 5^n$. Selection of L is a tradeoff between estimation accuracy and computational burden. For each particle state $\{\mathbf{x}_{0/0}^j : j = 1,2,\ldots,L\}$, assign weight $w_{0/0}^j = 1/L$.

2. *Time update*: For a measurement at time t_i, integrate the nonlinear dynamic equations for each particle to obtain

$$\boxed{\mathbf{x}^j_{i/i-1} = \mathbf{f}(\mathbf{x}^j_{i-1/i-1}, \mathbf{q}^j_{i,i-1})} \tag{11.8-4}$$

where each process noise vector $\mathbf{q}^j_{i,i-1}$ is randomly generated according to the known probability density function; for example, $N(\mathbf{0},\mathbf{Q}_{i,i-1})$ if the assumed distribution is Gaussian. Integration to the new time does not change weights (i.e., $w^j_{i/i-1} = w^j_{i-1/i-1}$) since they are conditioned on past measurements.

3. *Measurement update of probability*: Given measurement m-vector $\mathbf{y}_i$, compute the conditional probability density $p_{Y|X}(\mathbf{y}_i \mid \mathbf{x}^j_{i/i-1})$ for the assumed measurement model $\mathbf{y}_i = \mathbf{h}_i(\mathbf{x}_i) + \mathbf{r}_i$. For example, if the measurement errors $\mathbf{r}_i$ are $N(\mathbf{0},\mathbf{R}_i)$,

$$\boxed{p_{Y|X}(\mathbf{y}_i \mid \mathbf{x}^j_{i/i-1}) = \frac{1}{(2\pi)^{m/2}\,|\mathbf{R}_i|^{1/2}} \exp\left(-0.5(\mathbf{y}_i - \mathbf{h}_i(\mathbf{x}^j_{i/i-1}))^T \mathbf{R}_i^{-1} (\mathbf{y}_i - \mathbf{h}_i(\mathbf{x}^j_{i/i-1}))\right).} \tag{11.8-5}$$

Because $p_{Y|X}(\mathbf{y}_i|\mathbf{x}_i)$ is not a function of $\mathbf{y}_{i-1}, \mathbf{y}_{i-2},\ldots$, it can be shown using Bayes' rule that

$$p_{X|Y}(\mathbf{x}_i \mid \mathbf{y}_i, \mathbf{y}_{i-1},\ldots) = \frac{p_{Y|X}(\mathbf{y}_i \mid \mathbf{x}_i)\, p_{X|Y}(\mathbf{x}_i \mid \mathbf{y}_{i-1},\ldots)}{p_Y(\mathbf{y}_i \mid \mathbf{y}_{i-1},\ldots)}. \tag{11.8-6}$$

The weights $w^j_{i/i}$ are treated as conditional probabilities for each particle; that is, $w^j_{i/i} \propto \Pr\left\{\mathbf{x}^j_{i/i} = \mathbf{x}_i \mid \mathbf{y}_i, \mathbf{y}_{i-1},\ldots\right\}$ taking into account the spacing between points $\mathbf{x}^j_{i/i}$ and using actual values of $\mathbf{y}_i, \mathbf{y}_{i-1},\ldots$. This suggests that the posterior weights could be computed as

$$w^j_{i/i} = \frac{p_{Y|X}(\mathbf{y}_i \mid \mathbf{x}^j_{i/i-1})\, w^j_{i/i-1}}{p_Y(\mathbf{y}_i \mid \mathbf{y}_{i-1},\ldots)}.$$

The probability density function $p_Y(\mathbf{y}_i|\mathbf{y}_{i-1},\ldots)$ is not known, but it is the same for all particles. Thus the weights are first computed as

$$\boxed{w^j_{i/i} = p_{Y|X}(\mathbf{y}_i \mid \mathbf{x}^j_{i/i-1})\, w^j_{i/i-1}} \tag{11.8-7}$$

and then normalized:

$$\boxed{w^j_{i/i} \Leftarrow \frac{w^j_{i/i}}{\sum_{j=1}^{L} w^j_{i/i}}.} \tag{11.8-8}$$

This is a somewhat oversimplified description of weight updating. Practical algorithms also include an *importance density* in the calculations (see Ristic et al. 2004).

4. *Resample particles:* After processing a measurement, the weights for some particles will be small and others weights will be large. If no adjustments are made, weights for many particles will become negligible after processing

several measurements. To avoid this problem, the particles are resampled when

$$\eta = \frac{1}{L\sum_{j=1}^{L}(w_{i/i}^{j})^{2}} < W_{thresh} \tag{11.8-9}$$

Notice that if all $w_{i/i}^{j} = 1/L$, then $\eta = 1$. If all weights but one are equal to zero, then $\eta = 1/L$. Since resampling reduces accuracy of the particle distribution, W_{thresh} is set so that particles are only resampled when many weights are small. Resampling creates multiple new particles from the particles with large weights, and eliminates particles with small weights. It is graphically demonstrated in a later example. Several methods may be used, but *sequential resampling* is efficient and easily implemented as follows:

a. Form the cumulative sum of weights (CSW): $c_j = \sum_{k=1}^{j} w_{i/i}^{k}, \quad j = 1,\ldots,L.$
b. Select starting point u_1 from a uniform random distribution: $0 \le u_1 \le 1/L$.
c. For $j = 1{:}L$
    i. $u_j = u_1 + (j-1)/L$
    ii. Find index k such that $c_{k-1} < u_j \le c_k$
    iii. Set a new particle as $\mathbf{z}_{i/i}^{j} = \mathbf{x}_{i/i}^{k}$ and $w_{i/i}^{j} = 1/L$
d. End loop
e. Copy $\mathbf{x}_{i/i}^{j} = \mathbf{z}_{i/i}^{j}$ for $j = 1{:}L$

5. *Compute mean and covariance:* The sample mean is computed from the L particles as

$$\hat{\mathbf{x}}_{i/i} = \sum_{j=1}^{L} \mathbf{x}_{i/i}^{j} w_{i/i}^{j}. \tag{11.8-10}$$

If needed, the sample covariance is

$$\mathbf{P}_{i/i} = \sum_{j=1}^{L} w_{i/i}^{j}(\mathbf{x}_{i/i}^{j} - \hat{\mathbf{x}}_{i/i})(\mathbf{x}_{i/i}^{j} - \hat{\mathbf{x}}_{i/i})^{T}. \tag{11.8-11}$$

Steps 2–5 are repeated for each new measurement.

There are a number of practical problems in implementing particle filters. For example, when process noise is small, the same particles are repeatedly selected in resampling, leading to *sample impoverishment*. Methods used to overcome various problems have led to a number of different particle filter implementations. These options are discussed in Ristic et al. (2004) and Simon (2006, chapter 15). Particle filters are an active research area, and have been used in many different application areas. As with unscented filters, the benefits are most likely to be noticed when nonlinearities are large and state observability from available measurements is poor.

Example 11.7: Particle Filter for Scalar Nonlinear Model

The time update for scalar x_i is

$$x_i = x_{i-1}^{1.2} + q_{i,i-1}$$

where $q_{i,i-1}$ is uncorrelated $N(0,0.2^2)$ noise, and x_0 is $N(2,1)$. Measurement $y_1 = 1.8$ is modeled in the filter as $y_1 = x_1 + r_1$, where r_1 is $N(0,0.5^2)$. A 10-particle filter is initialized as shown in the top line of Figure 11.15. Subsequent lines show the steps of the processing as follows:

1. Initial $x_{0/0}^j$ values for 10-particles ($w_{0/0}^j = 1/10$).
2. Propagated particles $x_1^j = (x_0^j)^{1.2} + q_{1,0}^j$ at time t_1.
3. Measurement y_1.
4. Particle weights (vertical line length) after processing the measurement.
5. Resampled particles (weights $w_{1/0}^j = 1/10$).
6. Particles propagated to time t_2.

Figure 11.16 shows the cumulative sum of weights (CSW) after measurement processing. Also shown is the CSW before and after resampling. Notice that the weights for the last five particles on the right are very small after measurement processing, and thus these particles are replaced in the resampling. The first two particles on the left become six particles after resampling.

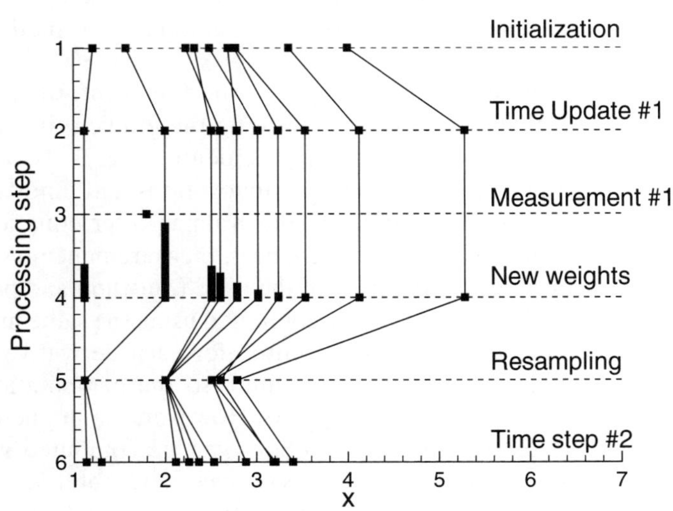

FIGURE 11.15: Particle values during processing.

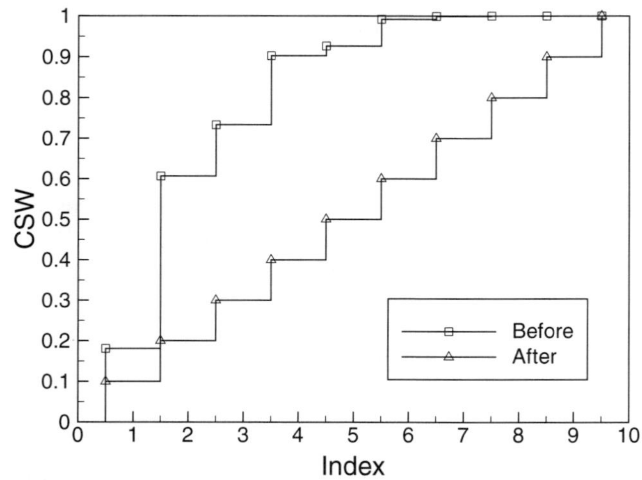

FIGURE 11.16: Cumulative sum-of-weights (CSW) before/after resampling.

11.9 SUMMARY

Model parameters such as state initial conditions, process noise variances, measurement noise variances, and dynamic model constants are often poorly known, and errors in these parameters can cause erroneous filter estimates. Because the likelihood function can be easily computed from Kalman filter innovations and innovation variances, it is possible to minimize model errors by determining the parameters values that maximize the likelihood. An iterative method called scoring—a generalization of the Gauss-Newton method—is used to determine the maximum likelihood estimate. Since the models are often nonlinear, it is sometimes necessary to constrain the scoring steps using a nonlinear optimization method such as Levenburg-Marquardt.

As with Gauss-Newton iterations, scoring uses the Fisher information matrix—the expected Hessian—that can be computed using the gradient of the log likelihood with respect to the unknown parameters, denoted as $\boldsymbol{\theta}$. That gradient is computed from partial derivatives of the filter innovations and innovation covariances with respect to $\boldsymbol{\theta}$. Thus derivatives of the Kalman filter equations must be computed with respect to process noise variances, measurement noise variances, and dynamic model constants. Derivatives of the state transition and process noise matrices are also required, and these can be computed using the same methods used to compute $\boldsymbol{\Phi}$ and $\mathbf{Q}$. The computations required for each derivative are slightly greater than those required for the Kalman filter, so the computational burden can be great when solving for many parameters. However, the gradient of the log likelihood with respect to state initial conditions can be computed with far less computation by using the sensitivity equations derived in Chapter 9.

Two examples demonstrate ML parameter identification. The first is a four-state model that includes two first-order Markov process models, and the estimated parameters include the feedback constants, process noise variances, and measurement noise variances. Using simulated measurements with randomly generated process and measurement noise, the one-sample Hessian, Fisher information matrix, and averaged 1000-sample Hessian were compared and shown to be approximately equal. When initialized at the true $\boldsymbol{\theta}$, scoring steps converged in three iterations regardless of the type of parameterization used in $\boldsymbol{\theta}$. The second example estimated 17 clock process noise PSD values for a 29-GPS satellite constellation using PR data. This demonstrated the practicality of implementing MLE for large-scale problems: the filter included 371 states, a variable number of pass-disposable station-satellite biases, and 103,505 scalar measurements.

When model parameters change with time, it is desirable that filters adapt to those changes. In many cases the level of disturbing "forces" are variable, and sample variances of filter innovations can be used to adjust the process noise covariance. A variety of simple methods have been developed for this purpose. The sample variances are either computed over a sliding window, or exponential weighting is used so that the response to measurement outliers is reduced.

This approach can work well when changes in system characteristics are slow, but it has a significant lag when system inputs change abruptly. A jump detection/estimation method based on hypothesis testing is usually a better approach in these cases, and a method based on stepwise regression is shown to work well either for off-line post-processing or for online real-time operation. It is designed to handle multiple jumps and has been used operationally to process IMU data from missile tests in which 40 or 50 jumps occur in a single flight. Accuracy of the post-processing approach was demonstrated using simulated data. The ability to detect multiple jumps allows a real-time, sliding window implementation to use a longer data window than other methods. An example based on simulated aircraft maneuvering demonstrated the real-time implementation.

When systems smoothly or abruptly transition between a finite number of models, it often better to operate multiple filters in parallel where each filter uses a different model. When all filter state vectors have the same size and definition, the composite state at any given time may be determined by weighted averaging of the filter states. The weighting should be based on the probability that each filter model is correct, so the likelihood function computed from the filter innovations is used. When tracking data spans are short or filter state models are different, selection of the state from the filter with the greatest likelihood can work well. A fire-control example demonstrated improved prediction accuracy when tracking a maneuvering tank. Another technique called interactive multiple model (IMM) is designed to track targets when the system may repeatedly transition between multiple models, and prior information on the mode transition probabilities is available.

In some cases it is necessary to enforce equality or inequality constraints on filter estimates. The methods employed are similar to those used for least-squares estimation as discussed in Chapter 7. The approaches for equality constraints include redefinition of variables to include the constraint, heavy weighting of the constraint as a pseudo-measurement, and Lagrange multipliers. Inequality constraints can also be enforced using the same methods used for least squares, but the filter response is randomly nonlinear.

It is sometimes important to limit the maximum errors of a linear combination of filter states for worst case assumptions about the process noise, measurement noise, and initial condition errors. The H_∞ or minimax filter is designed to minimize a weighted norm of state errors for worst case conditions. The cost function to be minimized is normalized by weighted quadratic sums involving the measurement noise, process noise, and initial condition errors. This game theoretic problem is solved by first finding the process noise and initial conditions that maximize the cost function, and then finding the state estimate that minimizes the cost function for the worst case conditions. The resulting filter equations are similar to the Kalman filter equations, but include an extra term that is multiplied by a cost function constraint parameter. This parameter and two weighting arrays are tuning variables for the H_∞ filter. Referenced examples demonstrate that the H_∞ filter can greatly limit worst case estimate errors compared with the Kalman filter, but it can also have larger errors in nominal conditions.

Two newer methods for handling model nonlinearities include the UKF and particle filter. The UKF does not linearize the model about the current estimate, as done in the EKF. Rather, carefully selected state perturbations are used in the nonlinear dynamic and/or measurement functions to allow computation of the conditional mean and covariance required by the filter. The effects of the nonlinearities are correctly modeled up through the third order of the Taylor series. Computations required by the UKF can be comparable to those of an EKF implemented using central difference numeric partial derivatives. An example of ballistic reentry demonstrates the improved accuracy of the UKF compared with the EKF.

Particle filters are a Monte Carlo approach to filtering. They use a cloud of randomly selected evaluation points and associated weights (the particles) to approximate the probability density function. Particle filters are potentially more accurate than the UKF, but require far more computation.

CHAPTER 12

EMPIRICAL MODELING

Previous chapters have emphasized use of models developed from first principles, or models motivated by physical considerations. Unfortunately physical information is unavailable or only partially available for some systems. Examples include systems that have no physical basis (e.g., economic), very complex systems that are difficult to model (e.g., chemical production processes, power plans), complex systems affected by many poorly understood inputs (e.g., weather, biological), and systems that are partially well modeled but the driving inputs are stochastic with unknown correlation characteristics (e.g., the maneuvering tank problem of Section 3.2.1). In these cases it is often necessary to develop estimation models using empirical methods.

There are two primary methods used for empirical model development. The first approach computes the power spectrum (*power spectral density* or *PSD*) of a measured output signal and then attempts to infer the differential or difference equations that will produce that spectrum when driven by white noise inputs. Since white noise has a uniform spectrum, the output spectrum is the magnitude squared of the system transfer function (Fourier transform of the system impulse response). The power spectrum defines the magnitude but not the phase response, so the computed impulse response function is non-unique, That is not a problem for real-valued signals because system poles and zeroes must occur in complex conjugate pairs. For a stable system poles must lie in the left-half complex plane, and for a causal system zeroes must also lie in the left-half plane. Hence the power spectrum and knowledge of system input signals (often assumed to be white noise) are sufficient to uniquely define a system transfer function.

The second approach assumes a discrete autoregressive (AR), autoregressive moving average (ARMA), or autoregressive moving average with exogenous inputs (ARMAX) model and uses discretely sampled output (and possibly input) measurements to compute parameters of the model. Then either the discrete model is used directly in the Kalman filter, or the power spectrum computed from the discrete model is used to infer differential equations that can be integrated in the Kalman filter. Both the power spectral and "ARMA" modeling approaches normally require

Advanced Kalman Filtering, Least-Squares and Modeling: A Practical Handbook, by Bruce P. Gibbs

discrete measurements at a fixed sampling interval, where the system is statistically stationary for the period of the measurements.

Empirical modeling directly or indirectly involves computation of power spectra. If you were disappointed to discover that estimation is somewhat more of an art than expected, you will be even more disappointed to learn that the computed power spectra depend on assumptions used to compute the spectra. There is no one perfect method. Hence the following sections describe the theory and assumptions used for the various methods, and explain caveats associated with the approaches. Since the focus of the book is modeling for least-squares estimation and Kalman filtering, we emphasize methods that are directed toward that goal rather than a general discussion of spectral analysis.

It may seem odd that this chapter is located at the end of the book rather than closer to Chapter 2. It will be seen that some methods used for discrete modeling are based on least-squares estimation, so it would have been premature to present this material before chapters on least squares.

Finally we note that empirical model development is an off-line process that generally does not require production software. We have included code for many of the methods in the associated software, but we generally recommend that you use standard packages such as the MATLAB System Identification Toolkit™ (trademark of The Mathworks, Inc.) or ADAPTx™ (trademark of Adaptics, Inc.) for empirical model building. Integrated packages allow interactive testing using well-validated techniques and direct display of plotted results. This is a great benefit when conducting the analysis. The supplied software is provided for readers desiring to experiment with the algorithms or to better understand them.

12.1 EXPLORATORY TIME SERIES ANALYSIS AND SYSTEM IDENTIFICATION

Analysis of time series data for purposes of model development is sometimes called *exploratory data analysis*. The term originated with John Tukey (1977) who wrote a book of that name. Tukey's approach was directed to understanding relationships between different measured quantities for the purposes of developing good regression models. Hence the approach was not directly intended for time series characterization, but the methods do apply. He advocated plotting the data in various different ways, transforming variables, removing trends, generating histograms, and using multiple variable regression analysis. He placed particular emphasis on data plotting as the eye can detect patterns that are sometimes difficult to find without prior knowledge. Plotting also allows easy detection of data outliers.

A later source for information on time series analysis methods and ARMA modeling is the book by Box et al. (2008). This is the fourth edition of a classic book and has been updated to include many new topics. The books by Brockwell and Davis (2006) and Shumway and Stoffer (2006) are also relevant.

Model development for dynamic systems is often called *system identification*. In practice this is usually done using discrete ARMAX models. Ljung provides a good summary of system identification methods and practical approaches in *The Control Handbook* (Levine 1996, section XII). Other more thorough treatments include Ljung and Glad (1994), Söderström and Stoica (1989), and Ljung (1999). The MATLAB implementation is described in Ljung (1994).

The methods used to develop colored noise models for the alongtrack and crosstrack tank accelerations of Section 3.2.1 demonstrate the general approach for determining characteristics of colored noise input to a physical model. The initial step involved plotting of the tank position data versus time, and x versus y. This allowed detection of data gaps and jumps due to changes in sensor configurations. Then the positions, sampled at 0.1 s intervals in a Cartesian frame, were numerically differenced to compute Cartesian velocity. Since numeric differencing exaggerates the effects of position measurement noise, the velocity values were smoothed by averaging a few points before and after each output value. Then the velocity values were again differenced to compute acceleration in the alongtrack and crosstrack directions. No additional averaging was applied at this step. The computed acceleration data were first plotted to determine general behavior, to allow editing of outliers, and to select "stationary" sections of the track for analysis. Next the mean alongtrack and crosstrack acceleration for each section were computed and subtracted from the data. No significant trends were noticed. (The importance of mean and trend removal will be explained later.) The power spectrum of the data for each section of interest was computed using the discrete Fourier transform. As expected most of the power was concentrated at the low frequencies, but significant peaks were observed and different tanks had different characteristics. The spectra gave a general indication of the appropriate Markov process model orders (one to three). A similar spectral analysis was also conducted using AR modeling, and the optimal AR model order was found to be much greater than three. This was partly due to the presence of measurement noise, but it also indicated that numerator terms of an ARMA model were probably important in accurately characterizing the acceleration. Finally, first-, second- and third-order Markov process acceleration models were implemented in a Kalman filter and the maximum likelihood estimation (MLE) approach of Section 11.1 was used to determine parameters of the perturbing acceleration models. The analysis used many of the methods described in this chapter.

Other examples that use methods of this chapter include modeling of tank gun tube flexure (Section 12.5), modeling of process disturbances in a power plant (Section 3.5), and temporal modeling of misalignments in an optical imaging instrument (Example 6.6).

12.2 SPECTRAL ANALYSIS BASED ON THE FOURIER TRANSFORM

This section discusses Fourier transforms and power spectra, and their use for inferring models. Most of the material can be found in standard texts such as Davenport and Root (1958), Cooper and McGillem (1967), Priestley (1981), Rabiner and Gold (1975), Oppenheim and Schafer (1975), Marple (1987), Kay (1988), Brockwell and Davis (2006), Shumway and Stoffer (2006), Box et al. (2008), and Papoulis and Pillai (2002). Collections of important papers on spectral analysis may be found in Rabiner and Rader (1972), Childers (1978), and Kesler (1986). It is not possible to thoroughly cover the extensive volume of material in a short section. Also the material will be well known to many readers. For those reasons we summarize the most relevant concepts and refer you to other references for more information. The subject is included here because assumptions of the theory have important implications when used to compute power spectra. To understand the limitations it is necessary to understand the assumptions. The focus is on methods used for model development, not a full discussion of spectral analysis.

It is instructive to explain the operation of early electronic spectrum analyzers to understand PSD. These analog devices passed an input signal through a band-pass filter having a narrow pass band and relatively sharp cutoff outside the pass band. The voltage output of the pass band filter was measured and the displayed PSD was computed as the voltage squared divided by the equivalent "rectangular" bandwidth of the band-pass filter. Hence it had units of voltage-squared per Hz or V^2s. The power spectrum over the desired frequency range was displayed on a cathode ray tube by sweeping the center frequency of the band-pass filter. The center frequency of the filter was not actually varied—the input signal was mixed (multiplied) with a local oscillator of variable frequency and the mixed signal was fed to a fixed-parameter band-pass filter. In operation the system behaved as if the center frequency of the filter was swept.

Notice that the displayed PSD is not a perfect measure of the true PSD for several reasons. First, the calculated PSD is a measure of the power passed through a band-pass filter having a non-ideal (not rectangular) transfer function. When a pure sinusoid is fed to the spectrum analyzer the display shows the transfer function of the band-pass filter rather than an ideal impulse at the center frequency of the sinusoid. Usually that pass band is small, so the display shows a fairly sharp spike. Second, the displayed PSD at a given frequency is an average of the power received during the effective integration time of the power meter. When measuring stochastic signals the displayed output at each frequency will vary from one display sweep to the next. Use of a long integration time or averaging of multiple sweeps is necessary to obtain a meaningful measurement of average power.

Modern digital spectrum analyzers use the *fast Fourier transform* (FFT) to accomplish the same function as analog devices, but the same problems (and others) are present when digitally computing power spectra of discretely sampled signals. To understand why, we start with the definition of the Fourier series and Fourier transform, and explain how they are used to compute power spectra. Although computation of power spectra using the Fourier transform is usually the first step in exploratory model development, it has a number of problems that are avoided or minimized when using alternate techniques. The FFT approach has the advantage that it does not rely on model structure assumptions and thus is less affected by structural prejudices of the analyst. However, other implementation issues associated with use of the FFT still require analyst decisions.

Computation of power spectra using the Fourier transform is called classical spectral analysis. Later approaches based on ARMA models are sometimes called "modern" spectral analysis, although most were first developed more than three decades ago. There are two general approaches for classical spectral analysis: the periodogram and the correlogram. The periodogram computes the magnitude-squared Fourier series coefficients while the correlogram computes the Fourier transform of a weighted sample autocorrelation function. Both methods suffer from a variety of problems generally referred to as "bias" and "variance." The finite time span and discrete sampling of the data cause systematic errors—or bias—in the computed spectra. The fact that the power spectrum of a stochastic signal is also a random variable requires that the periodogram be computed multiple times using independent time samples, and then the periodograms are averaged. While the mean periodogram for independent infinite-duration samples of a stationary process converges to the true power spectrum as the number of samples increases, the variance may not approach zero (see Davenport and Root 1958, p. 108 or Marple 1987, appendix 4.A). This is a fundamental limitation of the method. For a fixed-length data sample, the trade-off between bias and variance is controlled by the number of non-overlapping data segments used to compute separate periodograms before averaging. Non-equal weighting (windowing) of the data can also be used to minimize spectral leakage from narrowband signals to adjacent frequencies.

The correlogram approach does not require averaging of multiple spectra because the averaging is performed when computing the autocorrelation function. This approach is also subject to bias and variance problems. In this method the trade-off between bias and variance is controlled by the number of lags used to compute the autocorrelation function, and the weighting applied as a function of the autocorrelation lags.

Parameteric AR, moving average (MA), and ARMA modeling approaches tend to have smaller bias and variance, but uncertainty as to the true model order and structure causes other problems. For these reasons, both classical and ARMA spectral analysis techniques are useful when developing empirical models. The ARMA approaches are potentially more accurate and also have the benefit of directly producing a model.

The following sections discuss Fourier series and transforms and their use for power spectral analysis. Some derivations follow those of Cooper and McGillem (1967), or Davenport and Root (1958) because their developments are particularly clear.

12.2.1 Fourier Series for Periodic Functions

A continuous periodic function $x(t)$ that has within any one period $-T/2 \le t \le T/2$ a finite number of minima, maxima, and discontinuities can be expanded in an infinite series of trigonometric functions as

$$\boxed{x(t) = a_0 + \sum_{k=1}^{\infty}(a_k \cos k\omega_0 t + b_k \sin k\omega_0 t)} \tag{12.2-1}$$

where $\omega_0 = 2\pi/T$. The coefficients are determined as

$$\left.\begin{aligned} a_0 &= \frac{1}{T}\int_{-T/2}^{T/2} x(t)\,dt \\ a_k &= \frac{1}{T}\int_{-T/2}^{T/2} x(t)\cos k\omega_0 t\,dt \\ b_k &= \frac{1}{T}\int_{-T/2}^{T/2} x(t)\sin k\omega_0 t\,dt \end{aligned}\right\} k \neq 0 \quad (12.2\text{-}2)$$

The series converges to $x(t)$ for values of t at which $x(t)$ is continuous. Notice that the set of trigonometric functions are orthogonal over the period T:

$$\left.\begin{aligned} &\int_{-T/2}^{T/2} \cos i\omega_0 t \cos k\omega_0 t\,dt = 0 \\ &\int_{-T/2}^{T/2} \sin i\omega_0 t \sin k\omega_0 t\,dt = 0 \\ &\int_{-T/2}^{T/2} \sin i\omega_0 t \cos k\omega_0 t\,dt = 0 \end{aligned}\right\} \text{ integer } i^2 \neq k^2.$$

If the series is truncated after a finite number of terms, the truncated Fourier series is the least-squares approximation to $x(t)$.

Since $e^{jk\omega_0 t} = \cos k\omega_0 t + j\sin k\omega_0 t$ (where $j = \sqrt{-1}$), the series can also be written in complex form as

$$x(t) = \sum_{k=-\infty}^{\infty} c_k e^{jk\omega_0 t} \quad (12.2\text{-}3)$$

where $c_0 = a_0$, $c_k = \left(\sqrt{a_k^2 + b_k^2}\,/2\right)(\cos\theta_k + j\sin\theta_k)$ and $\theta_k = -\tan^{-1}(b_k/a_k)$ for $k \neq 0$. Notice that the harmonic terms are located at both positive and negative frequency with one-half the magnitude in each.

12.2.2 Fourier Transform of Continuous Energy Signals

While the Fourier series is used to model periodic functions, the Fourier integral transform may be used for nonperiodic functions. The conditions under which the Fourier integrals exist—known as the *Dirichlet conditions*—are the same as for the Fourier series: the function must be single-valued with a finite number of minima, maxima, and discontinuities, with the added requirement that $\int_{-\infty}^{\infty} |x(t)|\,dt < \infty$. The Fourier transform of a continuous time signal $x(t)$ for $-\infty < t < \infty$ is defined as

$$\mathscr{F}[x(t)] = X(j\omega) \triangleq \int_{-\infty}^{\infty} x(t)e^{-j\omega t}\,dt \quad (12.2\text{-}4)$$

where $\omega = 2\pi f$ is frequency in radians per second and f is frequency in cycles per second (Hz). The output $X(j\omega)$ has units of $V\cdot s$ if x is measured in volts and t is in seconds. Notice that the Fourier transform can be derived as the limiting case of the Fourier series as the period $T \to \infty$ and $\omega_0 \to d\omega$.

The inverse Fourier transform, defined as

$$\mathscr{F}^{-1}[X(j\omega)] \triangleq \frac{1}{2\pi}\int_{-\infty}^{\infty} X(j\omega)e^{j\omega t}\,d\omega = x(t), \quad (12.2\text{-}5)$$

can be used to recover the original signal. The functions $x(t)$ and $X(j\omega)$ are called Fourier transform pairs. Often a lower case letter is used for the time signal and an upper case letter is used to denote the forward Fourier transform. That convention is used in this chapter.

We first discuss the transform of finite *energy signals* for which $\int_{-\infty}^{\infty} x^2(t)\,dt < \infty$. For example, the rectangular "boxcar" function, $x(t) = 1$ for $|t| \le T$ and $x(t) = 0$ for $|t| > T$, has energy equal to $2T$ and Fourier transform

$$\begin{aligned} X(j\omega) &= \int_{-T}^{T} e^{-j\omega t}\,dt = -\frac{e^{-j\omega t}}{j\omega}\Bigg|_{-T}^{T} = 2\frac{\sin(\omega t)}{\omega}\Bigg|_{0}^{T} \\ &= 2T\frac{\sin(\omega T)}{\omega T} \end{aligned} . \qquad (12.2\text{-}6)$$

The function $\mathrm{sinc}(y) \triangleq \sin(y)/y$ occurs often in signal processing and is plotted in Figure 12.1. Notice that it oscillates about zero, has zero crossings at multiples of $y = \pi$ (or $\omega T = \pi$), and never decays to exactly zero (except at $|y| = \infty$). Using the sinc function it can be shown that the transform of a time-limited signal extends to infinite frequency and the inverse transform of a band-limited spectrum is infinite in time. In other words, it is impossible to have a signal that is both time and frequency limited.

Another signal of interest for spectral estimation is the time-limited ramp $x(t) = t$ for $|t| \le T$ and $x(t) = 0$ otherwise. This is useful when computing the effects of trends in the data. The Fourier transform is

$$\begin{aligned} X(j\omega) &= \int_{-T}^{T} t e^{-j\omega t}\,dt = -j\int_{-T}^{T} t\sin(\omega t)\,dt \\ &= -j\left(\frac{1}{\omega^2}\sin(\omega t) - \frac{t}{\omega}\cos(\omega t)\right)\Bigg|_{-T}^{T} \\ &= -2j\left(\frac{\sin(\omega T)}{\omega^2} - \frac{T\cos(\omega T)}{\omega}\right) \end{aligned} . \qquad (12.2\text{-}7)$$

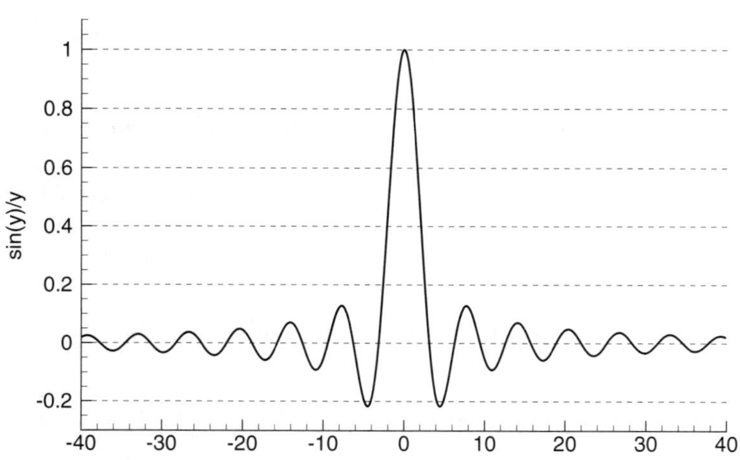

FIGURE 12.1: Sinc function (Fourier transform of rectangular pulse).

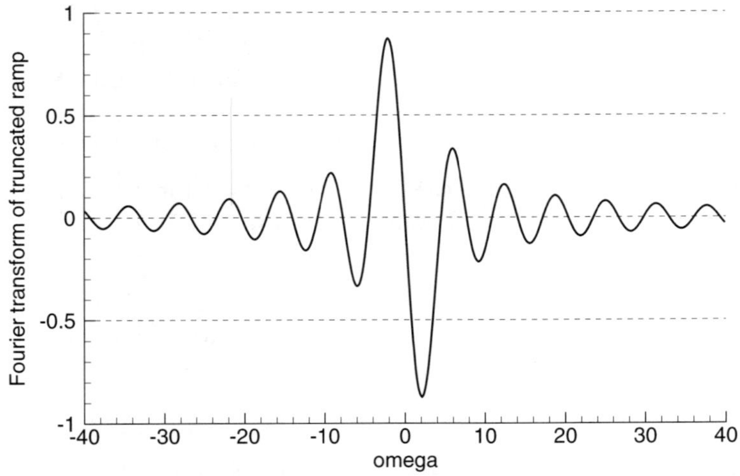

FIGURE 12.2: Fourier transform of time-limited ramp (imaginary component).

Figure 12.2 is a plot of the transform versus frequency for $T = 1$ Notice that transform is imaginary (no real component), it is an oscillatory odd function of ω with zero value at $\omega = 0$, and zeroes spaced at approximately $\omega = \pi$, except for the first zero crossing.

Several properties of the Fourier transform are important when computing and analyzing power spectra. First, the Fourier transform of the product of two energy signals, $f(t) = x(t)y(t)$, is equal to $1/2\pi$ times the convolution (denoted by *) of the Fourier transforms, that is

$$\boxed{\begin{aligned} F(j\omega) &= \frac{1}{2\pi} X(j\omega) * Y(j\omega) \\ &= \frac{1}{2\pi}\int_{-\infty}^{\infty} X(j\lambda)Y(j\omega - j\lambda)\,d\lambda \end{aligned}} \qquad (12.2\text{-}8)$$

where $X(j\omega)$, $Y(j\omega)$ and $F(j\omega)$ are, respectively, the Fourier transforms of $x(t)$, $y(t)$, and $f(t)$. This can be verified by taking the inverse transform

$$\begin{aligned} f(t) &= \frac{1}{2\pi}\int_{-\infty}^{\infty} F(j\omega)e^{j\omega t}d\omega \\ &= \frac{1}{(2\pi)^2}\int_{-\infty}^{\infty}\left(\int_{-\infty}^{\infty} X(j\lambda)Y(j\omega - j\lambda)\,d\lambda\right)e^{j\omega t}d\omega \end{aligned}$$

Now interchanging the order of integration and changing variables of integration,

$$\begin{aligned} f(t) &= \frac{1}{(2\pi)^2}\int_{-\infty}^{\infty} X(j\lambda)\left(\int_{-\infty}^{\infty} Y(j\omega - j\lambda)\,e^{j\omega t}d\omega\right)d\lambda \\ &= \frac{1}{(2\pi)^2}\int_{-\infty}^{\infty} X(j\lambda)\left(\int_{-\infty}^{\infty} Y(j\xi)\,e^{j(\xi+\lambda)t}d\xi\right)d\lambda \\ &= \frac{1}{(2\pi)^2}\int_{-\infty}^{\infty} X(j\lambda)e^{j\lambda t}\left(\int_{-\infty}^{\infty} Y(j\xi)\,e^{j\xi t}d\xi\right)d\lambda. \\ &= \frac{1}{2\pi}\left(\int_{-\infty}^{\infty} X(j\lambda)e^{j\lambda t}d\lambda\right)y(t) \\ &= x(t)\,y(t) \end{aligned} \qquad (12.2\text{-}9)$$

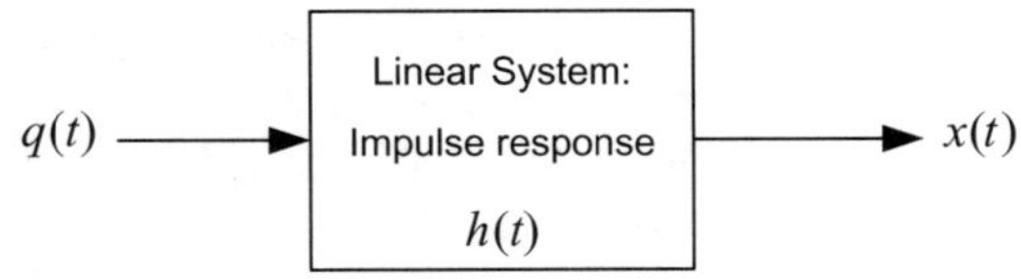

FIGURE 12.3: Linear system driven by input signal.

This convolution property is used when considering the Fourier transform of a time-limited sample of an infinite-duration signal. If the time-limiting *window function* is the rectangular boxcar defined above, the computed Fourier transform of the sample will be equal to the convolution of the true signal transform and the sinc function of equation (12.2-6).

The transform of signals convolved in the time domain is also of interest. It can be shown that the output $x(t)$ of a linear system driven by an input signal $q(t)$ (see Fig. 12.3) is equal to the convolution of the input signal and the system impulse response $h(t)$,

$$x(t) = \int_{-\infty}^{\infty} q(t)h(t-\lambda)\,d\lambda. \tag{12.2-10}$$

Using an approach similar to that above, we find that

$$X(j\omega) = Q(j\omega)H(j\omega) \tag{12.2-11}$$

where $H(j\omega)$ is the Fourier transform of $h(t)$. In other words, convolution in the time domain is equivalent to multiplication in the frequency domain, and the transform of the output of a linear system is equal to the product of transforms of the input signal and the system impulse response. This property allows computation of $H(j\omega)$ from power spectra, as explained later.

Another transform property of interest involves signal delays. If signal $f(t)$ is a delayed copy of $x(t)$, as $f(t) = x(t - \tau)$, the Fourier transform of the delayed signal is

$$\begin{aligned} F(j\omega) &= \int_{-\infty}^{\infty} x(t-\tau)e^{-j\omega t}dt = \int_{-\infty}^{\infty} x(\lambda)e^{-j\omega(\lambda+\tau)}d\lambda \\ &= e^{-j\omega\tau}\int_{-\infty}^{\infty} x(\lambda)e^{-j\omega\lambda}d\lambda = e^{-j\omega\tau}X(j\omega) \end{aligned}. \tag{12.2-12}$$

That is, the transform of $f(t)$ is the original transform multiplied by $e^{-j\omega\tau}$.

To define the spectral density using the Fourier transform, we start with the definition of energy:

$$\begin{aligned} E &\triangleq \int_{-\infty}^{\infty} (x(t))^2 dt \\ &= \frac{1}{2\pi}\int_{-\infty}^{\infty} x(t)\left(\int_{-\infty}^{\infty} X(j\omega)e^{j\omega t}d\omega\right)dt \\ &= \frac{1}{2\pi}\int_{-\infty}^{\infty} X(j\omega)\left(\int_{-\infty}^{\infty} x(t)e^{j\omega t}dt\right)d\omega \\ &= \frac{1}{2\pi}\int_{-\infty}^{\infty} X(j\omega)X(-j\omega)\,d\omega \end{aligned}. \tag{12.2-13}$$

If $x(t)$ is real (no imaginary component), it is easily shown from definition (eq. 12.2-4) that $X(-j\omega) = X^*(j\omega)$, the complex conjugate of $X(j\omega)$. Hence

$$\boxed{\begin{aligned} E &= \frac{1}{2\pi}\int_{-\infty}^{\infty} X(j\omega)X^{*}(j\omega)\,d\omega \\ &= \frac{1}{2\pi}\int_{-\infty}^{\infty} |X(j\omega)|^2\,d\omega \end{aligned}}. \tag{12.2-14}$$

In other words, the energy of a signal is equal to $1/2\pi$ times the integral over frequency of the magnitude squared of the Fourier transform. This is *Parseval's theorem*. The term that is integrated, $|X(j\omega)|^2$, is the energy density per unit of frequency and has units of magnitude squared per Hz.

The next section extends consideration to signals that are not limited in total energy.

12.2.3 Fourier Transform of Power Signals

Some signals have infinite time extent and energy. The ordinary Fourier transform is defined for signals that are absolutely integrable, that is $\int_{-T}^{T} |x(t)|\,dt < \infty$, so it cannot be directly used for infinite energy signals. Some of these signals can be handled by allowing the Fourier transform to contain impulses.

The average power of a signal is defined as

$$P \triangleq \lim_{T\to\infty} \frac{1}{2T}\int_{-T}^{T} [x(t)]^2\,dt, \tag{12.2-15}$$

and a *power signal* is one for which $0 < P < \infty$. For example $x(t) = \sin(\omega_0 t)$ is a power signal with power 1/2. Use of equation (12.2-4) to compute the Fourier transform of $\sin(\omega_0 t)$ yields

$$\begin{aligned} X(j\omega) &= \int_{-\infty}^{\infty} \sin\omega_0 t\; e^{-j\omega t}\,dt \\ &= -2j\int_{0}^{\infty} \sin\omega_0 t\,\sin\omega t\;dt \\ &= j\left(\frac{\sin(\omega_0+\omega)t}{(\omega_0+\omega)} - \frac{\sin(\omega_0-\omega)t}{(\omega_0-\omega)}\right)\Bigg|_0^{\infty} \end{aligned},$$

which is not defined. The Fourier transform of power signals must be redefined to avoid problems associated with convergence or infinite energy. One approach for some power signals takes a time limited sample of an infinite-duration signal, where the time window is allowed to approach infinity. A finite-duration sample of the signal $x(t)$ is defined as

$$\begin{aligned} x_T(t) &\triangleq x(t) \qquad |t| \le T < \infty \\ &\triangleq 0 \qquad\quad\;\; |t| > T \end{aligned} \tag{12.2-16}$$

and the corresponding Fourier transform is

$$X_T(j\omega) \triangleq \int_{-\infty}^{\infty} x_T(t)e^{-j\omega t}\,dt = \int_{-T}^{T} x(t)e^{-j\omega t}\,dt. \tag{12.2-17}$$

Hence the transform of $x_T(t) = \sin(\omega_0 t)$ for $|t| \le T$ is

$$X_T(j\omega) = jT\left(\frac{\sin(\omega_0+\omega)T}{(\omega_0+\omega)T} - \frac{\sin(\omega_0-\omega)T}{(\omega_0-\omega)T}\right). \tag{12.2-18}$$

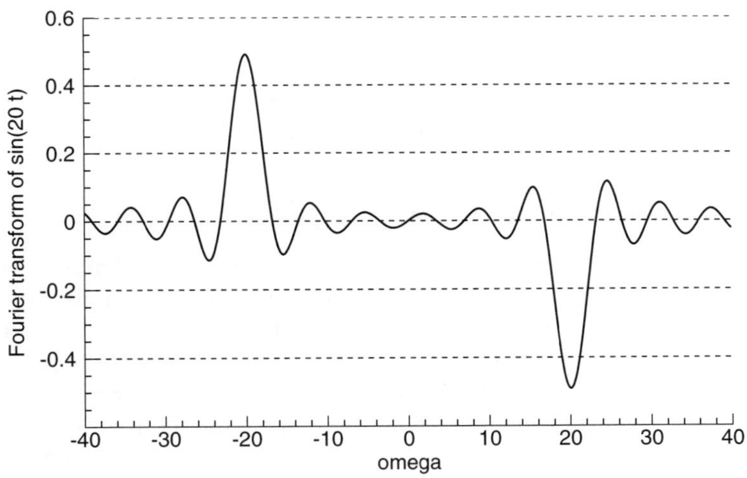

FIGURE 12.4: Fourier transform of energy signal sin(20t), $|t| \le 1$.

Notice that the transform is imaginary (no real component), and that sinc functions of opposite signs are centered at $\pm\omega_0$. The width of the sinc functions decreases with increasing T, but the magnitude increases linearly. Figure 12.4 shows the Fourier transform of the energy signal $x_T(t) = \sin(\omega_0 t)$ for $|t| \le 1$, where $\omega_0 = 20$.

Now as the time window T approaches ∞, the ω-width of the sinc functions approaches zero but the area $\int_{-\infty}^{\infty} T\,\text{sinc}(\omega T)\,d\omega$ remains constant since $\int_{-\infty}^{\infty} \text{sinc}(x)\,dx = \pi$. Hence in an integral sense $\text{sinc}(\omega T) \to \pi\delta(\omega)$ as $T \to \infty$ and the Fourier transform becomes

$$X(j\omega) = j\pi[\delta(\omega_0 + \omega) - \delta(\omega_0 - \omega)]. \qquad (12.2\text{-}19)$$

In other words, the transform is an imaginary Dirac impulse in the frequency domain with half the power located at frequency $-\omega_0$ and the other half at ω_0. This derivation is not rigorous, but it produces the correct answer. Notice that the inverse transform of equation (12.2-19) is indeed $\sin(\omega_0 t)$.

Another approach to computing Fourier transforms of power signals multiplies the signal by a suitable convergence factor such as $e^{-a|t|}$, and takes the limit as $a \to 0$. Fourier transforms of other power signals may be found in standard reference books, for example, Cooper and McGillem (1967).

Using a generalization of Parseval's theorem, the average power of a real-valued power signal $x_T(t)$ is computed from the Fourier transform as

$$\begin{aligned} P &= \lim_{T\to\infty} \frac{1}{2T}\int_{-T}^{T} [x(t)]^2 dt \\ &= \lim_{T\to\infty} \frac{1}{4\pi T}\int_{-\infty}^{\infty} X_T(j\omega)X_T(-j\omega)\,d\omega. \qquad (12.2\text{-}20) \\ &= \lim_{T\to\infty} \frac{1}{4\pi T}\int_{-\infty}^{\infty} |X_T(j\omega)|^2\,d\omega \end{aligned}$$

The PSD of $x(t)$ is computed from the Fourier transform as:

$$\boxed{S_x(\omega) \triangleq \lim_{T\to\infty} \frac{|X_T(j\omega)|^2}{2T}} \tag{12.2-21}$$

where $S_x(\omega)$ also has units of magnitude-squared per Hz.

12.2.4 Power Spectrum of Stochastic Signals

This chapter focuses on stochastic signals, since empirical modeling is most often used to determine characteristics of a system driven by white noise and possibly other known signals. The Fourier transform (eq. 12.2-4) is not defined for random power signals because the limit does not exist as the interval $T \to \infty$. This can be seen by considering the change in the Fourier transform as an increment of time dt is added to a previously calculated Fourier integral for interval $T < \infty$. Since the signal in the added time increment is random, the sum will not converge as the integration time span approaches infinity.

However, the PSD of a statistically stationary stochastic signal does converge and is defined as an expected value. As with power signals, we start with a time limited sample of an infinite-duration signal, and then extend the duration to infinity. The average power of $x_T(t)$ over the finite time window $|t| \le T$ is

$$\frac{1}{2T}\int_{-\infty}^{\infty}[x_T(t)]^2\,dt = \frac{1}{4\pi T}\int_{-\infty}^{\infty}|X_T(j\omega)|^2\,d\omega. \tag{12.2-22}$$

We now introduce three definitions for random processes. *Wide sense stationary* processes have autocorrelation functions and means that are constant functions of time. The time average of a stationary random process is generally a random variable, but for a *mean ergodic* process it is equal to the constant statistical (ensemble) average. *Autocorrelation ergodic* processes require that time and ensemble moments up to fourth order be equivalent. Unless specified otherwise, we use the term *ergodic* to imply *autocorrelation ergodic*.

If the signal $x_T(t)$ is mean ergodic, the average power from the left side of equation (12.2-22) will approach the mean-squared value as $T \to \infty$. The right side of equation (12.2-22) cannot be computed for $T \to \infty$ since the Fourier transform does not exist, but the expected value does exist. Hence we take the expected value of both sides:

$$E\left[\frac{1}{2T}\int_{-\infty}^{\infty}[x_T(t)]^2\,dt\right] = E\left[\frac{1}{4\pi T}\int_{-\infty}^{\infty}|X_T(j\omega)|^2\,d\omega\right]. \tag{12.2-23}$$

Under conditions that are satisfied for stable systems and physical signals [see Cooper and McGillem 1967, appendix D), the order of integration and expectation can be interchanged. Taking the limit as $T \to \infty$ yields

$$\begin{aligned}\lim_{T\to\infty}\left[\frac{1}{2T}\int_{-\infty}^{\infty}E[x_T^2(t)]\,dt\right] &= \lim_{T\to\infty}\left[\frac{1}{4\pi T}\int_{-\infty}^{\infty}E\left[|X_T(j\omega)|^2\right]d\omega\right]. \\ &= \frac{1}{2\pi}\int_{-\infty}^{\infty}\lim_{T\to\infty}\frac{E\left[|X_T(j\omega)|^2\right]}{2T}\,d\omega\end{aligned} \tag{12.2-24}$$

The integrand on the right is the PSD

$$S_x(\omega) = \lim_{T\to\infty} \frac{E\left[|X_T(j\omega)|^2\right]}{2T}, \tag{12.2-25}$$

and for a stationary process the time average of the mean-squared signal is equal to

$$E[x^2(t)] = \frac{1}{2\pi}\int_{-\infty}^{\infty} S_x(\omega)\,d\omega. \tag{12.2-26}$$

Equation (12.2-25) is the basic relationship allowing computation of the power spectrum from the Fourier transform. Notice that it involves two operations in addition to the Fourier transform. Because the Fourier transform of a stochastic signal is also a random variable, it is necessary to compute the expected value of the magnitude-squared. This requires repeated computation of the transform using different data spans, and averaging of the magnitude-squared transforms. Furthermore, the signal time span for which the transform is computed must be long so that the time-averaged power approaches the average power.

After computation of the PSD the system model is defined by finding a system transfer function $H(j\omega)$ (transform of the impulse response) such that

$$S_x(\omega) = |H(j\omega)|^2\, S_q(\omega) \tag{12.2-27}$$

where $S_q(\omega)$ is the PSD of the system input signal $q(t)$. In most cases where $q(t)$ is not directly measured, it is assumed to be $N(0,1)$ white noise so that $|H(j\omega)|^2 = S_x(\omega)$. In general $H(j\omega)$ is non-unique because $S_x(\omega)$ does not contain phase information. However, for real-valued signals the poles and zeroes of $H(j\omega)$ must occur in complex-conjugate pairs; that is, $H(j\omega)H^*(j\omega) = S_x(\omega)/S_q(\omega)$ where $S_x(\omega)$ and $S_q(\omega)$ are real. Further, the poles of $H(j\omega)$ must lie in the left-half complex plane for a stable system, and the zeroes must also lie in the left-half plane to minimize phase shifts for an assumed causal system. Thus $H(j\omega)$ can be uniquely determined using $S_x(\omega)$ and $S_q(\omega)$.

Finally we note that $S_x(\omega)$ can also be defined from the autocorrelation function. Since $|X_T(j\omega)|^2 = X_T(j\omega)X_T(-j\omega)$ we can write

$$\begin{aligned} S_x(\omega) &= \lim_{T\to\infty}\frac{1}{2T} E\left[\int_{-\infty}^{\infty} x_T(t)e^{j\omega t}dt \int_{-\infty}^{\infty} x_T(\lambda)e^{-j\omega\lambda}d\lambda\right] \\ &= \lim_{T\to\infty}\frac{1}{2T} E\left[\int_{-\infty}^{\infty}\left(\int_{-\infty}^{\infty} x_T(t)x_T(\lambda)e^{-j\omega(\lambda-t)}dt\right)d\lambda\right]. \\ &= \lim_{T\to\infty}\frac{1}{2T}\int_{-\infty}^{\infty}\left(\int_{-\infty}^{\infty} E[x_T(t)x_T(\lambda)]e^{-j\omega(\lambda-t)}dt\right)d\lambda \end{aligned} \tag{12.2-28}$$

Again the conditions for interchanging expectation and integration are satisfied for most signals of interest. Now the expected value of the term in brackets is the truncated autocorrelation function defined as

$$\begin{aligned} E[x_T(t)x_T(\lambda)] &= \rho_{xx}(t,\lambda) \qquad |t|,|\lambda| \le T \\ &= 0 \qquad\qquad\quad \text{otherwise} \end{aligned}. \tag{12.2-29}$$

Using the substitution $\tau = \lambda - t$ the PSD can be written as

$$\begin{aligned} S_x(\omega) &= \lim_{T\to\infty} \frac{1}{2T} \int_{-\infty}^{\infty} \left(\int_{-T}^{T} \rho_{xx}(t,t+\tau) e^{-j\omega\tau} dt \right) d\tau \\ &= \int_{-\infty}^{\infty} \left(\lim_{T\to\infty} \frac{1}{2T} \int_{-T}^{T} \rho_{xx}(t,t+\tau) dt \right) e^{-j\omega\tau} d\tau \end{aligned} \tag{12.2-30}$$

Hence the PSD is the Fourier transform of the time average of the autocorrelation function, that is,

$$S_x(\omega) = \mathcal{F}[\langle \rho_{xx}(t,t+\tau) \rangle] \tag{12.2-31}$$

where $\langle \cdot \rangle$ denotes the time average of the given random variable. For a stationary process the autocorrelation function is independent of t and the equation simplifies to

$$S_x(\omega) = \mathcal{F}[\rho_{xx}(\tau)]. \tag{12.2-32}$$

where $\rho_{xx}(-\tau) = \rho_{xx}(\tau)$ for a real-valued signal. Equation (12.2-32) is used in the correlogram method for computing the PSD from a sample autocorrelation function. The equation is also used as the basis for AR modeling methods discussed later in this chapter. Again notice that the computation involves time averaging and use of a long data span.

12.2.5 Time-Limiting Window Functions

Since all power spectra calculations for real data involve a finite data span, the impact of the time-limited window must be considered when computing power spectra. As noted previously, use of the simple rectangular (boxcar) window function implicitly convolves the true (infinite-time) Fourier transform by the sinc function when the transform of the time-limited data is computed. Notice that the side lobes (response for $\omega T > \pi$) of $\sin(\omega T)/\omega T$ only decay as $1/\omega T$, with the result that about 10% of the total power is contained in the side lobes. This causes spectral leakage of the true spectrum to frequencies offset from the actual. For example, the Fourier transform of a pure sinusoid at ω_0 is a pair of imaginary impulses at $\pm\omega_0$, but when the transform is applied to a finite data sample of the sinusoid, the result is the pair of sinc functions shown in Figure 12.4. This behavior is potentially a problem when power spectra contain pure sinusoids or narrowband signals, but it is less of an issue with wideband signals. Although some physical systems have resonances causing narrowband signals, most are well damped. Furthermore physical systems are seldom driven by sinusoids. Thus spectral leakage is usually not an issue.

A variety of nonrectangular window functions have been developed to minimize spectral leakage. Compared with the sinc function, they all have the effect of slightly widening the central lobe and greatly reducing the magnitude of the side lobes. The side lobe reduction is accomplished by tapering the edges of the rectangular window so that higher frequencies of the transform are attenuated. To apply a window function, multiply the signal $x(t)$ defined for $|t| \le T$ by the window function $w(t)$ to compute $y(t)$ over the same interval:

$$\begin{aligned} y(t) &= w(t)x(t) \qquad |t| \le T \\ &= 0 \qquad\qquad\quad\; |t| > T \end{aligned}. \tag{12.2-33}$$

TABLE 12.1: Common Discrete Window Functions

Window Function	$w(k)$ for $-N \le k \le N$	Peak Amplitude of Side Lobes (dB)	−6 dB Width of Central Lobe
Rectangular	1	−13.3	0.602/N
10% cosine taper	1 $\quad |k| \le 0.8N$ $0.5 + 0.5\cos(\theta) \quad 0.8N < |k| \le N$ $(\theta = (10|k| - 8N)\pi/2N$	−13.5	0.668/N
Hamming	$0.5 + 0.5\cos(\pi k/N)$	−31.5	1.000/N
Bartlett	$1 - |k|/N$	−35.7	0.886/N
Hamming	$0.54 + 0.46\cos(\pi k/N)$	−42.7	0.907/N
Kaiser	$\dfrac{I_0\left(\omega_a\sqrt{N^2-k^2}\right)}{I_0(\omega_a N)}$, I_0 is the modified zeroth order Bessel function of the first kind	−47.5 for $\omega_a N = 6.5$	1.012/N
Parzen (N even)	$2\left(1-\dfrac{|k|}{N}\right)^3 - \left(1-2\dfrac{|k|}{N}\right)^3 \quad |k| \le N/2$ $2\left(1-\dfrac{|k|}{N}\right)^3 \quad N/2 < |k| \le N$	−53.1	1.276/N
Blackman	$0.42 + 0.5\cos(\pi k/N) + 0.08\cos(2\pi k/N)$	−58.3	1.150/N

Then compute the transform of $y(t)$. The resulting transform will be the convolution of $X_T(j\omega)$ and $W_T(j\omega) = \mathscr{F}[w(t)]$. Since the area under $w^2(t)$ for $|t| \le T$ will not be $2T$, the power of the resulting transform should be scaled by

$$\frac{1}{\int_{-T}^{T} w^2(t)\,dt}$$

to obtain the power in $x(t)$.

Table 12.1 lists common window functions and their properties. Since window functions are typically used with discretely sampled data, the independent variable in the window equations is listed as k where $-N \le k \le N$; that is, there are $2N + 1$ discrete samples of the signal $x(k\Delta t)$ centered at $k = 0$. If the sample indices are numbered from 1 to $2N + 1$, the corresponding index should be changed to $j = k + N + 1$. Figure 12.5 shows the time response of several window functions for $N = 200$ and $\Delta t = 1$. Figure 12.6 shows the frequency response computed from the real component of a *discrete Fourier transform* (DFT)—defined in the next section—for windows centered at $t = 0$. The window data was "padded" with zeroes over the interval $-2000 \le k \le 2000$ so that the transform could be plotted at intervals smaller than the DTF spacing of $1/(2N\Delta t) = 0.0025$ Hz for the nominal window width. With padding the points can be plotted at intervals of 0.00025 Hz. It should be noted that zero padding does not actually improve DFT resolution: it only allows interpolation between normal frequency samples.

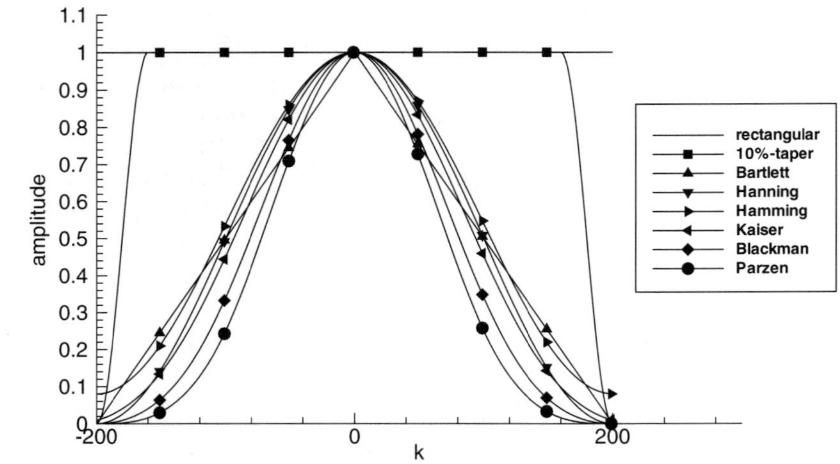

FIGURE 12.5: Time response of common window functions.

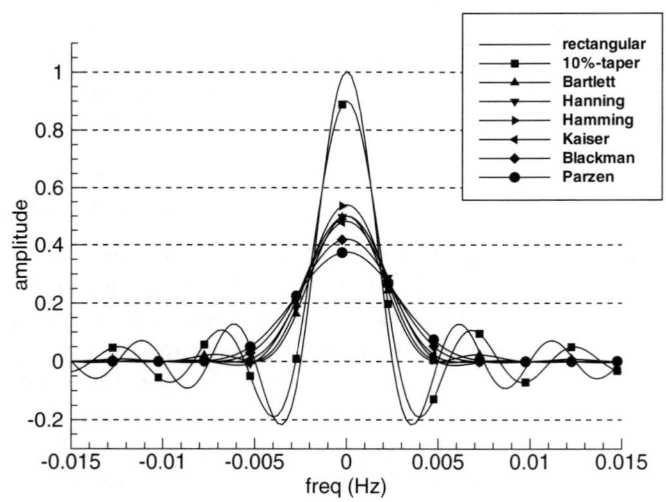

FIGURE 12.6: Frequency response of common window functions.

Other properties of common window functions are described in Rabiner and Gold (1975), Oppenheim and Schafer (1975), Kay (1988), Marple (1987), and Harris (1978). The Kaiser window function has the largest energy in the main lobe for a given peak side lobe amplitude, with the ω_a parameter used to trade off main lobe width and side lobe amplitude. (Side lobe reduction causes an increase in main lobe width.) Typical values are $4 < \omega_a N < 9$. Only the Bartlett and Parzen window functions have non-negative Fourier transforms for all frequencies. This is not important when using the periodogram, but can be important when the Fourier transform of the sample autocorrelation function is used to compute power spectra.

The selection of the best window function for spectral analysis depends on the types of signals appearing in the data. The primary reasons for using a window function are to minimize spectral leakage of a narrowband signal to adjacent frequencies, and to avoid the "ringing" effect of the sinc function at transients. As noted previously, windowing is not always necessary when attempting to model physical systems. If a window function is to be used, the Bartlett, Hamming, or Parzen windows are simple to implement and usually acceptable. Notice that application of a window function effectively smoothes the computed PSD because the output Fourier transform is the convolution of the data and window transforms.

12.2.6 Discrete Fourier Transform

Although discussion in the previous sections has assumed that the signal was continuous, most practical applications involve discrete sampling of a continuous signal. The effect of discrete sampling over a finite time span can be analyzed using two approaches. First, the discrete sample can be treated as the multiplication of a sampling impulse train with the continuous signal over the sampling interval, and the Fourier transform of an energy signal can be computed taking into account the convolution of the sampling function with the continuous signal. Alternatively the discrete values can be treated as samples of a periodic function, and the Fourier series can be used to compute power spectra. The effect on the resulting power spectra is best understood using both approaches.

The sampling function is defined as an impulse train,

$$g(t) = \sum_{n=-\infty}^{\infty} \delta(t - n\Delta t), \tag{12.2-34}$$

where Δt is the sampling interval. Hence the discretely sampled output function is

$$y(t) = g(t)x(t). \tag{12.2-35}$$

Of course the impulse function only has meaning when appearing within an integral, but this is temporarily ignored as integration will eventually be used.

To understand the effect of sampling on the transform of $x(t)$, the transform of $g(t)$ is needed. This can be computed by representing the periodic sampling function $g(t)$ in a Fourier series as

$$g(t) = \frac{1}{\Delta t}\sum_{n=-\infty}^{\infty} e^{jn\omega_0 t} \tag{12.2-36}$$

where $\omega_0 = 2\pi/\Delta t$. It can be verified using equations (12.2-2) and (12.2-34) that all Fourier series coefficients are equal to $1/\Delta t$. Then the Fourier transform of $g(t)$ is computed as

$$\begin{aligned} G(j\omega) &= \frac{1}{\Delta t}\sum_{n=-\infty}^{\infty} \mathscr{F}(e^{jn\omega_0 t}) = \frac{1}{\Delta t}\sum_{n=-\infty}^{\infty} \int_{-\infty}^{\infty} e^{jn\omega_0 t} e^{-j\omega t}\, dt \\ &= \frac{1}{\Delta t}\sum_{n=-\infty}^{\infty} \int_{-\infty}^{\infty} e^{j(n\omega_0 - \omega)t}\, dt = \frac{2\pi}{\Delta t}\sum_{n=-\infty}^{\infty} \delta(n\omega_0 - \omega) \end{aligned} \tag{12.2-37}$$

The last step can be verified by noting that the inverse transform of the impulse train is equal to equation (12.2-36):

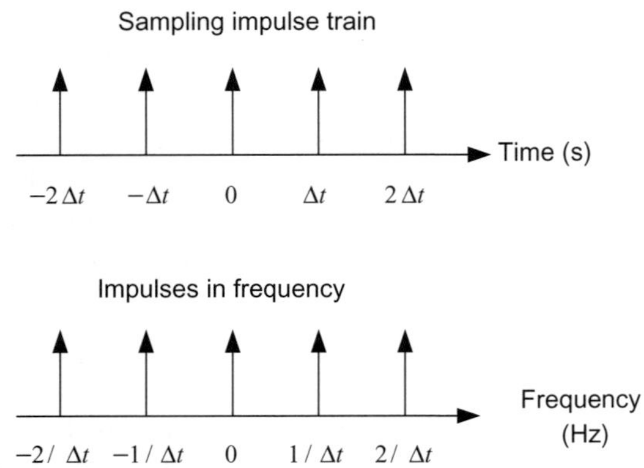

FIGURE 12.7: Impulse train in time and frequency domains.

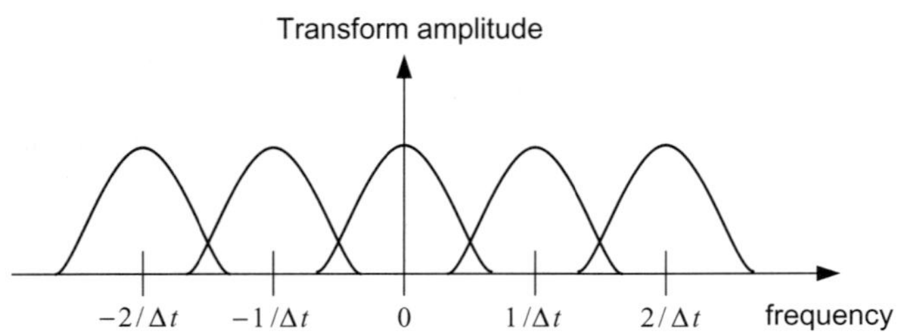

FIGURE 12.8: Power spectrum of discretely sampled low-pass spectrum.

$$\mathscr{F}^{-1}\left(\frac{2\pi}{\Delta t}\sum_{n=-\infty}^{\infty}\delta(n\omega_0-\omega)\right)=\frac{1}{\Delta t}\int_{-\infty}^{\infty}\sum_{n=-\infty}^{\infty}\delta(n\omega_0-\omega)e^{j\omega t}d\omega=\frac{1}{\Delta t}\sum_{n=-\infty}^{\infty}e^{jn\omega_0 t}.$$

Hence the Fourier transform of an impulse train in the time domain is an impulse train spaced at $\omega_0 = 2\pi/\Delta t$ in the frequency domain. Figure 12.7 shows the sampling function and the equivalent impulse train in the frequency domain.

Since multiplication in the time domain is convolution in the frequency domain, the original spectrum of $x(t)$ is repeated at $f_0 = 1/\Delta t$ Hz. This is graphically shown in Figure 12.8 for a typical low pass power spectrum. If the power spectrum of $x(t)$ has content above the *Nyquist frequency* $f_N = 1/2\Delta t$, the power spectrum of the discrete signal $y(t)$ for the repeated spectrum at $\Delta f = \pm 1/\Delta t$ will *alias* into the spectrum centered at $f = 0$. Thus the sampling frequency must be greater than the two times the highest frequencies in $x(t)$, or $x(t)$ must be band-limited before sampling to remove content above the Nyquist frequency.

Notice that for a finite data window of $2N + 1$ discretely sampled data points spaced at Δt, the sampled signal $y(k\Delta t)$ for $-N \le k \le N$ can be represented as

$$y(t) = g(t)w(t)x(t) \qquad |t| \le T = N\Delta t. \tag{12.2-38}$$

or

$$y(k\Delta t) = w(k\Delta t)x(k\Delta t) \qquad |k| \le N. \tag{12.2-39}$$

Hence the Fourier transform of $y(k\Delta t)$ is the convolution of the transform of the signal $x(k\Delta t)$, the window function $w(k\Delta t)$ and the sampling function $g(k\Delta t)$. When the rectangular window function is used, the transform of the sinc function has zero crossings at $f = 1/2T$ where $T = 2N\Delta t$ is the duration of the time window. At any other sampled frequencies the Fourier transform will be affected by the signal frequency content at nearby frequencies because of the side lobes of the sinc function. This suggests that a discrete approximation of the continuous Fourier transform—where the integral is replaced with a sum—can be defined to compute the same number of frequency samples as in the discrete time series:

$$\begin{aligned} Y_D(ji\omega_0) &= \Delta t \sum_{k=-N}^{N} y(k\Delta t)e^{-jik\omega_0\Delta t} \\ &= \Delta t \sum_{k=-N}^{N} y(k\Delta t)\,(\cos(\pi ik/N) - j\sin(\pi ik/N)) \qquad -N \le i \le N \end{aligned} \tag{12.2-40}$$

where $\omega_0 = 2\pi/2T = \pi/N\Delta t$. The DFT is sometimes defined without the multiplying Δt or with $1/2N$ replacing Δt (to be more consistent with the continuous power Fourier transform.) Notice that the DFT with the assumed rectangular window function does not introduce correlation at the sampled frequencies. If the input signal $x(k\Delta t)$ is white noise, the Fourier transform of $y(k\Delta t)$ will also be white uncorrelated noise. This is useful for some signal processing purposes as the DFT of a correlated signal plus white measurement noise can be treated as a statistically equivalent signal. When nonrectangular window functions are used, the discrete frequency samples will correlate white measurement noise present in $x(k\Delta t)$. However, this property is usually unimportant in modeling applications.

Notice that equation (12.2-40) involves repeated calculation of identical trigonometric functions at $-\pi N \le ik\pi/N \le \pi N$ and all values can be obtained from the sine and cosine evaluated at $N/2$ angles. Furthermore the same product ik is used multiple times. A much more efficient algorithm—the FFT—results from grouping the trigonometric evaluations and multiplications when N can be factored using small prime numbers—particularly 2. The FFT is well documented in many publications such as Press et al. (2007), Bergland (1969), Rabiner and Gold (1975), Singleton (1969), and *Programs for Digital Signal Processing* (1979). Singleton's method is the most flexible because it works with mixed-radix factors, not just two, and it is also very efficient. An updated version of the code is included in the library associated with this book. The FFT is not discussed further in this chapter because it is not critical to understanding spectral analysis. Note that FFT software is usually defined with shifted indices, $0 \le i \le 2N$ (or $0 \le i \le N$ for real signals) and $1 \le k \le 2N + 1$.

Although equation (12.2-40) was developed as a discrete form of the Fourier integral, the similarity to the Fourier series is obvious. Since the Fourier series is only used for periodic functions, it is apparent that the DFT implicitly assumes $2T$ periodicity of $y(k\Delta t)$. If $y(-N)$ is significantly different from $y(N)$, the power spectrum will have added high and low frequency content. Figure 12.2 shows the effect on the spectrum when linear trends are present. To avoid this problem, large trends should be subtracted from the data before applying the DFT. Use of a tapered window function can also minimize this problem, but the presence of a large bias in the data will cause the computed PSD to contain the spectrum of the window function centered at zero frequency. For this reason it is also advisable to remove large biases.

12.2.7 Periodogram Computation of Power Spectra

With that lengthy introduction we now show how the Fourier transform can be used to compute the power spectrum. Using definition equation (12.2-25) for the continuous PSD of a wide sense stationary process, the PSD for discrete data is defined as

$$S_x(\omega) = \lim_{N\to\infty} E\left[\frac{1}{2N+1}\left|\sum_{k=-N}^{N} x(k\Delta t)e^{-j\omega k}\right|^2\right], \tag{12.2-41}$$

where $\omega = 2\pi f$.

(Note that use of data array $x[k] = x(k\Delta t)$ for indices $-N \le k \le N$ is accommodated in MATLAB or Fortran 90/95/2003, but it may be convenient to use positive data indices $0 \le k \le 2N$ with $\exp(-j\omega(k+N))$ in some computer languages. Many FFT utilities require an even number of data values and the user normally sets elements $0 \le k \le N$ in the first half of the input data array and elements $-N+1 \le k \le -1$ in the second half. It is assumed that the data are periodic in $2N$ so $x[-N] = x[N]$.)

Given a single data set i over the time interval $-N\Delta t \le t \le N\Delta t$, the single sample estimate of the periodogram for the finite window is

$$\boxed{\hat{S}_{x\_i}(\omega) = \frac{1}{2N+1}\left|\sum_{k=-N}^{N} x(k\Delta t)e^{-j\omega k}\right|^2.} \tag{12.2-42}$$

The periodogram sample is the magnitude squared of the DFT. This is potentially evaluated at any frequency, but is normally limited to $\omega = \pi i / N\Delta t$ for $|i| \le N/2$. Given M sets of data each over the interval $-N\Delta t \le t \le N\Delta t$, the averaged estimate of the periodogram is

$$\boxed{\hat{S}_{x\_AV}(\omega) = \frac{1}{M}\sum_{i=1}^{M} \hat{S}_{x\_i}(\omega).} \tag{12.2-43}$$

It can be shown that averaging of independent periodograms is equivalent to convolving the true PSD with the Fourier transform of the Bartlett window (Kay 1988, appendix 4A). On the average, this produces a smooth version of the true PSD.

The preceding sections have discussed the many potential problems of using Fourier transforms to compute power spectra via the periodogram. To summarize:

1. The periodogram only converges to the true power spectrum for a wide sense stationary process as the time duration of the sample approaches infinity. Since this is not possible, the sample periodogram is the convolution of the true PSD and the Fourier transform of the window function. The side lobes of the window transform cause spectral leakage and bias in the PSD. This is a particular problem for the rectangular window function where side lobes of the sinc function decay slowly. Hence the presence of narrowband signals or transients cause spurious peaks in the PSD. This is usually (but not always) more an issue for signal processing applications than when modeling physical systems.

2. Use of window functions other than rectangular can reduce the leakage caused by narrowband signals, thus reducing bias in the PSD. However, any bias in the data will cause the resulting transform to also include the transform of the window function centered at zero frequency. Hence, large data biases should be removed before applying the window function.
3. The presence of a trend or discontinuities between the first and last data points will also result in spurious frequencies. Thus large ramp (or quadratic) trends should be removed.
4. For stochastic signals the sample periodogram is a noisy estimator of the true power spectrum, and the variance of the periodogram does not approach zero as the data window approaches infinity. Hence it is necessary to compute multiple periodograms and average them. If only a limited data sample is available, this requires breaking the sample into non-overlapping segments with consequently greater PSD bias and poorer resolution in each segment. In a method proposed by Welch (1967), multiple segments are used with tapered window functions, and the segments may overlap if available data is limited. Marple (1987, p. 155) states that Welch's method, implemented using the FFT, is the most frequently used periodogram estimation approach.
5. Discrete sampling will cause aliasing of the power spectrum if the signal bandwidth is greater than the Nyquist frequency before sampling. This is a problem with all spectral estimation methods.
6. Since bias in the PSD estimate is caused by narrowband signals, it may be beneficial to pre-whiten the data using a digital filter to remove approximately known PSD structure (Blackman and Tukey 1958). The output of the filter will have a flatter spectrum that will result in smaller spectral bias when a periodogram is computed.
7. The periodogram assumes a fixed sampling time interval, but this is not an inherent limitation.
8. Efficient implementations of the DFT only computes the spectrum at frequency intervals of $1/2N\Delta t$, but this is not an inherent limitation as zero padding can be used to produce output at other frequencies.

Despite the above problems, periodogram computation of power spectra is often the first step in empirical modeling as it does not depend on assumed model structure. However, it is a poor spectral estimator and should not be the last step in the analysis process.

A recommended procedure for computing and using a periodogram is:

1. Compute biases and trends in the data, and if they are large relative to the signal variance, remove them.
2. Either break a single data set into multiple non-overlapping segments (or possibly overlapping if using Welch's windowed method) or use multiple data sets. A minimum of 10 segments is desirable if the total number of data values is sufficiently large. It is often helpful to try different numbers of segments and compare the results.
3. Apply a window function to each segment (optional).
4. Compute the DFT and periodogram for each segment.

5. Average the periodograms.
6. Examine the averaged periodogram to determine the approximate location of dominant poles and zeroes. This may be difficult because the PSD will still be "noisy." If the output spectrum is relatively flat for assumed white noise inputs, it may be concluded that the system has no significant poles or zeroes. Likewise a simple low pass characteristic decaying at –20 dB/decade may indicate that the signal can be modeled as a first-order Markov process. For many systems further analysis using AR or ARMA models is required, but the periodogram provides information on the appropriate maximum model order and structure.

12.2.8 Blackman-Tukey (Correlogram) Computation of Power Spectra

Equation (12.2-32) showed that the PSD can also be computed as the Fourier transform of the autocorrelation function. Since the true autocorrelation function is unknown, a sample value is computed from a finite data set. A biased estimate of the autocorrelation function can be computed as

$$\hat{\rho}_{xx\_b}[k] = \frac{1}{N} \sum_{i=0}^{N-k-1} x_i \, x_{i+k} \qquad 0 \le k \le N-1 \tag{12.2-44}$$

where $x_i = x(i\Delta t)$ and we have modified the definition (compared to usage in previous sections) of the data time indices and N to be more consistent with typical computer storage. Since $\hat{\rho}_{xx}[-k] = \hat{\rho}_{xx}[k]$ for real-valued data, a total of $2N-1$ autocorrelation lags can be computed for N data samples. Notice that the divisor of equation (12.2-44) is fixed but the number of terms summed is smaller for large lags: this biases $\hat{\rho}_{xx\_b}[k]$ since it will be small for large lags. It can be shown (e.g., Marple 1987, appendix 4.A) that when the PSD is computed as $S_x(\omega) = \mathscr{F}(\hat{\rho}_{xx\_b}[k])$, the result is the same as when using the periodogram for the same sample. The linearly decreasing number of terms in $\hat{\rho}_{xx\_b}$ with lag number effectively applies a Bartlett window function to the true $\rho_{xx}[k]$, and this is equivalent to averaging periodograms computed with a rectangular window.

An *unbiased* estimate of the autocorrelation can be computed as

$$\hat{\rho}_{xx\_u}[k] = \frac{1}{N-k} \sum_{i=0}^{N-k-1} x_i \, x_{i+k} \tag{12.2-45}$$

since the divisor matches the number of terms summed. However, the power spectrum computed from this definition may have negative power, so the biased definition equation (12.2-44) is often preferred when all lags of the autocorrelation are to be used.

In the Blackman-Tukey (BT) method of PSD computation, the maximum lag used for the autocorrelation function is limited to be less than $N/10$. This minimizes randomness in the PSD caused by large variations in $\hat{\rho}_{xx}[k]$ for large lags. However, this rectangular window truncation of $\hat{\rho}_{xx}[k]$ is equivalent to sinc function convolution with the true PSD. For that reason it is important to apply a smooth window function to $\hat{\rho}_{xx}[k]$ before the Fourier transform is computed. The window must be normalized so that $w[0] = 1$. The Fourier transform of the selected window function should not be negative for any frequencies, so the Bartlett or Parzen windows are preferred. However, randomness in sample autocorrelation values may still cause the PSD to be negative at some frequencies. The BT power spectrum is formally defined as

$$S_{xx\_BT}(\omega) \triangleq \Delta t \sum_{k=-L}^{L} w[k]\hat{\rho}_{xx\_u}[k]e^{-j\omega k\Delta t} \qquad (12.2\text{-}46)$$

where $w[k]$ is the window function and $\hat{\rho}_{xx\_u}[k]$ is the unbiased autocorrelation function.

Kay (1988, section 4.6) compares the periodogram and BT methods on several signals that include closely spaced sinusoids with added white noise, broadband noise, and narrowband noise. The averaged periodogram is better able to resolve closely spaced sinusoids, but exhibits more variance. The spectral bias of the periodogram is smaller than for the BT method.

A recommended procedure for computing the BT PSD is:

1. Compute biases and trends in the data, and if they are large relative to the signal variance, remove them.
2. Compute the unbiased sample autocorrelation function using equation (12.2-45). Limit the maximum lag order to approximately $N/10$.
3. Apply a window function to the autocorrelation function.
4. Compute the Fourier transform of the windowed autocorrelation function as equation (12.2-46). The PSD is the magnitude of the Fourier transform.
5. Examine the PSD to determine the approximate location of dominant poles and zeroes. See the previous comments for the periodogram.

The following example demonstrates use of the periodogram and BT correlogram methods. The same example is also used for other spectral estimation algorithms so that performance can be compared.

Example 12.1: Fourth-Order ARMAX Model with Two Outputs and One Input

A fourth-order ARMAX model with two output signals and one sinusoidal input signal is used to evaluate the various spectral analysis methods. While not modeling any particular system, it has characteristics that may be found in physical systems. The model is defined in state space form because the ARMAX modeling approach to be used later directly computes a state space model. The simulated data from this model are not intended to rigorously evaluate all the different modeling and spectral analysis methods discussed in this chapter. The unbiased data have distinct narrowband noise components, a relatively high signal-to-noise ratio, and a medium-length span. With different conditions, the results for different spectral analysis approaches may be different from those shown here.

Figure 12.9 is a block diagram of the discrete system: q_1,q_2 are, respectively, white $N(0,1)$ and $N(0,8)$ process noises, r_1,r_2 are white $N(0,0.5^2)$ measurement noises, y_1,y_2 are the measurements, and $u_i = 0.8\sin(2\pi i\Delta t/25)$ is an exogenous "control" input. The time step $\Delta t = 1$. Notice that the two measurements are coupled because $c_1(y_1 - r_1)$ is added to y_2. Also notice that control u_i affects the

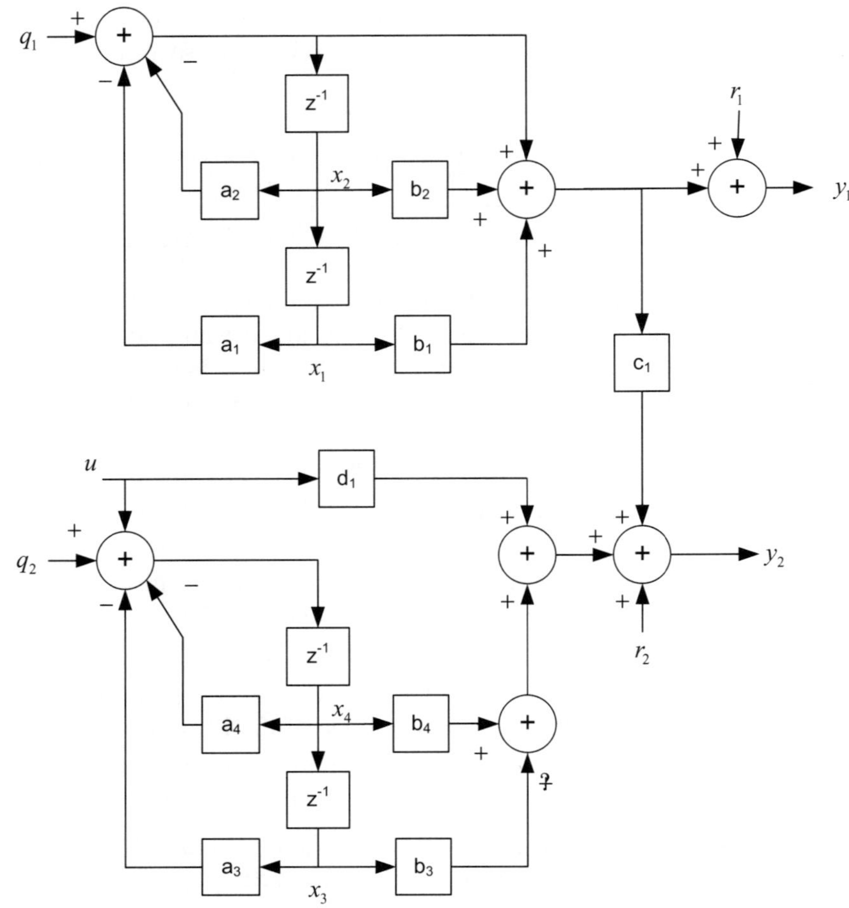

FIGURE 12.9: Example fourth-order ARMAX model.

dynamics of states x_3,x_4 and is also added to y_2. The various coefficients are set as follows

$$\begin{aligned} &a_1 = 0.81, \quad a_2 = -1.4562, \quad a_3 = 0.81, \quad a_4 = -0.5562 \\ &b_1 = 0.5, \quad b_2 = 1.0, \quad b_3 = 0.50, \quad b_4 = 1.0, \quad c_1 = 0.30, \quad d_1 = 0.20 \end{aligned}$$

In state space notation the discrete system is

$$\begin{bmatrix} y_1 \\ y_2 \end{bmatrix}_i = \begin{bmatrix} b_1 - a_1 & b_2 - a_2 & 0 & 0 \\ c_1(b_1 - a_1) & c_1(b_2 - a_2) & b_3 & b_4 \end{bmatrix} \begin{bmatrix} x_1 \\ x_2 \\ x_3 \\ x_4 \end{bmatrix}_i + \begin{bmatrix} 0 \\ d_1 \end{bmatrix} u_i + \begin{bmatrix} 0 & 1 & 0 & 0 \\ 0 & c_1 & 0 & 0 \end{bmatrix} \begin{bmatrix} 0 \\ q_2 \\ 0 \\ q_4 \end{bmatrix}_i + \begin{bmatrix} r_1 \\ r_2 \end{bmatrix}_i$$

$$\begin{bmatrix} x_1 \\ x_2 \\ x_3 \\ x_4 \end{bmatrix}_{i+1} = \begin{bmatrix} 0 & 1 & 0 & 0 \\ -a_1 & -a_2 & 0 & 0 \\ 0 & 0 & 0 & 1 \\ 0 & 0 & -a_3 & -a_4 \end{bmatrix} \begin{bmatrix} x_1 \\ x_2 \\ x_3 \\ x_4 \end{bmatrix}_i + \begin{bmatrix} 0 \\ 0 \\ 0 \\ 1 \end{bmatrix} u_i + \begin{bmatrix} 0 \\ q_2 \\ 0 \\ q_4 \end{bmatrix}_i$$

This is more compactly represented as

$$\begin{aligned}\mathbf{y}_i &= \mathbf{H}\mathbf{x}_i + \mathbf{A}u_i + \mathbf{B}\mathbf{q}_i + \mathbf{r}_i\\ \mathbf{x}_{i+1} &= \mathbf{\Phi}\mathbf{x}_i + \mathbf{G}u_i + \mathbf{q}_i\end{aligned}$$

where

$$\mathbf{y}_i = \begin{bmatrix} y_1 \\ y_2 \end{bmatrix}_i, \quad \mathbf{x}_i = \begin{bmatrix} x_1 \\ x_2 \\ x_3 \\ x_4 \end{bmatrix}_i, \quad \mathbf{r}_i = \begin{bmatrix} r_1 \\ r_2 \end{bmatrix}_i, \quad \mathbf{q}_i = \begin{bmatrix} 0 \\ q_2 \\ 0 \\ q_4 \end{bmatrix}_i,$$

$$\mathbf{H} = \begin{bmatrix} b_1 - a_1 & b_2 - a_2 & 0 & 0 \\ c_1(b_1 - a_1) & c_1(b_2 - a_2) & b_3 & b_4 \end{bmatrix}, \quad \mathbf{A} = \begin{bmatrix} 0 \\ d_1 \end{bmatrix}, \quad \mathbf{B} = \begin{bmatrix} 0 & 1 & 0 & 0 \\ 0 & c_1 & 0 & 0 \end{bmatrix}$$

$$\mathbf{\Phi} = \begin{bmatrix} 0 & 1 & 0 & 0 \\ -a_1 & -a_2 & 0 & 0 \\ 0 & 0 & 0 & 1 \\ 0 & 0 & -a_3 & -a_4 \end{bmatrix}, \quad \mathbf{G} = \begin{bmatrix} 0 \\ 0 \\ 0 \\ 1 \end{bmatrix},$$

and

$$E[\mathbf{r}_i\mathbf{r}_i^T] = \begin{bmatrix} 0.5^2 & 0 \\ 0 & 0.5^2 \end{bmatrix}, \quad E[\mathbf{q}_i\mathbf{q}_i^T] = \begin{bmatrix} 0 & 0 & 0 & 0 \\ 0 & 1 & 0 & 0 \\ 0 & 0 & 0 & 0 \\ 0 & 0 & 0 & 8 \end{bmatrix}.$$

This state space structure is the most general form of an ARMAX model. The eigenvalues (poles) of $\mathbf{\Phi}$ for this system are

$$\begin{aligned}\lambda_1 &= 0.72812 + j0.52901, \quad \lambda_2 = 0.72812 - j0.52901\\ \lambda_3 &= 0.27812 + j0.85595, \quad \lambda_4 = 0.27812 - j0.85595\end{aligned}$$

which correspond to resonance frequencies at 0.1 Hz (λ_1,λ_2), and 0.2 Hz (λ_3,λ_4). Both pairs of poles are located in the z-plane at a radius of 0.9 (0.1 from the unit circle) so that damping is light. The zeroes in this ARMAX model are far from the unit circle and thus do not create significant nulls in the spectrum. This is typical of many physical systems. However, this should not be interpreted to mean that zeroes of the ARMAX model can be eliminated. The zeroes are important in modeling system behavior.

Simulated measurements with random measurement and process noise were generated at 1 s intervals for 500 s. Figure 12.10a,b,c shows the simulated measurements and control. Notice that the frequency content of the y_1 measurement is much lower than that of y_2. The measurement noise standard deviation of 0.5 is small compared with the sample measurement standard deviations of 7.2 and 6.7, respectively, so the signal-to-noise ratios (23.2 and 22.5 dB) are quite high. Also notice that the control signal u_1 is small ($\sigma = 0.4$) compared with the measurement magnitudes, so the effect of the control is unlikely to be

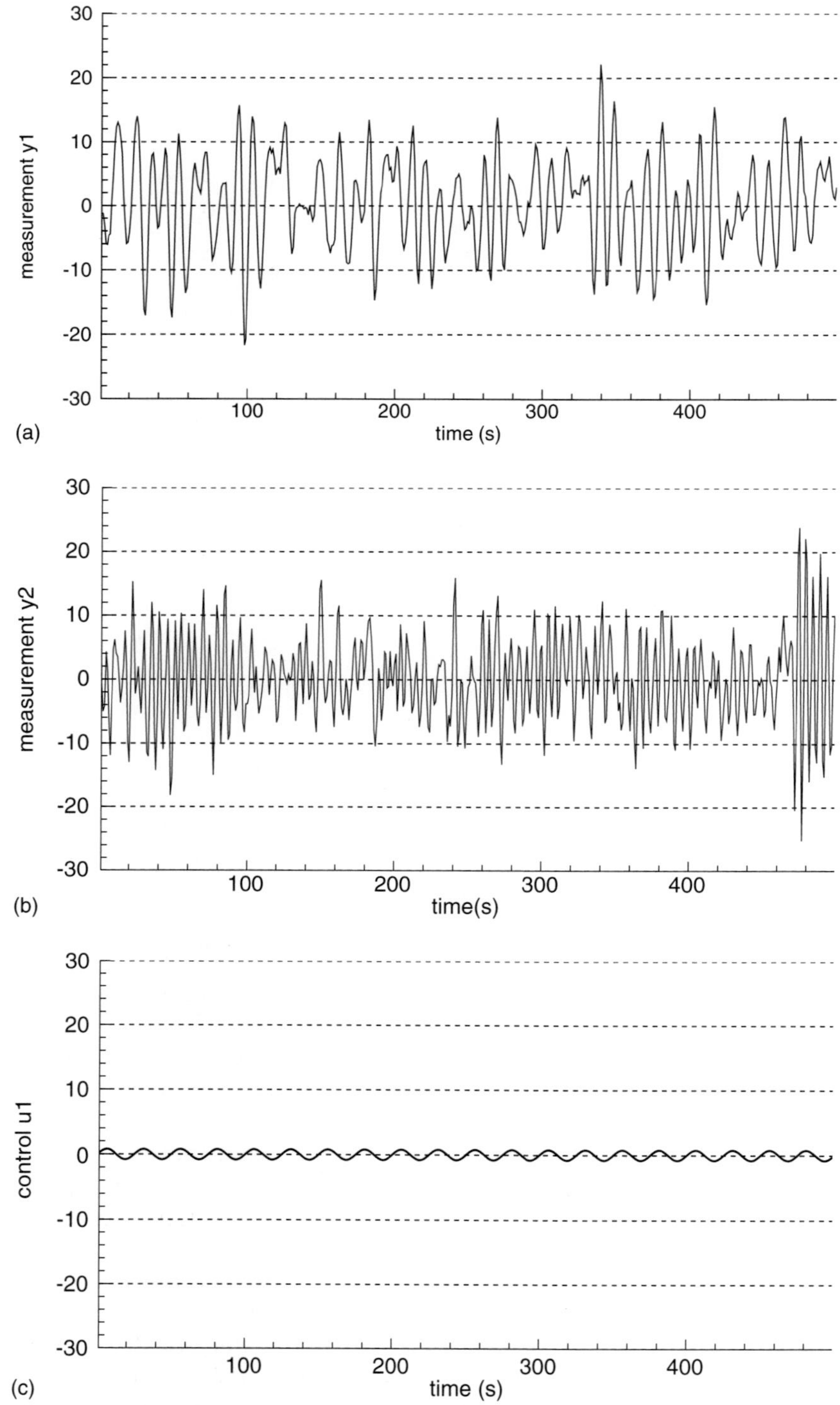

FIGURE 12.10: (a) ARMAX model measurement y_1. (b) ARMAX model measurement y_2. (c) ARMAX model measurement u_1.

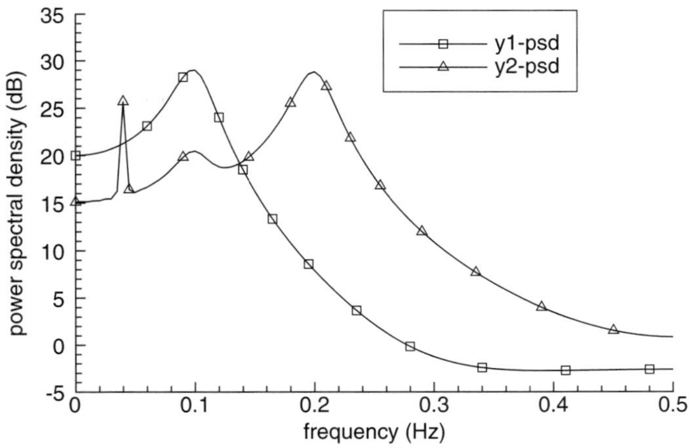

FIGURE 12.11: True PSD of simulated fourth-order ARMAX system.

detected using normal spectral analysis techniques. This will be important in later evaluation of an ARMAX modeling approach. The y_2 signal appears to change statistical characteristics near the end of the data span. This was not done intentionally. Other "nonstationary" behavior appears when using longer data spans, so statistically stationary data can appear nonstationary for short spans.

Figure 12.11 shows the true power spectrum for both measurements, computed as two times the diagonals of the 2×2 PSD matrix $\mathbf{S}_{yy}(\omega)$. The equations for computing the power spectrum from the state space model will be defined later in Section 12.5.6. The multiplier 2.0 maps all power to positive frequencies since power is evenly split between negative and positive frequencies. Notice the lightly damped peaks at 0.1 and 0.2 Hz. The effect of the 0.04 Hz control on y_2 is evident, but the total power is very small. The effect of random measurement noise is apparent as a $10\log_{10}(0.5^2/0.5\,\text{Hz}) = -3.0\,\text{dB}$ spectral "floor" for y_1, but systematic signals dominate the noise for y_2 and no floor appears.

Figure 12.12 shows the PSD computed as the average of five periodograms obtained by breaking the 500 s data into non-overlapping 100 s segments. A rectangular window function was used. Spectral peaks at 0.1 and 0.2 are readily observable, but the control signature at 0.04 Hz is barely noticeable because power is small and the variance of the spectrum is large. Figure 12.13 is a similar plot that averages ten periodograms of 50 s each. Here the spectrum is smoother. In Figure 12.14 the rectangular window is replaced with a Hamming window function: the spectral peaks are slightly more distinct, but the y_1 measurement noise does not appear as a floor. Finally Figure 12.15 shows the PSD when using the Welch approach that averages non-overlapping periodograms and applies a Bartlett triangular window. This has sharper peaks than any of the previous 10-sample periodograms. The control signal at 0.004 Hz is barely noticeable and the peak is slightly smaller than shown in the five-sample average of Figure 12.12.

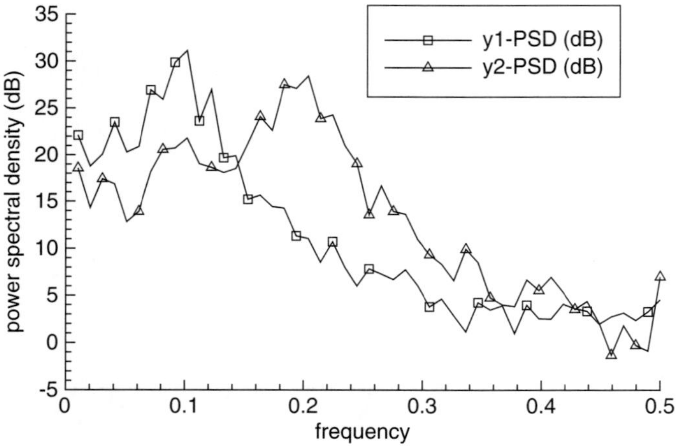

FIGURE 12.12: Five-sample average periodogram of fourth-order ARMAX system (rectangular window).

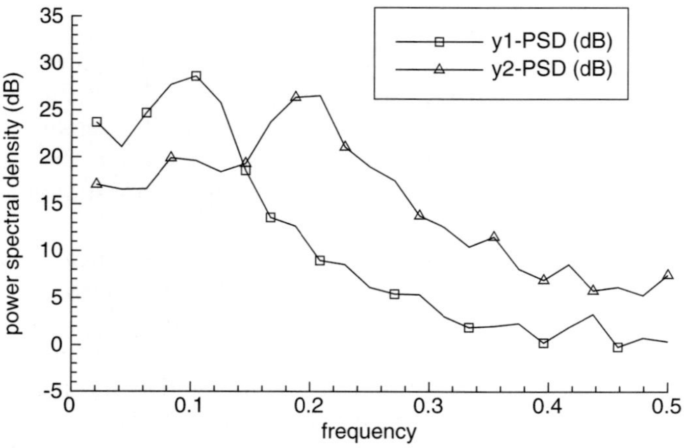

FIGURE 12.13: Ten-sample average periodogram of fourth-order ARMAX system (rectangular window).

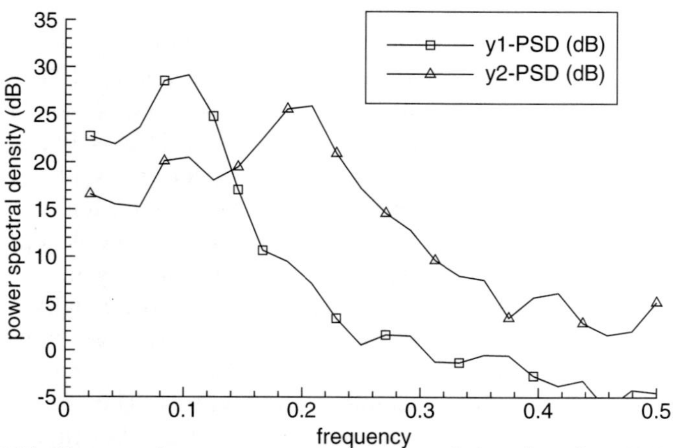

FIGURE 12.14: Ten-sample average periodogram of fourth-order ARMAX system (Hamming window).

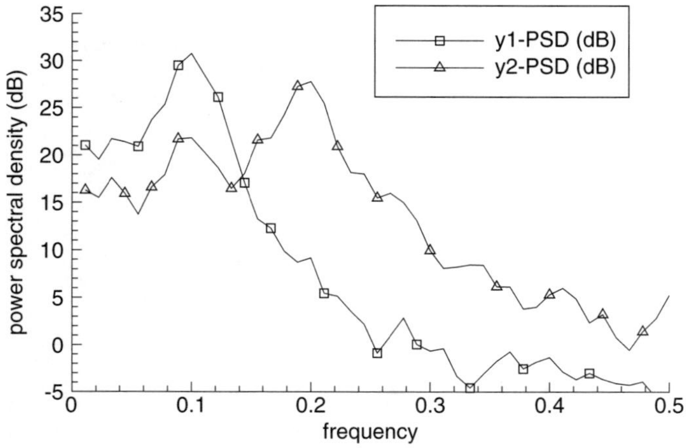

FIGURE 12.15: Nine 50% overlapping sample average (Welch) periodogram of fourth-order ARMAX system (Bartlett window).

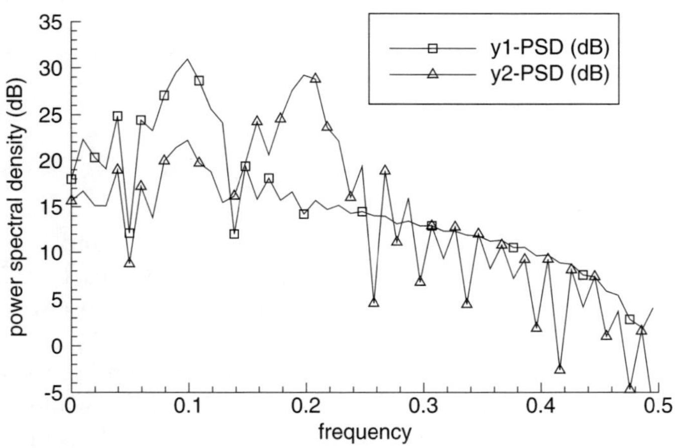

FIGURE 12.16: Blackman-Tukey PSD of fourth-order ARMAX system (nlag=n/10, Bartlett window).

Figure 12.16 shows the BT correlogram PSD when using the Bartlett window with the number of lags set to $N/10 = 50$, which is the recommended upper limit. Notice that the y_1 spectrum is fairly smooth but the y_2 spectrum is quite noisy. The control signal at 0.004 Hz appears in both measurements even though it is actually present only in y_2. Figure 12.17 is a similar plot using $N/15 = 33$ lags: the y_2 spectrum is much smoother but the control signal barely appears. Not shown is a plot with $N/5 = 100$ lags: the y_1 and y_2 spectrums are both so noisy that model identification is doubtful.

To summarize, the periodogram and correlogram approaches for computing the PSD are useful for preliminary evaluation of the model order, dominant poles, and general structure, but it would be difficult to develop accurate models

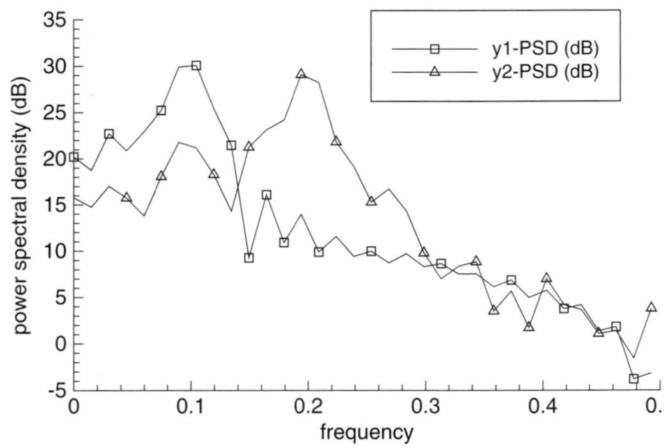

FIGURE 12.17: Blackman-Tukey PSD of fourth-order ARMAX system (nlag=n/15, Bartlett window).

from the PSD. Of the two basic approaches, the Welch periodogram using 50% overlapping segments with a Bartlett window is probably the best method, but it is barely adequate.

12.3 AUTOREGRESSIVE MODELING

The previous example demonstrated that the periodogram and BT correlogram methods for computing the PSD are not acceptable, by themselves, for empirical model development. This observation is generally true for most spectral analysis applications.

A better spectral analysis approach assumes that the time series of measured data is the output of a finite order linear system driven by white noise. Typical discrete models considered are AR, MA, ARMA, or ARMAX. Hence spectral analysis involves determination of the "ARMA" model parameters and the magnitude of the white process noise. For obvious reasons this is called a parametric approach.

Although ARMA or ARMAX processes are the most general, we first consider AR processes because many physical systems can be adequately represented as AR models. Discrete linear systems can be modeled as AR or MA processes of infinite order, although only finite-order models are useful. AR models are best suited to systems with spectral peaks (poles), while MA models are better suited to systems with spectral nulls (zeroes). Physical systems tend to have either low-pass or band-pass characteristics, with stop bands rather rare. Hence AR processes are the first choice for many systems. However, it is difficult for an AR process to model spectral nulls or additive white noise, so ARMA models are a better choice in these cases.

Recall that the output of a stationary AR process, y_i for $-\infty < i < \infty$, is represented as

$$y_i = -\sum_{k=1}^{n} \alpha_k y_{i-k} + q_i \tag{12.3-1}$$

or

$$\begin{bmatrix} y_{i-1} & y_{i-2} & \cdots & y_{i-n} \end{bmatrix} \begin{bmatrix} \alpha_1 \\ \alpha_2 \\ \vdots \\ \alpha_n \end{bmatrix} = q_i - y_i \tag{12.3-2}$$

where q_i is zero-mean white noise. This implies that $E[y_i] = 0$. If $E[y_i] \neq 0$ the mean should be removed before further analysis. In vector form the AR process is

$$\mathbf{y}_{i-1}^T \boldsymbol{\alpha} = q_i - y_i \tag{12.3-3}$$

where

$$\begin{aligned} \mathbf{y}_{i-1}^T &= [y_{i-1} \quad y_{i-2} \quad \cdots \quad y_{i-n}] \\ \boldsymbol{\alpha}^T &= [\alpha_1 \quad \alpha_2 \quad \cdots \quad \alpha_n] \end{aligned}.$$

Multiplying both sides of equation (12.3-3) by $\mathbf{y}_{i-1}$ and taking the expected value,

$$E[\mathbf{y}_{i-1}\mathbf{y}_{i-1}^T]\boldsymbol{\alpha} = E[\mathbf{y}_{i-1}(q_i - y_i)], \tag{12.3-4}$$

gives

$$\boxed{\begin{bmatrix} \rho_{yy}[0] & \rho_{yy}[-1] & \cdots & \rho_{yy}[-n+1] \\ \rho_{yy}[1] & \rho_{yy}[0] & \cdots & \rho_{yy}[-n+2] \\ \vdots & \vdots & \ddots & \vdots \\ \rho_{yy}[n-1] & \rho_{yy}[n-2] & \cdots & \rho_{yy}[0] \end{bmatrix} \boldsymbol{\alpha} = -\begin{bmatrix} \rho_{yy}[1] \\ \rho_{yy}[2] \\ \vdots \\ \rho_{yy}[n] \end{bmatrix}} \tag{12.3-5}$$

where $\rho_{yy}[j] = E[y_i y_{i+j}]$ is the autocorrelation function for a stationary process. Since $E[y_i] = 0$, $\rho_{yy}[j]$ is also the auto-covariance function. Equation (12.3-5) is called the *AR Yule-Walker(YW) normal equations* (Yule 1927; Walker 1931). Notice that the autocorrelation matrix multiplying $\boldsymbol{\alpha}$ is Toeplitz (each diagonal is constant). Further $\rho_{yy}[-j] = \rho_{yy}[j]$ for a real stationary process, or $\rho_{yy}[-j] = \rho_{yy}^*[j]$ for complex data, so only n autocorrelation lags are required and the matrix is symmetric (or Hermitian if the data are complex). Hence given $\rho_{yy}[j]$ for $0 \le j \le n - 1$, the $\boldsymbol{\alpha}$ coefficients can be computed by matrix inversion, which is acceptably fast for most problems. The noise power $\sigma_q^2 = E[q_i^2]$ can then be computed by multiplying equation (12.3-1) by y_i and taking the expected value to give

$$E[y_i^2] = -\sum_{k=1}^{n} \alpha_k E[y_i y_{i-k}] + E[y_i q_i],$$

or after rearranging,

$$\boxed{\sigma_q^2 = \rho_{yy}[0] + \sum_{k=1}^{n} \alpha_k \rho_{yy}[k].} \tag{12.3-6}$$

Vector $\boldsymbol{\alpha}$ can be calculated more efficiently using a recursive procedure developed by Levinson (1947) and Durbin (1960) for Toeplitz matrices. Ulrych and Bishop (1975) demonstrate how the a_j coefficients and σ_q^2 can be recursively computed without forming the autocorrelation matrix. Their method has another benefit that will be explained shortly.

After $\hat{\boldsymbol{\alpha}}$ is computed using the YW method or alternatives, the PSD is computed for positive frequencies as

$$\boxed{S_y(f) = \frac{2\Delta t\,\sigma_q^2}{\left|1 + \sum_{i=1}^{n} \alpha_i \,\mathrm{e}^{-j2\pi f i \Delta t}\right|^2}}. \tag{12.3-7}$$

Notice in equation (12.3-5) that the YW equations make no direct assumptions about $\rho_{yy}[j]$ for $|j| > n$, although these autocorrelation values are implicitly defined from the AR model. This contrasts with the BT correlogram where $\rho_{yy}[j] = 0$ is assumed for $|j| > n$. To minimize spectral leakage, the BT correlogram method tapers the autocorrelation function toward zero using a window function. This is neither necessary nor desirable with the YW method. Because the YW method does not use autocorrelation values for lags greater than the model order, it has the property of maximizing statistical entropy. This is explained in the next section (also see Ulrych and Bishop 1975 or Burg 1967, 1968).

12.3.1 Maximum Entropy Method (MEM)

The thermodynamic property entropy is a measure of the disorder of a system, where disorder or uncertainty is a measure of the number of possible states (such as quantized energy of individual particles) in which the system may be arranged for a given set of conditions (such as temperature and pressure) (e.g., see El-Saden 1965). Entropy of a gas tends to increase with temperature and decrease with pressure. Shannon (1948) defined information entropy as a measure of the number of possible arrangements of information (such as bits for a binary signal) in a message prior to receiving the message. More formally entropy H is the average information per time interval provided by a message, or $H = I_{total}/T$ (e.g., see Abramson 1963). When there are M possible event types, information for each type is $I_i = \log_2(1/\Pr\{i\})$, where $\Pr\{i\}$ is the probability of event i in a unit sample period. Hence when the probability of an event is small, knowledge of its occurrence provides much information. When a system is observed for a long period of time T so that multiple events of each type are detected, the expected number of type i events is $\Pr\{i\}T$. The total information about the system per unit of time is thus

$$H = \frac{I_{total}}{T} = -\sum_{i=1}^{M} \Pr\{i\} \log_2 \Pr\{i\}.$$

Entropy is a measure of the total uncertainty in all events. It is zero when the probability of one event is 1 and the probability of all other events is 0. Entropy is maximized when all events are equally likely. Smylie et al. (1973) showed that the entropy of a discrete stationary Gaussian process $y_1, y_2, \ldots, y_m, y_{m+1}$ can be computed from the definition of the multivariate Gaussian density function (eq. B2-17) and the covariance matrix. The result is

$$H = \frac{1}{2}\log\left(\det\begin{bmatrix} \rho_{yy}[0] & \rho_{yy}[1] & \cdots & \rho_{yy}[m] \\ \rho_{yy}[1] & \rho_{yy}[0] & \cdots & \rho_{yy}[m-1] \\ \vdots & \vdots & \ddots & \vdots \\ \rho_{yy}[m] & \rho_{yy}[m-1] & \cdots & \rho_{yy}[0] \end{bmatrix}\right). \tag{12.3-8}$$

Hence if $\rho_{yy}[0], \rho_{yy}[1], \ldots, \rho_{yy}[n-1]$ are exactly known for an AR model of order n, the goal of the MEM is to compute $\rho_{yy}[n], \rho_{yy}[n+1], \ldots, \rho_{yy}[m]$ such that entropy is maximized. Entropy is maximized with respect to $\rho_{yy}[n], \rho_{yy}[n+1], \ldots, \rho_{yy}[m]$ when the determinant of the autocorrelation matrix in equation (12.3-8) is zero for $m > n - 1$. That happens automatically in the YW method because $\rho_{yy}[n], \rho_{yy}[n+1], \ldots$ can be computed from $\alpha_1, \alpha_2, \ldots, \alpha_n$ using the AR definition (12.3-1); that is, $\rho_{yy}[n], \rho_{yy}[n+1], \ldots$ are a linear combination of $\rho_{yy}[0], \rho_{yy}[1], \ldots, \rho_{yy}[n-1]$. Hence the autocorrelation matrix for $m > n - 1$ is singular.

This interpretation of the AR-based power spectrum as a maximum entropy spectrum was introduced by Burg (1967, 1968), and his observations were responsible for renewing interest in the parametric AR approach. However, rather than using the YW method to compute the AR coefficients, Burg proposed an alternate procedure described in the next section.

The purpose of this discussion is to demonstrate that the YW method approach is optimal in a well-defined sense. Other optimization criteria can also be used to define algorithms for computing AR coefficients, and properties of alternate algorithms may be different from those of MEM. For example, the *maximum likelihood method* (MLM) (Capon 1969) is optimized to maximize the power in a narrow band about each computed frequency while minimizing power due to out-of-band frequencies. The Fourier transform of the MLM power spectrum does not generally match measured autocorrelation values. The MLM has less spectral resolution than MEM, but is statistically more stable because MEM is highly sensitive to the signal-to-noise ratio. Both MEM and MLM provide better spectral resolution than the periodogram or BT correlogram when using short data records. More information comparing MEM and MLE methods may be found in Kay (1988) and Marple (1987). The MEM is not the best method when the true system is ARMA or when measurement noise is present. Cadzow (1983) notes that there benefits to using an overdetermined number of YW equations to solve for ARMA parameters as a least-squares problem.

12.3.2 Burg MEM

We now discuss implementation of AR modeling methods. There are two issues to be addressed in practical implementations: the method used to compute a sample autocorrelation function and the selection of AR model order. The YW method typically computes the sample $\rho_{yy}[j]$ as the biased auto-covariance equation (12.2-44) (with the mean subtracted from the data) because the available data span is often small and the mean-squared error of $\hat{\rho}_{xx\_b}[k]$ is generally smaller than that of other estimators. Use of equation (12.2-44) also generates a positive semi-definite autocorrelation matrix.

An alternate method developed by Burg (1967) does not require initial computation of the autocorrelation function. Instead Burg recursively fits the higher order AR coefficients by summing the squared one-step prediction errors for one "prediction" filter operating forward in time and another "prediction" filter running backward in time. A prediction filter is defined as

$$\hat{y}_i = -\sum_{j=1}^{n} \alpha_i y_{i-j} \tag{12.3-9}$$

and $\tilde{y}_i = y_i - \hat{y}_i$ is the prediction error. When operating backward in time y_{i+j} replaces y_{i-j}. (Note: Burg states that a "prediction filter works equally well when reversed and run backward in time." This appears incorrect because $E[q_i y_{i+j}] \neq 0$ for an AR model. However, the same correlation coefficients apply in reverse time, so the statement is true.)

The n-step recursive procedure starts with step $j = 1$ to compute $\hat{\alpha}_1$ minimizing the variance of the prediction error; that is, it is a least-squares solution for α_1. This process continues for additional steps j and the computed $\hat{\alpha}_{j-1}, \hat{\alpha}_{j-2}, \ldots$ for past steps are recalculated using the Levinson recursion relationship, not by estimating all previous $\hat{\alpha}_k$ as part of the least-squares problem. Use of the recursion guarantees that all poles of the AR model are within the unit circle (i.e., the system is stable). Unlike the YW method, the Burg method effectively gives smaller weighting to data near the ends without assuming that the y_i values are zero outside the given data span. Ulrych and Bishop (1975) list code for an efficient implementation of Burg's method, including optimal order selection. (See MEMOPT.F and MEMPR.F in the software associated with this book.) Marple (1987) also provides code for this algorithm.

The second problem, selection of model order, is addressed in the next section.

12.3.3 Final Prediction Error (FPE) and Akaike Information Criteria (AIC)

The primary problem in early applications of MEM methods was selection of AR order. If the order is too low the spectrum lacks resolution, and if the order is too high the model fits noise. This problem is addressed using a method developed by Akaike (1969, 1970a, 1970b). The optimal model order is chosen as the one that minimizes the FPE for an AR process. The FPE is defined as

$$FPE = E[(y_i - \hat{y}_i)^2] \tag{12.3-10}$$

where $\hat{y}_i$ is a prediction of y_i based on past values $y_{i-1}, y_{i-2}, \ldots$. For the AR model,

$$\hat{y}_i = -\sum_{j=1}^{n} \hat{\alpha}_j y_{i-j} \tag{12.3-11}$$

where the $\hat{\alpha}_j$ are computed as the least mean-squared error solution for an AR model of order n. The difference of $\hat{\alpha}_j$ from the true α_j is designated as $\tilde{\alpha}_j = \alpha_j - \hat{\alpha}_j$, with $E[\tilde{\alpha}_j] = 0$. For purposes of the derivation, Akaike wanted the errors $\tilde{\alpha}_j$ to be independent of $y_{i-1}, y_{i-2}, \ldots, y_{i-n}$, so he assumed that the $\hat{\alpha}_j$ were computed from a different realization of the process. The alternate realizations, designated as $x_{i-1}, x_{i-2}, \ldots, x_{i-n}$, are assumed to have the same statistical properties as $y_{i-1}, y_{i-2}, \ldots, y_{i-n}$. Note that $E[\tilde{\alpha}_j \alpha_k] = 0$ because α_j is not a random variable, and $E[\tilde{\alpha}_j y_i] = 0$ because $\hat{\alpha}_j$ is computed from a different data set. Thus the FPE is computed as

$$
\begin{aligned}
FPE_n &= E\left[(y_i + \sum_{j=1}^{n} \hat{\alpha}_j y_{i-j})^2 \right] \\
&= E\left[(y_i + \sum_{j=1}^{n} \alpha_j y_{i-j} - \sum_{j=1}^{n} \tilde{\alpha}_j y_{i-j})^2 \right] \\
&= E\left[(y_i + \sum_{j=1}^{n} \alpha_j y_{i-j})^2 - 2\left(y_i \sum_{j=1}^{n} \tilde{\alpha}_j y_{i-j} + \sum_{j=1}^{n}\sum_{k=1}^{n} \tilde{\alpha}_k \alpha_j y_{i-k} y_{i-j} \right) + \left(\sum_{j=1}^{n} \tilde{\alpha}_j y_{i-j} \right)^2 \right]. \\
&= E\left[(y_i + \sum_{j=1}^{n} \alpha_j y_{i-j})^2 + \sum_{j=1}^{n}\sum_{k=1}^{n} \tilde{\alpha}_k \tilde{\alpha}_j y_{i-k} y_{i-j} \right] \\
&= S_n^2 + \sum_{j=1}^{n}\sum_{k=1}^{n} E[\tilde{\alpha}_j \tilde{\alpha}_j] E[y_{i-j} y_{i-k}]
\end{aligned}
\tag{12.3-12}
$$

The FPE consists of two terms: the minimum residual sum-of-squares,

$$S_n^2 = (y_i + \sum_{j=1}^{n} \alpha_j y_{i-j})^2,$$

and another term that includes the statistical deviation of $\hat{\alpha}_j$ from α_j. Generally S_n^2 decreases with increasing order n while the second term increases with n. For an n-th order fit, Akaike defines an efficient estimate of the FPE as

$$FPE_n = \left(1 + \frac{n+1}{m} \right) \sigma_q^2 \tag{12.3-13}$$

where m is the total number of measurements. Akaike showed that σ_q^2 can be estimated from S_n^2 as

$$\hat{\sigma}_q^2 = \left(\frac{m}{m-(n+1)} \right) S_n^2,$$

so the FPE estimate is

$$\boxed{FPE_n = \left(\frac{m+n+1}{m-(n+1)} \right) S_n^2 .} \tag{12.3-14}$$

This is evaluated for successfully higher orders up to a limit, and the optimal order is selected as the one with minimum FPE_n. This algorithm is included in Ulrych and Bishop's code, and also in software associated with this book.

Jones (1976) notes that the bias in the FPE of the YW method increases with AR order, which tends to cause selection of order lower than actual. The Burg FPE is unbiased for $n < m/2$, but is then biased downward. This can be corrected using a modified FPE calculation:

$$\boxed{FPE_n = \begin{cases} \dfrac{m+n+1}{m-n-1} S_n^2, & 1 < n \le m/2 \\ \dfrac{(m+n+1)\, S_n^2}{(m-n-1)(1.5-n/m)}, & m/2 < n \le m-1 \end{cases}.} \tag{12.3-15}$$

Although Akaike derived the FPE assuming that the $\hat{\alpha}_j$ were computed using a data set statistically independent of the data used to compute the FPE, in practice the same data are often used for both purposes. When $m \gg n$ the error in the FPE due to the correlation may be small, but the correlation can be significant for small data sets.

Based on information theoretic and entropy concepts, Akaike (1974b) later developed the more general AIC. The AIC uses the likelihood function as a more sensitive indicator of model error, where the model is not necessarily restricted to AR. The AIC is defined as

$$AIC_n = \sum_{i=1}^{m} (\ln(2\pi) + \ln|\mathbf{P}_{y\_i}(n)|) + 2n \tag{12.3-16}$$

where n is the number of free parameters in the model, m is the number of measurements, and $\mathbf{P}_{y\_i}(n)$ is the prediction error covariance of measurement i for the model with n parameters. The $\ln(2\pi)$ term is often ignored because it is fixed for a given number of measurements. As with the FPE, the model with smallest AIC is selected as optimal. Kay (1988) notes that the AIC may be a better indicator of model order than the FPE for short data records. However, others have noted that neither the FPE nor AIC work well for short data spans. Both metrics also tend to underestimate the order of harmonic processes, although that is not a serious problem when modeling physical systems. The AIC is often used to select the order of ARMA models, although the $2n$ component is sometimes replaced with a factor that increases faster because equation (12.3-16) tends to overestimate model order. Rissanen (1983) derived an alternate *minimum description length* metric that replaces the $2n$ term of the AIC with $n\ln(m)$.

12.3.4 Marple AR Spectral Analysis

A number of alternative AR modeling methods are available. Marple (1980) developed an AR spectral analysis approach that, as with Burg's method, operates one prediction filter forward in time and another backward in time. It is also has computational efficiency comparable to the Burg method. However, Marple computes the AR parameters as a least-squares problem without the Levinson recursion constraint, and the model order is determined using prediction error tolerances (rather than the FPE). In comparing this approach to the Burg and YW methods, Marple (and others) demonstrate that the algorithm reduces bias in frequency estimates of spectral components, reduces variance in frequency estimates over an ensemble of spectra, and does not "split" spectral lines when working with very narrow band signals.

The primary problem in using the algorithm comes in setting two tolerance inputs that control stopping of the internal model order calculations. One tolerance, "*tol1*" stops the iterations when the prediction error variance is smaller than "*tol1*" times the data variance. The second tolerance, "*tol2*", stops iterations when the change in prediction error variances is smaller than "*tol2*" times the current prediction error variance. Marple recommends that both tolerances be set in the range 0.01 to 0.001. This author found that the algorithm works best on the given ARMA example when *tol1* = 0.01 and *tol2* = 0.002, although *tol1* never terminated the iterations for this data.

Code for this algorithm is listed by Marple (1980, 1987). Subroutine MARPLE_SPECT.F, included in software associated with this book, computes power spectra using Marple's algorithm.

12.3.5 Summary of MEM Modeling Approaches

Experience with numerous examples has demonstrated the superiority of the YW, Burg, and Marple approaches compared with the periodogram and BT correlogram methods. However, there are still potential problems with MEM. Ulrych and Bishop (1975) make the following observations about the YW and Burg methods:

1. YW estimates of the AR coefficients are very sensitive to rounding errors.
2. Use of the biased autocorrelation function equation (12.2-44) assumes that $x_i = 0$ for $i > m$ (or $i > N - 1$ in the notation of equation (12.2-44)), which contradicts the principle of maximum entropy. The Burg method makes no such assumptions.
3. For short time series, the variance of the autocorrelation computed using the Burg MEM is greater than that of the YW estimate and does not decrease to zero with increasing lag. However, the resolution properties of the Burg method are much superior.
4. The variance of the Burg FPE is greater than that of the YW FPE in all cases and a cutoff of $n < m/2$ should be imposed on the Burg method. The YW FPE is biased toward lower orders for high-order AR processes.
5. The Burg method generally gives the correct FPE for AR processes, but tends to overestimate the order for ARMA processes. In these cases use of the first FPE minimum gives more consistent results than the global minimum. (This author has observed the problem.)
6. The width and peak of a MEM spectral line may have considerable variance because the estimate is actually a spectral density estimate.
7. Processes with sharp spectral lines often do not exhibit a clear minimum in the FPE, and the AR order should be limited to $n < m/2$.

Another problem of the Burg and YW methods, demonstrated by Marple (1987, section 8.6), is the tendency to split a sharp spectral peak into two peaks. Again this is not a serious problem when modeling physical systems. Marple's modified covariance method does not have this tendency.

Despite these problems, MEM is a very useful tool and this author often uses Burg's method when analyzing time series data. However, do not be surprised if it is necessary to restrict the model order range when using these approaches.

Example 12.2: Fourth-Order ARMAX Data Using AR Models

The same data of Example 12.1 is used to evaluate the MEM methods. The algorithms and code are those of Ulrych and Bishop, and Marple. Figure 12.18 shows the PSD computed using Burg's method. The minimum FPE AR model orders are 5 for y_1 and 16 for y_2. Notice that the PSD estimate for y_1 is a fairly close match to the true PSD of Figure 12.11, except that the 0.1 Hz peak is slightly higher and narrower. The total spectral power of y_1 is almost exactly equal to the actual variance of y_1. The PSD for y_2 is has many extra wiggles not present in the true PSD. The wiggles are due to over-fitting the fourth-order ARMA model with a 16th-order AR model. (More will be said about model order in a following paragraph.) The AR models missed the effect in y_2 of the exogenous "control" input at 0.04 Hz because this sinusoid is not included in the AR model, and total power of the signal is small. It would have been more accurate to use an autoregressive with exogenous inputs (ARX) model for this input signal, but the results are almost unchanged when the 0.04 Hz signal is removed from y_2. Again the integrated spectral power of y_2 is almost exactly equal to the actual variance.

Figure 12.19 shows the PSD computed using the YW method. Both the model orders and the PSD are essentially equal to those of the Burg MEM method. The model orders computed using Marple's method are 6 for y_1 and 16 for y_2, but the PSDs of all three methods appear nearly identical within the resolution of the plots.

To better understand why a 16th-order model (with all the wiggles) was chosen as optimal for the y_2 data, the simulation was executed ten times using different seeds for the random number generator. This was done both with and without the 0.04 Hz control input. In all cases the Burg and YW methods selected model orders from 6 to 9 for y_1. Marple's method usually chose an order of 10 or less, except for two cases that used 14 and 16. Since Marple's order selection algorithm is significantly different from the FPE, it is not surprising that it occa-

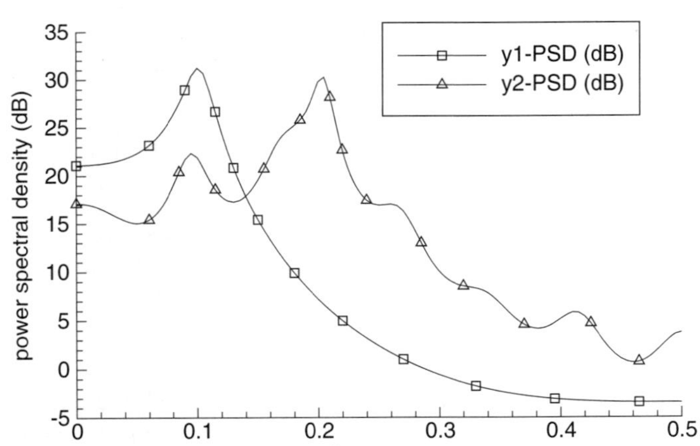

FIGURE 12.18: Burg MEM (orders 5, 16) PSD of fourth-order ARMAX system.

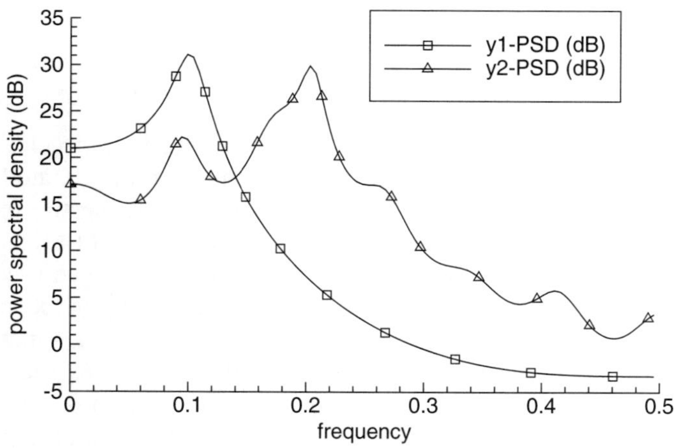

FIGURE 12.19: Yule-Walker MEM (orders 5, 16) PSD of fourth-order ARMAX system.

sionally selects a different model order. For approximately 60% of the cases, the selected y_2 model order used by all methods was in the range 5 to 10. As expected, the PSD plots for these cases are much smoother than shown above, and better match the idealized PSD of the simulation. Selected model orders for other cases were in the range of 12 to 17, with one exception: the minimum FPE of the Burg method for one particular seed occurred at order 51, with the minimums for the YW and Marple methods at orders 5 and 13, respectively. The PSD computed using the order 51 AR model bears little similarity to the true spectrum. When the 0.04 Hz control input was removed from the data, the selected orders were 7, 6, and 13 for the Burg, YW, and Marple methods, respectively. This shows that Burg's method with the FPE algorithm can occasionally be very sensitive to non-ideal signatures in the data. In most cases, however, the model order and PSD are nearly unchanged when the 0.04 Hz control input is removed.

To summarize, the AR methods work well for computing the PSD of the y_1 ARMAX process, but a combination of unmodeled effects (wide-band measurement noise, sinusoidal input, and zeroes in the ARMA transfer function) sometimes cause selection of an unrealistically high model order for y_2. All three methods exhibited this behavior, but Burg and Marple methods showed somewhat greater variability in model order selection.

12.4 ARMA MODELING

While AR modeling methods can be successful in computing power spectra for simple low-order systems, they tend to overestimate the true model order for many real systems, as demonstrated in Example 12.2. They also cannot model a pole-only system with additive white measurement noise such as

$$y_i = -\sum_{k=1}^{n} \alpha_k y_{i-k} + q_i$$

$$z_i = y_i + r_i$$

where z_i is the measurement and r_i is white noise. The tank acceleration models discussed in Section 3.2.1 demonstrate another limitation of AR models. Here the pole-only models were not completely successful in modeling the more dynamic Twister and Scout tank accelerations. When zeros were added to the second-order transfer functions, Kalman filter tracking of the motion improved greatly.

Although MA models can also be used for computing power spectra, we do not discuss them further because transfer function zeroes are a poor match to behavior of physical systems. The ARMA or ARMAX models are the most general form of discrete models, and they can be directly used in a Kalman filter, if desired. Many different methods have been developed to estimate parameters of ARMA models. We describe several of these but concentrate on one technique—*canonical variate analysis* (CVA)—because it is one of the most general. Furthermore, the experience of this author and others has demonstrated the superiority of CVA for some problems. Prior to presenting the CVA method (in Section 12.5), we describe the ARMA model and briefly summarize a few methods for determining parameters.

12.4.1 ARMA Parameter Estimation

An ARMA process is defined as

$$y_i = -\sum_{k=1}^{n} \alpha_k y_{i-k} + \sum_{k=0}^{l} \beta_k q_{i-k}. \tag{12.4-1}$$

Multiplying both sides of equation (12.4-1) by y_{i-j} and taking the expected value for the assumed stationary process yields

$$\rho_{yy}[j] = -\sum_{k=1}^{n} \alpha_k \rho_{yy}[j-k] + \sum_{k=0}^{l} \beta_k \rho_{qy}[j-k] \tag{12.4-2}$$

where $\rho_{qy}[k] = E[q_i y_{i-k}]$. Since $\rho_{qy}[j-k] = 0$ for $j > k$, equation (12.4-2) may be written in matrix form as

$$\begin{bmatrix} \rho_{yy}[l] & \rho_{yy}[l-1] & \cdots & \rho_{yy}[l-n] \\ \rho_{yy}[l+1] & \rho_{yy}[l] & \cdots & \rho_{yy}[l-n+1] \\ \vdots & \vdots & \ddots & \vdots \\ \rho_{yy}[l+n] & \rho_{yy}[l+n-1] & \cdots & \rho_{yy}[l] \end{bmatrix} \begin{bmatrix} \alpha_1 \\ \alpha_2 \\ \vdots \\ \alpha_n \end{bmatrix} = - \begin{bmatrix} \rho_{yy}[l+1] \\ \rho_{yy}[l+2] \\ \vdots \\ \rho_{yy}[l+n] \end{bmatrix}, \tag{12.4-3}$$

These are the *ARMA YW normal equations*. Given the autocorrelation values $\rho_{yy}[l]$, $\rho_{yy}[l+1], \ldots, \rho_{yy}[l+n]$, equation (12.4-3) may be solved directly for the AR parameters $\alpha_1, \alpha_2, \ldots, \alpha_n$. A modified version of the Levinson-Durbin recursion can be used to obtain the solution (see Marple 1987, section 8.5).

Solution for the MA coefficients is more difficult. Equation (12.4-2) may be rewritten using the causal system impulse response function, $h[k]$, which defines the system output due to an impulse at $k = 0$ and is, of course, a function of $\boldsymbol{\alpha}$ and $\boldsymbol{\beta}$. Using $h[k]$, the system response to the noise sequence q_i is

$$y_i = \sum_{k=0}^{\infty} h[k] q_{i-k}. \tag{12.4-4}$$

This is used to compute

$$\rho_{yq}[j] = \begin{cases} 0 & \text{for } j > 0 \\ \sigma_q^2 h[0] & \text{for } j = 0 \\ \sigma_q^2\, h[-j] & \text{for } j < 0 \end{cases} \tag{12.4-5}$$

where again $\sigma_q^2 = E[q_i^2]$. Thus

$$\rho_{yy}[j] = \begin{cases} \rho_{yy}[-j] & \text{for } j < 0 \\ -\sum_{k=1}^{n} \alpha_k \rho_{yy}[j-l] + \sigma_q^2 \sum_{k=j}^{l} \beta_k\, h[k-j] & \text{for } 0 \le j \le l. \\ -\sum_{k=1}^{n} \alpha_k \rho_{yy}[j-l] & \text{for } j > l \end{cases} \tag{12.4-6}$$

Because the β_k parameters are convolved with the impulse response $h[k]$, the relationship between the autocorrelation $\rho_{yy}[j]$ and β_k is nonlinear. Hence a direct solution for the MA parameters is not possible. Of course iterative algorithms can be used. General information on ARMA modeling methods may be found in Marple (1987), Kay (1988), Brockwell and Davis (1991), Shumway and Stoffer (2006), Box et al. (2008), Ljung (1999, 2002) and Levine (1996, chapter 58).

Methods used for ARMA parameter estimation include the following:

1. AR and MA parameters can both be computed as a single nonlinear problem using an approximate maximum likelihood or minimum variance approach. The highly nonlinear equations must be solved iteratively; for example, using Newton iterations. This approach requires a good initial estimate of the parameters and the order, so an alternate technique is also required. The iterations may not always converge or may not converge to the global minimum. When convergence is achieved for different orders, the proper AR and MA orders are determined using the AIC.
2. The AR and MA parameters are computed as separate problems. Usually a sample autocorrelation function is first computed and used in the ARMA (extended) YW equations (12.4-3) to solve for the $\boldsymbol{\alpha}$ parameters. The data signal is then passed through an inverse filter that removes the effects of the AR model poles. The output can be modeled as an MA process, and several approximate techniques can be used to compute the MA coefficients. One method fits a high-order AR model to the signal and then computes the MA model that best matches the AR model. These separate solution techniques have several problems. The results depend heavily on the assumed model orders, and the model orders should not be larger than the true orders. Use of the ARMA YW equations requires knowledge of the MA order, or at least an upper limit on the order. Further the variance of AR estimates derived from the ARMA YW equations is generally poor where the signal PSD is

small. Somewhat better results may be obtained using an overdetermined set of correlation equations, where the AR parameters are computed as the solution to a least-squares problem.

3. Estimate a high-order AR model and convert it to a lower order ARMA model. This works because ARMA models can be expanded as infinite-order AR models; for example, the z-transform of the finite-order (l,n) ARMA model, $B_l(z)/C_n(z)$, is equivalent to the transform of the infinite-order AR model $1/A_\infty(z)$. The MA parameters are determined by first solving simultaneous linear equations, and then the AR parameters of the ARMA model are computed as a convolution using the MA parameters. Marple (1987, section 6.4) provides details of the method. The approach requires knowledge of the model orders, so different AR and MA orders must be tested and compared using the AIC. Since this method starts with an AR model, the ARMA coefficient estimates may be poor because the variance of the AR estimates can be high.
4. Use Cadzow's (1980, 1983) "high-performance spectral estimator." This is an example of newer techniques that utilize singular value decomposition (SVD). The approach starts by forming a sample nonsymmetric, overdetermined autocorrelation matrix where the model order is greater than the true AR or MA orders. The correlation matrix is rectangular because the number of future lags is large, and nonsymmetric because autocorrelations for past and future lags are not numerically equal. The rank of the autocorrelation matrix is determined using SVD, and the matrix is reconstructed using the non-negligible singular values. The AR parameters are computed using the reconstructed autocorrelation matrix, and MA parameters are computed using the estimated AR parameters and the reconstructed autocorrelation matrix. Prasad and Hari (1987) note that better results are obtained when the reconstructed autocorrelation matrix is forced to be Toeplitz. They also note that the performance of the CVA method is a significant improvement over Cadzow's method when resolving sinusoidal signals in white noise.
5. Iteratively select orthogonal states that maximally reduce the squared measurement prediction error at each step (Korenberg 1985, 1988, 1989). This method also computes past-past and past-future correlation matrices. An orthogonal decomposition of the past-past correlation matrix is computed, and an initial state vector is selected as the orthogonal component with maximum effect on reducing the squared prediction error. The orthogonal state is then deleted from the candidate set and the process is repeated until there is little change in the prediction error. Wu et al. (1999) compare this approach with CVA for nonlinear ARMA models.

12.5 CANONICAL VARIATE ANALYSIS

CVA was first proposed by Akaike (1974a, 1975, 1976), and greatly extended by Larimore (1983, 1987, 1988, 1990a,b, 1992a, 1999, 2004). It differs from many other ARMA methods in directly producing a state-space model appropriate for use in a

Kalman filter. Like Cadzow's method, it depends on the SVD of past-future autocorrelations, but the matrices are formed differently. Also AIC is used when determining CVA model order. This author became convinced of the advantages of CVA when attempting to predict future flexure motion of a tank gun tube as the tank traversed rough terrain. Although the gun tube has well-defined flexure modes and resonances, the quasi-random effects of the terrain and interaction with the tank chassis, suspension, and gun controller make prediction of future motion during projectile transit quite difficult. Several techniques—including maximum likelihood parameter estimation as described in Chapter 11—were tested and CVA was clearly superior (Gibbs and Cantor 1986).

A method called Numerical Subspace State Space System ID (N4SID) is similar to CVA and is implemented in the MATLAB System Identification Toolkit. Larimore (1999) compared some features of the ADAPTx CVA implementation with N4SID and other packages. The advantages of ADAPTx are particularly evident when inputs are nonwhite and/or feedback from outputs to inputs (such as control feedback) is present. Dahlen et al. (1998) studied CVA and the three N4SID algorithms described by Van Overschee and De Moor (1994), and have shown (using simulated data) that the N4SID algorithms can produce covariance sequences that are not positive semi-definite. This could be a reliability issue for N4SID in critical applications. Dahlen (2001) also analyzed the CVA algorithm and concluded that it will always produce positive-semi-definite covariances.

The following description of CVA is based on the various papers by Larimore, Akaike, Wu et al. (1999), and Prasad and Hari (1987). This development mostly documents the original (1983) algorithm of Larimore that was rediscovered by Deistler et al. (1995). For large samples the algorithm asymptotically produces unbiased and minimum variance parameter estimates when the only inputs are white noise (see Bauer and Ljung 2002; Bauer 2005). When colored noise inputs and/or feedback from outputs to inputs are present, the CVA algorithm should be modified to include an initial step that removes the effects of future inputs on future outputs. This is done in ADAPTx by initially fitting an ARX model, and then using the computed model to correct the future outputs. Larimore (1996a,b) demonstrates the large sample behavior of the modified algorithm using simulated data. Large sample theory and the new CVA implementation are developed in Larimore (2004). Chiuso (2007) discusses the asymptotic equivalence of the SSARX algorithm of Jansson (2003) and a "Future Corrected-Canonical Correlation Analysis" (FC-CCA) algorithm that appears to be Larimore's method.

In addition to complicating parameter estimation algorithms, control feedback has another effect that adversely affects system identification: control signals are designed to suppress certain frequency bands of the system, and this reduces observability of model parameters. The issues are discussed in Ljung et al. (1974), Ljung (1999, chapter 13), Åström and Eykhoff (1971), and Isermann (1980). In the power plant modeling problem of Chapter 3, the plant was first allowed to reach steady-state conditions using control feedback. Then it was operated open loop with symmetric pulsing of individual controls. This provided the needed pulse response without causing instability or creating problems associated with control feedback.

12.5.1 CVA Derivation and Overview

A state-space model is not required for the CVA derivation, but the implementation of Larimore is based on the most general form of an ARMAX model:

$$\begin{aligned} \mathbf{y}_i &= \mathbf{H}\mathbf{x}_i + \mathbf{A}\,\mathbf{u}_i + \mathbf{B}\mathbf{q}_i + \mathbf{r}_i \\ \mathbf{x}_{i+1} &= \mathbf{\Phi}\mathbf{x}_i + \mathbf{G}\mathbf{u}_i + \mathbf{q}_i \end{aligned}. \tag{12.5-1}$$

This is the same model used in Example 12.1. As before, $\mathbf{y}_i$ is the m-element vector of measurements at time t_i, $\mathbf{x}_i$ is the n-element state vector, $\mathbf{u}_i$ is an l-element exogenous input (control) vector that may or may not be present, $\mathbf{q}_i$ is $N(\mathbf{0},\mathbf{Q})$ white process noise, $\mathbf{r}_i$ is $N(\mathbf{0},\mathbf{R})$ white measurement noise, and $E[\mathbf{q}_i\mathbf{r}_i^T]=\mathbf{0}$. The available time span is defined by $1 \le i \le M$.

Although $\mathbf{q}_i$ and $\mathbf{r}_i$ are assumed zero mean, $\mathbf{u}_i$ may be biased and trended. If this is true, $\mathbf{y}_i$ and $\mathbf{x}_i$ may also have biases and trends. There are three possible approaches for handling biases and trends, which we generically refer to as "biases." They may be subtracted from the measurements and controls before processing with CVA. This is a safe approach but requires extra work. Also the correlations between biases and measurements will not be modeled in CVA. Assuming that biases appear in $\mathbf{u}_i$, the second approach simply uses CVA unaltered. CVA will detect the correlation between bias and measurements, but it has a tendency to overestimate model order in this case. The third approach models the biases directly in the CVA algorithm, as done in ADAPTx. This requires extensive enhancements to the code and may still result in overestimation of model order. For purposes of the discussion to follow, we assume that biases have been removed from the measurements and controls so that all sample correlation matrices are actually covariances.

The goal of CVA is to find the simplest (canonical) relationship between past values of measurements $\mathbf{y}_{i-1}, \mathbf{y}_{i-2},\ldots$ and controls $\mathbf{u}_{i-1}, \mathbf{u}_{i-2},\ldots$ $\mathbf{u}_i$ that allows accurate prediction of future measurements $\mathbf{y}_i, \mathbf{y}_{i+1},\ldots$. This is stated more formally by first defining past and future input and output vectors as follows:

$$\mathbf{p}_i \triangleq \begin{bmatrix} \mathbf{y}_{i-1}^T & \mathbf{u}_{i-1}^T & \mathbf{y}_{i-2}^T & \mathbf{u}_{i-2}^T & \cdots & \mathbf{y}_{i-L-1}^T & \mathbf{u}_{i-L-1}^T \end{bmatrix}^T \tag{12.5-2}$$

$$\mathbf{f}_i \triangleq \begin{bmatrix} \mathbf{y}_{i+L}^T & \mathbf{y}_{i+L-1}^T & \cdots & \mathbf{y}_i^T \end{bmatrix}^T. \tag{12.5-3}$$

The $\mathbf{p}_i$ and $\mathbf{f}_i$ vectors shift at every time update. The number of lags carried in the vectors is denoted as L. Notice that the "past" vector includes both output measurements and input controls while "future" only includes measurements. The total length of the past and future vectors are $L(m + l)$ and Lm, respectively. L must be chosen so that Lm is greater than the true state order. For simplicity it is assumed that L is the same for past and future vectors. However, Larimore notes that when the total number of data points M is small and l/m is significant (e.g., >0.2), it may be beneficial to use more lags for the future than past so that weighting of future is closer to that of the past. This is less of a problem when L and m are both large.

CVA attempts to find a finite-span linear weighting of the past (matrix $\mathbf{F}$) that predicts the future as $\hat{\mathbf{f}}_i = \mathbf{F}\,\mathbf{p}_i$. The use of a finite-past time span may seem inconsistent with the infinite-impulse-response (IIR) state-space model equation (12.5-1), but Larimore (1987) shows that for bounded conditional expectations of past and future on a finite time window, the optimal predictor of $\mathbf{y}_i$ can be approximated arbitrarily closely by a state affine system.

The prediction matrix $\mathbf{F}$ is selected to minimize the normalized squared prediction error

$$\varepsilon^2 = (\mathbf{f}_i - \hat{\mathbf{f}}_i)^T \mathbf{R}_{ff}^{-1} (\mathbf{f}_i - \hat{\mathbf{f}}_i), \tag{12.5-4}$$

where the weighting is the inverse of the correlation matrix $\mathbf{R}_{ff} = E\left[\mathbf{f}_i\mathbf{f}_i^T\right]$. The prediction matrix $\mathbf{F}$ should also represent the "simplest" model consistent with the data so that it is modeling the actual system rather than measurement noise. The minimum variance prediction of $\mathbf{f}_i$ is obtained by taking the expected value of ε^2 and rearranging using the trace operator:

$$\begin{aligned} E[\varepsilon^2] &= E\left[tr\left\{\mathbf{R}_{ff}^{-1}(\mathbf{f}_i - \mathbf{F}\mathbf{p}_i)(\mathbf{f}_i - \mathbf{F}\mathbf{p}_i)^T\right\}\right] \\ &= tr\left\{\mathbf{R}_{ff}^{-1}(\mathbf{R}_{ff} - \mathbf{F}\mathbf{R}_{pf} - \mathbf{R}_{fp}\mathbf{F}^T + \mathbf{F}\mathbf{R}_{pp}\mathbf{F}^T\right\} \end{aligned}$$

where $\mathbf{R}_{fp} = E\left[\mathbf{f}_i\mathbf{p}_i^T\right]$ and $\mathbf{R}_{pp} = E\left[\mathbf{p}_i\mathbf{p}_i^T\right]$. $E[\varepsilon^2]$ is minimized with $\mathbf{F} = \mathbf{R}_{fp}\mathbf{R}_{pp}^{-1}$ or

$$\boxed{\hat{\mathbf{f}}_i = \mathbf{R}_{fp}\mathbf{R}_{pp}^{-1}\,\mathbf{p}_i\,.} \tag{12.5-5}$$

CVA decomposes this prediction in two steps. A canonical state n-element vector is first defined as a nonsingular transformation on the past,

$$\boxed{\mathbf{x}_i = \mathbf{J}\,\mathbf{p}_i\,,} \tag{12.5-6}$$

where

$$E[\mathbf{x}_i\mathbf{x}_i^T] = \mathbf{J}\mathbf{R}_{pp}\mathbf{J}^T = \mathbf{I}_{n\times n} \tag{12.5-7}$$

and the maximum possible state dimension is $n = L(m + l)$. Hence $\mathbf{J}$ must be determined so that the correlation of $\mathbf{x}_i$ is the identity matrix. Similarly, the future vector is mapped using a nonsingular transformation to another canonical vector $\mathbf{d}_i$ with similar properties:

$$\boxed{\mathbf{d}_i = \mathbf{L}\mathbf{f}_i\,,\quad E[\mathbf{d}_i\mathbf{d}_i^T] = \mathbf{L}\mathbf{R}_{ff}\mathbf{L}^T = \mathbf{I}_{mL\times mL}} \tag{12.5-8}$$

where matrix $\mathbf{L}$ must also be determined. An additional constraint is also imposed on $\mathbf{J}$ and $\mathbf{L}$: the correlation matrix between $\mathbf{x}_i$ and $\mathbf{d}_i$ must be diagonal:

$$\mathbf{R}_{xd} = E[\mathbf{x}_i\mathbf{d}_i^T] = \mathbf{J}\mathbf{R}_{pf}\mathbf{L}^T = \mathbf{\Lambda} \tag{12.5-9}$$

where

$$\mathbf{\Lambda} = \begin{bmatrix} \lambda_1 & 0 & 0 & 0 \\ 0 & \lambda_2 & 0 & 0 \\ \vdots & \vdots & \ddots & \vdots \\ 0 & 0 & \cdots & \lambda_n \\ \vdots & \vdots & \ddots & \vdots \\ 0 & 0 & \cdots & 0 \end{bmatrix}$$

and $\lambda_1 \geq \lambda_2 \geq \cdots \geq \lambda_n$. The λ_j are called canonical correlations. $\mathbf{R}_{xd} = \mathbf{\Lambda}$ is not square because past and future vectors are of different lengths when controls are present.

Since the transformations **J** and **L** are square and nonsingular, we can write $\mathbf{p}_i = \mathbf{J}^{-1}\mathbf{x}_i$ and $\mathbf{f}_i = \mathbf{L}^{-1}\mathbf{d}_i$. This leads to

$$\mathbf{R}_{pp} = \mathbf{J}^{-1}\mathbf{J}^{-T}, \quad \mathbf{R}_{ff} = \mathbf{L}^{-1}\mathbf{L}^{-T} \tag{12.5-10}$$

$$\mathbf{R}_{pf} = \mathbf{J}^{-1}\mathbf{\Lambda}\mathbf{L}^{-T}, \quad \mathbf{R}_{fp} = \mathbf{L}^{-1}\mathbf{\Lambda}^T\mathbf{J}^{-T}. \tag{12.5-11}$$

Hence from equation (12.5-5) the optimal prediction of the future is

$$\boxed{\hat{\mathbf{f}}_i = \mathbf{R}_{fp}\mathbf{R}_{pp}^{-1}\mathbf{p}_i = \mathbf{L}^{-1}\mathbf{\Lambda}^T\mathbf{J}\mathbf{p}_i = \mathbf{L}^{-1}\mathbf{\Lambda}^T\mathbf{x}_i} \tag{12.5-12}$$

and the expected prediction error variance is

$$\begin{aligned} E[\varepsilon^2] &= tr\{\mathbf{R}_{ff}^{-1}(\mathbf{R}_{ff} - \mathbf{R}_{fp}\mathbf{R}_{pp}^{-1}\mathbf{R}_{pf}\} \\ &= tr\{\mathbf{I} - \mathbf{L}^T\mathbf{L}\mathbf{L}^{-1}\mathbf{\Lambda}^T\mathbf{J}^{-T}\mathbf{J}^T\mathbf{J}\mathbf{J}^{-1}\mathbf{\Lambda}\mathbf{L}^{-T}\} \\ &= Lm - tr\{\mathbf{L}^T\mathbf{\Lambda}^T\mathbf{\Lambda}\mathbf{L}^{-T}\} \\ &= Lm - \sum_{k=1}^{Lm}\lambda_k^2 \end{aligned} \quad . \tag{12.5-13}$$

Thus the minimum prediction error variance is obtained by maximizing the canonical correlations between past and future. Since the correlation function must eventually decay with increasing lag for a Markov process, the canonical correlations λ_k become negligibly small for sufficiently large k. Thus the order ($n \le Lm$) of the canonical model required to achieve a given prediction error variance is determined by comparing the magnitude of λ_k^2 to the sum of previous terms. Model reduction is desirable because model orders greater than the true system order will cause predictions of the future to respond to measurement noise in $\mathbf{y}_i$ rather than system behavior. Model order reduction can be implemented without reducing the number of lags used to compute $\mathbf{R}_{ff}$ and $\mathbf{R}_{pp}$. However, the **J** and **L** arrays become rectangular, so matrix inverses in the previous equations must be replaced with pseudo-inverses. These details are discussed below after first explaining computation of the canonical prediction arrays.

Computation of **J** and **L** arrays that satisfy the above requirements can be accomplished by first factoring $\mathbf{R}_{pp} = \mathbf{R}_{pp}^{1/2}\mathbf{R}_{pp}^{T/2}$ and $\mathbf{R}_{ff} = \mathbf{R}_{ff}^{1/2}\mathbf{R}_{ff}^{T/2}$. Cholesky factorization could be used, but because the correlation matrices are samples computed from a finite-length data set, they may not necessarily be positive definite. Hence it is safer to use the SVD (or eigen decomposition) to factor $\mathbf{R}_{pp} = \mathbf{U}_{pp}\mathbf{S}_{pp}\mathbf{V}_{pp}^T$ where $\mathbf{U}_{pp} = \mathbf{V}_{pp}$ because $\mathbf{R}_{pp}$ is symmetric. Then compute $\mathbf{R}_{pp}^{-1/2} = \mathbf{U}_{pp}\mathbf{S}_{pp}^{-1/2}\mathbf{V}_{pp}^T$ as a pseudo-inverse where numerically insignificant singular values are ignored. (That is, replace singular values where $S_k < \eta S_1$ with $S_k^{-1} = 0$. η is typically 10^{-12} when using double precision.) The same procedure is used for $\mathbf{R}_{ff}^{-1/2}$. Then the product $\mathbf{R}_{pp}^{-1/2}\mathbf{R}_{pf}\mathbf{R}_{ff}^{-T/2}$ is formed and factored using the SVD as:

$$\boxed{\mathbf{R}_{pp}^{-1/2}\mathbf{R}_{pf}\mathbf{R}_{ff}^{-T/2} = \mathbf{U}\mathbf{S}\mathbf{V}^T} \tag{12.5-14}$$

where $\mathbf{U}^T\mathbf{U} = \mathbf{I}$, $\mathbf{V}^T\mathbf{V} = \mathbf{I}$, and $\mathbf{S} = \mathbf{\Lambda}$ is diagonal. A "full" SVD must be used so that **U** is square and includes the nullspace. Notice that the dimensions of **U**, **S**, and $\mathbf{R}_{pp}$ are, respectively, $L(m + l) \times L(m + l)$, $L(m + l) \times Lm$, and $Lm \times Lm$. The rank of $\mathbf{R}_{pp}^{-1/2}\mathbf{R}_{pf}\mathbf{R}_{ff}^{-T/2}$ will be less than or equal to Lm.

Next compute

$$\boxed{\mathbf{J} = \mathbf{U}^T \mathbf{R}_{pp}^{-1/2},} \tag{12.5-15}$$

which is a square $L(M + l) \times L(M + l)$ array. All rows of **J** below Lm represent the nullspace of $\mathbf{R}_{pf}$ and thus there will be at most Lm useful canonical states.

Matrix **L** could be computed as $\mathbf{L} = \mathbf{V}^T \mathbf{R}_{ff}^{-1/2}$, but it is not actually used in subsequent calculations so it can be ignored. It easily verified that **J** and **L** as defined here satisfy the conditions of equations (12.5-7) to (12.5-9). Of more use than **L** is the array that maps the canonical state $\mathbf{x}_i$ to the future. From (12.5-12),

$$\boxed{\hat{\mathbf{f}}_i = (\mathbf{L}^{-1}\mathbf{S}^T)\mathbf{x}_i = (\mathbf{R}_{ff}^{1/2}\mathbf{V}\mathbf{S}^T)\mathbf{x}_i.} \tag{12.5-16}$$

It is desirable that the order of **x** not be greater than that required to have statistically significant reductions in the prediction error ε^2. One method for truncating the model order simply eliminates singular values S_k that are smaller than a specified threshold of S_1. (This approach is similar to Cadzow's.) From equation (12.5-13) this is equivalent to specifying the relative change in ε^2. Then **J** and **L** are computed using the reduced-order **S**. This is not a particularly robust approach because it does not directly indicate the statistical significance of adding canonical states, that is, it does not indicate whether adding states will cause over-fitting of measurement noise.

A better method for selecting model order uses the AIC calculated as equation (12.3-16) for the prediction. The order with minimal AIC is selected as optimal. Let that be designated as n_{opt}. The unneeded rows (row index > n_{opt}) of **J** and columns of $\mathbf{R}_{ff}^{1/2}\mathbf{V}\mathbf{S}^T$ are then eliminated. Using this reduced-order definition of the canonical $\mathbf{x}_i$, various arrays in the state-space model equation (12.5-1) are computed from the sample correlation matrices and **J**. This is explained in greater detail in Section 12.5.5. Finally, the PSD of $\mathbf{y}_i$ is computed from the state-space model. This is explained in Section 12.5.6.

12.5.2 Summary of CVA Steps

The major steps in CVA are:

1. Remove biases and trends in measurements and controls (optional).
2. Form the sample past-past, past-future, and future-future correlation matrices. We use $\mathbf{\Sigma}$ to designate the sample estimates of the correlation matrices (i.e., $E[\mathbf{\Sigma}_{pp}] = \mathbf{R}_{pp}$, $E[\mathbf{\Sigma}_{pf}] = \mathbf{R}_{pf}$, and $E[\mathbf{\Sigma}_{ff}] = \mathbf{R}_{ff}$).
3. Factor $\mathbf{\Sigma}_{pp} = \mathbf{\Sigma}_{pp}^{1/2}\mathbf{\Sigma}_{pp}^{T/2}$ and $\mathbf{\Sigma}_{ff} = \mathbf{\Sigma}_{ff}^{1/2}\mathbf{\Sigma}_{ff}^{T/2}$ using the SVD. Determine the rank of $\mathbf{\Sigma}_{pp}$ and $\mathbf{\Sigma}_{ff}$. Compute $\mathbf{\Sigma}_{pp}^{-1/2}$ and $\mathbf{\Sigma}_{ff}^{-T/2}$ as pseudo-inverses if they are not full rank.
4. Form $\mathbf{\Sigma}_{pp}^{-1/2}\mathbf{\Sigma}_{pf}\mathbf{\Sigma}_{ff}^{-T/2}$ and factor as $\mathbf{U}\mathbf{S}\mathbf{V}^T$ using the SVD. The rank of **S** (determined as the number of normalized singular values (S_k/S_1) above a given threshold such as 10^{-12}) is used to determine the maximum order of the canonical state **x**. Compute $\mathbf{J} = \mathbf{U}^T\mathbf{\Sigma}_{pp}^{-1/2}$ and $\mathbf{\Sigma}_{ff}^{1/2}\mathbf{V}\mathbf{S}^T$, and truncate the number of rows of **J** and columns of $\mathbf{\Sigma}_{ff}^{1/2}\mathbf{V}\mathbf{S}^T$ to match the order of **x**.

5. Compute the AIC versus model order using the algorithm described in Section 12.5.4. The optimal state order is that with the minimum AIC.
6. Compute the state-space model from the correlation matrices as described in Section 12.5.5.
7. Compute the power spectrum from the state-space model using equations in Section 12.5.6.

The accuracy of the CVA model depends somewhat on the number of lags used to compute the sample correlation matrices. It will be demonstrated later that use of too many lags degrades the computed model. Larimore suggested using the AIC to determine the appropriate number of lags for step 2 above. Unfortunately the implementation used in our CVA.F module tends to select a lag number that may be smaller than ideal. For that reason it is advisable to test several different lags. This is done in CVA.F by setting the MINLAG (minimum) and NLAG (maximum) variables.

CVA is a difficult algorithm to implement in software because of alternating measurement and control terms for multiple lags in the past and future vectors. Use of variable array indexing and array operators, available in MATLAB and Fortran 90/95/2003, greatly simplify the implementation.

The following sections describe several CVA steps in greater detail.

12.5.3 Sample Correlation Matrices

The sample correlation matrices are computed as

$$\boxed{\boldsymbol{\Sigma}_{ff} = \frac{1}{M-2L+1}\sum_{k=L+1}^{M-L+1} \mathbf{f}_k \mathbf{f}_k^T,} \tag{12.5-17}$$

$$\boxed{\boldsymbol{\Sigma}_{fp} = \frac{1}{M-2L+1}\sum_{k=L+1}^{M-L+1} \mathbf{f}_k \mathbf{p}_k^T,} \tag{12.5-18}$$

and

$$\boxed{\boldsymbol{\Sigma}_{pp} = \frac{1}{M-2L+1}\sum_{k=L+1}^{M-L+1} \mathbf{p}_k \mathbf{p}_k^T.} \tag{12.5-19}$$

Notice that the sample correlations are unbiased because a fixed number of samples is used for each correlation element. However, different data samples are used for different correlation elements, so the correlation may not necessarily be positive definite. For CVA to work properly it is important that all correlation sums use the same time index range.

The computations required for the correlation matrices are significant. For example, $\boldsymbol{\Sigma}_{pp}$ requires $L(L+1)(m+l)^2(M-2L+1)/2$ flops even when taking advantage of symmetry. Larimore derived an alternate method that can reduce computation of this Toeplitz matrix by an order of magnitude or more. First, $2L$ sample correlation partitions of $(m+l)\times(m+l)$ elements each are computed as

$$\mathbf{C}(j) = \frac{1}{M-2L+1}\sum_{k=1}^{M-j}\begin{bmatrix}\mathbf{y}_{k+j}\\ \mathbf{u}_{k+j}\end{bmatrix}\begin{bmatrix}\mathbf{y}_k^T & \mathbf{u}_k^T\end{bmatrix}, \qquad 0 \le j \le 2L-1. \tag{12.5-20}$$

Correlation partitions for negative lags are computed as $\mathbf{C}(-j) = \mathbf{C}^T(j)$. An approximate version of $\boldsymbol{\Sigma}_{pp}$ can be obtained using the $\mathbf{C}(j)$ partitions, where the error is due to data end effects. The $\boldsymbol{\Sigma}_{ff}$ and $\boldsymbol{\Sigma}_{fp}$ matrices can be similarly approximated by deleting rows and columns of $\mathbf{C}(j)$ partitions corresponding to future controls. Corrections for beginning and end effects are then applied. These corrections are difficult to describe, so for brevity we omit them. They are, however, used in subroutine CVACORR1 included in the CVA software associated with this book. In the past it was desirable to use algorithmic tricks such as this to reduce computations, but the speed of current computers has made this much less important. Even when implemented using the brute-force equations (12.5-17) to (12.5-19), correlation matrix computation times are usually acceptable for moderate sized problems.

12.5.4 Order Selection Using the AIC

Equation (12.3-16) defines the AIC in terms of the covariance of the measurement prediction error, and the estimation degrees-of-freedom. For our assumed state-space model equation (12.5-1), the one-step predicted measurement conditioned on knowledge of $\mathbf{x}_i$ and $\mathbf{u}_i$ is

$$\hat{\mathbf{y}}_i = [\boldsymbol{\Sigma}_{yu} \quad \boldsymbol{\Sigma}_{yx}] \begin{bmatrix} \boldsymbol{\Sigma}_{uu} & \boldsymbol{\Sigma}_{ux} \\ \boldsymbol{\Sigma}_{xu} & \boldsymbol{\Sigma}_{xx} \end{bmatrix}^{-1} \begin{bmatrix} \mathbf{u}_i \\ \mathbf{x}_i \end{bmatrix}, \tag{12.5-21}$$

or simply $\hat{\mathbf{y}}_i = \boldsymbol{\Sigma}_{yx}\boldsymbol{\Sigma}_{xx}^{-1}\mathbf{x}_i$ when direct feed of control-to-measurement is absent ($\mathbf{A} = \mathbf{0}$). Hence in the general case the prediction error (residual) is

$$\begin{aligned} \tilde{\mathbf{y}}_i &= \mathbf{y}_i - [\boldsymbol{\Sigma}_{yu} \quad \boldsymbol{\Sigma}_{yx}] \begin{bmatrix} \boldsymbol{\Sigma}_{uu} & \boldsymbol{\Sigma}_{ux} \\ \boldsymbol{\Sigma}_{xu} & \boldsymbol{\Sigma}_{xx} \end{bmatrix}^{-1} \begin{bmatrix} \mathbf{u}_i \\ \mathbf{x}_i \end{bmatrix} \\ &= \mathbf{y}_i - [\boldsymbol{\Sigma}_{yu} \quad \boldsymbol{\Sigma}_{yp}\mathbf{J}_n^T] \begin{bmatrix} \boldsymbol{\Sigma}_{uu} & \boldsymbol{\Sigma}_{up}\mathbf{J}_n^T \\ \mathbf{J}_n\boldsymbol{\Sigma}_{pu} & \mathbf{I} \end{bmatrix}^{-1} \begin{bmatrix} \mathbf{u}_i \\ \mathbf{J}_n\mathbf{p}_i \end{bmatrix} \end{aligned}. \tag{12.5-22}$$

We use the notation $\mathbf{J}_n = [\mathbf{I}_{n\times n} \quad \mathbf{0}_{n\times(Lm+Ll-n)}]\mathbf{J}$ to indicate that only the first n rows of the total $\mathbf{J}$ are used when evaluating the AIC for state model order n. The covariances are obtained as

$$\boxed{\boldsymbol{\Sigma}_{yp} = E[\mathbf{y}_i\mathbf{p}_i^T] \quad (\text{from first } m \text{ rows of } \boldsymbol{\Sigma}_{fp})} \tag{12.5-23}$$

$$\boxed{\boldsymbol{\Sigma}_{up} = E[\mathbf{u}_i\mathbf{p}_i^T] = \frac{1}{M-2L+1}\sum_{k=L+1}^{M-L+1} \mathbf{u}_k\mathbf{p}_k} \tag{12.5-24}$$

$$\boxed{\boldsymbol{\Sigma}_{yy} = E[\mathbf{y}_i\mathbf{y}_i^T] \quad (\text{from } (1:m, 1:m) \text{ partition of } \boldsymbol{\Sigma}_{ff})} \tag{12.5-25}$$

$$\boxed{\boldsymbol{\Sigma}_{yu} = E[\mathbf{y}_i\mathbf{u}_i^T] = \frac{1}{M-2L+1}\sum_{k=L+1}^{M-L+1} \mathbf{y}_k\mathbf{u}_k} \tag{12.5-26}$$

$$\boxed{\boldsymbol{\Sigma}_{uu} = E[\mathbf{u}_i\mathbf{u}_i^T] = \frac{1}{M-2L+1}\sum_{k=L+1}^{M-L+1} \mathbf{u}_k\mathbf{u}_k.} \tag{12.5-27}$$

The covariance of the residual $\tilde{\mathbf{y}}_i$ is then

$$\begin{aligned}\mathbf{\Sigma}_{yres\_yres} &\triangleq E[\tilde{\mathbf{y}}_i\tilde{\mathbf{y}}_i^T] \\ &= \mathbf{\Sigma}_{yy} - [\mathbf{\Sigma}_{yu} \quad \mathbf{\Sigma}_{yx}]\begin{bmatrix}\mathbf{\Sigma}_{uu} & \mathbf{\Sigma}_{ux} \\ \mathbf{\Sigma}_{xu} & \mathbf{\Sigma}_{xx}\end{bmatrix}^{-1}\begin{bmatrix}\mathbf{\Sigma}_{uy} \\ \mathbf{\Sigma}_{xy}\end{bmatrix} \\ &= \mathbf{\Sigma}_{yy} - [\mathbf{\Sigma}_{yu} \quad \mathbf{\Sigma}_{yp}\mathbf{J}_n^T]\begin{bmatrix}\mathbf{\Sigma}_{uu} & \mathbf{\Sigma}_{up}\mathbf{J}_n^T \\ \mathbf{J}_n\mathbf{\Sigma}_{pu} & \mathbf{I}_{n\times n}\end{bmatrix}^{-1}\begin{bmatrix}\mathbf{\Sigma}_{uy} \\ \mathbf{J}_n\mathbf{\Sigma}_{py}\end{bmatrix}\end{aligned} . \quad (12.5\text{-}28)$$

This equation can be directly used in the AIC calculations, but because the AIC must be evaluated for all possible model orders, it efficient to first calculate

$$\begin{aligned}\mathbf{\Sigma}_{yx} &= \mathbf{\Sigma}_{yp}\mathbf{J}^T \triangleq [\bar{\mathbf{y}}_1 \quad \bar{\mathbf{y}}_2 \quad \cdots \quad \bar{\mathbf{y}}_{Lm}] \\ \mathbf{\Sigma}_{ux} &= \mathbf{\Sigma}_{up}\mathbf{J}^T \triangleq [\bar{\mathbf{u}}_1 \quad \bar{\mathbf{u}}_2 \quad \cdots \quad \bar{\mathbf{u}}_{Lm}]\end{aligned} . \quad (12.5\text{-}29)$$

Larimore (1990b, 2002) showed that Σ_{yres_yres} can be recursively calculated for $n = 1, 2, \ldots, L(m + l)$ using partitioned matrix inversion, thus greatly reducing computations. However, the time savings is usually negligible for moderate sized problems.

The AIC calculation also requires the model degrees-of-freedom. The arrays in the state-space model have the following number of elements:

$$\begin{aligned}&\mathbf{H}: mn, \quad \mathbf{A}: ml, \quad \mathbf{B}: mn, \quad \mathbf{R}: m(m+1)/2 \\ &\mathbf{\Phi}: n^2, \quad \mathbf{G}: nl, \quad \mathbf{Q}: n(n+1)/2\end{aligned} .$$

However, constraints on the canonical variables, equations (12.5-7) to (12.5-9), reduce the free variables. Larimore shows that the degrees-of-freedom are $n(2m + l) + m(m + 1)/2 + ml$. The ml term is deleted when direct control feed-through is not allowed.

For each candidate model order n, $1 \le n \le Lm$, the AIC is evaluated as

$$\begin{aligned}\mathbf{\Sigma}_{yres\_yres} &= \mathbf{\Sigma}_{yy} - \sum_{k=1}^{n}\bar{\mathbf{y}}_k\bar{\mathbf{y}}_k^T \\ &\quad - \left(\mathbf{\Sigma}_{yu} - \sum_{k=1}^{n}\bar{\mathbf{y}}_k\bar{\mathbf{u}}_k^T\right)\left(\mathbf{\Sigma}_{uu} - \sum_{k=1}^{n}\bar{\mathbf{u}}_k\bar{\mathbf{u}}_k^T\right)^{-1}\left(\mathbf{\Sigma}_{yu} - \sum_{k=1}^{n}\bar{\mathbf{y}}_k\bar{\mathbf{u}}_k^T\right)^T\end{aligned} \quad (12.5\text{-}30)$$

$$\begin{aligned}AIC_n &= (M - 2L + 1)(m + m\ln(2\pi) + \ln|\mathbf{\Sigma}_{yres\_yres}|) \\ &\quad + 2n(2m + l) + 2ml + m(m + 1)\end{aligned} \quad (12.5\text{-}31)$$

where $\bar{\mathbf{y}}_k$ and $\bar{\mathbf{u}}_k$ are the column vectors of $\mathbf{\Sigma}_{yx}$ and $\mathbf{\Sigma}_{ux}$, respectively. If controls $\mathbf{u}$ are not present ($l = 0$) or direct feed-through is not permitted ($\mathbf{A}_c = \mathbf{0}$),

$$\mathbf{\Sigma}_{yres\_yres} = \mathbf{\Sigma}_{yy} - \sum_{k=1}^{n}\bar{\mathbf{y}}_k\bar{\mathbf{y}}_k^T \quad (12.5\text{-}32)$$

and the $2ml$ term is deleted from AIC_n.

It should be noted that $\mathbf{\Sigma}_{yres\_yres}$ can be calculated using alternate methods, including implementation of the state-space model in a Kalman filter. The degrees-of-freedom may also be adjusted depending on model assumptions.

12.5.5 State-Space Model

To compute the state space arrays for the "minimum AIC" model order n_{opt}, we first write equation (12.5-1) as

$$\begin{bmatrix} \mathbf{y}_i \\ \mathbf{x}_{i+1} \end{bmatrix} = \begin{bmatrix} \mathbf{A} & \mathbf{H} \\ \mathbf{G} & \mathbf{\Phi} \end{bmatrix} \begin{bmatrix} \mathbf{u}_i \\ \mathbf{x}_i \end{bmatrix} + \begin{bmatrix} \mathbf{B} \\ \mathbf{I} \end{bmatrix} \mathbf{q}_i + \begin{bmatrix} \mathbf{r}_i \\ \mathbf{0} \end{bmatrix}. \tag{12.5-33}$$

(This section assumes that n_{opt} has been previously determined, and thus notation indicating order is dropped.) Post-multiplying both sides by $[\mathbf{u}_i^T \quad \mathbf{x}_i^T]$ and taking the expected value gives

$$E\left[\begin{bmatrix} \mathbf{y}_i \\ \mathbf{x}_{i+1} \end{bmatrix} [\mathbf{u}_i^T \quad \mathbf{x}_i^T]\right] = \begin{bmatrix} \mathbf{A} & \mathbf{H} \\ \mathbf{G} & \mathbf{\Phi} \end{bmatrix} E\left[\begin{bmatrix} \mathbf{u}_i \\ \mathbf{x}_i \end{bmatrix} [\mathbf{u}_i^T \quad \mathbf{x}_i^T]\right]. \tag{12.5-34}$$

Using $\mathbf{x}_i = \mathbf{J}\mathbf{p}_i$ (where $\mathbf{J}$ means $\mathbf{J}_{n\_opt}$ of the previous section), the last term on the right can be written as

$$\boxed{E\left[\begin{bmatrix} \mathbf{u}_i \\ \mathbf{x}_i \end{bmatrix} [\mathbf{u}_i^T \quad \mathbf{x}_i^T]\right] = \begin{bmatrix} \mathbf{\Sigma}_{uu} & \mathbf{\Sigma}_{ux} \\ \mathbf{\Sigma}_{xu} & \mathbf{\Sigma}_{xx} \end{bmatrix} = \begin{bmatrix} \mathbf{\Sigma}_{uu} & \mathbf{\Sigma}_{up}\mathbf{J}^T \\ \mathbf{J}\mathbf{\Sigma}_{pu} & \mathbf{I} \end{bmatrix}.} \tag{12.5-35}$$

Thus

$$\boxed{\begin{aligned} \begin{bmatrix} \mathbf{A} & \mathbf{H} \\ \mathbf{G} & \mathbf{\Phi} \end{bmatrix} &= E\left[\begin{bmatrix} \mathbf{y}_i \\ \mathbf{x}_{i+1} \end{bmatrix} [\mathbf{u}_i^T \quad \mathbf{x}_i^T]\right] \begin{bmatrix} \mathbf{\Sigma}_{uu} & \mathbf{\Sigma}_{up}\mathbf{J}^T \\ \mathbf{J}\mathbf{\Sigma}_{pu} & \mathbf{I} \end{bmatrix}^{-1} \\ &= \begin{bmatrix} \mathbf{\Sigma}_{yu} & \mathbf{\Sigma}_{yp}\mathbf{J}^T \\ \mathbf{J}E[\mathbf{p}_{i+1}\mathbf{u}_i^T] & \mathbf{J}E[\mathbf{p}_{i+1}\mathbf{p}_i^T]\mathbf{J}^T \end{bmatrix} \begin{bmatrix} \mathbf{\Sigma}_{uu} & \mathbf{\Sigma}_{up}\mathbf{J}^T \\ \mathbf{J}\mathbf{\Sigma}_{pu} & \mathbf{I} \end{bmatrix}^{-1} \end{aligned}} \tag{12.5-36}$$

The two expectations in the lower rows can be computed as

$$\boxed{E[\mathbf{p}_{i+1}\mathbf{p}_i^T] = \begin{bmatrix} E[\mathbf{y}_i\mathbf{p}_i^T] \\ E[\mathbf{u}_i\mathbf{p}_i^T] \\ E[\mathbf{y}_{i-1}\mathbf{p}_i^T] \\ \vdots \\ E[\mathbf{u}_{i-L+1}\mathbf{p}_i^T] \end{bmatrix} = \begin{bmatrix} \mathbf{\Sigma}_{yp} \\ \mathbf{\Sigma}_{up} \\ \text{first } (L-1)(m+l) \text{ rows of } \mathbf{\Sigma}_{pp} \end{bmatrix}} \tag{12.5-37}$$

and

$$\boxed{E[\mathbf{p}_{i+1}\mathbf{u}_i^T] = \begin{bmatrix} E[\mathbf{y}_i\mathbf{u}_i^T] \\ E[\mathbf{u}_i\mathbf{u}_i^T] \\ E[\mathbf{y}_{i-1}\mathbf{u}_i^T] \\ \vdots \\ E[\mathbf{u}_{i-L+1}\mathbf{u}_i^T] \end{bmatrix} = \begin{bmatrix} \mathbf{\Sigma}_{yu} \\ \mathbf{\Sigma}_{uu} \\ (1{:}(L-1)(m+l), m+1 : m+l) \text{ partition of } \mathbf{\Sigma}_{pp} \end{bmatrix}.} \tag{12.5-38}$$

If it is assumed that the control does not directly feed the measurements, $\mathbf{A} = \mathbf{0}$ can be set in the above and following equations.

To determine **Q** and **R**, compute the covariance of the left and right sides of equation (12.5-33):

$$E\left[\begin{bmatrix}\mathbf{y}_i\\ \mathbf{x}_{i+1}\end{bmatrix}\begin{bmatrix}\mathbf{y}_i^T & \mathbf{x}_{i+1}^T\end{bmatrix}\right]=\begin{bmatrix}\mathbf{A} & \mathbf{H}\\ \mathbf{G} & \mathbf{\Phi}\end{bmatrix}\begin{bmatrix}\mathbf{\Sigma}_{uu} & \mathbf{\Sigma}_{up}\mathbf{J}^T\\ \mathbf{J}\mathbf{\Sigma}_{pu} & \mathbf{I}\end{bmatrix}\begin{bmatrix}\mathbf{A} & \mathbf{H}\\ \mathbf{G} & \mathbf{\Phi}\end{bmatrix}^T+\begin{bmatrix}\mathbf{B}\\ \mathbf{I}\end{bmatrix}\mathbf{Q}\begin{bmatrix}\mathbf{B}^T & \mathbf{I}\end{bmatrix}+\begin{bmatrix}\mathbf{R} & \mathbf{0}\\ \mathbf{0} & \mathbf{0}\end{bmatrix} \tag{12.5-39}$$

where we have used the white, independent properties of $\mathbf{q}_i$ and $\mathbf{r}_i$. Notice that the sample value of $E[\mathbf{x}_{i+1}\mathbf{x}_{i+1}^T]=\mathbf{J}E[\mathbf{p}_{i+1}\mathbf{p}_{i+1}^T]\mathbf{J}^T$ will be approximately but not exactly equal to **I**. Both $E[\mathbf{p}_{i+1}\mathbf{p}_{i+1}^T]$ and $E[\mathbf{y}_i\mathbf{p}_{i+1}^T]$ are computed from partitions of previously computed arrays. Now letting

$$\begin{bmatrix}\mathbf{S}_{11} & \mathbf{S}_{12}\\ \mathbf{S}_{21} & \mathbf{S}_{22}\end{bmatrix}\triangleq\begin{bmatrix}\mathbf{\Sigma}_{yy} & E[\mathbf{y}_i\mathbf{p}_{i+1}^T]\mathbf{J}^T\\ \mathbf{J}E[\mathbf{p}_{i+1}\mathbf{y}_i^T] & \mathbf{J}E[\mathbf{p}_{i+1}\mathbf{p}_{i+1}^T]\mathbf{J}^T\end{bmatrix}-\begin{bmatrix}\mathbf{A} & \mathbf{H}\\ \mathbf{G} & \mathbf{\Phi}\end{bmatrix}\begin{bmatrix}\mathbf{\Sigma}_{uu} & \mathbf{\Sigma}_{up}\mathbf{J}^T\\ \mathbf{J}\mathbf{\Sigma}_{pu} & \mathbf{I}\end{bmatrix}\begin{bmatrix}\mathbf{A} & \mathbf{H}\\ \mathbf{G} & \mathbf{\Phi}\end{bmatrix}^T \tag{12.5-40}$$

we obtain

$$\begin{aligned}\mathbf{Q}&=\mathbf{S}_{22}\\ \mathbf{B}&=\mathbf{S}_{12}\mathbf{S}_{22}^{\#}\\ \mathbf{R}&=\mathbf{S}_{11}-\mathbf{B}\mathbf{S}_{21}\end{aligned} \tag{12.5-41}$$

where $\mathbf{S}_{22}^{\#}$ denotes the pseudo-inverse of $\mathbf{S}_{22}$ since **Q** may be singular. Note that matrix **R** may lose precision when it small compared with $\mathbf{\Sigma}_{yy}$ because of the two subtractions. Larimore developed an innovations version of CVA to better handle this case.

Notice that when the **B** matrix in equation (12.5-33) is nonzero, the measurement and state update errors are correlated, so it is necessary to use the correlated Kalman filter equation (8.2-5).

12.5.6 Measurement Power Spectrum Using the State-Space Model

Computation of the state space model may be the last step if the goal is to develop a model for use in a Kalman filter, but for validation purposes it is still good practice to compute the PSD. From equation (12.2-25), the power spectrum of a stochastic signal is the expected value of the magnitude-squared Fourier transform divided by the time window. For finite-time window measurements $\mathbf{y}_i=\mathbf{y}(i\Delta t)$, $1\le i\le M$, this is

$$\mathbf{S}_{yy}(\omega)=\frac{1}{M\Delta t}E[\mathcal{F}[\mathbf{y}(i\Delta t]\mathcal{F}^*[\mathbf{y}(i\Delta t)]]. \tag{12.5-42}$$

This can be evaluated using the z-transform of real-valued data $\mathbf{y}(i\Delta t)$,

$$\mathbf{y}(z)=\sum_{i=1}^{M}\mathbf{y}(i\Delta t)z^{-i}=\frac{1}{\Delta t}\mathcal{F}[\mathbf{y}(i\Delta t)] \tag{12.5-43}$$

where $z^{-1}=e^{-j\omega\Delta t}$ is the unit delay operator. Although we previously used an upper case letter for the transform of a signal, we are now using a lower case here to avoid confusion with vector/matrix conventions on case. The distinction between the data $\mathbf{y}(i\Delta t)$ and the transform $\mathbf{y}(z)$ is indicated by the argument. Thus the PSD can be written as

$$\begin{aligned}\mathbf{S}_{yy}(\omega) &= \frac{\Delta t}{M}E\left[\mathbf{y}(z)\mathbf{y}(z^{-1})\right] \\ &= \frac{\Delta t}{M}E\left[\mathbf{y}(e^{j\omega\Delta t})\mathbf{y}(e^{-j\omega\Delta t})\right]\end{aligned}. \tag{12.5-44}$$

We compute $\mathbf{y}(z)$ by taking the z-transform of the ARMAX model (12.5-1). The z-transform of the lower equation is

$$(z\mathbf{I}-\mathbf{\Phi})\mathbf{x}(z) = \mathbf{G}\mathbf{u}(z)+\mathbf{q}(z).$$

This equation is then solved for $\mathbf{x}(z)$ and substituted in the equation for $\mathbf{y}_i$ to obtain

$$\begin{aligned}\mathbf{y}(z) &= \mathbf{H}\mathbf{x}(z)+\mathbf{A}\,\mathbf{u}(z)+\mathbf{B}\mathbf{q}(z)+\mathbf{r}(z) \\ &= \mathbf{H}\left((z\mathbf{I}-\mathbf{\Phi})^{-1}(\mathbf{G}\mathbf{u}(z)+\mathbf{q}(z))\right)+\mathbf{A}\,\mathbf{u}(z)+\mathbf{B}\mathbf{q}(z)+\mathbf{r}(z) \\ &= \left(\mathbf{H}(z\mathbf{I}-\mathbf{\Phi})^{-1}+\mathbf{B}\right)\mathbf{q}(z)+\left(\mathbf{H}(z\mathbf{I}-\mathbf{\Phi})^{-1}\mathbf{G}+\mathbf{A}\right)\mathbf{u}(z)+\mathbf{r}(z)\end{aligned}. \tag{12.5-45}$$

The PSD is then computed as

$$\boxed{\begin{aligned}\mathbf{S}_{yy}(\omega) &= \frac{\Delta t}{M}E\left[\mathbf{y}(e^{j\omega\Delta t})\mathbf{y}(e^{-j\omega\Delta t})\right] \\ &= \left[\mathbf{H}(e^{j\omega\Delta t}\mathbf{I}-\mathbf{\Phi})^{-1}+\mathbf{B}\right]\mathbf{Q}\left[\mathbf{H}(e^{-j\omega\Delta t}\mathbf{I}-\mathbf{\Phi})^{-1}+\mathbf{B}\right]^T+\mathbf{R} \\ &\quad +\Delta t\left(\mathbf{H}(z\mathbf{I}-\mathbf{\Phi})^{-1}\mathbf{G}+\mathbf{A}\right)\mathbf{u}(z)\,\mathbf{u}(-z)\left(\mathbf{H}(z\mathbf{I}-\mathbf{\Phi})^{-1}\mathbf{G}+\mathbf{A}\right)^T/M\end{aligned}} \tag{12.5-46}$$

where

$$\begin{aligned}\mathbf{Q} &\triangleq E\left[\mathbf{q}_i\mathbf{q}_i^T\right] = \frac{\Delta t}{M}E\left[\mathbf{q}(z)\mathbf{q}^T(-z)\right] \\ \mathbf{R} &\triangleq E\left[\mathbf{r}_i\mathbf{r}_i^T\right] = \frac{\Delta t}{M}E\left[\mathbf{r}(z)\mathbf{r}^T(-z)\right].\end{aligned}$$

The z-transform of the actual sample signal $\mathbf{u}_i$ is used since the autocorrelation function is not generally known. $\mathbf{S}_{yy}(\omega)$ is defined for $-\pi/\Delta t \le \omega \le \pi/\Delta t$, so it should be multiplied by 2.0 (except for $\omega = 0$) when computing total PSD versus frequency.

Example 12.3: CVA on ARMAX(4,4) Model of Example 12.1

We again use the same ARMAX example to evaluate CVA performance. Figures 12.20 and 12.21 show the PSD computed by CVA when using 3 and 20 lags, respectively, for the sample correlation matrices. The PSD deviates less than 1 dB from Figure 12.21 when using 30 lags. Three lags provides the minimum AIC for the autoregressive (not ARMAX) model when attempting to automate selection of the optimal lag. Although three lags works, it seems somewhat low.

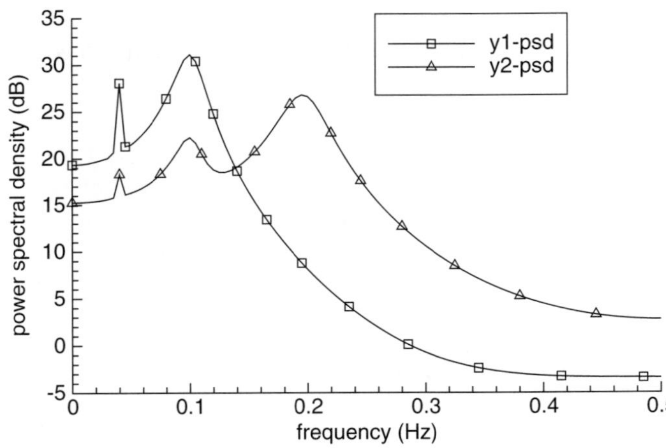

FIGURE 12.20: CVA-computed PSD for ARMAX(4,4) model using 3 lags.

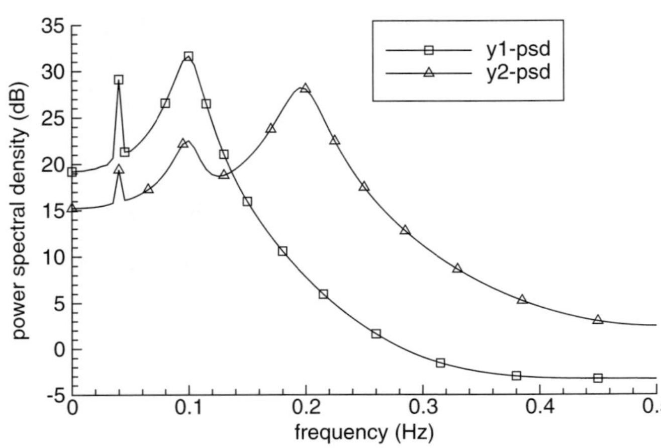

FIGURE 12.21: CVA-computed PSD for ARMAX(4,4) model using 20 lags.

Figure 12.22 shows that the PSD of $\mathbf{y}_2$ loses much detail when the correlation functions are computed using 40 lags. Thus there is danger when using far more lags than necessary. However, in all cases the minimum AIC is correctly obtained for a fourth-order ARMAX state model.

Notice that there is little difference between Figures 12.20 and 12.21, and they are both similar to the true spectrum of Figure 12.11. The raised spectral floor of $\mathbf{y}_2$ and −3 dB floor of $\mathbf{y}_1$ at high frequencies is properly modeled. The effect of the small 0.04 Hz control signal is properly detected in $\mathbf{y}_2$, but it is also incorrectly detected in $\mathbf{y}_1$. Apparently the randomness of the short data span allows for small correlation between $\mathbf{y}_1$ and $\mathbf{u}_1$. Other differences include a slightly larger $\mathbf{y}_2$ peak at 0.1 Hz and slightly smaller peak at 0.2 Hz in the CVA-computed PSD. The integrated power near these two frequencies is approximately the same for the CVA and true PSDs, so the difference in peak values is not significant. The eigenvalues of the CVA-computed $\boldsymbol{\Phi}$ are

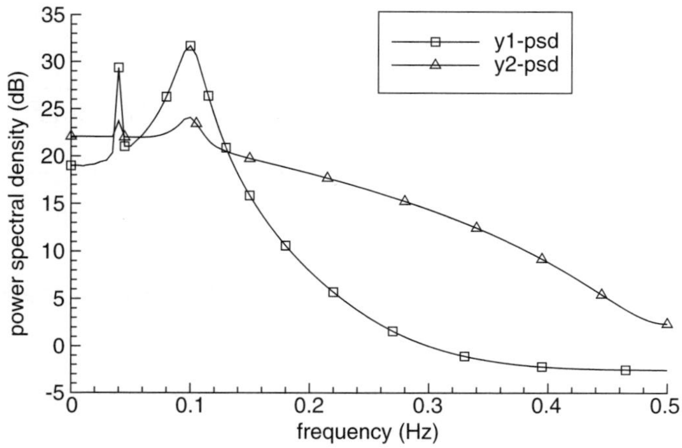

FIGURE 12.22: CVA-computed PSD for ARMAX(4,4) model using 40 lags.

$$\begin{aligned}
&3 \text{ lags: } \lambda_{1,2} = 0.74537 \pm j\ 0.54767, \quad \lambda_{3,4} = 0.28707 \pm j\ 0.83112\\
&20 \text{ lags: } \lambda_{1,2} = 0.75283 \pm j\ 0.54791, \quad \lambda_{3,4} = 0.28950 \pm j\ 0.84288.\\
&30 \text{ lags: } \lambda_{1,2} = 0.74693 \pm j\ 0.54786, \quad \lambda_{3,4} = 0.27589 \pm j\ 0.81481
\end{aligned}$$

These are fairly close to the simulation true values listed in Example 12.1. The computed measurement noise standard deviations from the **R** matrix are 0.117 and 0.017 for y_1 and y_2, respectively. These are significantly smaller than the true values of 0.5 for both measurements. The inability to accurately determine the measurement noise is not surprising (for any method) because the signal-to-noise ratio is high and the y_2 PSD is well above the measurement noise floor at all frequencies. Furthermore the presence of direct feed-through of process noise to the measurements makes separation of process and measurement noise effects difficult when **R** is small. Larimore developed an innovations form of CVA to better handle the small **R** case, but we did not use that method for this example. The loss of precision in the two subtractions required to form **R** did not cause negative variances, although **R** is nearly singular. Considering the difficulty in detecting the effects of measurement noise that is much smaller than the signal, it would be prudent to increase the diagonals of **R** when implementing the model in a Kalman filter.

To summarize, this CVA implementation accurately modeled the important characteristics of the high signal-to-noise, medium-length span ARMAX data containing a small-magnitude control signal. This may not happen in all cases, but experience on other problems has demonstrated that CVA usually works better than alternative techniques. It should be noted that this version of CVA did not include the modifications required to properly handle time-correlated inputs. Hence it is not surprising that CVA incorrectly determined that some of the 0.04 Hz control input signal appears in $\mathbf{y}_1$. Further testing without the control signal produced nearly the same $\mathbf{\Phi}$, **Q**, **H**, **R** as listed above, and the PSD matched Figure 12.21 within 0.35 dB except near 0.04 Hz. Hence the CVA implementation error did not have a significant effect on the general model identification. However, this good behavior cannot be guaranteed in cases when correlated inputs are larger or when feedback from output to input is present. In these cases an initial ARX modeling step should be added, and the effect of future inputs on future outputs should be removed.

12.6 CONVERSION FROM DISCRETE TO CONTINUOUS MODELS

As mentioned, a discrete ARMA or ARMAX model may be directly used in a Kalman filter that takes into account modeled correlations between measurement and process noise (eq. 8.2-5). In some cases the ARMA/ARMAX model may have been created using measurements at a fixed sampling interval, but the intended filter application may have to handle measurement dropouts or outlying measurements that must be edited. This is not a real problem because Kalman filters using ARMA/ARMAX models can perform the time update and then bypass measurement processing. However, in cases where the operational sampling interval is irregular, or measurement types may change from one sampling interval to another, or the discrete model only applies to a small part (e.g., driving forces) of a continuous nonlinear system, direct use of a discrete model is a problem. In these cases the discrete model should usually be converted to a continuous model that can be integrated in the Kalman filter to match actual time steps between measurements.

Conversion of a multivariate discrete model to a continuous model is nontrivial. In simple low-order cases it may be possible to examine the PSD computed from the discrete model and visually determine the location and of dominant poles. Each pole will result in an asymptotic PSD drop of –20 dB/decade, and each zero will cause a +20 dB/decade asymptotic increase. Since physical systems tend to have low-pass characteristics, typical measurement PSDs are constant up to the first pole "break frequency," and then they drop asymptotically at –20 dB/decade for a single pole or –40 dB/decade for a pair of complex poles. Complex pairs of poles also cause spectral peaks, where the magnitude rise above the asymptotes is determined by the damping ratio ζ in quadratic Laplace transform equations of the form $s^2 + 2\zeta\omega_0 s + \omega_0^2 = 0$. The relationship between damping ratio and PSD rise near the pole for low-pass second-order systems is listed in Table 12.2.

TABLE 12.2: Under-Damped Second-Order Pole Peaks

Damping Ratio ζ	Peak Rise (dB)	Peak Location (ω)
0.05	+20	0.997 ω_0
0.10	+14	0.990 ω_0
0.20	+8	0.959 ω_0
0.30	+5	0.906 ω_0
0.50	+1.2	0.707 ω_0
0.70	+0.002	0.141 ω_0
0.707	0	—

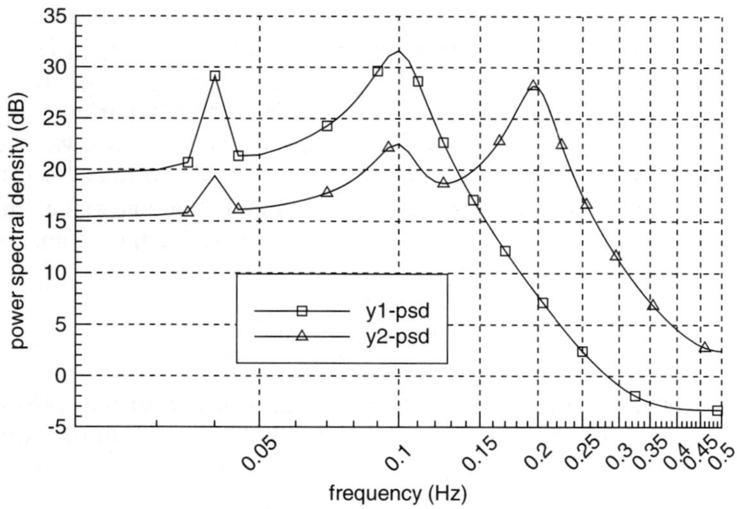

FIGURE 12.23: CVA-computed PSD for ARMAX(4,4) model using 20 lags (log scale).

Transfer function plots for second-order systems may be found in many engineering textbooks, such as Close (1966, p. 319).

Identification of poles and zeroes from power spectral plots is best done when the PSD is plotted on log-log scales. Figure 12.23 is a copy of the Figure 12.21 PSD for the "20-lag" CVA model, but it is plotted on a log-log scale. The lowest frequencies down to 0.002 Hz are not plotted because the more important structure is in the displayed frequency range. Notice that the PSD is nearly constant at low frequencies, indicating that poles and zeroes are either not located near $f = 0$ or they nearly cancel. The lightly damped poles at 0.1 and 0.2 Hz are readily detected, and it is also noted that the 0.2 Hz pole does not appear in y_1. In the range 0.20 to 0.25 Hz, the y_1 PSD falls at −40 dB/decade, which indicates that the system is second order at high frequencies. It is more difficult to determine asymptotic behavior for y_2 because the 0.5 Hz Nyquist frequency is only 2.5 times greater than the 0.20 Hz pole frequency. However, it appears that the asymptotic high-frequency behavior is also approximately −40 dB/decade. Since two second-order pole peaks appear at lower frequencies, two zeroes must also lie in the 0.05 to 0.2 Hz range for the drop above 0.3 Hz to appear second order. The measurement noise floor of −2 dB is also noticeable in the y_1 PSD.

To summarize, we can determine solely from the PSD plots that the continuous system is fourth order, has lightly damped second-order poles at 0.1 and 0.2 Hz, has no poles or zeroes at low frequencies, and probably none above 0.25 Hz. There are probably two zeroes in the 0.05 to 0.2 Hz range that affect y_2, and none that affect y_1. The effect of the exogenous input at 0.04 Hz appears prominently in y_1 and to a lesser extent in y_2. The magnitude of the measurement noise in y_1 is about −3 dB. This information is sufficient to develop a continuous model of the system, but the development is not a simple process and it depends on accurate computation of the PSD. Of the methods tested in this chapter, only CVA produced an accurate power spectrum, and most of the above model information is available directly from the CVA model.

Because the pole peaks are so prominent in this example, it is an easier test of PSD-based model development than many real systems. It is often difficult to locate poles and zeroes from PSD plots when working with real, noisy data. If an approximate continuous model structure can be defined, then the ML parameter identification method of Chapter 11 can be used to refine the model. This was the method used to model tank acceleration in Section 3.2.

Another approach for converting a discrete model to a continuous model uses the bilinear transform. This is an approximate mapping of the z-transform to the Laplace transform:

$$z = e^{s\Delta t} = e^{s\Delta t/2} / e^{-s\Delta t/2} \approx \frac{1+s\Delta t/2}{1-s\Delta t/2}. \tag{12.6-1}$$

This approximation works reasonably well up to one-fourth of the sampling frequency (one-half of the Nyquist frequency), but then the nonlinearity of the frequency mapping becomes severe. The mapping from discrete frequency ω_d to continuous frequency ω_c is

$$\omega_c = \frac{2}{\Delta t}\tan(\omega_d \Delta t/2).$$

Thus at $\omega_d = 2\pi/4\Delta t$ (half of the Nyquist frequency) the frequency ratio is $\omega_c / \omega_d = 1.27$. Provided that most spectral power is located below $f_d = 1/4\Delta t$ Hz, the bilinear transform is a potentially reasonable approach for converting a discrete state-space model to a continuous model. However, even for low-order systems this can be difficult. Consider the z-transform of the discrete dynamic model

$$\mathbf{x}_{i+1} = \mathbf{\Phi}\mathbf{x}_i + \mathbf{G}\mathbf{u}_i + \mathbf{q}_i, \tag{12.6-2}$$

which is

$$z\,\mathbf{x}(z) = \mathbf{\Phi}\mathbf{x}(z) + \mathbf{G}\mathbf{u}(z) + \mathbf{q}(z). \tag{12.6-3}$$

Now using the bilinear transform we have

$$\frac{1+s\Delta t/2}{1-s\Delta t/2}\mathbf{x}(z) = \mathbf{\Phi}\mathbf{x}(z) + \mathbf{G}\mathbf{u}(z) + \mathbf{q}(z)$$

or

$$[(1+s\Delta t/2)\mathbf{I} - (1-s\Delta t/2)\mathbf{\Phi}]\mathbf{x}(z) = (1-s\Delta t/2)(\mathbf{G}\mathbf{u}(z) + \mathbf{q}(z)). \tag{12.6-4}$$

After rearrangement this is

$$s\,\mathbf{x}(z) = (2/\Delta t)(\mathbf{I}+\mathbf{\Phi})^{-1}(\mathbf{\Phi}-\mathbf{I})\mathbf{x}(z) + (2/\Delta t - s)(\mathbf{I}+\mathbf{\Phi})^{-1}(\mathbf{G}\mathbf{u}(z) + \mathbf{q}(z)). \tag{12.6-5}$$

The homogenous part of this equation involving $(2/\Delta t)(\mathbf{I} + \mathbf{\Phi})^{-1}(\mathbf{\Phi} - \mathbf{I})$ is straightforward, but the term $(2/\Delta t - s)(\mathbf{G}\mathbf{u}(z) + \mathbf{q}(z))$ is a problem because it involves both the values and time derivatives of $\mathbf{u}_i$ and $\mathbf{q}_i$. If the system only responds to low frequencies ($2\pi f \ll 2/\Delta t$), it may be acceptable to assume $(2/\Delta t - s) \approx 2/\Delta t$ so that the differential equation becomes

$$\boxed{\dot{\mathbf{x}}(t) \cong (2/\Delta t)(\mathbf{I}+\mathbf{\Phi})^{-1}(\mathbf{\Phi}-\mathbf{I})\mathbf{x}(t) + (2/\Delta t)(\mathbf{I}+\mathbf{\Phi})^{-1}(\mathbf{G}\mathbf{u}(t) + \mathbf{q}(t)).} \tag{12.6-6}$$

This is an approximation that may or may not be valid. If used, output of a stochastic simulation based on the model should be compared with actual data to verify accuracy of the approximation.

When possible the discrete model should be used directly in the Kalman filter so that conversion to a continuous model is unnecessary. Unfortunately this is not always an option.

12.7 SUMMARY

In many cases it is not practical to develop first-principle models of systems. This may happen because the required information is not available, the system is too complex, the system is not physical, or driving noises for a physical system are stochastic. It is therefore necessary to develop empirical models of these systems or subsystems for use in a Kalman filter.

Initial modeling efforts often attempt to infer the transfer function of the system by comparing the PSD of the output signals with either the PSD of the inputs when measured, or assumed white noise when input measurements are unavailable. Equivalently the autocorrelation functions of the output and input may also be used. Since the output signal of a linear system is the convolution of the input signal with the system impulse response, the output Fourier or Laplace transform is the product of the transforms of the input signal and the system impulse response. Furthermore, the power spectrum is the squared magnitude of the Fourier transform, so knowledge of the PSD is sufficient to infer the magnitude of the Fourier transform. The phase can also be determined for a stable, causal system since system poles and zeroes must both lie in the left-half complex plane.

Most signals of interest for Kalman filter applications are stochastic, since they include both measurement noise and driving process noise. The power spectrum of a stochastic signal is defined as the expected value of the time-averaged squared magnitude of the Fourier transform. It is also equal to the Fourier transform of the autocorrelation function. These two definitions lead to two classical approaches for computing the power spectrum: the periodogram and correlogram.

The periodogram computes the Fourier transform of multiple discrete samples of the signal, and then averages the squared magnitudes. It is usually beneficial to multiply the time series data by a tapered window function before computing the transform. Without tapered windowing the effects of the "rectangular" finite-time data sampling can lead to spectral "leakage" from actual signal frequencies to computed frequencies. Discrete data sampling also leads to fold-over of power above the Nyquist frequency to frequencies below. For reasons explained in Section 12.2, the periodogram-computed PSD will exhibit spectral bias and variance. When only a limited time series of measurements is available, variance is reduced by increasing the number of segments into which the data is partitioned for computing the Fourier transform. However, that reduces the number of data points for each transform, thus degrading resolution and increasing bias. Use of a window function also reduces variance but degrades resolution. A method introduced by Welch combines overlapping segments and window functions, and is generally regarded as the best periodogram method.

The BT correlogram computes the unbiased sample autocorrelation function of the signal, and then multiples the function by a Bartlett (triangular) window function. The PSD is computed as the Fourier transform of the modified autocorrelation function. Unfortunately both periodogram and correlogram methods suffer badly from bias and variance. Either method (preferably Welch's periodogram) may be used for initial evaluation of time series data, but neither is satisfactory by itself for model building.

Newer methods directly or indirectly use the sample autocorrelation function to compute coefficients of an AR model. In the YW method, the autocorrelation samples are used to form an autocorrelation matrix that is inverted to solve for the AR coefficients. An alternate method due to Burg operates one prediction filter on the data forward in time, and another operating backward in time. The squared prediction errors are summed and the AR coefficients are computed as the solution to a least-squares problem. Another method due to Marple is similar. These methods are often called MEMs because they assume that autocorrelations for lags greater than the AR order are equally probable. The optimal AR order can be determined using Akaike's FPE. In practice these methods usually work well (within limitations of the AR model) when sharp narrowband spectral peaks and additive measurement noise are not present. Burg's method tends to work better than the YW with sinusoidal signals, but it may still split the peaks. Marple's method is less likely than the other methods to have frequency bias, spectral variance or line splitting, but order determination is dependent on correct selection of two prediction error tolerances.

Real systems are usually better characterized as ARMA or ARMAX models than as AR models. This is true even when sharp nulls in the spectrum do not exist. Unfortunately linear approaches used for AR modeling cannot be directly applied to ARMA models. Several approximate ARMA modeling approaches are available. It has been this author's experience that CVA, as developed by Akaike and Larimore, often works better than alternate methods, particularly for multivariate inputs and outputs. CVA starts by computing past-past, past-future, and future-future sample autocorrelation matrices. The normalized correlation matrix between past and future is factored using SVD, and a canonical state vector is defined from the SVD factors. Because the singular values are ordered from largest to smallest, each state added to the canonical vector reduces the prediction error variance by an amount that can be predicted. The AIC is used to determine the optimal number of states (order of the ARMAX model). The matrices of a canonical state space model are constructed from autocorrelation matrices and the SVD decomposition. This model can be directly implemented in a Kalman filter, or the power spectrum computed from the state-space model can be used to develop a continuous system model that is integrated in a discrete Kalman filter.

Simulated data from an ARMAX(4,4) model is used to compare the different methods, and it is shown that the CVA PSD most accurately matches the true PSD for this example system. However, to properly handle cases containing nonwhite inputs or feedback from output to input, the CVA algorithm should be modified as described by Larimore (2004).

Finally it is noted that operational filter implementations may have to process measurements at irregular time intervals, or different measurement types may be available at different times, or the discrete model only applies to a small part of a continuous nonlinear model. In these cases it is necessary to either infer a continuous model from computed power spectra, or to convert a discrete ARMA/ARMAX model to a continuous model. An example demonstrated the links between PSD behavior and model structure. An ARMA/ARMAX model can be converted to a continuous model using the bilinear transform, but this is an approximation that may or may not be realistic.

APPENDIX A

SUMMARY OF VECTOR/MATRIX OPERATIONS

This Appendix summarizes properties of vector and matrices, and vector/matrix operations that are often used in estimation. Further information may be found in most books on estimation or linear algebra; for example, Golub and Van Loan (1996), DeRusso et al. (1965), and Stewart (1988).

A.1 DEFINITION

A.1.1 Vectors

A *vector* is a linear collection of elements. We use a lower case bold letter to denote vectors, which by default are assumed to be column vectors. For example,

$$\mathbf{a} = \begin{bmatrix} a_1 \\ a_2 \\ a_3 \end{bmatrix}$$

is a three-element column vector. A row vector is (obviously) defined with elements in a row; for example,

$$\mathbf{a} = \begin{bmatrix} a_1 & a_2 & a_3 \end{bmatrix}.$$

A vector is called *unit* or *normalized* when the sum of elements squared is equal to 1: $\sum_{i=1}^{n} a_i^2 = 1$ for an n-element vector $\mathbf{a}$.

A.1.2 Matrices

A *matrix* is a two-dimensional collection of elements. We use bold upper case letters to denote matrices. For example,

---

Advanced Kalman Filtering, Least-Squares and Modeling: A Practical Handbook, by Bruce P. Gibbs

$$\mathbf{A} = \begin{bmatrix} A_{11} & A_{12} & A_{13} \\ A_{21} & A_{22} & A_{23} \end{bmatrix}$$

is a matrix with two rows and three columns, or a 2×3 matrix. Individual elements are labeled with the first subscript indicating the row, and the second indicating the column. An n-element column vector may be considered an $n \times 1$ matrix, and an n-element row vector may be considered a $1 \times n$ matrix.

A.1.2.1 Symmetric Matrix A square matrix is symmetric if elements with interchanged subscripts are equal: $A_{ji} = A_{ij}$. For example,

$$\mathbf{A} = \begin{bmatrix} 2 & 5 & 9 \\ 5 & 1 & 3 \\ 9 & 3 & 4 \end{bmatrix}$$

is symmetric.

A.1.2.2 Hermitian Matrix A square complex matrix is *Hermitian* if elements with interchanged subscripts are equal to the complex conjugate of each other: $A_{ji} = A_{ij}^*$.

A.1.2.3 Toeplitz Matrix A square matrix is *Toeplitz* if all elements along the upper left to lower right diagonals are equal: $A_{i,j} = A_{i-1,j-1}$. For example,

$$\mathbf{A} = \begin{bmatrix} 1 & 2 & -1 & 4 \\ 3 & 1 & 2 & -1 \\ 5 & 3 & 1 & 2 \\ -2 & 5 & 3 & 1 \end{bmatrix}$$

is Toeplitz.

A.1.2.4 Identity Matrix A square matrix that is all zero except for ones along the main diagonal is the identity matrix, denoted as $\mathbf{I}$. For example,

$$\begin{bmatrix} 1 & 0 & 0 & 0 \\ 0 & 1 & 0 & 0 \\ 0 & 0 & 1 & 0 \\ 0 & 0 & 0 & 1 \end{bmatrix}$$

is a 4×4 identity matrix. Often a subscript is added to indicate the dimension, as $\mathbf{I}_n$ is an $n \times n$ identity matrix.

A.1.2.5 Triangular Matrix All elements of a lower triangular matrix above the main diagonal are zero. All elements of an upper triangular matrix below the main diagonal are zero. For example,

$$\begin{bmatrix} 1 & 7 & 4 & -4 \\ 0 & 2 & 6 & 1 \\ 0 & 0 & 3 & 9 \\ 0 & 0 & 0 & 5 \end{bmatrix}$$

is upper triangular.

A.2 ELEMENTARY VECTOR/MATRIX OPERATIONS

A.2.1 Transpose

The *transpose* of a matrix, denoted with superscript T, is formed by interchanging row and column elements: $\mathbf{B} = \mathbf{A}^T$ is the transpose of $\mathbf{A}$ where $B_{ji} = A_{ij}$. For example,

$$\mathbf{B} = \begin{bmatrix} A_{11} & A_{12} & A_{13} \\ A_{21} & A_{22} & A_{23} \end{bmatrix}^T = \begin{bmatrix} A_{11} & A_{21} \\ A_{12} & A_{22} \\ A_{13} & A_{23} \end{bmatrix}.$$

If the matrix is complex, the complex conjugate transpose is denoted as $\mathbf{A}^H = (\mathbf{A}^*)^T$. The matrix is Hermitian if $\mathbf{A} = \mathbf{A}^H$.

A.2.2 Addition

Two or more vectors or matrices of the same dimensions may be added or subtracted by adding/subtracting individual elements. For example, if

$$\mathbf{A} = \begin{bmatrix} 1 & 2 & 3 \\ 4 & 5 & 6 \end{bmatrix}, \quad \mathbf{B} = \begin{bmatrix} 3 & 7 & 5 \\ 2 & 1 & -2 \end{bmatrix}$$

then

$$\mathbf{C} = \mathbf{A} + \mathbf{B} = \begin{bmatrix} 4 & 9 & 8 \\ 6 & 6 & 4 \end{bmatrix}.$$

Matrix addition is commutative; that is, $\mathbf{C} = \mathbf{A} + \mathbf{B} = \mathbf{B} + \mathbf{A}$.

A.2.3 Inner (Dot) Product of Vectors

The *dot product* or *inner product* of two vectors of equal size is the sum of the products of corresponding elements. If vectors $\mathbf{a}$ and $\mathbf{b}$ both contain m elements, the dot product is a scalar:

$$\langle \mathbf{a}, \mathbf{b} \rangle = \mathbf{a} \bullet \mathbf{b} = \mathbf{a}^T \mathbf{b} = \sum_{i=1}^{m} a_i b_i.$$

If $\langle \mathbf{a}, \mathbf{b} \rangle = 0$, the vectors are *orthogonal*.

A.2.4 Outer Product of Vectors

The *outer product* of two vectors (of possibly unequal sizes) is a matrix of products of corresponding vector elements. If vectors **a** and **b** contain m- and n-elements, respectively, then the outer product is an $m \times n$ matrix:

$$\mathbf{C} = \mathbf{a} \otimes \mathbf{b} = \mathbf{a}\mathbf{b}^T$$

or $C_{ij} = a_i b_j$.

A.2.5 Multiplication

Two matrices, where the column dimension of the first (m) is equal to the row dimension of the second, may be multiplied by forming the dot product of the rows of the first matrix and the columns of the second; that is, $C_{ij} = \sum_{k=1}^{m} A_{ik} B_{kj}$. For example, if

$$\mathbf{A} = \begin{bmatrix} 1 & 2 & 3 \\ 4 & 5 & 6 \end{bmatrix}, \quad \mathbf{B} = \begin{bmatrix} 3 & 2 \\ 0 & 1 \\ 5 & -2 \end{bmatrix}$$

then

$$\mathbf{C} = \mathbf{AB} = \begin{bmatrix} 18 & -2 \\ 42 & 1 \end{bmatrix}.$$

Matrix multiplication is not commutative; that is, $\mathbf{C} = \mathbf{AB} \neq \mathbf{BA}$. A matrix multiplying or multiplied by the identity **I** is unchanged; that is, $\mathbf{AI} = \mathbf{IA} = \mathbf{A}$.

The transpose of the product of two matrices is the reversed product of the transposes: $(\mathbf{AB})^T = \mathbf{B}^T\mathbf{A}^T$.

Vector-matrix multiplication is defined as for matrix-matrix multiplication. If matrix **A** is $m \times n$ and vector **x** has m-elements, $\mathbf{y} = \mathbf{x}^T\mathbf{A}$ or

$$y_j = \sum_{i=1}^{m} x_i A_{ij} \quad \text{for } j = 1, 2, \ldots, n$$

is an n-element row vector. If vector **x** has n elements, $\mathbf{y} = \mathbf{Ax}$ is an m-element column vector.

A.3 MATRIX FUNCTIONS

A.3.1 Matrix Inverse

A square matrix that multiplies another square matrix to produce the identity matrix is called the inverse, and is denoted by a superscript −1; that is, if $\mathbf{B} = \mathbf{A}^{-1}$, then $\mathbf{AB} = \mathbf{BA} = \mathbf{I}$. Just as scalar division by zero is not defined, a matrix is called *indeterminate* if the inverse does not exist. The matrix inverse may be computed by various methods. Two popular methods for general square matrices are Gauss-Jordan elimination with pivoting, and LU decomposition followed by inversion of the LU factors (see Press et al. 2007, chapter 2). Inversion based on cofactors and determinants (explained below) is also used for small matrices. For symmetric matrices, inversion based on Cholesky factorization is recommended.

The inverse of the product of two matrices is the reversed product of the inverses: $(\mathbf{AB})^{-1} = \mathbf{B}^{-1}\mathbf{A}^{-1}$. Nonsquare matrices generally do not have an inverse, but left or right inverses can be defined; for example, for $m \times n$ matrix $\mathbf{A}$, $((\mathbf{A}^T\mathbf{A})^{-1}\mathbf{A}^T)\mathbf{A} = \mathbf{I}_n$, so $(\mathbf{A}^T\mathbf{A})^{-1}\mathbf{A}^T$ is a left inverse provided that $(\mathbf{A}^T\mathbf{A})^{-1}$ exists, and $\mathbf{A}(\mathbf{A}^T(\mathbf{A}\mathbf{A}^T)^{-1}) = \mathbf{I}_m$ so $\mathbf{A}^T(\mathbf{A}\mathbf{A}^T)^{-1}$ is a right inverse provided that $(\mathbf{A}\mathbf{A}^T)^{-1}$ exists.

A square matrix is called *orthogonal* when $\mathbf{A}^T\mathbf{A} = \mathbf{A}\mathbf{A}^T = \mathbf{I}$. Thus the transpose is also the inverse: $\mathbf{A}^{-1} = \mathbf{A}^T$. If rectangular matrix $\mathbf{A}$ is $m \times n$, it is called *column orthogonal* when $\mathbf{A}^T\mathbf{A} = \mathbf{I}$ since the columns are orthonormal. This is only possible when $m \geq n$. If $\mathbf{A}\mathbf{A}^T = \mathbf{I}$ for $m \leq n$, matrix $\mathbf{A}$ is called *row orthogonal* because the rows are orthonormal.

A square symmetric matrix must be positive definite for it to be invertible. A *symmetric positive definite* matrix is a square symmetric matrix for which $\mathbf{x}^T\mathbf{A}\mathbf{x} > 0$ for all nonzero vectors $\mathbf{x}$. A symmetric *positive semi-definite* or *non-negative definite* matrix is one for which $\mathbf{x}^T\mathbf{A}\mathbf{x} \geq 0$.

A.3.2 Partitioned Matrix Inversion

It is often helpful to compute the inverse of a matrix in partitions. For example, consider the inverse of

$$\begin{bmatrix} \mathbf{A} & \mathbf{B} \\ \mathbf{C} & \mathbf{D} \end{bmatrix}$$

where the four bold letters indicate smaller matrices. We express the inverse as

$$\begin{bmatrix} \mathbf{E} & \mathbf{F} \\ \mathbf{G} & \mathbf{H} \end{bmatrix}$$

and write

$$\begin{bmatrix} \mathbf{A} & \mathbf{B} \\ \mathbf{C} & \mathbf{D} \end{bmatrix}\begin{bmatrix} \mathbf{E} & \mathbf{F} \\ \mathbf{G} & \mathbf{H} \end{bmatrix} = \begin{bmatrix} \mathbf{I} & \mathbf{0} \\ \mathbf{0} & \mathbf{I} \end{bmatrix}$$

or

$$\mathbf{AE} + \mathbf{BG} = \mathbf{I} \tag{A3-1}$$

$$\mathbf{AF} + \mathbf{BH} = \mathbf{0} \tag{A3-2}$$

$$\mathbf{CE} + \mathbf{DG} = \mathbf{0} \tag{A3-3}$$

$$\mathbf{CF} + \mathbf{DH} = \mathbf{I}. \tag{A3-4}$$

Using equations (A3-2), (A3-4), (A3-1), and (A3-3) in that order, we obtain:

$$\begin{aligned}
\mathbf{F} &= -\mathbf{A}^{-1}\mathbf{BH} \\
\mathbf{H} &= (\mathbf{D} - \mathbf{CA}^{-1}\mathbf{B})^{-1} \\
\mathbf{E} &= \mathbf{A}^{-1}(\mathbf{I} - \mathbf{BG}) \quad \text{(intermediate)} \\
\mathbf{G} &= -(\mathbf{D} - \mathbf{CA}^{-1}\mathbf{B})^{-1}\mathbf{CA}^{-1} \\
&= -\mathbf{HCA}^{-1} \\
\mathbf{E} &= \mathbf{A}^{-1} + \mathbf{A}^{-1}\mathbf{BHCA}^{-1}
\end{aligned} \tag{A3-5}$$

Alternately, using (A3-3), (A3-1), (A3-4), and (A3-2), we obtain

$$\begin{aligned}
\mathbf{G} &= -\mathbf{D}^{-1}\mathbf{C}\mathbf{E} \\
\mathbf{E} &= (\mathbf{A}-\mathbf{B}\mathbf{D}^{-1}\mathbf{C})^{-1} \\
\mathbf{H} &= \mathbf{D}^{-1}(\mathbf{I}-\mathbf{C}\mathbf{F}) \\
\mathbf{F} &= -(\mathbf{A}-\mathbf{B}\mathbf{D}^{-1}\mathbf{C})^{-1}\mathbf{B}\mathbf{D}^{-1} \\
&= -\mathbf{E}\mathbf{B}\mathbf{D}^{-1} \\
\mathbf{H} &= \mathbf{D}^{-1}+\mathbf{D}^{-1}\mathbf{C}\mathbf{E}\mathbf{B}\mathbf{D}^{-1}
\end{aligned} \tag{A3-6}$$

Thus the partitioned inverse can be written in two forms:

$$\begin{bmatrix} \mathbf{A} & \mathbf{B} \\ \mathbf{C} & \mathbf{D} \end{bmatrix}^{-1} = \begin{bmatrix} \mathbf{A}^{-1}+\mathbf{A}^{-1}\mathbf{B}(\mathbf{D}-\mathbf{C}\mathbf{A}^{-1}\mathbf{B})^{-1}\mathbf{C}\mathbf{A}^{-1} & -\mathbf{A}^{-1}\mathbf{B}(\mathbf{D}-\mathbf{C}\mathbf{A}^{-1}\mathbf{B})^{-1} \\ -(\mathbf{D}-\mathbf{C}\mathbf{A}^{-1}\mathbf{B})^{-1}\mathbf{C}\mathbf{A}^{-1} & (\mathbf{D}-\mathbf{C}\mathbf{A}^{-1}\mathbf{B})^{-1} \end{bmatrix} \tag{A3-7}$$

or

$$\begin{bmatrix} \mathbf{A} & \mathbf{B} \\ \mathbf{C} & \mathbf{D} \end{bmatrix}^{-1} = \begin{bmatrix} (\mathbf{A}-\mathbf{B}\mathbf{D}^{-1}\mathbf{C})^{-1} & -(\mathbf{A}-\mathbf{B}\mathbf{D}^{-1}\mathbf{C})^{-1}\mathbf{B}\mathbf{D}^{-1} \\ -\mathbf{D}^{-1}\mathbf{C}(\mathbf{A}-\mathbf{B}\mathbf{D}^{-1}\mathbf{C})^{-1} & \mathbf{D}^{-1}+\mathbf{D}^{-1}\mathbf{C}(\mathbf{A}-\mathbf{B}\mathbf{D}^{-1}\mathbf{C})^{-1}\mathbf{B}\mathbf{D}^{-1} \end{bmatrix}. \tag{A3-8}$$

If the matrix to be inverted is symmetric:

$$\begin{bmatrix} \mathbf{A} & \mathbf{B} \\ \mathbf{B}^T & \mathbf{D} \end{bmatrix}^{-1} = \begin{bmatrix} \mathbf{A}^{-1}+\mathbf{A}^{-1}\mathbf{B}(\mathbf{D}-\mathbf{B}^T\mathbf{A}^{-1}\mathbf{B})^{-1}\mathbf{B}^T\mathbf{A}^{-1} & -\mathbf{A}^{-1}\mathbf{B}(\mathbf{D}-\mathbf{B}^T\mathbf{A}^{-1}\mathbf{B})^{-1} \\ -(\mathbf{D}-\mathbf{B}^T\mathbf{A}^{-1}\mathbf{B})^{-1}\mathbf{B}^T\mathbf{A}^{-1} & (\mathbf{D}-\mathbf{B}^T\mathbf{A}^{-1}\mathbf{B})^{-1} \end{bmatrix} \tag{A3-9}$$

or

$$\begin{bmatrix} \mathbf{A} & \mathbf{B} \\ \mathbf{B}^T & \mathbf{D} \end{bmatrix}^{-1} = \begin{bmatrix} (\mathbf{A}-\mathbf{B}\mathbf{D}^{-1}\mathbf{B}^T)^{-1} & -(\mathbf{A}-\mathbf{B}\mathbf{D}^{-1}\mathbf{B}^T)^{-1}\mathbf{B}\mathbf{D}^{-1} \\ -\mathbf{D}^{-1}\mathbf{B}^T(\mathbf{A}-\mathbf{B}\mathbf{D}^{-1}\mathbf{B}^T)^{-1} & \mathbf{D}^{-1}+\mathbf{D}^{-1}\mathbf{B}^T(\mathbf{A}-\mathbf{B}\mathbf{D}^{-1}\mathbf{B}^T)^{-1}\mathbf{B}\mathbf{D}^{-1} \end{bmatrix} \tag{A3-10}$$

where **A** and **D** are also symmetric.

A.3.3 Matrix Inversion Identity

The two equivalent expressions for the partitioned inverse suggest a matrix equivalency that is the link between batch least squares and recursive least squares (and Kalman filtering). This formula and variations on it have been attributed to various people (e.g., Woodbury, Ho, Sherman, Morrison), but the relationship has undoubtedly been discovered and rediscovered many times.

From the lower right corner of equations (A3-7) and (A3-8):

$$(\mathbf{D}-\mathbf{C}\mathbf{A}^{-1}\mathbf{B})^{-1} = \mathbf{D}^{-1}+\mathbf{D}^{-1}\mathbf{C}(\mathbf{A}-\mathbf{B}\mathbf{D}^{-1}\mathbf{C})^{-1}\mathbf{B}\mathbf{D}^{-1}. \tag{A3-11}$$

In the symmetric case when $\mathbf{C} = \mathbf{B}^T$:

$$(\mathbf{D}-\mathbf{B}^T\mathbf{A}^{-1}\mathbf{B})^{-1} = \mathbf{D}^{-1}+\mathbf{D}^{-1}\mathbf{B}^T(\mathbf{A}-\mathbf{B}\mathbf{D}^{-1}\mathbf{B}^T)^{-1}\mathbf{B}\mathbf{D}^{-1}, \tag{A3-12}$$

or by changing the sign of **A**:

$$\boxed{(\mathbf{D}+\mathbf{B}^T\mathbf{A}^{-1}\mathbf{B})^{-1}=\mathbf{D}^{-1}-\mathbf{D}^{-1}\mathbf{B}^T(\mathbf{A}+\mathbf{B}\mathbf{D}^{-1}\mathbf{B}^T)^{-1}\mathbf{B}\mathbf{D}^{-1}.} \quad \text{(A3-13)}$$

Equation (A3-13) is the connection between the measurement update of Bayesian least squares and the Kalman filter.

If $\mathbf{D}=\mathbf{I}$ and $\mathbf{C}=\mathbf{B}^T$:

$$\boxed{(\mathbf{I}-\mathbf{B}^T\mathbf{A}^{-1}\mathbf{B})^{-1}=\mathbf{I}+\mathbf{B}^T(\mathbf{A}-\mathbf{B}\mathbf{B}^T)^{-1}\mathbf{B}.} \quad \text{(A3-14)}$$

or

$$\boxed{(\mathbf{I}+\mathbf{B}^T\mathbf{A}^{-1}\mathbf{B})^{-1}=\mathbf{I}-\mathbf{B}^T(\mathbf{A}+\mathbf{B}\mathbf{B}^T)^{-1}\mathbf{B}.} \quad \text{(A3-15)}$$

If **b** is a row vector and a is scalar:

$$(\mathbf{D}-\mathbf{b}^T\mathbf{b}/a)^{-1}=\mathbf{D}^{-1}+(\mathbf{D}^{-1}\mathbf{b}^T)(\mathbf{b}\mathbf{D}^{-1})/(a-\mathbf{b}\mathbf{D}^{-1}\mathbf{b}^T) \quad \text{(A3-16)}$$

or

$$(\mathbf{D}+\mathbf{b}^T\mathbf{b}/a)^{-1}=\mathbf{D}^{-1}-(\mathbf{D}^{-1}\mathbf{b}^T)(\mathbf{b}\mathbf{D}^{-1})/(a+\mathbf{b}\mathbf{D}^{-1}\mathbf{b}^T). \quad \text{(A3-17)}$$

A.3.4 Determinant

The determinant of a square matrix is a measure of scale change when the matrix is viewed as a linear transformation. When the determinant of a matrix is zero, the matrix is *indeterminate* or *singular*, and cannot be inverted. The *rank* of matrix $|\mathbf{A}|$ is the largest square array in **A** that has nonzero determinant.

The determinant of matrix **A** is denoted as det(**A**) or $|\mathbf{A}|$. Laplace's method for computing determinants uses *cofactors*, where a cofactor of a given matrix element ij is $C_{ij}=(-1)^{i+j}|\mathbf{M}_{ij}|$ and $\mathbf{M}_{ij}$, called the *minor* of ij, is the matrix formed by deleting the i row and j column of matrix **A**. For 2×2 matrix

$$\begin{bmatrix} A_{11} & A_{12} \\ A_{21} & A_{22} \end{bmatrix},$$

the cofactors are

$$C_{11}=A_{22},\quad C_{12}=-A_{21},\quad C_{21}=-A_{12},\quad C_{22}=A_{11}.$$

The determinant is the sum of the products of matrix elements and cofactors for any row or column. Thus

$$\begin{aligned}|\mathbf{A}|&=A_{11}C_{11}+A_{12}C_{12}=A_{11}C_{11}+A_{21}C_{21}=A_{21}C_{21}+A_{22}C_{22}=A_{12}C_{12}+A_{22}C_{22}\\&=A_{11}A_{22}-A_{12}A_{21}\end{aligned}.$$

For a 3×3 matrix **A**,

$$\begin{bmatrix} A_{11} & A_{12} & A_{13} \\ A_{21} & A_{22} & A_{23} \\ A_{31} & A_{32} & A_{33} \end{bmatrix},$$

the cofactors are

$$C_{11} = A_{22}A_{33} - A_{23}A_{32}, \quad C_{21} = -(A_{12}A_{33} - A_{13}A_{32}), \quad C_{31} = A_{12}A_{23} - A_{13}A_{22}, \quad \ldots$$

so using the first column,

$$\begin{aligned} |\mathbf{A}| &= A_{11}C_{11} + A_{21}C_{21} + A_{31}C_{31} \\ &= A_{11}(A_{22}A_{33} - A_{23}A_{32}) - A_{21}(A_{12}A_{33} - A_{13}A_{32}) + A_{31}(A_{12}A_{23} - A_{13}A_{22}) \end{aligned}.$$

A matrix inverse can be computed from the determinant and cofactors as

$$\boxed{\mathbf{A}^{-1} = \frac{1}{|\mathbf{A}|}[\mathbf{C}]^T = \frac{1}{|\mathbf{A}|}\begin{bmatrix} C_{11} & C_{21} & \cdots & C_{n1} \\ C_{12} & C_{22} & \cdots & C_{n2} \\ \vdots & \vdots & \ddots & \vdots \\ C_{1n} & C_{2n} & \cdots & C_{nn} \end{bmatrix}}, \tag{A3-18}$$

where $[\mathbf{C}]$ is the matrix of cofactors for $\mathbf{A}$.

Computation of determinants using cofactors is cumbersome, and is seldom used for dimensions greater than three. The determinant is more easily computed by factoring $\mathbf{A} = \mathbf{LU}$ using Crout reduction, where $\mathbf{L}$ is unit lower triangular and $\mathbf{U}$ is upper triangular. Since the determinant of the product of matrices is equal to the product of determinants,

$$|\mathbf{AB}| = |\mathbf{A}||\mathbf{B}|, \tag{A3-19}$$

we have $|\mathbf{A}| = |\mathbf{L}||\mathbf{U}|$, which is simply equal to the product of the diagonals of $\mathbf{U}$ because $|\mathbf{L}| = 1$. A number of other methods may also be used to compute determinants. Often the determinant is a bi-product of matrix inversion algorithms.

A.3.5 Matrix Trace

The *trace* of a square matrix is the sum of the diagonal elements. This property is often useful in least-squares or minimum variance estimation, as the sum of squared elements in an n-vector can be written as

$$\boxed{\sum_{i=1}^{n} a_i^2 = \mathbf{a}^T\mathbf{a} = tr\left[\mathbf{a}\mathbf{a}^T\right].} \tag{A3-20}$$

This rearrangement of vector order often allows solutions for minimum variance problems: it has been used repeatedly in previous chapters.

Since the trace only involves diagonal elements, $tr(\mathbf{A}^T) = tr(\mathbf{A})$. Also,

$$\boxed{tr(\mathbf{A}+\mathbf{B}) = tr(\mathbf{A}) + tr(\mathbf{B})} \tag{A3-21}$$

and for scalar c,

$$\boxed{tr(c\mathbf{A}) = c \cdot tr(\mathbf{A}).} \tag{A3-22}$$

Unlike the determinant, the trace of the matrix products is not the product of traces. If $\mathbf{A}$ is an $n \times m$ matrix and $\mathbf{B}$ is an $m \times n$ matrix,

$$\begin{aligned} tr(\mathbf{AB}) &= \sum_{i=1}^{n}\sum_{j=1}^{m} A_{ij}B_{ji} = \sum_{j=1}^{m}\sum_{i=1}^{n} B_{ji}A_{ij} \\ &= tr(\mathbf{BA}) \end{aligned}. \tag{A3-23}$$

However, this commutative property only works for pairs of matrices, or interchange of "halves" of the matrix product:

$$\begin{aligned} tr(\mathbf{ABC}) &= tr(\mathbf{C}(\mathbf{AB})) = tr(\mathbf{BCA}) \\ &\neq tr(\mathbf{ACB}) \\ &\neq tr(\mathbf{CBA}) \end{aligned}. \tag{A3-24}$$

When the three individual matrices are square and symmetric, any permutation works:

$$tr(\mathbf{ABC}) = tr(\mathbf{C}^T\mathbf{B}^T\mathbf{A}^T) = tr(\mathbf{CBA}). \tag{A3-25}$$

This permutation does not work with four or more symmetric matrices.

Permutation can be used to express the weighted quadratic form $\mathbf{a}^T\mathbf{Wa}$ as

$$\begin{aligned} \mathbf{a}^T\mathbf{Wa} &= \mathbf{a}^T\mathbf{W}^{T/2}\mathbf{W}^{1/2}\mathbf{a} = tr(\mathbf{W}^{1/2}\mathbf{aa}^T\mathbf{W}^{T/2}) = tr(\mathbf{W}^{T/2}\mathbf{W}^{1/2}\mathbf{aa}^T) \\ &= tr(\mathbf{W}(\mathbf{aa}^T)) \end{aligned}, \tag{A3-26}$$

where $\mathbf{a}$ is a vector and matrix $\mathbf{W} = \mathbf{W}^{T/2}\mathbf{W}^{1/2}$ is symmetric.

For a transformation of the form $\mathbf{T}^{-1}\mathbf{AT}$ (called a similarity transformation), the trace is unchanged:

$$tr(\mathbf{T}^{-1}\mathbf{AT}) = tr(\mathbf{TT}^{-1}\mathbf{A}) = tr(\mathbf{A}). \tag{A3-27}$$

A.3.6 Derivatives of Matrix Functions

The derivative of matrix $\mathbf{A}$ with respect to scalar variable t is $\partial\mathbf{A}/\partial t$, which is a matrix with the same dimensions as $\mathbf{A}$. The derivative of matrix product $\mathbf{BA}$ with respect to t can be written as

$$\frac{\partial(\mathbf{BA})}{\partial t} = \frac{\partial(\mathbf{BA})}{\partial \mathbf{A}}\frac{\partial \mathbf{A}}{\partial t}, \tag{A3-28}$$

where $\partial(\mathbf{BA})/\partial\mathbf{A}$ is a four-dimensional variable:

$$\frac{\partial(\mathbf{BA})}{\partial A_{11}}, \frac{\partial(\mathbf{BA})}{\partial A_{12}}, \frac{\partial(\mathbf{BA})}{\partial A_{13}}, \ldots.$$

Thus matrix derivatives for a function of $\mathbf{A}$, $\mathbf{C}(\mathbf{A})$, can be written using $\partial\mathbf{C}/\partial\mathbf{A}$ if the four dimensions are properly handled.

The derivative of the inverse of a matrix is obtained by differentiating $\mathbf{AB} = \mathbf{I}$, which gives $(d\mathbf{A})\mathbf{B} + \mathbf{A}(d\mathbf{B}) = \mathbf{0}$. Thus

$$d(\mathbf{A}^{-1}) = d\mathbf{B} = -\mathbf{A}^{-1}(d\mathbf{A})\mathbf{A}^{-1}$$

is two-dimensional, but has perturbations with respect to all elements of $d\mathbf{A}$; that is, each i, j element $(d\mathbf{B})_{ij}$ is a matrix of derivatives for all perturbations $d\mathbf{A}$. Thus

$$\boxed{\frac{\partial(\mathbf{A}^{-1})}{\partial t} = -\mathbf{A}^{-1}\left(\frac{\partial \mathbf{A}}{\partial t}\right)\mathbf{A}^{-1}.} \tag{A3-29}$$

The derivative of the trace of matrix products is computed by examining the derivative with respect to an individual element:

$$\frac{\partial tr(\mathbf{AB})}{\partial B_{ij}} = \frac{\partial}{\partial B_{ij}}\left(\sum_{k=1}^{m}\sum_{l=1}^{m} A_{kl}B_{lk}\right) = A_{ji}.$$

Thus

$$\boxed{\frac{\partial tr(\mathbf{AB})}{\partial \mathbf{B}} = \mathbf{A}^T.} \tag{A3-30}$$

For products of three matrices,

$$\boxed{\frac{\partial tr(\mathbf{ABC})}{\partial \mathbf{B}} = \frac{\partial tr(\mathbf{CAB})}{\partial \mathbf{B}} = (\mathbf{CA})^T.} \tag{A3-31}$$

The derivative of the determinant is computed by rearranging equation (A3-18) as

$$|\mathbf{A}|\mathbf{I} = \mathbf{A}[\mathbf{C}]^T. \tag{A3-32}$$

Thus for any diagonal element $i = 1, 2, \ldots, n$,

$$|\mathbf{A}| = \sum_{k=1}^{n} A_{ik}C_{ik}.$$

The partial derivative of $|\mathbf{A}|$ with respect to any $\mathbf{A}$ element in row i is

$$\frac{\partial|\mathbf{A}|}{\partial A_{ij}} = \frac{\partial}{\partial A_{ij}}\left(\sum_{k=1}^{n} A_{ik}C_{ik}\right) = C_{ij},$$

where $\partial C_{ik}/\partial A_{ij} = 0$ from cofactor definitions. Hence

$$\boxed{\frac{\partial|\mathbf{A}|}{\partial \mathbf{A}} = [\mathbf{C}] = |\mathbf{A}|\mathbf{A}^{-T}.} \tag{A3-33}$$

By a similar development

$$\boxed{\frac{\partial|\mathbf{A}|}{\partial t} = |\mathbf{A}|\, tr\left(\mathbf{A}^{-1}\frac{\partial \mathbf{A}}{\partial t}\right).} \tag{A3-34}$$

A.3.7 Norms

Norm of vectors or matrices is often useful when analyzing growth of numerical errors. The Hölder p-norms for vectors are defined as

$$\boxed{\|\mathbf{x}\|_p = \left(\sum_{i=1}^{n}|x_i|^p\right)^{1/p}.} \tag{A3-35}$$

The most frequently used p-norms are

$$\|\mathbf{x}\|_1 = \sum_{i=1}^{n} |x_i|, \quad \|\mathbf{x}\|_2 = \left(\sum_{i=1}^{n} x_i^2\right)^{1/2}, \quad \|\mathbf{x}\|_\infty = \max_i |x_i|. \tag{A3-36}$$

The l_2-norm $\|\mathbf{x}\|_2$ is also called the Euclidian or root-sum-squared norm.

An l_2-like norm can be defined for matrices by treating the matrix elements as a vector. This leads to the *Frobenius norm*

$$\|\mathbf{A}\|_F = \left(\sum_{i=1}^{m}\sum_{j=1}^{n} A_{ij}^2\right)^{1/2}. \tag{A3-37}$$

Induced matrix norms measure the ability of matrix $\mathbf{A}$ to modify the magnitude of a vector; that is,

$$\|\mathbf{A}\| = \max_{\|\mathbf{x}\|=1}\left(\frac{\|\mathbf{Ax}\|}{\|\mathbf{x}\|}\right).$$

The l_1-norm is $\|\mathbf{A}\|_1 = \max_j \|\mathbf{a}_{:j}\|_1$ where $\mathbf{a}_{:j}$ is the j-th column of $\mathbf{A}$, and the l_∞-norm is $\|\mathbf{A}\|_\infty = \max_i \|\mathbf{a}_{i:}\|_1$ where $\mathbf{a}_{i:}$ is the i-th row of $\mathbf{A}$. It is more difficult to compute an l_2-norm based on this definition than a Frobenius norm. $\|\mathbf{A}\|_2$ is equal to the square root of the maximum eigenvalue of $\mathbf{A}^T\mathbf{A}$—or equivalently the largest singular value of $\mathbf{A}$. These terms are defined in Section A.4.

Norms of matrix products obey inequality conditions:

$$\|\mathbf{Ax}\| \le \|\mathbf{A}\|\|\mathbf{x}\| \quad \text{or} \quad \|\mathbf{AB}\| \le \|\mathbf{A}\|\|\mathbf{B}\|. \tag{A3-38}$$

The l_2 and Frobenius matrix norms are unchanged by orthogonal transformations, that is,

$$\left.\begin{aligned}\|\mathbf{AB}\|_2 &= \|\mathbf{A}\|_2\\ \|\mathbf{BA}\|_2 &= \|\mathbf{A}\|_2\end{aligned}\right\} \quad \text{if} \quad \mathbf{B}^T\mathbf{B} = \mathbf{I}. \tag{A3-39}$$

A.4 MATRIX TRANSFORMATIONS AND FACTORIZATION

A.4.1 LU Decomposition

LU decomposition has been mentioned previously. Crout reduction is used to factor square matrix $\mathbf{A} = \mathbf{LU}$ where $\mathbf{L}$ is unit lower triangular and $\mathbf{U}$ is upper triangular. This is often used for matrix inversion or when repeatedly solving equations of the form $\mathbf{Ax} = \mathbf{y}$ for $\mathbf{x}$. The equation $\mathbf{Lz} = \mathbf{y}$ is first solved for $\mathbf{z}$ using forward substitution, and then $\mathbf{Ux} = \mathbf{z}$ is solved for $\mathbf{x}$ using backward substitution.

A.4.2 Cholesky Factorization

Cholesky factorization is essentially LU decomposition for symmetric matrices. Usually it implies $\mathbf{A} = \mathbf{LL}^T$ where $\mathbf{L}$ is lower triangular, but it is sometimes used to indicate $\mathbf{A} = \mathbf{L}^T\mathbf{L}$. Cholesky factorization for symmetric matrices is much more efficient and accurate than LU decomposition.

A.4.3 Similarity Transformation

A similarity transformation on square matrix $\mathbf{A}$ is one of the form $\mathbf{B} = \mathbf{T}^{-1}\mathbf{AT}$. Since $\mathbf{TB} = \mathbf{AT}$, matrices $\mathbf{A}$ and $\mathbf{B}$ are called similar. Similarity transformations can be used to reduce a given matrix to an equivalent canonical form; for example, eigen decomposition and singular value decomposition discussed below.

A.4.4 Eigen Decomposition

The vectors $\mathbf{x}_i$ for which $\mathbf{Ax}_i = \lambda_i\mathbf{x}_i$, where λ_i are scalars, are called the eigenvectors of square matrix $\mathbf{A}$. For an $n \times n$ matrix $\mathbf{A}$, there will be n eigenvectors $\mathbf{x}_i$ with corresponding eigenvalues λ_i. The λ_i may be complex (occurring in complex conjugate pairs) even when $\mathbf{A}$ is real, and some eigenvalues may be repeated. The eigenvectors are the directions that are invariant with pre-multiplication by $\mathbf{A}$.

The eigenvector/eigenvalue relationship can be written in matrix form as

$$\boxed{\mathbf{AM} = \mathbf{M\Lambda}} \tag{A4-1}$$

or

$$\boxed{\mathbf{A} = \mathbf{M\Lambda M}^{-1}} \tag{A4-2}$$

where

$$\mathbf{M} = [\mathbf{x}_1 \quad \mathbf{x}_2 \quad \cdots \quad \mathbf{x}_n], \quad \mathbf{\Lambda} = \begin{bmatrix} \lambda_1 & 0 & \cdots & 0 \\ 0 & \lambda_2 & \cdots & 0 \\ \vdots & \vdots & \ddots & \vdots \\ 0 & 0 & \cdots & \lambda_n \end{bmatrix}.$$

Matrix $\mathbf{M}$ is called the modal matrix and λ_i are the eigenvalues. The eigenvalues are the roots of the *characteristic polynomial* $p(s) = |s\mathbf{I} - \mathbf{A}|$, and they define the spectral response of the linear system $\dot{\mathbf{x}}(t) = \mathbf{A}\,\mathbf{x}(t)$ when $\mathbf{x}(t)$ is the system state vector (not eigenvectors). Eigen decomposition is a similarity transformation, and thus

$$\boxed{|\mathbf{A}| = \lambda_1\lambda_2 \ldots \lambda_n.} \tag{A4-3}$$

Also

$$\boxed{tr(\mathbf{A}) = \lambda_1 + \lambda_2 + \ldots + \lambda_n.} \tag{A4-4}$$

When real $\mathbf{A}$ is symmetric and nonsingular, the λ_i are all real, and the eigenvectors are distinct and orthogonal. Thus $\mathbf{M}^{-1} = \mathbf{M}^T$ and $\mathbf{A} = \mathbf{M\Lambda M}^T$.

Eigenvectors and eigenvalues are computed in LAPACK using either generalized QR decomposition or a divide-and-conquer approach.

A.4.5 Singular Value Decomposition (SVD)

While eigen decomposition is used for square matrices, SVD is used for either square or rectangular matrices. The SVD of $m \times n$ matrix $\mathbf{A}$ is

$$\boxed{\mathbf{A} = \mathbf{USV}^T} \tag{A4-5}$$

where **U** is an $m \times m$ orthogonal matrix ($\mathbf{U}\mathbf{U}^T = \mathbf{I}_m$), **V** is an $n \times n$ orthogonal matrix ($\mathbf{V}\mathbf{V}^T = \mathbf{I}_n$), and **S** is an $m \times n$ upper diagonal matrix of singular values. In least-squares problems $m > n$ is typical, so

$$\mathbf{A} = [\mathbf{u}_1 \;\; \ldots \;\; \mathbf{u}_n \;\; \mathbf{u}_{n+1} \;\; \cdots \;\; \mathbf{u}_m] \begin{bmatrix} S_1 & 0 & \cdots & 0 \\ 0 & S_2 & \cdots & 0 \\ \vdots & \vdots & \ddots & \vdots \\ 0 & 0 & \cdots & S_n \\ \vdots & \vdots & \ddots & \vdots \\ 0 & 0 & \cdots & 0 \end{bmatrix} \begin{bmatrix} \mathbf{v}_1^T \\ \mathbf{v}_2^T \\ \\ \mathbf{v}_n^T \end{bmatrix}$$

where $\mathbf{u}_i$ are the left singular vectors and $\mathbf{v}_i$ are the right singular vectors. The left singular vectors $\mathbf{u}_{n+1}, \mathbf{u}_{n+2}, \ldots, \mathbf{u}_m$ are the nullspace of **A** as they multiply zeroes in **S**. The above SVD is called "full," but some SVD utilities have the option to omit computation of $\mathbf{u}_{n+1}, \mathbf{u}_{n+2}, \ldots, \mathbf{u}_m$. These are called "thin" SVDs.

When **A** is a real symmetric matrix formed as $\mathbf{A} = \mathbf{H}^T\mathbf{H}$ (such as the information matrix in least-squares estimation), the eigenvalues of **A** are the squares of the singular values of **H**, and the eigenvectors are the right singular vectors $\mathbf{v}_i$. This is seen from

$$\mathbf{A} = \mathbf{H}^T\mathbf{H} = (\mathbf{V}\mathbf{S}^T\mathbf{U}^T)(\mathbf{U}\mathbf{S}\mathbf{V}^T) = \mathbf{V}(\mathbf{S}^T\mathbf{S})\mathbf{V}^T.$$

The accuracy of the SVD factorization is much greater than that for eigen decomposition of $\mathbf{A} = \mathbf{H}^T\mathbf{H}$ because the "squaring" operation doubles the condition number (defined below). This doubles sensitivity to numerical round-off errors.

A.4.6 Pseudo-Inverse

When **A** is rectangular or singular, **A** does not have an inverse. However, Penrose (1955) defined a *pseudo-inverse* $\mathbf{A}^{\#}$ uniquely determined by four properties:

$$\boxed{\begin{aligned} \mathbf{A}\mathbf{A}^{\#}\mathbf{A} &= \mathbf{A} \\ \mathbf{A}^{\#}\mathbf{A}\mathbf{A}^{\#} &= \mathbf{A}^{\#} \\ (\mathbf{A}\mathbf{A}^{\#})^T &= \mathbf{A}\mathbf{A}^{\#} \\ (\mathbf{A}^{\#}\mathbf{A})^T &= \mathbf{A}^{\#}\mathbf{A} \end{aligned}} \tag{A4-6}$$

This Moore-Penrose pseudo-inverse is sometimes used in least-squares problems when there is insufficient measurement information to obtain a unique solution. A pseudo-inverse for the least-squares problem based on measurement equation $\mathbf{y} = \mathbf{H}\mathbf{x} + \mathbf{r}$ can be written using the SVD of $\mathbf{H} = \mathbf{U}\mathbf{S}\mathbf{V}^T$. Then the normal equation least-squares solution $\hat{\mathbf{x}} = (\mathbf{H}^T\mathbf{H})^{-1}\mathbf{H}^T\mathbf{y}$ is obtained using the pseudo-inverse of **H**,

$$\begin{aligned} \mathbf{H}^{\#} &= (\mathbf{V}\mathbf{S}^T\mathbf{U}^T\mathbf{U}\mathbf{S}\mathbf{V}^T)^{-1}\mathbf{V}\mathbf{S}^T\mathbf{U}^T \\ &= \mathbf{V}(\mathbf{S}^T\mathbf{S})^{-1}\mathbf{V}^T\mathbf{V}\mathbf{S}^T\mathbf{U}^T \\ &= \mathbf{V}(\mathbf{S}^T\mathbf{S})^{-1}\mathbf{S}^T\mathbf{U}^T \\ &= \mathbf{V}[\mathbf{S}_1^{\#} \;\; \mathbf{0}]\mathbf{U}^T \end{aligned},$$

where $\mathbf{S}_1^{\#}$ is the pseudo-inverse of the nonzero square portion of **S**. For singular values that are exactly zero, they are replaced with zero in the same location when forming $\mathbf{S}_1^{\#}$. Thus

$$\hat{\mathbf{x}} = \mathbf{V}\begin{bmatrix}\mathbf{S}_1^{\#} & \mathbf{0}\end{bmatrix}\mathbf{U}^T\mathbf{y}.$$

This shows that a pseudo-inverse can be computed even when $\mathbf{H}^T\mathbf{H}$ is singular. The pseudo-inverse provides the minimal norm solution for a *rank-deficient* (rank < min(m,n)) **H** matrix.

A.4.7 Condition Number

The *condition number* of square matrix **A** is a measure of sensitivity of errors in $\mathbf{A}^{-1}$ to perturbations in **A**. This is used to analyze growth of numerical errors. The condition number is denoted as

$$\boxed{\kappa_p(\mathbf{A}) = \|\mathbf{A}\|_p \|\mathbf{A}^{-1}\|_p} \tag{A4-7}$$

when using the an l_p induced matrix norm for **A**. Because it is generally not convenient to compute condition numbers by inverting a matrix, they are most often computed from the singular values of matrix **A**. Decomposing $\mathbf{A} = \mathbf{U}\mathbf{S}\mathbf{V}^T$ where **U** and **V** are orthogonal and S_i are the singular values in **S**, then

$$\boxed{\kappa_2(\mathbf{A}) = \frac{\max_i(S_i)}{\min_i(S_i)}.} \tag{A4-8}$$

APPENDIX B

PROBABILITY AND RANDOM VARIABLES

This Appendix provides a short summary of probability, random variables, and stochastic processes as applied to estimation. Further information may be found in most books on estimation, probability or signal processing; for example, Papoulis and Pillai (2002), Parzen (1960), Fisz (1963), Jazwinski (1970), Levine (1996, Chapter 34), Åström (1970), Whalen (1971), Davenport and Root (1958), Bar-Shalom et al. (2001), Simon (2006), Maybeck (1979), and Anderson and Moore (1979).

B.1 PROBABILITY

B.1.1 Definitions

Probability theory was originally developed to analyze the unpredictability of gambling. More generally, probability is used to describe the average relative occurrence of random events. Events are considered random if the outcome cannot be precisely predicted. For example, the throw of a die results in six possible outcomes, and the outcome can only be predicted as an average fraction of the total number of throws.

The *event set* or *sample space* Ω contains samples representing all possible outcomes of an "experiment." The exposed face of the die defines the outcome of a die toss, and we use the number of dots to represent the face; for example, $\Omega = \{1,2,3,4,5,6\}$. Since the outcome must be one of these six possibilities, the sample space Ω is a *certain event*, and the probability of a certain event is defined as one. More generally the sample space is represented as $\Omega = \{A_1, A_2, \ldots, A_n\}$ where A_i are the sample outcomes. Certain (measurable) subsets of the sample space are called *events*. For example, even and odd event subsets can be defined for a single die toss,

$$S_1 = \{1,3,5\}, \quad S_2 = \{2,4,6\},$$

or two-outcome events for two die tosses can be defined as

Advanced Kalman Filtering, Least-Squares and Modeling: A Practical Handbook, by Bruce P. Gibbs

$$S_1 = \{1 \cap 1, 1 \cap 2, 1 \cap 3, 1 \cap 4, 1 \cap 5, 1 \cap 6\}$$
$$S_2 = \{2 \cap 2, 2 \cap 3, 2 \cap 4, 2 \cap 5, 2 \cap 6\}$$
$$\vdots$$

where $\cap$ means *intersection* of events ($A \cap B = AB = A$ and B). These subsets are considered events for purposes of computing probabilities. Additional information on set theory, Borel fields, and probability rules governing subsets may be found in listed references.

The axiomatic definition of probability is based on three postulates:

1. The probability of event A is non-negative: $\Pr(A) \geq 0$.
2. The probability of the certain event (Ω) is one: $\Pr(\Omega) = 1$.
3. The probability that either of mutually exclusive events A and B will occur is the sum of the probabilities: $\Pr(A \cup B) = \Pr(A) + \Pr(B)$ where $\cup$ means the *union* of events ($A \cup B = A + B = A$ or B). Mutually exclusive events are those having no elements in common so that $A \cap B = \phi$ and ϕ is the empty set.

Probabilities are defined for given events. Using these axioms it can be shown that $\Pr(\phi) = 0$, $\Pr(\bar{A}) = 1 - \Pr(A)$, where $\bar{A}$ is the complement of A, and

$$\Pr(A_1 \cup A_2 \cup \ldots \cup A_n) \leq \sum_{i=1}^{n} \Pr(A_i) \qquad \text{(B1-1)}$$

where the A_i may or may not be mutually exclusive events.

Use of the above rules leads to an empirical definition based on the relative frequency of occurrence of an event:

$$\Pr(A) = \lim_{n \to \infty} \frac{n_A}{n} \qquad \text{(B1-2)}$$

where n is the total number of trials and n_A is the number of occurrences of event A. This relationship does not allow precise measurement because the number of trials in an experiment is always finite.

B.1.2 Joint and Conditional Probability, and Independence

Most applications of probability theory involve two or more events, and thus it is necessary to compute the probability that multiple events occur. That probability depends on whether the events are independent or correlated. Events are considered *independent* when the outcome of one does not impact the probability of the other. For example, the second toss of a die is independent of the result of the first toss. However, when receiving an English message, the probability of receiving letter "u" increases when the previous letter is "q," so the events are somewhat dependent.

When two events are independent, the *joint probability* of events is equal to the product of the individual probabilities:

$$\Pr\{A \cap B\} = \Pr\{AB\} = \Pr\{A\}\Pr\{B\}.$$

When tossing two unbiased coins, there are four equally probable outcomes: $\Omega = \{hh, ht, th, tt\}$. The probability of obtaining two "heads" is $\Pr\{hh\} = \Pr\{h\}\Pr\{h\} = 0.25$, so the events are independent. Three or more events may be independent in pairs but not necessarily independent as a group.

When events depend on each other, *conditional probability* is used to specify the probability of one event given knowledge of the other event. The conditional probability of event A given knowledge that event B has occurred is written as $\Pr\{A \mid B\}$. The joint probability of both events is

$$\begin{aligned}\Pr\{AB\} &= \Pr\{A \mid B\}\Pr\{B\} \\ &= \Pr\{B \mid A\}\Pr\{A\}\end{aligned}.$$

Thus

$$\Pr\{A \mid B\} = \frac{\Pr\{B \mid A\}\Pr\{A\}}{\Pr\{B\}}$$

provided that $\Pr\{B\} > 0$. This is one form of Bayes' theorem for conditional probabilities. It can be expanded to three or more events with appropriate groupings. Conditional probabilities of mutually exclusive events are additive.

Conditional probabilities of multiple events are used to derive many estimation algorithms.

B.2 RANDOM VARIABLE

Probabilities are defined for events. Estimators use measured variables to infer the values of other system variables, where measured and unmeasured variables are random in the sense that they cannot be perfectly predicted. Random variables are the link between measured variables and probability space.

A random variable, denoted by x, is a function $X(\omega)$ that assigns a real value to every event ω of an "experiment." The two main requirements on function $X(\omega)$ are (1) the set $\{X(\omega) \le x\}$ must be an event for every x, and (2) $\Pr\{X = -\infty\} = \Pr\{X = \infty\} = 0$. There is also a technical requirement that $X(\omega)$ be a measurable function, and this is always true for piecewise continuous $X(\omega)$. (See Grimmett and Stirzaker 2001, chapter 2.)

Random variables may have discrete values, or may be continuous functions of ω. For example, the number of dots on the face of a die may be regarded as function $X(\omega)$ of the experiment of tossing the die. This is a discrete random variable. In estimation applications, random variables include measured variables, measurement noise, system state variables, and process noise. These variables are usually continuous.

B.2.1 Distribution and Density Functions

Since $\{X \le x\}$ is an event, $\Pr\{X \le x\}$ can be computed. This is called the *probability distribution function*:

$$F_X(x) \triangleq \Pr\{X \le x\} \tag{B2-1}$$

defined for $-\infty < x < \infty$. From definitions it is seen that $F_X(-\infty) = 0$, $F_X(\infty) = 1$, and $F_X(x + \Delta x) \geq F_X(x)$ for any (all) $\Delta x > 0$.

The derivative of the distribution function is the *probability density function*

$$p_X(x) \triangleq \lim_{\Delta x \to 0} \frac{F_X(x+\Delta x) - F_X(x)}{\Delta x} = \frac{dF_X(x)}{dx} \tag{B2-2}$$

provided that $dF_X(x)/dx$ exists. Again from definitions we note that $p_X(-\infty) = p_X(\infty) = 0$ and $\int_{-\infty}^{\infty} p_X(x)\,dx = 1$. If $F_X(x)$ is discontinuous, $p_X(x)$ contains impulses.

When working with pairs of random variables, the joint distribution function is

$$F_{X,Y}(x, y) \triangleq \Pr\{(X \leq x) \cap (Y \leq y)\}. \tag{B2-3}$$

This can be differentiated to obtain a joint density function

$$p_{X,Y}(x, y) \triangleq \frac{\partial^2}{\partial x \partial y} F_{X,Y}(x, y). \tag{B2-4}$$

Since $F_X(x) = F_{X,Y}(x,\infty)$,

$$p_X(x) = \int_{-\infty}^{\infty} p_{X,Y}(x, y)\,dy. \tag{B2-5}$$

Multiple random variables that are independent and have the same distribution function are often called *independent, identically distributed* (*i.i.d*).

B.2.2 Bayes' Theorem for Density Functions

Bayes' theorem can also be used for density functions of jointly distributed random variables. The joint probability density function of X and Y is

$$p_{X,Y}(x, y) = p_{X|Y}(x \mid y) p_Y(y) = p_{Y|X}(y \mid x) p_X(x) \tag{B2-6}$$

Where

$p_{X|Y}(x|y)$ is the conditional probability density function of X given Y = y,

$p_{Y|X}(y|x)$ is the conditional probability density function of Y given X = x, and

$p_X(x)$ and $p_Y(y)$ are unconditional (marginal) probability densities.

Hence the conditional density of X given y can be computed from the conditional probability density of Y given x:

$$p_{X|Y}(x \mid y) = \frac{p_{Y|X}(y \mid x) p_X(x)}{p_Y(y)} \tag{B2-7}$$

provided that $p_Y(y) > 0$. This gives the *a posteriori* probability density of X given the *a priori* probability density $p_X(x)$. The denominator of (B2-7) can be written as

$$p_Y(y) = \int_{-\infty}^{\infty} p_{Y|X}(y \mid x) p_X(x)\,dx \tag{B2-8}$$

to obtain

$$p_{X|Y}(x \mid y) = \frac{p_{Y|X}(y \mid x) p_X(x)}{\int_{-\infty}^{\infty} p_{Y|X}(y \mid x) p_X(x) dx}. \tag{B2-9}$$

This is Bayes' theorem. Since the integrand is equal to the numerator, $p_Y(y)$ can be viewed as a normalization factor that makes

$$\int_{-\infty}^{\infty} p_{X|Y}(x \mid y) dx = 1.$$

For vector random variables the vector form of (B2-7) is

$$p_{\mathbf{X}|\mathbf{Y}}(\mathbf{x} \mid \mathbf{y}) = \frac{p_{\mathbf{Y}|\mathbf{X}}(\mathbf{y} \mid \mathbf{x}) p_{\mathbf{X}}(\mathbf{x})}{p_{\mathbf{Y}}(\mathbf{y})} \tag{B2-10}$$

where $p_{\mathbf{X}|\mathbf{Y}}(\mathbf{x} \mid \mathbf{y}) = p_{X_1,X_2,\ldots X_n|Y_1,Y_2,\ldots Y_n}(x_1, x_2, \ldots, x_n \mid y_1, y_2, \ldots, y_m)$ and other vector probability densities are defined accordingly.

B.2.3 Moments of Random Variables

It is often easier to derive estimation algorithms using first- and second-order moments (mean and covariances) of random variables rather than probability densities. The mean or expected value of a random variable is a "center-of-mass" integral,

$$\bar{x} \triangleq E[x] = \int_{-\infty}^{\infty} x\, p_X(x) dx, \tag{B2-11}$$

and the variance is

$$\sigma_x^2 \triangleq E[(x-\bar{x})^2] = \int_{-\infty}^{\infty} (x-\bar{x})^2 p_X(x) dx. \tag{B2-12}$$

For vector random variables the mean is

$$\bar{\mathbf{x}} \triangleq E[\mathbf{x}] = \int_{-\infty}^{\infty} \mathbf{x}\, p_{\mathbf{X}}(\mathbf{x}) d\mathbf{x}. \tag{B2-13}$$

where the vector integral is defined as

$$\int_{-\infty}^{\infty} \mathbf{x}\, p_{\mathbf{X}}(\mathbf{x}) d\mathbf{x} = \int_{-\infty}^{\infty} \cdots \int_{-\infty}^{\infty} \begin{bmatrix} x_1 \\ x_2 \\ \vdots \\ x_n \end{bmatrix} p_{\mathbf{X}}(x_1, x_2, \ldots, x_n) dx_1, \ldots, dx_n.$$

The conditional mean is

$$E[\mathbf{x} \mid \mathbf{y}] = \int_{-\infty}^{\infty} \mathbf{x}\, p_{\mathbf{X}|\mathbf{Y}}(\mathbf{x} \mid \mathbf{y}) d\mathbf{x} \tag{B2-14}$$

Notice that $E[\mathbf{x} \mid \mathbf{y}]$ is a random variable, conditioned on $\mathbf{y}$, that has expected value

$$\begin{aligned} E[E[\mathbf{x} \mid \mathbf{y}]] &= \int_{-\infty}^{\infty} \left(\int_{-\infty}^{\infty} \mathbf{x}\, p_{\mathbf{X}|\mathbf{Y}}(\mathbf{x} \mid \mathbf{y}) d\mathbf{x} \right) p_{\mathbf{Y}}(\mathbf{y}) d\mathbf{y} \\ &= \int_{-\infty}^{\infty} \mathbf{x}\, p_{\mathbf{X}}(\mathbf{x}) d\mathbf{x} \\ &= \bar{\mathbf{x}} \end{aligned} \tag{B2-15}$$

The conditional mean plays a fundamental role in estimation and advanced treatment of probability theory.

The covariance of **x** is

$$E\left[(\mathbf{x}-\bar{\mathbf{x}})(\mathbf{x}-\bar{\mathbf{x}})^T\right]=\int_{-\infty}^{\infty}(\mathbf{x}-\bar{\mathbf{x}})(\mathbf{x}-\bar{\mathbf{x}})^T p_{\mathbf{X}}(\mathbf{x})\,d\mathbf{x}. \tag{B2-16}$$

Expectation is a linear operation.

B.2.4 Gaussian Distribution

For a single continuous random variable x with mean $\bar{x}$ and nonzero standard deviation σ_x, the *Gaussian* or *normal* probability density is

$$p_X(x)=\frac{1}{\sqrt{2\pi}\sigma_x}\exp\left(-(x-\bar{x})^2/(2\sigma_x^2)\right). \tag{B2-17}$$

The Gaussian distribution function is

$$\begin{aligned}F_X(x)&=\frac{1}{\sqrt{2\pi}\sigma_x}\int_{-\infty}^{x}\exp\left(-(\lambda-\bar{x})^2/(2\sigma_x^2)\right)d\lambda\\&=\begin{cases}\frac{1}{2}+\frac{1}{2}\operatorname{erf}\left(\frac{x-\bar{x}}{\sqrt{2}\sigma_x}\right) & \text{for x} \geq \bar{x}\\ \frac{1}{2}-\frac{1}{2}\operatorname{erf}\left(\frac{\bar{x}-x}{\sqrt{2}\sigma_x}\right) & \text{for x} < \bar{x}\end{cases}\end{aligned} \tag{B2-18}$$

where $\operatorname{erf}\left((x-\bar{x})/(\sqrt{2}\sigma_x)\right)$ is the error function.

The joint probability density for an m-vector of Gaussian random variables with a common mean $\bar{\mathbf{x}}$ and covariance $\mathbf{P}_{xx}$ is equal to

$$p_{\mathbf{X}}(\mathbf{x})=\frac{1}{(2\pi)^{m/2}\left|\mathbf{P}_{xx}\right|^{1/2}}\exp\left(-\frac{1}{2}(\mathbf{x}-\bar{\mathbf{x}})^T\mathbf{P}_{xx}^{-1}(\mathbf{x}-\bar{\mathbf{x}})\right) \tag{B2-19}$$

where $|\mathbf{P}_{xx}|$ is the determinant of $\mathbf{P}_{xx}$.

B.2.5 Chi-Squared Distribution

The sum of m independent, squared normalized $N(0,\sigma_i^2)$ Gaussian variables,

$$\chi^2=\sum_{i=1}^{m}(x_i/\sigma_i)^2, \tag{B2-20}$$

is a Chi-squared-distributed random variable having m degrees-of-freedom. The cost function of an optimally weighted least-squares problem is a χ^2-distributed variable (or 1/2 of that value depending on the definition of cost). The χ^2 probability density is

$$p_X(x)=\frac{x^{m/2-1}}{2^{m/2}\Gamma(m/2)}\exp(-x/2) \qquad x\geq 0 \tag{B2-21}$$

where Γ is the gamma function. The distribution function is

$$F_X(x) = P\left(\frac{m}{2}, \frac{x}{2}\right) \qquad x \geq 0 \tag{B2-22}$$

where P is the incomplete gamma function (see Press et al. 2006, section 6.2). The mean of a χ^2 distributed variable with m degrees-of-freedom is m, and the variance is $2m$.

B.3 STOCHASTIC PROCESSES

While random variables are rules (functions) for assigning outcomes of experiments to numbers, a stochastic process $\{X(\tau): -\infty < \tau < \infty\}$ is the numerical value assigned to every experimental outcome ω by the real function $X(t,\omega)$. For each ω there is a whole function $X(t)$ for $-\infty < t < \infty$, and elemental ω values may be associated with different points in time. $X(t)$ is a random variable that changes with time (or possibly space.) If time t is real and continuous, the stochastic process is either a discrete or continuous random process, depending on whether the values $X(t)$ are discrete or continuous. It time is discrete, $X(t_i)$ is either a discrete or continuous random sequence. Stochastic processes may be a function of space rather than time, but the term usually implies that the independent variable is time.

Probability density and distribution functions are defined for stochastic processes by allowing the densities and distributions to be functions of time. A stochastic process may have a mean, variance, and an autocorrelation function, defined as

$$\rho_{xx}(t,\lambda) \triangleq E[X(t)X(\lambda)]. \tag{B3-1}$$

Wide sense stationary processes have means and autocorrelation functions that are invariant with time shifts:

$$E[X(t)] = \text{constant}, \qquad \text{for all } t \tag{B3-2}$$

$$\rho_{xx}(\tau) = E[X(t)X(t+\tau)], \quad \text{for all t and } \tau. \tag{B3-3}$$

If probability density functions and joint probability density functions for different time samples are invariant with time shifts, the process is *strictly stationary*.

The time average of a stationary random process is generally a random variable, but for a *mean ergodic* process it is equal to the constant statistical (ensemble) average. The time and ensemble moments up to fourth order are equivalent for an *autocorrelation ergodic* process.

The power spectral density (PSD) of a wide-sense stationary stochastic process is the Fourier transform of the time average of the autocorrelation function:

$$\begin{aligned} S_x(\omega) &= \lim_{T\to\infty} \frac{1}{2T} \int_{-\infty}^{\infty} \left(\int_{-T}^{T} \rho_{xx}(t,t+\tau) e^{-j\omega\tau} dt \right) d\tau \\ &= \int_{-\infty}^{\infty} \left(\lim_{T\to\infty} \frac{1}{2T} \int_{-T}^{T} \rho_{xx}(t,t+\tau)\, dt \right) e^{-j\omega\tau} d\tau \end{aligned} . \tag{B3-4}$$

Stochastic processes $X(t)$ and $Y(t)$ are called

1. *mutually independent* (independent in pairs) if $F_{X,Y}[X(t_1),Y(t_2)] = F_X[X(t_1)]$ $F_Y[Y(t_2)]$ or equivalently $p_{X,Y}[X(t_1),Y(t_2)] = p_X[X(t_1)]p_Y[Y(t_2)]$,
2. *orthogonal* if $E[X(t_1)Y(t_2)] = 0$, and
3. *uncorrelated* if $E[(X(t_1)-\bar{X})(Y(t_2)-\bar{Y})] = 0$

for all pairs t_1 and t_2. Note that independence in pairs is insufficient to prove independence in general for stochastic processes. More generally, *independence* means that every possible joint probability density is equal to the products of the marginal densities; for example,

$$p_{X,Y,...}[X(t_1),Y(t_2),X(t_3),Y(t_4),...]=p_X[X(t_1)]p_Y[Y(t_2)]p_X[X(t_3)]p_Y[Y(t_4)]....$$

If the stochastic processes are Gaussian, they are independent if and only if $E[X(t_1)Y(t_2)] = E[X(t_1)]E[Y(t_2)]$ for all pairs t_1 and t_2.

B.3.1 Wiener or Brownian Motion Process

A *Wiener* or *Brownian motion* process is often modeled as:

$$\dot{x}(t)=w(t) \tag{B3-5}$$

where $w(t)$ is white noise. The variance of $x(t)$ grows linearly with time:

$$\sigma_x^2(t_2)-\sigma_x^2(t_1)=W_s\cdot(t_2-t_1) \tag{B3-6}$$

where W_s is the constant-with-frequency PSD of $w(t)$; that is, $E[w(t)w(\tau)] = W_s\delta(t-\tau)$. Equation (B3-6) is often used as one condition defining a Wiener process. Equation (B3-5) is a differential model of the process that has the desired integral properties. The discretely sampled integral of Brownian motion is a *random walk* sequence.

B.3.2 Markov Process

A continuous-time Markov process is a stochastic process in which the probability distribution of any future state depends only on the current state, not on the past history leading to that state. A continuous *first-order Markov process* is one in which the probability distribution of the scalar output depends on only one point immediately in the past:

$$F_X[x(t_k)\,|\,x(t_{k-1}),\ldots,x(t_1)]=F_X[x(t_k)\,|\,x(t_{k-1})] \tag{B3-7}$$

for every $t_1 < t_2 < \ldots < t_k$. The time-invariant stochastic differential equation

$$\dot{x}(t)=-\frac{1}{\tau}x(t)+w(t) \tag{B3-8}$$

models a *first-order Markov process* when $w(t)$ is white noise. This process has a "low-pass" characteristic, because the PSD of $x(t)$ is nearly constant up to frequency $1/2\pi\tau$, and then asymptotically falls off at −20 dB/decade. Most low-order Markov processes have low-pass or band-pass characteristics, and thus are often referred to as colored noise models. Colored noise processes are used to model time-correlated system inputs, or inputs that are quasi-random and otherwise difficult to model.

It should be noted that white noise has infinite power, so inclusion of white noise input in a differential equation is not physically realistic. However, the above model for a Markov process is commonly used in engineering applications because discrete samples of the continuous process have the desired properties.

B.3.3 Differential and Integral Equations with White Noise Inputs

Stochastic processes are usually modeled as differential or integral equations containing white noise inputs. The presence of infinite-power white noise in differential equations presents problems because integrals or derivatives of functions $f[w(t),t]$ are not uniquely defined when $w(t)$ is white noise.

To understand the problem, suppose that we have samples of a stochastic process $f[w(t)]$ where $f[\cdot]$ is a piecewise continuous function and $w(t)$ is white noise. If we try to express the time derivative as

$$\frac{df}{dt} = \lim_{h \to 0} \frac{f[w(t+h)] - f[w(t)]}{h},$$

we see that the limit does not exist because samples of $w(t_i + h)$ for different h are uncorrelated. There is also a problem when trying to express

$$\int_0^T f[w(\tau)]d\tau = \lim_{n \to \infty} \left(\frac{T}{n}\right) \sum_{i=0}^{n-1} (\alpha f[w(iT/n)] + (1-\alpha) f[w((i+1)T/n)]).$$

for $0 \le \alpha \le 1$. Notice that different results are obtained for different values of α, and again the limit does not exist.

We are primarily interested in convergence in expected value of integrals involving random processes. However, we first define two terms. A stochastic process $x(t)$ is *mean square continuous* at t if

$$\lim_{h \to 0} E\left[|x(t+h) - x(t)|^2\right] = 0 \tag{B3-9}$$

for all t and $t + h$ within a specified region. Notice that the requirement involves the expected value. Functions are mean square continuous if their autocorrelation functions at zero time shift are continuous. White noise is not mean square continuous.

A stochastic process $x(t)$ is said to be *mean square integrable* over $[0,T]$ if

$$\lim_{h \to 0} E\left[\left| h \sum_{i=0}^{T/h} x(\lambda_i) - \int_0^T x(\tau)d\tau \right|^2\right] = 0 \tag{B3-10}$$

where $ih \le \lambda_i < (i+1)h$. For the limit to exist, the condition

$$E\left[\left| h \sum_{i=0}^{T/h} x(\lambda_i) \right|^2\right] < \infty$$

must be satisfied for all $h > 0$ and

$$E\left[\left| \int_0^T x(\tau)d\tau \right|^2\right] < \infty. \tag{B3-11}$$

Actually, the step sizes can be variable rather than fixed at h, provided that the maximum step approaches zero in the limit. Functions $x(t)$ having autocorrelation functions that are Riemann integrable over the specified time shifts are mean square integrable. This requirement again excludes white noise.

To evaluate integrals involving white noise, Itô and Stratonovich developed two different definitions of stochastic integrals that express the differential equation

$$\frac{dx(t)}{dt} = f[x(t)] + g[x(t)]w(t)$$

as

$$dx(t) = f[x(t)]\,dt + g[x(t)]\,d\beta(t) \tag{B3-12}$$

where $w(t) \sim d\beta(t)/dt$ and $d\beta(t)$ is a process of independent Brownian motions. Then the integral of equation (B3-12) can be defined to have a mean-squared limit.

For mean square integrable functions, the operations of expectation and integration can be interchanged. That is,

$$E\left[\int_a^b x(t)\,dt\right] = \int_a^b E[x(t)]\,dt \tag{B3-13}$$

where the limits a and b may be finite or infinite. Furthermore, for mean square continuous functions $g[x(t)]$, expectations of stochastic integrals can generally be evaluated as ordinary integrals.

Jazwinski (1970, chapter 3) provides an extended discussion of differential and integral equations involving white noise and Brownian processes. This includes interpretations of stochastic integrals by Itô and Stratonovich. Åström (1970, chapter 3), Schweppe (1973, section 4.2), Levine (1996, chapters 34, 60), Davenport and Root (1958, section 4.7), and Cooper and McGillem (1967, appendix D) also provide more information on the issues.

BIBLIOGRAPHY

Abramson, N. (1963) *Information Theory and Coding*. New York: McGraw-Hill.

Agee, W.S. and R.H. Turner (1972) Triangular decomposition of a positive definite matrix plus a symmetric dyad with applications to Kalman filtering. White Sands Missile Range Technical Report No. 38.

Akaike, H. (1969) Fitting autoregressive models for prediction. *Annals of the Institute of Statistical Mathematics*, 21: 243–247.

Akaike, H. (1970a) Statistical prediction identification, *Annals of the Institute of Statistical Mathematics*, 22: 203–217.

Akaike, H. (1970b) On a semi-automatic power spectrum procedure. *Proceedings of the 3rd Hawaii International Conference on System Sciences*, Honolulu, HI, pp. 974–977, January 14–16.

Akaike, H. (1971) Autoregressive model fitting for control. *Annals of the Institute of Statistical Mathematics*, 23: 163–180.

Akaike, H. (1972) Information theory and an extension of the maximum likelihood principle. *2nd International Symposium on Information Theory*, Armenia, USSR. Budapest: Akadémiai Kiadó.

Akaike, H. (1974a) Stochastic theory of minimal realization. *IEEE Transactions on Automatic Control*, AC-19: 667–674.

Akaike, H. (1974b) A new look at the statistical model identification. *IEEE Transactions on Automatic Control*, AC-19: 716–723; reprinted in *Modern Spectral Analysis* by IEEE Press, 1978.

Akaike, H. (1975) Markovian representation of stochastic processes by canonical variables. *SIAM Journal on Control*, 13: 162–173.

Akaike, H. (1976) Canonical correlation analysis of time series and the use of an information criterion. In *System Identification Advances and Case Studies*, pp. 27–96, R.K. Mehra and D.G. Lainiotis (Eds.). New York: Academic Press.

Aldrich, G.T. (1971) *Final Technical Report for KAlman Program for Positioning Aircraft (KAPPA)*. Riverdale, MD: Wolf R&D Corp.

Anderson, J.D. (2005) *Introduction to Flight*, 5th edition. New York: McGraw-Hill.

Anderson, B.D.O. and J.B. Moore (1979) *Optimal Filtering*. Englewood Cliffs, NJ: Prentice-Hall; republished by Dover, 2005.

Advanced Kalman Filtering, Least-Squares and Modeling: A Practical Handbook, by Bruce P. Gibbs

Anderson, M.P. and W.W. Woessner (1992) *Applied Groundwater Modeling*. San Diego, CA: Academic Press.

Anderson, E., Z. Bai, C. Bischof, S. Blackford, J. Demmel, J. Dongarra, J. Du Croz, A. Greenbaum, S. Hammarling, A. McKenney, and D. Sorensen (1999) *LAPACK Users' Guide*, 3rd edition. Philadelphia: SIAM.

Andrews, A. (1968) A square root formulation of the Kalman covariance equations. *AIAA Journal*, 6: 1165–1166.

Arnold, W.F. III and A.J. Laub (1984) Generalized eigenproblem algorithms and software for algebraic riccati equations. *Proceedings of the IEEE*, 72: 1746–1754.

Åström, K.J. (1970) *Introduction to Stochastic Control Theory*. New York: Academic Press.

Åström, K.J. (1980) Maximum likelihood and prediction error methods. *Automatica*, 16(5): 551–574.

Åström, K.J. and P. Eykhoff (1971) System identification, a survey, preprints 2nd IFAC Symposium on Identification, Prague. *Automatica*, 7: 123–162.

Astronomical Almanac (2009) US Naval Observatory/Nautical Almanac Office. http://asa.usno.navy.mil/

Athans, M. and P. Falb (1966) *Optimal Control*. New York: McGraw-Hill.

Avallone, E.A. and T. Baumeister III (1996) *Marks' Standard Handbook for Mechanical Engineers*, 10th edition. New York: McGraw-Hill.

Balchen, J.G. and K.I. Mummé (1988) *Process Control Structures and Applications*. New York: Van Nostrand Reinhold.

Balchen, J.G., D. Ljundquist, and S. Strand (1988) Predictive control based on state space models. *American Control Conference*, Atlanta, GA, June 1988.

Banavar, R.N. and J.L. Speyer (1991) A linear-quadratic game approach to estimation and smoothing. *Proceedings of the IEEE Automatic Control Conference*, Brighton, UK, pp. 2818–2822, December 1991.

Bar-Shalom, Y. and T.E. Fortmann (1988) *Tracking and Data Association*. San Diego, CA: Academic Press.

Bar-Shalom, Y., X. Rong Li, and T. Kirubarajan (2001) *Estimation with Applications to Tracking and Navigation*. New York: Wiley.

Barrett, R., M. Berry, T.F. Chan, J. Demmel, J. M. Donato, J. Dongarra, V. Eijkhout, R. Pozo, C. Romine, and H. Van der Vorst (1993) *Templates for the Solution of Linear Systems: Building Blocks for Iterative Methods*. Philadelphia: SIAM. http://www.netlib.org/templates/templates.html

Basseville, M. (1988) Detecting changes in signals and systems—A survey. *Automatica*, 24: 309–326.

Basseville, M. and A. Benveniste (Eds.) (1986) *Detection of Abrupt Changes in Signals and Dynamical Systems*. Berlin: Springer-Verlag.

Bate, R.R., D.D. Mueller, and J.E. White (1971) *Fundamentals of Astrodynamics*. New York: Dover.

Battin, R.H. (1964) *Astronautical Guidance*. New York: McGraw-Hill.

Bauer, D. (2005) Comparing the CCA subspace method to pseudo maximum likelihood methods in the case of no exogenous inputs. *Journal of Time Series Analysis*, 26(5): 631–668.

Bauer, D. and L. Ljung (2002) Some facts about the choice of the weighting matrices in Larimore type of subspace algorithms. *Automatica*, 38: 763–773.

Bellantoni, J.F. and K.W. Dodge (1967) A square-root formulation of the Kalman-Schmidt filter. *AIAA Journal*, 5: 1309–1314.

Bennett, S. (1993) *A History of Control Engineering, 1930–1955. IEE Control Engineering Series 47*. London: Peter Peregrinus.

Bergland, G.D. (1969) A guided tour of the Fast Fourier Transform. *IEEE Spectrum*, 6: 41–52; reprinted in *Digital Signal Processing*, IEEE Press, 1972.

Bierman, G.J. (1974) Sequential square root filtering and smoothing of discrete linear systems. *Automatica*, 10: 147–158.

Bierman, G.J. (1977a) An application of the square root information filter to large-scale linear interconnected systems. *IEEE Transactions on Automatic Control*, AC-22(6): 989–991.

Bierman, G.J. (1977b) *Factorization Methods for Discrete Sequential Estimation*. New York: Academic Press; republished by Dover, 2006.

Bierman, G.J. (1983) A new computationally efficient fixed-interval, discrete-time smoother. *Automatica*, 19(5): 503–511.

Bierman, G.J. (1988) A reformulation of the Rauch-Tung-Striebel discrete time fixed interval smoother. *Proceedings of the 27th Conference on Decision and Control*, Austin, TX, pp. 840–844, December 1988.

Bierman, G.J. and K.H. Bierman (1984) Estimation subroutine library user guide–Preliminary. Factorized Estimation Applications, Report 81584. http://netlib.org/a/esl.tgz

Bierman, G.J. and M.W. Nead (1978) A parameter estimation subroutine package (ESP). NASA Jet Propulsion Laboratory, #77-26 Rev 2, October 1978.

Bierman, G.J. and C.L. Thornton (1977) Numerical comparison of Kalman filter algorithms: Orbit determination case study. *Automatica*, 13: 23–35.

Bierman, G.J., M.R. Belzer, J.S. Vandergraft, and D.W. Porter (1990) Maximum likelihood estimation using square root information filters. *IEEE Transactions on Automatic Control*, 35(12): 1293–1298.

Bingham, C., M.D. Godfrey, and J. Tukey (1967) Modern techniques of power spectrum estimation. *IEEE Transactions on Audio and Electroacoustics*, AU-15: 56–66; reprinted in *Modern Spectrum Analysis*, IEEE Press, 1978.

Björck, Å (1996) *Numerical Methods for Least Squares Problems*. Philadelphia: SIAM.

Blackman, S.S. (1986) *Multiple-Target Tracking with Radar Applications*. Norwood, MA: Artech House.

Blackman, S. and R. Popoli (1999) *Design and Analysis of Modern Tracking Systems*. Norwood, MA: Artech House.

Bode, H.W. and C.E. Shannon (1950) A simplified derivation of linear least square smoothing and prediction theory. *Proceedings of the IRE*, 38: 417–426.

Boggs, P.T., R.H. Byrd, and R.B. Schnabel (1985) *Numerical Optimization 1984*. Philadelphia: SIAM.

Box, G.E.P., G.M. Jenkins, and G.C. Reinsel (2008) *Time Series Analysis Forecasting and Control*, 4th edition. San Francisco, CA: Holden-Day.

Bozic, S.M. (1994) *Digital and Kalman Filtering*. Portsmouth, NH: Butterworth-Heinemann.

Britting, K.R. (1971) *Inertial Navigation Systems Analysis*. New York: Wiley-Interscience.

Brockwell, P.J. and R.A. Davis (2006) *Time Series: Theory and Methods*, 2nd edition. New York: Springer-Verlag.

Bronzino, J. (2006) *The Biomedical Engineering Handbook*. Boca Raton, FL: CRC Press.

Brouwer, D. and G. Clemence (1961) *Methods of Celestial Mechanics*. New York: Academic Press.

Bryson, A.E. and M. Frazier (1963) Smoothing for Linear and Nonlinear Dynamic Systems, Technical Report ASD-TDR-63-119, Wright Patterson Air Force Base, Ohio, pp. 353–364.

Bryson, A.E. and L.J. Henrikson (1968) Estimation using sampled data containing sequentially correlated noise. *Journal of Spacecraft and Rockets*, 5: 662–665; reprinted in *Kalman Filtering: Theory and Application*, IEEE Press, 1985.

Bucy, R.S. and P.D. Joseph (1968) *Filtering for Stochastic Processes with Applications to Guidance*, 2nd edition. New York: John Wiley; reprinted by AMS Chelsea Publishing, 2005.

Bühler, W.K (1981) *Gauss, A Bibliographical Study*. Berlin: Springer-Verlag.

Burg, J.P. (1967) Maximum entropy spectral analysis. *Proceedings of the 37th Meeting of Society of Exploration Geophysicists*, Oklahoma City: OK; reprinted in *Modern Spectrum Analysis*, IEEE Press, 1978.

Burg, J.P. (1968) A new analysis technique for time series data. *NATO Advanced Study Institute on Signal Processing with Emphasis on Underwater Acoustics*, Enchede, Holland, August 1968; reprinted in *Modern Spectrum Analysis*, IEEE Press, 1978.

Byers, R. (1987) Solving the algebraic Riccati equation with the matrix sign. *Linear Algebra and Its Applications*, 85: 267–279.

Cadzow, J.A. (1980) High performance spectral estimation—A new ARMA method. *IEEE Transactions on Acoustics, Speech, and Signal Processing*, ASSP-28: 524–529.

Cadzow, J.A. (1983) ARMA time series modeling: An effective method. *IEEE Transactions on Aerospace and Electronic Systems*, AES-19(1): 49–58.

Caglayan, A.K. and R.E. Lancraft (1983) Reinitialization Issues in Fault Tolerant Systems. *Proceedings of the American Control Conference*, New York, NY, pp. 952–955, June 1983.

Campbell, M.E. and B. Udrea (2002) Collision avoidance in satellite clusters *Proceedings of the American Control Conference*, Anchorage, AK, pp. 1686–1692, May 2002.

Capon, J. (1969) High Resolution frequency-wavenumber spectrum analysis. *Proceedings of the IEEE*, 57: 1408–1418.

Carlson, N.A. (1973) Fast triangular formulation of the square-root filter. *AIAA Journal*, 11(9): 1259–1265.

Carr, J.L. (2001) Integrating equations of motion as an optimization problem. *2001 Flight Mechanics Symposium*, NASA/GSFC, NASA CP-2001-209986, June 2001

Carr, J.L., B.P. Gibbs, H. Mandani, and N. Allen (2008) INR performance of the GOES-13 spacecraft measured in orbit. *AAS GN&C Conference*, AAS 08–077, January 2008

Chang, C.B., M. Athans, and R. Whiting (1977) On the state and parameter estimation for maneuvering reentry vehicles. *IEEE Transactions on Automatic Control*, 22(1): 99–105.

Chien, T.T. (1972) *An Adaptive Technique for a Redundant Sensor Navigation System*. Cambridge, MA: Draper Labs.

Childers, D.G. (Ed.) (1978) *Modern Spectrum Analysis*. New York: IEEE Press.

Chin, M.M., C.C. Goad, and T.V. Martin (1972) *GEODYN System Description*. Riverdale, MD: Wolf Research and Development Corporation.

Chiuso, A. (2007) Some insights on the choice of the future horizon in closed-loop CCA-type Subspace Algorithms. *Proceedings of 2007 American Control Conference*, New York, July 2007.

Chiuso, A. (2010) On the asymptotic properties of closed loop CCA-type subspace algorithms: Equivalence results and choice of the future horizon. *IEEE Transactions on Automatic Control*, to appear.

Chobotov, V.A. (2002) *Orbital Mechanics*. Reston, VA: AIAA.

Cline, A.K., C.B. Moler, G.W. Stewart, and J.H. Wilkinson (1979) An estimate for the condition number of a matrix. *SIAM Journal on Numerical Analysis*, 8: 368–375.

Close, C.M. (1966) *The Analysis of Linear Circuits*, New York: Harcourt, Brace and World.

Close, C.M., D.K. Frederick, and J.C. Newell (2001) *Modeling and Analysis of Dynamic Systems*, 3rd edition. New York: John Wiley.

Collins, F.G. and E.C. Knox (1994a) Parameters of Nocilla gas/surface interaction model from measured accommodation coefficients. *AIAA Journal*, 32(4): 765–773.

Collins, F.G. and E.C. Knox (1994b) Determination of wall boundary conditions for high-speed-ratio direct simulation Monte Carlo calculations. *Journal of Science Research*, 31(6): 965–970.

Conte, S.D. (1965) *Elementary Numerical Analysis*. New York: McGraw-Hill.

Cook, S.R. and M.A. Hoffbauer (1997) Nocilla model parameters obtained from forces exerted on surfaces by molecular beams. *Journal of Science Research*, 34(3): 379–383.

Cooley, J.W., P.A.W. Lewis, and P.D. Welch (1967) Historical notes on the Fast Fourier Transform. *IEEE Transactions on Audio and Electroacoustics*. AU-15: 76–79; reprinted in *Digital Signal Processing*, IEEE Press, 1972.

Cooper, G.R. and C.D. McGillem (1967) *Methods of Signal and System Analysis*. New York: Holt, Rinehart and Winston.

Cotton, K.C. and P. Schofield (1970) Analysis of Changes in Performance Characteristics of Steam Turbines. *ASME Winter Annual Meeting*, GER-2561, New York, November 1970.

Cramér, H. (1946) *Mathematical Methods of Statistics*. Princeton, NJ: Princeton University Press.

Cressie, N. (1991) *Statistics for Spatial Data*. New York: John Wiley.

Cruz, J.Y., M.D. Menn, W.A. Feess, and V. Nuth (2005) GPS constellation navigation: L-band measurement only. *Institute of Navigation NTM*, San Diego, CA: 792–798, January 2005.

Curkendall, D.W. (1978) Algorithms for isolating worst case systematic data errors. *Journal of Guidance and Control*, 1(1): 56–62.

Dahlen A. (2001) Identification of stochastic systems: Subspace methods and covariance extension, Doctoral Thesis, Royal Institute of Technology, Stockholm, Sweden.

Dahlen, A., A. Liindquist, and J. Mari (1998) Experimental evidence showing that stochastic subspace identification methods may fail. *Systems and Control Letters*, 34: 303–312.

Davenport, W.B. and W.L. Root (1958) *Random Signals and Noise*. New York: McGraw-Hill.

Deistler, M., K. Peternell, and W. Scherrer (1995) Consistency and relative efficiency of subspace methods. *Automatica*, 31: 1865–1875.

Denman, E. and A. Beavers (1976) The matrix sign function and computations in systems. *Applied Mathematics and Computation*, 2: 63–94.

Denham, W.F. and S. Pines (1966) Sequential estimation when measurement function nonlinearity is comparable to measurement error. *AIAA Journal*, 4: 1071–1076.

Dennis, J.E. and R.B. Schnabel (1983) *Numerical Methods for Unconstrained Optimization and Nonlinear Equations*, Philadelphia: Prentice-Hall; republished by SIAM, 1996.

Dennis, J.E., D.M. Gay, and R.E. Welsch (1981a) An Adaptive Nonlinear Least-Squares Algorithm. *TOMS*, 7: 348–368.

Dennis, J.E., D.M. Gay, and R.E. Welsch (1981b) Algorithm 573 N2LSOL-An Adaptive Nonlinear Least-Squares Algorithm [E4]. *TOMS*, 7: 369–383.

Department of Defense World Geodetic System 1984, Its Definition and Relationships With Local Geodetic Systems, Third Edition (2004) National Geospatial-Intelligence Agency TR8350.2, June 2004. http://earth-info.nga.mil/GandG/publications/tr8350.2/tr8350_2.html.

DeRusso, P.M., R.J. Roy, and C.M. Close (1965) *State Variables for Engineers*. New York: John Wiley.

Dixon, W.J. (Ed.) (1975) *BMDP Biomedical Computer Programs*. Berkeley, CA: University of California Press.

Dongarra, J. and H. Van Der Vorst (1993) Performance of various computers using standard sparse linear equations solving techniques. In *Computer Benchmarks*, pp. 177–188, J. Dongarra and W. Gentzsch (Eds.). New York: Elsevier Science Publishers B.V.

Dongarra, J., C. Moler, J. Bunch, and G. Stewart (1979) *LINPACK Users' Guide*. Philadelphia: SIAM.

Dongarra, J., I. Duff, D. Sorenson and H. Van Der Vorst (1991) *Solving Linear Systems on Vector and Shared Memory Computers*. Philadelphia: SIAM.

Doob, J.L. (1953) *Stochastic Processes*. New York: John Wiley.

Draper, N. and H. Smith (1998) *Applied Regression Analysis*, 3rd edition. New York: Wiley-Interscience.

Durbin, J. (1960) The fitting of time series models. *Review of the International Institute of Statistics*, 28: 233–244.

Dyer, P. and S. McReynolds (1969) Extension of square-root smoothing to include process noise. *Journal of Optimization Theory and Applications*, 3(6): 444–459.

Dynamics of the Airframe (1952) Northrup Aircraft, BU AER Report AE-61-4II (September 1952); US Defense Documentation Center, Defense Technical Information Center, AD 11366.

Efroymson, M.A. (1960) Multiple regression analysis. In *Mathematical Methods for Digital Computers*, A. Ralston and H.S. Wilf (Eds.). New York: John Wiley.

Eklund, K. and I. Gustavsson (1973) Identification of drum boiler dynamics. *3rd IFAC Symposium*, Hague/Delft, June 1973.

El-Saden, M.R. (1965) *Engineering Thermodynamics*. Princeton, NJ: Van Nostrand.

Escobal, P.R. (1976) *Methods of Orbit Determination*. Malabar, FL: Krieger.

Fasol, K.H. and H.P. Jörgl (1980) Principles of model building and identification. *Automatica*, 16: 505–518.

Favier, G. and A. Smolders (1984) Adaptive smoother-predictors for tracking maneuvering targets. *Proceedings of the 23rd Conference on Decision and Control*, Las Vegas, NV, pp. 831–836, December 1984.

Finn, J.D. (1974) *A General Model for Multivariate Analysis*. New York: Holt, Rinehart and Winston.

Fisher, R.A. (1918) The correlation between relatives on the supposition of Mendelian inheritance. *Philosophical Transactions of the Royal Society Edinburgh*, 52: 399–433.

Fisher, R.A. (1922) On the mathematical foundations of theoretical statistics. *Philosophical Transactions of the Royal Society London*, 222: 309–368.

Fisher, R.A. (1925a) Theory of Statistical Estimation. *Proceedings of the Cambridge Philosophical Society*, 22: 700–725.

Fisher, R.A. (1925b) *Statistical Methods for Research Workers*; Edinburgh: Oliver and Boyd; republished by Oxford University Press, 1990.

Fisher, R.A. (1934) Two new properties of mathematical likelihood. *Proceedings of the Royal Society London*, 144: 285–307.

Fisher, R.A. (1935) The logic of inductive inference. *Journal of Royal Statistical Society*, 98: 39–82.

Fisz, M. (1963) *Probability Theory and Mathematical Statistics*. New York: John Wiley.

Fitzgerald, R.T. (1971) Divergence of the Kalman filter. *IEEE Transactions on Automatic Control*, AC-16: 736–747; reprinted in *Kalman Filtering: Theory and Application*, IEEE Press, 1985.

Forsythe, G.E. and C.B. Moler (1967) *Computer Solution of Linear Algebraic Systems*. Englewood Cliffs, NJ: Prentice-Hall.

Forsythe, G.E., M.A. Malcolm and C.B. Moler (1977) *Computer Methods for Mathematical Computations*. Englewood Cliffs, NJ: Prentice-Hall.

Francis, J.G.F. (1961) The QR transformation, part I. *Computer Journal*, 4: 265–271.

Francis, J.G.F. (1962) The QR transformation, part II. *Computer Journal*, 4: 332–345.

Fraser, D.C. and J.E. Potter (1969) The optimum linear smoother as a combination of two optimum linear filters. *IEEE Transactions on Automatic Control*, AC-14: 387–390; reprinted in *Kalman Filtering: Theory and Application*, IEEE Press, 1985.

Friedland, B. (1969) Treatment of bias in recursive filtering. *IEEE Transactions on Automatic Control*, AC-14: 359–367; reprinted in *Kalman Filtering: Theory and Application*, IEEE Press, 1985.

Friedland, B. (1979) Maximum-likelihood estimation of a process with random transitions (failures). *IEEE Transactions on Automatic Control*, AC-24(6): 932–937.

Friedland, B. (1981) Multidimensional maximum likelihood failure detection and estimation. *IEEE Transactions on Automatic Control*, AC-26(2): 567–570.

Friedland B. and S.M. Grabousky (1982) Estimating sudden changes of biases in linear dynamic systems. *IEEE Transactions on Automatic Control*, AC-27(1): 237–240.

Furnival, G.M. and R. W. Wilson (1971) All possible regressions with less computations. *Technometrics*, 13: 403–408.

Gauss, K.F. (1809) *Theory of the Motion of Heavenly Bodies Moving about the Sun in Conic Sections*. Göttingen, Germany; republished by Dover, 1963.

Gelb, A. (Ed.) (1974) *Applied Optimal Estimation*. Cambridge, MA: MIT Press.

Gibbs, B.P. (1977) Results of the PLACE/MARAD L-Band satellite navigation experiment. *IEEE EASCON Proceedings*, Alexandria, VA, September 1977.

Gibbs, B.P. (1978) *Precision Recursive Estimator for Ephemeris Refinement (PREFER) Math Description*. Greenbelt, MD: Business and Technological Systems.

Gibbs B.P. (1979) Precision recursive estimator for ephemeris refinement. *GSFC Flight Mechanics/Estimation Theory Symposium*, Greenbelt, MD, October 1979.

Gibbs, B.P. (1981) Geomagnetic modeling by optimal recursive filtering. *Fourth Scientific Assembly, International Association of Geomagnetism and Aeronomy*, Edinburgh, July 1981.

Gibbs, B.P. (1982) *Precision Recursive Estimator for Ephemeris Refinement (PREFER) Mathematical Description*. Greenbelt, MD: Business and Technological Systems.

Gibbs, B.P. (1992) Jump detection/estimation using stepwise regression. *IEEE Transactions on Aerospace and Electronic Systems*, 28(4): 1105–1118.

Gibbs, B.P. (1995) *Application of Nonlinear Model-Based Predictive Control to Power Plants*. Columbia, MD: Coleman Research Corp.

Gibbs, B.P. (1998) *Hydrological Fusion and Control Tool (HydroFACT) User's Guide*. New Carrollton, MD: Fusion and Control Technology.

Gibbs, B.P. (2008) GOES image navigation and registration. *SatMagazine*; www.satmagazine.com, July/Aug. 2008.

Gibbs, B.P. and J.L. Cantor (1986) *Final Report: Stochastic Modeling for Improved Weapon Performance*. Greenbelt, MD: Business and Technological Systems.

Gibbs, B.P. and B.T. Fang (1975) *IMP-9 (H) Attitude Study*. Riverdale, MD: Wolf R&D Corporation.

Gibbs, B.P. and W.E. Hatch (1973) *Final Report on Real-Time Attitude Estimator for the Atmospheric Explorer Satellite*. Riverdale, MD: Wolf R&D Corporation.

Gibbs, B.P. and D.W. Porter (1980) Development of an adaptive algorithm for predicting tank motion. *Proceedings of the 19th IEEE Conference on Decision and Control*, Albuquerque, NM, December 1980.

Gibbs, B.P. and D.S. Weber (1992a) Nonlinear model-based predictive control for fossil power plants. *1992 American Automatic Control Conference*, Chicago, IL, June 1992.

Gibbs, B.P. and D.S. Weber (1992b) Parameter identification for nonlinear model predictive control of fossil power plants. *35th Power Instrumentation Symposium*, Instrument Society of America and Electric Power Research Institute, Orlando, FL, June 1992.

Gibbs, B.P., D.R. Haley, B.T. Fang (1975) *IMP-8 (J) Attitude Study–Final Report*. Riverdale, MD: Wolf R&D Corporation.

Gibbs, B.P., D.S. Weber, and D.W. Porter (1991) Application of nonlinear model-predictive control to fossil power plants. *Proceedings of the 30th IEEE Conference on Decision and Control*, Brighton, GB, December 1991.

Gibbs, B.P., D.W. Porter and W.E. Yancey (1998) Apparatus and method for fusing diverse data. Patent #5,729,451, March 1998.

Gibbs, B.P., J. Carr, D.R. Uetrecht, and C. Sayal (2008a) Analysis of GOES-13 orbit and attitude determination. *SpaceOps 2008*, Heidelberg: AIAA-2008-3222, May 2008.

Gibbs, B.P., D.R. Uetrecht, and C. Sayal (2008b) GOES-13 propulsion model. *SpaceOps 2008*, Heidelberg: AIAA-2008-3434, May 2008.

Gill, P.E., W. Murray and M.H. Wright (1991) *Numerical Linear Algebra and Optimization*, Vol. 1. Redwood City, CA: Addison-Wesley.

Goddard Trajectory Determination System (GTDS) Mathematical Theory (1989) NASA/GSFC FDD/552-89/001, July 1989.

Godfrey, K.R. (1980) Correlation methods. *Automatica*, 16(5): 527–534.

GOES N-Q Orbit and Attitude Tracking System (OATS) Software Maintenance Manual (2007) ISI-NQ-OATS-0003 Rev C, Columbia, MD: Integral Systems.

Goldstein, H. (1950) *Classical Mechanics*. Reading, MA: Addison-Wesley.

Golub, G.H. and C.F. Van Loan (1996) *Matrix Computations*, 3rd edition. Baltimore, MD: John Hopkins University Press.

Gordon, N.J., D.J. Salmond, and A.F.M. Smith (1993) Novel approach to nonlinear/non-Gaussian Bayesian state estimation. *IEE Proceedings F in Radar and Signal Processing*, 140(2): 107–113.

Greco, C. and V. Marchis (1982) A steam power plant identification. *IASTED*, 113–117.

Grewal, M.S. and A.P. Andrews (2001) *Kalman Filtering Theory and Practice Using MATLAB*, 2nd edition. New York: Wiley-Interscience.

Grewal, M.S., L.R. Weill, and A.P. Andrews (2001) *Global Positioning Systems, Inertial Navigation and Integration*. New York: Wiley-Interscience.

Grimmett, G.R. and D.R. Stirzaker (2001) *Probability and Random Processes*, 3rd edition. Oxford, GB: Oxford University Press.

Guler, I., H. Hardalac, and M. Kaymaz (2002) Comparison of FFT and adaptive ARMA methods in transcranial Doppler signals recorded from the cerebral vessels. *Computers in Biology and Medicine*, 32(6): 445–453.

Gupta, N.K. and R.K. Mehra (1974) Computational aspects of maximum likelihood estimation and reduction in sensitivity function calculations. *IEEE Transactions on Automatic Control*, AC-19: 774–783.

Gura, A. and A.B. Bierman (1971) On computational efficiency of filtering algorithms *Automatica*, 7: 299–314.

Hampel, F.R., E.M. Ronchetti, P.J. Rousseeuw, and W.A. Stahel (1986) *Robust Statistics—The Approach Based on Influence Functions*, New York: John Wiley; republished in paperback, 2005.

Harbaugh, A.W. and M.G. McDonald (1996) User's documentation for MODFLOW-96, an update to the U.S. Geological Survey modular finite-difference ground-water flow model, U.S. Geological Survey Open-File Report 96-485.

Harris, F.J. (1978) On the use of windows for harmonic analysis with the discrete Fourier transform. *Proceedings of the IEEE*, 66: 51–83.

Harvey, A.C. (1990) *Forecasting, Structural Time Series Models and the Kalman Filter*. Cambridge, UK: Cambridge University Press.

Hayes, M.H. (1996) *Statistical Digital Signal Processing and Modeling*. New York: John Wiley.

Ho, Y.C. and R.C.K. Lee (1964) A Bayesian approach to problems in stochastic estimation and control. *IEEE Transactions on Automatic Control*, AC-9: 333–339; reprinted in *Kalman Filtering: Theory and Application*, IEEE Press, 1985.

Hoaglin, D.C., F. Mosteller, and J.W. Tukey (1983) *Understanding Robust and Exploratory Data Analysis*. New York: John Wiley; republished, 2000.

Hocking, R.R. (1983) Developments in linear regression methodology: 1959–1982. *Technometrics*, 25(3).

Housner, G.W. and D.E. Hudson (1959) *Applied Mechanics, Volume II, Dynamics*. Toronto: Van Nostrand.

Huber, P.S. (1981) *Robust Statistics*, New York: John Wiley; republished in paperback, 2004.

Huyakorn, P.S. and G.F. Pinder (1983) *Computational Methods in Subsurface Flow*. San Diego, CA: Academic press.

Ipakchi, A., D. Skedzielewski, B. Wells, and S.M. Divakaruni (1989) Process integration and information systems for fossil plant automation: R & D plan for EPRI. *Conference on Power Plant Control and Automation*.

Isermann, R. (1980) Practical aspects of process identification. *Automatica*, 16(5): 575–587.

Isermann, R. (1984) Process fault detection based on modeling and estimation methods—A survey. *Automatica*, 20: 387–404.

Jansson, M. (2003), Subspace identification and ARX modeling, *Proceedings ofIFAC Symposium on System Identification*, Rotterdam, August 2003.

Jazwinski, A.H. (1969) Adaptive filtering. *Automatica*, 5: 475–485; reprinted in *Kalman Filtering: Theory and Application*, IEEE Press, 1985.

Jazwinski, A.H. (1970) *Stochastic Processes and Filtering Theory*. New York: Academic Press; republished by Dover, 2007.

Johnson, G. (1969) A deterministic theory of estimation and control. *IEEE Transactions on Automatic Control*, 14(4): 380–384.

Jones, R.H. (1976) Autoregression order selection. *Geophysics*, 41: 771–773.

Julier, S.J. (2002) The scaled unscented transform. *Proceeedings of the American Control Conference*, Anchorage, AK, May 2002.

Julier, S.J. and J.K. Uhlmann (1997) A new extension of the Kalman filter to nonlinear systems. *International Symposium on Aerospace/Defense Sensing, Simulation, and Controls*, Orlando, FL, pp. 182–193, April 1997.

Julier, S.J. and J.K. Uhlmann (2002) Comment on "A new method for the nonlinear transformation of means and covariances in filters and estimators (authors' reply)," *IEEE Transactions on Automatic Control*, 47(8): 1408–1409.

Julier, S.J. and J.K. Uhlmann, (2004a) Unscented Filtering and Nonlinear Estimation. *Proc. IEEE*, 92(3): 401–422.

Julier, S.J. and J.K. Uhlmann, (2004b) Corrections to 'Unscented Filtering and Nonlinear Estimation. *Proc. IEEE*, 92(12): 1958–1958.

Julier, S.J., J.K. Uhlmann, and H.F. Durrant-Whyte (1995) A new approach for filtering nonlinear systems. *American Control Conference*, Seattle, WA, pp. 1628–1632, June 1995.

Julier, S.J., J.K. Uhlmann, and H.F. Durrant-Whyte (2000) A new method for the nonlinear transformation of means and covariances in filters and estimators. *IEEE Transactions on Automatic Control*, 45(3): 477–482.

Kailath, T. (1968) An innovation approach to least-squares estimation—part I: Linear filtering in additive white noise. *IEEE Transactions on Automatic Control*, AC-13(6): 646–655; reprinted in *Kalman Filtering: Theory and Application*, IEEE Press, 1985.

Kailath, T. (1975) Supplement to "A survey of data smoothing." *Automatica*, 11: 109–111.

Kailath, T. (Ed.) (1977) *Linear Least Squares Estimation*. Stroudsburg, PA: Dowden, Hutchinson & Ross.

Kailath, T. and P. Frost (1968) An innovation approach to least-squares estimation—part II: Linear smoothing in additive white noise. *IEEE Transactions on Automatic Control*, AC-13(6): 655–660; reprinted in *Kalman Filtering: Theory and Application*, IEEE Press, 1985.

Kailath, T., A. Sayed, and B. Hassibi (2000) *Linear Estimation*. Upper Saddle River, NJ: Prentice-Hall.

Kalman, R.E. (1959) On the general theory of control systems. *IRE Transactions on Automatic Control*, 4(3): 110–111.

Kalman, R.E. (1960) A new approach to linear filtering and prediction problems. *ASME Journal of Basic Engineering*, 82(1): 35–45; reprinted in *Kalman Filtering: Theory and Application*, IEEE Press, 1985.

Kalman, R.E. (1961) New Methods and Results in Linear Filtering and Prediction Theory. *ASME Journal of Basic Engineering*, Series D, 83.

Kalman, R.E. and J.E. Bertram (1958) General synthesis procedure for computer control of single and multi-loop linear systems. *Transactions of AIEE*, 77(II): 602–609.

Kalman, R.E. and J.E. Bertram (1959) Control system analysis and design via the second method of Lyapunov: (I) continuous-time systems, (II) discrete time systems. *IRE Transactions on Automatic Control*, 4(3): 112.

Kalman, R.E. and R.S. Bucy (1961) New results in linear filtering and prediction theory. *ASME Journal of Basic Engineering*, 83: 95–108; reprinted in *Kalman Filtering: Theory and Application*, IEEE Press, 1985.

Kaminski, P.G., A.E. Bryson and S. Schmidt (1971) Discrete square root filtering: A survey of current techniques. *IEEE Transactions on Automatic Control*, AC-16: 727–735; reprinted in *Kalman Filtering: Theory and Application*, IEEE Press, 1985.

Kaminski, P.G. and A.E. Bryson (1972) Discrete square root smoothing. *Proceedings of the 1972 AIAA Guidance and Control Conference*, Paper #72–877.

Kaplan, E.D. and C.J. Hegarty (2006) *Understanding GPS Principles and Applications*, 2nd edition. Boston, MA: Artech House.

Kaula, W.M. (1966) *Theory of Satellite Geodesy*. New York: Blaisdell Publishing; republished by Dover, 2000.

Kay, S.M. (1988) *Modern Spectral Estimation, Theory and Application*. Englewood Cliffs, NJ: Prentice-Hall.

Kay, S.M. and S.L. Marple (1981) Spectrum analysis—a modern perspective. *Proceedings of IEEE*, 69: 1380–1419.

Kearton, W.J. (1951) *Steam Turbines Theory and Practice*. London: Pitman & Sons.

Kelley, C.T. (1995) *Iterative Methods for Linear and Nonlinear Equations*. Philadelphia: SIAM.

Kenney, C. and C.J. Laub (1988). Controllability and stability radii for companion form systems. *Mathematics of Control, Signals, and Systems*, 1: 239–256.

Kepler, J. (1619) *Harmonies of the World*.

Kerr, T.H. (1983) A conservative view of the GLR failure and event detection approach. *IEEE Transactions on Information Theory*, 29(6).

Kesler, S.B. (Ed.) (1986) *Modern Spectrum Analysis, II*. New York: IEEE Press.

Kolmogorov, A.N. (1941) Interpolation und extrapolation von stationaren zufalligen kolgen. *Bulletin of Academy of Science U.S.S.R, Series Math*, 5: 3–14.

Korenberg, M.J. (1985) Orthogonal identification of nonlinear difference equation models. In *Proceedings 28th Midwest Symposium on Circuits and Systems*, pp. 90–95, Louisville, KY.

Korenberg, M.J. (1988) Identifying nonlinear difference equation and functional expansion representations: the fast orthogonal algorithm. *Annals of Biomedical Engineering*, 16: 123–142.

Korenberg, M.J. (1989) A robust orthogonal algorithm for system identification and time-series analysis. *Biological Cybernetics*, 60: 267–276.

Koskela, P.E. (1967) *Astrodynamic Analysis for the Advanced Orbit/Ephemeris Subsystem*. Newport Beach, CA: Aeronutronic Publication U-4180, Philco-Ford Corporation.

Kwan, H.W. and J.H. Anderson (1970) A mathematical model of a 200 MW boiler. *International Journal of Control*, 12.

Larimore, W.E. (1983) System identification, reduced-order filtering and modeling via canonical variate analysis. *Proceedings of the 1983 American Control Conference*, New York: IEEE Press pp. 445–451, June 1983.

Larimore, W.E. (1987) Identification of nonlinear systems using canonical variate analysis. *Proceedings of the 26th IEEE Conference, Decision and Control*, Los Angeles, CA, pp. 1694–1699, December 1987.

Larimore, W.E. (1988) Generalized canonical variate analysis of nonlinear systems. *Proceedings of the 27th IEEE Conference, Decision and Control*, Austin, TX, pp. 1720–1725, December 1987.

Larimore, W.E. (1990a) Canonical variate analysis in identification, filtering and adaptive control. *Proceedings of the 29th IEEE Conference, Decision and Control*, Honolulu, HI, pp. 596–604, December 1990.

Larimore, W.E. (1990b) Identification and filtering of nonlinear systems using canonical variate analysis. *Proceedings of the 29th IEEE Conference Decision and Control*, Honolulu, HI, pp. 635–640, December 1990.

Larimore, W.E. (1990c) Order-recursive factorization of the pseudoinverse of a covariance matrix. *IEEE Transactions on Automatic Control*, 35: 1299–1303.

Larimore, W.E. (1992a) *ADAPTx Automated System Identification Software Users Manual*. McLean, VA: Adaptics.

Larimore, W.E. (1992b) Identification and filtering of nonlinear systems using canonical variate analysis. In *Nonlinear Modeling and Forecasting*, pp. 283–303, M. Casdagli and S. Eubank (Eds.). Reading, MA: Addison-Wesley.

Larimore, W.E. (1996a) Statistical optimality and canonical variate analysis system identification. *Signal Processing*, 52: 131–144.

Larimore, W.E. (1996b) Optimal order selection and efficiency of canonical variate analysis system identification. *Proceedings of the13th IFAC World Congress*, Vol. 1. San Francisco, CA, pp. I: 151–156, July 1996.

Larimore, W.E. (1999) Automated multivariable system identification and industrial applications. *Proceedings of the 1999 American Control Conference*, San Diego, CA, June 1999.

Larimore, W.E. (2002) Reply to Comment on Order-recursive factorization of the pseudoinverse of a covariance matrix. *IEEE Transactions on Automatic Control*, 47: 1953–1957

Larimore, W.E. (2004) Large sample efficiency for ADAPTx subspace system identification with unknown feedback. *IFAC Dynamics and Control of Process Systems (DYCOPS)*, Boston, MA, July 2004.

Laub, A.J. (1979) A Schur method for solving algebraic Riccati equations. *IEEE Transactions on Automatic Control*, 24: 913–921.

Lawson, C.L. and R.J. Hanson (1974) *Solving Least Squares Problems*. Philadelphia: Prentice-Hall; republished by SIAM, 1995.

Lee, R.C.K. (1964) *Optimal Estimation, Identification and Control*. Cambridge, MA: MIT Press.

Legendre, A.M. (1806) *Nouvelles méthodes pour la détermination des orbites des comètes*.

Levine, W. (Ed.) (1996) *Control System Handbook*. Boca Raton, FL: CRC Press.

Levinson, H. (1947) The Wiener RMS (root mean square) error criterion in filter design and prediction. *Journal of Mathematical Physics*, 25: 261–278.

Ljung, L. (1994) *The System Identification Toolbox: The Manual*, 4th edition. Natick, MA: The Mathworks.

Ljung, L. (1999) *System Identification-Theory For the User*, 2nd edition. Upper Saddle River, NJ: Prentice Hall.

Ljung, L. (2002) Prediction error estimation methods. *Circuits, Systems and Signal Processing*, 21(2): 11–21.

Ljung, L. and T. Glad (1994) *Modeling of Dynamic Systems*. Englewood Cliffs, NJ: Prentice Hall.

Ljung, L. and T. Söderström (1983) *Theory and Practice of Recursive Identification*. Cambridge, MA: MIT Press.

Ljung, L., I. Gustavsson, and T. Söderström (1974) Identification of linear multivariable systems under linear feedback control. *IEEE Transactions on Automatic Control*, AC-19: 836–841.

LSQR-Algorithm 583, collected algorithms from ACM (1982) *ACM Transactions on Mathematical Software*, 8(2): 195–209.

Lu, S., K.H. Ju, and K.K. Chon (2001) A new algorithm for linear and nonlinear ARMA model parameter estimation using affine geometry (and application to blood flow/pressure data). *IEEE Transactions on Biomedical Engineering*, 48(10): 1116–1124.

Kulikova, M.V. (2009) Maximum likelihood estimation via the extended covariance and combined square-root filters. *Mathematics and Computers in Simulation*, 79(5): 1641–1657.

Makowski, A., W.S. Levine, and M.S. Asher (1984) The non-linear MMSE filter for partially observed systems driven by non-Gaussian white noise, with applications to failure estimation. *Proceedings of the 23rd IEEE Conference on Decision and Control*, Las Vegas, NV, December 1984.

Markelov, G.N. (2007) Plume impingement analysis for Aeolus spacecraft and gas/surface interaction models. *Journal of Spacecraft and Rockets*, 44(3): 607–618.

Maronna, R., D. Martin, and V. Yohai (2006) *Robust Statistics: Theory and Methods*. Hoboken, NJ: Wiley Interscience.

Marple, S.L. (1980) A New Autogregressive Spectral Analysis Algorithm. *IEEE Transactions on Acoustics, Speech, and Signal Processing*, ASSP-28: 441–454.

Marple, S.L. (1987) *Digital Spectral Analysis with Applications*. Englewood Cliffs, NJ: Prentice-Hall.

Marsaglia, G. and Tsang (1983) A fast, easily implemented method for sampling from decreasing or symmetric unimodal density functions. *SIAM Journal on Scientific Computing*, 5.

Masada, G.Y. and D.N. Wormley (1982) Evaluation of lumped heat exchanger dynamic models. New York: ASME, paper 82-WA/DSC-16, pp. 1–10.

Maybeck, P.S. (1979) *Stochastic Models, Estimation, and Control*, Vol. 1. New York: Academic Press; reprinted Navtech Book and Software, 1994.

Maybeck, P.S. (1982a) *Stochastic Models, Estimation, and Control*, Vol. 2. New York: Academic Press; reprinted Navtech Book and Software, 1994.

Maybeck, P.S. (1982b), *Stochastic Models, Estimation, and Control*, Vol. 3. New York: Academic Press; reprinted Navtech Book and Software, 1994.

McCarthy, D.D. (Ed.) (1996) *IERS Conventions (1996), IERS Technical Note 21*. Observatoire de Paris: International Earth Rotation Service, Observatoire de Paris.

McCarthy, D.D. and G. Petit (Eds.) (2004) *IERS Conventions (2003), IERS Technical Note 32*. Bundesamt für Kartographie und Geodäsie: International Earth Rotation Service.

McGee, L.A. and S.F. Schmidt (1985) Discovery of the Kalman filter as a practical tool for aerospace and industry. NASA Technical Memorandum 86847.

Meditch, J. (1969) *Stochastic Optimal Linear Estimation and Control*. New York: McGraw-Hill.

Meditch, J. (1973) A survey of data smoothing for linear and nonlinear dynamic systems. *Automatica*, 9(2): 151–162; reprinted in *Kalman Filtering: Theory and Application*, IEEE Press, 1985.

Meeus, J. (2005) *Astronomical Algorithms*. Richmond, VA: Willmann-Bell.

Mehra, R. (1970) On the identification of variances and adaptive Kalman filtering. *IEEE Transactions on Automatic Control*, AC-15(2): 175–184.

Melzer, S.M. (1978) The extension of Friedland's technique for decoupling state and bias estimates to fixed interval smoothing. Aerospace Corporation, Technical Memo ATM-0078(3901-03)-16, May 1978

Mendel, J. (1971) Computation requirements for a discrete Kalman filter. *IEEE Transactions on Automatic Control*, AC-16: 748–758; reprinted in *Kalman Filtering: Theory and Application*, IEEE Press, 1985.

Mendel, J. (1987) *Lessons in Digital Estimation Theory*. Englewood Cliffs, NJ: Prentice-Hall.

Misra, P. and P. Enge (2001) *Global Positioning System–Signals, Measurements and Performance*. Lincoln, MA: Ganga-Jamuna Press.

Moler, C. and C. Van Loan (2003) Nineteen dubious ways to compute the exponential of a matrix, twenty-five years later. *SIAM Review*, 45(1): 3–49.

Moore, R.L. and F.C. Schweppe (1973) Model identification for adaptive control of nuclear power plants. *Automatica*, 9: 309–318.

Morari, M., C.E. Garcia and D.M. Prett (1988) Model predictive control: Theory and practice. *American Control Conference*, Atlanta, GA, June 1988.

Moré, J.J. (1977) The Levenberg-Marquardt algorithm: Implementation and theory. In *Lecture Notes in Mathematics (1977)*, 630: 105–116, G.A. Watson. (Ed.). Berlin: Springer-Verlag.

Morf, M., B. Levy, and T. Kailaith (1978) Square-root algorithms for the continuous-time linear least squares estimation problem. *IEEE Transactions on Automatic Control*, AC-23(5): 907–911.

Morgan, R.C., M.S. Asher, D.W. Porter, and W.S. Levine (1985) Maximum likelihood detection and estimation of jumps in dynamic systems. *Proceedings of the 8th MIT/ONR Workshop on C^3 Systems*, Cambridge, MA, December 1985.

Mosteller, F. and J.W. Tukey (1977) *Data Analysis and Regression*. Reading, MA: Addison-Wesley.

NAG Numerical Libraries, Numerical Algorithms Group, 256 Banbury Road, Oxford OX27DE, UK. http://www.nag.co.uk/numeric/numerical_libraries.asp

Nakamura, H. and H. Akaike (1981) Statistical identification for optimal control of supercritical thermal power plants. *Automatica*, 17(1): 143–155.

Nakamura, H., M. Ochida, T. Kitami, and H. Akaike (1979) Application of an optimal control system to a supercritical thermal power plant. *Control of Power Systems Conference and Exhibition*, IEEE Conference Records, CH1377-1.

Newton, I. (1687) *Mathematical Principles of Natural Philosophy*.

Oppenheim, A.V. and R.W Schafer (1975) *Digital Signal Processing*. London: Prentice-Hall.

Paige, C.C. and M.A. Saunders (1982) LSQR: An algorithm for sparse linear equations and sparse least squares. *ACM Transactions on Mathematical Software*, 8(1): 43–71.

Papoulis, A. and S.U. Pillai (2002) *Probability, Random Variables and Stochastic Processes*, 4th edition. New York: McGraw-Hill.

Parkinson, B.W. and J.J. Jr. Spilker (1996) *Global Positioning System: Theory and Applications*, Vol. 1. Washington: AIAA.

Parlett, B.N. (1980) *The Symmetric Eigenvalue Problem*. Englewood Cliffs, NJ: Prentice-Hall.

Parzen, E. (1960) *Modern Probability Theory and Its Applications*. New York: John Wiley.

Parzen, E. (1974) Some recent advances in time series modeling. *IEEE Transactions on Automatic Control*, AC-19: 723–730; reprinted in *Modern Spectral Analysis*, IEEE Press, 1978.

Penrose, R. (1955) A generalized inverse for matrices. *Proceedings of the Cambridge Philosophical Society*, 51: 406–413.

Picone, J.M., A.E. Hedin, S.L. Coffey, J. Lean, D.P. Drob, H. Neal, D.J. Melendez-Alvira, R.R Meier, and J.Y. Mariska (1997) The Naval Research Laboratory program on empirical models of the neutral upper atmosphere. *Astrodynamics 1997, Proceedings of the AAS/AIAA Astrodynamics Conference*, Sun Valley, ID, pp. 515–527. August 1997.

Porter, D.W., B. Gibbs, W. Jones, P. Huyakorn, L. Hamm, and G. Flach (1997) Data fusion modeling for groundwater systems *Journal of Contaminant Hydrology*, 42: 303–335.

Porter, D.W., D.R. Haley, W.S. Levine, C.J. Vahlberg, and B.P. Gibbs (1980) Estimation of Kalman filter model parameters from an ensemble of tests. *Proceedings of Flight Mechanics/Estimation Theory Symposium*, Greenbelt, MD, October 1980.

Porter, D.W., B.P. Gibbs, J.S. Vandergraft, W.E. Hatch, P.W. Kelsey, P. Hoekstra, B. Hoekstra, M. Blohm, D. Phillips, and R. Bates (1993) Waste site characterization using data fusion workstation at Hanford and Savannah River DOE sites. *Environmental Remediation (ER) 93*.

Porter, D.W., J.S. Vandergraft, B. Beck, and B.P. Gibbs (1994) Data fusion workstation. Patent #5,321,613, June 1994.

Potter, J.E. (1966) Matrix quadratic solutions. *SIAM Journal on Applied Mathematics*, 14: 496–501.

Prasad, S. and K.V.S. Hari (1987) Improved ARMA spectral estimation using the canonical variate method. *IEEE Transactions on Acoustics, Speech, and Signal Processing*, ASSP-35(6): 900–903.

Press, W.H., S.A. Teukolsky, W.T. Vetterling, and B.P. Flannery (1992) *Numerical Recipes in Fortran*, 2nd edition. New York: Cambridge University Press.

Press, W.H., S.A. Teukolsky, W.T. Vetterling, and B.P. Flannery (1996) *Numerical Recipes in Fortran 90*. New York: Cambridge University Press.

Press, W.H., S.A. Teukolsky, W.T. Vetterling, and B.P. Flannery (2007) *Numerical Recipes*, 3rd edition. New York: Cambridge University Press.

Prett, D.M. and C.E. Garcia (1988) *Fundamental Process Control*. Boston: Butterworths.

Priestley, M.B. (1981) *Spectral Analysis and Time Series*. New York: Academic Press.

Programs for Digital Signal Processing (1979) Digital Signal Processing Committee (Ed.), IEEE Acoustics, Speech and Signal Processing Society. New York: IEEE Press.

Prvan, T. and M.R. Osborne (1988) A square-root fixed-interval discrete-time smoother. *Journal of Australian Mathematical Society, Ser. B*, 30: 57–68.

Rabiner, L.R. and B. Gold (1975) *Theory and Application of Digital Signal Processing*. Englewood Cliffs, NJ: Prentice-Hall.

Rabiner, L.R. and C.M. Rader (Ed.) (1972) *Digital Signal Processing*. New York: IEEE Press.

Rake, H. (1980) Step response and frequency response methods. *Automatica*, 16(5): 519–526.

Rao, C.R. (1945) Information and accuracy obtainable in the estimation of statistical parameters. *Bulletin of the Calcutta Mathematical Society*, 37: 81–91.

Rao, C.R. (1968) *Linear Statistical Inference and its Applications*. New York: John Wiley.

Rauch, H.E., F. Tung, and C.T. Striebel (1965) Maximum likelihood estimates of linear dynamic systems. *AIAA Journal*, 3: 1445–1450; reprinted in *Kalman Filtering: Theory and Application*, IEEE Press, 1985.

RETRAN-02-A Program for Transient Thermal-Hydraulic Analysis of Complex Fluid Flow Systems (1981) Energy Incorporated, NP-1850, EPRI Research Project 889, May 1981

Rissanen, J.A. (1983) A universal prior for the integers and estimation by minimum description length. *Annals of Statistics*, 11: 417–431.

Ristic, B., S. Arulampalam, N. Gordon (2004) *Beyond the Kalman Filter: Particle Filters for Tracking Applications*. Boston, MA: Artech House Radar Library.

Roweis, S. and Z. Ghahramani (1999) A Unifying Review of Linear Gaussian Models. *Neural Computation*, 11(2): 305–345.

Sabersky, R.H. and A.J. Acosta (1964) *Fluid Flow—A First Course in Fluid Dynamics*. New York: Macmillan.

Sage, A.P. (1968) *Optimum Systems Control*. Englewood Cliffs, NJ: Prentice-Hall.

Salisbury, J.K. (1974) *Steam Turbines and Their Cycles*. Malabar, FL: Krieger.

Schlee, F.H., C.J. Standish and N.F. Toda (1967) Divergence in the Kalman filter. *AIAA Journal*, 5: 1114–1120; reprinted in *Kalman Filtering: Theory and Application*, IEEE Press, 1985.

Schmidt, S.F. (1966) Applications of state-space methods to navigation problems. In *Advances in Control Systems*, 3: 293–340, C.T. Leondes (Ed.). New York: Academic Press.

Schmidt, S.F. (1970) Computational techniques in Kalman filtering. *Theory and Application of Kalman Filtering*, Advisory Group for Aerospace Research and Development, AGARDograph 139 (AD 704 306), February 1970.

Schmidt, S.F. (1972) Precision navigation for approach and landing operations. *Proceedings of the Joint Automatic Control Conference*, Stanford University, Stanford, CA, pp. 455–463, June 1972.

Schmidt, S.F. (1981) The Kalman filter: Its recognition and development for aerospace applications. *AIAA Journal of Guidance Control*, 4(1): 4–7.

Schweppe, F.C. (1973), *Uncertain Dynamic Systems*. Englewood Cliffs, NJ: Prentice-Hall.

Seidelmann, P.K. (2006) *Explanatory Supplement to the Astronomical Almanac*. Sausalito, CA: University Science Books.

Shannon, C.E. (1948) A mathematical theory of communication. *Bell System Technical Journal*, 27: 379–423.

Shen, X. and L. Deng (1995) Discrete H-infinity filtering design with application to speech enhancement. *Proceedings of the IEEE International Conference on Acoustics, Speech, and Signal Processing*, Detroit, MI, pp. 1504–1507, May 1995.

Shen, X. and L. Deng (1997) Game theory approach to H-infinity filter design. *IEEE Transactions on Signal Processing*, 45(4): 1092–1095.

Shen, X. and L. Deng (1999) A dynamic system approach to Speech enhancement using the H-inf filtering algorithm. *IEEE Transactions on Speech and Audio Processing*, 7: 391–399.

Shen, X., L. Deng, and A. Yasmin (1966) H-infinity filtering for speech enhancement. *Proceedings of the International Conference on Spoken Language Processing*, Philadelphia, PA, pp. 873–876, October 1996.

Shumway, R.H. and D.S. Stoffer (2006) *Time Series Analysis and Its Applications with R Examples*, 2nd edition. New York: Springer-Verlag.

Shuster, M.D., D.W. Porter (1984) Efficient estimation of initial-condition parameters for partially observable initial conditions *Proceedings of the 23rd IEEE Conference on Decision and Control*, Las Vegas, NV, pp. 1272–1275, December 1984.

Shuster, M.D., D.W. Porter, and B.P. Gibbs (1983) *Maximum Likelihood Estimation Techniques for Trident Missile System Parameters*. Greenbelt, MD: Business and Technological Systems.

Siegel, A.F. (1982) Robust regression using repeated medians. *Biometrica*, 69: 242–244.

Sima, V. (2005) Computational experience in solving algebraic Riccati equations. *Proceedings of the 44th IEEE Conference on Decision and Control*, Seville, Spain, pp. 7982–7987, December 2005.

Simon, H. (1988) Incomplete LU preconditioners for conjugate-gradient-type iterative methods. *Society of Petroleum Engineers Reservoir Engineering*, February: 302–306.

Simon, D. (2006) *Optimal State Estimation: Kalman, H Infinity, and Nonlinear Approaches*. Hoboken, NJ: Wiley-Interscience.

Singer, J.G. (1991) *Combustion–Fossil Power*, Windsor, CT: Combustion Engineering.

Singleton, R.C. (1969) An algorithm for computing the mixed radix Fast Fourier Transform. *IEEE Transactions on Audio and Electroacoustics*, AU-17: 93–103; reprinted in *Digital Signal Processing*, IEEE Press, 1972.

Singleton, R.C. (1979) Mixed radix Fast Fourier Transform. In *Programs for Digital Signal Processing*, Digital Signal Processing Committee (Ed.). New York: IEEE Press.

Smylie, D.E., G.K.C. Clarke, and T.J. Ulrych (1973) Analysis of irregularities in the earth's rotation. *Methods in Computational Physics*, 13: 391–420. New York: Academic Press.

Söderström, T. and P. Stoica (1989) *System Identification*. London: Prentice-Hall Int.

Sorenson, H.W. (1966) Kalman filtering techniques. *Advances in Control Systems Theory and Applications*, 3: 219–292; reprinted in *Kalman Filtering: Theory and Application*, IEEE Press, 1985.

Sorenson, H.W. (1970) Least-squares estimation: From Gauss to Kalman, *IEEE Spectrum*, 7: 63–68; reprinted in *Kalman Filtering: Theory and Application*, IEEE Press, 1985.

Sorenson, H.W. (Ed.) (1985) *Kalman Filtering: Theory and Application*. New York: IEEE Press.

Spencer, R.C., K. C. Cotton, and C.N. Cannon (1974) A method for predicting the performance of steam turbine-generators…16,500 KW and larger, revised July 1974, Paper GER2007C, Schenectady, NY: General Electric Company.

Stengel, R.F. (1994) *Optimal Control and Estimation*. Mineola, NY: Dover.

Stengel, R.F. (2004) *Flight Dynamics*. Princeton, NJ: Princeton University Press.

Stewart, G.W. (1998) *Matrix Algorithms*, Vol. 1. Philadelphia: SIAM.

Stoer, J. and R. Bulirsch (1980), *Introduction to Numerical Analysis*. New York: Springer-Verlag.

Strejc, V. (1980) Least Squares Parameter Estimation. *Automatica*, 16(5): 535–550.

Stultz, S.C. and J.B. Kitto (Ed.) (2007) *Steam–Its Generation and Use*, 41st edition. Barberton, OH: Babcock & Wilcox.

Swerling, P. (1959) First order error propagation in a stagewise smoothing procedure for satellite observations. *Journal of Astronautical Science*, 6: 46–52; reprinted in *Kalman Filtering: Theory and Application*, IEEE Press, 1985.

Tanenbaum, M. (1977) Private communication to B. Gibbs, NSWC/Dahlgren Laboratory, December 1977.

Tapley, B.D., B.E. Schutz, and G.H. Born (2004) *Statistical Orbit Determination*. Amsterdam: Elsevier Academic.

Thomas, L.C. (1993) *Heat Transfer—Professional Version*. Englewood Cliffs, NJ: Prentice-Hall.

Thornton, C.L. (1976) Triangular covariance factorizations for Kalman filtering, Ph.D. Thesis, University of California at LA, School of Engineering, NASA/JPL Technical Memo 33-798.

Thornton, C.L. and G.J. Bierman (1975) Gram-Schmidt algorithms for covariance propagation. *Proceedings of the 1975 IEEE Conference on Decision and Control*, Houston, TX, pp. 489–498, December 1975.

Thornton, C.L. and G.J. Bierman (1976) A numerical comparison of discrete Kalman filtering algorithms: An orbit determination case study. JPL Tech. Memo 33-771, Pasadena, CA.

Thornton, C.L. and G.J. Bierman (1978) Filtering and error analysis via the UDU^T covariance factorization. *IEEE Transactions on Automatic Control*, 23(5): 901–907.

Tukey, J.W. (1977) *Exploratory Data Analysis*. Reading, MA: Addison-Wesley.

Tyssø, A. (1981) Modeling and parameter estimation of a ship boiler. *Automatica*, 17(1).

Tyssø, A. and J.C. Brembo (1976) The design of a multivariate control system for a ship boiler. *Automatica*, 12(3): 211–224.

Tyssø, A. and J.C. Brembo (1978) Installation and operation of a multivariate ship boiler control system. *Automatica*, 14: 213–221.

Ulrych, T.J. and T.N. Bishop (1975) Maximum entropy spectral analysis and autoregressive decomposition. *Reviews of Geophysics and Space Physics*, 13: 183–200; reprinted in *Modern Spectrum Analysis*, IEEE Press, 1978.

Vallado, D.A. and W.D. McClain (2001) *Fundamentals of Astrodynamics and Applications*. El Segundo, CA: Microcosm Press.

Van der Merwe, R. and E. Wan (2001) The square-root unscented Kalman filter for state and parameter estimation. *IEEE International Conference on Acoustics, Speech, and Signal Processing*, pp. 3461–3464, May.

Van der Vorst, H. (1989) High performance preconditioning. *SIAM Journal on Scientific and Statistical Computing*, 10: 1174–1185.

Van der Vorst, H. (1990) The convergent behavior of preconditioned CG and CG-S in the presence of rounding errors. In *Preconditioned Conjugate Gradient Methods*, O. Axelsson and L.Y. Kolotilina (Eds.). Berlin: Springer-Verlag.

Van der Vorst, H. (1992) Bi-CGSTAB: A fast and smoothly converging variant of Bi-CG for the solution of nonsymmetric linear systems. *SIAM Journal on Scientific and Statistical Computing*, 13: 631–644.

Van Overschee, P. and B. De Moor (1994) A unifying theorem for three subspace system identification algorithms. *American Control Conference*, Baltimore, MD, pp. 1645–1649.

Van Trees, H.L. (1968) *Detection, Estimation and Modulation Theory—Part 1*. New York: John Wiley.

Van Wylen, G.J. and R.E. Sonntag (1986) *Fundamentals of Classical Thermodynamics*, 3rd edition. New York: John Wiley.

Vandergraft, J.S. (1983) *Introduction to Numerical Analysis*. San Diego CA: Academic Press.

Vandergraft, J.S. (1987) Efficient optimization methods for maximum likelihood parameter estimation. *Proceedings of the 24th IEEE Conference on Decision and Control*, Ft. Lauderdale, FL, December 1987.

Vandergraft, J.S. (1991) *An Overview of Smoothers*. Columbia, MD: Coleman Research Corporation.

Verhaegen, M. and P. Van Dooren (1986) Numerical aspects of different Kalman filter implementations. *IEEE Transactions on Automatic Control*, AC-31(10): 907–917.

Vinsome, P.K.W. (1976) Orthomin, an iterative method for solving sparse sets of simultaneous linear equations. *Society of Petroleum Engineers AIME*, SPE 5729.

Walker, G. (1931) On periodicity in series of related terms. *Proceedings of the Royal Society of London, Series A*, 131: 518–532.

Wampler, R.H. (1969) An evaluation of linear least squares computer programs. *National Bureau of Standards Journal of Research, Series B Math and Science*, 73B: 59–90.

Warwick, K. (1987) Optimal observers for ARMA models. *International Journal of Control*, 46(5): 1493–1503.

Weber, D.S. (1991) Simplified once-through supercritical model development—An eleventh order model. Unpublished.

Welch, P.D. (1967) The use of fast Fourier transform for the estimation of power spectra: A method based on time averaging over short, modified periodograms. *IEEE Transactions on Audio and Electroacoustics*, AU-15: 70–73; reprinted in *Digital Signal Processing*, IEEE Press, 1972, and *Modern Spectrum Analysis*, IEEE Press, 1978.

Wertz, J.R. (Ed.) (1978) *Spacecraft Attitude Determination and Control*. Dordrecht: Kluwer Academic Publishing.

Whalen, A.D. (1971) *Detection of Signals in Noise*. New York: Academic Press.

White, F.M. (1988) *Heat and Mass Transfer*. Reading, MA: Addison-Wesley.

Whittle, P. (1954) On stationary processes in the plane. *Biometrica*, 41: 434–449.

Widnall, W.S. (1973) Optimal filtering and smoothing simulation results for Cirus inertial and precision ranging data. *AIAA Guidance and Control Conference*, paper 73–872, Key Biscayne, FL, August 1973.

Wiener, N. (1949) *Extrapolation, Interpolation and Smoothing of Stationary Time Series*. New York: John Wiley.

Wilkinson, J.H. (1965) *The Algebraic Eigenvalue Problem*. London: Oxford University Press.

Wilkinson, J.H. and C. Reinsch (1971) *Handbook for Automatic Computation*, Vol. 2: *Linear Algebra*. New York: Springer-Verlag.

Willsky, A.S. (1976) A survey of design methods for failure detection in dynamical systems. *Automatica*, 12: 601–611.

Willsky, A.S. and H.L. Jones (1974) A generalized likelihood ratio approach to state estimation in linear systems subject to abrupt changes. *Proceedings of the IEEE Conference on Decision and Control*, Phoenix, AZ, November 1974.

Willsky, A.S. and H.L. Jones (1976) A generalized likelihood ratio approach to the detection and estimation of jumps in linear systems. *IEEE Transactions on Automatic Control*, 21(1): 108–112.

Wishner, R.P., J.A. Tabaczynski, and M. Athans (1968) On the estimation of the state of noisy nonlinear multivariable systems. *IFAC Symposium on Multivariable Control System*, Dusseldorf, Germany, October 1968.

Wishner, R.P., J.A. Tabaczynski, and M. Athans (1969) A comparison of three nonlinear filters. *Automatica*, 5: 487–496.

Wolberg, J. (2006) *Data Analysis Using the Method of Least Squares–Extracting the Most Information from Experiments*. Berlin: Springer-Verlag.

Wonham, W.M. (1968) On a matrix Ricatti equation of stochastic control. *SIAM Journal of Control and Optimization*, 6: 681–697.

Woodburn, J. and S. Tanygin (2002) Position covariance visualization. *AIAA/AAS Astrodynamics Specialist Conference*, paper 2002–4832, Monterey, CA, August 2002.

Wu, Y., M. Sun, D. Krieger, and R.J. Sclabassi (1999) Comparison of orthogonal search and canonical variate analysis for the identification of neurobiological systems. *Annals of Biomedical Engineering*, 27(5): 592–606.

Yancey, W.E. (1994) *A Random Field Model for Spatial Continuity*. Columbia, MD: Coleman Research Corporation.

Yule, G.U. (1927) On the method of investigating periodicities in disturbed series with special reference to Wolfer's sunspot number. *Philosophical Transactions of the Royal Society London, Series A*, 226: 267–298.

Zadeh, L.A. and J.R. Ragazzini (1950) An extension of Wiener's theory of prediction. *Journal of Applied Physics*, 21: 645–655.

Zarchan, P. and H. Musoff (2005) *Fundamentals of Kalman Filtering: A Practical Approach*, 2nd edition. Reston. VA: AIAA.

INDEX

Some entries in this index refer to the Wiley ftp site for this book, which can be found at ftp://ftp.wiley.com/public/sci_tech_med/least_squares/.

A

Accelerometer, 75–80
ADAPTx™, 494, 535
Akaike final prediction error (FPE), 526–528
Akaike information criteria (AIC), 528, 541, 542
Attitude determination, 53, 62, 96, 97

B

Bayes' theorem, 117, 118, 571–573
Bierman, Gerald, 295, 362, 389–411
Bilinear transform, 550
Bode, Hendrik, 7, 335
Bucy, Richard, 1, 7, 14

C

Canonical variate analysis (CVA)—*see* Empirical modeling, ARMA/ARMAX, canonical variate analysis (CVA)
CHAMP spacecraft, 95, 96, 280–282
Chebyshev polynomial, 23, 178, 179, 231
Convolution, 20, 21, 500, 501, 510, 512, 514
Cost function
  Bayesian, 116, 222, 225, 226
  least-squares, 3, 102, 103, 158
  maximum *a posteriori* (MAP), 133
  maximum likelihood (ML), 128
  minimum mean squared error (MMSE), 121, 122
  unweighted least-squares, 102, 103
  weighted least-squares, 108
Cost gradient, 103, 104, 262
Cramér-Rao bound, 129, 270
Curve fitting, 22, 101, 107

D

Delta function
  Dirac, 21
  Kronecker, 290
Dirichlet conditions, 498
Domain
  covariance, 259, 390
  frequency, 6, 335, 501, 503, 510
  information, 259, 390, 404, 417
  time, 6, 335, 501, 510

E

Eigenvalue, 48, 207–211, 566
Eigenvector, 207–211, 566
Empirical modeling
  AR models, 522, 523
    Burg MEM, 525, 526
    Marple, 528, 529
    maximum entropy method (MEM), 524, 525
    MEM comparison, 529–531

Advanced Kalman Filtering, Least-Squares and Modeling: A Practical Handbook, by Bruce P. Gibbs

Empirical modeling (*Continued*)
 Yule-Walker AR normal equation, 523
 Yule-Walker MEM, 523, 524
 ARMA/ARMAX
 canonical variate analysis (CVA), 534–548
 parameter estimation, 532–534
 Yule-Walker normal equations, 532
 via spectral analysis—*see* Spectral analysis
Error analysis, 60, 208–221, 370–384. *See also* Least-squares, error analysis; Kalman (discrete-time) filter, error analysis
Error function, 126, 576
Estimate
 a posteriori, 9, 109, 116, 126, 133, 291
 a priori, 9, 115, 116, 120, 123, 291
Estimation
 Bayesian, 115–121, 213, 216
 best linear unbiased (BLUE), 109
 filtering, 6–8, 289
 Kalman filter—*see* Kalman filter
 least-squares—*see* Least-squares
 maximum *a posteriori* (MAP), 126, 133–137, 347, 348
 maximum likelihood (ML), 126, 128–132, 463
 maximum likelihood parameter identification, 71, 432–449
 minimum mean squared error (MMSE), 121–124
 minimum variance—*see* Estimation, minimum mean squared error (MMSE)
 parameter, 3–5, 432–449
 prediction, 8, 332. *See also* Least-squares, prediction
 robust, 282, 283
 smoothing, 6, 8, 352–370, 401–403, 408–410
Estimation Subroutine Library/Package
 COV2UD, 393
 description, 390
 RANK1, 400
 RINCON, 144
 TDHHT, 407
 THHC, 390, 408
 U2SIG, 393
 UD2COV, 393
 UDMEAS, 396
 WGSG, 398
Euler angles, 29, 30
Exploratory data analysis, 494

F

Fault detection—*see* Jump detection/estimation
Filter
 band-pass, 39–42
 finite impulse response, 6
 H-infinity, 471–474
 high-pass, 34
 infinite impulse response, 6
 interactive multiple model (IMM), 464–468
 Kalman—*see* Kalman (discrete-time) filter
 Kalman-Bucy—*see* Kalman-Bucy (continuous-time) filter
 low-pass, 33, 34, 41, 42
 particle, 485–490
 square root—*see* Kalman filter (discrete-time), factored
 square-root information (SRIF)—*see* Kalman (discrete-time), filter, factored, square root information (SRIF)
 Wiener, 6, 332–340
Fisher, Ronald, 5
Fourier series, 24, 497, 498
Fourier transform
 discrete (DFT), 509–512
 of energy signals, 498–502
 fast (FFT), 496, 511
 of power signals, 502–504
 of stochastic signals, 34, 504–506, 575
Friedland, Bernard, 321, 438, 455

G

Gauss, Carl Friedrich, 4, 5, 102, 193
Gauss-Jordan elimination, 149, 151, 153–156, 175–183
Generalized likelihood ratio (GLR) test, 451–454, 456, 458, 459
Geostationary Operational Environmental Satellites (GOES)
 GOES-13, 25, 63, 64, 96, 97, 202–206
 GOES-14, 206, 207
Global Positioning System (GPS), 1, 91–95, 97, 98, 369, 370
Gyroscope, 77

H

Helmert, Friedrich, 5

J

Jazwinski, Andrew, 349, 351, 449
Jump detection/estimation, 74, 450–461

K

Kalman, Rudolf, 1, 7, 14, 292, 344
Kalman (discrete-time) filter
  adaptive, 449–470
  augmented state, 306–308, 353, 356, 381–384
  bias states, 295
  bias-free/bias-restoring, 321–325, 452–456
  consider parameter, 325–328
  constrained, 471
  correlated measurement and process noise, 303–305
  covariance matrix, 291, 292
  covariance symmetrization, 294, 355
  delay states, 306–308
  divergence, 345, 349
  equations, 290–294
  error analysis, 370–384, 403, 404, 411, 412
  extended (EKF), 344–349
  factored
    Andrews, 389
    Bellantoni and Dodge, 389
    Bierman, 389, 392–416
    Carlson, 389
    Dyer and McReynolds, 389
    Schmidt, 389
    square-root information filter (SRIF), 164, 404–408, 417–426
    SRIF data equation, 404–412, 425
    U-D, 392–404, 412–416
  innovations, 311–314
  iterated-extended, 349–352
  iterated linear filter-smoother, 349–352
  Joseph-stabilized, 294, 391, 413–416
  likelihood function, 314
  linearized, 98, 344, 345
  measurement editing, 312
  measurement preprocessing, 385
  measurement update, 292, 294, 295, 439–440
  multiple model, 69–72, 461–470
  nuisance parameter, 325
  numerical errors, 390–392, 412–416
  reduced-order, 371, 377–381
  residual derivatives, 374, 375
  robust, 471–474
  Schmidt consider, 325–328, 359
  steady-state, 298, 299, 328–331, 339–340
  time update, 291, 438
  time-correlated measurement noise, 305–311
  unscented KF, 474–485
Kalman-Bucy (continuous-time) filter
  derivation, 314–318, 337–340
  steady-state, 319–321, 328–331, 338–340
Kepler orbit elements, 81
Kolmogorov, Andrey, 5, 6
Krylov space method, 169 -172

L

Lagrange multiplier, 250, 251
LAPACK
  BLAS, 182, 183
  DGEESX, 330
  DGESDD, 176, 182, 183
  DGESVD, 176, 182, 183
  DGGLSE, 249, 254, 255
  DGGSVD, 254, 255
  DPOTF2, 151
  matrix condition number, 144
Laplace, Pierre-Simon, 5
Laplace transform, 34, 38, 46, 47
Larimore, Wallace, 534–536, 540, 542, 544, 547
Least-squares
  analysis of variance, 238, 239
  batch, 4, 8
  Bayesian, 115–121, 207, 208, 213
  best linear unbiased estimate (BLUE), 109
  CGNR algorithm, 170, 171, 174–183, 426
  Cholesky factor solution, 149–152, 175–183
  confidence bounds, 208–212
  consider parameters, 215–219
  constrained
    active set method, 256, 257
    constraint types, 249–257
  covariance matrix, 109, 208, 216
  error analysis, 212–219
  Gauss-Jordan solution, 149, 175–183
  generalized, 109
  information matrix, 109, 129, 208
  iterative methods, 167–174
  LSQR algorithm, 170–172, 174–183
  measurement editing, 283–285
  measurement fit residual, 198–206, 214, 215, 219, 220, 229–232

Least-squares (*Continued*)
measurement prediction residual, 196, 201, 203–206, 214, 215, 219, 220, 229–232
measurement preprocessing, 285–286
method, 4, 5, 101–104
model order, 225–237, 242–244
model validation, 194–207
nonlinear, 259–263
nonlinear solution
backtracking step, 265–268, 277–281
conjugate gradient, 169, 264
Gauss-Newton step, 260–263, 265–273, 432
Levenburg-Marquardt, 266, 268, 269, 275–282
Newton step, 260, 272–274
NL2SOL, 266, 279–280
steepest descent, 264
stopping criteria, 269–271
trust region, 265
normal equations, 104, 109, 145–156
numerical errors, 142–144, 153–156, 165, 175–181
optimal data span, 244, 245
partitioned solution, 147, 162
polynomial model, 152–156, 174–183, 197–201
prediction
examples, 201–207, 231–232
measurement residual equations, 214, 215, 219, 220
model evaluation, 196, 197, 226, 227
pseudo-inverse, 186–190
QR solution, 157–165, 175–183
recursive, 7, 257–259
robust estimation, 282–285
solution validation, 194–197
stepwise regression, 239–244
SVD solution, 165–167, 175–183
weighted, 108–114
Legendre, Adrien-Marie, 4, 102
Likelihood function, 128, 314
Linearization, 18, 260, 344, 345
LINPACK, 144, 176

M

Maritime Administration, 62, 63
Markov, Andrey, 5
Markov process—*see* Stochastic process, Markov
Markov random field, 422–424
MATLAB®
CARE, 330
DARE, 330
EXPM1, 47
EXPM2, 47
EXPM3, 49
matrix-vector notation, 10
N4SID, 535
System Identification Toolkit™, 494, 535
Matrix
condition number, 142–144, 568
correlation, 109, 523
covariance, 109, 208, 291, 292, 574
decomposition
Cholesky, 61, 149, 565
eigen, 127, 566
generalized singular value (GSVD), 253
LU (Crout), 149, 565
QR, 157
Schur, 48, 329, 330
singular value (SVD), 106, 165, 566, 567
derivative, 563, 564
determinant, 561, 562
Fisher information, 109, 129, 441–444
Hermitian, 556
Hessian, 104, 260, 273, 443–447
inverse
definition, 558, 559
derivative, 564
Gauss-Jordan elimination, 149, 558
partitioned, 559, 560
inversion identity, 120, 560, 561
norms
Frobenius, 140, 565
induced matrix, 140, 565
operations, 557, 558
orthogonal, 559
positive definite, 559
positive semi-definite, 559
pseudo-inverse, 109, 141, 185–190, 567, 568
rank, 561
rank-1 update, 395, 399, 400
singular values, 106, 141, 567
sparse array storage, 167
state process noise covariance—*see* State process noise covariance
state transition—*see* State transition matrix
Toeplitz, 556
trace, 562, 563
transformation
Givens, 158, 159

Householder, 159–161
modified Gram-Schmidt (MGS), 162–164
modified weighted Gram-Schmidt (MWGS), 398
orthogonal, 140, 157, 567
similarity, 48, 566
transpose, 557
triangular, 556, 557
Vandermonde, 153
Model
aircraft, 28–31, 73
basis function, 14, 22–25
continuous-time, 17, 18
deterministic, 2, 3
discrete-time
autoregressive (AR), 14–16
autoregressive integrated moving average (ARIMA), 14–16
autoregressive moving average (ARMA), 14–16
autoregressive moving average with exogenous inputs (ARMAX), 14–17, 536
definition, 14, 15, Appendix D on ftp.wiley.com
moving average (MA), 14–16
discrete-to-continuous conversion, 548–551
empirical, 493–495
errors, 212–220
errors and system biases, 62–64
first-principles, 25–31
forward, 2, 3
hydrological flow, 419–427
inertial navigation system errors, 74–80, 377–380, 412–416, 457–459
inverse, 2–4
linear regression, 42, 43
measurement, 2, 3, 7, 58–60, 92–95
parametric, 14
power plant, 100, Section 3.5 on ftp.wiley.com
reduced-order, 44, 72, 371, 372, 377–381
satellite, 80–90
ship, 27
state space, 16, 543–545
stochastic, 3, 5–7. *See also* Stochastic process
tank gun-tube flexure, 535
tank motion, 69–72, 468–470
time-invariant, 3

N

NASA, 7, 62, 63, 89, 206, 325, 390
Navigation, 63
Noise
colored, 33, 70. *See also* Stochastic process, Markov
correlated measurement-process, 303–305
measurement, 3, 59–61
process, 2, 3, 7, 13, 17–18, 20–21
white, 7, 17, 493, 511, 536, 576–578
Nonrandom variable—*see* Parameter, fixed-but-unknown
Normal equation—*see* Least-squares, normal equation
Nyquist frequency, 510

O

Observablility, 7, 105, 143, 145, 183–185
Orbit determination, 4, 62, 63, 80–98, 147, 280–282, 369, 417, 418
Orthogonal polynomial, 23, 178, 179, 231
Orthogonal projection, 124, 125
Orthogonality principle, 124, 292

P

Parameter
consider (Schmidt-type), 325–328, 373–376, 379
fixed-but-unknown, 105, 125, 126, 237
unadjusted analyze (consider), 215, 216, 373–380
Parseval's theorem, 502
Power spectral density (PSD), 6, 16, 21, 493–497, 504–506, 512–524, 528, 544–548, 577
Power spectrum—*see* Power spectral density
Probability
conditional, 118, 573
density
a posteriori, 118
a priori, 118
Chi-squared, 111, 112, 574, 575
function (PDF), 5, 126, 572
Gaussian (normal), 5, 126, 127, 574
joint, 118, 126, 127, 574, 576
Rayleigh, 134, 135
skewed, 347, 348
distribution function, 126, 571, 572
independent, 570, 571
joint, 118, 570, 571
marginal, 118, 572

Q

QR decomposition—*see* Matrix, decomposition, QR
Quaternion, 31

R

Radar, 74, 480–482
Ragazzini, John, 7
Random process—*see* Stochastic process
Random variable
  conditional mean, 119, 573
  correlation function, 6, 505, 506, 514, 575
  definition, 571
  independent, 571, 572, 575
  independent, identically-distributed, 111, 572
  mean (expectation), 573
  notation, 11, 571
  orthogonal, 124, 575
  uncorrelated, 124, 575
  variance, 573
Recursive processing, 4, 6, 257–259, 337
Riccati equation, 58, 317, 328

S

Schmidt, Stanley, 7, 325, 344, 397
Scoring, 432, 441–443
Shannon, Claude, 7, 335, 524
Simplex method, 264
Simulating stochastic systems, 60, 61
Singular value decomposition—*see* Matrix, decomposition, singular value (SVD)
Smoother
  Dyer-McReynolds covariance, 390, 410
  fixed-interval, 357–370
  fixed-lag, 356
  fixed-point, 352–355
  forward-backward filters, 357–359
  modified Bryson-Frazier, 359, 361, 362
  pass-disposable biases, 366–370
  Rauch-Tung-Striebel, 359–361, 363–370
  square root information, 390, 408–412
  U-D RTS, 401–403
Spectral analysis
  AR models—*see* Empirical modeling, AR
  ARMA/ARMAX models—*see* Empirical modeling, ARMA/ARMAX
  Blackman-Tukey correlogram, 514, 515, 521
  methods, 496, 497
  periodogram, 512–514, 519–521
  spectrum analyzer, 496
  stochastic signals, 504–506
  Welch periodogram, 513, 519, 521
  window functions, 506–509
Spectrum—*see* Power spectral density
Square-root information filter (SRIF)—*see* Kalman filter (discrete-time), factored, square-root information (SRIF)
State process noise covariance
  computation, 57, 58
  definition, 20, 21
State transition matrix
  computation, 45–56
    hump, 54
    integration, 48, 437
    inverse Laplace transform, 47
    matrix decomposition methods, 48, 49
    matrix exponential, 46
    numeric derivatives, 50–52, 437
    Padé series expansion, 47
    partitioned combinations of methods, 52
    scaling and squaring, 49, 437, 438
    Taylor series expansion, 46, 435, 436
  definition, 19
Stochastic process
  autocorrelation, 33, 505, 575
  Brownian motion (Wiener process), 31, 576
  correlation function, 6, 505, 506, 514, 575
  definition, 575
  ergodic, 504, 575
  Markov, 33–42, 71, 576
    first-order, 33–36
    second-order, 36–42
  independent, 575
  mean square continuous, 577
  mean square integrable, 577
  orthogonal, 575
  random walk, 31, 32, 35, 41, 42, 576
  strictly stationary, 575
  uncorrelated, 575
  wide-sense stationary, 504, 575
Subroutine library—*see* ftp.wiley.com
  COVAR_ANAL, 376
  CVA, 540
  MARPLE_SPECT, 529
  MEMOPT, 526
  MLE_PARAMETERID, 444
  OPTL_BACK, 265

OPTL_LM, 266
PHIU_M, 413, 416
SINV, 151
SMULT, 439
SMULT_BLOCK, 439
SPECT, 529
TDHHT_M, 416
THHC_M, 416
WGS_M, 398, 413, 416
Superposition, 18
Swerling, Peter, 7
System
causal, 16, 335, 493
continuous-time, 13
controllable, 329
damping, 36–42, 548
detectable, 329
discrete-time, 14, 290
homogeneous solution, 19, 20
minimum phase, 16
nonlinear, 17, 259, 344
observable, 145, 184, 185, 329
particular solution, 19, 20
poles, 16, 336, 493, 505, 522, 548–550
stable, 16, 329, 493
time-invariant, 18, 20
time-varying, 20, 314
zeroes, 16, 493, 505, 522, 548–550
system identification, 494, 495, 535

T

Taylor series, 18, 20, 46, 55–57, 260, 435
Thornton, Catherine, 389, 392, 397, 398, 404
Time series analysis, 494
Tracking
aircraft, 63, 73, 74, 364–366, 459–461
ballistic reentry, 480–485
falling body, 271–274
maneuvering target, 69–74, 363–366, 468–470
ship, 67, 68, 274–280
tank, 69–72, 363, 364, 468–470
target, 8
target motion analysis, 67
Tukey, John, 494

U

Unscented transformation, 475–477

V

Vector norms
Euclidian, 105, 140, 565
Hölder p-norms, 139, 140, 564

W

Wavelets, 24
Wiener, Norbert, 5, 6, 332
Wiener-Hopf equation, 6, 316, 333, 334

Z

Zadeh, Lofti, 7
z-transform, 15, 16, 534, 544, 545, 550